THE CHEMISTRY AND BIOLOGY OF BETA-LACTAMS

THE CHEMISTRY AND BIOLOGY OF BETA-LACTAMS

Zerong Wang, PhD

AAP | APPLE ACADEMIC PRESS

First edition published 2023

Apple Academic Press Inc.
1265 Goldenrod Circle, NE,
Palm Bay, FL 32905 USA

760 Laurentian Drive, Unit 19,
Burlington, ON L7N 0A4, CANADA

CRC Press
6000 Broken Sound Parkway NW,
Suite 300, Boca Raton, FL 33487-2742 USA

4 Park Square, Milton Park,
Abingdon, Oxon, OX14 4RN UK

Apple Academic Press exclusively co-publishes with CRC Press, an imprint of Taylor & Francis Group, LLC

Library and Archives Canada Cataloguing in Publication

Title: The chemistry and biology of Beta-Lactams / Zerong Wang, PhD.
Names: Wang, Zerong (Daniel Zerong), author.
Description: First edition. | Includes bibliographical references and index.
Identifiers: Canadiana (print) 20220443939 | Canadiana (ebook) 20220443971 | ISBN 9781774911716 (hardcover) | ISBN 9781774911723 (softcover) | ISBN 9781003330288 (ebook)
Subjects: LCSH: Beta lactamases.
Classification: LCC QP609.B46 W36 2023 | DDC 579.3/5—dc23

Library of Congress Cataloging-in-Publication Data

CIP data on file with US Library of Congress

ISBN: 978-1-77491-171-6 (hbk)
ISBN: 978-1-77491-172-3 (pbk)
ISBN: 978-1-00333-028-8 (ebk)

About the Author

Zerong Wang, PhD
Professor of Chemistry, College of Science and Engineering, University of Houston-Clear Lake, Texas, USA

Zerong Wang, PhD, is a Full Professor at the University of Houston-Clear Lake, Texas. Prior to that, he worked at the Institute for Biological Sciences of the National Research Council of Canada for several years. Through his career, the author has gained specific training and expertise in organic chemistry, particularly in physical and organic chemistry and other subdisciplines that include photochemistry, spectroscopies, carbohydrate chemistry, sulfur chemistry, nucleosides and heterocycles, material science, reaction methodology, computational chemistry, among other. Dr. Wang has developed research projects relating to sulfur chemistry, computational chemistry, nucleoside analogs, heterocycle chemistry, materials science, and macromolecules (pillarene, calix[n]arene, and melamine-based dendrimers, etc.) and has received 22 research grants, including from NSF-MRI, NSF-STEM, Welch Research Grant, Welch Departmental Research Grant, and University of Houston-Clear Lake's Faculty Research and Support Fund (FRSF) Grants.

The author has developed two compendiums in organic chemistry: *Comprehensive Organic Named Reactions, with Detailed Mechanism Discussions and Updated Experimental Procedures* (3 volumes) (Wiley, 2009) and *Encyclopedia of Physical Organic Chemistry* (6 volumes) (Wiley, 2017), the PROSE Award winner in 2018.

While conducting research activities, the author also teaches courses for both graduate and undergraduate students. To date, the author has taught courses on General Chemistry, General Chemistry Laboratory, Analytical Chemistry, Quantitative Chemical Analysis, Forensic Chemistry, Organic Chemistry, Organic Chemistry Laboratory, Advanced Organic Chemistry, Physical Organic Chemistry, Synthetic Organic Chemistry, Organometallic Chemistry, Biochemistry, Biochemistry Laboratory, Polymer Chemistry, Introduction to Chemical Engineering, Nutrition and Diet Chemistry, Green Chemistry, Introduction to NMR Spectroscopy, Chemistry Seminar, Graduate Research, and Chemistry for Non-Science Majors.

Dr. Wang earned his BS degree in Chemistry from Lanzhou University, PR China, and earned his MS and PhD degrees from the Shanghai Institute of Organic Chemistry, Chinese Academy of Sciences. He conducted his postdoctoral research at the Department of Chemistry, University of California Berkeley and York University (Canada).

Contents

Abbreviations

ACE	angiotensin-converting enzyme
ADAC	acyclic diaminocarbenes
AMR	access antimicrobial resistance
APAF-1	apoptotic protease activating factor 1
ARDB	antibiotic resistance genes database
ARDs	antibiotic resistance determinants
ARGO	antibiotic resistance genes online
ARO	antibiotic resistance ontology
BHT	butylated hydroxytoluene
BLAD	beta-lactamase database
BLAST	basic local alignment search tool
BLDB	beta-lactamase database
BTPP	*tert*-butylimino-tri(pyrrolidino)phosphorane
CAACs	cyclic (alkyl)(amino)carbenes
CAL-B	*Candida Antarctica* lipase B
CAN	ceric ammonium nitrate
CARD	comprehensive antibiotic resistance database
CBMAR	comprehensive β-lactamase molecular annotation resource
CDC	centers for disease control and prevention
CDD	conserved domains database
CMPI	2-chloro-1-methylpyridinium iodide
CODLE	comprehensive online database of β-lactamase enzymes
COG	clusters of orthologous groups of proteins
CSI	chlorosulfonyl isocyanate
DABCO	1,4-diazabicyclo[2.2.2]octane
DEAD	diethyl azodicarboxylate
DIAD	diisopropyl azodicarboxylate
DIPEA	diisopropylethylamine
DMDO	dimethyldioxirane
DME	1,2-dimethoxyethane
DMPU	*N,N'*-dimethylpropyleneurea
EARS-Net	European Antimicrobial Resistance Surveillance Network
ECDC	European Center for Disease Prevention and Control
EDCI	1-ethyl-3-(3-dimethylaminopropyl)carbodiimide

EDTA	ethylenediaminetetraacetic acid
ESBL	extended-spectrum beta-lactamases
FARME	functional antibiotic resistant metagenomic element
FDA	Food and Drug Administration
GES	Guiana-extended-spectrum
HCMV	human cytomegalovirus
HIV-1 PR	human immunodeficiency virus type 1 proteinase
HMMs	hidden Markov models
HWE	Horner-Wadsworth-Emmons
IBC-Cl	isobutyl carbonochloridate
IJSB	international journal of systematic bacteriology
IJSEM	International Journal of Systematic and Evolutionary Microbiology
IMDs	indwelling medical devices
IMP	imipenemase
KEGG	Kyoto Encyclopedia of Genes and Genomes
KOTMS	potassium trimethylsilanolate
KPC	*Kiebsiella pneumoniae* carbapenemase
LacED	lactamase engineering database
LDL	low-density lipoproteins
MBLs	metallo-β-lactamases
MCR	multi-component reaction
MDR	multi-drug resistant
MIC	minimal inhibitory concentration
MRIs	mobile resistance integrons
MRSA	methicillin-resistant *staphylococcus aureus*
NAAA	*N*-acylethanolamine acid amidase
NCA	*N*-carboxyanhydride
NCBI	National Center for Biotechnology Information
NCS	*N*-chlorosuccinimide
NDARO	national database of antibiotic-resistant organisms
NDM	New Delhi MBL
NHC	*N*-heterocyclic carbene
NMI	*N*-methylimidazole
OBHA·HCl	*O*-benzylhydroxylamine hydrochloride
PASs	poly(amido-saccharide)s
PATRIC	pathosystems resource integration center
PBPs	penicillin-binding proteins
PBS	phosphate-buffered saline

PCM	pairwise comparative modeling
PDB	protein data bank
PGA	penicillin G acylase
PHE	Public Health Emergency
PivCl	pivaloyl chloride
PPY	4-(pyrrolidin-1-yl)pyridine
PS	proton sponge
Qnr	quinolone resistance genes
RAC	repository of antibiotic resistance cassettes
RAST	rapid annotation using subsystem technology
ROS	reactive oxygen species
SDS	sodium dodecyl sulfate
SET	single electron transfer
SHVED	SHV β-lactamase engineering database
SPSS	solid phase supported synthesis
TB	tuberculosis
TBAF	tetrabutylammonium fluoride
TCAI	trichloroacetyl isocyanate
TCT	2,4,6-trichloro-1,3,5-triazine
TEMPO	tetramethylpiperidin-1-yl)oxyl
TMEDA	tetramethylethylenediamine
TMSOTf	trifluoromethanesulfonate
TsCl	*p*-toluenesulfonyl chloride
TTN	total turnover number
USDA	United States Department of Agriculture
UTI	urinary tract infection
VCD	vibrational circular dichroism
VIM	Verona integron-encoded MBL
VLDL	very-low-density lipoproteins
WGS	whole genome shotgun
WHO	World Health Organization
XDR	extensively drug-resistant

Acknowledgments

Writing this book series was harder than I expected, even though I had already been through this process for my three-volume book *Comprehensive Organic Named Reactions, with Detailed Mechanism Discussions and Updated Experimental Procedures* (2009, ISBN: 978-0-471-70450-8).

I'm eternally grateful to my wife, Xi Liu, and my daughter, Izellah, who have taken care of me so that I could focus on this book series in my spare time. It would not have been possible to complete this monograph and the four-volume book series of "AMINO ACIDS: Insights and Roles in Heterocyclic Chemistry" without their long-time support.

A very special thanks to our librarians in the Newman Library of the University of Houston-Clear Lake, who helped me locate necessary references in a timely manner.

Finally, I thank my colleagues and friends who have provided me with endless guidance.

Preface

β-lactams are four-membered cyclic amides with high chemical reactivity due to their ring strain and that have been commonly applied as antibiotics in the treatment of bacterial infections. The contents collected in the four chapters of this book provide the most updated information about β-lactams relating to both the pharmaceutical industry and synthetic chemistry.

Chapter 1 collects more than seven types of β-lactam related antibiotics, including 40 penicillin group of antibiotics of four generations, 78 cephalosporin antibiotics of five generations, 52 carbapenem antibiotics, and 46 monobactam antibiotics. These antibiotics have been organized in different tables that also contain their chemical structures, CAS numbers, IUPAC names, and major biological activities. In addition to these popular β-lactams antibiotics, other β-lactam related antibiotics, including clavam antibiotics (e.g., clavamycins), penem antibiotics (e.g., faropenem, sulopenem), carbapenem antibiotics (e.g., OA-6129D, OA-6129E), carbacephem antibiotics (e.g., carbacefaclor), oxacephem antibiotics (e.g., moxalactam), and penam sulfone antibiotics (e.g., tazobactam) have also been collected in this chapter.

Chapter 2 focuses on the antibiotic activities of β-lactams, and contains: (a) the nomenclature of bacteria and the classification of bacteria based on the Gram stain; (b) the antibacterial mechanism of β-lactams; (c) the self-defense mechanism of bacteria via the formation of biofilm as well as conversion of themselves into the *L*-forms, and the secretion of β-lactamases to deactivate the β-lactam antibiotics; (d) the Ambler and Bush-Jacoby-Medeiros systems to classify the β-lactamases; and (e) more than 20 different β-lactamase databases.

Chapter 3 concentrates on the syntheses of β-lactams, including the two major synthetic strategies of cycloaddition and intramolecular cyclization. Four cycloaddition methods have been summarized in this chapter, including the Staudinger's [2+2] cycloaddition involving imines and ketenes, the Wolff rearrangement of α-diazoketones into ketenes to undergo the subsequent Staudinger cycloaddition, the [2+2] cycloadditions between alkenes and isocyanates, and the Kinugasa reaction involving nitrones and alkynes. For comparison, the intramolecular cyclizations to give β-lactams include lactamization of β-amino acid derivatives, intramolecular amidation, intramolecular substitutive cyclization with either nitrogen or carbon nucleophiles, intramolecular Michael addition, and electrocyclic ring closure reactions, photolysis of α,β-unsaturated amides, carbonylation of allylamines, Ugi-multicomponent reactions, ring enlargement of aziridines, and ring contraction reactions. In addition to these many classified synthetic methods for β-lactams, there are also other synthetic methods such as the enzymatic synthesis of β-lactams which has existed in nature prior to the discovery of any of the above methods.

Chapter 4 explores the chemical reactivities of β-lactams that include the hydrolysis, alcoholysis, aminolysis, and thiolysis of β-lactams into the corresponding β-amino acids, β-amino esters, β-amino amides, and β-amino thioesters. Hydrazinolysis of β-lactams and reduction of the β-lactams with hydride to give β-amino alcohols (or aldehydes, azetidines), cleavage of the C3-C4 bond or N1-C4 bond of β-lactams are other reactivities of β-lactams. Moreover, substitutions taking place at C3, C4, or N1 of β-lactams have also been collected in this chapter, as well as the ring-opening polymerization of β-lactams to yield poly(β-amides) or poly(β-peptides) that are generally known as nylon 3.

The author wishes that this book provides readers with valuable information regarding antibiotic activities as well as the chemical reactivities of β-lactams.

CHAPTER 1

β-Lactams and Antibiotics

1.1 BRIEF INTRODUCTION

β-Lactams are a special group of organic molecules that originate from β-amino acids. When a carboxylic acid reacts with an amine, an amide is generated. Intramolecular amidation (or cyclization) of amino acids leads to the formation of cyclic amides, generally known as lactams. As a result, the intramolecular cyclization of carboxylic acid containing an amino group at the β-, γ-, or δ-position would yield the β-, γ-, or δ-lactam, respectively. The tautomers of the corresponding lactams are known as lactims, owing to the presence of an imino functionality (C=N bond), as shown in Figure 1.1. According to the IUPAC nomenclature system, azetidine is the base name of a four-membered cyclic molecule with one nitrogen atom on the ring, so that β-lactam is generally known as azetidin-2-one, where the "one" indicates the presence of the carbonyl group. Likewise, the base names of five- and six-membered cyclic molecules with one nitrogen atom are pyrrolidine and piperidine, respectively. Due to the high ring strain, α-amino acids are generally known to undergo intermolecular amidation to form peptides, instead of intramolecular cyclization to form α-lactams with a highly unstable three-membered ring. In contrast, β-amino acids are often found to undergo intramolecular amidation to form β-lactams, although stability and reactivity of β-lactams are still largely related to the high strain of the four-membered rings. The high ring strain of β-lactams can be explained in two aspects: (a) the α- and β-carbon atoms in sp^3 hybridization require a bond angle of 109.5° and the carbonyl carbon atom and the nitrogen atom in sp^2 hybridization prefer a typical bond angle of 120°, but all these bond angles are confined roughly to 90°; (b) the bond strength has been weakened as the maximum bond strength can only be reached at the desired bond angle due to the perfect interaction of the corresponding atomic orbitals. Different from other simple heterocycles originating from α-amino

The Chemistry and Biology of Beta-Lactams. Zerong Wang, PhD

acids, such as hydantoins, thiohydantoins, 2,5-diketopiperazines, α-amino acid *N*-carboxyanhydrides (NCAs), sydnones, azlactones, and oxazolidin-5-ones, β-lactams-containing molecules have demonstrated many critical biological activities, primarily as antibiotics, largely owing to the relative instability and high reactivity of the four-membered ring, as described below.

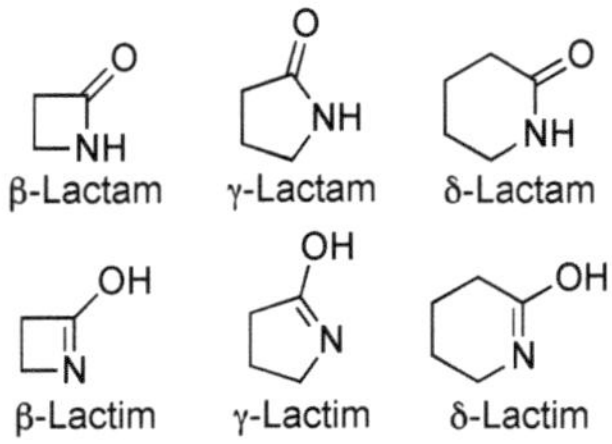

FIGURE 1.1 The simple structure of β-, γ-, and δ-lactams and the corresponding lactims

While the first synthetic β-lactam was reported as early as 1907 by Hermann Staudinger from the [2+2] cycloaddition between diphenylketene and a Schiff base [1], the importance of β-lactams was initially realized after Alexander Fleming's discovery and isolation of powerful antibacterial "penicillin" from cultures of *Penicillium notatum* in 1929 [2]. In his commendable paper, Fleming instructed how to filter the active agent from the culturing broth of the mold and stated that the antibacterial activity could be preserved for a longer time after neutralization which would otherwise vanish after 10–14 days at room temperature. This compound is soluble in alcohol, but not in ether and chloroform, and is particularly effective for pyogenic cocci and the diphtheria group of bacilli. Penicillin can be destroyed in an autoclave at 115°C for 20 minutes but can withstand boiling water for a few minutes.

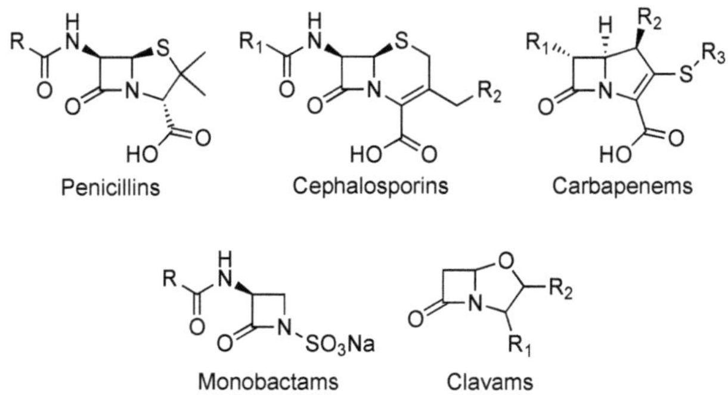

FIGURE 1.2 The five groups of antibacterial β-lactams

Soon after the introduction of penicillin in medical treatment in 1942 [3], antibacterial resistance emerges, thus many β-lactam-containing antibacterial agents have been developed, which can be divided into five groups: penicillins, cephalosporins, monobactams, carbapenems, and clavams, as displayed in Figure 1.2 [4, 5]. Clearly, three out of these five types of β-lactams still contain an α-amino group, including penicillins, cephalosporins, and monobactams. Currently, because of their effectiveness in controlling both Gram-negative and Gram-positive bacteria, β-lactams have become one of the three largest antibiotic classes, with the other two classes being macrolides and fluoroquinolones [6].

1.2 PENICILLIN ANTIBIOTICS

Penicillins, generally known as penams, are the acyl derivatives of (2*S*,5*R*,6*R*)-6-amino-3,3-dimethyl-7-oxo-4-thia-1-azabicyclo[3.2.0]heptane-2-carboxylic acid (i.e., 6β-amino-penicillanic acid, 6-APA), whereas the "penicillin" coined by Fleming in 1928 refers to benzylpenicillin or penicillin G, that was extracted from *Penicillium rubens*, a common fungus of an indoor environment. While most penicillins in clinical uses are semi-synthesized from naturally existing penicillins, two natural penicillin derivatives, including penicillin G and penicillin V are still in clinical use, where the penicillin G is for intravenous or intramuscular use and penicillin V is taken by mouth. Penicillins are known to treat odontogenic infections, acute otitis media (middle ear infection), helicobacter pylori infection, sinusitis, life-threatening cervicofacial infection, otosyphilis, bacterial endocarditis and related infections, perianal streptococcal dermatitis, syphilis, among others [7]. More infections that can be cured by penicillin G can be found at: https://en.wikipedia.org/wiki/Penicillin. Penicillin V has similar medical usages but is more tolerable for stomach acid.

Nowadays penicillin is commonly used to refer to any β-lactam antimicrobial agent that contains a thiazolidine ring fused to the β-lactam ring, regardless of being a natural or semi-synthesized compound. Due to the development of drug resistance, more, and more penicillin derivatives have been synthesized and added to the arsenal of penicillin antibiotics. So far, a total of four generations of penicillin derivatives have been developed and used in medical treatments, as summarized in Table 1.1. The penicillins of the first generation are sensitive to β-lactamases, which cause antibacterial resistance. In contrast, the penicillins in the second generation are resistant to β-lactamases. However, both generations of penicillins have a narrow spectrum of antibacterial activities. Further development of penicillins leads to the third and fourth generations of penicillins with an extended spectrum of antibacterial activities. The penicillins in the third generation are aminopenicillins, and the penicillins in the fourth generation include carboxypenicillins and ureidopenicillins. Besides these penicillin derivatives, there are other penicillin-related derivatives, such as mecillinam, pivmecillinam, and sulbenicillin.

TABLE 1.1 The List of Penicillin Derivatives

Name	CAS No.	Structure	Full Name	Major Uses
1st Generation				
Penicillin G [8, 9] (Benzylpenicillin)	61-33-6	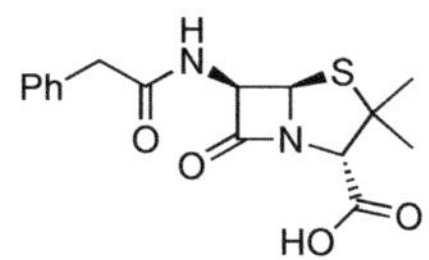	(2*S*,5*R*,6*R*)-3,3-Dimethyl-7-oxo-6-(2-phenylacetamido)-4-thia-1-azabicyclo[3.2.0] heptane-2-carboxylic acid	The first-line drug for *syphilis* in all its stages. High intrinsic activity against *streptococci* and *pneumococci*
Penicillin G sodium	69-57-8	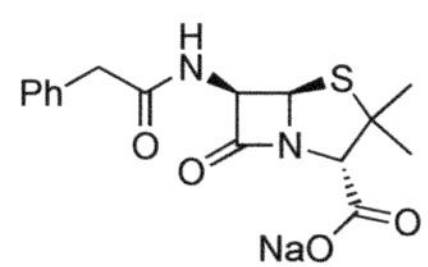	Sodium (2*S*,5*R*,6*R*)-3,3-dimethyl-7-oxo-6-(2-phenylacetamido)-4-thia-1-azabicyclo[3.2.0] heptane-2-carboxylate	The first-line drug for *syphilis* in all its stages. High intrinsic activity against *streptococci* and *pneumococci*
Penicillin G potassium	113-98-4	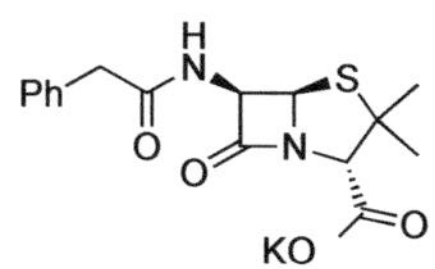	Potassium (2*S*,5*R*,6*R*)-3,3-dimethyl-7-oxo-6-(2-phenylacetamido)-4-thia-1-azabicyclo[3.2.0] heptane-2-carboxylate	The first-line drug for *syphilis* in all its stages. High intrinsic activity against *streptococci* and *pneumococci*
Penicillin F [10]	118-53-6	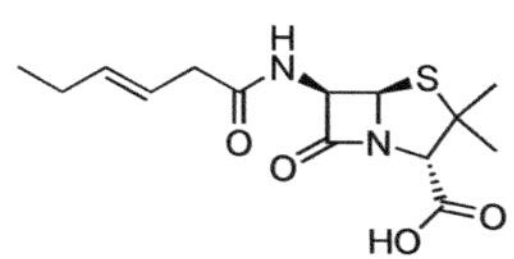	(2*S*,5*R*,6*R*)-6-((*E*)-Hex-3-enamido)-3,3-dimethyl-7-oxo-4-thia-1-azabicyclo[3.2.0] heptane-2-carboxylic acid	Effective against Gram-positive bacteria
Penicillin K [10]	525-97-3	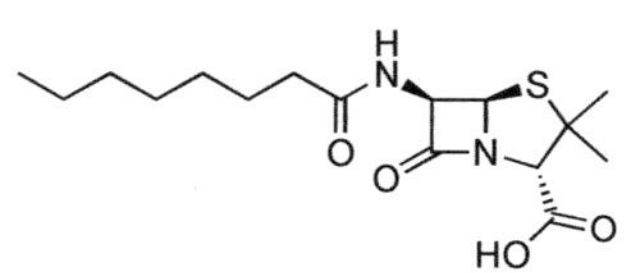	(2*S*,5*R*,6*R*)-3,3-Dimethyl-6-octanamido-7-oxo-4-thia-1-azabicyclo[3.2.0] heptane-2-carboxylic acid	Effective against Gram-positive bacteria

Name	CAS No.	Structure	Full Name	Major Uses
Penicillin X [10]	525-91-7	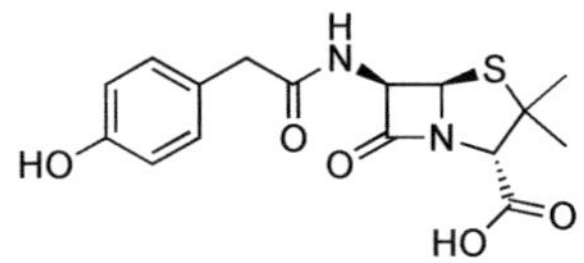	(2*S*,5*R*,6*R*)-6-(2-(4-Hydroxyphenyl) acetamido)-3,3-dimethyl-7-oxo-4-thia-1-azabicyclo[3.2.0]heptane-2-carboxylic acid	Effective against Gram-positive bacteria
Azidocillin [9]	17243-38-8	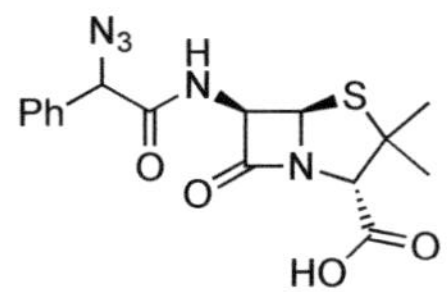	(2*S*,5*R*,6*R*)-6-(2-Azido-2-phenylacetamido)-3,3-dimethyl-7-oxo-4-thia-1-azabicyclo[3.2.0] heptane-2-carboxylic acid	High intrinsic activity against *streptococci* and *pneumococci*
Clometocillin [11] Clometacillin	1926-49-4	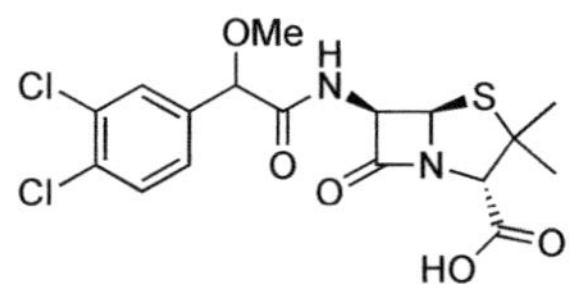	(2*S*,5*R*,6*R*)-6-(2-(3,4-Dichlorophenyl)-2-methoxyacetamido)-3,3-dimethyl-7-oxo-4-thia-1-azabicyclo[3.2.0] heptane-2-carboxylic acid	More effective than benzylpenicillin and phenoxyalkylpenicillins against pneumococcal septicemia
Penamecillin [12]	983-85-7	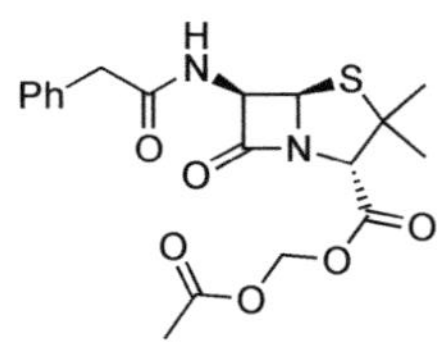	Acetoxymethyl (2*S*,5*R*,6*R*)-3,3-Dimethyl-7-oxo-6-(2-phenylacetamido)-4-thia-1-azabicyclo[3.2.0] heptane-2-carboxylate	Administered orally to infants with mild respiratory infections
Penicillin V [9]	87-08-1	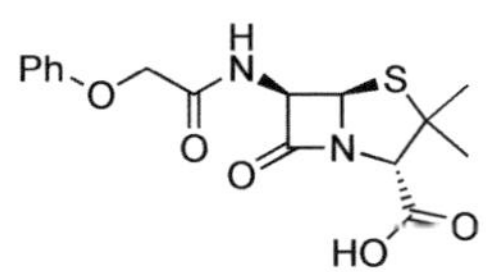	(2*S*,5*R*,6*R*)-3,3-Dimethyl-7-oxo-6-(2-phenoxyacetamido)-4-thia-1-azabicyclo[3.2.0] heptane-2-carboxylic acid	High intrinsic activity against *streptococci* and *pneumococci*

TABLE 1.1 *(Continued)*

Name	CAS No.	Structure	Full Name	Major Uses
Pheneticillin [13]	147-55-7	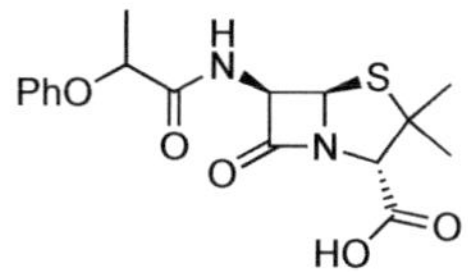	(2*S*,5*R*,6*R*)-3,3-Dimethyl-7-oxo-6-(2-phenoxypropanamido)-4-thia-1-azabicyclo[3.2.0]heptane-2-carboxylic acid	Treatment of severe streptococcal infections
Propicillin [9]	551-27-9	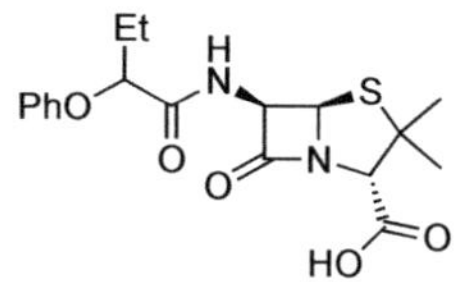	(2*S*,5*R*,6*R*)-3,3-Dimethyl-7-oxo-6-(2-phenoxybutanamido)-4-thia-1-azabicyclo[3.2.0]heptane-2-carboxylic acid	High intrinsic activity against *streptococci* and *pneumococci*
Benzathine benzylpenicillin [14]	1538-09-6	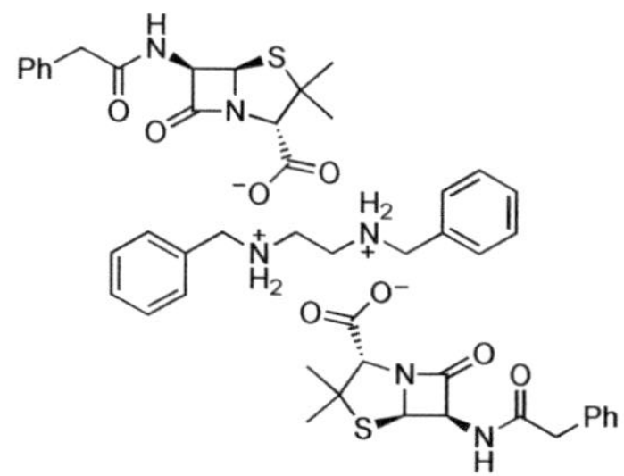	*N*[1],*N*[2]-Dibenzylethane-1,2-diaminium Bis-(2*S*,5*R*,6*R*)-3,3-dimethyl-7-oxo-6-(2-phenylacetamido)-4-thia-1-azabicyclo[3.2.0]heptane-2-carboxylate	Treatment of syphilis
2nd Generation				
Cloxacillin [8]	61-72-3	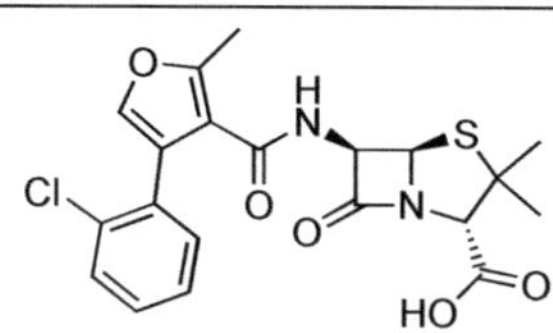	(2*S*,5*R*,6*R*)-6-(4-(2-Chlorophenyl)-2-methylfuran-3-carboxamido)-3,3-dimethyl-7-oxo-4-thia-1-azabicyclo[3.2.0]heptane-2-carboxylic acid	Against *Staphylococci* producing β-lactamase

Name	CAS No.	Structure	Full Name	Major Uses
Dicloxacillin [15]	3116-76-5	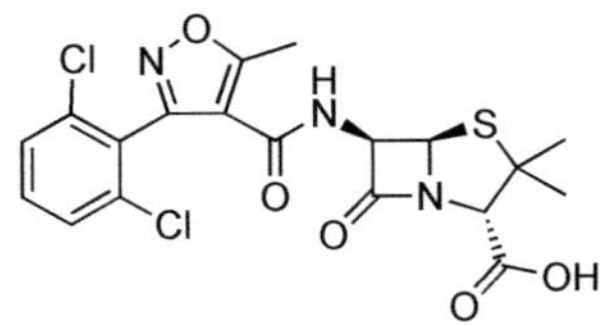	(2*S*,5*R*,6*R*)-6-(3-(2,6-Dichlorophenyl)-5-methylisoxazole-4-carboxamido)-3,3-dimethyl-7-oxo-4-thia-1-azabicyclo[3.2.0] heptane-2-carboxylic acid	Against *Staphylococci*
Flucloxacillin [8]	5250-39-5	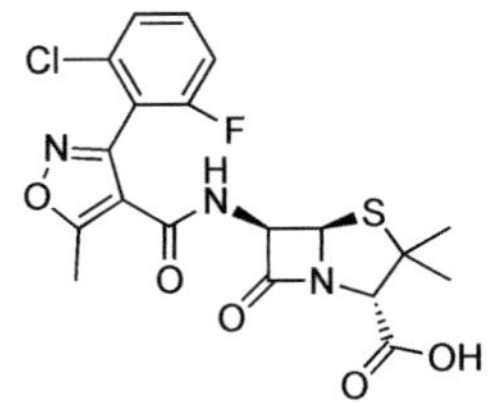	(2*S*,5*R*,6*R*)-6-(3-(2-Chloro-6-fluorophenyl)-5-methylisoxazole-4-carboxamido)-3,3-dimethyl-7-oxo-4-thia-1-azabicyclo[3.2.0] heptane-2-carboxylic acid	Used to treat infections caused by susceptible Gram-positive bacteria, including β-lactamase-producing *Staphylococcus aureus*
Methicillin [8] Meticillin Methycillin	61-32-5	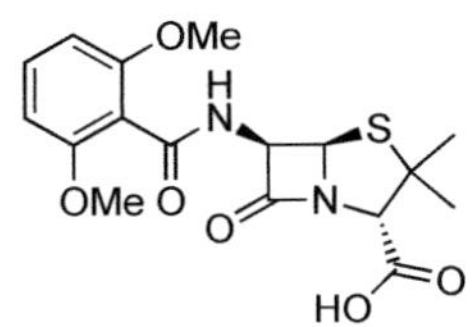	(2*S*,5*R*,6*R*)-6-(2,6-Dimethoxybenzamido)-3,3-dimethyl-7-oxo-4-thia-1-azabicyclo[3.2.0] heptane-2-carboxylic acid	Used to treat infections caused by *Staphylococcus*, but largely replaced by flucloxacillin and dicloxacillin
Nafcillin [16]	147-52-4	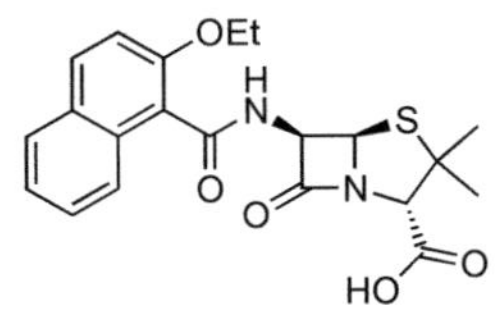	(2*S*,5*R*,6*R*)-6-(2-Ethoxy-1-naphthamido)-3,3-dimethyl-7-oxo-4-thia-1-azabicyclo[3.2.0] heptane-2-carboxylic acid	Against *Staphylococcal* species

TABLE 1.1 *(Continued)*

Name	CAS No.	Structure	Full Name	Major Uses
Oxacillin [17]	66-79-5	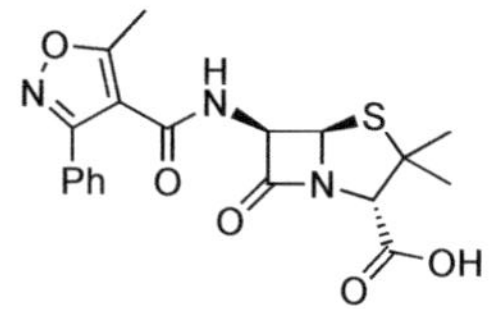	(2*S*,5*R*,6*R*)-3,3-Dimethyl-6-(5-methyl-3-phenylisoxazole-4-carboxamido)-7-oxo-4-thia-1-azabicyclo[3.2.0]heptane-2-carboxylic acid	Used to treat many different types of infections caused by *Staphylococcus*
3rd Generation				
Amoxicillin [8]	26787-78-0	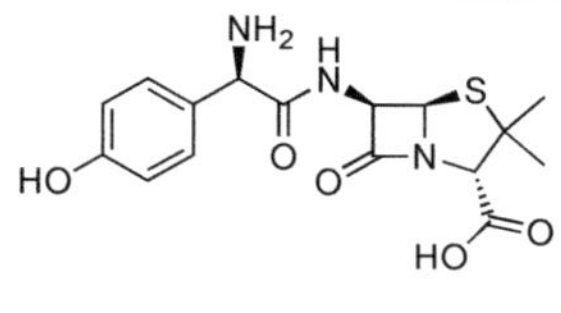	(2*S*,5*R*,6*R*)-6-((*R*)-2-Amino-2-(4-hydroxyphenyl)acetamido)-3,3-dimethyl-7-oxo-4-thia-1-azabicyclo[3.2.0]heptane-2-carboxylic acid	Effective against *Streptococcus*, *Bacillus subtilis*, *Enterococcus*, *Haemophilus*, *Helicobacter*, and *Moraxella*
Ampicillin [8]	69-53-4	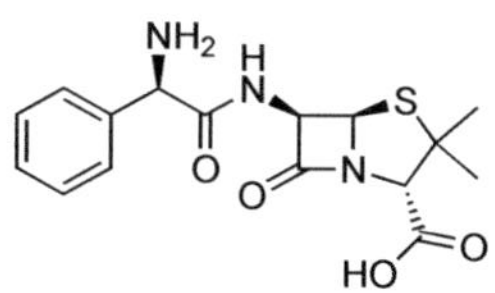	(2*S*,5*R*,6*R*)-6-((*R*)-2-Amino-2-phenylacetamido)-3,3-dimethyl-7-oxo-4-thia-1-azabicyclo[3.2.0]heptane-2-carboxylic acid	Effective against many gram-negative species
Bacampicillin [18]	37661-08-8	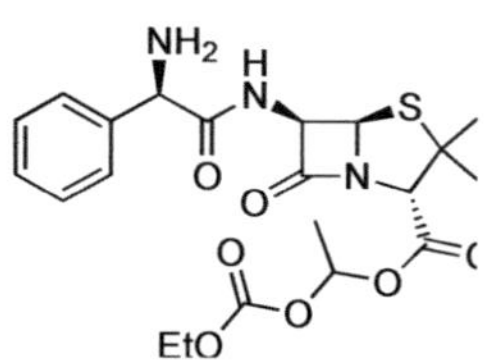	1-((Ethoxycarbonyl)oxy)ethyl (2*S*,5*R*,6*R*)-6-((*R*)-2-amino-2-phenylacetamido)-3,3-dimethyl-7-oxo-4-thia-1-azabicyclo[3.2.0]heptane-2-carboxylate	Against Gram-positive and gram-negative bacteria

Name	CAS No.	Structure	Full Name	Major Uses
Cyclacillin [19] ciclacillin	3485-14-1	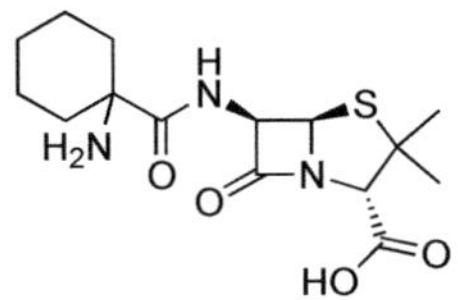	(2*S*,5*R*,6*R*)-6-(1-Aminocyclohexane-1-carboxamido)-3,3-dimethyl-7-oxo-4-thia-1-azabicyclo[3.2.0] heptane-2-carboxylic acid	Similar to ampicillin
Epicillin [20]	26774-90-3	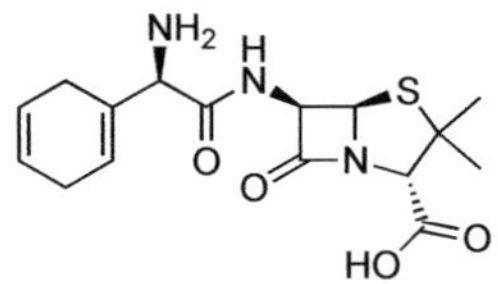	(2*S*,5*R*,6*R*)-6-((*R*)-2-Amino-2-(cyclohexa-1,4-dien-1-yl) acetamido)-3,3-dimethyl-7-oxo-4-thia-1-azabicyclo[3.2.0]heptane-2-carboxylic acid	Similar to ampicillin
Hetacillin [20]	3511-16-8	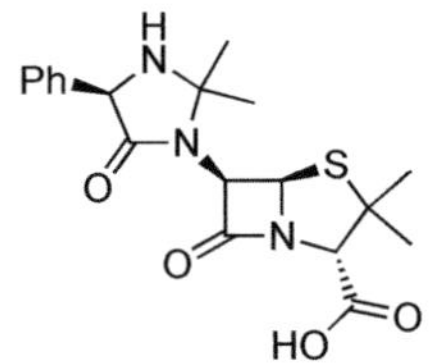	(2*S*,5*R*,6*R*)-6-((*R*)-2,2-Dimethyl-5-oxo-4-phenylimidazolidin-1-yl)-3,3-dimethyl-7-oxo-4-thia-1-azabicyclo[3.2.0] heptane-2-carboxylic acid	Similar to ampicillin
Metampicillin [21]	6489-97-0	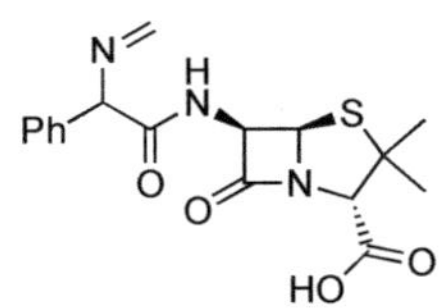	(2*S*,5*R*,6*R*)-3,3-Dimethyl-6-(2-(methyleneamino)-2-phenylacetamido)-7-oxo-4-thia-1-azabicyclo[3.2.0] heptane-2-carboxylic acid	Similar to ampicillin
Pivampicillin [22]	33817-20-8	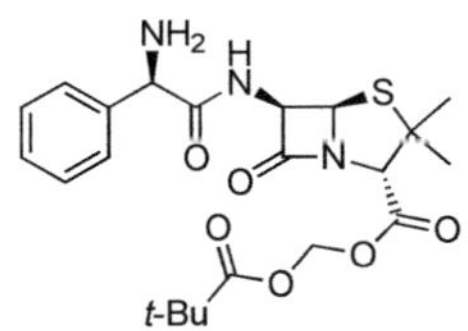	(Pivaloyloxy)methyl (2*S*,5*R*,6*R*)-6-((*R*)-2-Amino-2-phenylacetamido)-3,3-dimethyl-7-oxo-4-thia-1-azabicyclo[3.2.0] heptane-2-carboxylate	Similar to ampicillin

TABLE 1.1 *(Continued)*

Name	CAS No.	Structure	Full Name	Major Uses
Talampicillin [23]	47747-56-8	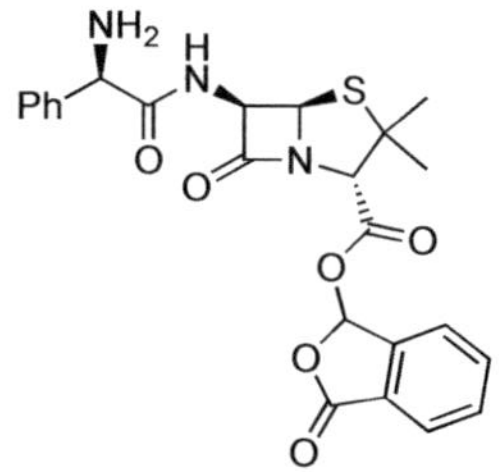	3-Oxo-1,3-dihydroisobenzofuran-1-yl (2*S*,5*R*,6*R*)-6-((*R*)-2-Amino-2-phenylacetamido)-3,3-dimethyl-7-oxo-4-thia-1-azabicyclo[3.2.0]heptane-2-carboxylate	Broad-spectrum antibacterial activity against Gram-positive and gram-negative bacteria
4th Generation				
Azlocillin [24]	37091-66-0	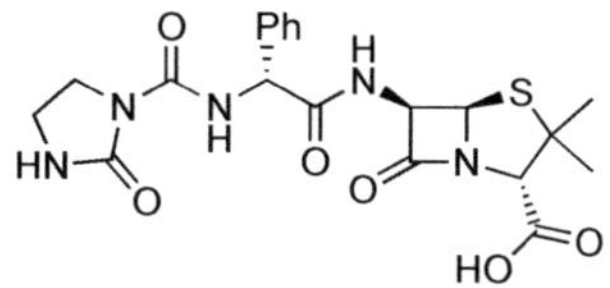	(2*S*,5*R*,6*R*)-3,3-Dimethyl-7-oxo-6-((*R*)-2-(2-oxoimidazolidine-1-carboxamido)-2-phenylacetamido)-4-thia-1-azabicyclo[3.2.0]heptane-2-carboxylic acid	Effective against a broad spectrum of bacteria, including *Pseudomonas aeruginosa* and *enterococci*
Mezlocillin [24]	51481-65-3	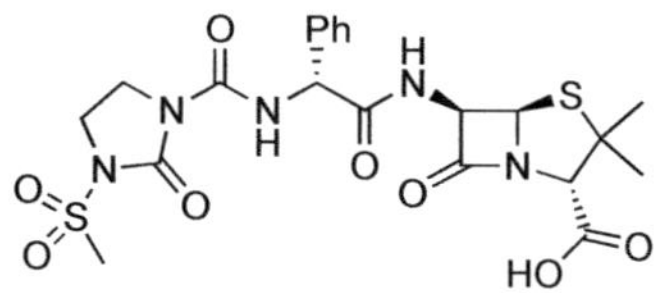	(2*S*,5*R*,6*R*)-3,3-Dimethyl-6-((*R*)-2-(3-(methylsulfonyl)-2-oxoimidazolidine-1-carboxamido)-2-phenylacetamido)-7-oxo-4-thia-1-azabicyclo[3.2.0]heptane-2-carboxylic acid	Effective against a broad spectrum of bacteria, including *Pseudomonas aeruginosa* and *enterococci*
Piperacillin [8]	59703-84-3	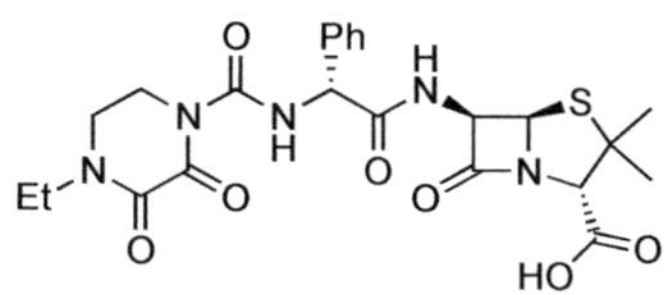	(2*S*,5*R*,6*R*)-6-((*R*)-2-(4-Ethyl-2,3-dioxopiperazine-1-carboxamido)-2-phenylacetamido)-3,3-dimethyl-7-oxo-4-thia-1-azabicyclo[3.2.0]heptane-2-carboxylic acid	Often used in combination with tazobactam

Name	CAS No.	Structure	Full Name	Major Uses
Carbenicillin [25]	4697-36-3	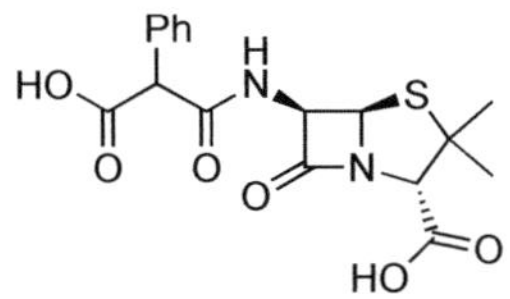	(2*S*,5*R*,6*R*)-6-(2-Carboxy-2-phenylacetamido)-3,3-dimethyl-7-oxo-4-thia-1-azabicyclo[3.2.0]heptane-2-carboxylic acid	Competitive reversible inhibitors of enkephalinase
Carfecillin [25]	27025-49-6	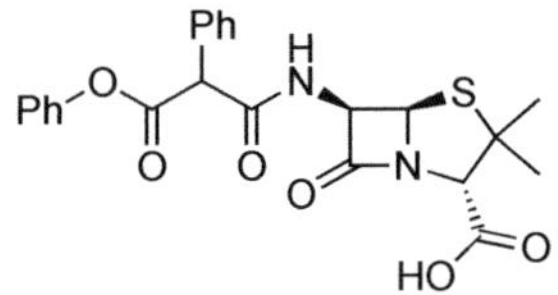	(2*S*,5*R*,6*R*)-3,3-Dimethyl-7-oxo-6-(3-oxo-3-phenoxy-2-phenylpropanamido)-4-thia-1-azabicyclo[3.2.0]heptane-2-carboxylic acid	Competitive reversible inhibitors of enkephalinase
Carindacillin [26]	26605-69-6	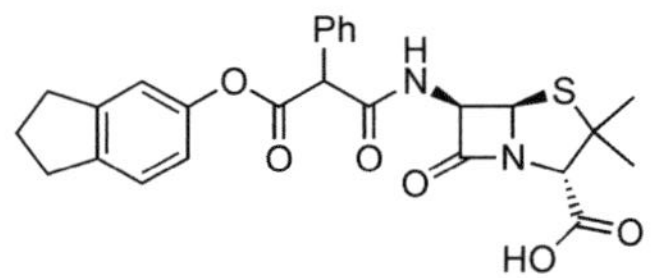	(2*S*,5*R*,6*R*)-6-(3-((2,3-Dihydro-1H-inden-5-yl)oxy)-3-oxo-2-phenylpropanamido)-3,3-dimethyl-7-oxo-4-thia-1-azabicyclo[3.2.0]heptane-2-carboxylic acid	Treatment of acute and chronic infections of the upper and lower urinary tract and in asymptomatic bacteriuria
Temocillin [10]	66148-78-5	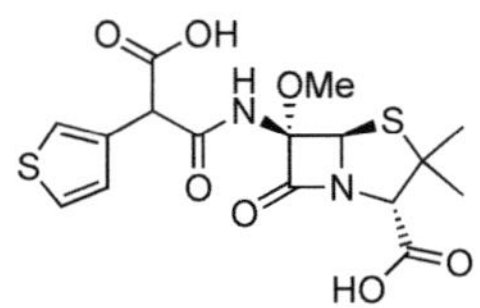	(2*S*,5*R*,6*S*)-6-(2-Carboxy-2-(thiophen-3-yl)acetamido)-6-methoxy-3,3-dimethyl-7-oxo-4-thia-1-azabicyclo[3.2.0]heptane-2-carboxylic acid	Only active against *Enterobacteriaceae*
Ticarcillin [8]	34787-01-4	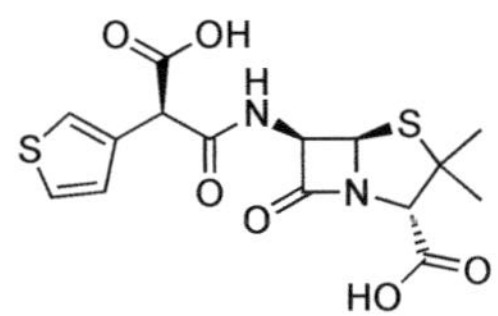	(2*S*,5*R*,6*R*)-6-((*R*)-2-Carboxy-2-(thiophen-3-yl)acetamido)-3,3-dimethyl-7-oxo-4-thia-1-azabicyclo[3.2.0]heptane-2-carboxylic acid	An IV antibiotic for the treatment of gram-negative bacteria, particularly *Pseudomonas aeruginosa*

TABLE 1.1 *(Continued)*

Name	CAS No.	Structure	Full Name	Major Uses
Others				
Amdinocillin [27] Mecillinam	32887-01-7	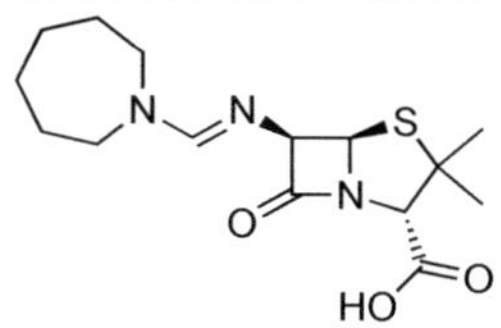	(2*S*,5*R*,6*R*)-6-(((*E*)-Azepan-1-ylmethylene)amino)-3,3-dimethyl-7-oxo-4-thia-1-azabicyclo[3.2.0] heptane-2-carboxylic acid	mainly for the treatment of uncomplicated urinary tract infections
Pivmecillinam [28]	32886-97-8	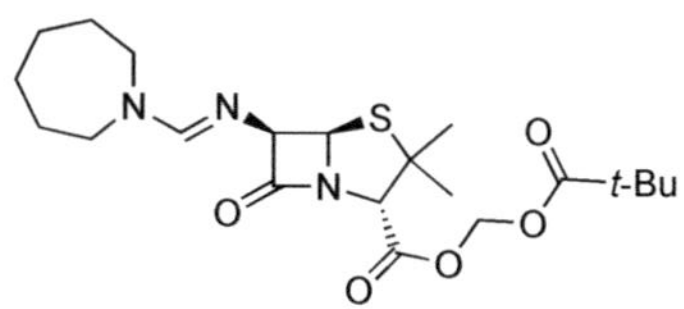	(Pivaloyloxy)methyl (2*S*,5*R*,6*R*)-6-(((*E*)-Azepan-1-ylmethylene) amino)-3,3-dimethyl-7-oxo-4-thia-1-azabicyclo[3.2.0] heptane-2-carboxylate	Against gram-negative organisms such as *Escherichia coli* and other *Enterobacteriaceae*
Sulbenicillin [29]	41744-40-5	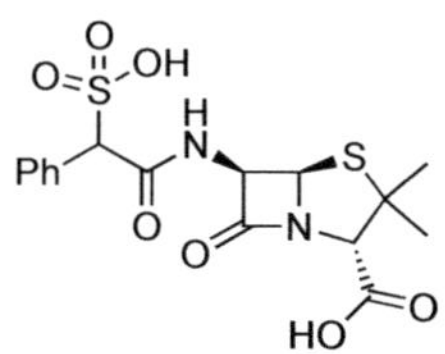	(2*S*,5*R*,6*R*)-3,3-Dimethyl-7-oxo-6-(2-phenyl-2-sulfoacetamido)-4-thia-1-azabicyclo[3.2.0] heptane-2-carboxylic acid	Against *Pseudomonas aeruginosa*
FR-900318 [30]	83670-99-9	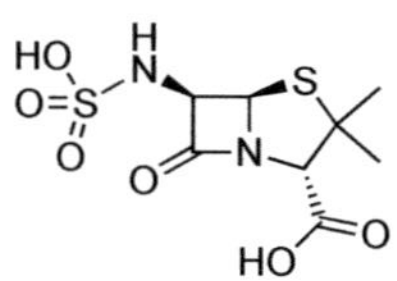	(2*S*,5*R*,6*R*)-3,3-Dimethyl-7-oxo-6-(sulfoamino)-4-thia-1-azabicyclo[3.2.0] heptane-2-carboxylic acid	Against β-lactamases from *Citrobacter freundii*, *Escherichia coli*, *Serratia marcescens*, and common penicillinase

On the other hand, most penicillins are the semi-synthetic derivatives of 6-aminopenicillic acid, which are classified into six groups, i.e., Group 1 (Penicillin G and its long-acting parenteral forms), Group 2 (orally absorbed penicillins, e.g., Penicillin V), Group 3 (antistaphylococcal penicillins, e.g., meticillin, flucloxacillin), Group 4 (extended-spectrum penicillins, e.g., amoxicillin), Group 5 (antipseudomonal penicillins, e.g., ticarcillin, piperacillin) and Group 6 (β-lactamase-resistant penicillins) [3].

1.3 CEPHALOSPORIN ANTIBIOTICS

Cephalosporins are another major class of bactericidal β-lactam antibiotics and are chemically characterized by a β-lactam fused to a dihydrothiazine ring [31]. They are derived from the mold *Acremonium* (previously known as *Cephalosporium*), including cephalosporin N, cephalosporin C and cephalosporin P. Many cephalosporins are semi-synthetic derivatives of Cephalosporin C, the parent antibiotic of this class. This initial substance was isolated in 1945 from a *Cephalosporium acremonium* strain by Guiseppe Brotzu at the Instituto d'Igiene in Cagliari Sardinia and published in 1948 [32, 33], of which the structure was elucidated by Abraham and Newton in 1961 [34]. Although Cephalosporin C itself has weak antibacterial activity, it is more stable toward acid than penicillin; in addition, it is not likely affected by penicillinases, and is not cross-allergenic with penicillins [35]. Therefore, cephalosporins are the largest class of β-lactam antibiotics, for the inhibition of enzymes in the cell wall of susceptible bacteria as well as disrupting the synthesis of bacteria cells. So far, a total of five generations of cephalosporins have been developed, as listed in Table 1.2. The classification of these five generations of cephalosporins is based on the types of bacteria. For example, the "first-generation cephalosporins are potentially useful for the prevention and treatment of mild-to-moderate infections caused by Gram-positive organisms, especially *Staphylococcus aureus* and *Streptococcus pyogenes*" [36]. In addition, the first-generation cephalosporins are also effective in the treatment of infections caused by some Gram-negative bacteria and are generally prescribed for the treatment of ear infections, pneumonia, skin, and soft tissue infections, strep throat, urinary tract infections (UTI) and cystitis (bladder infection). While cephalosporins of the second generation are effective against both Gram-positive and Gram-negative bacteria, they are a little less effective against Gram-positive bacteria compared to the first-generation cephalosporins. Cephalosporins of the second generation are often used to treat respiratory infections, such as bronchitis or pneumonia. In addition, these cephalosporins are used to treat ear infections, gonorrhea, *Haemophilus influenzae* infection, meningitis, sepsis, sinus infections and UTI. The cephalosporins of the third generation are more effective against Gram-negative bacteria than the first two generations of cephalosporins and can attain high concentrations in the cerebrospinal fluid whereas the cephalosporins of the first two generations do not readily cross the blood-brain barrier. Cephalosporins of the third generation are usually reserved for more severe infections, such as gonorrhea, Lyme disease, meningitis, pneumonia, sepsis, skin and soft tissue infections, and UTIs. Compared to the first three generations of cephalosporins, the cephalosporins in the fourth and fifth generations are generally reserved for more severe infections or for patients with weakened immune systems. Particularly, the fifth generation cephalosporins are used for multi-drug resistant (MDR) *Staphylococcus aureus* (MRSA), for which penicillin antibiotics are usually ineffective.

1.4 CARBAPENEM ANTIBIOTICS

Compared to penicillin and cephalosporin antibiotics, carbapenems were discovered much later by two groups from Beecham Research Laboratories and Merck Sharp and Dohme Research Laboratories. The initial study from the former group identified three compounds from a culture of *Streptomyces olivaceus* [80–82], which were named as MM 4550, MM 13902, and MM 17880, respectively; whereas the research group at the Merck Sharp and Dohme Research Laboratories isolated thienamycin from *Streptomyces cattleya* [83, 84]. The unique nuclear structure within carbapenem antibiotics differs from the penam nucleus of the penicillins for having a carbon atom to replace the sulfur at position 1 and an unsaturated C=C bond between carbon atoms 2 and 3 in the 5-membered ring, i.e., with a core of *(R)*-1-azabicyclo[3.2.0]hept-2-en-7-one. Coincidently, the isolation of olivanic acid from *Streptomyces olivaceus* also led to the discovery of clavulanic acid [81].

Carbapenems are highly effective antibiotic agents commonly used for the treatment of severe or high-risk bacterial infections. As a result, they are usually reserved for known or suspected multi-drug resistant (MDR) bacterial infections. Similar to penicillins and cephalosporins, carbapenems kill bacteria by binding to penicillin-binding proteins and inhibit the synthesis of the bacterial cell wall, whilst with a broader spectrum of activity with respect to most cephalosporins and penicillins. Furthermore, carbapenems are typically unaffected by emerging antibiotic resistance, even to other β-lactams. However, the application of carbapenems faces the challenge arising from the occurrence of infections caused by carbapenem-resistant bacteria (such as *Klebsiella pneumoniae* and other carbapenem-resistant *Enterobacteriaceae*) [82].

TABLE 1.2 The List of Cephalosporins

Name	CAS No.	Structure	Full Name	Major Uses
1st Generation				
Cefacetrile [37]	10206-21-0	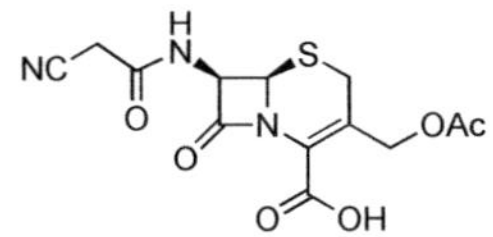	(6*R*,7*R*)-3-(Acetoxymethyl)-7-(2-cyanoacetamido)-8-oxo-5-thia-1-azabicyclo[4.2.0]oct-2-ene-2-carboxylic acid	Against Gram-positive bacteria, e.g., *Staphylococci*, and *Streptococci*, and Gram-negative *E. coli*, *Klebsiella*, and *Salmonella* spp.
Cefadroxil [8]	66592-87-8	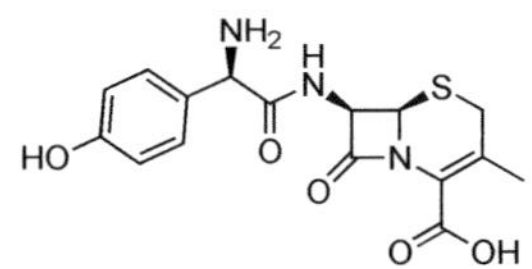	(6*R*,7*R*)-7-((*R*)-2-Amino-2-(4-hydroxyphenyl)acetamido)-3-methyl-8-oxo-5-thia-1-azabicyclo[4.2.0]oct-2-ene-2-carboxylic acid	Highly effective against Gram-positive and Gram-negative bacterial infections (e.g., *E. coli*, *S. aureus*, and *S. pneumoniae*)
Cefalexin [8]	15686-71-2	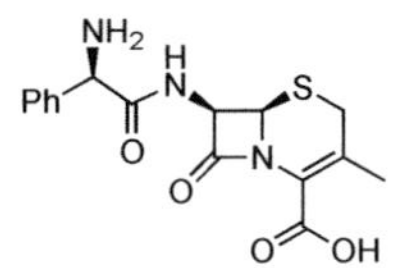	(6*R*,7*R*)-7-((*R*)-2-Amino-2-phenylacetamido)-3-methyl-8-oxo-5-thia-1-azabicyclo[4.2.0]oct-2-ene-2-carboxylic acid	Inhibits bacterial peptidoglycan biosynthesis
Cefaloglycin [38]	3577-01-3	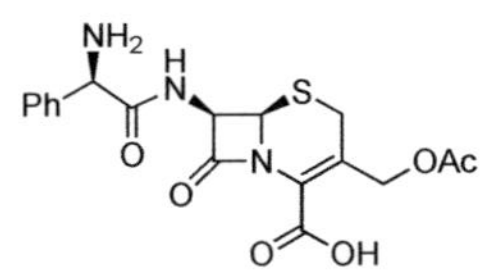	(6*R*,7*R*)-3-(Acetoxymethyl)-7-((*R*)-2-amino-2-phenylacetamido)-8-oxo-5-thia-1-azabicyclo[4.2.0]oct-2-ene-2-carboxylic acid	Binds to and inactivates penicillin-binding proteins located on the inner membrane of the bacterial cell wall, results in the weakening of the bacterial cell wall and causes cell lysis.

TABLE 1.2 *(Continued)*

Name	CAS No.	Structure	Full Name	Major Uses
Cefalonium [39]	5575-21-3	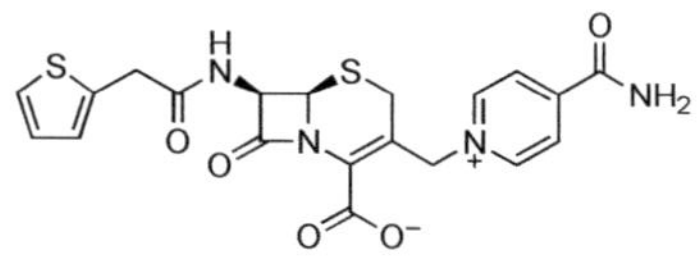	(6*R*,7*R*)-3-((4-Carbamoylpyridin-1-ium-1-yl)methyl)-8-oxo-7-(2-(thiophen-2-yl)acetamido)-5-thia-1-azabicyclo[4.2.0]oct-2-ene-2-carboxylate	Treatment of *S. aureus* mastitis infection
Cefaloridine [40]	50-59-9	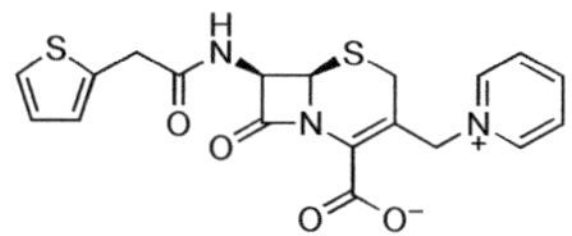	(6*R*,7*R*)-8-Oxo-3-(pyridin-1-ium-1-ylmethyl)-7-(2-(thiophen-2-yl)acetamido)-5-thia-1-azabicyclo[4.2.0]oct-2-ene-2-carboxylate	Treatment of tuberculosis and other mycobacteriosis
Cefalotin [38]	153-61-7	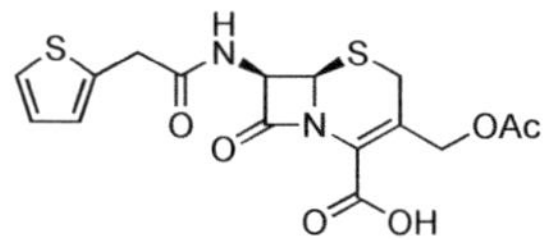	(6*R*,7*R*)-3-(Acetoxymethyl)-8-oxo-7-(2-(thiophen-2-yl)acetamido)-5-thia-1-azabicyclo[4.2.0]oct-2-ene-2-carboxylic acid	Intravenous antimicrobial agent
Cefapirin [41]	21593-23-7	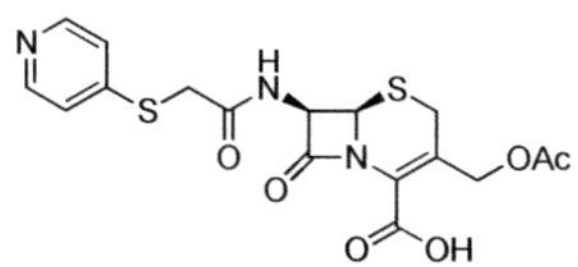	(6*R*,7*R*)-3-(Acetoxymethyl)-8-oxo-7-(2-(pyridin-4-ylthio)acetamido)-5-thia-1-azabicyclo[4.2.0]oct-2-ene-2-carboxylic acid	Veterinary cephalosporins used for recurrent and economically relevant infections
Cefatrizine [42]	51627-14-6	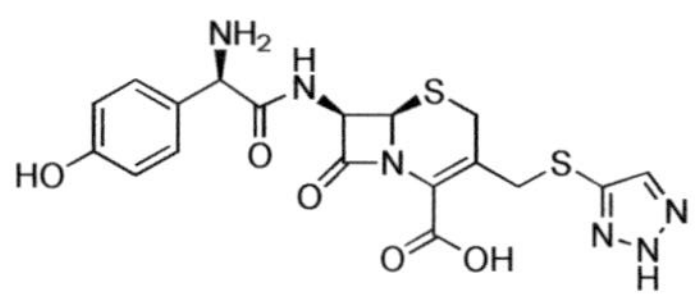	(6*R*,7*R*)-3-(((2*H*-1,2,3-Triazol-4-yl)thio)methyl)-7-((*R*)-2-amino-2-(4-hydroxyphenyl)acetamido)-8-oxo-5-thia-1-azabicyclo[4.2.0]oct-2-ene-2-carboxylic acid	Effective in infections from *S. pneumoniae*, β-hemolytic *Streptococcus*, *S. aureus*, *E. coli*, *P. mirabilis*, *Klebsiella spp.*

Name	CAS No.	Structure	Full Name	Major Uses
Cefazaflur [43]	52123-49-6	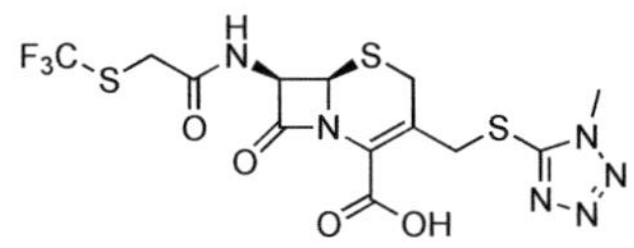	(6*R*,7*R*)-3-(((1-methyl-1*H*-tetrazol-5-yl)thio)methyl)-8-oxo-7-(2-((trifluoromethyl)thio)acetamido)-5-thia-1-azabicyclo[4.2.0]oct-2-ene-2-carboxylic acid	Inhibition of majority of Gram-positive cocci at concentrations <1 μg/mL except for *Enterococci*.
Cefazedone [9]	56187-47-4	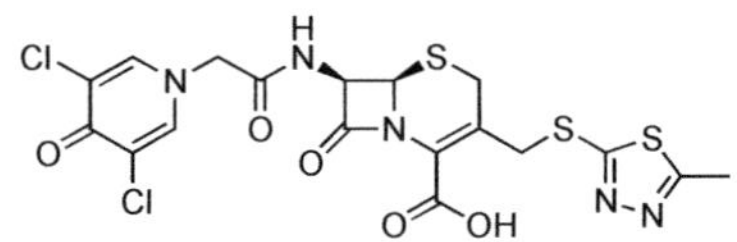	(6*R*,7*R*)-7-(2-(3,5-Dichloro-4-oxopyridin-1(4*H*)-yl)acetamido)-3-(((5-methyl-1,3,4-thiadiazol-2-yl)thio)methyl)-8-oxo-5-thia-1-azabicyclo[4.2.0]oct-2-ene-2-carboxylic acid	Against *streptococci* and *staphylococci* (including penicillin-G-resistant strains)
Cefazolin [8]	25953-19-9	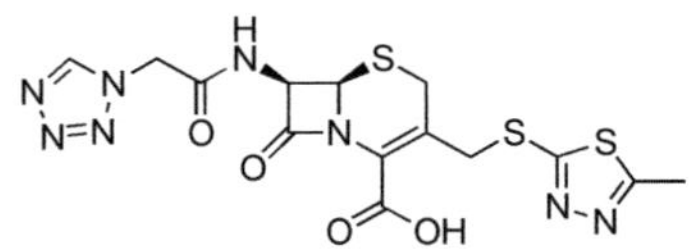	(6*R*,7*R*)-7-(2-(1*H*-tetrazol-1-yl)acetamido)-3-(((5-methyl-1,3,4-thiadiazol-2-yl)thio)methyl)-8-oxo-5-thia-1-azabicyclo[4.2.0]oct-2-ene-2-carboxylic acid	Against *Staphylococcus aureus*
Cefradine [44]	38821-53-3	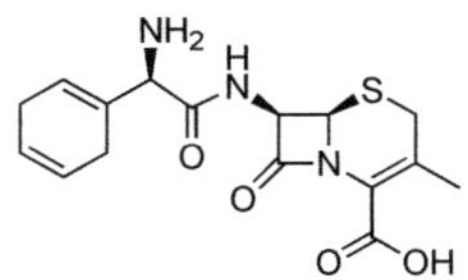	(6*R*,7*R*)-7-((*R*)-2-Amino-2-(cyclohexa-1,4-dien-1-yl)acetamido)-3-methyl-8-oxo-5-thia-1-azabicyclo[4.2.0]oct-2-ene-2-carboxylic acid	Against Gram-positive bacteria but low activity against Gram-negative organisms
Cefroxadine [45]	51762-05-1	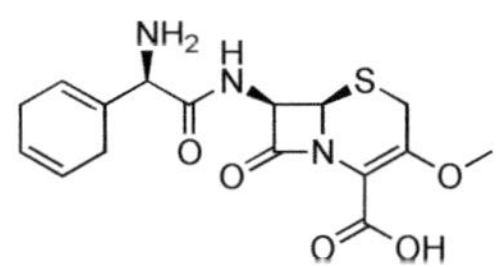	(6*R*,7*R*)-7-((*R*)-2-Amino-2-(cyclohexa-1,4-dien-1-yl)acetamido)-3-methoxy-8-oxo-5-thia-1-azabicyclo[4.2.0]oct-2-ene-2-carboxylic acid	Twice effectiveness against *Escherichia coli* and *Klebsiella pneumoniae* than cefalexin.

TABLE 1.2 *(Continued)*

Name	CAS No.	Structure	Full Name	Major Uses
Ceftezole [46]	26973-24-0	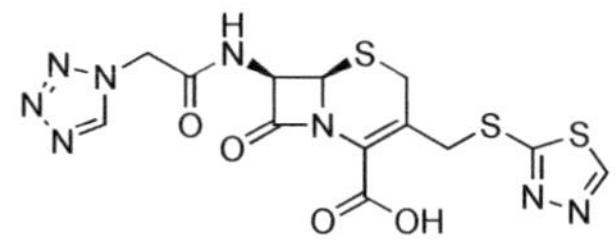	(6*R*,7*R*)-3-(((1,3,4-Thiadiazol-2-yl)thio)methyl)-7-(2-(1*H*-tetrazol-1-yl)acetamido)-8-oxo-5-thia-1-azabicyclo[4.2.0]oct-2-ene-2-carboxylic acid	Against Gram-positive and negative bacteria, but inactive against *Proteus vulgaris IAM-1025* and *Pseudomonas aeruginosa IAM-1095*.
2nd Generation				
Cefaclor [38]	53994-73-3	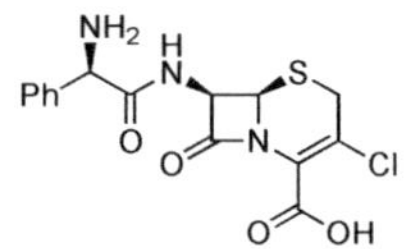	(6*R*,7*R*)-7-((*R*)-2-Amino-2-phenylacetamido)-3-chloro-8-oxo-5-thia-1-azabicyclo[4.2.0]oct-2-ene-2-carboxylic acid	Treat some infections caused by bacteria such as pneumonia
Cefamandole [47]	34444-01-4	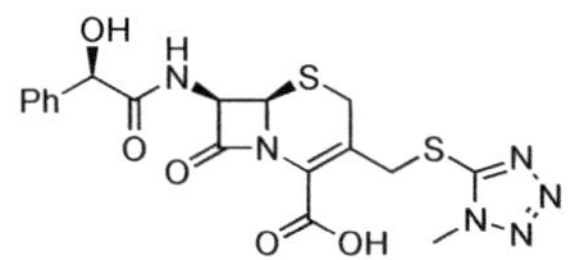	(6*R*,7*R*)-7-((*R*)-2-Hydroxy-2-phenylacetamido)-3-(((1-methyl-1*H*-tetrazol-5-yl)thio)methyl)-8-oxo-5-thia-1-azabicyclo[4.2.0]oct-2-ene-2-carboxylic acid	Against H. influenzae and the *Enterobacteriaceae* resulting from an increased stability to β-lactamases.
Cefbuperazone [48]	76610-84-9	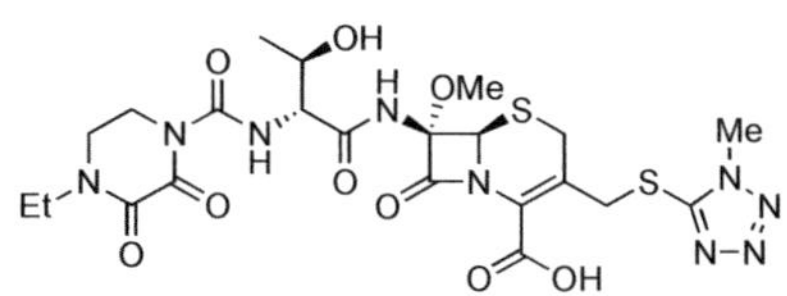	(6*R*,7*S*)-7-((2*R*,3*R*)-2-(4-Ethyl-2,3-dioxopiperazine-1-carboxamido)-3-hydroxybutanamido)-7-methoxy-3-(((1-methyl-1*H*-tetrazol-5-yl)thio)methyl)-8-oxo-5-thia-1-azabicyclo[4.2.0]oct-2-ene-2-carboxylic acid	A useful antibiotic in biliary infection

Name	CAS No.	Structure	Full Name	Major Uses
Cefmetazole [49]	56796-20-4	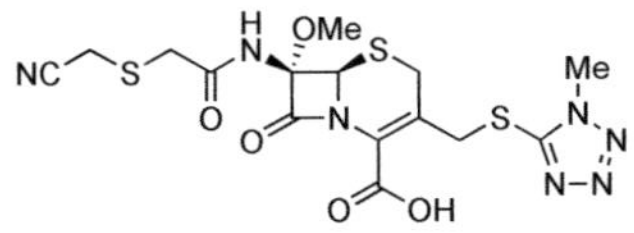	(6*R*,7*S*)-7-(2-((Cyanomethyl)thio)acetamido)-7-methoxy-3-(((1-methyl-1*H*-tetrazol-5-yl)thio)methyl)-8-oxo-5-thia-1-azabicyclo[4.2.0]oct-2-ene-2-carboxylic acid	Active against *B. fragilis*. broader spectrum and stronger activity against *E. coli* and *Klebsiella spp.*
Cefminox [50]	84305-41-9	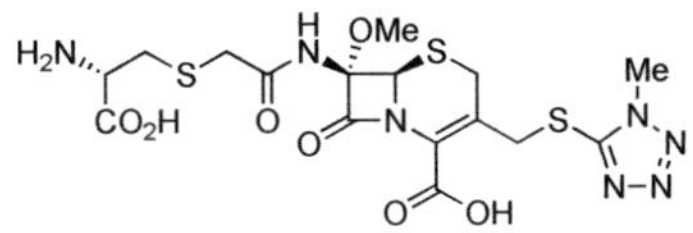	(6*R*,7*S*)-7-(2-(((*R*)-2-Amino-2-carboxyethyl)thio)acetamido)-7-methoxy-3-(((1-methyl-1*H*-tetrazol-5-yl)thio)methyl)-8-oxo-5-thia-1-azabicyclo[4.2.0]oct-2-ene-2-carboxylic acid	A dual agonist of IP and PPARγ that significantly inhibits PASMC proliferation by up-regulation of PTEN and cAMP, with potential for treatment of PAH
Cefonicid [51]	61270-58-4	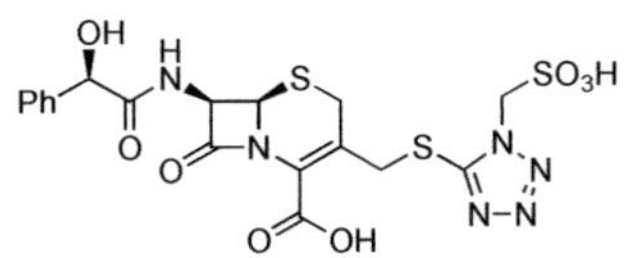	(6*R*,7*R*)-7-((*R*)-2-Hydroxy-2-phenylacetamido)-8-oxo-3-(((1-(sulfomethyl)-1*H*-tetrazol-5-yl)thio)methyl)-5-thia-1-azabicyclo[4.2.0]oct-2-ene-2-carboxylic acid	Inhibition of carnitine/carnitine antiport catalyzed by the reconstituted transporter
Ceforanide [52]	60925-61-3	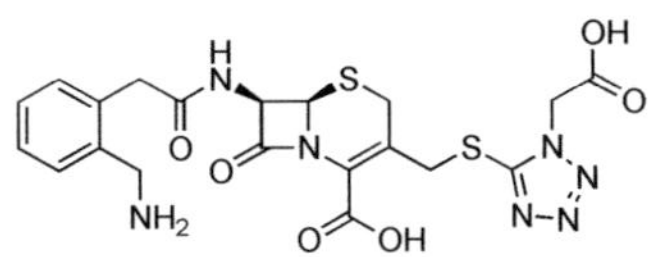	(6*R*,7*R*)-7-(2-(2-(Aminomethyl)phenyl)acetamido)-3-(((1-(carboxymethyl)-1*H*-tetrazol-5-yl)thio)methyl)-8-oxo-5-thia-1-azabicyclo[4.2.0]oct-2-ene-2-carboxylic acid	More active than cefazolin against *Enterobacteriaceae*, *E. coli*, *K. pneumoniae*, *Salmonella*, and *P. mirabilis*
Cefotetan [52]	69712-56-7	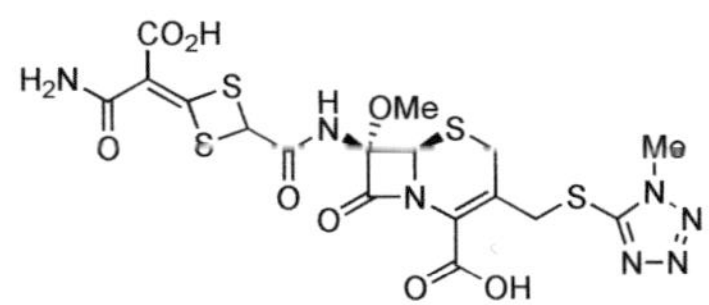	(6*R*,7*S*)-7-(4-(2-Amino-1-carboxy-2-oxoethylidene)-1,3-dithietane-2-carboxamido)-7-methoxy-3-(((1-methyl-1*H*-tetrazol-5-yl)thio)methyl)-8-oxo-5-thia-1-azabicyclo[4.2.0]oct-2-ene-2-carboxylic acid	Half as active as cefoxitin against Gram-positive bacteria but 3 to 100 times more effective against Gram-negative organisms.

TABLE 1.2 *(Continued)*

Name	CAS No.	Structure	Full Name	Major Uses
Cefoxitin [8]	35607-66-0	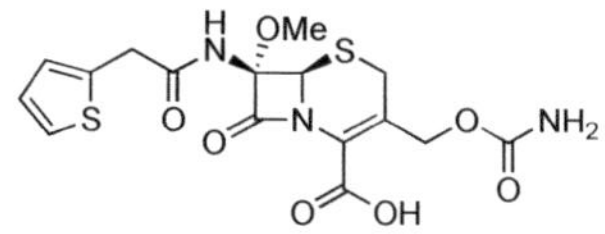	(6*R*,7*S*)-3-((Carbamoyloxy)methyl)-7-methoxy-8-oxo-7-(2-(thiophen-2-yl)acetamido)-5-thia-1-azabicyclo[4.2.0]oct-2-ene-2-carboxylic acid	Against the indole-positive *Proteus* group
Cefprozil [38]	92665-29-7	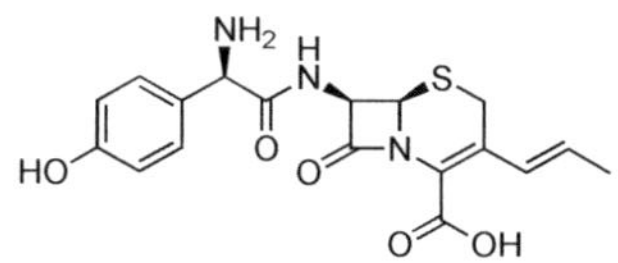	(6*R*,7*R*)-7-((*R*)-2-Amino-2-(4-hydroxyphenyl)acetamido)-8-oxo-3-((*E*)-prop-1-en-1-yl)-5-thia-1-azabicyclo[4.2.0]oct-2-ene-2-carboxylic acid	Used to treat bronchitis, ear infections, skin infections, and other bacterial infections.
Cefuroxime [53]	55268-75-2	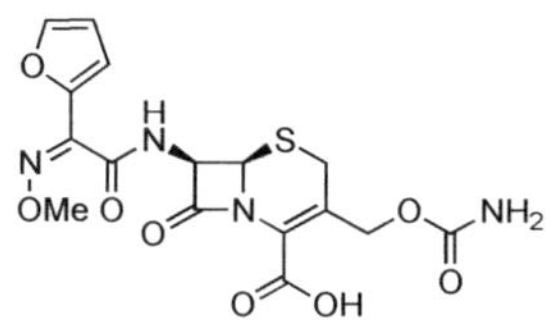	(6*R*,7*R*)-3-((Carbamoyloxy)methyl)-7-((*Z*)-2-(furan-2-yl)-2-(methoxyimino)acetamido)-8-oxo-5-thia-1-azabicyclo[4.2.0]oct-2-ene-2-carboxylic acid	Against *H. influenzae* and the *Enterobacteriaceae* resulting from an increased stability to β-lactamases.
Cefuroxime Axetil [54]	64544-07-6	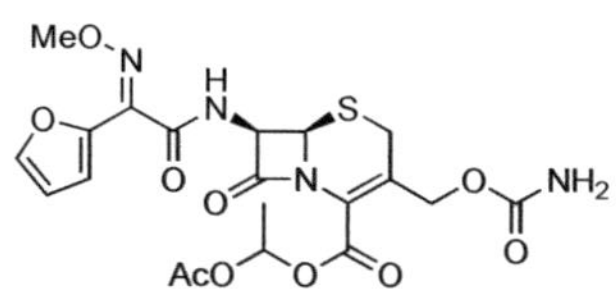	1-Acetoxyethyl (6*R*,7*R*)-3-((Carbamoyloxy)methyl)-7-((*E*)-2-(furan-2-yl)-2-(methoxyimino)acetamido)-8-oxo-5-thia-1-azabicyclo[4.2.0]oct-2-ene-2-carboxylate	Broad-spectrum antibiotic
Cefuzonam [55]	82219-78-1	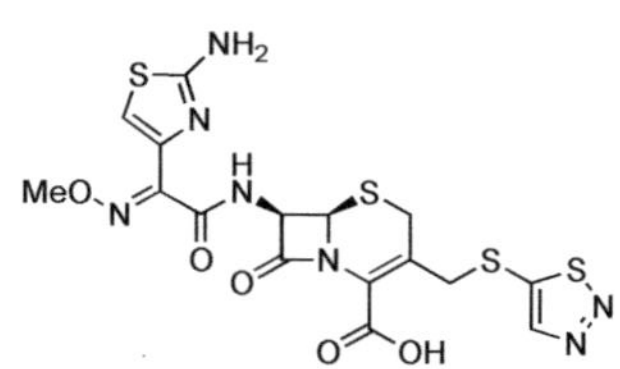	(6*R*,7*R*)-3-(((1,2,3-Thiadiazol-5-yl)thio)methyl)-7-((*E*)-2-(2-aminothiazol-4-yl)-2-(methoxyimino)acetamido)-8-oxo-5-thia-1-azabicyclo[4.2.0]oct-2-ene-2-carboxylic acid	Against Gram-positive bacteria and Gram-negative bacteria except *Pseudomonas aeruginosa*.

Name	CAS No.	Structure	Full Name	Major Uses
3rd Generation				
Cefcapene [56]	135889-00-8	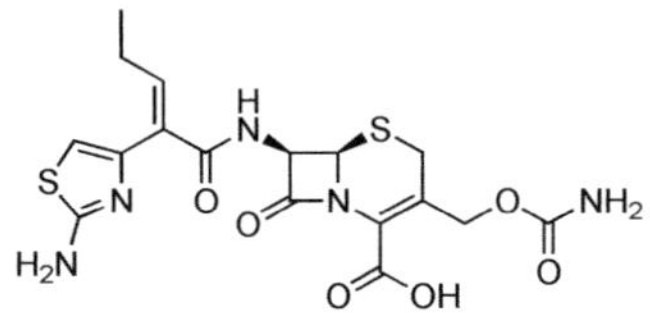	(6*R*,7*R*)-7-((E)-2-(2-Aminothiazol-4-yl)pent-2-enamido)-3-((carbamoyloxy)methyl)-8-oxo-5-thia-1-azabicyclo[4.2.0]oct-2-ene-2-carboxylic acid	Antimicrobial activity against *S. pyogene*
Cefdaloxime [57]	80195-36-4	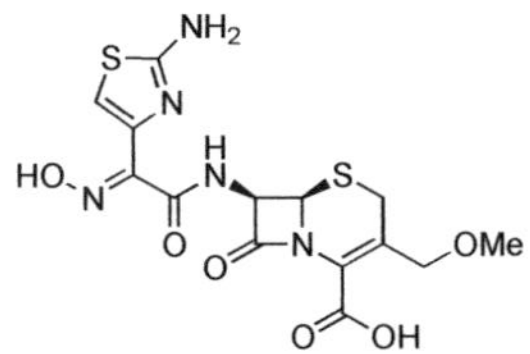	(6*R*,7*R*)-7-((*E*)-2-(2-Aminothiazol-4-yl)-2-(hydroxyimino)acetamido)-3-(methoxymethyl)-8-oxo-5-thia-1-azabicyclo[4.2.0]oct-2-ene-2-carboxylic acid	Against *Enterococcus faecalis*
Cefdinir [38]	91832-40-5	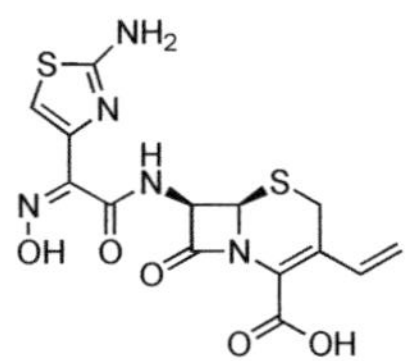	(6*R*,7*R*)-7-((*Z*)-2-(2-Aminothiazol-4-yl)-2-(hydroxyimino)acetamido)-8-oxo-3-vinyl-5-thia-1-azabicyclo[4.2.0]oct-2-ene-2-carboxylic acid	Used for soft tissue infections and respiratory tract infections
Cefditoren [58]	104145-95-1	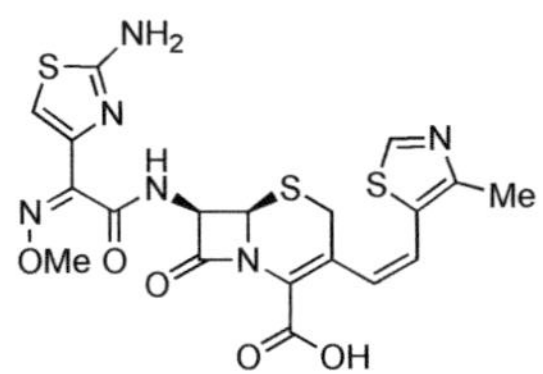	(6*R*,7*R*)-7-((*Z*)-2-(2-Aminothiazol-4-yl)-2-(methoxyimino)acetamido)-3-((Z)-2-(4-methylthiazol-5-yl)vinyl)-8-oxo-5-thia-1-azabicyclo[4.2.0]oct-2-ene-2-carboxylic acid	Broad-spectrum against many Gram-negative and Gram-positive aerobes with high stability by many common β-lactamases

TABLE 1.2 *(Continued)*

Name	CAS No.	Structure	Full Name	Major Uses
Cefetamet [59]	65052-63-3	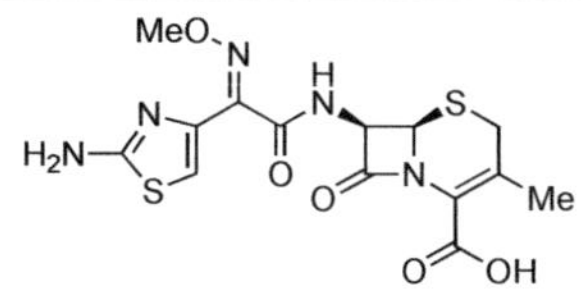	(6*R*,7*R*)-7-((*E*)-2-(2-Aminothiazol-4-yl)-2-(methoxyimino)acetamido)-3-methyl-8-oxo-5-thia-1-azabicyclo[4.2.0]oct-2-ene-2-carboxylic acid	Highly active against enterobacterial strains isolated from community-acquired urinary tract infections
Cefetamet Pivoxil Hydrochloride [38]	111696-23-2	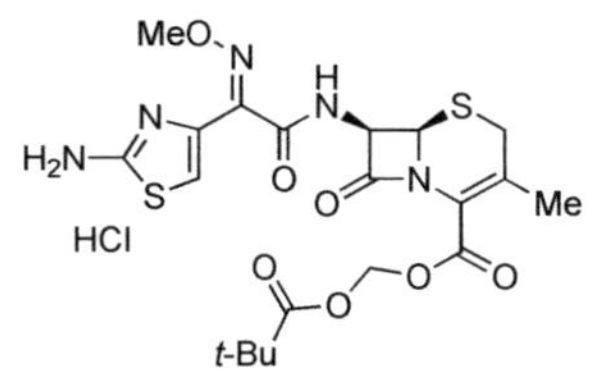	(Pivaloyloxy)methyl (6*R*,7*R*)-7-((*E*)-2-(2-aminothiazol-4-yl)-2-(methoxyimino)acetamido)-3-methyl-8-oxo-5-thia-1-azabicyclo[4.2.0]oct-2-ene-2-carboxylate hydrochloride	Effective in the treatment for sorts of infections caused by sensitive bacteria
Cefixime [8]	79350-37-1	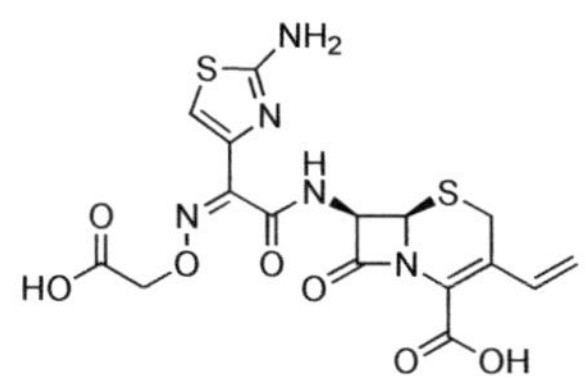	(6*R*,7*R*)-7-((*Z*)-2-(2-Aminothiazol-4-yl)-2-((carboxymethoxy)imino)acetamido)-8-oxo-3-vinyl-5-thia-1-azabicyclo[4.2.0]oct-2-ene-2-carboxylic acid	Used to treat bacterial infections of the ear, urinary tract, and upper respiratory tract by Gram-positive and Gram-negative bacteria
Cefmenoxime [49]	65085-01-0	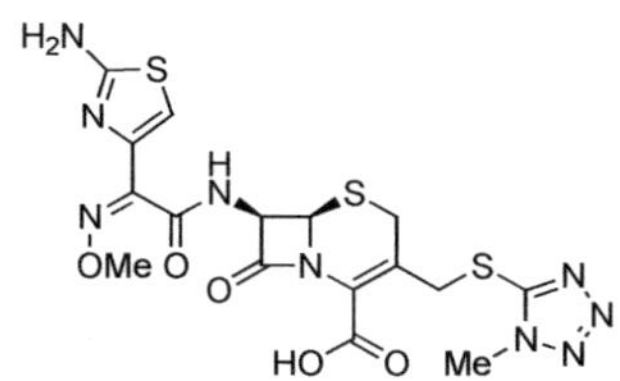	(6*R*,7*R*)-7-((*Z*)-2-(2-Aminothiazol-4-yl)-2-(methoxyimino)acetamido)-3-(((1-methyl-1*H*-tetrazol-5-yl)thio)methyl)-8-oxo-5-thia-1-azabicyclo[4.2.0]oct-2-ene-2-carboxylic acid	Broad antibacterial spectrum, high degree of β-lactamase stability

Name	CAS No.	Structure	Full Name	Major Uses
Cefodizime [44]	69739-16-8	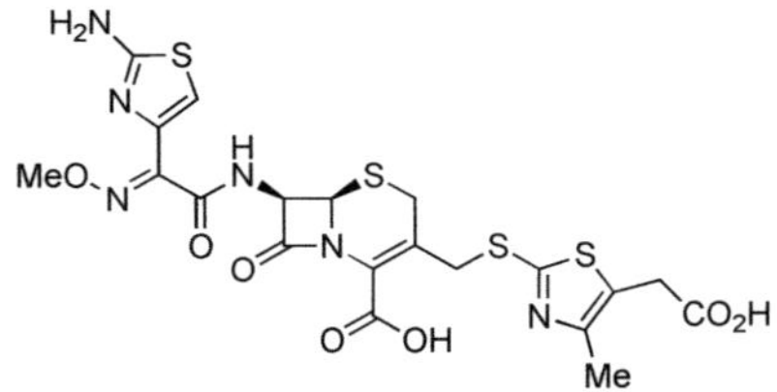	(6*R*,7*R*)-7-((*E*)-2-(2-Aminothiazol-4-yl)-2-(methoxyimino)acetamido)-3-(((5-(carboxymethyl)-4-methylthiazol-2-yl)thio)methyl)-8-oxo-5-thia-1-azabicyclo[4.2.0]oct-2-ene-2-carboxylic acid	Against extended Gram-negative spectrum, but may be less active against some Gram-positive bacteria, e.g., *Staphylococcus aureus*
Cefoperazone [38]	62893-19-0	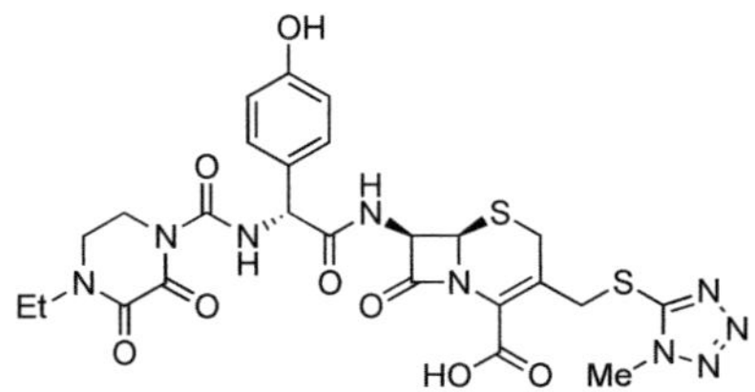	(6*R*,7*R*)-7-((*R*)-2-(4-Ethyl-2,3-dioxopiperazine-1-carboxamido)-2-(4-hydroxyphenyl)acetamido)-3-(((1-methyl-1*H*-tetrazol-5-yl)thio)methyl)-8-oxo-5-thia-1-azabicyclo[4.2.0]oct-2-ene-2-carboxylic acid	Inhibition of rMrp2-mediated [^{3}H]E217βG uptake
Cefotaxime [8]	63527-52-6	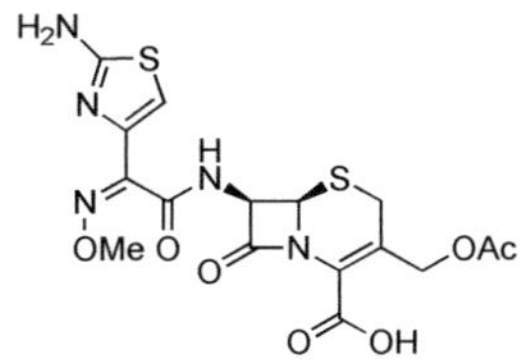	(6*R*,7*R*)-3-(Acetoxymethyl)-7-((*Z*)-2-(2-aminothiazol-4-yl)-2-(methoxyimino)acetamido)-8-oxo-5-thia-1-azabicyclo[4.2.0]oct-2-ene-2-carboxylic acid	Against *Staphylococcus aureus*
Cefotiam [52]	61622-34-2	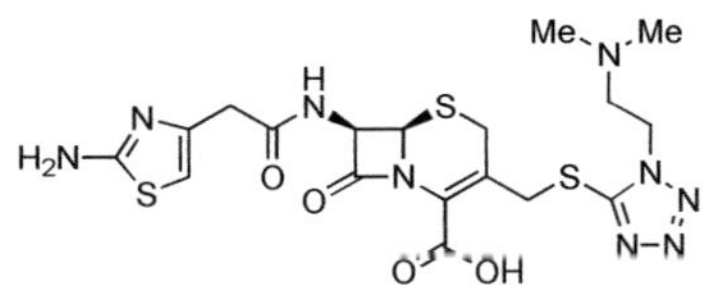	(6*R*,7*R*)-7-(2-(2-Aminothiazol-4-yl)acetamido)-3-(((1-(2-(dimethylamino)ethyl)-1*H*-tetrazol-5-yl)thio)methyl)-8-oxo-5-thia-1-azabicyclo[4.2.0]oct-2-ene-2-carboxylic acid	*In vitro* activity equal to cefazolin or cephalothin against Gram-positive organisms, and 10 times more potent against *H. influenzae* and *Enterobacteriaceae*

TABLE 1.2 *(Continued)*

Name	CAS No.	Structure	Full Name	Major Uses
Cefovecin [60]	234096-34-5	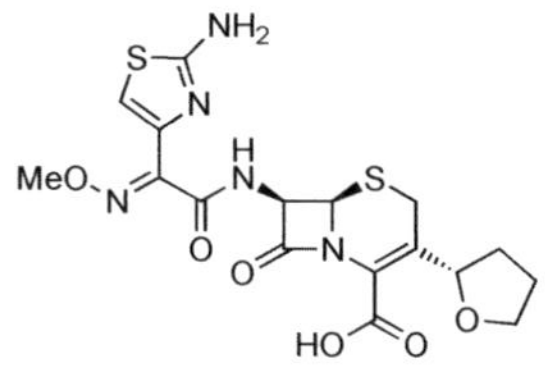	(6*R*,7*R*)-7-((*E*)-2-(2-aminothiazol-4-yl)-2-(methoxyimino)acetamido)-8-oxo-3-((S)-tetrahydrofuran-2-yl)-5-thia-1-azabicyclo[4.2.0]oct-2-ene-2-carboxylic acid	Treatment of skin and urinary infections
Cefpimizole [61]	84880-03-5	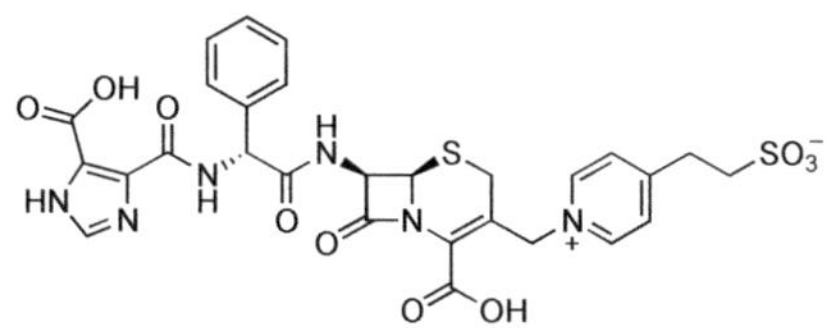	2-(1-(((6*R*,7*R*)-2-Carboxy-7-((*R*)-2-(5-carboxy-1*H*-imidazole-4-carboxamido)-2-phenylacetamido)-8-oxo-5-thia-1-azabicyclo[4.2.0]oct-2-en-3-yl)methyl)pyridin-1-ium-4-yl)ethane-1-sulfonate	Against *Proteus vulgaris* and Enterobacter species
Cefpiramide [38]	70797-11-4	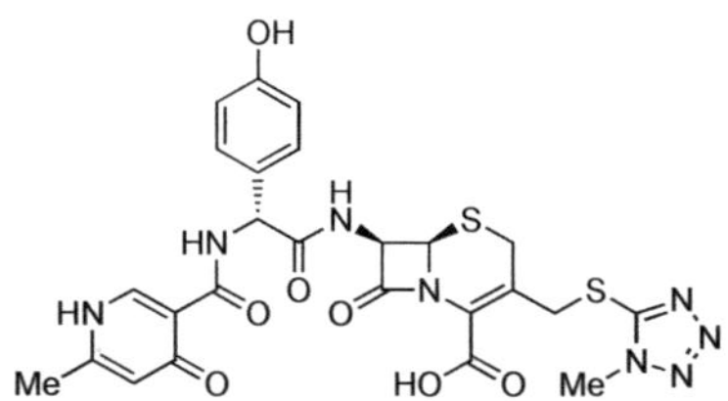	(6*R*,7*R*)-7-((*R*)-2-(4-Hydroxyphenyl)-2-(6-methyl-4-oxo-1,4-dihydropyridine-3-carboxamido)acetamido)-3-(((1-methyl-1*H*-tetrazol-5-yl)thio)methyl)-8-oxo-5-thia-1-azabicyclo[4.2.0]oct-2-ene-2-carboxylic acid	Active against *Acinetobacter spp.* and *Pseudomonas spp*. It inhibits many *Pseudomonas aeruginosa*
Cefpodoxime [38]	80210-62-4	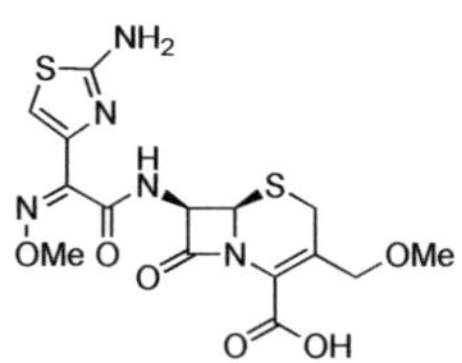	(6*R*,7*R*)-7-((*Z*)-2-(2-Aminothiazol-4-yl)-2-(methoxyimino)acetamido)-3-(methoxymethyl)-8-oxo-5-thia-1-azabicyclo[4.2.0]oct-2-ene-2-carboxylic acid	Active against most Gram-positive and Gram-negative organisms. It can be used to treat acute otitis media, *pharyngitis*, *sinusitis*, and *gonorrhea*.

Name	CAS No.	Structure	Full Name	Major Uses
Cefsulodin [38]	52152-93-9	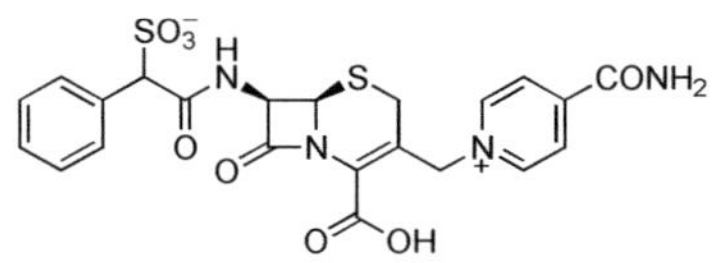	2-(((6*R*,7*R*)-3-((4-Carbamoylpyridin-1-ium-1-yl)methyl)-2-carboxy-8-oxo-5-thia-1-azabicyclo[4.2.0]oct-2-en-7-yl)amino)-2-oxo-1-phenylethane-1-sulfonate	Inhibition of many *Pseudomonas aeruginosa*
Ceftazidime [8]	72558-82-8	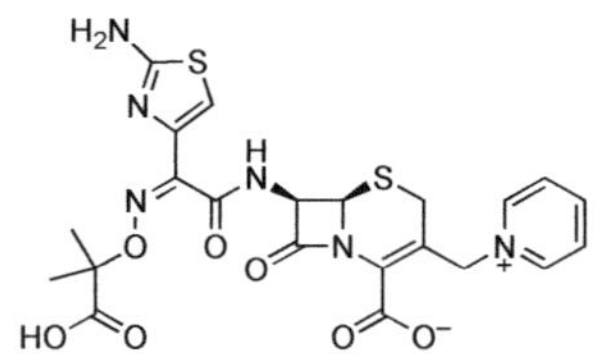	(6*R*,7*R*)-7-((*Z*)-2-(2-Aminothiazol-4-yl)-2-(((2-carboxypropan-2-yl)oxy)imino)acetamido)-8-oxo-3-(pyridin-1-ium-1-ylmethyl)-5-thia-1-azabicyclo[4.2.0]oct-2-ene-2-carboxylate	Against Enterobacteriaceae, *Haemophilus influenzae*, *Neisseria gonorrhoeae*, and Lancefield group A β-hemolytic Streptococci
Cefteram [62]	82547-58-8	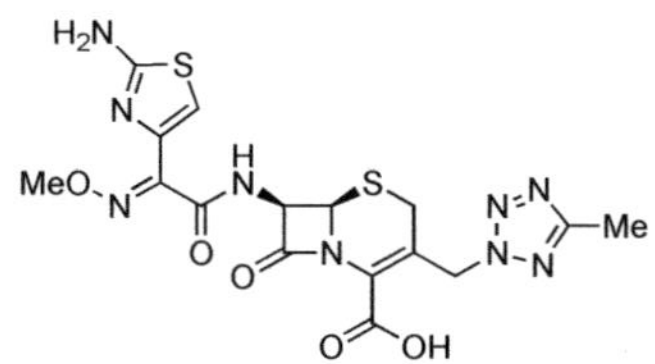	(6*R*,7*R*)-7-((*E*)-2-(2-Aminothiazol-4-yl)-2-(methoxyimino)acetamido)-3-((5-methyl-2*H*-tetrazol-2-yl)methyl)-8-oxo-5-thia-1-azabicyclo[4.2.0]oct-2-ene-2-carboxylic acid	Excellent antibacterial activity against methicillin-susceptible *S. aureus* and *S. pyogenes*
Ceftibuten [38]	97519-39-6	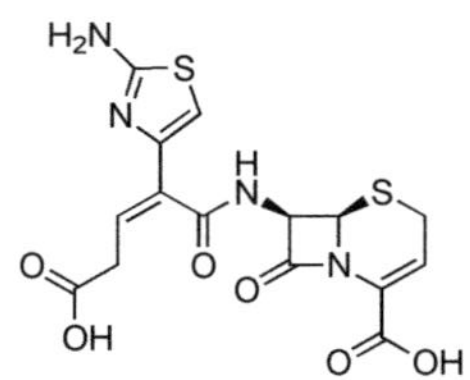	(6*R*,7*R*)-7-((*Z*)-2-(2-Aminothiazol-4-yl)-4-carboxybut-2-enamido)-8-oxo-5-thia-1-azabicyclo[4.2.0]oct-2-ene-2-carboxylic acid	Against *H. influenzae* and *B. catarrhalis*. Used to treat acute bacterial exacerbations of chronic bronchitis (ABECB), acute bacterial otitis media, *pharyngitis*, and *tonsilitis*

TABLE 1.2 *(Continued)*

Name	CAS No.	Structure	Full Name	Major Uses
Ceftiofur [41]	80370-57-6	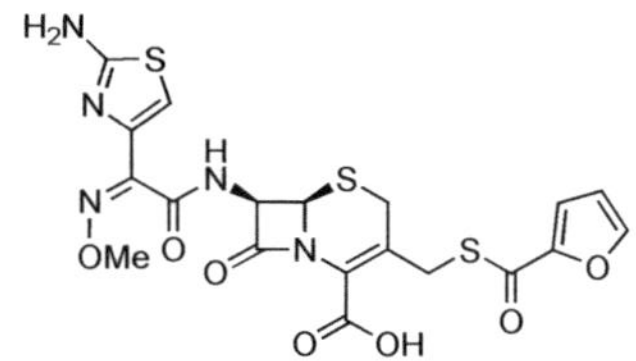	(6*R*,7*R*)-7-((*Z*)-2-(2-Aminothiazol-4-yl)-2-(methoxyimino)acetamido)-3-(((furan-2-carbonyl)thio)methyl)-8-oxo-5-thia-1-azabicyclo[4.2.0]oct-2-ene-2-carboxylic acid	Veterinary cephalosporins used for recurrent and economically relevant infections
Ceftiolene [63]	77360-52-2	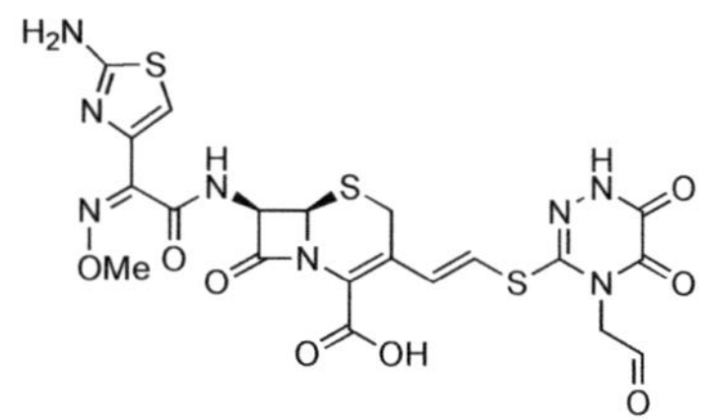	(6*R*,7*R*)-7-((*Z*)-2-(2-Aminothiazol-4-yl)-2-(methoxyimino)acetamido)-3-((*E*)-2-((5,6-dioxo-4-(2-oxoethyl)-1,4,5,6-tetrahydro-1,2,4-triazin-3-yl)thio)vinyl)-8-oxo-5-thia-1-azabicyclo[4.2.0]oct-2-ene-2-carboxylic acid	Broad antibacterial spectrum similar to cefotaxime or ceftriaxone
Ceftizoxime [38]	68401-81-0	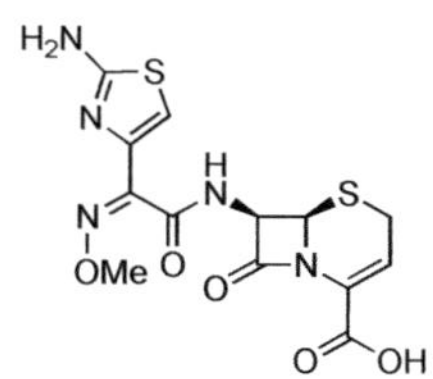	(6*R*,7*R*)-7-((*Z*)-2-(2-Aminothiazol-4-yl)-2-(methoxyimino)acetamido)-8-oxo-5-thia-1-azabicyclo[4.2.0]oct-2-ene-2-carboxylic acid	For the treatment of infections due to susceptible strains of microorganisms
Ceftriaxone [8]	73384-59-5	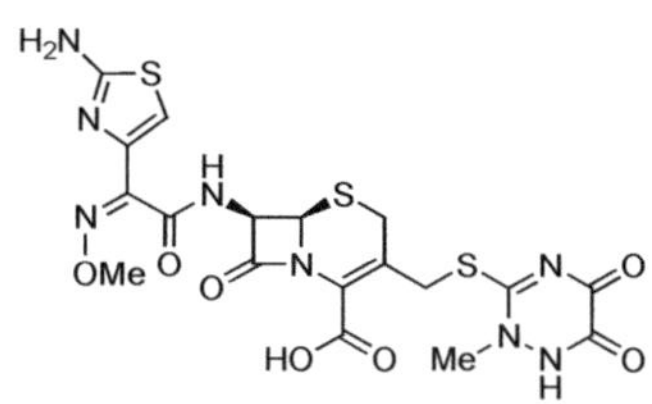	(6*R*,7*R*)-7-((*Z*)-2-(2-Aminothiazol-4-yl)-2-(methoxyimino)acetamido)-3-(((2-methyl-5,6-dioxo-1,2,5,6-tetrahydro-1,2,4-triazin-3-yl)thio)methyl)-8-oxo-5-thia-1-azabicyclo[4.2.0]oct-2-ene-2-carboxylic acid	Against Gram-positive bacteria as well as many Gram-negative bacteria.

Name	CAS No.	Structure	Full Name	Major Uses
4th Generation				
Cefclidine [64]	105239-91-6	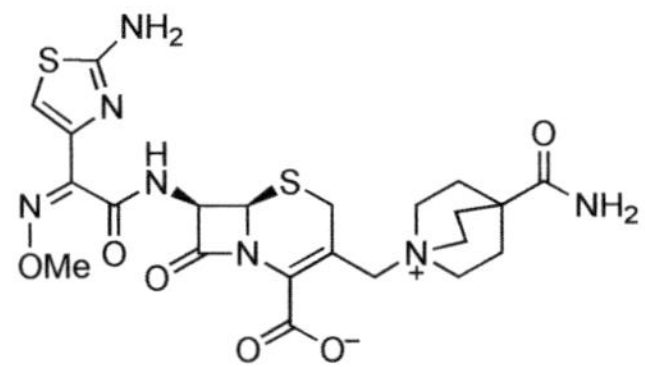	(6*R*,7*R*)-7-((*Z*)-2-(2-Aminothiazol-4-yl)-2-(methoxyimino)acetamido)-3-((4-carbamoylquinuclidin-1-ium-1-yl)methyl)-8-oxo-5-thia-1-azabicyclo[4.2.0]oct-2-ene-2-carboxylate	Highly effective against Gram-negative organisms, e.g., *P. aeruginosa*, *Enterobacteriaceae*, *Haemophilus*, *Proteus*, *Neisseria*
Cefepime [8]	88040-23-7	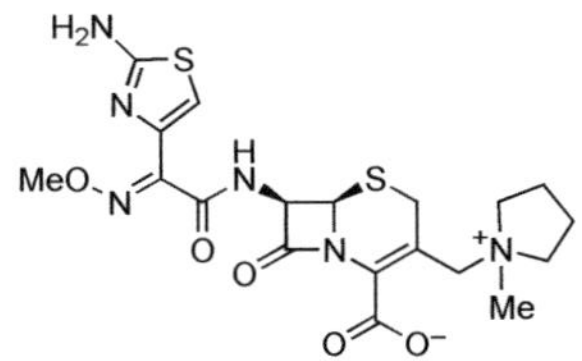	(6*R*,7*R*)-7-((*E*)-2-(2-Aminothiazol-4-yl)-2-(methoxyimino)acetamido)-3-((1-methylpyrrolidin-1-ium-1-yl)methyl)-8-oxo-5-thia-1-azabicyclo[4.2.0]oct-2-ene-2-carboxylate	Highly effective against *Pseudomonas aeruginosa*, *Staphylococcus aureus*, *Streptococcus pneumoniae*, and certain *Enterobacteriaceae*
Cefiderocol [65]	1225208-94-5	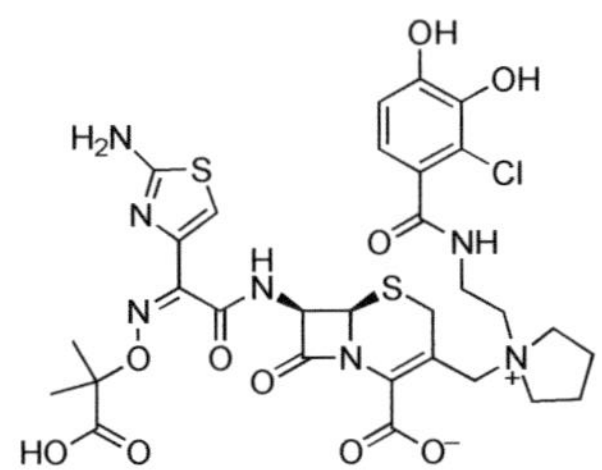	(6*R*,7*R*)-7-((*Z*)-2-(2-Aminothiazol-4-yl)-2-(((2-carboxypropan-2-yl)oxy)imino)acetamido)-3-((1-(2-(2-chloro-3,4-dihydroxybenzamido)ethyl)pyrrolidin-1-ium-1-yl)methyl)-8-oxo-5-thia-1-azabicyclo[4.2.0]oct-2-ene-2-carboxylate	Against overall MDR-GNB

TABLE 1.2 *(Continued)*

Name	CAS No.	Structure	Full Name	Major Uses
Cefluprenam [64]	116853-25-9	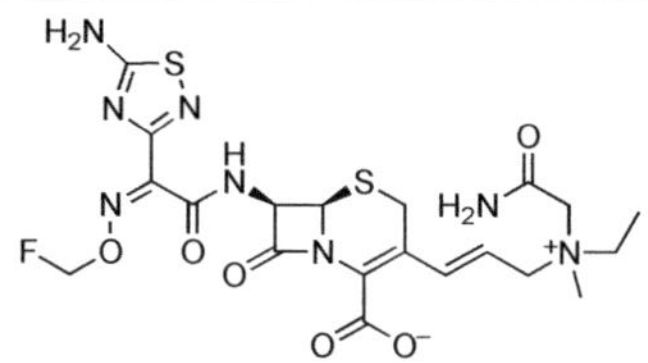	(6*R*,7*R*)-7-((*Z*)-2-(5-Amino-1,2,4-thiadiazol-3-yl)-2-((fluoromethoxy)-imino) acetamido)-3-((*E*)-3-((2-amino-2-oxoethyl)(ethyl)(methyl) ammonio)-prop-1-en-1-yl)-8-oxo-5-thia-1-azabicyclo[4.2.0] oct-2-ene-2-carboxylate	Highly effective against Gram-negative organisms, e.g., *P. aeruginosa*, *Enterobacteriaceae*, *Haemophilus*, *Proteus*, *Neisseria*
Cefoselis [38]	122841-10-5	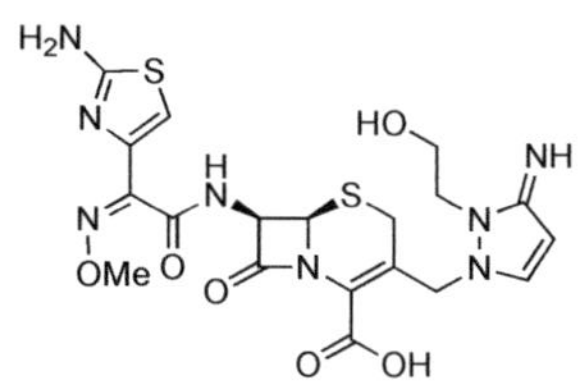	(6*R*,7*R*)-7-((*Z*)-2-(2-Aminothiazol-4-yl)-2-(methoxyimino) acetamido)-3-((2-(2-hydroxyethyl)-3-imino-2,3-dihydro-1*H*-pyrazol-1-yl) methyl)-8-oxo-5-thia-1-azabicyclo[4.2.0] oct-2-ene-2-carboxylic acid	Against *Hemophilus influenzae* and *Moraxella catarrhalis*
Cefozopran [64]	113359-04-9	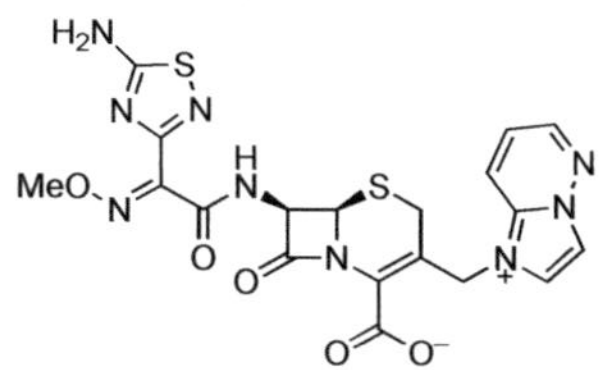	(6*R*,7*R*)-7-((*E*)-2-(5-Amino-1,2,4-thiadiazol-3-yl)-2-(methoxyimino) acetamido)-3-(imidazo[1,2-b] pyridazin-1-ium-1-ylmethyl)-8-oxo-5-thia-1-azabicyclo[4.2.0] oct-2-ene-2-carboxylate	Highly effective against Gram-negative organisms, e.g., *P. aeruginosa*, *Enterobacteriaceae*, *Haemophilus*, *Proteus*, *Neisseria*
Cefpirome [8]	84957-29-9	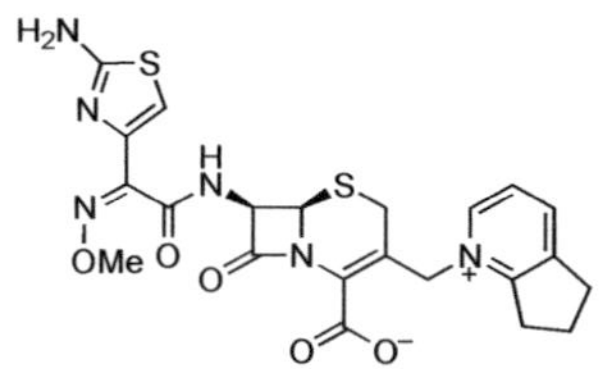	(6*R*,7*R*)-7-((*Z*)-2-(2-Aminothiazol-4-yl)-2-(methoxyimino)acetamido)-3-((6,7-dihydro-5*H*-cyclopenta[b] pyridin-1-ium-1-yl)methyl)-8-oxo-5-thia-1-azabicyclo[4.2.0] oct-2-ene-2-carboxylate	Against extended-spectrum cephalosporin- and Sme-1-containing carbapenem-resistant clinical strains

Name	CAS No.	Structure	Full Name	Major Uses
Cefquinome [64]	84957-30-2	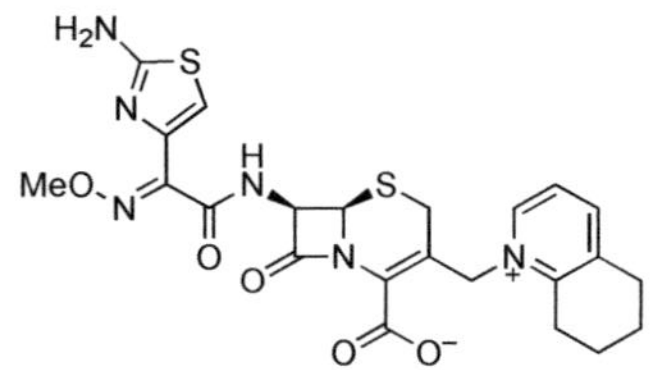	(6*R*,7*R*)-7-((*E*)-2-(2-Aminothiazol-4-yl)-2-(methoxyimino) acetamido)-8-oxo-3-((5,6,7,8-tetrahydroquinolin-1-ium-1-yl) methyl)-5-thia-1-azabicyclo[4.2.0] oct-2-ene-2-carboxylate	Highly effective against Gram-negative organisms, e.g., *P. aeruginosa*, *Enterobacteriaceae*, *Haemophilus*, *Proteus*, *Neisseria*
5th Generation				
Ceftobiprole [8]	209467-52-7	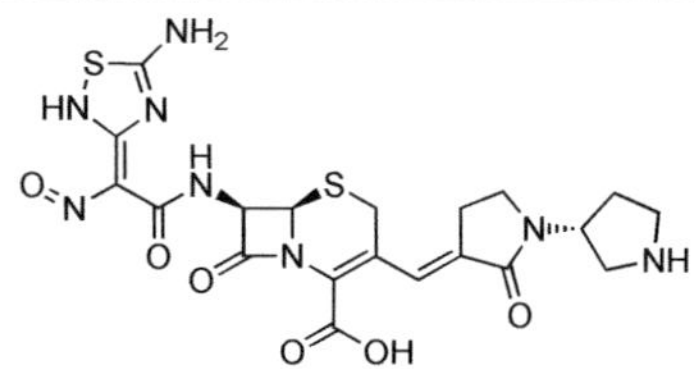	(6*R*,7*R*)-7-((*Z*)-2-(5-Amino-1,2,4-thiadiazol-3(2*H*)-ylidene)-2-nitrosoacetamido)-8-oxo-3-((*E*)-((*R*)-2-oxo-[1,3'-bipyrrolidin]-3-ylidene) methyl)-5-thia-1-azabicyclo[4.2.0] oct-2-ene-2-carboxylic acid	Highly effective against MRSA, penicillin-resistant *Streptococcus pneumoniae*, *Pseudomonas aeruginosa*, and certain *enterococci*
Ceftaroline Fosamil [8]	229016-73-3	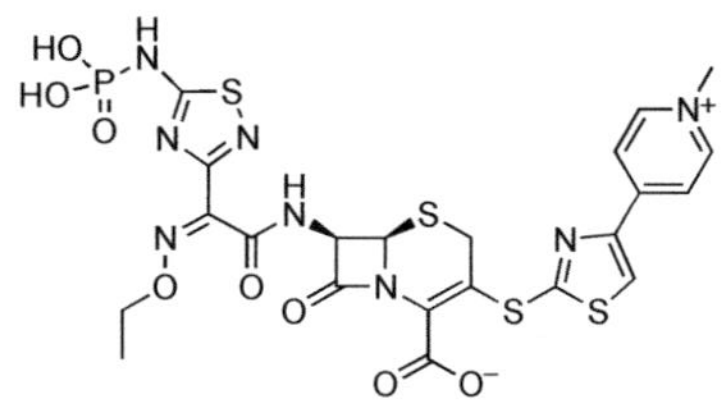	(6*R*,7*R*)-7-((*Z*)-2-(Ethoxyimino)-2-(5-(phosphonoamino)-1,2,4-thiadiazol-3-yl) acetamido)-3-((4-(1-methylpyridin-1-ium-4-yl)thiazol-2-yl)thio)-8-oxo-5-thia-1-azabicyclo[4.2.0] oct-2-ene-2-carboxylate	Against MRSA and broad-spectrum activity against Gram-positive bacteria
Ceftolozane [66]	689293-68-3	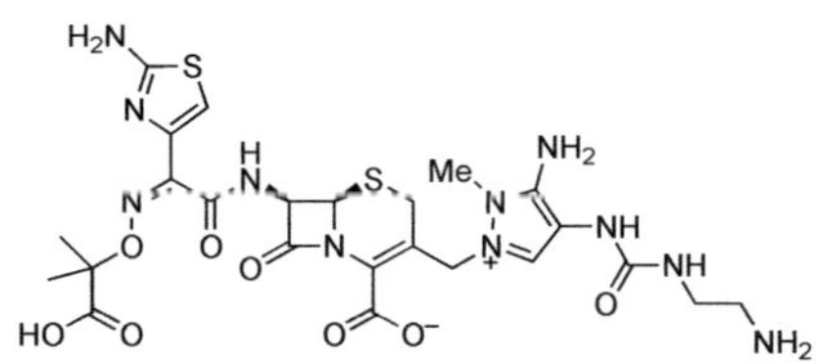	(6*R*,7*R*)-3-((5-Amino-4-(3-(2-aminoethyl)ureido)-1-methyl-1*H*-pyrazol-2-ium-2-yl) methyl)-7-((*Z*)-2-(2-aminothiazol-4-yl)-2-(((2-carboxypropan-2-yl) oxy)imino)acetamido)-8-oxo-5-thia-1-azabicyclo[4.2.0]oct-2-ene-2-carboxylate	Broad-spectrum activity, including against drug- and multidrug-resistant *Pseudomonas aeruginosa*, along with potent antipseudomonal activity

TABLE 1.2 *(Continued)*

Name	CAS No.	Structure	Full Name	Major Uses
Others				
Cefaloram [67]	859-07-4	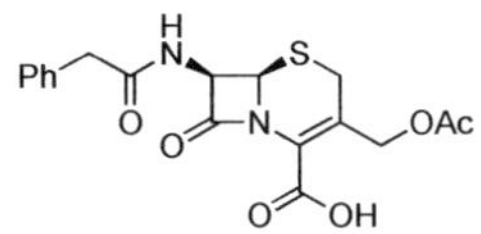	(6*R*,7*R*)-3-(Acetoxymethyl)-8-oxo-7-(2-phenylacetamido)-5-thia-1-azabicyclo[4.2.0]oct-2-ene-2-carboxylic acid	Against *Haemophilus influenzae* and *Streptococcus pneumoniae*
Cefaparole [68]	51627-20-4	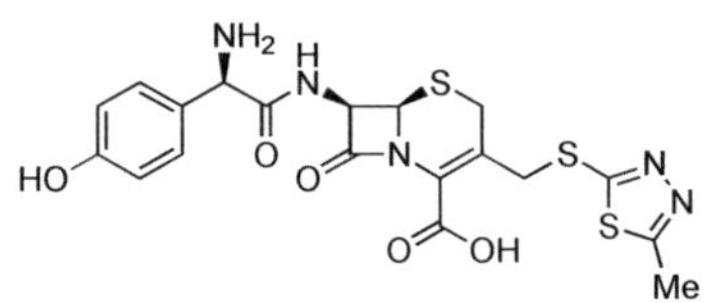	(6*R*,7*R*)-7-((*R*)-2-Amino-2-(4-hydroxyphenyl)acetamido)-3-(((5-methyl-1,3,4-thiadiazol-2-yl)thio)methyl)-8-oxo-5-thia-1-azabicyclo[4.2.0]oct-2-ene-2-carboxylic acid	Antipseudomonal activity
Cefcanel [69]	41952-52-7	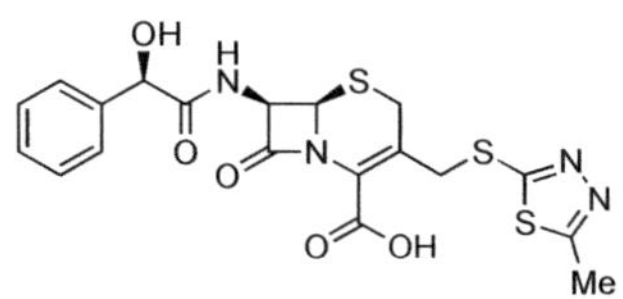	(6*R*,7*R*)-7-((*R*)-2-Hydroxy-2-phenylacetamido)-3-(((5-methyl-1,3,4-thiadiazol-2-yl)thio)methyl)-8-oxo-5-thia-1-azabicyclo[4.2.0]oct-2-ene-2-carboxylic acid	Against *Escherichia coli, Klebsiella aerogenes*, and *Proteus mirabilis*
Cefedrolor [70]	57847-69-5	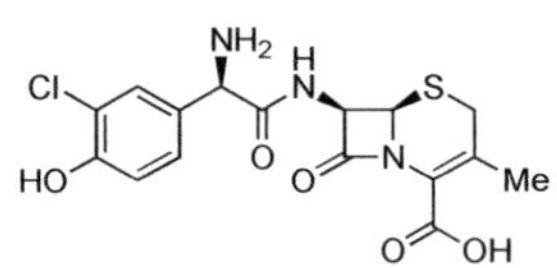	(6*R*,7*R*)-7-((*R*)-2-Amino-2-(3-chloro-4-hydroxyphenyl)acetamido)-3-methyl-8-oxo-5-thia-1-azabicyclo[4.2.0]oct-2-ene-2-carboxylic acid	*In vitro* antibacterial activity similar to that of ampicillin and cephalexin
Cefempidone [71]	103238-57-9	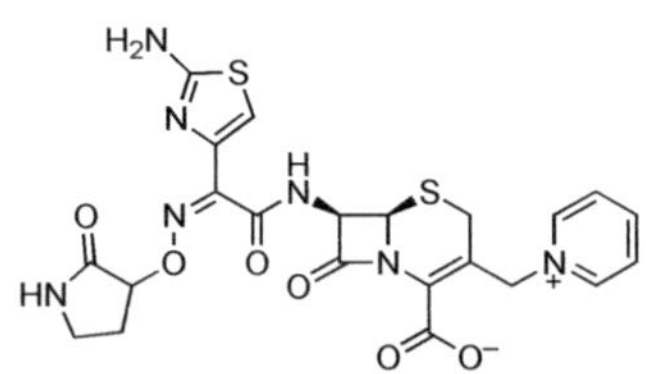	(6*R*,7*R*)-7-((*Z*)-2-(2-Aminothiazol-4-yl)-2-(((2-oxopyrrolidin-3-yl)oxy)imino)acetamido)-8-oxo-3-(pyridin-1-ium-1-ylmethyl)-5-thia-1-azabicyclo[4.2.0]oct-2-ene-2-carboxylate	Against multi-drug resistant bacteria

Name	CAS No.	Structure	Full Name	Major Uses
Cefetrizole [72]	65307-12-2	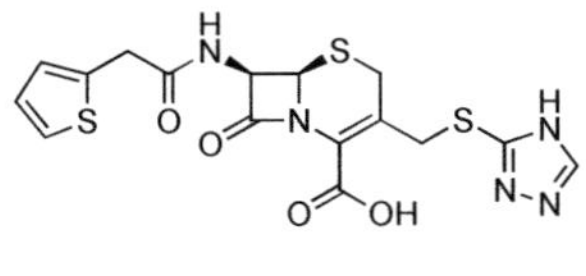	(6*R*,7*R*)-3-(((4*H*-1,2,4-triazol-3-yl)thio)methyl)-8-oxo-7-(2-(thiophen-2-yl) acetamido)-5-thia-1-azabicyclo[4.2.0] oct-2-ene-2-carboxylic acid	Against penicillin-resistant *Staphylococcus*
Cefivitril [73]	66474-36-0	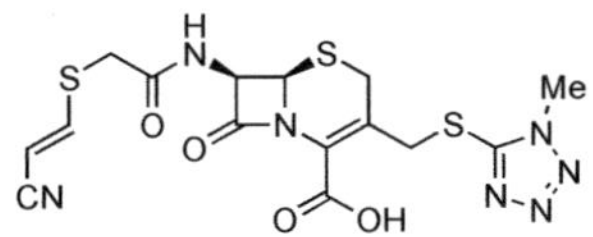	(6*R*,7*R*)-7-(2-(((*E*)-2-Cyanovinyl) thio)acetamido)-3-(((1-methyl-1*H*-tetrazol-5-yl)thio)methyl)-8-oxo-5-thia-1-azabicyclo[4.2.0] oct-2-ene-2-carboxylic acid	Wide *in-vitro* activity on Gram-positive and Gram-negative bacteria
Cefmatilen [74]	140128-74-1	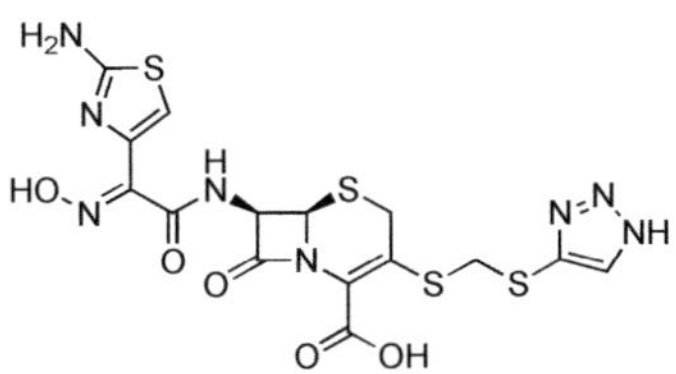	(6*R*,7*R*)-3-((((1*H*-1,2,3-triazol-4-yl) thio)methyl)thio)-7-((*E*)-2-(2-aminothiazol-4-yl)-2-(hydroxyimino) acetamido)-8-oxo-5-thia-1-azabicyclo[4.2.0]oct-2-ene-2-carboxylic acid	A cephalosporin antibiotic for oral use
Cefmepidium Chloride [75]	107452-79-9	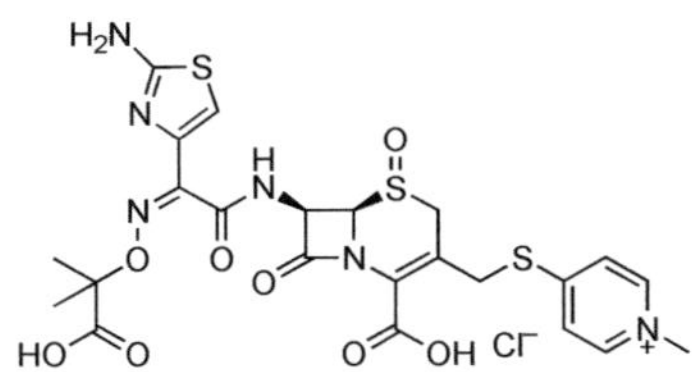	4-((((5*R*,6*R*,7*R*)-7-((*Z*)-2-(2-Aminothiazol-4-yl)-2-(((2-carboxypropan-2-yl)oxy)imino) acetamido)-2-carboxy-5-oxido-8-oxo-5-thia-1-azabicyclo[4.2.0] oct-2-en-3-yl)methyl)thio)-1-methylpyridin-1-ium chloride	Treatment of infection
Cefoxazole [76]	36920-48-6	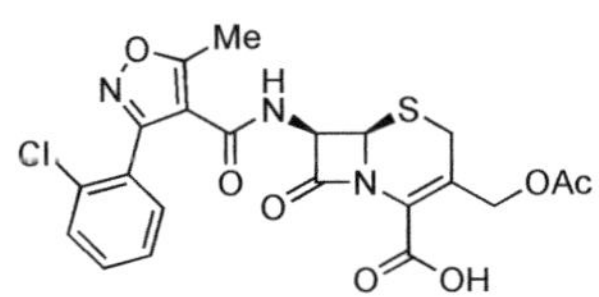	(6*R*,7*R*)-3-(Acetoxymethyl)-7-(3-(2-chlorophenyl)-5-mcthylisoxazolc-4-carboxamido)-8-oxo-5-thia-1-azabicyclo[4.2.0] oct-2-ene-2-carboxylic acid	Treatment of bovine mastitis.

TABLE 1.2 *(Continued)*

Name	CAS No.	Structure	Full Name	Major Uses
Cefrotil [75]	52231-20-6	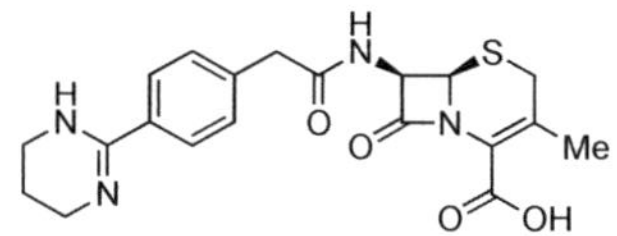	(6*R*,7*R*)-3-Methyl-8-oxo-7-(2-(4-(1,4,5,6-tetrahydropyrimidin-2-yl) phenyl)acetamido)-5-thia-1-azabicyclo[4.2.0]oct-2-ene-2-carboxylic acid	Treatment of infection
Cefsumide [77]	54818-11-0	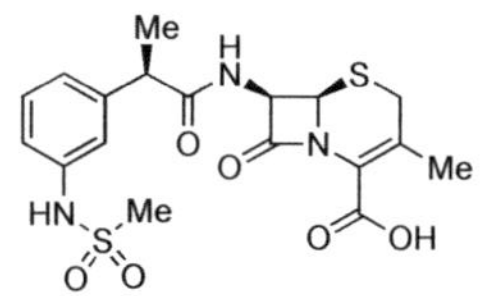	(6*R*,7*R*)-3-Methyl-7-((*R*)-2-(3-(methylsulfonamido)phenyl) propanamido)-8-oxo-5-thia-1-azabicyclo[4.2.0]oct-2-ene-2-carboxylic acid	Against *Klebsiella pneumonia* and *Proteus mirabilis* strains
Ceftioxide [78]	71048-88-9	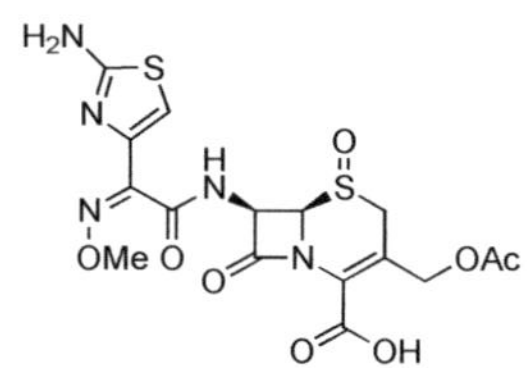	(5*R*,6*R*,7*R*)-3-(Acetoxymethyl)-7-((Z)-2-(2-aminothiazol-4-yl)-2-(methoxyimino) acetamido)-8-oxo-5-thia-1-azabicyclo[4.2.0]oct-2-ene-2-carboxylic acid 5-oxide	Against *Klebsiella aeruginosa* 1082 E and *Enterobacter cloacae*
Cefuracetime [75]	39685-31-9	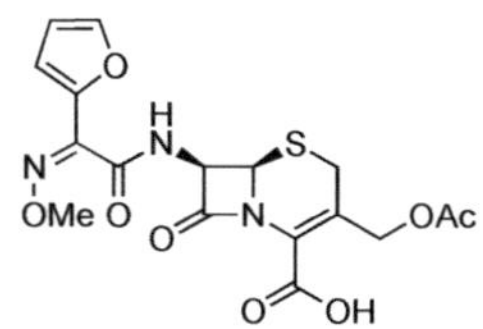	(6*R*,7*R*)-3-(Acetoxymethyl)-7-((Z)-2-(furan-2-yl)-2-(methoxyimino) acetamido)-8-oxo-5-thia-1-azabicyclo[4.2.0]oct-2-ene-2-carboxylic acid	Treatment of infection
Nitrocefin [79]	41906-86-9	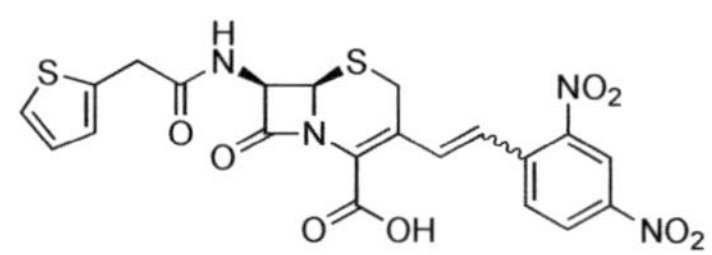	(6*R*,7*R*)-3-(2,4-Dinitrostyryl)-8-oxo-7-(2-(thiophen-2-yl) acetamido)-5-thia-1-azabicyclo[4.2.0] oct-2-ene-2-carboxylic acid	Not an antibiotic of clinical use, but used as a chromogenic substrate in nitrocefin assay

These carbapenems can be organized into several groups, such as olivanic acids, asparenomycins, carpetimycins, pluracidomycins, PS series, thienamycins, and others, based on where they were initially isolated. For example, olivanic acids were isolated from *Streptomyces olivaceus* [80–82], the asparenomycins were isolated from the cultures of *Streptomyces tokunonesis* and *Streptomyces argenteolus*, carpetimycins were isolated from a filtrate of *Streptomyces spp.* KC-6643; pluracidomycin A1 (SF 2103A) was isolated from a strain of *Streptomyces sulfonofaciens*. Similarly, the PS series were isolated from *Streptomyces cremeus subsp. auratilis* admixed with olivanic acids and epithenamycins. Most carbapenem antibiotics with clear biological activities are listed in Table 1.3.

1.5 MONOBACTAM ANTIBIOTICS

Compared to the above three types of β-lactam antibiotics with fused bicyclic heterocycle cores, all naturally occurring monobactams have a 2-oxoazetidine-1-sulfonic acid moiety with variation in the 3β or 3α side chain. The monobactam name was coined by Sykes et al. for the single cyclic (so for "mono") β-lactam [120]. In contrast to penicillins and cephalosporins that are extracted from fungi, monobactam antibiotics are produced by actinomycetes, which also produce nocardicins [121], cephamycins [122], clavulanic acid [81], and carbapenems [112, 123]. Nocardicins were isolated in 1976 as the first group of examples for this type of β-lactams [121]. While these naturally occurring monobactams did not exhibit decent antimicrobial activity, modification of the side-chain provides the potential to improve such biological property. The simple monocyclic structures of such β-lactams with potent antimicrobial activity have prompted great impetus for the study of structural features necessary for the recognition and activation of β-lactams. Extensive studies on the structure-activity relationship indicate that the most surprising aspects of the monobactams are the presence of a sulfamic acid group. Similar to other types of β-lactam antibiotics, monobactams are open to side-chain modification, and can potentially be tuned to be effective against Gram-positive or Gram-negative bacteria or being broad spectrum in character. In terms of the extremes of the spectrum that can be achieved, the monobactams might surpass their classic counterparts, although currently only a few monobactams are in practical use, such as aztreonam [124]. Nevertheless, all these β-lactam antibiotics including penicillins, cephalosporins, etc., share a common mechanism for antibiotic activity, i.e., interfering with the bacterial cell wall synthesis [125].

TABLE 1.3 The List of Carbapenems

Name	CAS No.	Structure	Full Name	Major Uses
Asparenomycin A [85–87]	76466-24-5	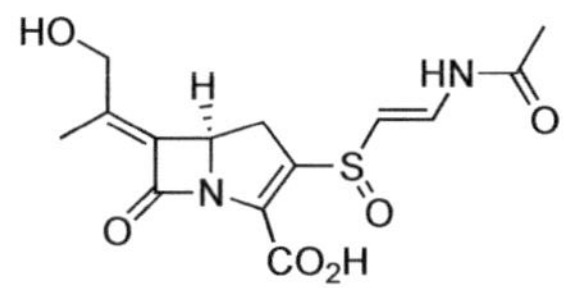	(5*R*,*E*)-3-(((*E*)-2-Acetamidovinyl) sulfinyl)-6-(1-hydroxypropan-2-ylidene)-7-oxo-1-azabicyclo[3.2.0] hept-2-ene-2-carboxylic acid	Broad-spectrum antibacterial agent and potent β-lactamase inhibitor.
Asparenomycin B [86, 87]	79366-72-6	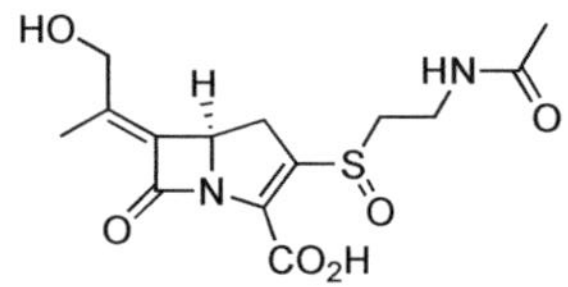	(5*R*,*E*)-3-((2-Acetamidoethyl) sulfinyl)-6-(1-hydroxypropan-2-ylidene)-7-oxo-1-azabicyclo[3.2.0] hept-2-ene-2-carboxylic acid	Inhibited a wide range of β-lactamases, including both cephalosporinases, and penicillinases, usually at conc. <3 μM
Asparenomycin C [86, 87]	81018-71-5	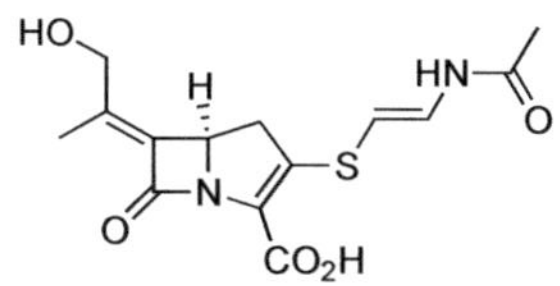	(*R*,*E*)-3-(((*E*)-2-Acetamidovinyl) thio)-6-(1-hydroxypropan-2-ylidene)-7-oxo-1-azabicyclo[3.2.0] hept-2-ene-2-carboxylic acid	Inhibited a wide range of β-lactamases, including both cephalosporinases, and penicillinases, usually at conc. <3 μM
Biapenem [8]	120410-24-4	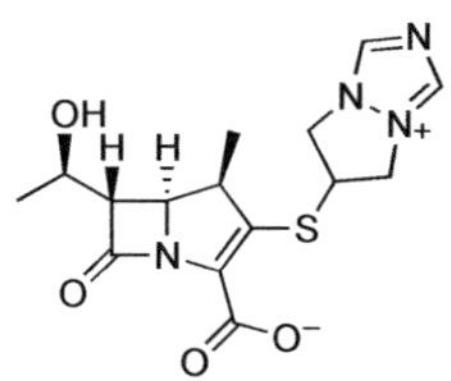	(4*R*,5*S*,6*S*)-3-((6,7-Dihydro-5*H*-pyrazolo[1,2-a][1,2,4] triazol-4-ium-6-yl)thio)-6-((*R*)-1-hydroxyethyl)-4-methyl-7-oxo-1-azabicyclo[3.2.0] hept-2-ene-2-carboxylate	Against the family of *Enterobacteriaceae*, including ceftazidime-resistant strains

Name	CAS No.	Structure	Full Name	Major Uses
BMS-181139 [88]	135918-08-0	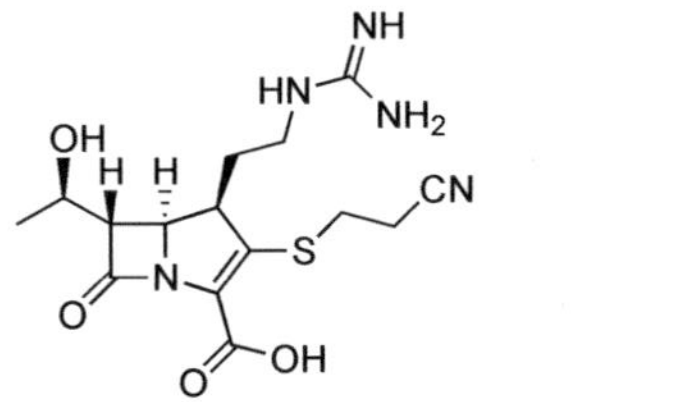	(4*R*,5*S*,6*S*)-3-((2-Cyanoethyl) thio)-4-(2-guanidinoethyl)-6-((*R*)-1-hydroxyethyl)-7-oxo-1-azabicyclo[3.2.0] hept-2-ene-2-carboxylic acid	Against species of *Pseudomonas* and the related genera *Alteromonas* and *Burkholderia*
BO-2727 [89] L 739428	149882-71-3	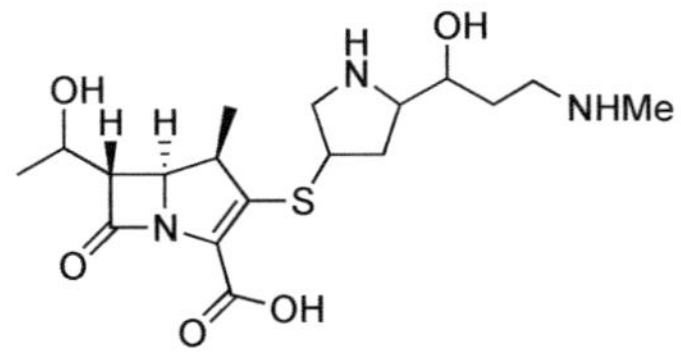	(4*R*,5*S*,6*S*)-3-((5-(1-Hydroxy-3-(methylamino)propyl) pyrrolidin-3-yl)thio)-6-(1-hydroxyethyl)-4-methyl-7-oxo-1-azabicyclo[3.2.0] hept-2-ene-2-carboxylic acid	Against *Staphylococcus aureus Smith*, *Escherichia coli ML4707*, and *P. aeruginosa GN11189*
C-19393 E_5 [90] PA 31088II	83310-72-9	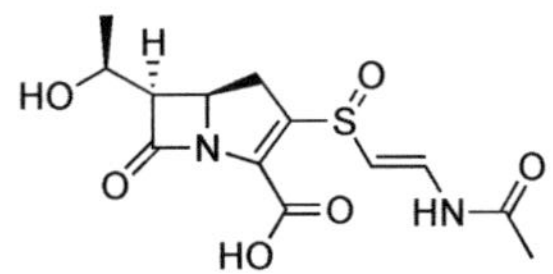	(5*R*,6*R*)-3-(((*E*)-2-Acetamidovinyl) sulfinyl)-6-((*S*)-1-hydroxyethyl)-7-oxo-1-azabicyclo[3.2.0]hept-2-ene-2-carboxylic acid	Shows broad antimicrobial spectrum and strong inhibitory activity against β-lactamases.
Carpetimycin A [91]	76025-73-5	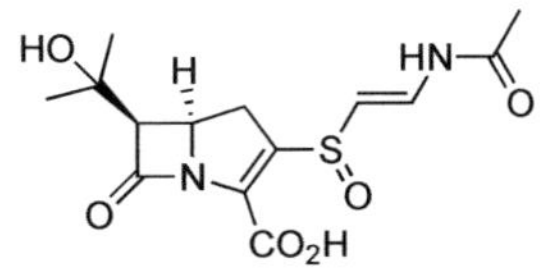	(5*R*,6*R*)-3-(((*E*)-2-Acetamidovinyl) sulfinyl)-6-(2-hydroxypropan-2-yl)-7-oxo-1-azabicyclo[3.2.0] hept-2-ene-2-carboxylic acid	Broad spectra and potent activity against Gram-pos. and Gram-neg. bacteria, including various anaerobes
Carpetimycin B [91]	76094-36-5	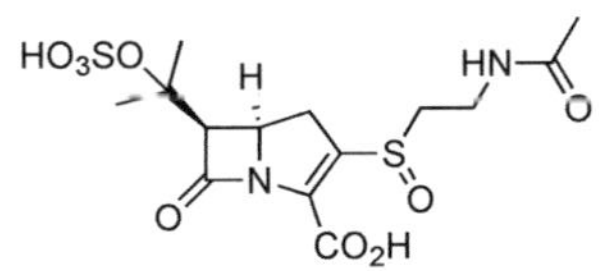	(5*R*,6*R*)-3-(((*E*)-2-Acetamidovinyl)sulfinyl)-7-oxo-6-(2-(sulfooxy) propan-2-yl)-1-azabicyclo[3.2.0] hept-2-ene-2-carboxylic acid	Broad spectra and potent activity against Gram-pos. and Gram-neg. bacteria, including various anaerobes

TABLE 1.3 *(Continued)*

Name	CAS No.	Structure	Full Name	Major Uses
CS-834 [92]	157542-49-9	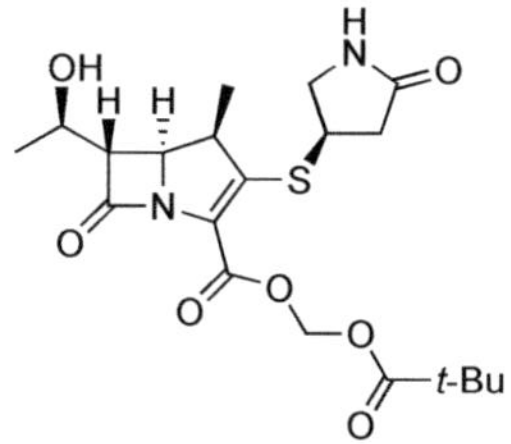	(Pivaloyloxy)methyl (4*R*,5*S*,6*S*)-6-((*R*)-1-Hydroxyethyl)-4-methyl-7-oxo-3-((5-oxopyrrolidin-3-yl)thio)-1-azabicyclo[3.2.0]hept-2-ene-2-carboxylate	Against Gram-positive and Gram-negative species, such as *Staphylococcus aureus*, *Escherichia coli*, and *Klebsiella pneumoniae*
Doripenem [8]	148016-81-3	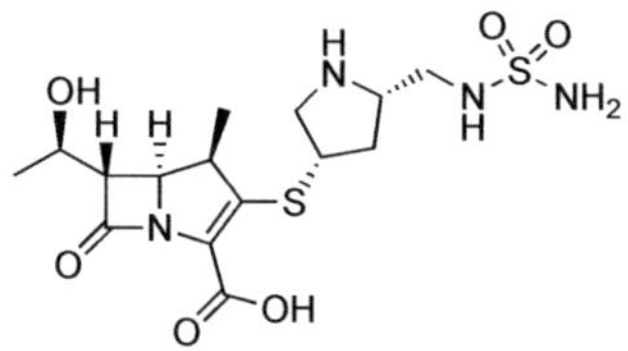	(4*R*,5*S*,6*S*)-6-((*R*)-1-Hydroxyethyl)-4-methyl-7-oxo-3-(((3*S*,5*S*)-5-((sulfamoylamino)methyl)yrrolidine-3-yl)thio)-1-azabicyclo[3.2.0]hept-2-ene-2-carboxylic acid	Against both Gram-positive and Gram-negative bacteria, more active against *Pseudomonas aeruginosa* than other carbapenems
DZ-2640 [93]	184633-88-3	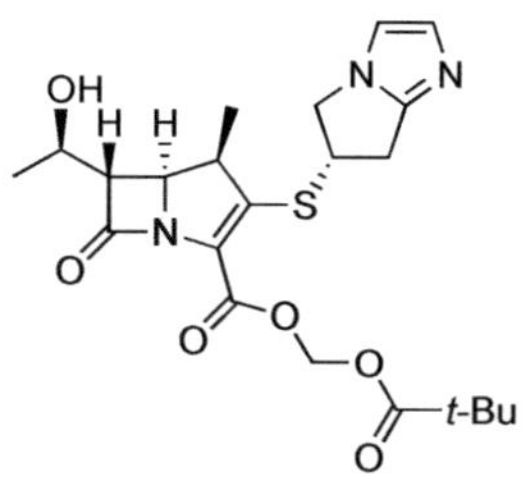	(Pivaloyloxy)methyl (4*R*,5*S*,6*S*)-3-(((*S*)-6,7-dihydro-5*H*-pyrrolo[1,2-a]imidazol-6-yl)thio)-6-((*R*)-1-hydroxyethyl)-4-methyl-7-oxo-1-azabicyclo[3.2.0]hept-2-ene-2-carboxylate	Oral carbapenem prodrug of DU-6681

Name	CAS No.	Structure	Full Name	Major Uses
DU-6681 [94]	129951-17-3	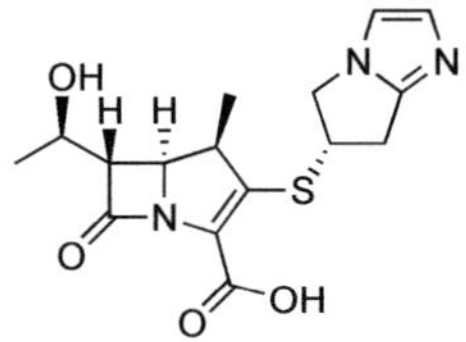	(4*R*,5*S*,6*S*)-3-(((*S*)-6,7-Dihydro-5*H*-pyrrolo[1,2-a]imidazol-6-yl)thio)-6-((*R*)-1-hydroxyethyl)-4-methyl-7-oxo-1-azabicyclo[3.2.0]hept-2-ene-2-carboxylic acid	Against *S. aureus*, *S. pneumoniae*, *Escherichia coli*, and *H. influenzae*.
E-1010 [95, 96] ER-35786	186319-97-1	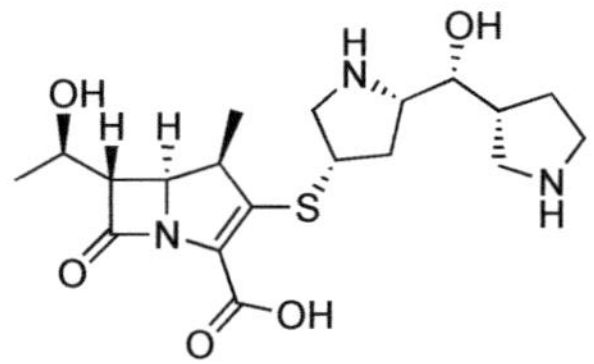	(4*R*,5*S*,6*S*)-3-(((3*S*,5*S*)-5-((*R*)-hydroxy((*R*)-pyrrolidin-3-yl)methyl)pyrrolidin-3-yl)thio)-6-((*R*)-1-hydroxyethyl)-4-methyl-7-oxo-1-azabicyclo[3.2.0]hept-2-ene-2-carboxylic acid	Against a broad spectrum of bacteria and useful in the treatment of anaerobic obstetric/Gynecol. infections
Epithienamycin A [97] MM 22380 [84]	63582-78-5	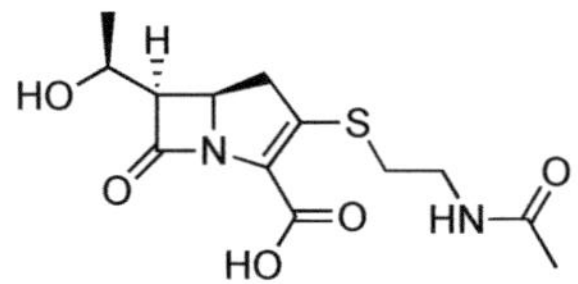	(5*R*,6*R*)-3-((2-Acetamidoethyl)thio)-6-((*S*)-1-hydroxyethyl)-7-oxo-1-azabicyclo[3.2.0]hept-2-ene-2-carboxylic acid	Are cell wall active antibodies structurally related to N-acetylthienamycin, showing activity *in vitro* vs. a broad spectrum of bacterial species
Epithienamycin B [97]	65376-20-7	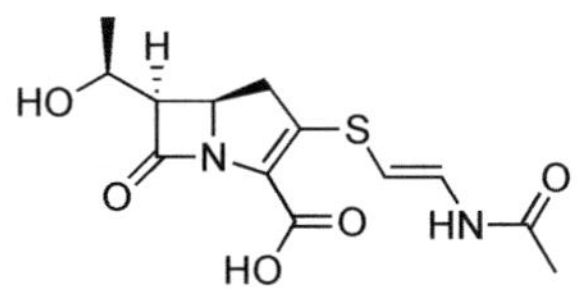	(5*R*,6*R*)-3-(((*E*)-2-Acetamidovinyl)thio)-6-((*S*)-1-hydroxyethyl)-7-oxo-1-azabicyclo[3.2.0]hept-2-ene-2-carboxylic acid	Are cell wall active antibodies structurally related to N-acetylthienamycin, showing activity *in vitro* vs. a broad spectrum of bacterial species

TABLE 1.3 *(Continued)*

Name	CAS No.	Structure	Full Name	Major Uses
Epithienamycin C [97] MM 22381 [84, 98]	63599-16-6 63701-32-6	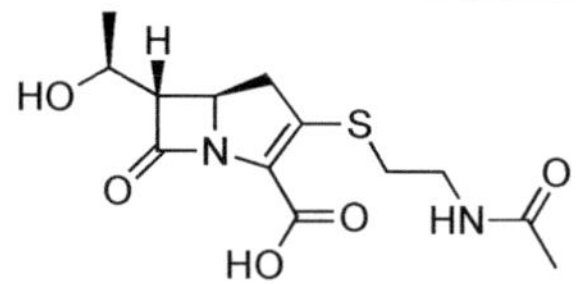	(5*R*,6*S*)-3-((2-Acetamidoethyl)thio)-6-((*S*)-1-hydroxyethyl)-7-oxo-1-azabicyclo[3.2.0]hept-2-ene-2-carboxylic acid	Broad-spectrum activity against Gram-positive and Gram-negative bacteria
Epithienamycin D [97] MM 22383 [84, 98]	65322-98-7 68738-07-8	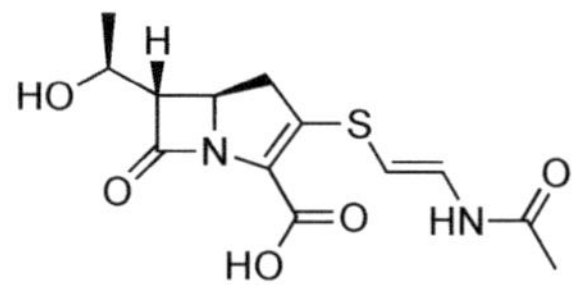	(5*R*,6*S*)-3-(((*E*)-2-Acetamidovinyl)thio)-6-((*S*)-1-hydroxyethyl)-7-oxo-1-azabicyclo[3.2.0]hept-2-ene-2-carboxylic acid	Broad-spectrum activity against Gram-positive and Gram-negative bacteria
Ertapenem [99]	153832-46-3	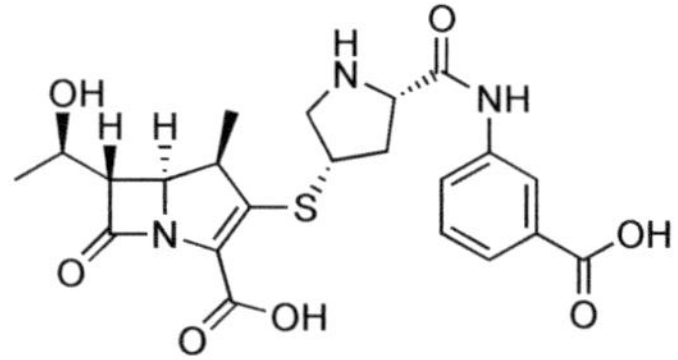	(4*R*,5*S*,6*S*)-3-(((3*S*,5*S*)-5-((3-Carboxyphenyl)carbamoyl)yrrolidine-3-yl)thio)-6-((*R*)-1-hydroxyethyl)-4-methyl-7-oxo-1-azabicyclo[3.2.0]hept-2-ene-2-carboxylic acid	The first line against ESBL-producing Gram-negative bacteria, and wide spectrum of activity against community-acquired pathogens infections
Imipenem [100] MK 0787 [84]	64221-86-9	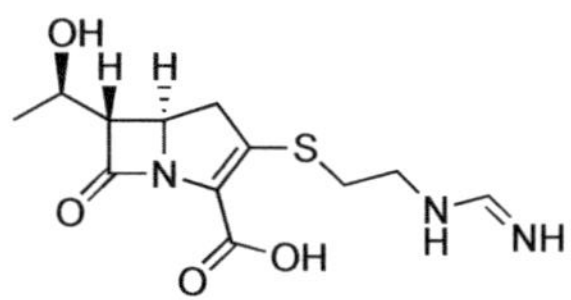	(5*R*,6*S*)-3-((2-Formimidamidoethyl)thio)-6-((*R*)-1-hydroxyethyl)-7-oxo-1-azabicyclo[3.2.0]hept-2-ene-2-carboxylic acid	The first carbapenem discovered, highly potent, broad-spectrum antimicrobial agent with a strong safety profile

Name	CAS No.	Structure	Full Name	Major Uses
J-110441 [101]	251949-13-0	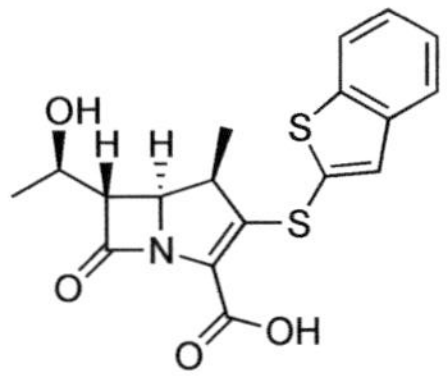	(4*R*,5*S*,6*S*)-3-(Benzo[b] thiophen-2-ylthio)-6-((*R*)-1-hydroxyethyl)-4-methyl-7-oxo-1-azabicyclo[3.2.0] hept-2-ene-2-carboxylic acid	Against *Bacteroides fragilis*, *Stenotrophomonas maltophilia*, *Bacillus cereus* and TEM-type class A serine β-lactamase and chromosomal class C serine β-lactamase
Lenapenem [102]	149951-16-6	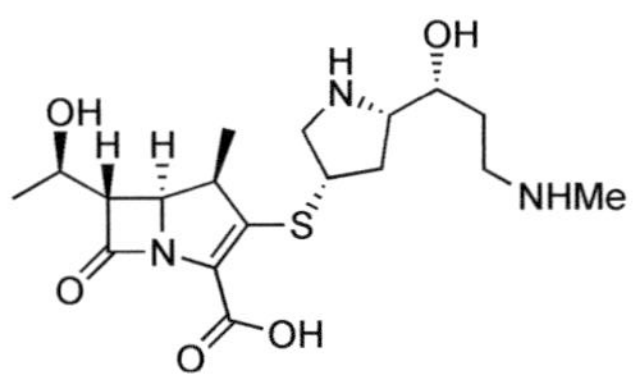	(4*R*,5*S*,6*S*)-3-(((3*S*,5*S*)-5-((*R*)-1-Hydroxy-3-(methylamino) propyl) epacianne-3-yl)thio)-6-((*R*)-1-hydroxyethyl)-4-methyl-7-oxo-1-azabicyclo[3.2.0] hept-2-ene-2-carboxylic acid	Against Gram-positive and Gram-negative bacteria, including some imipenem- and/ or meropenem-resistant *Pseudomonas aeruginosa* strains
ME 1036 [103, 104] CP5609	432038-96-5	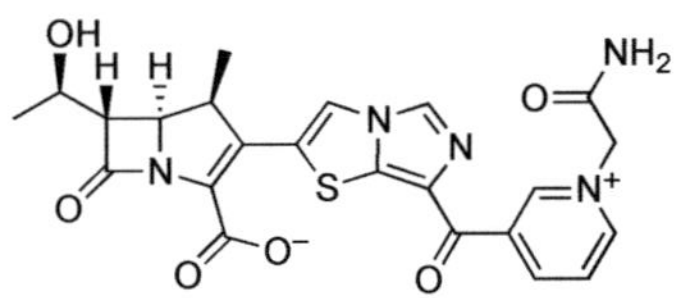	(4S,5*R*,6*S*)-3-(7-(1-(2-Amino-2-oxoethyl) epacian-1-ium-3-carbonyl)imidazo[5,1-b] thiazol-2-yl)-6-((*R*)-1-hydroxyethyl)-4-methyl-7-oxo-1-azabicyclo[3.2.0] hept-2-ene-2-carboxylate	Against community-acquired pneumonia isolates causing serious invasive infections, including MRSA
Meropenem [105] SM-7338	96036-03-2	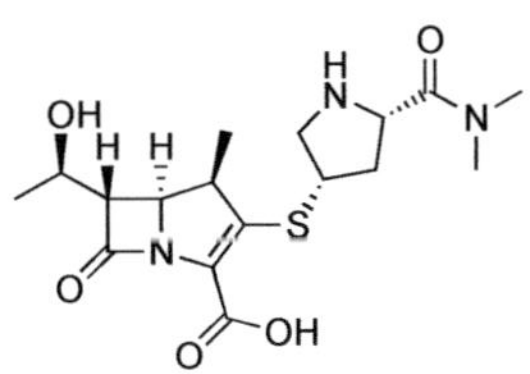	(4*R*,5*S*,6*S*)-3-(((3S,5S)-5-(Dimethylcarbamoyl) epacianne-3-yl)thio)-6-((*R*)-1-hydroxyethyl)-4-methyl-7-oxo-1-azabicyclo[3.2.0] hept-2-ene-2-carboxylic acid	Excellent bactericidal activity *in vitro* against almost all clinically significant aerobes and anaerobes.

TABLE 1.3 *(Continued)*

Name	CAS No.	Structure	Full Name	Major Uses
MM 4550 [98]	12795-21-0	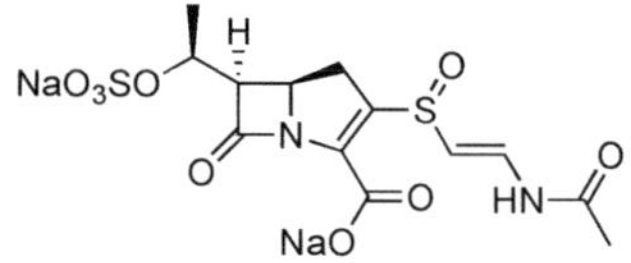	Sodium (5*R*,6*R*)-3-(((*E*)-2-Acetamidovinyl)sulfinyl)-7-oxo-6-((*S*)-1-(sulfonatooxy)ethyl)-1-azabicyclo[3.2.0]hept-2-ene-2-carboxylate	Much less effective than MM 13902
MM 13902 [98] Epithienamycin E	57459-82-2	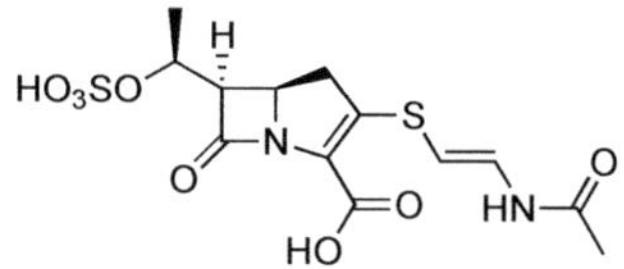	(5*R*,6*S*)-3-(((*E*)-2-Acetamidovinyl)thio)-7-oxo-6-(1-(sulfooxy)ethyl)-1-azabicyclo[3.2.0]hept-2-ene-2-carboxylic acid	Potent antibiotics with MIC values against Gram-positive and Gram-negative bacteria in the range 0.1–3.1 μg/mL
MM 17880 [98]	61036-81-5	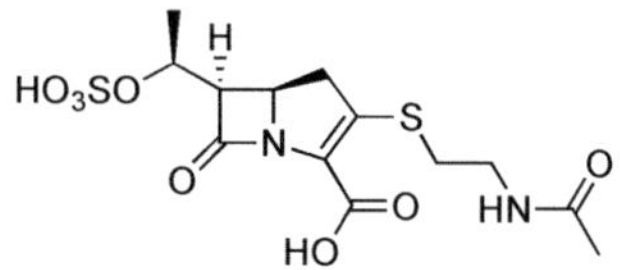	(5*R*,6*R*)-3-((2-Acetamidoethyl)thio)-7-oxo-6-((*S*)-1-(sulfooxy)ethyl)-1-azabicyclo[3.2.0]hept-2-ene-2-carboxylic acid	Potent antibiotics with MIC values against Gram-positive and Gram-negative bacteria in the range 0.1-3.1 μg/mL
OA-6129A [106]	82475-11-4	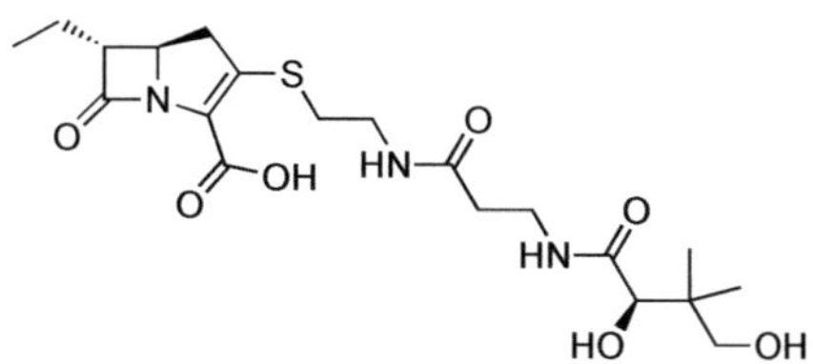	(5*R*,6*R*)-3-((2-(3-((*R*)-2,4-Dihydroxy-3,3-dimethylbutanamido)propanamido)ethyl)thio)-6-ethyl-7-oxo-1-azabicyclo[3.2.0]hept-2-ene-2-carboxylic acid	Wide spectrum of antimicrobial activity against Gram-positive and Gram-negative bacteria and potent inhibitory effect on β-lactamases.

Name	CAS No.	Structure	Full Name	Major Uses
OA-6129B1 [106]	82510-13-2	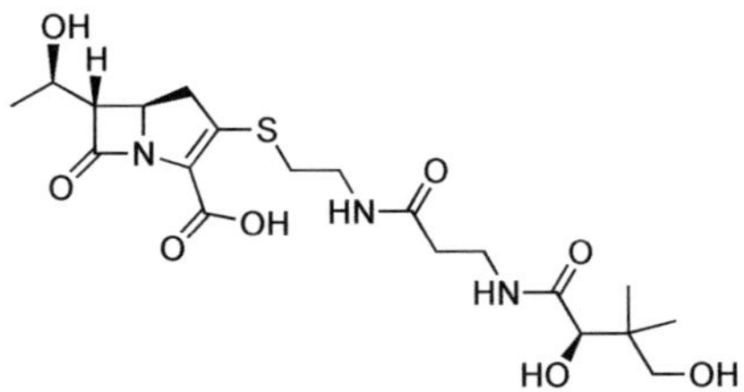	(5*R*,6*S*)-3-((2-(3-((*R*)-2,4-Dihydroxy-3,3-dimethylbutanamido)propanamido)ethyl)thio)-6-((*R*)-1-hydroxyethyl)-7-oxo-1-azabicyclo[3.2.0]hept-2-ene-2-carboxylic acid	Wide spectrum of antimicrobial activity against Gram-positive and Gram-negative bacteria and potent inhibitory effect on β-lactamases.
OA-6129B2 [106]	82475-10-3	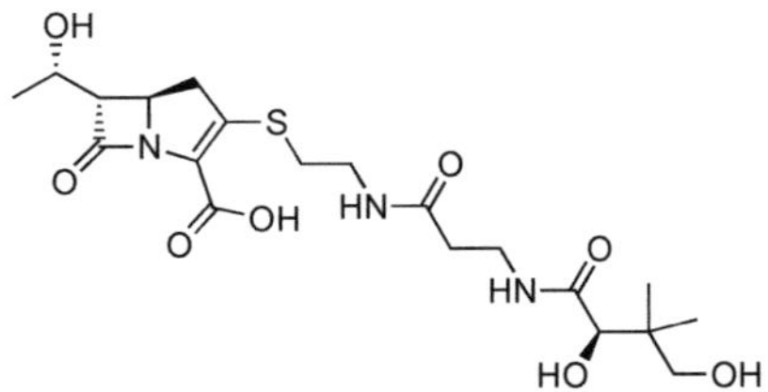	(5*R*,6*S*)-3-((2-(3-((*R*)-2,4-Dihydroxy-3,3-dimethylbutanamido)propanamido)ethyl)thio)-6-((*S*)-1-hydroxyethyl)-7-oxo-1-azabicyclo[3.2.0]hept-2-ene-2-carboxylic acid	Wide spectrum of antimicrobial activity against Gram-positive and Gram-negative bacteria and potent inhibitory effect on β-lactamases.
OA-6129C [106]	82475-09-0	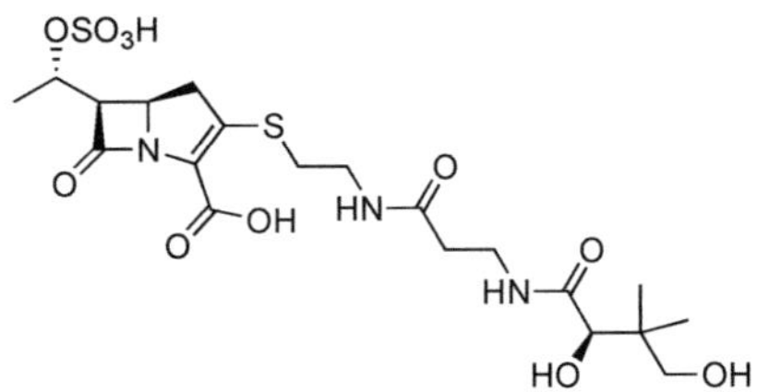	(5*R*,6*R*)-3-((2-(3-((*R*)-2,4-Dihydroxy-3,3-dimethylbutanamido)propanamido)ethyl)thio)-7-oxo-6-((*S*)-1-(sulfooxy)ethyl)-1-azabicyclo[3.2.0]hept-2-ene-2-carboxylic acid	Wide spectrum of antimicrobial activity against Gram-positive and Gram-negative bacteria and potent inhibitory effect on β-lactamases.
Olivanic acid [107]	64761-66-6	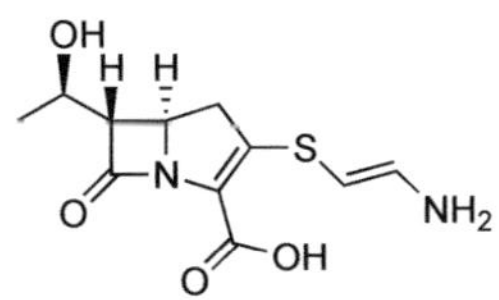	(5*R*,6*S*)-3-(((*E*)-2-Aminovinyl)thio)-6-((*R*)-1-hydroxyethyl)-7-oxo-1-azabicyclo[3.2.0]hept-2-ene-2-carboxylic acid	Against Gram-positive and Gram-negative organisms, including β-lactamase-producing strains

TABLE 1.3 *(Continued)*

Name	CAS No.	Structure	Full Name	Major Uses
Panipenem [103]	87726-17-8	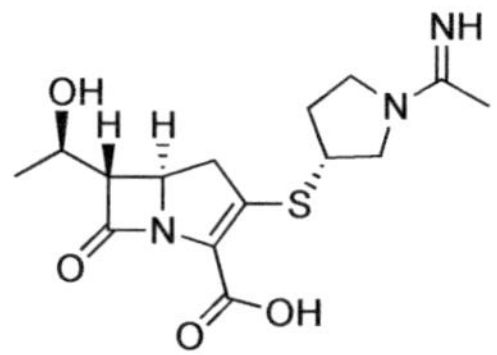	(5*R*,6*S*)-6-((*R*)-1-Hydroxyethyl)-3-(((*R*)-1-(1-iminoethyl) epacianne-3-yl)thio)-7-oxo-1-azabicyclo[3.2.0]hept-2-ene-2-carboxylic acid	Against *Enterobacteriaceae* and common respiratory tract pathogens
Pluracidomycin A1 SF2103A [108]	82189-02-4	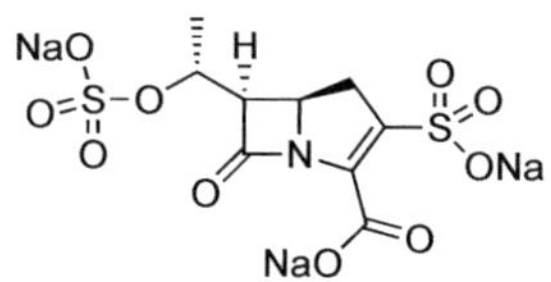	Sodium (5*R*,6*R*)-7-Oxo-3-sulfonato-6-((*R*)-1-(sulfonatooxy) ethyl)-1-azabicyclo[3.2.0] hept-2-ene-2-carboxylate	A strong inhibitor of various types of β-lactamase
Pluracidomycin A2 [109] PA 41746A2	92755-79-8	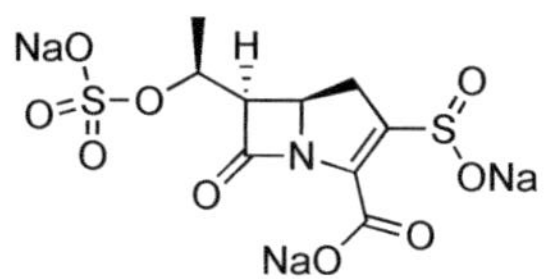	Sodium (5*R*,6*R*)-7-Oxo-3-sulfinato-6-((*S*)-1-(sulfonatooxy) ethyl)-1-azabicyclo[3.2.0] hept-2-ene-2-carboxylate	The first sulfinic acid was identified as a microbial metabolite. Broad antibacterial and β-lactamase activity
Pluracidomycin B [110]	82188-59-8	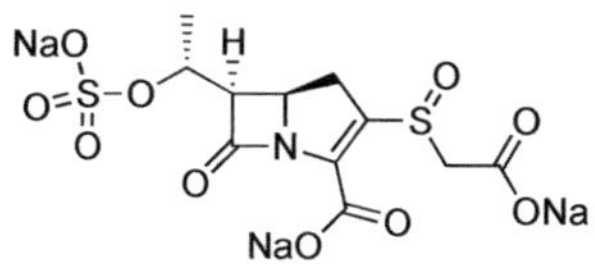	Sodium (5*R*,6*R*)-3-((Carboxylatomethyl)sulfinyl)-7-oxo-6-((*R*)-1-(sulfonatooxy) ethyl)-1-azabicyclo[3.2.0] hept-2-ene-2-carboxylate	Effective against *Escherichia coli NIHJ JC-2*, *Escherichia coli 377*, *Klebsiella sp. 363*
Pluracidomycin C1 [109]	82188-60-1	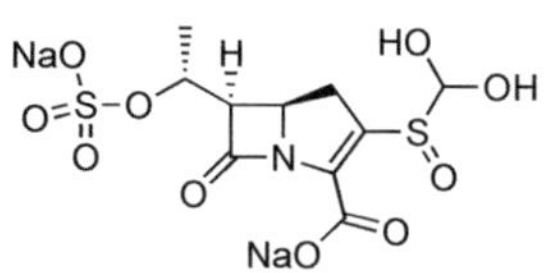	Sodium (5*R*,6*R*)-3-((Dihydroxymethyl)sulfinyl)-7-oxo-6-((*R*)-1-(sulfonatooxy) ethyl)-1-azabicyclo[3.2.0] hept-2-ene-2-carboxylate	Broad antibacterial and β-lactamase inhibitory activities

Name	CAS No.	Structure	Full Name	Major Uses
Pluracidomycin C2 [109]	92842-68-7	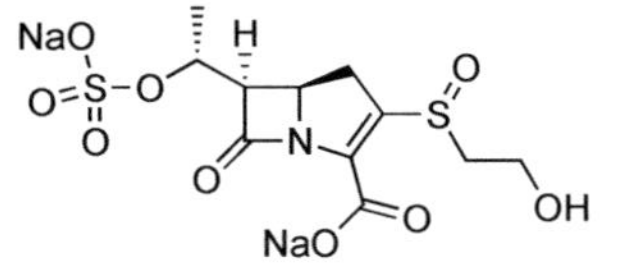	Sodium (5*R*,6*R*)-3-((2-hydroxyethyl)sulfinyl)-7-oxo-6-((*R*)-1-(sulfonatooxy) ethyl)-1-azabicyclo[3.2.0] hept-2-ene-2-carboxylate	Precursor of pluracidomycin B
Pluracidomycin C3 [109]	92934-62-8	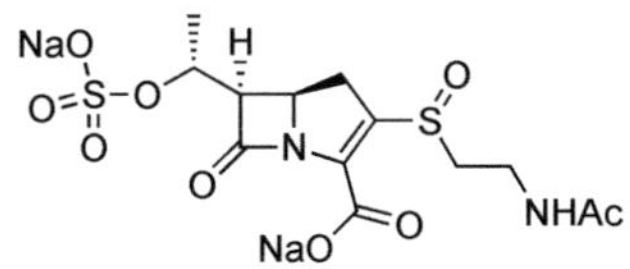	Sodium (5*R*,6*R*)-3-((2-acetamidoethyl)sulfinyl)-7-oxo-6-((*R*)-1-(sulfonatooxy) ethyl)-1-azabicyclo[3.2.0] hept-2-ene-2-carboxylate	Broad antibacterial and β-lactamase inhibitory activities
Pluracidomycin D [109]	86992-80-5	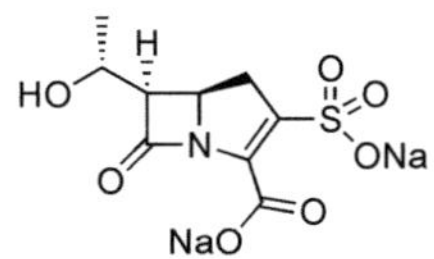	Sodium (5*R*,6*R*)-6-((*R*)-1-hydroxyethyl)-7-oxo-3-sulfonato-1-azabicyclo[3.2.0] hept-2-ene-2-carboxylate	Broad antibacterial and β-lactamase inhibitory activities
PS-5 [111, 112] MM 22744 [84]	67007-79-8	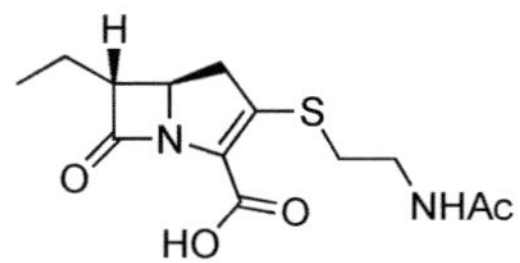	(5*R*,6*R*)-3-((2-acetamidoethyl) thio)-6-ethyl-7-oxo-1-azabicyclo[3.2.0]hept-2-ene-2-carboxylic acid	No activity specified
PS-6 [111]	72615-19-1	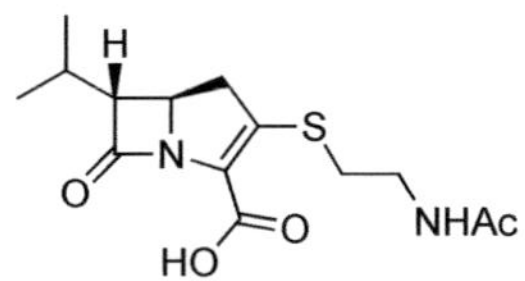	(5*R*,6*R*)-3-((2-Acetamidoethyl) thio)-6-isopropyl-7-oxo-1-azabicyclo[3.2.0]hept-2-ene-2-carboxylic acid	No activity specified

TABLE 1.3 *(Continued)*

Name	CAS No.	Structure	Full Name	Major Uses
PS-7 [111]	72615-18-0	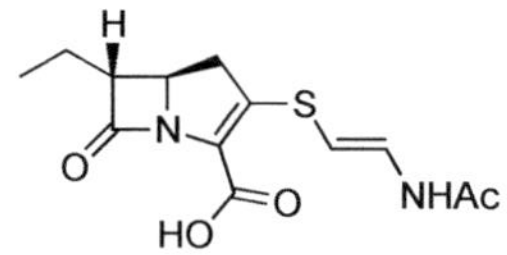	(5*R*,6*R*)-3-(((*E*)-2-Acetamidovinyl)thio)-6-ethyl-7-oxo-1-azabicyclo[3.2.0]hept-2-ene-2-carboxylic acid	No activity specified
PS-8 [111]	69242-34-8	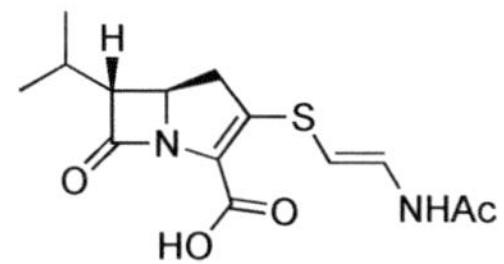	(5*R*,6*R*)-3-(((*E*)-2-Acetamidovinyl)thio)-6-isopropyl-7-oxo-1-azabicyclo[3.2.0]hept-2-ene-2-carboxylic acid	No activity specified
Razupenem [103] PTZ 601 PZ 601 SM 216601 SMP 601	426253-04-5	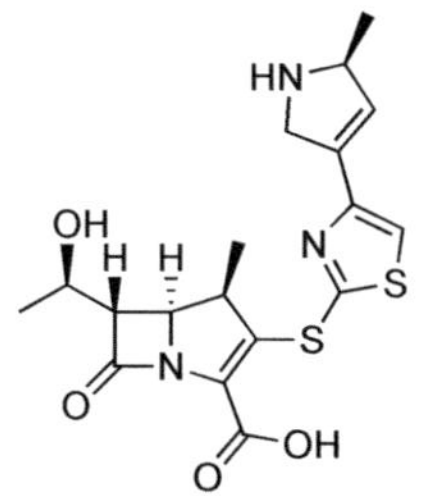	(4*R*,5*S*,6*S*)-6-((*R*)-1-Hydroxyethyl)-4-methyl-3-((4-((*S*)-5-methyl-2,5-dihydro-1*H*-pyrrol-3-yl)thiazol-2-yl)thio)-7-oxo-1-azabicyclo[3.2.0]hept-2-ene-2-carboxylic acid	Against ESBL-producers, with a synergistic activity with amikacin or ciprofloxacin against *B. epacian* and *S. marcescens*
S-4661 [113]	2114982-73-7	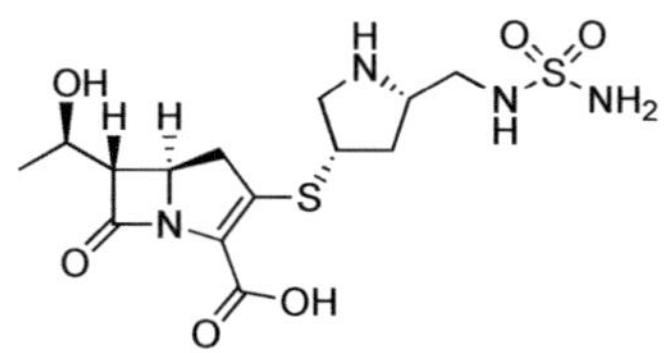	(5*R*,6*S*)-6-((*R*)-1-Hydroxyethyl)-7-oxo-3-(((3*S*,5*S*)-5-((sulfamoylamino)methyl)pyrrolidin-3-yl)thio)-1-azabicyclo[3.2.0]hept-2-ene-2-carboxylic acid	Possesses a potent and well-balanced activity against respiratory pathogens. 2 times more active than imipenem, meropenem, and biapenem against Gram-positive bacteria.

Name	CAS No.	Structure	Full Name	Major Uses
Sanfetrinem [103]	156769-21-0	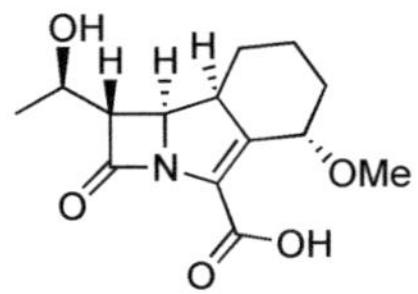	(1*S*,5*S*,8a*S*,8b*R*)-1-((*R*)-1-Hydroxyethyl)-5-methoxy-2-oxo-1,2,5,6,7,8,8a,8b-octahydroazeto[2,1-a] isoindole-4-carboxylic acid	Against *P. vulgaris* and *K. oxytoca*
SM-17466 [114]	161681-09-0	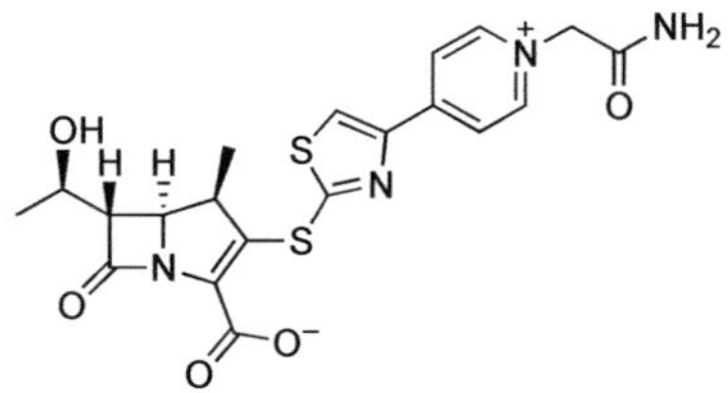	(4*R*,5*S*,6*S*)-3-((4-(1-(2-Amino-2-oxoethyl)pyridin-1-ium-4-yl)thiazol-2-yl) thio)-6-((*R*)-1-hydroxyethyl)-4-methyl-7-oxo-1-azabicyclo[3.2.0] hept-2-ene-2-carboxylate	Against methicillin-resistant *Staphylococcus aureus*
SM-197436 [115]	475263-06-0	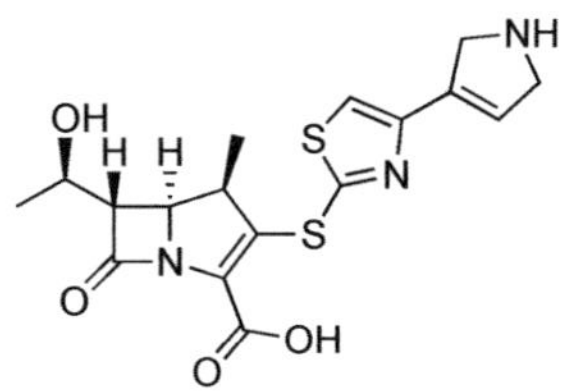	(4*R*,5*S*,6*S*)-3-((4-(2,5-Dihydro-1*H*-pyrrol-3-yl)thiazol-2-yl) thio)-6-((*R*)-1-hydroxyethyl)-4-methyl-7-oxo-1-azabicyclo[3.2.0] hept-2-ene-2-carboxylic acid	Against methicillin-resistant *Staphylococcus aureus* (MRSA) and *Staphylococcus epidermidis* (MRSE)
SM-232721 [115]	426253-09-0	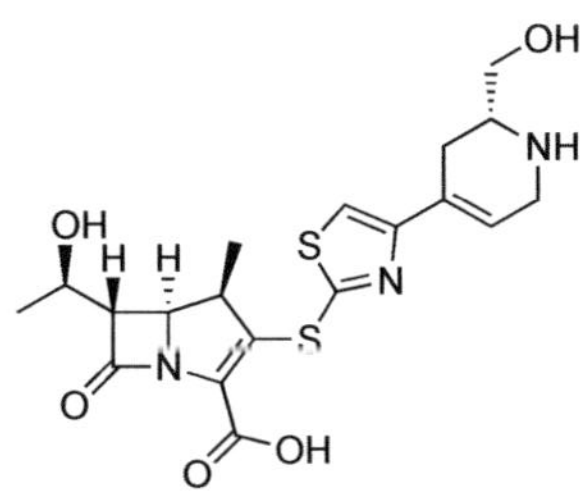	(4*R*,5*S*,6*S*)-6-((*R*)-1-Hydroxyethyl)-3-((4-((*R*)-2-(hydroxymethyl)-1,2,3,6-tetrahydropyridin-4-yl) thiazol-2-yl)thio)-4-methyl-7-oxo-1-azabicyclo[3.2.0]hept-2-ene-2-carboxylic acid	Against methicillin-resistant *Staphylococcus aureus* (MRSA) and *Staphylococcus epidermidis* (MRSE)

TABLE 1.3 *(Continued)*

Name	CAS No.	Structure	Full Name	Major Uses
SM-232724 [115]	426253-11-4	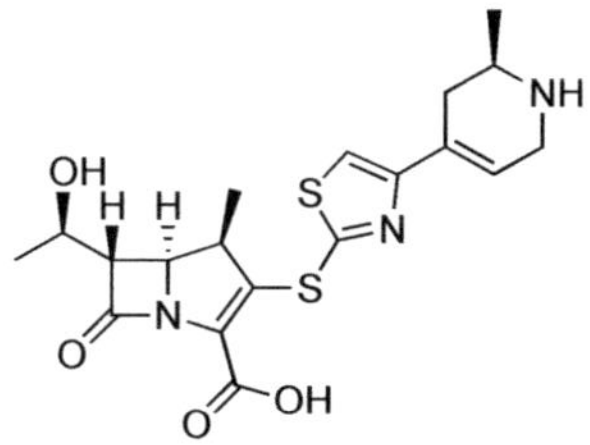	(4*R*,5*S*,6*S*)-6-((*R*)-1-Hydroxyethyl)-4-methyl-3-((4-((*R*)-2-methyl-1,2,3,6-tetrahydropyridin-4-yl)thiazol-2-yl)thio)-7-oxo-1-azabicyclo[3.2.0]hept-2-ene-2-carboxylic acid	Against methicillin-resistant *Staphylococcus aureus* (MRSA) and *Staphylococcus epidermidis* (MRSE)
Sulopenem [116]	120788-07-0	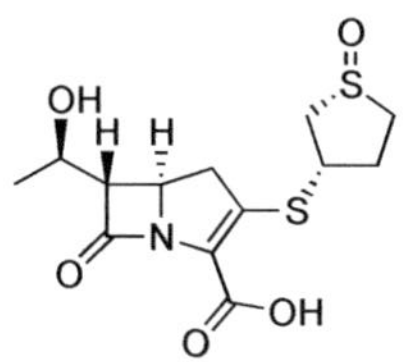	(5*R*,6*S*)-6-((*R*)-1-Hydroxyethyl)-3-(((1*R*,3*S*)-1-oxidotetrahydrothiophen-3-yl)thio)-7-oxo-1-azabicyclo[3.2.0]hept-2-ene-2-carboxylic acid	Against *Bacteroides fragilis*, *Bacteroides thetaiotaomicron*, *Prevotella* intermedia, *Porphyromonas gingivalis*, *Fusobacterium spp.*, and *Pepto-streptococcus spp.*
Tebipenem pivoxil [117]	161715-24-8	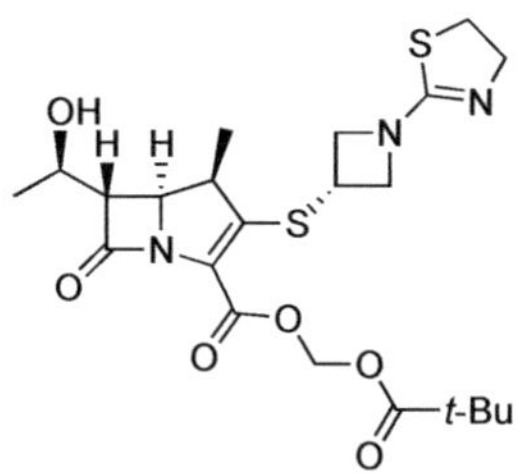	(Pivaloyloxy)methyl (4*R*,5*S*,6*S*)-3-((1-(4,5-Dihydrothiazol-2-yl)azetidin-3-yl)thio)-6-((*R*)-1-hydroxyethyl)-4-methyl-7-oxo-1-azabicyclo[3.2.0]hept-2-ene-2-carboxylate	Broad-spectrum activity against Gram-positive and Gram-negative bacteria including ESBL-producing Gram-negative bacteria

Name	CAS No.	Structure	Full Name	Major Uses
Thienamycin [118]	59995-64-1	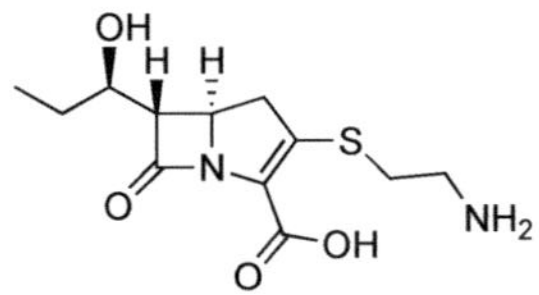	(5*R*,6*S*)-3-((2-Aminoethyl) thio)-6-((*R*)-1-hydroxypropyl)-7-oxo-1-azabicyclo[3.2.0]hept-2-ene-2-carboxylic acid	The first representative of carbapenems produced by *Streptomyces cattleya* with excellent potency and antibacterial spectrum
Tomopenem [119] RO-4908463 CS-023	222400-20-6	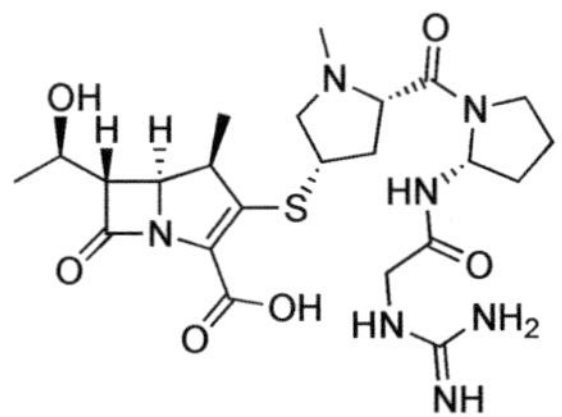	(4*R*,5*S*,6*S*)-3-(((3*S*,5*S*)-5-((*R*)-2-(2-Guanidinoacetamido) pyrrolidine-1-carbonyl)-1-methylpyrrolidin-3-yl) thio)-6-((*R*)-1-hydroxyethyl)-4-methyl-7-oxo-1-azabicyclo[3.2.0] hept-2-ene-2-carboxylic acid	Against methicillin-resistant *Staphylococcus aureus* and *Pseudomonas aeruginosa*

TABLE 1.4 The List of Monobactams

Name	CAS No.	Structure	Full Name	Major Uses
Azthreonam [126] Aztreonam SQ 26776	78110-38-0	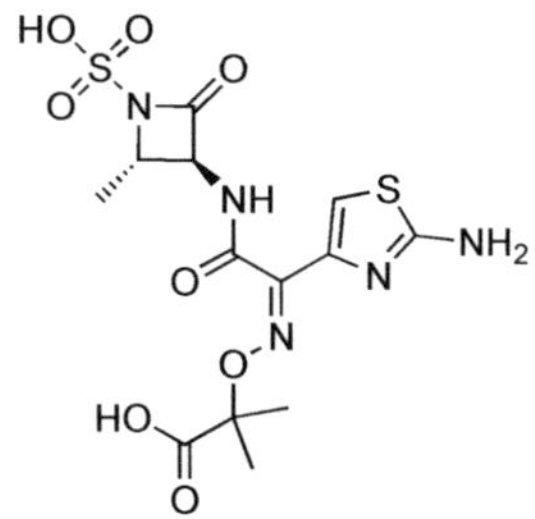	2-((((*Z*)-1-(2-Aminothiazol-4-yl)-2-(((2*S*,3*S*)-2-methyl-4-oxo-1-sulfoazetidin-3-yl) amino)-2-oxoethylidene) amino)oxy)-2-methylpropanoic acid	Treatment of acute bacterial skin and skin structure infections
BAL19764 [127, 128] PTX 2416 SYN 2416	404824-79-9	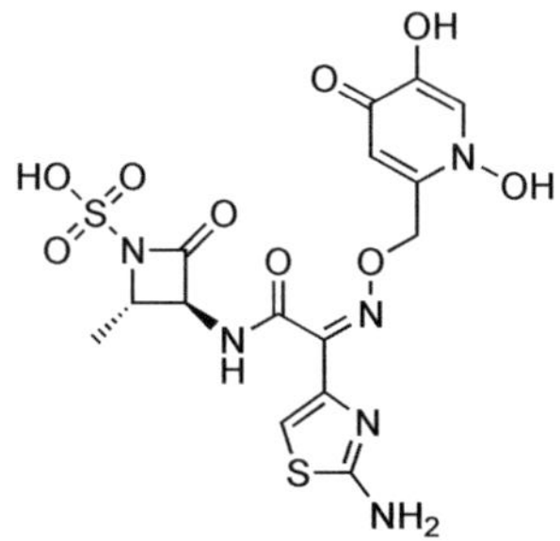	(2*S*,3*S*)-3-((*Z*)-2-(2-Aminothiazol-4-yl)-2-(((1,5-dihydroxy-4-oxo-1,4-dihydropyridin-2-yl) methoxy)imino) acetamido)-2-methyl-4-oxoazetidine-1-sulfonic acid	Against Gram-negative bacilli that express multiple β-Lactamases
BAL29880 [127, 129]	941285-92-3	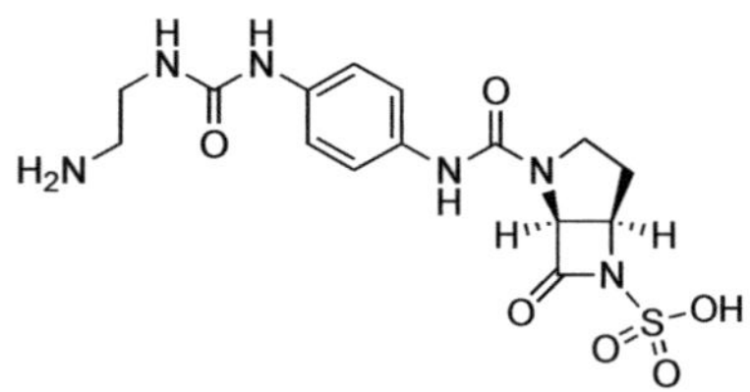	(*1S*,*5R*)-2-((4-(3-(2-Aminoethyl)ureido)phenyl) carbamoyl)-7-oxo-2,6-diazabicyclo[3.2.0]heptane-6-sulfonic acid	β-Lactamase inhibitor

Name	CAS No.	Structure	Full Name	Major Uses
BAL30072 [130]	941285-15-0	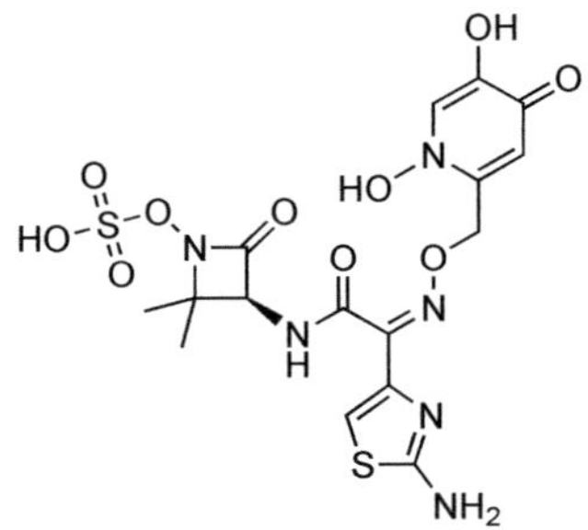	(*S*,*Z*)-3-(2-(2-Aminothiazol-4-yl)-2-(((1,5-dihydroxy-4-oxo-1,4-dihydropyridin-2-yl) methoxy)imino) acetamido)-2,2-dimethyl-4-oxoazetidin-1-yl hydrogen sulfate	Treatment of severe infections caused by multidrug-resistant Gram-negative bacteria
BMS-180680 [131] PA-1806	142654-34-0	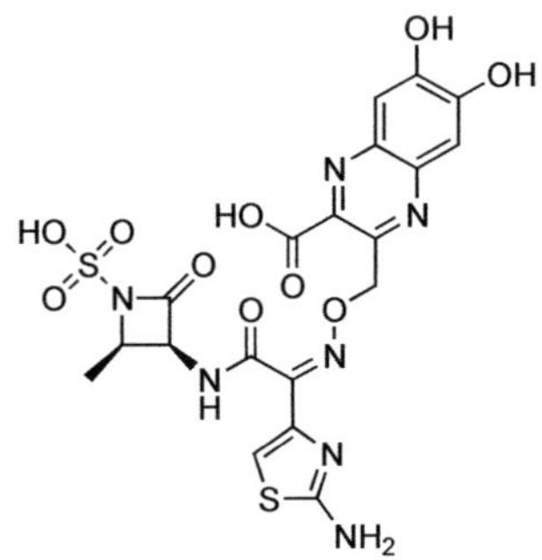	3-(((((*Z*)-1-(2-Aminothiazol-4-yl)-2-(((2*S*,3*R*)-2-methyl-4-oxo-1-sulfoazetidin-3-yl) amino)-2-oxoethylidene) amino)oxy)methyl)-6,7-dihydroxyquinoxaline-2-carboxylic acid	Potent activity against *Pseudomonas aeruginosa*, *Burkholderia cepacia*, *Stenotrophomonas maltophilia* and other Gram-negative organisms

TABLE 1.4 *(Continued)*

Name	CAS No.	Structure	Full Name	Major Uses
BO-1165 [132]	89426-64-2	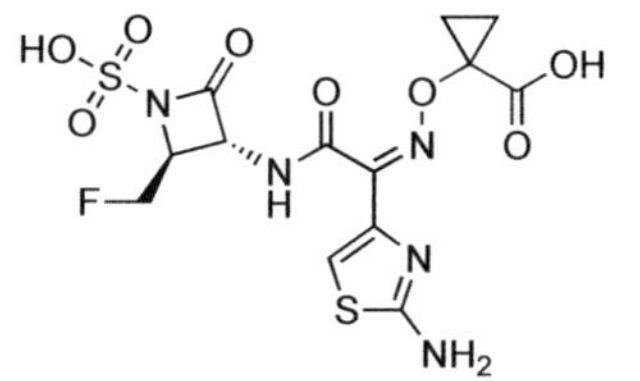	1-((((*Z*)-1-(2-Aminothiazol-4-yl)-2-(((2*R*,3*S*)-2-(fluoromethyl)-4-oxo-1-sulfoazetidin-3-yl)amino)-2-oxoethylidene)amino)oxy)cyclopropane-1-carboxylic acid	Against the majority of *Escherichia coli*, *Klebsiella pneumoniae*, *Klebsiella oxytoca*, *Citrobacter diversus*, *Aeromonas hydrophila*, *Proteus mirabilis*, *Proteus vulgaris*, *Providencia rettgeri*, *Providencia stuartii*, *Yersinia enterocolitica*, *etc.*
BO-1166 [133]	104266-31-1	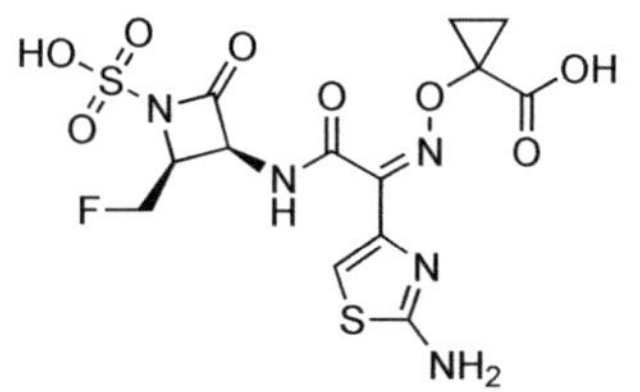	1-((((*Z*)-1-(2-Aminothiazol-4-yl)-2-(((2*S*,3*S*)-2-(fluoromethyl)-4-oxo-1-sulfoazetidin-3-yl)amino)-2-oxoethylidene)amino)oxy)cyclopropane-1-carboxylic acid	Strong antibacterial activity against Gram-negative bacteria including *Pseudomonas aeruginosa* and good stability to various β-lactamases
Carumonam [134] RO17-2301	87638-04-8	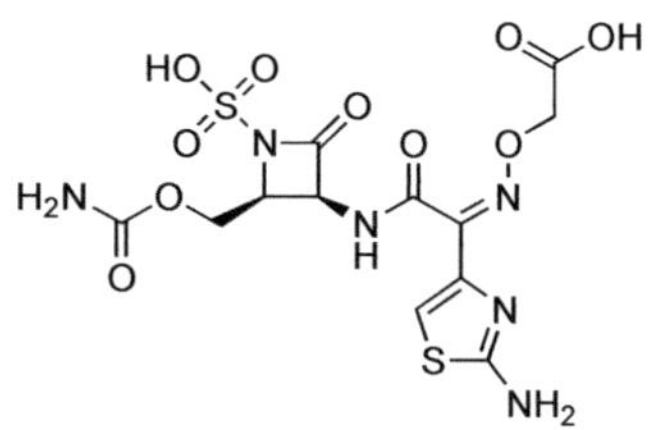	2-((((*Z*)-1-(2-Aminothiazol-4-yl)-2-(((2*S*,3*S*)-2-((carbamoyloxy)methyl)-4-oxo-1-sulfoazetidin-3-yl)amino)-2-oxoethylidene)amino)oxy)acetic acid	Treating bacterial infection caused by bacteria producing *Klebsiella pneumoniae* carbapenemase (KPC) enzymes

Name	CAS No.	Structure	Full Name	Major Uses
EM5210 [135] SQ 26445	79720-09-5	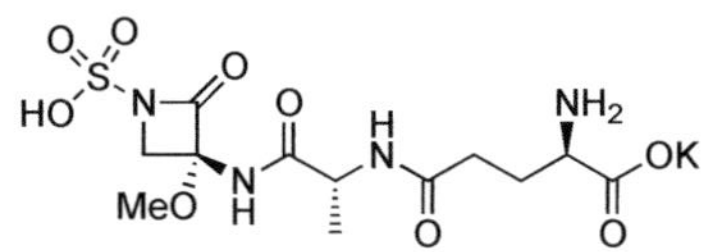	Potassium N^5-((*S*)-1-(((*R*)-3-Methoxy-2-oxo-1-sulfoazetidin-3-yl)amino)-1-oxopropan-2-yl)-*D*-glutaminate	Moderate antibacterial activity against a variety of strains, particularly some members of the Enterobacteriaceae
SQ 26700 [136, 137]	81919-28-0	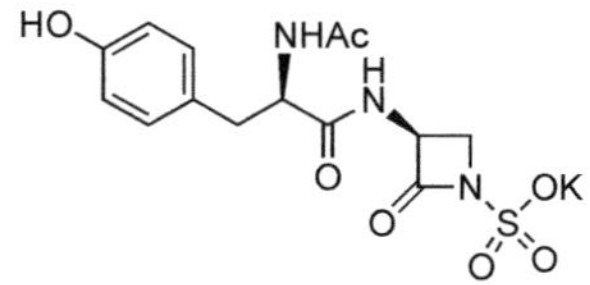	Potassium (*S*)-3-((*R*)-2-Acetamido-3-(4-hydroxyphenyl)propanamido)-2-oxoazetidine-1-sulfonate	Weak antibacterial activity but highly stable to hydrolysis by β-lactamases
SQ 26823 [136, 137]	79720-13-1	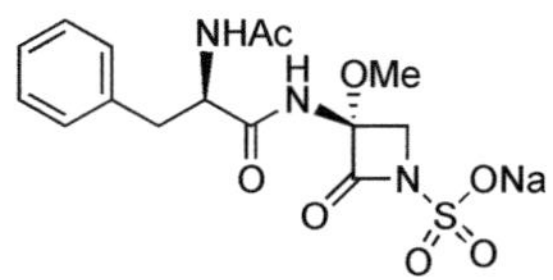	Sodium (*R*)-3-((*R*)-2-Acetamido-3-phenylpropanamido)-3-methoxy-2-oxoazetidine-1-sulfonate	Weak antibacterial activity but highly stable to hydrolysis by β-lactamases
SQ 26875 [136, 137]	79720-14-2	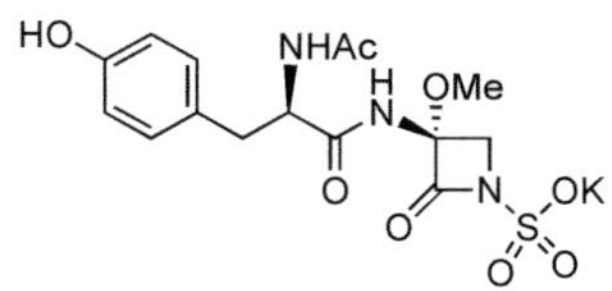	Potassium (*R*)-3-((*R*)-2-Acetamido-3-(4-hydroxyphenyl)propanamido)-3-methoxy-2-oxoazetidine-1-sulfonate	Weak antibacterial activity but highly stable to hydrolysis by β-lactamases
SQ 26970 [136, 137]	79720-16-4	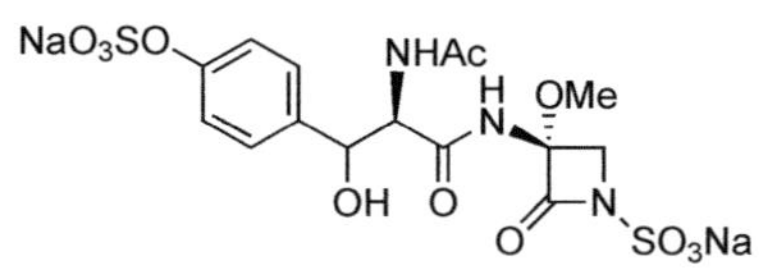	Sodium 4-((2*R*)-2-Acetamido-1-hydroxy-3-(((*R*)-3-methoxy-2-oxo-1-sulfonatoazetidin-3-yl)amino)-3-oxopropyl)phenyl sulfate	Weak antibacterial activity but highly stable to hydrolysis by β-lactamases

TABLE 1.4 *(Continued)*

Name	CAS No.	Structure	Full Name	Major Uses
SQ 26812 [136, 137] M138	79720-17-5	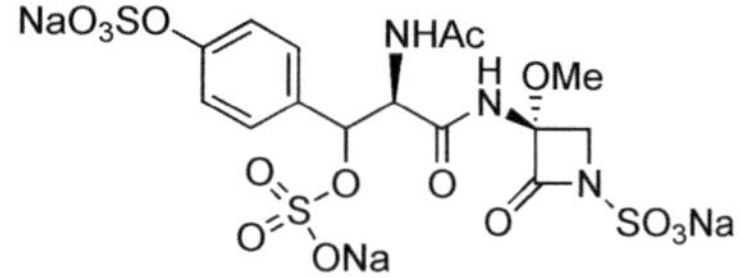	Sodium 4-((2R)-2-Acetamido-3-(((*R*)-3-methoxy-2-oxo-1-sulfonatoazetidin-3-yl) amino)-3-oxo-1-(sulfonatooxy)propyl) phenyl sulfate	Weak antibacterial activity but highly stable to hydrolysis by β-lactamases
Ezetimibe [138] Ezedoc Ezentia Ezetrol Sch 58235	163222-33-1	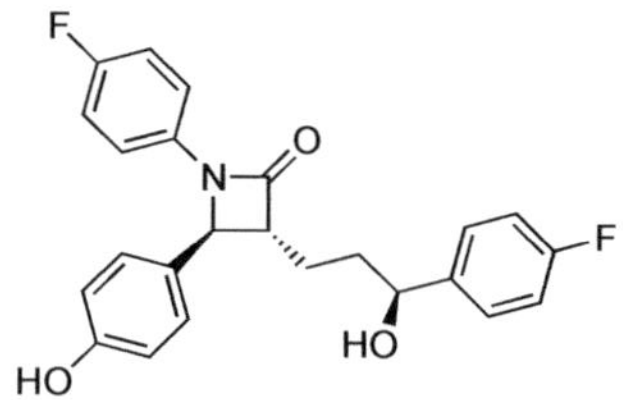	(*3R,4S*)-1-(4-Fluorophenyl)-3-((*S*)-3-(4-fluorophenyl)-3-hydroxypropyl)-4-(4-hydroxyphenyl) azetidin-2-one	First-in-class cholesterol absorption inhibitor
Formadicin C TAN-585C [139, 140] PA 42702A	99124-79-5	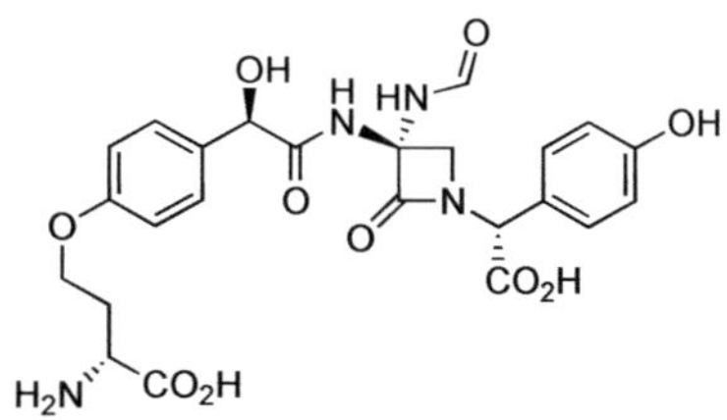	O-(4-((*R*)-2-(((*S*)-1-((*R*)-Carboxy(4-hydroxyphenyl) methyl)-3-formamido-2-oxoazetidin-3-yl)amino)-1-hydroxy-2-oxoethyl) phenyl)-*D*-homoserine	Highly active against some species of *Pseudomonas*, *Proteus*, and *Alcaligenes*. Resistant to β-lactamase and effective against Gram-negative bacteria

Name	CAS No.	Structure	Full Name	Major Uses
Gloximonam [10, 141] SQ 82531	90850-05-8	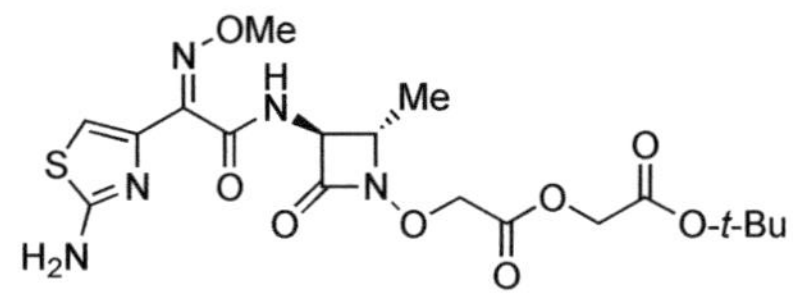	2-(*tert*-Butoxy)-2-Oxoethyl 2-(((2*S*,3*S*)-3-((*Z*)-2-(2-aminothiazol-4-yl)-2-(methoxyimino) acetamido)-2-methyl-4-oxoazetidin-1-yl)oxy) acetate	Antibacterial agent, prodrug of oximonam to improve the oral availability
Isosulfazecin [142]	77912-79-9	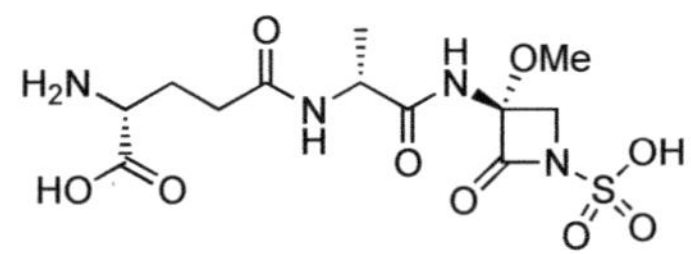	N^5-((*R*)-1-(((*R*)-3-Methoxy-2-oxo-1-sulfoazetidin-3-yl) amino)-1-oxopropan-2-yl)-*D*-glutamine	Weakly active against Gram-positive and Gram-negative bacteria and strongly active against mutants hypersensitive to β-lactam antibiotics.
LYS228 [143]	1810051-96-7	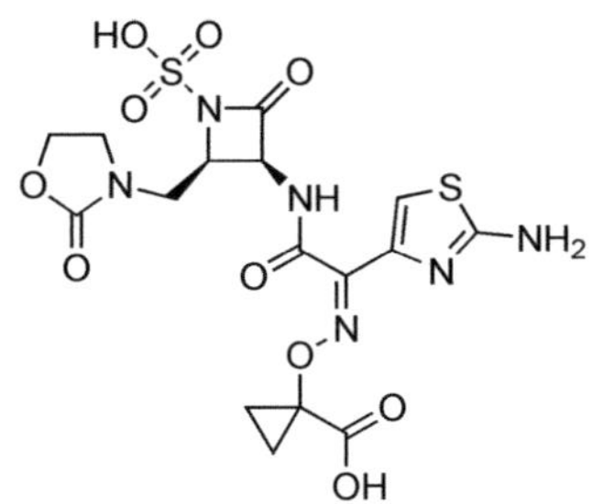	1-((((*Z*)-1-(2-Aminothiazol-4-yl)-2-oxo-2-(((3*S*,4*R*)-2-oxo-4-((2-oxooxazolidin-3-yl) methyl)-1-sulfoazetidin-3-yl)amino)ethylidene) amino)oxy)cyclopropane-1-carboxylic acid	Against *Escherichia coli* and *K. pneumoniae* strains, as well as inactivating isolates encoding TEM-1, SHV-1, CTX-M-14, CTX-M-15, KPC-2, KPC-3, and NDM-1 β-lactamases.

TABLE 1.4 *(Continued)*

Name	CAS No.	Structure	Full Name	Major Uses
MM 42842 [144]	113784-28-4	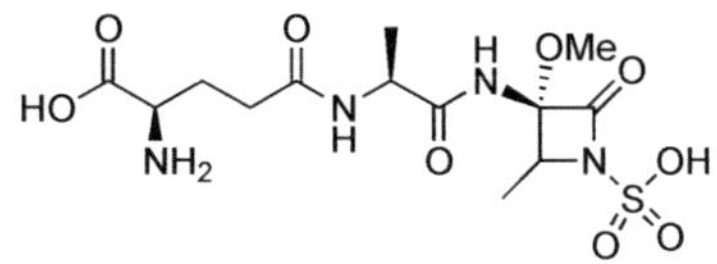	N^5-((2*S*)-1-(((3*S*)-3-Methoxy-2-methyl-4-oxo-1-sulfoazetidin-3-yl) amino)-1-oxopropan-2-yl)-*D*-glutamine	With activity similar to MM 42842
Monamphilectine A [145]	1256166-32-1	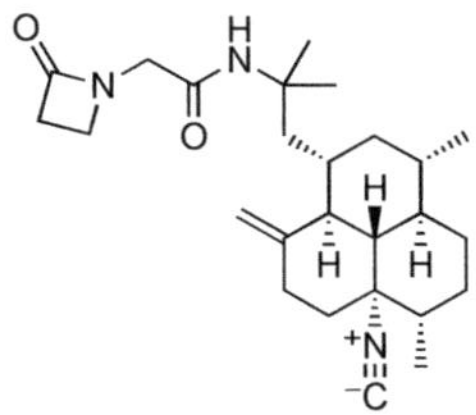	N-(1-((1*S*,3*S*,3a*R*,3a1*S*,6*S*,6a*S*,9a*S*)-6a-Isocyano-3,6-dimethyl-9-methylenedodecahydro-1*H*-phenalen-1-yl)-2-methylpropan-2-yl)-2-(2-oxoazetidin-1-yl)acetamide	Potent antimalarial activity
Nocardicin A [52, 146]	39391-39-4	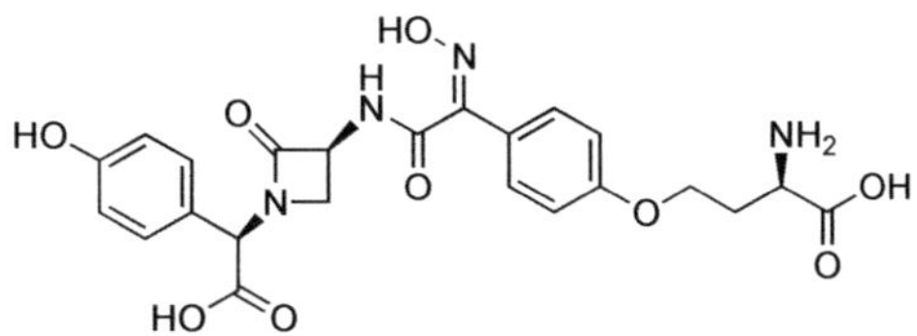	O-(4-((*Z*)-2-(((*S*)-1-((*R*)-Carboxy(4-hydroxyphenyl) methyl)-2-oxoazetidin-3-yl) amino)-1-(hydroxyimino)-2-oxoethyl) phenyl)-*D*-homoserine	Against bacterial infection, enhancement of humoral antibody formation, and anticancer activity in experimental animals
Nocardicin B [52, 146]	60134-71-6	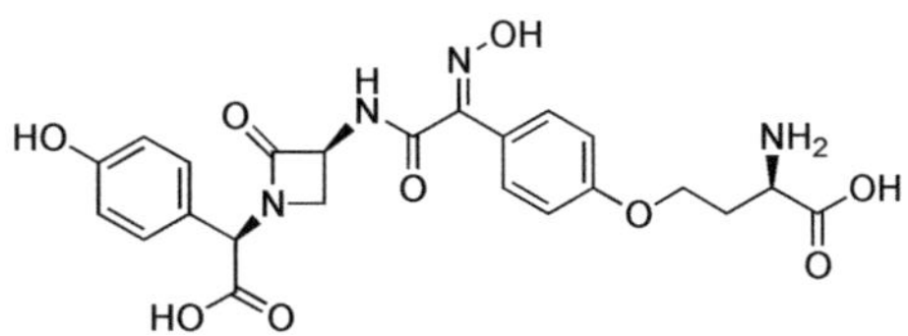	O-(4-((*E*)-2-(((*S*)-1-((*R*)-Carboxy(4-hydroxyphenyl) methyl)-2-oxoazetidin-3-yl) amino)-1-(hydroxyimino)-2-oxoethyl) phenyl)-*D*-homoserine	Against bacterial infection, enhancement of humoral antibody formation, and anticancer activity in experimental animals

Name	CAS No.	Structure	Full Name	Major Uses
Nocardicin C [52, 146]	99297-17-3	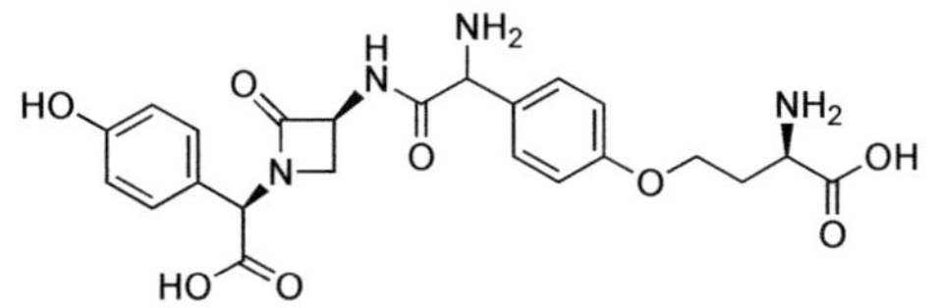	O-(4-(1-Amino-2-(((*S*)-1-((*R*)-Carboxy(4-hydroxyphenyl) methyl)-2-oxoazetidin-3-yl)amino)-2-oxoethyl) phenyl)-*D*-homoserine	Against bacterial infection, enhancement of humoral antibody formation, and anticancer activity in experimental animals
Nocardicin D [52, 146]	61425-17-0	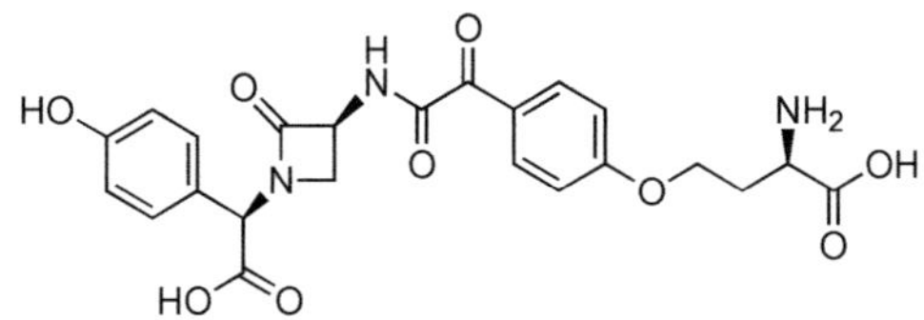	O-(4-(2-(((*S*)-1-((*R*)-Carboxy(4-hydroxyphenyl) methyl)-2-oxoazetidin-3-yl)amino)-2-oxoacetyl) phenyl)-*D*-homoserine	Against bacterial infection, enhancement of humoral antibody formation, and anticancer activity in experimental animals
Nocardicin E [52, 146]	63555-59-9	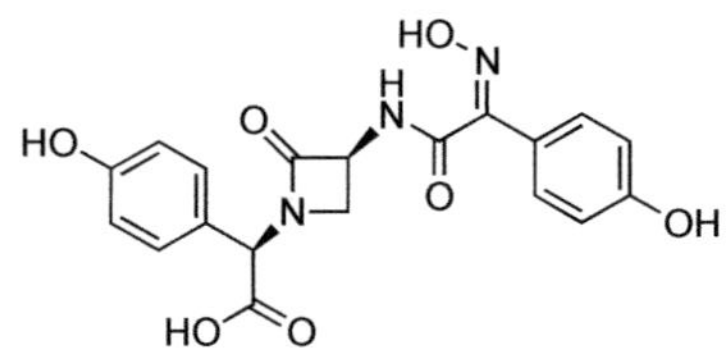	(*R*)-2-((*S*)-3-((*Z*)-2-(Hydroxyimino)-2-(4-hydroxyphenyl) acetamido)-2-oxoazetidin-1-yl)-2-(4-hydroxyphenyl) acetic acid	Against bacterial infection, enhancement of humoral antibody formation, and anticancer activity in experimental animals
Nocardicin F [52, 146]	63598-46-9	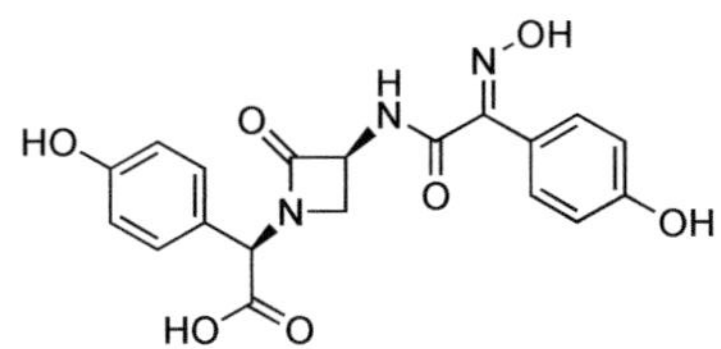	(*R*)-2-((*S*)-3-((*E*)-2-(Hydroxyimino)-2-(4-hydroxyphenyl) acetamido)-2-oxoazetidin-1-yl)-2-(4-hydroxyphenyl) acetic acid	Against bacterial infection, enhancement of humoral antibody formation, and anticancer activity in experimental animals

TABLE 1.4 *(Continued)*

Name	CAS No.	Structure	Full Name	Major Uses
Nocardicin G [52, 146]	65309-11-7	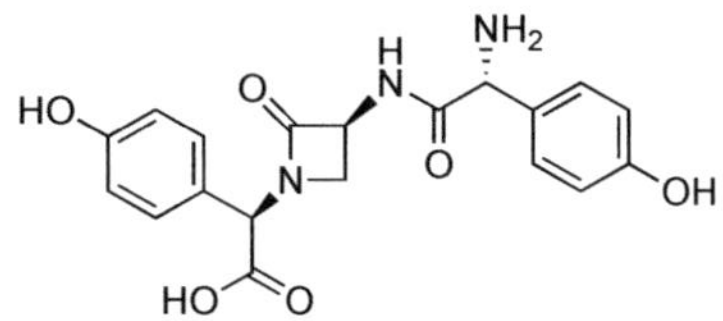	(*R*)-2-((*S*)-3-((*R*)-2-Amino-2-(4-hydroxyphenyl) acetamido)-2-oxoazetidin-1-yl)-2-(4-hydroxyphenyl) acetic acid	Against bacterial infection, enhancement of humoral antibody formation, and anticancer activity in experimental animals
Oximonam [147, 148] SQ 82291	90898-90-1	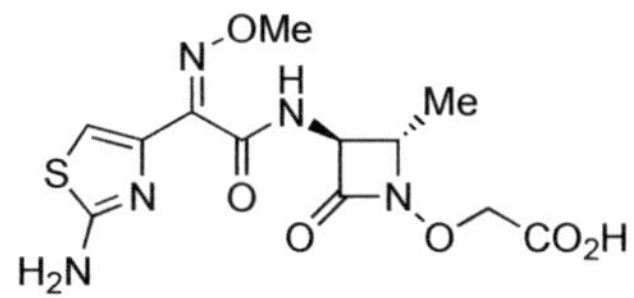	2-(((2*S*,3*S*)-3-((*Z*)-2-(2-Aminothiazol-4-yl)-2-(methoxyimino) acetamido)-2-methyl-4-oxoazetidin-1-yl)oxy)acetic acid	Good activity against different bacteria of the family *Enterobacteriaceae* and *Haemophilus influenzae*
Pachystermine A	6156-99-6 [149]	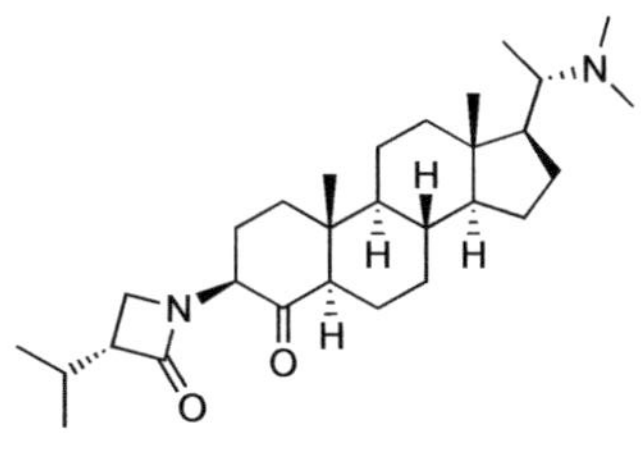	(*R*)-1-((3*S*,5*R*,8*S*,9*S*,10*R*,13*S*,14*S*,17*S*)-17-((*S*)-1-(Dimethylamino) ethyl)-10,13-dimethyl-4-oxohexadecahydro-1*H*-cyclopenta[a] phenanthren-3-yl)-3-isopropylazetidin-2-one	Inhibiting breast cancer invasion and potential antimetastasis therapeutic agent
PB-5266 A [150, 151]	108065-94-7	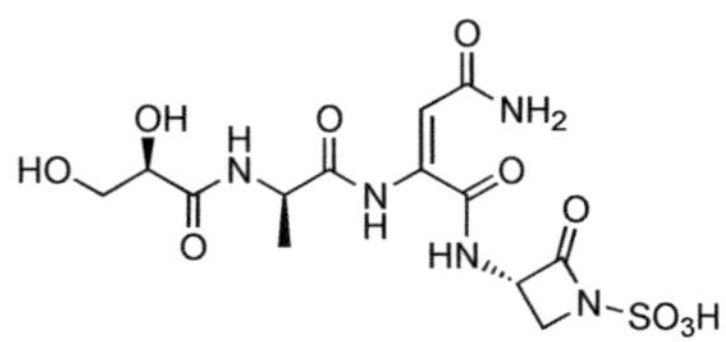	(*S*)-3-((*E*)-4-Amino-2-((*R*)-2-((*R*)-2,3-dihydroxypropanamido) propanamido)-4-oxobut-2-enamido)-2-oxoazetidine-1-sulfonic acid	Exhibited weak antibacterial activity against a mutant of *Escherichia coli* sensitive to β-lactam antibiotics.

Name	CAS No.	Structure	Full Name	Major Uses
PB-5266 B [150, 151]	108065-95-8	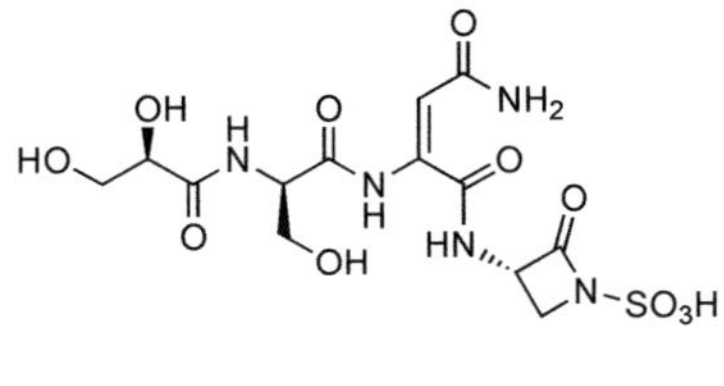	(*S*)-3-((*E*)-4-Amino-2-((*R*)-2-((*R*)-2,3-dihydroxypropanamido)-3-hydroxypropanamido)-4-oxobut-2-enamido)-2-oxoazetidine-1-sulfonic acid	Exhibited weak antibacterial activity against a mutant of *Escherichia coli* sensitive to β-lactam antibiotics.
PB-5266 C [150, 151]	108065-96-9	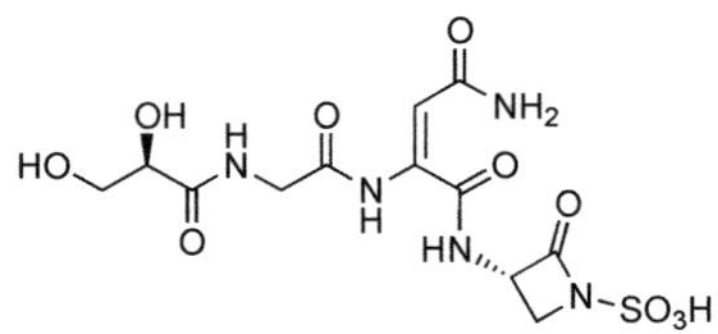	(*S*)-3-((*E*)-4-Amino-2-(2-((*R*)-2,3-dihydroxypropanamido) acetamido)-4-oxobut-2-enamido)-2-oxoazetidine-1-sulfonic acid	Exhibited weak antibacterial activity against a mutant of *Escherichia coli* sensitive to β-lactam antibiotics.
Ro 48-1256 [152]	197702-36-6	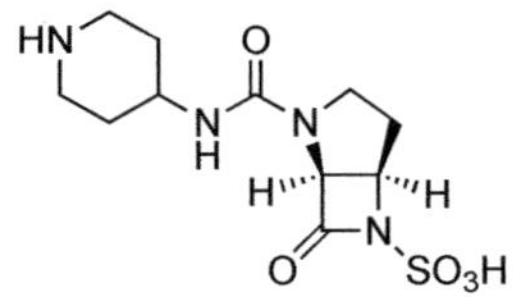	(*1S,5R*)-7-Oxo-2-(Piperidin-4-ylcarbamoyl)-2,6-diazabicyclo[3.2.0] heptane-6-sulfonic acid	Inhibitor of class C β-lactamases, without significant antibacterial activity of its own, can be used with ceftazidime or piperacillin to overcome derepression of AmpC.
RU 44790 [153]	93608-14-1	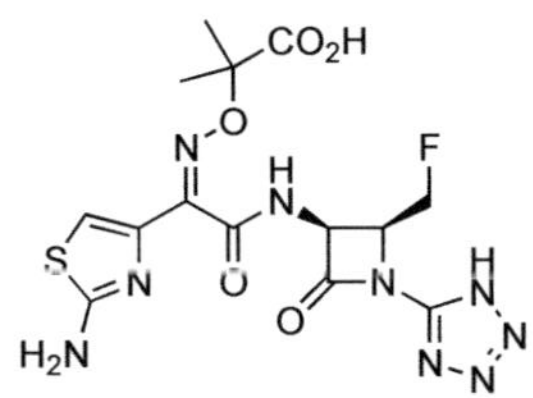	2-((((*Z*)-1-(2-Aminothiazol-4-yl)-2-(((*2S,3S*)-2-(fluoromethyl)-4-oxo-1-(1*H*-tetrazol-5-yl) azetidin-3-yl)amino)-2-oxoethylidene)amino)oxy)-2-methylpropanoic acid	Against Gram-negative bacteria

TABLE 1.4 *(Continued)*

Name	CAS No.	Structure	Full Name	Major Uses
SQ 26180 [154]	80542-73-0	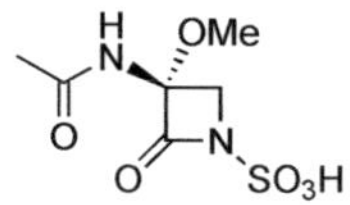	(*R*)-3-Acetamido-3-methoxy-2-oxoazetidine-1-sulfonic acid	Inactivation of both R61 and R39 serine *DD*-peptidases
SQ 26324 [154]	80904-76-3	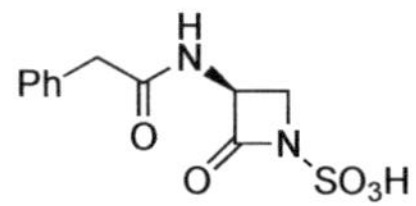	(*S*)-2-Oxo-3-(2-Phenylacetamido)azetidine-1-sulfonic acid	Inactivation of both R61 and R39 serine *DD*-peptidases
SQ 26917 [155]	82691-17-6	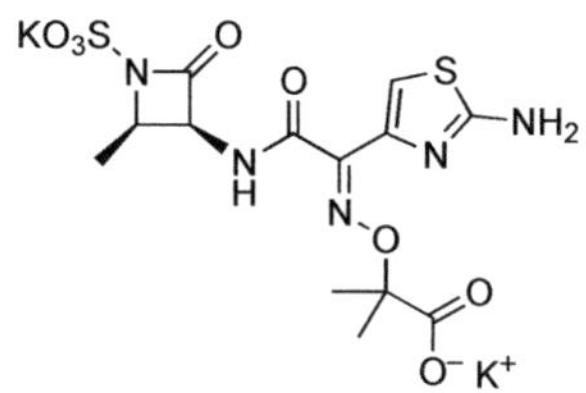	Potassium 2-((((1*E*,2*Z*)-1-(2-Aminothiazol-4-yl)-2-(((2R,3S)-2-methyl-4-oxo-1-sulfoazetidin-3-yl)imino)-2-oxidoethylidene)amino)oxy)-2-methylpropanoate	Against *Pseudomonas aeruginosa*, but no activity against Gram-positive isolates or anaerobic bacteria
SQ 28332 [156] PB 5582A	94992-64-0	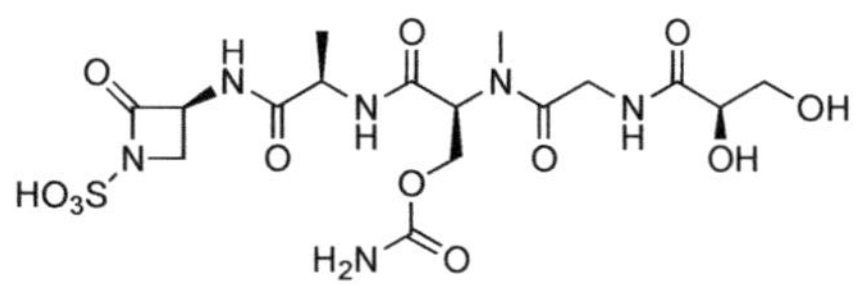	(*S*)-3-((*R*)-2-((*S*)-3-(Carbamoyloxy)-2-(2-((*R*)-2,3-dihydroxypropanamido)-N-methylacetamido)propanamido)propanamido)-2-oxoazetidine-1-sulfonic acid	Against Gram-positive bacteria, esp. *Staphylococcus aureus* and *Streptococcus pneumonia*

Name	CAS No.	Structure	Full Name	Major Uses
Pirazmonam [157] SQ 83360	108319-07-9	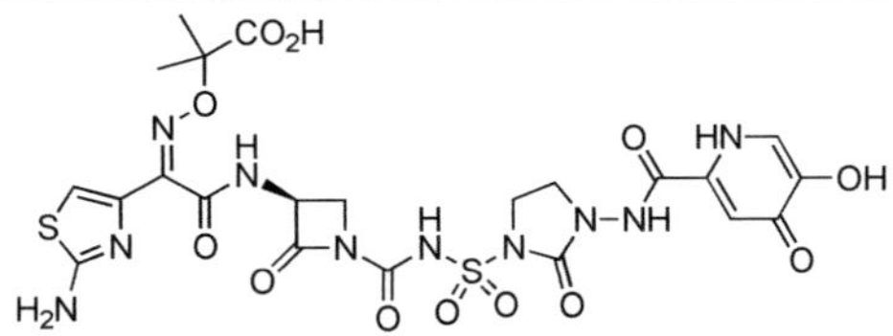	(*S,Z*)-2-(((1-(2-Aminothiazol-4-yl)-2-((1-(((3-(5-hydroxy-4-oxo-1,4-dihydropyridine-2-carboxamido)-2-oxoimidazolidin-1-yl)sulfonyl)carbamoyl)-2-oxoazetidin-3-yl)amino)-2-oxoethylidene)amino)oxy)-2-methylpropanoic acid	Against *P. aeruginosa* and stable to metallo-β-lactamases
Sulfazecin [158]	77912-79-9	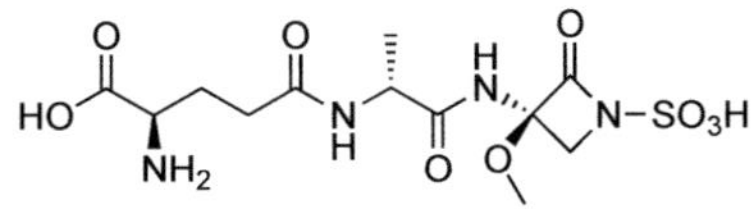	N[5]-((*R*)-1-(((*R*)-3-Methoxy-2-oxo-1-sulfoazetidin-3-yl)amino)-1-oxopropan-2-yl)-*D*-glutamine	Against multi-drug resistant pathogens
Syn2190 [159]	214963-75-4	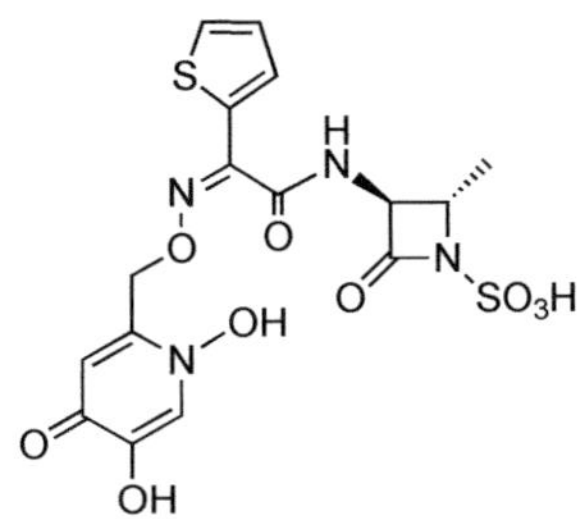	(*2S,3S*)-3-((*E*)-2-(((1,5-Dihydroxy-4-oxo-1,4-dihydropyridin-2-yl)methoxy)imino)-2-(thiophen-2-yl)acetamido)-2-methyl-4-oxoazetidine-1-sulfonic acid	β-lactamase inhibitor, synergist of ceftazidime and cefpirome, good for urinary tract infection with cephalosporin-resistant *P. aeruginosa*
Tabtoxin [160]	40957-90-2	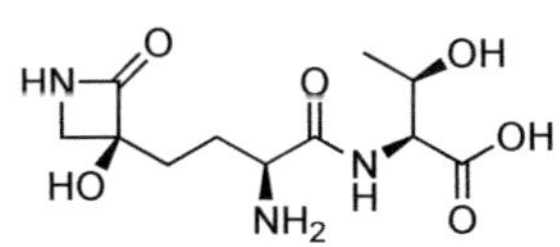	((*S*)-2-Amino-4-((*S*)-3-Hydroxy-2-oxoazetidin-3-yl)butanoyl)-*L*-threonine	Irreversibly inhibits glutamine synthetase

TABLE 1.4 *(Continued)*

Name	CAS No.	Structure	Full Name	Major Uses
TAN-585D [140]	99124-80-8	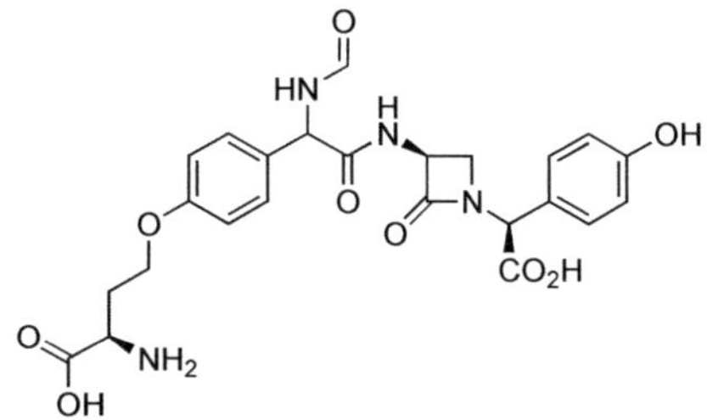	O-(4-(2-(((*S*)-1-((*S*)-Carboxy(4-hydroxyphenyl) methyl)-2-oxoazetidin-3-yl)amino)-1-formamido-2-oxoethyl) phenyl)-*D*-homoserine	Resistant to β-lactamase and effective against Gram-negative bacteria, including *Proteus vulgaris*.
TAN-850 [161]	111243-20-0	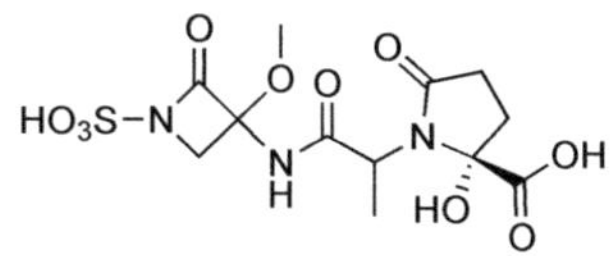	(2*R*)-2-Hydroxy-1-(1-((3-methoxy-2-oxo-1-sulfoazetidin-3-yl) amino)-1-oxopropan-2-yl)-5-oxopyrrolidine-2-carboxylic acid	Weak antibacterial activity and was highly stable to hydrolysis by β-lactamases, inactivates cephalosporinase from *Enterobacter cloaceae*
Tigemonam [162] SQ 30836	102507-71-1	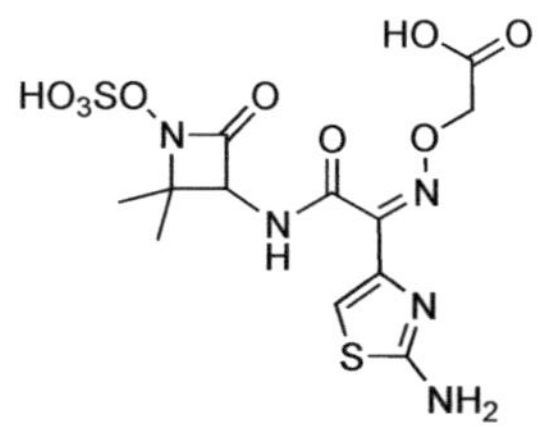	(*Z*)-2-(((1-(2-Aminothiazol-4-yl)-2-((2,2-dimethyl-4-oxo-1-(sulfooxy) azetidin-3-yl)amino)-2-oxoethylidene)amino)oxy) acetic acid	Against *Enterobacteriaceae*, *Haemophilus influenzae*, and *Neisseria gonorrhoeae*

Most known monobactam antibiotics are collected in Table 1.4. These monobactams have demonstrated a relatively high level of β-lactam ring reactivity to achieve good antibacterial activity by a not obviously equipped (fused ring) structure. Presumably, the activation of β-lactam is caused by the strong electronegativity of the sulfamate moiety in lieu of the ring strain effects. Other functional groups of high electronegativity such as phosphonates [-OP(OCH_3)OH, less effective], *N*-aminosulfonyl amides (-CONSONR_2' effective) and O-sulfates (-OSO_3H, very effective) can also be used, in addition to the sulfamate moiety [52].

1.6 CLAVAM ANTIBIOTICS

Compared to the other four classes of β-lactam antibiotics, the number of clavam antibiotics is very limited. Currently, the only notable member of this class is clavulanic acid (i.e., (2*R*,5*R*,*Z*)-3-(2-hydroxyethylidene)-7-oxo-4-oxa-1-azabi-cyclo[3.2.0]heptane-2-carboxylic acid, [CAS 58001-44-8], Figure 1.3), from which this class of compounds receives their names. Structurally, clavams are similar to penams, with oxygen substituted for the sulfur so they are also known as oxapenams. Clavulanic acid was initially isolated from *Streptomyces clavuligerus*, which is available as the potassium salt for oral and intravenous use in combination with amoxicillin [163]. Since only a few compounds of this kind are known as of today, the structures of these compounds are collectively displayed in Figures 1.3 and 1.4, instead of being organized in a table like the other four β-lactam classes. Among these clavams, (-)-2-(2-hydroxyethyl)clavam (i.e., (3*S*,5*S*)-3-(2-hydroxyethyl)-4-oxa-1-azabicyclo-[3.2.0]heptan-7-one, [CAS 79416-52-7], Figure 1.3) was isolated from *Streptomyces antibioticus* [164], 2-(3-alaninyl)clavam (i.e., clavalanine, Ro 22-5417, (*S*)-2-amino-3-((3*S*,5*S*)-7-oxo-4-oxa-1-azabicyclo[3.2.0]heptan-3-yl)propanoic acid, [CAS 74758-63-7], Figure 1.3) [165], as well as 2-hydroxy-methylclavam (i.e., (3*R*,5*S*)-3-(hydroxymethyl)-4-oxa-1-azabicyclo[3.2.0]-heptan-7-one, [CAS 66036-39-3], Figure 1.3), 2-formyloxymethylclavam (i.e., ((3*R*,5*S*)-7-oxo-4-oxa-1-azabicyclo-[3.2.0]heptan-3-yl)methyl formate, [CAS 64341-63-5], Figure 1.3), and clavam-2-carboxylic acid (i.e., (3*R*,5*S*)-7-oxo-4-oxa-1-azabicyclo[3.2.0]heptane-3-carboxylic acid, [CAS 212268-81-0], Figure 1.3) [166] were isolated from *Streptomyces clavuligerus*. Ro 22-5417 has demonstrated antimicrobial activities against a variety of bacteria and fungi [167]. Clavam antibiotics CA-146A (i.e., (3*R*)-2-((*S*)-2-amino-3-

methylbutanamido)-3-hydroxy-3-((3*R*,5*S*)-7-oxo-4-oxa-1-azabicyclo[3.2.0]heptan-3-yl)propanoic acid, [CAS 103059-95-6], Figure 1.3) andCA-146B(i.e.,(3*R*)-2-((*S*)-2-aminopropanamido)-3-hydroxy-3-((3*R*,5*S*)-7-oxo-4-oxa-1-azabicyclo[3.2.0]-heptan-3-yl)propanoic acid, [CAS 108147-53-1], Figure 1.3) were isolated from *Streptomyces lavendulae* CA-146, showing activity against microorganisms, e.g., *Candida albicans* [168]. Similarly, clavam antibiotic CA-31 (i.e., (3*R*)-2-(2-amino-3-hydroxy-3-((5*S*)-7-oxo-4-oxa-1-azabicyclo[3.2.0]heptan-3-yl)propanamido)-3-hydroxy-3-((3*R*,5*S*)-7-oxo-4-oxa-1-azabicyclo[3.2.0]heptan-3-yl)propanoic acid, [CAS 103059-93-4], Figure 1.3) was isolated from the culture of *Streptomyces platensis* CA-31, which also inhibits microorganisms, e.g., *Candida albicans* [169].

A series of clavam antibiotics known as clavamycins were isolated from two variants of *Streptomyces hygroscopicus,* NRRL 15846 and NRRL 15879. Examples of these clavamycins include clavamycin A (i.e., (3*R*)-2-((3*S*)-2-amino-3-hydroxy-3-((3*R*,5*S*)-7-oxo-4-oxa-1-azabicyclo [3.2.0]-heptan-3-yl)propanamido)-3-hydroxy-3-((3*R*,5*S*)-7-oxo-4-oxa-1-azabicyclo[3.2.0]-heptan-3-yl)propanoic acid, [CAS 103059-93-4], Figure 1.4), clavamycin B (i.e., 5-amino-2-((3*R*)-2-amino-3-hydroxy-3-((3*R*,5*S*)-7-oxo-4-oxa-1-azabicyclo[3.2.0]-heptan-3-yl)propanamido)-3,4-dihydroxypentanoic acid, [CAS 103080-36-0], Figure 1.4), clavamycin C (i.e., (3*R*)-2-(2,5-diamino-3,4-dihydroxypentanamido)-3-hydroxy-3-((3*R*,5*S*)-7-oxo-4-oxa-1-azabi-cyclo[3.2.0]heptan-3-yl)propanoic acid, [CAS 103059-94-5], Figure 1.4), clavamycin D (i.e., (3*R*)-2-((*S*)-2-amino-3-methylbutanamido)-3-hydroxy-3-((3R,5S)-7-oxo-4-oxa-1-azabicyclo[3.2.0] heptan-3-yl)propanoic acid, [CAS 103059-95-6], Figure 1.4), clavamycin E (i.e., (3*R*)-2-((*S*)-2-aminopropanamido)-3-hydroxy-3-((3*R*,5*S*)-7-oxo-4-oxa-1-azabicyclo-[3.2.0]heptan-3-yl)propanoic acid, [CAS 108147-53-1], Figure 1.4), and clavamycin F (i.e., (3*R*)-2-((*S*)-5-acetamido-2-aminopentanamido)-3-hydroxy-3-((3*R*,5*S*)-7-oxo-4-oxa-1-azabicyclo[3.2.0]heptan-3-yl)propanoic acid, [CAS 103059-97-8], Figure 1.4). Among these clavamycins, clavamycin A and D are potent inhibitors of *Candida* spp. including *Candida albicans*, and are highly active against *Trichophyton quinckeanum* and *Neurospora crassa* [170, 171].

Ignoring the difference in stereochemistry due to the unspecified stereostructures of some clavam antibiotics, clavam G0069A should be identical to clavamycin D, G0069B-3 should be 2-hydroxymethylclavam and clavam G0069C should be clavamycin E. In addition, clavam G0069A has antibiotic activity against cancer cell lines, with ED_{50} at ~0.03-0.6 ppm and weak antimicrobial activity against yeast and filamentous fungi [172]. Also, the structure of valclavam (might be Tu 1718z, CAS number 98359-78-5) is somewhat ambiguous, as it has been shown as either 2-(2-amino-3-methylbutanamido)-3-hydroxy-4-((3*S*,5*S*)-7-oxo-4-oxa-1-azabicyclo[3.2.0] heptan-3-yl)butanoic acid [173], or 3-((*S*)-2-amino-3-methylbutanamido)-2-hydroxy-4-((3*S*,5*S*)-7-oxo-4-oxa-1-azabicyclo[3.2.0]heptan-3-yl)butanoic acid, with antibacterial and antifungal activity [174]. Thus, the structures of these clavam antibiotics are not provided.

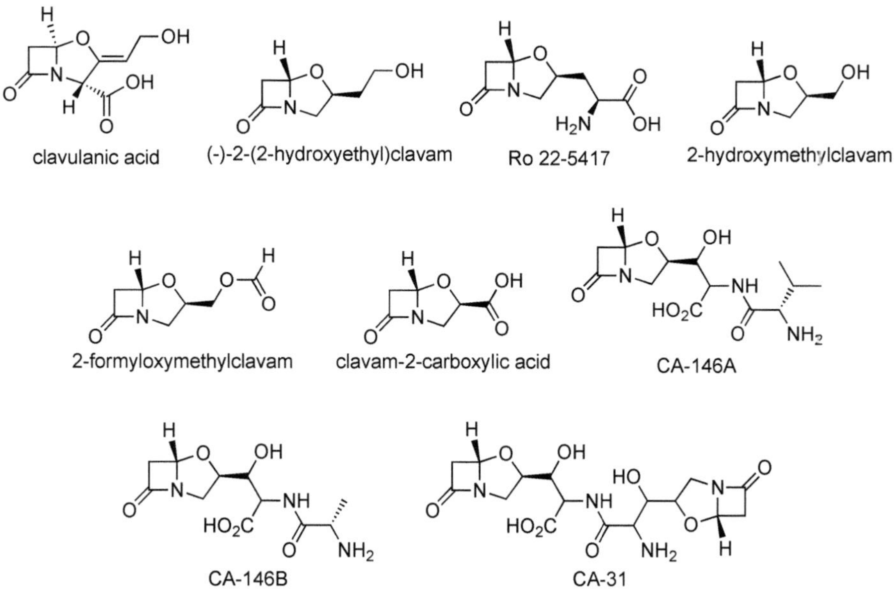

FIGURE 1.3 The list of clavam antibiotics

1.7 MISCELLANEOUS Β-LACTAM ANTIBIOTICS

In addition to these five classes of β-lactams, i.e., penicillins, cephalosporins, monobactams, carbapenems, and clavams, some other antibiotics also bear a β-lactam moiety. In fact, all β-lactam-containing antibiotics have been classified into different classes, according to the core structure of β-lactams. For example, penicillins are penam antibiotics, cephalosporins belong to cephem antibiotics, and clavams are oxapenam antibiotics. All these related core structures of β-lactam antibiotics are displayed in Figure 1.5 [163, 175].

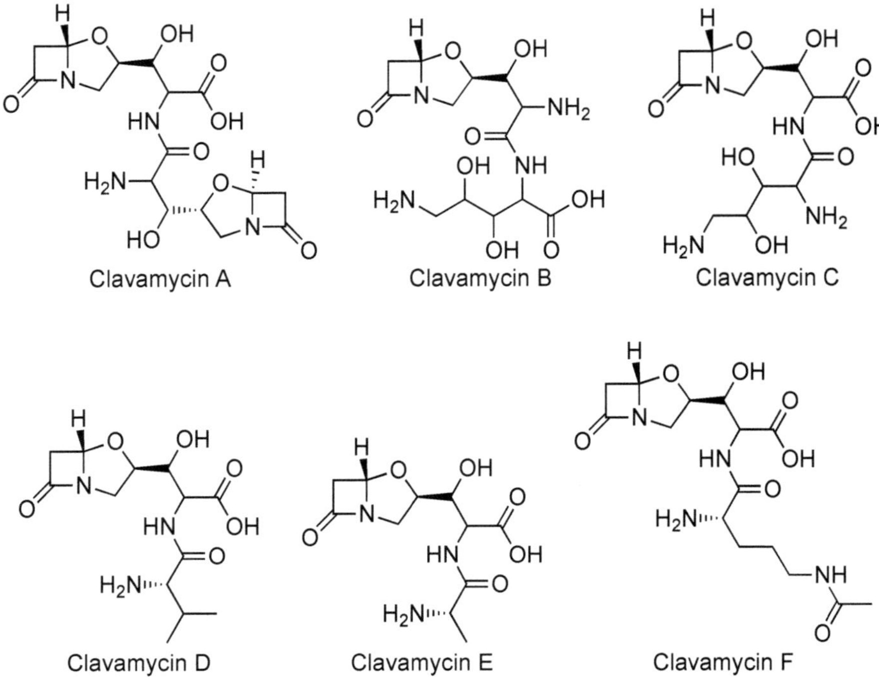

FIGURE 1.4 The list of clavamycin antiobiotics

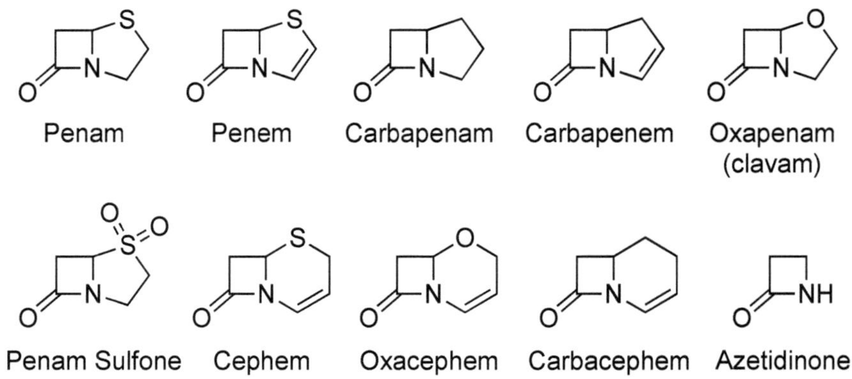

FIGURE 1.5 The core structures of β-lactam antibiotics

In addition to the above five major classes of β-lactam antibiotics, the representative penem antibiotics are BRL 42175, CGP 31608, CP-65,207, faropenem (SY5555 or SUN 5555, Faropenem sodium), FCE22891, HRE 664, MEN 10700, Ritipenem (FCE22101), Sch 29482, Sch 34343, sulopenem (CP-70,429), SYN-1012 and TMA-230. They are described in detail below.

While BRL 42715 (i.e., (*R*,*Z*)-6-((1-methyl-1*H*-1,2,3-triazol-4-yl)methylene)-7-oxo-4-thia-1-azabicyclo[3.2.0]hept-2-ene-2-carboxylic acid, [CAS 108413-41-8], Figure 1.6) alone possesses no apparent antibacterial activity except for methicillin-sensitive *Staphylococci*, it is a β-lactamase inhibitor with an intrinsic activity superior to that of clavulanic acid, sulbactam, and tazobactam. What is more, it can inactivate the plasmid-mediated OXA-1 β-lactamase and the chromosomally mediated enzymes of *Enterobacter*, *Citrobacter*, *Pseudomonas*, *Serratia*, and *Morganella* species, for which other inhibitors are even less effective. Particularly, BRL 42715 is able to strengthen the antibacterial properties of a wide range of β-lactamase-susceptible β-lactam antibiotics, including the third-generation cephalosporins [176]. CGP 31608 (i.e., (5*R*,6*S*)-3-(aminomethyl)-6-((*R*)-1-hydroxyethyl)-7-oxo-4-thia-1-azabicyclo[3.2.0]hept-2-ene-2-carboxylic acid, [CAS 107740-67-0], Figure 1.6), as a semi-synthetic antibiotic, has antibacterial activity against *Enterobacteriaceae*, *Pseudomonas aeruginosa*, *Haemophilus influenzae*, *Neisseria spp*., *Bacteroides spp*., *Clostridium spp*., *Staphylococci*, and *Streptococcus pneumoniae*, with a minimal inhibitory concentration (MIC) between 0.25 and 8 μg/mL against 90% of the 641 strains from the above family [177]. One of the most important penem antibiotics might be faropenem (i.e., (5*R*,6*S*)-6-((*R*)-1-hydroxyethyl)-7-oxo-3-((*R*)-tetrahydrofuran-2-yl)-4-thia-1-azabicyclo[3.2.0]hept-2-ene-2-carboxylic acid, [CAS 106560-14-9]), which has appeared in more than 600 entries via the SciFinder® search and demonstrated a broad-spectrum *in vitro* antimicrobial activity against many Gram-positive and Gram-negative aerobes and anaerobes, in addition to its resistance for hydrolysis by nearly all β-lactamases. On the other hand, faropenem is not very active against methicillin-resistant *Staphylococcus aureus* (MRSA), vancomycin-resistant *Enterococcus faecium*, *Pseudomonas aeruginosa* or *Stenotrophomonas maltophilia* [178]. Based on the treatment of over 1000 pediatric and adult patients with faropenem and referring 10 antimicrobials with CLSI microdilution method, faropenem showed excellent antibacterial activity against methicillin-susceptible *Staphylococcus* and *Streptococcus* [179]. The corresponding sodium salt of faropenem is known as SUN5555 (or SY5555) [180, 181]. CP-65,207 (i.e., (5*R*,6*S*)-6-((*R*)-1-hydroxyethyl)-3-((1-oxidotetrahydrothiophen-3-yl)thio)-7-oxo-4-thia-1-azabicyclo[3.2.0]hept-2-ene-2-carboxylic acid, [CAS 120788-07-0] and [CAS 128443-48-1]) is a 1:1 mixture of two stereoisomers varying at the side chain moiety, with a broad-spectrum of antimicrobial activity [182, 183]. FCE 22891 (i.e., acetoxymethyl (5*R*,6*S*)-3-((carbamoyloxy)methyl)-6-((*R*)-1-hydroxyethyl)-7-oxo-4-thia-1-azabicyclo-[3.2.0]hept-2-ene-2-carboxylate, [CAS 87238-52-6], Figure 1.6) is the oral prodrug of FCE 22101, a new broad-spectrum penem antibiotic [184]. HRE 664 (i.e., (5*R*,6*S*)-3-(4-carbamoylbenzyl)-6-((*R*)-1-hydroxyethyl)-7-oxo-

4-thia-1-azabicyclo[3.2.0]-hept-2-ene-2-carboxylic acid, Figure 1.6) displays potent *in vitro* antibacterial activity against a broad spectrum of clinically relevant bacterial strains including Gram-negative and Gram-positive aerobes and anaerobes. In addition, it is extremely stable against β-lactamases and is also active against methicillin-resistant *Staphylococci* [185]. MEN 10700 (i.e., (5*R*,6*S*)-3-(((2-amino-2-oxoethyl)(methyl)amino)methyl)-6-((*R*)-1-hydroxyethyl)-7-oxo-4-thia-1-azabicyclo[3.2.0]hept-2-ene-2-carboxylic acid, [CAS 195874-55-6], Figure 1.6), is a penem β-lactam antibiotic developed at Menarini with a broad-spectrum activity against Gram-positive and Gram-negative bacteria, with a potency greater than that of imipenem [186]. Ritipenem, also known as FCE22101 (i.e., (5*R*,6*S*)-3-((carbamoyloxy)methyl)-6-((*R*)-1-hydroxyethyl)-7-oxo-4-thia-1-azabicyclo[3.2.0]-hept-2-ene-2-carboxylic acid, [CAS 84845-57-8], Figure 1.6), is a novel oral penem antibiotic with a broad antibacterial spectrum and an even stronger agent against staphylococcal than other oral cephalosporins tested [187]. Sch 29482 (i.e., (5*R*,6*S*)-3-(ethylthio)-6-((*R*)-1-hydroxyethyl)-7-oxo-4-thia-1-azabicyclo[3.2.0]hept-2-ene-2-carboxylic acid, [CAS 77646-83-4], Figure 1.6) is a relatively old penem antibiotic with high activity against Gram-positive and Gram-negative organisms excluding *Pseudomonas*. It is stable to β-lactamases and can be absorbed orally [188]. Although Sch 34343 (i.e., (5*R*,6*S*)-3-((2-(carbamoyloxy)ethyl)thio)-6-((*R*)-1-hydroxyethyl)-7-oxo-4-thia-1-azabicyclo[3.2.0]hept-2-ene-2-carboxylic acid, [CAS 95415-91-1], Figure 1.6) is inactive against *Pseudomonas aeruginosa*, it is comparable to or superior to the third generation cephalosporins against *Enterobacteriaceae* [189].

Sulopenem (CP 70429, BCP17608, i.e., (5*R*,6*S*)-6-((*R*)-1-hydroxyethyl)-3-(((3*S*)-1-oxidotetrahydrothiophen-3-yl)thio)-7-oxo-4-thia-1-azabicyclo[3.2.0]hept-2-ene-2-carboxylic acid, [CAS 120788-07-0], Figure 1.7) has broad-spectrum activities against Gram-positive and Gram-negative bacteria, equivalent to or somewhat superior to imipenem in terms of inhibiting methicillin-sensitive *Staphylococcus aureus*, *Staphylococcus epidermidis*, *Enterococcus faecalis*, *Streptococcus pyogenes* and *Streptococcus pneumoniae* [190]. In contrast, CP 81054 (i.e., (5*R*,6*S*)-6-((*R*)-1-hydroxyethyl)-3-(((3*R*)-1-oxidotetrahydrothiophen-3-yl)thio)-7-oxo-4-thia-1-azabicyclo[3.2.0]hept-2-ene-2-carboxylic acid, [CAS 128443-48-1], Figure 1.7) is less active [182]. A 1:1 mixture of these two antibiotics are collectively called CP 65,207, of which the CP 70429 is identical to CP 65207-S whereas CP 81054 is CP 65207-R [183]. SYN-1012 (i.e., (*R*,*Z*)-6-((*E*)-3-(1-methyl-1*H*-1,2,3-triazol-4-yl)allylidene)-7-oxo-4-thia-1-azabicyclo[3.2.0]hept-2-ene-2-carboxylic acid, [CAS 763907-42-2], Figure 1.7), as indicated by its name, is a synthesized penem antibiotic, with β-lactamase inhibitory activity comparable to that of BRL-42715 [191]. TMA-230 (i.e., the acetoxymethyl ester prodrug of AMA-3176, acetoxymethyl (5*R*,6*S*)-6-((*R*)-1-hydroxyethyl)-7-oxo-3-(pyridin-3-yl)-4-thia-1-azabicyclo[3.2.0]hept-2-ene-2-carboxylate, [CAS 113141-80-3], Figure 1.7) [192], has activity similar to faropenem, ritipenem, sulopenem, Sch29482 and Sch34343 [193].

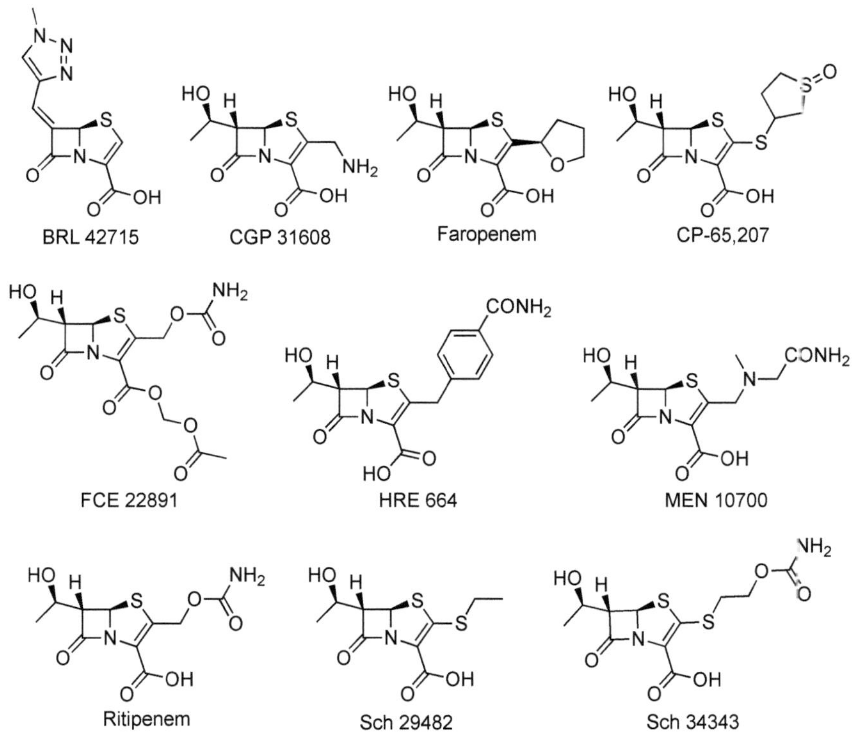

FIGURE 1.6 The list of penem antibiotics

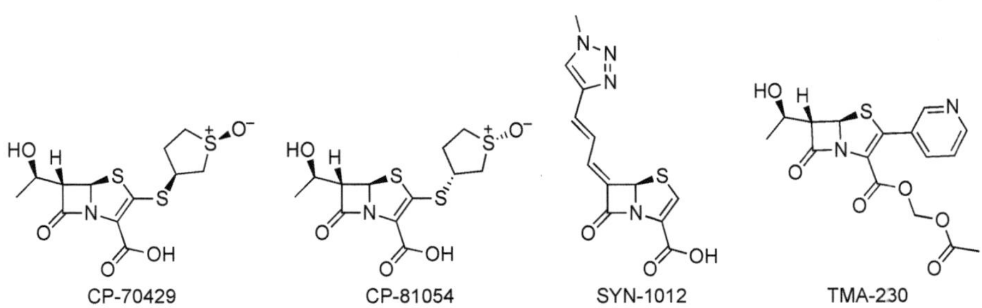

FIGURE 1.7 More examples of penem antibiotics

Compared to the carbapenem class of β-lactams, much less carbapenam β-lactams have been isolated and characterized. As of today, only three carbapenam β-lactams have been isolated from *Streptomyces sp.* OA-6129 according to SciFinder®, which are OA-6129D (i.e., (2*S*,3*S*,5*R*,6*R*)-3-((2-(3-((*R*)-2,4-dihydroxy-3,3-dimethylbutanamido)propanamido)ethyl) thio)-6-(1-hydroxy-ethyl)-7-oxo-1-azabicyclo[3.2.0]heptane-2-carboxylic acid, [CAS 85414-25-1], Figure 1.8), OA-6129E (i.e., (2*S*, 3*S*,5*R*,6*R*)-3-((2-(3-((*R*)-2,4-dihydroxy-3,3-dimethylbutanamido)propanamido)-ethyl)thio)-6-(2-hydroxypropan-2-yl)-7-oxo-1-azabicyclo[3.2.0]heptane-2-carbo-xylic acid, [CAS 85414-26-2], Figure 1.8) and No. 17927D substance (3-((2-acetamidoethyl)thio)-6-(1-hydroxyethyl)-7-oxo-1-azabicyclo[3.2.0]heptane-2-carboxylic acid, [CAS 74660-91-6], Figure 1.8) [194]. While the sodium salt of substance 17927D has neither antimicrobial activity against Gram-positive and Gram-negative bacteria nor inhibiting activity for β-lactamase [195], the *p*-nitrobenzyl esters of OA-6129D and OA-6219E showed extremely weak antimicrobial activity against *Comamonas terrigena* in a disc-agar diffusion assay [194]. This might be the reason that carbapenam β-lactams are poorly developed, although the "irrational and indiscriminate use of carbapenam antibiotics has resulted in the emergence of extremely drug resistant bugs." [196] Particularly, the carbapenam resistance among the *Enterobacteriaceae* has enormously increased in the last decade, for which carbapenam-resistant *Klebsiella pneumoniae* accounts for about 26–44% of the deaths. In addition, more, and more evidences has demonstrated that carbapenams will be converted into carbapenem β-lactams under enzymatic catalysis [197], so that carbapenams often exist as the intermediate compounds in bacteria culture. It is reported that carbapenam synthetase and β-lactam synthetase (β-LS) are evolutionarily linked to a common ancestor, i.e., asparagine synthetase B, all these three enzymes catalyze substrate acyl-adenylation and nucleophilic acyl substitution [198], where the carbapenam synthetase is an ATP/Mg^{2+}-dependent enzyme that catalyzes the formation of the β-lactam ring and is also involved in the biosynthesis of clavulanic acid [199].

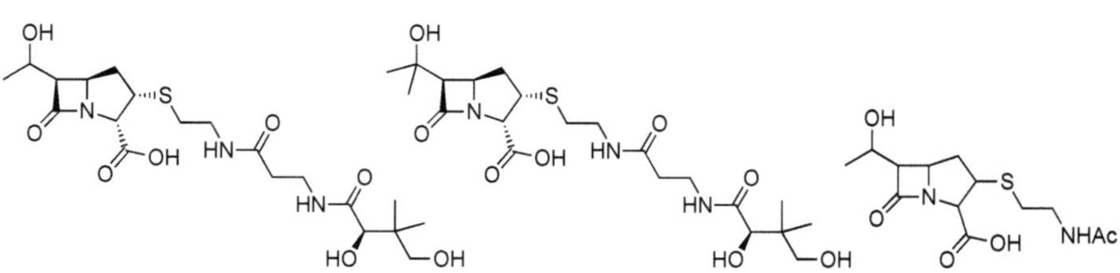

OA-6129D OA-6129E 17927D

FIGURE 1.8 The list of carbapenam β-lactam antibiotics

Similarly, carbacephem class of β-lactams are also less developed. Loracarbef, also known as LY163892 [200], KT-3777 [201], or carbacefaclor [202] (i.e., (6*R*,7*S*)-7-((*R*)-2-amino-2-phenylacetamido)-3-chloro-8-oxo-1-azabicyclo[4.2.0]-oct-2-ene-2-carbo-xylic acid, [CAS 76470-66-1], Figure 1.9), is the first clinically available carbacephem antibiotic [203]. It is active *in vitro* against pathogens associated with acute maxillary sinusitis [204], as well as skin infections, otitis media, sinusitis, bronchopulmonary infections, and UTIs [205]. Two other β-lactams of the carbacephem class are KT-3715 [CAS 72345-56-3] [206, 207], and KT-3767 [CAS 76497-77-3] [206–212]. According to SciFinder®, these two compounds show the same chemical structure; however, KT-3767 is for a structure of absolute stereochemistry, i.e., (6R,7S)-7-((Z)-2-(2-aminothiazol-4-yl)-2-(methoxyimino)-acetamido)-8-oxo-1-azabicyclo[4.2.0]oct-2-ene-2-carboxylic acid (Figure 1.9), whereas KT-3715 represents a structure of relative stereochemistry. KT-3767 has demonstrated an antimicrobial activity comparable to that of the 3rd generation cephalosporins whereas KT-3715 is inferior to KT-3767, bearing half the potency of antimicrobial activity of KT-3767 [207]. Other related derivatives include KT-3919 (i.e., (4R,6R,7S)-7-((Z)-2-(2-aminothiazol-4-yl)-2-(methoxyimino)acetamido)-4-hydroxy-8-oxo-1-azabicyclo[4.2.0]oct-2-ene-2-carboxylic acid, [CAS 78919-10-5], Figure 1.9) [208], KT-3937 (i.e., (4S,6R,7S)-7-((Z)-2-(2-aminothiazol-4-yl)-2-(methoxyimino)acetamido)-4-hydroxy-8-oxo-1-azabicyclo[4.2.0]oct-2-ene-2-carboxylic acid, [CAS 79083-76-4]) [209, 210], which is also coded as KT-3933 [208], KT-4380 (i.e., (6R,7S)-7-((S)-2-(2-aminothiazol-4-yl)-2-(3,4-dihydroxybenzamido)-acetamido)-8-oxo-1-azabicyclo[4.2.0]oct-2-ene-2-carboxylic acid, [CAS 86833-76-3], Figure 1.9) [211], KT-4697 (i.e., (6R,7S)-7-((S)-2-(2-aminothiazol-4-yl)-2-(5-hydroxy-4-oxo-1,4-dihydropyridine-2-carboxamido) acetamido)-8-oxo-1-azabicyclo[4.2.0]-oct-2-ene-2-carboxylic acid, [CAS 93129-06-7], Figure 1.9) [211], and KT-4788 (i.e., 4-(((6*R*,7*S*)-7-((*S*)-2-(2-aminothiazol-4-yl)-2-(5-hydroxy-4-oxo-1,4-dihydropyridine-2-carboxamido)acetamido)-2-carboxy-8-oxo-1-azabicyclo[4.2.0]oct-2-en-3-yl) thio)-1-methylpyridin-1-ium, Figure 1.9) [211]. Among these compounds, KT-3919 demonstrated high activity against *Escherichia coli* GN2411-5, *Klebsiella pneumoniae* 8045, *Serratia marcescens* T-26 and *Proteus mirabilis* 1287 [208]; KT-4380 was strongly active against a broad spectrum of Gram-negative bacteria including *Pseudomonas sp.* [210]; KT-4697 showed exceptionally strong activity against *Pseudomonas aeruginosa* as well as Gram-negative bacteria and KT-4788 was the highly active agent against Gram-positive and Gram-negative strains including *Pseudomonas aeruginosa*

[211]. One more carbacephem β-lactam is BP-102, showing potent bactericidal activity against Gram-positive cocci. "The MIC_{90} values were 0.25 μg/mL, 2 μg/mL and 4 μg/mL for MSSA, MRSA, and methicillin-resistant coagulase-negative staphylococci, respectively. BP-102 also exhibited potent bactericidal activity against both MSSA and MRSA in the neutropenic murine thigh-infection model" [213].

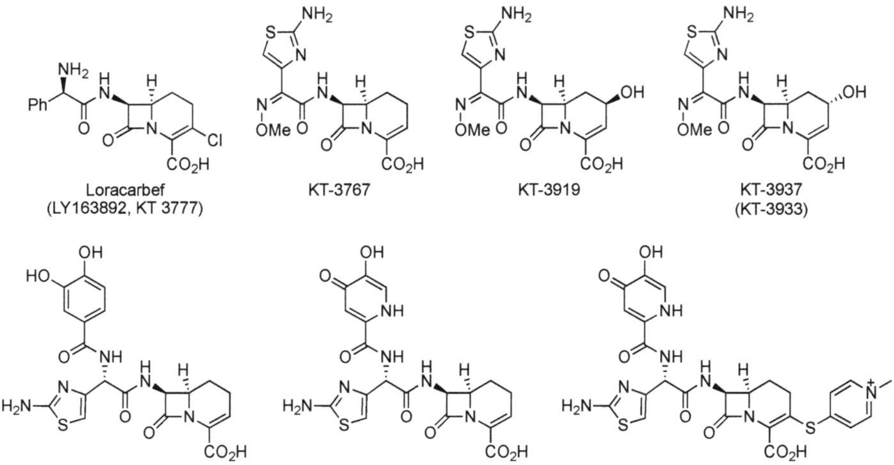

FIGURE 1.9 The list of carbacephem class of β-lactam antibiotics

While the oxacephem class of β-lactams has also been less developed, each individual oxacephem β-lactam has received great attention. For example, Flomoxef, also known as FMOX or 6315-S (i.e., (6*R*,7*R*)-7-(2-((difluoromethyl)thio)acetamido)-3-(((1-(2-hydroxyethyl)-1*H*-tetrazol-5-yl)thio)-methyl)-7-methoxy-8-oxo-5-oxa-1-azabicyclo-[4.2.0]oct-2-ene-2-carboxylic acid, [CAS 99665-00-6], Figure 1.10), reaches over 1000 entries in SciFinder®, and Moxalactam attains nearly 3000 entries. Flomoxef is potentially effective against β-lactamase-producing *Enterobacteriaceae* [214], particularly the ESBL-producing *Enterobacteriaceae*, providing a clearer distinction between ESBL and pAmpC-producing *Enterobacteriaceae* compared to cefmetazole [215]. In addition, it is used as the primary treatment for upper UTIs caused by extended-spectrum β-lactamase-producing *Escherichia coli* [216]. The less developed oxacephem L-656,575, also known as OCP-9-176 (i.e., (4*R*,6*S*,7*R*)-7-((*Z*)-2-(2-aminothiazol-4-yl)-2-(((2-carboxypropan-2-yl)oxy)imino)-acetamido)-4-methyl-3-(((1-methylpyridin-1-ium-4-yl)thio)methyl)-8-oxo-5-oxa-1-azabicyclo[4.2.0]oct-2-ene-2-carboxylate, [CAS 106036-65-1], Figure 1.10), has an *in vitro* activity against clinical isolates of *Enterobacter* species, methicillin-susceptible *Staphylococcus aureus* and *Staphylococcus epidermidis*, and multiply-resistant *Pseudomonas aeruginosa* that is superior to ceftazidime [217], but is roughly 2-fold less active than its cephem analog (e.g., ME 1228) against *Escherichia coli*, *Klebsiella pneumoniae*, *Klebsiella oxytoca*, *Providencia stuartii*, *Proteus vulgaris*, and *Serratia marcescens* [218]. Compared to L-656,575, the extensively studied moxalactam (i.e., (6*R*,7*R*)-7-(2-carboxy-2-(4-hydroxyphenyl)acetamido)-7-methoxy-3-(((1-methyl-1*H*-tetrazol-5-yl)thio)-methyl)-8-oxo-5-oxa-1-azabicyclo[4.2.0]oct-2-ene-2-carboxylic acid, [CAS 64952-97-2], Figure 1.10) has gained many acronyms, such as Lamoxactam [219], Latamoxef [220], Moxam [221], and Oxa-cephem [222]. Its disodium salt form has been known as 6059S [223], antibiotic 6059S [224], disodium latamoxef [225], Festamoxin [226], Latamoxef sodium [227], LY 12735, LY 127935 [228], Moxalactam disodium [229], or Shiomarin [230]. Moxalactam has good activity against ESBLs-producing and non-ESBLs-producing *Enterobacteriaceae* and anaerobia [231], and is especially good for Gram-negative bacillary infections [232], so it is a suitable alternative for the treatment of meningitis due to Gram-negative enteric bacilli [233].

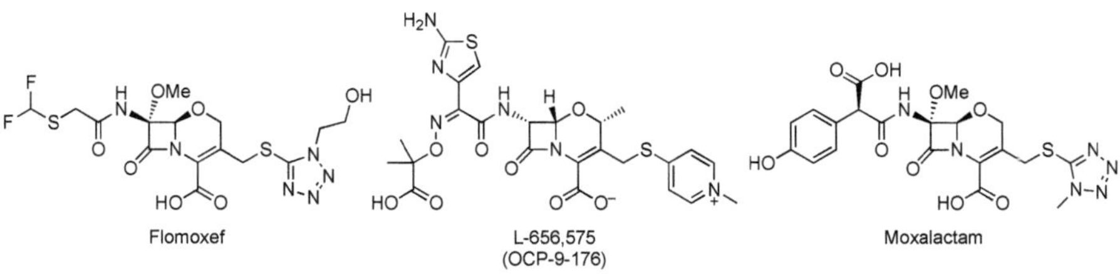

FIGURE 1.10 The oxacephem class of β-lactam antibiotics

The last class of β-lactam antibiotics are penam sulfones, which include LN1-255, PSR-3-226, PSR-4-155, PSR-4-157, Ro 48-1220, SA1-204 (SA-1-204), SA2-13, SA3-53 and tazobactam. PSR-3-226 (i.e., (*2S,3R*,5*R*)-3-((2-amino-2-oxoethoxy)carbonyl)-3-methyl-7-oxo-4-thia-1-azabicyclo[3.2.0]heptane-2-carbo-xylic acid 4,4-dioxide, [CAS 1427351-15-2], Figure 1.11) has been identified as a good lead for further structure-based inhibitor optimization owing to its *trans*-enamine conformation of high potential against KPC-2 β-lactamase [234]. PSR-4-155 (i.e., (*2S,3R*,5*R*)-3-(((3-carboxypropanoyl)oxy)methyl)-3-methyl-7-oxo-4-thia-1-azabicyclo[3.2.0]heptane-2-carboxylic acid 4,4-dioxide, [CAS 2242829-47-4], Figure 1.11) [235], SA2-13 (i.e., (*2S,3R*,5*R*)-3-(((4-carboxybutanoyl)oxy)-methyl)-3-methyl-7-oxo-4-thia-1-azabicyclo[3.2.0]heptane-2-carboxylic acid 4,4-

dioxide, [CAS 914988-69-5], Figure 1.11) [235] and PSR-4-157 (i.e., (*2S,3R,5R*)-3-(((5-carboxypen-tanoyl)oxy)methyl)-3-methyl-7-oxo-4-thia-1-azabicyclo[3.2.0]-heptane-2-carboxylic acid 4,4-dioxide, [CAS 2242829-45-2], Figure 1.11) are the analogs with different carbon numbers in the C2 diacid side chains. They have been developed to see the effect of chain length on the formation of *trans*-enamine conformation with respect to PSR-3-226, in order to develop more β-lactamase inhibitors, in addition to the only three β-lactamase inhibitors used clinically (i.e., clavulanate, sulbactam, and tazobactam). The results indicate that the length and composition of the C2 side chain linker profoundly affect the formation of the *trans*-enamine intermediate of the penam sulfones, where SA2-13 offers promise against IR SHV β-lactamases that possess the K234R substitution [235]. Ro 48-1220 (i.e., (2*S*,3*S*,5*R*)-3-((*Z*)-2-cyanovinyl)-3-methyl-7-oxo-4-thia-1-azabicyclo[3.2.0]-heptane-2-carboxylic acid 4,4-dioxide, [CAS 180323-80-2], Figure 1.11) generally inhibits β-lactamases that belong to groups 1, 2b and 2be [236]. The combination of Apalcillin and Ro 48-1220 is highly potent against β-lactamase-producing *Moraxella catarrhalis*, *Haemophilus influenzae* and *Neisseria gonorrhoeae* [237].

Likewise, LN1-255 (i.e., (*2S,3R,5R,Z*)-3-((2-(3,4-dihydroxyphenyl) acetoxy)-methyl)-3-methyl-7-oxo-6-(pyridin-2-ylmethylene)-4-thia-1-azabicyclo[3.2.0]hep-tane-2-carboxylic acid 4,4-dioxide, [CAS 748756-08-3], Figure 1.11) [238], SA1-204 (i.e., (*2S,3R,5R,Z*)-3-methyl-7-oxo-3-((2-phenylacetoxy) methyl)-6-(pyridin-2-ylmethylene)-4-thia-1-azabicyclo[3.2.0]-heptane-2-carboxylic acid 4,4-dioxide, [CAS 736123-32-3], Figure 1.11) [239] and SA3-53 (i.e., (*2S,3R,5R,Z*)-3-((((2-aminoethyl)carbamoyl)oxy)methyl)-3-methyl-7-oxo-6-(pyridin-2-ylmethylene)-4-thia-1-azabicyclo[3.2.0]heptane-2-carboxylic acid 4,4-dioxide, [CAS 1204432-71-2], Figure 1.11) [240] penam sulfones have been developed as potential inhibitors for class A serine β-lactamases [241], and possibly for class D β-lactamases as well [239]. Compared to these penam sulfones, tazobactam (also known as tazobactamum and YTR-830H, i.e., (2*S*,3*S*,5*R*)-3-((1*H*-1,2,3-triazol-1-yl)methyl)-3-methyl-7-oxo-4-thia-1-azabicyclo[3.2.0]heptane-2-carboxylic acid 4,4-dioxide, [CAS 89786-04-9], Figure 1.11) [242] has been extensively studied, which results in more than 18000 entries from SciFinder®. As one of the only three clinically used inhibitors for β-lactamases, tazobactam shows a strong synergistic effect with piperacillin to inhibit penicillinase- and oxyiminocephalosporinase-producing strains (e.g., *Proteus vulgaris*) and a moderate synergistic effect against cephalosporinase-producing strains. Its activity is comparable with that of clavulanic acid but stronger than that of sulbactam. In addition, tazobactam is able to inactivate intracellular β-lactamase in *Proteus vulgaris* and *Morganella morganii* [243]. Similarly, the combination of ceftolozane and tazobactam leads to an antibacterial agent with potent activity against *Pseudomonas aeruginosa* and many other Gram-negative pathogens, including extended-spectrum β-lactamase (ESBL)-producing *Enterobacteriaceae*. Such synergistic effects may be attributed partially to the ESBL-inhibitory activity of tazobactam, without adverse effect on the activity of other antibacterial agents given concurrently [244].

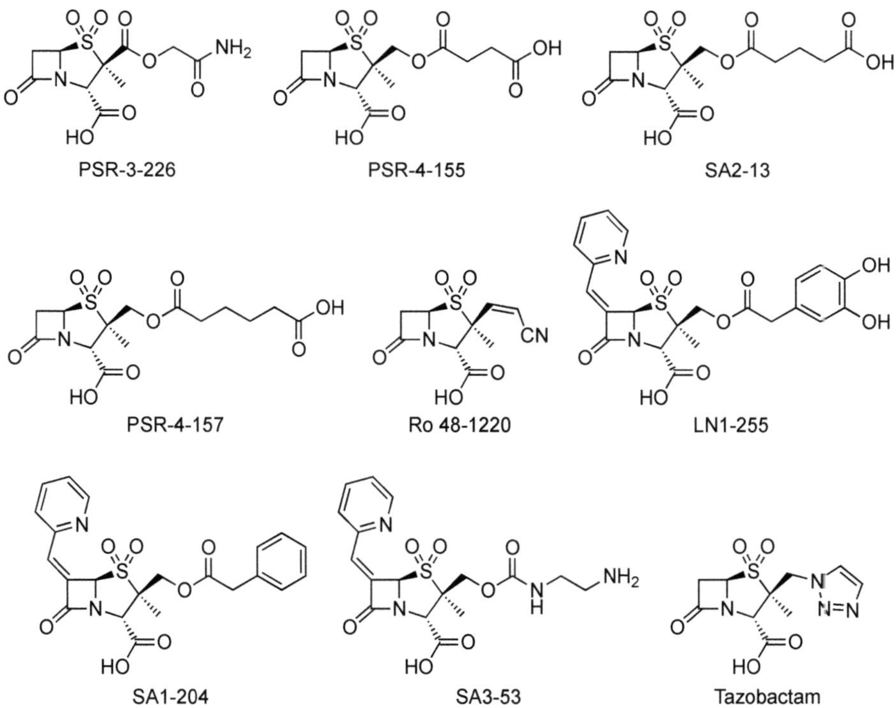

FIGURE 1.11 The list of penam sulfone antibiotics

KEYWORDS

- **penicillin**
- **cephalosporin**
- **monobactam**
- **carbapenem**
- **monobactam**
- **clavam antibiotic**
- **penem antibiotic**
- **carbapenam**
- **carbacephem**
- **oxacephem**
- **penam sulfone**

REFERENCES

1. Staudinger, H., (1907). Contribution to our knowledge of the ketenes; First paper. Diphenylketene. *Justus Liebigs Annalen der Chemie, 356*(1,2), 51–123. doi: 10.1002/jlac.19073560106.
2. Fleming, A., (1929). The antibacterial action of cultures of a penicillium, with special reference to their use in the isolation of B. influenzae. *British Journal of Experimental Pathology, 10*, 226–236.
3. Török, M. E., Moran, E., & Cooke, F. J., (2017). *Oxford Handbook of Infectious Diseases and Microbiology* (2nd edn.). Oxford University Press: Oxford, UK.
4. Ariza, A., Mayorga, C., Fernández, T. D., Barbero, N., Martín-Serrano, A., Pérez-Sala, D., Sánchez-Gómez, F. J., et al., (2015). Hypersensitivity reactions to β-lactams: Relevance of hapten-protein conjugates. *Journal of Investigational Allergology and Clinical Immunology, 25*(1), 12–25.
5. Krajnc, A., Lang, P. A., Panduwawala, T. D., Brem, J., & Schofield, C. J., (2019). Will morphing boron-based inhibitors beat the β-lactamases? *Current Opinion in Chemical Biology, 50*, 101–110. doi: 10.1016/j.cbpa.2019.03.001.
6. Fisher, J. F., Meroueh, S. O., & Mobashery, S., (2005). Bacterial resistance to β-lactam antibiotics: Compelling opportunism, compelling opportunity. *Chemical Reviews (Washington, DC, United States), 105*(2), 395–424. doi: 10.1021/cr030102i.
7. Parker, J. N., & Parker, P. M., (2004). *Penicillin: A Medical Dictionary, Bibliography, and Annotated Research Guide to Internet References*. ICON Group International, Inc.: San Diego, CA 92122 USA.
8. Purich, D. L., (2017). *The Inhibitor Index, A Desk Reference on Enzyme Inhibitors, Receptor Antagonists, Drugs, Toxins, Poisons & Therapeutic Leads*. CRC Press: Boca Raton, FL 33487-2742.
9. Grabe, M., Bjerklund-Johansen, T. E., Botto, H., Wullt, B., Çek, M., Naber, K. G., Pickard, R. S., et al., (2012). *Guidelines on Urological Infections*. European Association of Urology.
10. Page, M. G. P., (2012). Chapter 3: Beta-lactam antibiotics. In: Dougherty, T. J., & Pucci, M. J., (eds.), *Antibiotic Discovery and Development* (pp. 79–117). Springer: New York.
11. Rudzit, E. A., Pleshakov, M. G., Radkevich, T. P., Kulikova, D. A., Avidon, V. V., Kozlova, S. P., & Terekhina, A. I., (1974). Clometacillin. Synthesis and antibacterial properties. *Antibiotiki (Moscow), 19*(10), 887–890.
12. Ivady, G., Baranyai, P., Tuma, L., & Inkei, J., (1972). Clinico-pharmacological results on the efficacy of an oral penicillin preparation, penamecillin. *Acta Paediatrica Academiae Scientiarum Hungaricae, 13*(3), 215–219.
13. Dijkmans, A. C., Kweekel, D. M., Van, D. J. T., Van, E. M. J., Kamerling, I. M. C., & Burggraaf, J., (2019). The oral pheneticillin absorption test: An accurate method to identify patients with inadequate oral pheneticillin absorption. *Antibiotics, 8*, 119/1–119/4. doi: 10.3390/antibiotics8030119.
14. Organization, W. H., (1992). Essential drugs. *WHO Drug Information, 6*(1), 19–37.
15. Cunha, B. A., (2015). *Antibiotic Essentials*. Jaypee Brothers Medical Publishers Pvt. Ltd.: New Delhi.
16. Le, J., Dam, Q., Schweizer, M., Thienphrapa, W., Nizet, V., & Sakoulas, G., (2016). Effects of vancomycin versus nafcillin in enhancing killing of methicillin-susceptible *Staphylococcus aureus* causing bacteremia by human cathelicidin LL-37. *European*

Journal of Clinical Microbiology & Infectious Diseases, 35(9), 1441–1447. doi: 10.1007/s10096-016-2682-0.
17. Yourassowsky, E., & Monsieur, R., (1970). Paradoxical effect of sodium chloride on oxacillin activity with regard to sensitive or resistant strains of *Staphylococcus aureus*. *Pathologie et Biologie, 18*(19, 20), 819–822.
18. Otsuki, M., Nishino, T., Nakazawa, S., Isagai, K., & Ikeda, Y., (1979). Bacteriological studies on bacampicillin, a new orally active ampicillin ester. *Chemotherapy (Tokyo), 27*(Suppl. 4), 7–16. doi: 10.11250/chemotherapy1953.27.Supplement4_7.
19. Warren, G. H., (1976). Cyclacillin: Microbiological and pharmacological properties and use in chemotherapy of infection – a critical appraisal. *Chemotherapy, 22*, 154–182. doi: 10.1159/000221924.
20. Brogden, R. N., & Avery, G. S., (1972). New antibiotics: Epicillin, minocycline and spectinomycin a summary of their antibacterial activity, pharmacokinetic properties and therapeutic efficacy. *Drugs, 3*(5, 6), 314–330.
21. Sutherland, R., Elson, S., & Croydon, E. A. P., (1972). Metampicillin. Antibacterial activity and absorption and excretion in man. *Chemotherapy, 17*(3), 145–160. doi: 10.1159/000220849.
22. Ehrnebo, M., Nilsson, S. O., & Boréus, L. O., (1979). Pharmacokinetics of ampicillin and its prodrugs bacampicillin and pivampicillin in man. *Journal of Pharmacokinetics and Biopharmaceutics, 7*, 429–451. doi: 10.1007/BF01062386.
23. Tachibana, A., Ohtani, K., Komiya, M., Naito, S., Shimizu, C., & Yano, K., (1978). Talampicillin hydrochloride: Comparison with amoxicillin and ampicillin on the antibacterial activity and pharmacokinetics. *Japanese Journal of Antibiotics, 31*(3), 166–176.
24. Verbist, L., (1979). Comparison of the activities of the new ureidopenicillins piperacillin, mezlocillin, azlocillin, and bay k 4999 against gram-negative organisms. *Antimicrobial Agents and Chemotherapy, 16*(2), 115–119. doi: 10.1128/aac.16.2.115.
25. Livingston, S. J., Smith, H. J., Sewell, R. D. E., & Ahmed, S., (1992). Novel inhibitors of enkephalin-degrading enzymes. IV: Structure-activity relationships within the penicillins as enkephalinase inhibitors. *Journal of Enzyme Inhibition, 6*(2), 165–173. doi: 10.3109/14756369209040748.
26. Germinale, T., & Giglio, C., (1976). Carindacillin in the therapy of urinary tract infections. In: Williams, J. D., & Geddes, A. M., (eds.), *Penicillins and Cephalosporins. Chemotherapy* (Vol. 5, pp. 99–101). Springer: Boston, MA.
27. Thulin, E., Sundqvist, M., & Andersson, D. I., (2015). Amdinocillin (mecillinam) resistance mutations in clinical isolates and laboratory-selected mutants of *Escherichia coli. Antimicrobial Agents and Chemotherapy, 59*(3), 1718–1727. doi: 10.1128/AAC.04819-14.
28. Graninger, W., (2003). Pivmecillinam-therapy of choice for lower urinary tract infection. *International Journal of Antimicrobial Agents, 22*(Suppl. 2), S73–S78. doi: 10.1016/s0924-8579(03)00235-8.
29. Nagatomo, H., Kakizaki, M., & Tsuchiya, K., (1974). Combines action of sulbenicillin and gentamicin. *Journal of Antibiotics, 27*(3), 199–205. doi: 10.7164/antibiotics.27.199.
30. Yamashita, M., Hashimoto, S., Ezaki, M., Iwami, M., Komori, T., Kohsaka, M., & Imanaka, H., (1983). FR-900318, a novel penicillin with β-lactamase inhibitory activity. *Journal of Antibiotics, 36*(12), 1774–1776. doi: 10.7164/antibiotics.36.1774.

31. Georgopapadakou, N. H., & Sykes, R. B., (1983). Chapter 12. Bacterial enzymes interacting with β-lactam antibiotics. In: Demain, A. L., & Solomon, N. A. (eds.), *Antibiotics Containing the Beta-Lactam Structure. Part II* (Vol. 67, pp. 1–77). Part of the book series: Handbook of Experimental Pharmacology. Springer: Verlag.
32. White, R. J., (2012). Chapter 1. The early history of antibiotic discovery: Empiricism ruled. In: Dougherty, T. J., & Pucci, M. J., (eds.), *Antibiotic Discovery and Development* (pp. 3–31). Springer: New York.
33. Brotzu, G., (1948). Research on a new antibiotic. *Works of the Instituto D'Igiene du Cagliari,* 1–11.
34. Abraham, E. P., & Newton, G. G. F., (1961). The structure of cephalosporin C. *Biochemical Journal, 79*(2), 377–393. doi: 10.1042/bj0790377.
35. Considine, G. D., & Kulik, P. H., (2008). In: *Van Nostrand's Scientific Encyclopedia* (10th edn., Vol. 1, pp. 263, 264). Wiley-Interscience.
36. Prober, C. G., (1998). Cephalosporins: An update. *Pediatrics in Review, 19*(4), 118–127. doi: 10.1542/pir.19-4-118.
37. The European Agency for the Evaluation of Medicinal Products (1998). Committee for veterinary medicinal products: Cefacetrile. *Veterinary Medicines and Information Technology, EMEA/MRL/499/98-FINAL.*
38. BOC Sciences, *Antibacterial Inhibitors (Inhibitors, Agonists and Modulators).* https://www.bocsci.com/tag/antibacterial-62.html (accessed on 05 March 2022).
39. Demon, D., Ludwig, C., Breyne, K., Guede, D., Doerner, J. C., Froyman, R., & Meyer, E., (2012). The intramammary efficacy of first generation cephalosporins against *Staphylococcus aureus* mastitis in mice. *Veterinary Microbiology, 160*(1,2), 141–150. doi: 10.1016/j.vetmic.2012.05.017.
40. Kratky, M., & Vinsova, J., (2013). Antimycobacterial activity of quaternary pyridinium salts and pyridinium N-oxides: Review. *Current Pharmaceutical Design, 19*(7), 1343–1355. doi: 10.2174/1381612138048057l1.
41. Ribeiro, A. R., Sures, B., & Schmidt, T. C., (2018). Ecotoxicity of the two veterinarian antibiotics ceftiofur and cefapirin before and after photo-transformation. *Science of the Total Environment, 619–620*, 866–873. doi: 10.1016/j.scitotenv.2017.11.109.
42. Ueda, Y., (1976). Clinical and experimental studies on cefatriazone. *Chemotherapy (Tokyo), 9*, 1661–1984.
43. Aswapokee, N., & Neu, H. C., (1979). *In vitro* activity and β-lactamase stability of cefazaflur compared with those of β-lactamase-stable cephalosporins. *Antimicrobial Agents and Chemotherapy, 15*(3), 444–446. doi: 10.1128/AAC.15.3.444.
44. Dewick, P. M., (2002). Chapter 7. Peptides, proteins, and other amino acid derivatives. In: *Medicinal Natural Products: A Biosynthetic Approach* (2nd edn.). John Wiley & Sons, Ltd.
45. Yasuda, K., Kurashige, S., & Mitsuhashi, S., (1980). An orally active cephalosporin. *Antimicrobial Agents and Chemotherapy, 18*(1), 105–110. doi: 10.1128/AAC.18.1.105.
46. Nishida, M., Murakawa, T., Kamimura, T., Okada, N., Sakamoto, H., Fukada, S., Nakamoto, S., et al., (1976). *In vitro* and *in vivo* evaluation of ceftezole, a new cephalosporin derivative. *Antimicrobial Agents and Chemotherapy, 10*(1), 1–13. doi: 10.1128/AAC.10.1.1.
47. Adams, H. G., Stilwell, G. A., & Turck, M., (1976). *In vitro* evaluation of cefoxitin and cefamandole. *Antimicrobial Agents and Chemotherapy, 9*(6), 1019–1024. doi: 10.1128/AAC.9.6.1019.

48. Kameyama, J., Tsukamoto, M., Asano, T., Senda, H., Amemiya, K., Murakami, N., & Kaneda, N., (1989). Experimental studies on the pharmacokinetics and therapeutic effect of cefbuperazone in biliary infection. *Current Medical Research and Opinion, 11*(9), 576–584. doi: 10.1185/03007998909112674.
49. Uri, J. V., & Actor, P., (1983). Chapter 17. Therapeutic application of β-lactam antibiotics. In: Demain, A. L., & Solomon, N. A., (eds.), *Antibiotics Containing the Beta-Lactam Structure, Part II*. Springer: Verlag.
50. Xia, J., Yang, L., Dong, L., Niu, M., Zhang, S., Yang, Z., Wumaier, G., et al., (2018). Cefminox, a dual agonist of prostacyclin receptor and peroxisome proliferator-activated receptor-gamma identified by virtual screening, has therapeutic efficacy against hypoxia-induced pulmonary hypertension in rats. *Frontiers in Pharmacology, 9*, 134/1–134/11. doi: 10.3389/fphar.2018.00134.
51. Pochini, L., Galluccio, M., Scumaci, D., Giangregorio, N., Tonazzi, A., Palmieri, F., & Indiveri, C., (2008). Interaction of β-lactam antibiotics with the mitochondrial carnitine/acylcarnitine transporter. *Chemico-Biological Interactions, 173*(3), 187–194. doi: 10.1016/j.cbi.2008.03.003.
52. Hoover, J. R. E., (1983). Chapter 14, β-lactam antibiotics: Structure-activity relationships. In: Demain, A. L., & Solomon, N. A., (eds.), *Antibiotics Containing the Beta-Lactam Structure* (pp. 119–245). *Part II*, Springer: Verlag.
53. Brogden, R. N., Heel, R. C., Speight, T. M., & Avery, G. S., (1979). Cefuroxime: A review of its antibacterial activity, pharmacological properties and therapeutic use. *Drugs, 17*(4), 233–266. doi: 10.2165/00003495-197917040-00001.
54. Lalge, R., Thipsay, P., Shankar, V. K., Maurya, A., Pimparade, M., Bandari, S., Zhang, F., et al., (2019). Preparation and evaluation of cefuroxime axetil gastro-retentive floating drug delivery system via hot melt extrusion technology. *International Journal of Pharmaceutics (Amsterdam, Netherlands), 566*, 520–531. doi: 10.1016/j.ijpharm.2019.06.021.
55. Yoshida, T., Tateda, E., Hiramatsu, K., & Yokota, T., (1996). *In vitro* antibacterial activity of a new parenteral penem, sulopenem. *Japanese Journal of Antibiotics, 49*(4), 324–337.
56. Namba, E., Okumura, R., Chiba, M., Hoshino, K., & Tateda, K., (2013). Antimicrobial and rapid bactericidal activities of sitafloxacin and other agents against *Streptococcus pyogenes*. *Japanese Journal of Antibiotics, 66*(5), 293–304.
57. Buschelman, B. J., Jones, R. N., & Bale, M. J., (1994). Effects of blood medium supplements on activities of newer cephalosporins tested against enterococci. *Journal of Clinical Microbiology, 32*(2), 565–567. doi: 10.1128/jcm.32.2.565-567.1994.
58. Guay, D. R. P., (2001). Review of cefditoren, an advanced-generation, broad-spectrum oral cephalosporin. *Clinical Therapeutics, The International Peer-Reviewed Journal of Drug Therapy, 23*(12), 1924–1937. doi: 10.1016/S0149-2918(01)80147-8.
59. Sofianou, D., Tsakris, A., Nikolaidis, P., Kotis, E., & Tourkantonis, A., (1996). The activity of cefetamet against enterobacterial isolates from community-acquired urinary tract infections. *Journal of Antimicrobial Chemotherapy, 38*(2), 320.
60. Laposy, C. B., Silva, D. K. M., & Pessoto, J. N., (2014). Laboratory evaluation of the use of cefovecin in healthy dogs. *Semina: Ciencias Agrarias, 35*(1), 365–370.
61. Tsuji, A., Maniatis, A., Bertram, M. A., & Young, L. S., (1985). *In vitro* activity of BMY-28142 in comparison with those of other 1-lactam antimicrobial agents. *Antimicrobial Agents and Chemotherapy, 27*(4), 515–519. doi: 10.1128/AAC.27.4.515.

62. Yamaguchi, K., Ohno, A., Takahashi, S., Hayashi, M., Yamanaka, K., Hirakata, Y., & Mitsuyama, J., (1998). *In vitro* antibacterial activities of cefteram and other beta-lactam agents against recent clinical isolates. *Japanese Journal of Antibiotics, 51*(1), 11–25.
63. Williamson, R., Gutmann, L., Kitzis, M. D., & Acar, J. F., (1984). An evaluation of the bacteriolytic and biochemical properties of ceftiolene (42980RP). *Journal of Antimicrobial Chemotherapy, 14*(6), 581–593. doi: 10.1093/jac/14.6.581.
64. Gupta, S. K., Agarwal, R., & Srivastava, S., (2014). *Textbook on Clinical Ocular Pharmacology and Therapeutics* (1st edn., p. 110). Jaypee Brothers Medical Pub.
65. Jean, S. S., Gould, I. M., Lee, W. S., & Hsueh, P. R., (2019). International society of antimicrobial chemotherapy (ISAC), new drugs for multidrug-resistant gram-negative organisms: Time for stewardship. *Drugs, 79*(7), 705–714. doi: 10.1007/s40265-019-01112-1.
66. Chellat, M. F., Raguz, L., & Riedl, R., (2016). Targeting antibiotic resistance. *Angewandte Chemie, International Edition, 55*(23), 6600–6626. doi: 10.1002/anie.201506818.
67. Soren, O., Rineh, A., Silva, D. G., Cai, Y., Howlin, R. P., Allan, R. N., Feelisch, M., et al., (2020). Cephalosporin nitric oxide-donor prodrug DEA-C3D disperses biofilms formed by clinical cystic fibrosis isolates of *Pseudomonas aeruginosa Journal of Antimicrobial Chemotherapy, 75*(1), 117–125. doi: 10.1093/jac/dkz378.
68. Yamada, H., Jimpo, K., Tobiki, H., Komatsu, T., Noguchi, H., Irie, K., & Nakagome, T., (1983). New broad-spectrum cephalosporins with antipseudomonal activity. I. Synthesis and antibacterial activity of 7β-[*D*-2-[(4-hydroxy-1,5-naphthyridine-3-carbonylamino)- and (4-hydroxypyridine-3-carbonylamino)]-2-(4-hydroxyphenyl)acetamido]cephalosporins. *Journal of Antibiotics, 36*(5), 522–531. doi: 10.7164/antibiotics.36.522.
69. Bergan, T., & Da Fonseca, J., (1993). Antibacterial activity of cefcanel. *Chemotherapy (Basel, Switzerland), 39*(2), 96–104. doi: 10.1159/000239109.
70. Gooding, P. G., Berman, E., Rubinfeld, J., & Weber, A., (1974). Preliminary studies on new semisynthetic derivatives of ampicillin and cephalexin. In: Daikos, G. K., (ed.) *Prog. Chemother. (Antibacterial, Antiviral, Antineoplast.), Proc. Int. Congr. Chemother., 8th* (Vol. 2, pp. 167–172).
71. Qin, Y. Y., Chen, X. M., & Wei, L. P., (2012). Distribution characteristics of lower airway *Acinetobacter baumannii* infection and susceptible factors for drug resistance and multi-drug resistant bacteria. *Guangdong Yixue, 33*(18), 2812–2815.
72. Ferrer, S. C., & Colome, R. J., (1977). *Derivatives of 7-Aminocephalosporanic Acid.* Span., ES 447765 A1.
73. Bottarelli, E., Lucidi, E., & Geroldi, S., (1986). *In vitro* activity of a new cephalosporin (cefivitril) on bacteria from farm animals. I. Comparison with ampicillin and dicloxacillin. *Rivista di Zootecnia e Veterinaria, 14*(4), 210–215.
74. Takeuchi, M., Harada, M., Nagata, M., Watanabe, A., & Nakamura, K., (2001). Immunological properties of S-1090, cefmatilen hydrochloride hydrate. *The Journal of Toxicological Sciences, 26*(Suppl 1), 231–242.
75. Marban, C., Metz-Boutigue, M. H., Lavalle, P., Schaaf, P., & Haikel, Y., (2018). *New δ-Configured Cateslytin Peptide for Use in Treatment of Infection.* Eur. Pat. Appl., EP 3323422 A1.
76. Bogan, J., (1977). Cephoxazole. *Drugs of Today, 13*(1), 38–40.
77. Nishida, M., Murakawa, T., Kamimura, T., Okada, N., Sakamoto, H., Fukada, S., Nakamoto, S., et al., (1976). Laboratory evaluation of FR10612, a new oral cephalosporin derivative. *Journal of Antibiotics, 29*(4), 444–459. doi: 10.7164/antibiotics.29.444.

78. Duerckheimer, W., Klesel, N., Limbert, M., Schrinner, E., Seeger, K., & Seliger, H., (1980). HR 109, a highly active cephalosporin (S)-sulfoxide. *Special Publication – Royal Society of Chemistry, 38(Recent Adv. Chem. β-Lactam Antibiot.)*, 46–56.
79. Rehman, M. T., Faheem, M., & Khan, A. U., (2013). Insignificant β-lactamase activity of human serum albumin: No panic to nonmicrobial-based drug resistance. *Letters in Applied Microbiology, 57*(4), 325–329. doi: 10.1111/lam.12116.
80. Corbett, D. F., Eglington, A. J., & Howarth, T. T., (1977). Structure elucidation of MM 17880, a new fused β-lactam antibiotic isolated from *Streptomyces olivaceus*; a mild β-lactam degradation reaction. *Journal of the Chemical Society, Chemical Communications,* (24), 953, 954. doi: 10.1039/C39770000953.
81. Brown, A. G., Butterworth, D., Cole, M., Hanscomb, G., Hood, J. D., Reading, C., & Rolinson, G. N., (1976). Naturally-occurring beta-lactamase inhibitors with antibacterial activity. *The Journal of Antibiotics, 29*(6), 668, 669. doi: 10.7164/antibiotics.29.668.
82. Kattan, J. N., Villegas, M. V., & Quinn, J. P., (2008). New developments in carbapenems. *Clinical Microbiology and Infection, 14*, 1102–1111. doi: 10.1111/j.1469-0691.2008.02101.x.
83. Kahan, J. S., Kahan, F. M., Goegelman, R., Currie, S. A., Jackson, M., Stapley, E. O., Miller, T. W., et al., (1979). Thienamycin, a new beta-lactam antibiotic. I. Discovery, taxonomy, isolation and physical properties. *The Journal of Antibiotics, 32*(1), 1–12. doi: 10.7164/antibiotics.32.1.
84. Wise, R., (1986). *In vitro* and pharmacokinetic properties of the carbapenems. *Antimicrobial Agents and Chemotherapy, 30*(3), 343–349. doi: 10.1128/AAC.30.3.343.
85. Bouthillier, G., Mastalerz, H., Menard, M., Fung-Tomc, J., & Gradelski, E., (1992). The synthesis, antibacterial, and β-lactamase inhibitory activity of a novel asparenomycin analog. *Journal of Antibiotics, 45*(2), 240–245. doi: 10.7164/antibiotics.45.240.
86. Murakami, K., Doi, M., & Yoshida, T., (1982). Asparenomycins A, B and C, new carbapenem antibiotics. V. Inhibition of β-lactamases. *Journal of Antibiotics, 35*(1), 39–45. doi: 10.7164/antibiotics.35.39.
87. Kimura, Y., Motokawa, K., Nagata, H., Kameda, Y., Matsuura, S., Mayama, M., & Yoshida, T., (1982). Asparenomycins A, B and C, new carbapenem antibiotics. IV. Antibacterial activity. *Journal of Antibiotics, 35*(1), 32–38. doi: 10.7164/antibiotics.35.32.
88. Kessler, R. E., Fung-Tomc, J., Kolek, B., Minassian, B., Huczko, E., Gradelski, E., & Bonner, D. P., (1995). *In vitro* activity of BMS-181139, a new carbapenem with potent antipseudomonal activity. *Antimicrobial Agents and Chemotherapy, 39*(2), 380–385. doi: 10.1128/AAC.39.2.380.
89. Inoue, K., Hamana, Y., & Mitsuhashi, S., (1995). *In vitro* antibacterial activity and β-lactamase stability of a new carbapenem, BO-2727. *Antimicrobial Agents and Chemotherapy, 39*(10), 2331–2336. doi: 10.1128/AAC.39.10.2331.
90. Harada, S., Nozaki, Y., Shinagawa, S., & Kitano, K., (1982). C-19393 E5, a new carbapenem antibiotic. Fermentation, isolation and structure. *Journal of Antibiotics, 35*(8), 957–962. doi: 10.7164/antibiotics.35.957.
91. Kobayashi, F., Saino, Y., Koshi, T., Hattori, Y., Nakayama, M., Iwasaki, A., Mori, T., & Mitsuhashi, S., (1982). Antimicrobial and β-lactamase inhibitory activities of carpetimycins A and B, new carbapenem antibiotics. *Antimicrobial Agents and Chemotherapy, 21*(4), 536–544. doi: 10.1128/aac.21.4.536.
92. Van, O. M. L., (1999). CS-834 sankyo. *Current Opinion in Anti-Infective Investigational Drugs, 1*(1), 70–73.

93. Tanaka, M., Kato, K., Hakusui, H., Murakami, Y., Sato, K., Ito, Y., & Kawamoto, K., (2000). Pharmacokinetics and safety of ascending single doses of DZ-2640, a new oral carbapenem antibiotic, administered to healthy Japanese subjects. *Antimicrobial Agents and Chemotherapy, 44*(3), 578–582. doi: 10.1128/AAC.44.3.578-582.2000.
94. Tanaka, M., Hohmura, M., Nishi, T., Sato, K., & Hayakawa, I., (1997). Antimicrobial activity of DU-6681a, a parent compound of novel oral carbapenem DZ-2640. *Antimicrobial Agents and Chemotherapy, 41*(6), 1260–1268. doi: 10.1128/AAC.41.6.1260.
95. Yin, X. H., Mikamo, H., Hayasaki, Y., & Tamaya, T., (2001). *In vitro* and *in vivo* antibacterial activity of E1010, a new injectable carbapenem. *Current Therapeutic Research, 62*(8), 566–572. doi: 10.1016/S0011-393X(01)80062-X.
96. Köhler, T., Michea-Hamzehpour, M., Epp, S. F., & Pechere, J. C., (1999). Carbapenem activities against *Pseudomonas aeruginosa*: Respective contributions of OprD and efflux systems. *Antimicrobial Agents and Chemotherapy, 43*(2), 424–427. doi: 10.1128/AAC.43.2.424.
97. Stapley, E. O., Cassidy, P. J., Tunac, J., Monaghan, R. L., Jackson, M., Hernandez, S., Zimmerman, S. B., et al., (1981). Epithienamycins – novel β-lactams related to thienamycin. I. Production and antibacterial activity. *Journal of Antibiotics, 34*(6), 628–636. doi: 10.7164/antibiotics.34.628.
98. Basker, M. J., Boon, R. J., & Hunter, P. A., (1980). Comparative antibacterial properties *in vitro* of seven olivanic acid derivatives: MM 4550, MM 13902, MM 17880, MM 22380, MM 22381, MM 22382 and MM 22383. *Journal of Antibiotics (Tokyo), 33*(8), 878–884. doi: 10.7164/antibiotics.33.878.
99. El-Gamal, M. I., Brahim, I., Hisham, N., Aladdin, R., Mohammed, H., & Bahaaeldin, A., (2017). Recent updates of carbapenem antibiotics. *European Journal of Medicinal Chemistry, 131*, 185–195. doi: 10.1016/j.ejmech.2017.03.022.
100. Rodloff, A. C., Goldstein, E. J. C., & Torres, A., (2006). Two decades of imipenem therapy. *Journal of Antimicrobial Chemotherapy, 58*, 916–929. doi: 10.1093/jac/dkl354.
101. Nagano, R., Adachi, Y., Imamura, H., Yamada, K., Hashizume, T., & Morishima, H., (1999). Carbapenem derivatives as potential inhibitors of various β-lactamases, including class B metallo-β-lactamases. *Antimicrobial Agents and Chemotherapy, 43*(10), 2497–2503. doi: 10.1128/AAC.43.10.2497.
102. Nakagawa, S., Hashizume, T., Matsuda, K., Sanada, M., Okamoto, O., Fukatsu, H., & Tanaka, N., (1993). *In vitro* activity of a new carbapenem antibiotic, BO-2727, with potent antipseudomonal activity. *Antimicrobial Agents and Chemotherapy, 37*(12), 2756–2759 doi: 10.1128/AAC.37.12.2756.
103. Bassetti, M., Ginocchio, F., & Mikulska, M., (2011). New treatment options against gram-negative organisms. *Critical Care, 15*(2), 215/1–215/9. doi: 10.1186/cc9997.
104. Kurazono, M., Ida, T., Yamada, K., Hirai, Y., Maruyama, T., Shitara, E., & Yonezawa, M., (2004). *In vitro* activities of ME1036(CP5609), a novel parenteral carbapenem, against methicillin-resistant staphylococci. *Antimicrobial Agents and Chemotherapy, 48*(8), 2831–2837. doi: 10.1128/AAC.48.8.2831-2837.2004.
105. Edwards, J. R., (1995). Meropenem: A microbiological overview. *Journal of Antimicrobial Chemotherapy, 36*(suppl. A), 1–17. doi: 10.1093/jac/36.suppl_A.1.
106. Yoshioka, T., Kojima, I., Isshiki, K., Watanabe, A., Shimauchi, Y., Okabe, M., Fukagawa, Y., & Ishikura, T., (1983). Structures of OA-6129A, B1, B2, and C, new carbapenem

antibiotics produced by *Streptomyces sp*. OA-6129. *Journal of Antibiotics, 36*(11), 1473–1482. doi: 10.7164/antibiotics.36.1473.

107. Basker, M. J., Boon, R. J., Box, S. J., Brown, A. G., Davis, P., Ponsford, R. J., Southgate, R., & Spear, S. R., (1983). Comparative biological properties of some synthetic olivanic acid analogs. *Journal of Antibiotics, 36*(10), 1357–1371. doi: 10.7164/antibiotics.32.287.
108. Yamaguchi, A., Hirata, T., & Sawai, T., (1984). Novel carbapenem derivative SF2103A: Studies on the mode of β-lactamase inactivation. *Antimicrobial Agents and Chemotherapy, 25*(3), 348–353. doi: 10.1128/AAC.25.3.348.
109. Tsuji, N., Kobayashi, M., Terui, Y., Matsumoto, K., Takahashi, Y., & Kondo, E., (1985). Pluracidomycin A2, A new carbapenem bearing a sulfinic acid, and other minor pluracidomycins. *Journal of Antibiotics, 38*(2), 270–274. doi: 10.7164/antibiotics.38.270.
110. Tsuji, N., Nagashima, K., Kobayashi, M., Terui, Y., Matsumoto, K., & Kondo, E., (1982). The structures of pluracidomycins, new carbapenem antibiotics. *Journal of Antibiotics, 35*(4), 536–540. doi: 10.7164/antibiotics.35.536.
111. Fukagawa, Y., Okabe, M., Azuma, S., Kojima, I., Ishikura, T., & Kubo, K., (1984). Studies on the biosynthesis of carbapenem antibiotics. I. Biosynthetic significance of the OA-6129 group of carbepenem compounds as the direct precursors for PS-5, epithienamycins A and C and MM 17880. *Journal of Antibiotics, 37*(11), 1388–1393. doi: 10.7164/antibiotics.37.1388.
112. Okamura, K., Hirata, S., Okumura, Y., Fukagawa, Y., Shimauchi, Y., Kouno, K., & Ishikura, T., (1978). PS-5, a new beta-lactam antibiotic from *Streptomyces*. *The Journal of Antibiotics, 31*(5), 480–482. doi: 10.7164/antibiotics.31.480.
113. Watanabe, A., Takahashi, H., Kikuchi, T., Kobayashi, T., Gomi, K., Fujimura, S. Tokue, Y., & Nukiwa, T., (2000). Comparative *in vitro* activity of S-4661, A new parenteral carbapenem, and other antimicrobial agents against respiratory pathogens. *Chemotherapy (Basel), 46*(3), 184–187. doi: 10.1159/000007276.
114. Sumita, Y., Nouda, H., Kanazawa, K., & Fukasawa, M., (1995). Antimicrobial activity of SM-17466, a novel carbapenem antibiotic with potent activity against methicillin-resistant *Staphylococcus aureus*. *Antimicrobial Agents and Chemotherapy, 39*(4), 910–916. doi: 10.1128/AAC.39.4.910.
115. Ueda, Y., & Sunagawa, M., (2003). *In vitro* and *in vivo* activities of novel 2-(thiazol-2-ylthio)-1β-methylcarbapenems with potent activities against multiresistant gram-positive bacteria. *Antimicrobial Agents and Chemotherapy, 47*(8), 2471–2480. doi: 10.1128/AAC.47.8.2471-2480.2003.
116. Watanabe, K., Kato, N., Tanaka-Bandoh, K., Tanaka, Y., Kato, H., & Ueno, K., (1996). *In vitro* activities of sulopenem, a new parenteral penem, against anaerobes. *Japanese Journal of Antibiotics, 49*(4), 367–376.
117. Thamlikitkul, V., Lorchirachoonkul, N., & Tiengrim, S., (2014). *In vitro* and *in vivo* activity of tebipenem against ESBL-producing *E. coli*. *Journal of the Medical Association of Thailand, 97*(12), 1259–1268.
118. Kahan, F. M., Kropp, H., Sundelof, J. G., & Birnbaum, J., (1983). Thienamycin: Development of imipenen-cilastatin. *Journal of Antimicrobial Chemotherapy, 12*(Suppl D.), 1–35. doi: 10.1093/jac/12.suppl_d.1.
119. Koga, T., Masuda, N., Kakuta, M., Namba, E., Sugihara, C., & Fukuoka, T., (2008). Potent *in vitro* activity of tomopenem (CS-023) against methicillin-resistant *Staphylococcus aureus* and *Pseudomonas aeruginosa*. *Antimicrobial Agents and Chemotherapy, 52*(8), 2849–2854. doi: 10.1128/AAC.00413–08.

120. Sykes, R. B., Cimarusti, C. M., Bonner, D. P., Bush, K., Floyd, D. M., Georgopapadakou, N. H., Koster, W. H., et al., (1981). Monocyclic β-lactam antibiotics produced by bacteria. *Nature (London, United Kingdom), 291*(5815), 489–491. doi: 10.1038/291489a0.
121. Aoki, H., Sakai, H., Kohsaka, M., Konomi, T., Hosoda, J., Kubochi, Y., Iguchi, E., & Imanaka, H., (1976). Nocardicin A, a new monocyclic β-lactam antibiotic. I. Discovery, isolation and characterization. *Journal of Antibiotics, 29*(5), 492–500. doi: 10.7164/antibiotics.29.492.
122. Nagarajan, R., Boeck, L. D., Gorman, M., Hamill, R. L., Higgens, C. E., Hoehn, M. M., Stark, W. M., & Whitney, J. G., (1971). β-Lactam antibiotics from *Streptomyces*. *Journal of the American Chemical Society, 93*(9), 2308–2310. doi: 10.1021/ja00738a035.
123. Cassidy, P. J., Albers-Schonberg, G., Goegelman, R. T., Miller, T., Arison, B., Stapley, E. O., & Birnbaum, J., (1981). Epithienamycins. II. Isolation and structure assignment. *The Journal of Antibiotics, 34*(6), 637–648. doi: 10.7164/antibiotics.34.637.
124. Barbee, L. A., Soge, O. O., Ocbamichael, N., LeClair, A., & Golden, M. R., (2021). Single-arm open-label clinical trial of two grams of aztreonam for the treatment of *Neisseria gonorrhoeae*. *Antimicrobial Agents and Chemotherapy, 65*(1), e01739/1–e01739/8. doi: 10.1128/AAC.01739-20.
125. Sykes, R. B., & Bonner, D. P., (1985). Discovery and development of the monobactams. *Reviews of Infectious Diseases, 7*(Suppl. 4), S579–S593.
126. O'Riordan, W., McManus, A., Teras, J., Poromanski, I., Cruz-Saldariagga, M., Quintas, M., Lawrence, L., et al., (2018). A comparison of the efficacy and safety of intravenous followed by oral delaflxacin with vancomycin plus aztreonam for the treatment of acute bacterial skin and skin structure infections: A phase 3, multinational, double-blind, randomized study. *Clinical Infectious Diseases, 67*(5), 657–666. doi: 10.1093/cid/ciy165.
127. Worthington, R. J., & Melander, C., (2013). Overcoming resistance to β-lactam antibiotics. *Journal of Organic Chemistry, 78*(9), 4207–4213. doi: 10.1021/jo400236f.
128. Page, M. G. P., Dantier, C., Desarbre, E., Gaucher, B., Gebhardt, K., & Schmitt-Hoffmann, A., (2011). *In vitro* and *in vivo* properties of BAL30376, a β-lactam and dual β-lactamase inhibitor combination with enhanced activity against gram-negative bacilli that express multiple β-lactamases. *Antimicrobial Agents and Chemotherapy, 55*(4), 1510–1519. doi: 10.1128/AAC.01370-10.
129. Papp-Wallace, K. M., Mallo, S., Bethel, C. R., Taracila, M. A., Hujer, A. M., Fernandez, A., Gatta, J. A., et al., (2014). A kinetic analysis of the inhibition of FOX-4 β-lactamase, a plasmid-mediated AmpC cephalosporinase, by monocyclic β-lactams and carbapenems. *The Journal of Antimicrobial Chemotherapy, 69*(3), 682–690. doi: 10.1093/jac/dkt434.
130. Paech, F., Messner, S., Spickermann, J., Wind, M., Schmitt-Hoffmann, A. H., Witschi, A. T., Howell, B. A., et al., (2017). Mechanisms of hepatotoxicity associated with the monocyclic β-lactam antibiotic BAL30072. *Archives of Toxicology, 91*(11), 3647–3662. doi: 10.1007/s00204-017-1994-x.
131. Edlund, C., (1999). BMS-180680 Bristol-Myers Squibb. *Current Opinion in Anti-Infective Investigational Drugs, 1*(1), 96–100.
132. Neu, H. C., & Chin, N. X., (1987). *In vitro* activity and β-lactamase stability of a new monobactam, BO-1165. *Antimicrobial Agents and Chemotherapy, 31*(4), 505–511. doi: 10.1128/AAC.31.4.505.
133. Matsuda, K., Nakagawa, S., Nakano, F., Inoue, M., & Mitsuhashi, S., (1987). Structure-activity relations of 4-fluoromethyl monobactams. *Journal of Antimicrobial Chemotherapy, 19*(6), 753–760. doi: 10.1093/jac/19.6.753.

134. Rodny, O., (2014). *Therapeutic Uses of Tigemonam and Carumonam for Treatment of Bacterial Infections*. PCT Int. Appl., WO 2014071155 A1.
135. Liu, W. C., Parker, W. L., Wells, J. S., Principe, P. A., Trejo, W. H., Bonner, D. P., & Sykes, R. B., (1982). EM5210(SQ 26,445), a novel monobactam. In: Periti, P., & Gialdroni, G. G., (eds.), *Curr. Chemother. Immunother., Proc. Int. Congr. Chemother.* (Vol. 12th, pp. 328, 329).
136. Parker, W. L., & Rathnum, M. L., (1982). EM5400, a family of monobactam antibiotics produced by *Agrobacterium* radiobacter. II. Isolation and structure determination. *Journal of Antibiotics, 35*(3), 300–305. doi: 10.7164/antibiotics.35.300.
137. Wells, J. S., Trejo, W. H., Principe, P. A., Bush, K., Georgopapadakou, N., Bonner, D. P., & Sykes, R. B., (1982). EM5400, a family of monobactam antibiotics produced by *Agrobacterium* radiobacter. I. Taxonomy, fermentation and biological properties. *Journal of Antibiotics, 35*(3), 295–299. doi: 10.7164/antibiotics.35.295.
138. Mahajan, V. R., Basarkar, G. D., & Upasani, C. D., (2019). Development of self micro emulsifying drug delivery system of ezetimibe by spray drying technology: Characterization, *in-vitro* and *in-vivo* evaluation. *Journal of Pharmaceutical and Scientific Innovation, 8*(2), 64–78. doi: 10.7897/2277-4572.082130.
139. Katayama, N., Nozaki, Y., Okonogi, K., Ono, H., Harada, S., & Okazaki, H., (1985). Formadicins, new monocyclic β-lactam antibiotics of bacterial origin. I. Taxonomy, fermentation and biological activities. *Journal of Antibiotics, 38*(9), 1117–1127. doi: 10.7164/antibiotics.38.1117.
140. Harada, S., Tsuboya, S., & Ono, H., (1986). Antibiotics TAN-585C and D. *Jpn. Kokai Tokkyo Koho*. JP 61025488 A.
141. Breuer, H., (1985). Oral monobactams. Chemistry of [2-oxo-1-azetidinyl)oxy]acetic acids and prodrug esters thereof: An overview. In: Ishigami, J., (ed.), *Recent Adv. Chemother., Proc. Int. Congr. Chemother.* (pp. 1353, 1354). University of Tokyo Press.
142. Kintaka, K., Haibara, K., Asai, M., & Imada, A., (1981). Isosulfazecin, a new β-lactam antibiotic, produced by an acidophilic pseudomonad. Fermentation, isolation and characterization. *Journal of Antibiotics, 34*(9), 1081–1089. doi: 10.7164/antibiotics.34.1081.
143. Blais, J., Lopez, S., Li, C., Ruzin, A., Ranjitkar, S., Dean, C. R., Leeds, J. A., et al., (2018). *In vitro* activity of LYS228, a novel monobactam antibiotic, against multidrug-resistant *Enterobacteriaceae*. *Antimicrobial Agents and Chemotherapy, 62*(10), e00552-18/1–e00552-18/10. doi: 10.1128/aac.00552-18.
144. Box, S. J., Brown, A. G., Gilpin, M. L., Gwynn, M. N., & Spear, S. R., (1988). MM 42842, a new member of the monobactam family produced by *Pseudomonas cocovenenans*. II. Production, isolation and properties of MM 42842. *Journal of Antibiotics, 41*(1), 7–12. doi: 10.7164/antibiotics.41.7.
145. Aviles, E., & Rodriguez, A. D., (2010). Monamphilectine A, a potent antimalarial β-lactam from marine sponge *Hymeniacidon sp*: Isolation, structure, semisynthesis, and bioactivity. *Organic Letters, 12*(22), 5290–5293. doi: 10.1021/ol102351z.
146. Kamiya, T., Aoki, H., & Mine, Y., (1982). Nocardicins. In: Morin, R. B., & Gorman, M., (eds.) *Chemistry and Biology of β-lactam Antibiotic* (Vol. 2, pp. 165–226). Academic Press: New York.
147. Andreotti, D., Biondi, S., & Di Modugno, E., (2003). Chapter fourteen, β-lactam antibiotics. In: Abraham, D. J., (ed.) *Burger's Medicinal Chemistry and Drug Discovery, 6th Edition, Volume 5: Chemotherapeutic Agents* (pp. 607–736). Wiley-Interscience.

148. Garcia-Rodriguez, J. A., Garcia-Sanchez, J. E., Plata, A., & Trujillano, I., (1986). The *in-vitro* activity of SQ 82,291, a new monobactam, in comparison with aztreonam. *The Journal of Antimicrobial Chemotherapy, 17*(3), 303–307. doi: 10.1093/jac/17.3.303.
149. Zhai, H. Y., Zhao, C., Zhang, N., Jin, M. N., Tang, S. A., Qin, N., Kong, D. X., & Duan, H. Q., (2012). Alkaloids from *Pachysandra terminalis* inhibit breast cancer invasion and have potential for development as antimetastasis therapeutic agents. *Journal of Natural Products, 75*(7), 1305–1311. doi: 10.1021/np300207c.
150. Kato, T., Hinoo, H., Terui, Y., Nishikawa, J., Nakagawa, Y., Ikenishi, Y., & Shoji, J., (1987). PB-5266 A, B and C, new monobactams. II. Physicochemical properties and chemical structures. *Journal of Antibiotics, 40*(2), 139–144. doi: 10.7164/antibiotics.40.139.
151. Kato, T., Hinoo, H., Shoji, J., Matsumoto, K., Tanimoto, T., Hattori, T., Hirooka, K., & Kondo, E., (1987). PB-5266 A, B and C, new monobactams. I. Taxonomy, fermentation and isolation. *Journal of Antibiotics, 40*(2), 135–138. doi: 10.7164/antibiotics.40.135.
152. Livermore, D. M., & Chen, H. Y., (1997). Potentiation of β-lactams against *Pseudomonas aeruginosa* strains by Ro 48-1256, a bridged monobactam inhibitor of AmpC β-lactamases. *Journal of Antimicrobial Chemotherapy, 40*(3), 335–343. doi: 10.1093/jac/40.3.335.
153. Chantot, J. F., Klich, M., Teutsch, G., Bryskier, A., Collette, P., Markus, A., & Seibert, G., (1992). Antibacterial activity of RU44790, a new *N*-tetrazolyl monocyclic β-lactam. *Antimicrobial Agents and Chemotherapy, 36*(8), 1756–1763. doi: 10.1128/AAC.36.8.1756.
154. Frere, J. M., Klein, D., Kelly, J. A., & Ghuysen, J. M., (1984). Interaction between monobactams and model *D*-alanyl-*D*-alanine-cleaving peptidases. *FEMS Microbiology Letters, 21*(2), 213–217. doi: 10.1016/0378-1097(84)90297-0.
155. Neu, H. C., & Labthavikul, P., (1983). *In vitro* activity and β-lactamase stability of a monobactam, SQ 26,917, compared with those of aztreonam and other agents. *Antimicrobial Agents and Chemotherapy, 24*(2), 227–232. doi: 10.1128/AAC.24.2.227.
156. Shionogi and Co., Ltd., Japan, (1984), β-Lactam antibiotic PB-5582-A. *Jpn. Kokai Tokkyo Koho*, JP 59151895 A. (1984). β-lactam antibiotic PB-5582-A. *Jpn. Kokai Tokkyo Koho*. JP 59151895 A.
157. Murphy-Benenato, K. E., Bhagunde, P. R., Chen, A., Davis, H. E., Durand-Reville, T. F., Ehmann, D. E., Galullo, V., et al., (2015). Discovery of efficacious *Pseudomonas aeruginosa*-targeted siderophore-conjugated monocarbams by application of a semi-mechanistic pharmacokinetic/pharmacodynamic model. *Journal of Medicinal Chemistry, 58*(5), 2195–2205. doi: 10.1021/jm501506f.
158. Braga, D., & Lackner, G., (2017). One ring to fight them all: The sulfazecin story. *Cell Chemical Biology, 24*(1), 1, 2. doi: 10.1016/j.chembiol.2017.01.001.
159. Nishida, K., Kunugita, C., Uji, T., Higashitani, F., Hyodo, A., Unemi, N., Maiti, S. N., et al., (1999). *In vitro* and *in vivo* activities of Syn2190, a novel β-lactamase inhibitor. *Antimicrobial Agents and Chemotherapy, 43*(8), 1895–1900. doi: 10.1128/AAC.43.8.1895.
160. Baik, H. S., Gu, J. G., & Jun, H. K., (1989). Studies on the microbiological assay method for tabtoxin produced in *Pseudomonas syringae* pv. tabaci. *Misaengmul Hakhoechi, 27*(3), 310–315.
161. Katayama, N., Koyama, K., Nozaki, Y., Harada, S., & Ono, H., (1987). A new sulfazecin-type β-lactam antibiotic, TAN-850. *Takeda Kenkyushoho, 46*, 62–68.

162. Tanaka, S. K., Summerill, R. A. S., Minassian, B. F., Bush, K., Visnic, D. A., Bonner, D. P., & Sykes, R. B., (1987). *In vitro* evaluation of tigemonam, a novel oral monobactam. *Antimicrobial Agents and Chemotherapy, 31*(2), 219–225. doi: 10.1128/AAC.31.2.219.
163. Bush, K., (2010). Chapter 15. Other β-lactam antibiotics. In: Finch, R. G., Greenwood, D., Norrby, S. R., & Whitley, R. J., (eds.), *Antibiotic and Chemotherapy: Anti-Infective Agents and Their Use in Therapy* (9th edn., pp. 226–244). Saunders/Elsevier: Edinburgh.
164. Wanning, M., Zähner, H., Krone, B., & Zeeck, A., (1981). Ein neues antifungisches β-lactam-antibioticum der clavam-reihe. *Tetrahedron Letters, 22*(27), 2539–2540. doi: 10.1016/S0040-4039(01)90514-3.
165. Evans, R. H. Jr., Ax, H., Jacoby, A., Williams, T. H., Jenkins, E., & Scannell, J. P., (1983). Ro 22-5417, a new clavam antibiotic from *Streptomyces clavuligerus*. II. Fermentation, isolation and structure. *Journal of Antibiotics, 36*(3), 213–216. doi: 10.7164/antibiotics.36.213.
166. Brown, D., Evans, J. R., & Fletton, R. A., (1979). Structures of three novel β-lactams isolated from *Streptomyces clavuligerus*. *Journal of the Chemical Society, Chemical Communications,* (6), 282, 283. doi: 10.1039/C39790000282.
167. Kellett, M., Pruess, D., & Scannell, J. P., (1980). 2-(3-*Alaninyl*)*clavam antibiotic*. United States Patent, US 4202819A.
168. Kawamura, Y., Kondo, E., Matsumoto, K., Tsuji, N., & Tawara, K., (1987). Clavam antibiotics CA-146A and CA-146B. *Jpn. Kokai Tokkyo Koho*. JP 62010088 A.
169. Kawamura, Y., Kondo, E., Matsumoto, K., Tsuji, N., & Tawara, K., (1987). Clavam antibiotic CA-31. *Jpn. Kokai Tokkyo Koho*. JP 62010089 A.
170. King, H. D., Langhaerig, J., & Sanglier, J. J., (1986). Clavamycins, new clavam antibiotics from two variants of *Streptomyces hygroscopicus*. I. Taxonomy of the producing organisms, fermentation, and biological activities. *Journal of Antibiotics, 39*(4), 510–515. doi: 10.7164/antibiotics.39.510.
171. Naegeli, H. U., Loosli, H. R., & Nussbaumer, A., (1986). Clavamycins, new clavam antibiotics from two variants of *Streptomyces hygroscopicus*. II. Isolation and structures of clavamycins A, B and C from *Streptomyces hygroscopicus* NRRL 15846, and of clavamycins D, E and F from *Streptomyces hygroscopicus* NRRL 15879. *Journal of Antibiotics, 39*(4), 516–524. doi: 10.7164/antibiotics.39.516.
172. Zhang, W., Qi, C., Tian, J., Xie, M., Chen, W., & Sun, J., (1991). New clavam antibiotics G0069. *Zhongguo Kangshengsu Zazhi, 16*(1), 1–12.
173. Baldwin, J. E., Claridge, T. D. W., Goh, K. C., Keeping, J. W., & Schofield, C. J., (1993). Revised structures for Tue 1718B and valclavam. *Tetrahedron Letters, 34*(35), 5645–5648. doi: 10.1016/S0040-4039(00)73905-0.
174. Peter, H., Rabenhorst, J., Roehl, F., & Zaehner, H., (1985). Valclavam: An antifungal β-lactam antibiotic. *Recent Adv. Chemother., Proc. Int. Congr. Chemother., 14th* (antimicrobial sect. 1), 237–238.
175. Alcaide, B., Aragoncillo, C., & Almendros, P., (2008). 2.02 Cephalosporins. In: Katritzky, A. R., Ramsden, C. A., Scriven, E. F. V., & Taylor, R. J. K., (eds.), *Comprehensive Heterocyclic Chemistry* (IIIrd edn., Vol. 2, pp. 111–169). Elsevier. Four-membered Heterocycles.
176. Matagne, A., Ledent, P., Monnaie, D., Felici, A., Jamin, M., Raquet, X., Galleni, M., et al., (1995). Kinetic study of interaction between BRL 42715, β-lactamases, and δ-alanyl-*D*-alanine peptidases. *Antimicrobial Agents and Chemotherapy, 39*(1), 227–231. doi: 10.1128/AAC.39.1.227.

177. Wise, R., Andrews, J. M., & Piddock, L. J. V., (1987). *In vitro* activity of CGP 31608, a new penem. *Antimicrobial Agents and Chemotherapy, 31*(2), 267–273. doi: 10.1128/AAC.31.2.267.
178. Schurek, K. N., Wiebe, R., Karlowsky, J. A., Rubinstein, E., Hoban, D. J., & Zhanel, G. G., (2007). Faropenem: Review of a new oral penem. *Expert Review of Anti-Infective Therapy, 5*(2), 185–198. doi: 10.1586/14787210.5.2.185.
179. Kobayashi, I., Nakagawa, N., Urabe, E., Matsuzaki, K., Sato, Y., & Sunakawa, K., (2007). *In vitro* activity of faropenem against 1,014 clinical isolates. *Kagaku Ryoho no Ryoiki, 23*(11), 1763–1776.
180. Nishino, T., Maeda, Y., Ohtsu, E., Koizuka, S., Nishihara, T., Adachi, H., Okamoto, K., & Ishiguro, M., (1989). Studies on penem antibiotics. II. *In vitro* activity of SUN5555, a new oral penem. *The Journal of Antibiotics, 42*(6), 977–988. doi: 10.7164/antibiotics.42.977.
181. Cormican, M. G., & Jones, R. N., (1995). Evaluation of the *in-vitro* activity of furopenem (SY5555 or SUN5555) against respiratory tract pathogens and β-lactamase producing bacteria. *Journal of Antimicrobial Chemotherapy, 35*(4), 535–539. doi: 10.1093/jac/35.4.535.
182. Gootz, T., Girard, D., Schelkley, W., Tensfeldt, T., Foulds, G., Kellogg, M., Stam, J., et al., (1990). Pharmacokinetic studies in animals of a new parenteral penem CP-65,207 and its oral prodrug ester. *Journal of Antibiotics, 43*(4), 422–432. doi: 10.7164/antibiotics.43.422.
183. Foulds, G., Knirsch, A. K., Lazar, J. D., Tensfeldt, T. G., & Gerber, N., (1991). Pharmacokinetics of the penem CP-65,207 and its separate stereoisomers in humans. *Antimicrobial Agents and Chemotherapy, 35*(4), 665–671. doi: 10.1128/AAC.35.4.665.
184. Saathoff, A., Lode, H., Hampel, B., Deppermann, K. M., Borner, K., & Koeppe, P., (1990). Pharmacokinetics of FCE 22891, a new oral penem. *Antimicrobial Agents and Chemotherapy, 34*(6), 1001–1006. doi: 10.1128/AAC.34.6.1001.
185. Seibert, G., Isert, D., Klesel, N., Limbert, M., Pries, A., Schrinner, E., Cooke, M., et al., (1987). HRE 664, a new parenteral penem. I. Antibacterial activity *in vitro*. *Journal of Antibiotics, 40*(5), 660–667. doi: 10.7164/antibiotics.40.660.
186. Ferrari, L., Altamura, M., De Luca, M., & Ceruti, T., (2002). MEN 10700, a new penem antibiotic: *In vitro* antibacterial activity on clinical isolates. *Chemotherapy (Basel, Switzerland), 48*(1), 15–20. doi: 10.1159/000048582.
187. Tateda, E., Asada, K., Hiramatsu, K., & Yokota, T., (1995). *In vitro* antibacterial activity of ritipenem acoxil. *Nippon Kagaku Ryoho Gakkai Zasshi, 43*(Suppl. 3), 21–32. doi: 10.11250/chemotherapy1995.43.Supplement3_21.
188. Ganguly, A. K., Girijavallabhan, V. M., McCombie, S., Pinto, P., Rizvi, R., Jeffrey, P. D., & Lin, S., (1982). Synthesis of Sch 29482 – a novel penem antibiotic. *Journal of Antimicrobial Chemotherapy, 9*(Suppl. C), 1–6. doi: 10.1093/jac/9.suppl_C.1.
189. Giamarellou, H., Tsagarakis, J., Daourou, A., Grammatikou, M., & Avlami, A., (1985). Comparative *in vitro* activity of SCH 34343 with 14 antibiotics against 100 gram-negative blood isolates. *Recent Adv. Chemother., Proc. Int. Congr. Chemother., 14th (Antimicrobial Sect. 2)*, 1226, 1227.
190. Nagashima, M., Goto, S., Matsunaga, T., Shimohira, H., & Ogawa, M., (1996). *In vitro* and *in vivo* activities of sulopenem compared with those of imipenem and cephalosporins. *Japanese Journal of Antibiotics, 49*(4), 303–323.
191. Phillips, O. A., Czajkowski, D. P., Spevak, P., Singh, M. P., Hanehara-Kunugita, C., Hyodo, A., Micetich, R. G., & Maiti, S. N., (1997). Syn-1012: A new β-lactamase

inhibitor of penem skeleton. *Journal of Antibiotics, 50*(4), 350–356. doi: 10.7164/antibiotics.50.350.

192. Jefson, M. R., Hecker, S. J., & Dirlam, J. P., (1994). Chapter 12. Antibacterial agents. In: Hagmann, W. K., Lee, J. C., McCall, J. M., Plattner, J. J., Robertson, D. W., & Venuti, M. C., (eds.) *Annual Reports in Medicinal Chemistry* (Vol. 29, pp. 113–122). Academic Press, Inc.: San Diego.
193. Okamoto, K., Gotoh, N., & Nishino, T., (2001). *Pseudomonas aeruginosa* reveals high intrinsic resistance to penem antibiotics: Penem resistance mechanisms and their interplay. *Antimicrobial Agents and Chemotherapy, 45*(7), 1964–1971. doi: 10.1128/AAC.45.7.1964-1971.2001.
194. Yoshioka, T., Watanabe, A., Kojima, I., Shimauchi, Y., Okabe, M., Fukagawa, Y., & Ishikura, T., (1984). Structures of OA-6129D and E, new carbapenam antibiotics. *Journal of Antibiotics, 37*(3), 211–217. doi: 10.7164/antibiotics.37.211.
195. Haneishi, T., Nakajima, M., Serizawa, N., Inukai, M., Takiguchi, Y., Arai, M., Satoh, S., et al., (1983). A new carbapenam no. 17927 D substance. *Journal of Antibiotics, 36*(11), 1581–1584. doi: 10.7164/antibiotics.36.1581.
196. Tipparthi, S. K., Akula, S., Manderwad, G., & Rajkumar, H. R. V., (2019). Prevalence of carbapenam resistant *Klebsiella pneumoniae* in a tertiary care hospital, Hyderabad, Telangana State. *International Journal of Microbiology Research, 11*(8), 1694–1696.
197. Li, R., Stapon, A., Blanchfield, J. T., & Townsend, C. A., (2000). Three unusual reactions mediate carbapenem and carbapenam biosynthesis. *Journal of the American Chemical Society, 122*(38), 9296, 9297. doi: 10.1021/ja001723l.
198. Raber, M. L., Arnett, S. O., & Townsend, C. A., (2009). A conserved tyrosyl-glutamyl catalytic dyad in evolutionarily linked enzymes: Carbapenam synthetase and β-lactam synthetase. *Biochemistry, 48*(22), 4959–4971. doi: 10.1021/bi900432n.
199. Miller, M. T., Gerratana, B., Stapon, A., Townsend, C. A., & Rosenzweig, A. C., (2003). Crystal structure of carbapenam synthetase (CarA). *Journal of Biological Chemistry, 278*(42), 40996–41002. doi: 10.1074/jbc.M307901200.
200. Balatsouras, D. G., Korres, S., Rallis, E., Eliopoulos, P., & Ferekidis, E., (2005). Twice-daily dosing of loracarbef 15 μg/kg versus 30 μg/kg in the treatment of children with acute sinusitis. *Drugs Under Experimental and Clinical Research, 31*(Suppl.), 1–5.
201. Shetler, T., Bendele, A., Buening, M., Clemens, J., Colbert, W., Deldar, A., Helton, D., McGrath, J., Shannon, H., Turk, J., et al., (1993). General pharmacology of Loracarbef in animals. *Arzneimittel-Forschung, 43*(1), 60–70.
202. Matsukuma, I., Yoshiye, S., Mochida, K., Hashimoto, Y., Sato, K., Okachi, R., & Hirata, T., (1989). Synthesis and biological evaluation of 3-chloro-1-carbacephem compounds. *Chemical & Pharmaceutical Bulletin, 37*(5), 1239–1244. doi: 10.1248/cpb.37.1239.
203. Van, G. A. J. A., Janssen, A. J. H. M., Van, N. M., Verbrugh, H. A., & Diepersloot, R. J. A., (1995). *In vitro* activity of Loracarbef against common respiratory pathogens compared with amoxicillin-clavulanic acid: A multi-center study in the Netherlands. *International Journal of Antimicrobial Agents, 5*(3), 183–186. doi: 10.1016/0924-8579(94)00062-Y.
204. Stenquist, M., Olen, L., Jannert, M., Naslund, L., & Zeckel, M. L., (1996). Penetration of Loracarbef into the maxillary sinus: A pharmacokinetic assessment. *Clinical Therapeutics, 18*(2), 273–284. doi: 10.1016/S0149-2918(96)80008-7.
205. Doern, G., (1992). *In vitro* activity of Loracarbef and effects of susceptibility test methods. *The American Journal of Medicine, 92*(6A), 7S–15S. doi: 10.1016/0002-9343(92)90601-7.

206. Ogasa, T., Saito, H., Hashimoto, Y., Sato, K., & Hirata, T., (1989). Synthesis and biological evaluation of optically active 3-H-1-carbacephem compounds. *Chemical & Pharmaceutical Bulletin, 37*(2), 315–321. doi: 10.1248/cpb.37.315.
207. Hashimoto, Y., Takasawa, S., Ogasa, T., Saito, H., Hirata, T., & Kimura, K., (1984). Enzymic synthesis of optically pure 1-carbacephem compounds. *Annals of the New York Academy of Sciences, 434*(Enzyme Eng.), 206–209. doi: 10.1111/j.1749-6632.1984.tb29828.x.
208. Saito, H., Suzuki, F., & Hirata, T., (1989). Synthesis of 1,2-disubstituted-1-carbacephem. *Chemical & Pharmaceutical Bulletin, 37*(9), 2298–2302. doi: 10.1248/cpb.37.2298.
209. Mochida, K., & Hirata, T., (1988). Synthetic studies on 1-carbacephem antibiotics: New synthetic approach to 3H-carbacephems. *Chemical & Pharmaceutical Bulletin, 36*(9), 3642–3645. doi: 10.1248/cpb.36.3642.
210. Mochida, K., Shiraki, C., Yamasaki, M., Hirata, T., Sato, K., & Okachi, R., (1987). Aminothiazolylglycyl derivatives of carbacephems. I. Synthesis and antibacterial activity of novel carbacephems with substituted aminothiazolyl groups. *Journal of Antibiotics, 40*(1), 14–21. doi: 10.7164/antibiotics.40.14.
211. Mochida, K., Ono, Y., Yamasaki, M., Shiraki, C., Hirata, T., Sato, K., & Okachi, R., (1987). Aminothiazolylglycyl derivatives of carbacephem antibiotics. II. Synthesis and antibacterial activity of novel aminothiazolyl cephem compounds with hydroxypyridone moiety. *Journal of Antibiotics, 40*(2), 182–189. doi: 10.7164/antibiotics.40.182.
212. Takaya, T., Takasugi, H., Masugi, T., Chiba, T., Kochi, H., Takano, T., & Nakano, H., (1981). Studies of β-lactam antibiotics. 1. Structure-activity relations of sodium 7β-[(*Z*)-2-(2-amino-4-thiazolyl)-2-(methoxyimino)acetamido]-3-cephem-4-carboxylate (ceftizoxime) and its related compounds. *Nippon Kagaku Kaishi,* (5), 785–804. doi: 10.1246/nikkashi.1981.785.
213. Trampuz, A., & Zimmerli, W., (2006). Antimicrobial agents in orthopedic surgery: Prophylaxis and treatment. *Drugs, 66*(8), 1089–1105. doi: 10.2165/00003495-200666080-00005.
214. Lee, C. H., Chen, I. L., Li, C. C., & Chien, C. C., (2018). Relation between flomoxef minimum inhibitory concentrations and clinical outcomes of patients treated with flomoxef for *Enterobacteriaceae* bacteremia. *Infection and Drug Resistance, 11*, 2471–2480. doi: 10.2147/IDR.S185670.
215. Matsumura, Y., Yamamoto, M., Nagao, M., Tanaka, M., Takakura, S., & Ichiyama, S., (2016). *In vitro* activities and detection performances of cefmetazole and flomoxef for extended-spectrum β-lactamase and plasmid-mediated AmpC β-lactamase-producing *Enterobacteriaceae*. *Diagnostic Microbiology and Infectious Disease, 84*(4), 322–327. doi: 10.1016/j.diagmicrobio.2015.12.001.
216. Horie, A., Nariai, A., Katou, F., Abe, Y., Saito, Y., Koike, D., Hirade, T., et al., (2019). Increased community-acquired upper urinary tract infections caused by extended-spectrum beta-lactamase-producing *Escherichia coli* in children and the efficacy of flomoxef and cefmetazole. *Clinical and Experimental Nephrology, 23*(11), 1306–1314. doi: 10.1007/s10157-019-01775-w.
217. Weissberger, B., Abruzzo, G. K., Fromtling, R. A., Valiant, M. E., Shungu, D. L., & Gadebusch, H. H., (1988). L-656,575(OCP-9-176): A novel oxacephem. *in vitro* activity against aerobic and anaerobic clinical bacterial isolates. *Journal of Antibiotics, 41*(8), 1130–1136. doi: 10.7164/antibiotics.41.1130.

218. Gu, J. W., Chin, N. X., & Neu, H. C., (1991). *In vitro* activity of an oxycephem OCP 9-176 compared with its sulfur analog and other β-lactams. *Diagnostic Microbiology and Infectious Disease, 14*(2), 135–140. doi: 10.1016/0732-8893(91)90048-K.
219. Klech, H., & Fiser, I., (1983). Pharmacokinetics, *in vitro* activity, and clinical tests of lamoxactam. *Schweizerische Rundschau fuer Medizin/Praxis, 72*(11), 369–378.
220. Raetzel, G., Harnoss, B. M., Gortz, G., Haring, R., & Rodloff, A., (1986). Systemic antibiotic prophylaxis with metronidazole in elective colonic and rectal surgery. Results of a clinical controlled study and a critical literature review. *Arzneimittel-Forschung, 36*(6), 976–980.
221. Polk, R. E., (1982). Moxalactam (Moxam, Eli Lilly). *Drug Intelligence & Clinical Pharmacy, 16*(2), 104–112. doi: 10.1177/106002808201600203.
222. Van, K. B., Van, W. L. J., & Rutgers, A., (1980). Comparative susceptibility of *Pseudomonas aeruginosa* to four cephalosporins. *Journal of Antimicrobial Chemotherapy, 6*(5), 674–676. doi: 10.1093/jac/6.5.674.
223. Barry, A. L., Thornsberry, C., & Jones, R. N., (1980). *In vitro* evaluation of LY127935(6059S) compared with cefotaxime, eight other β-lactams and two aminoglycosides. *Journal of Antimicrobial Chemotherapy, 6*(6), 775–784. doi: 10.1093/jac/6.6.775.
224. Harada, M., Matsumoto, M., & Takeuchi, M., (1980). Immunological properties of a newly synthesized β-lactam antibiotic agent 6059-S. *Chemotherapy (Tokyo), 28*(Suppl. 7), 1191–1201. doi: 10.11250/chemotherapy1953.28.Supplement7_1191.
225. Fujita, T., Kimoto, H., & Koshiro, A., (1986). Stability of disodium latamoxef in aqueous solution and in the admixtures of 5% glucose solution. *Yakugaku Zasshi, 106*(1), 68–74. doi: 10.1248/yakushi1947.106.1_68.
226. Asif, M., (2015). Development of antitubercular agents: Potential for β-lactum based compounds. *International Journal of Medical Therapeutics, 1*(1), 1–14.
227. Nakanishi, K., Masukawa, T., Madada, M., & Nadai, T., (1994). Improvement of the rectal bioavailability of latamoxef sodium by adjuvants following administration of a suppository. *Biological & Pharmaceutical Bulletin, 17*(11), 1496–1500. doi: 10.1248/bpb.17.1496.
228. Brook, I., (1980). *In vitro* activity of Moxalactam (LY-127935) against anaerobic micro-organisms. *The Journal of Antimicrobial Chemotherapy, 6*(6), 676–679. doi: 10.1093/jac/6.5.676.
229. Bornstein, M., Lo, A. Y., Thomas, P. N., & Wilham, W. L., (1982). Moxalactam disodium compatibility with intramuscular and intravenous diluents. *American Journal of Hospital Pharmacy, 39*(9), 1495–1498.
230. Gu, H. Y., (2002). Trace determination of shiomarin using a hanging copper/amalgam drop electrode. *Journal of Pharmaceutical and Biomedical Analysis, 29*(1, 2), 341–346. doi: 10.1016/S0731-7085(02)00062-6.
231. Quan, J., Wang, Y., Ji, J., Wang, Y., Wang, H., Xu, Y., & Yu, Y., (2016). Activity of Moxalactam against *Enterobacteriaceae* and anaerobia *in vitro*. *Zhonghua Yixue Zazhi (Beijing, China), 96*(18), 1459–1464. doi: 10.3760/cma.j.issn.0376-2491.2016.18.015.
232. Livingston, W. K., Elliott, A. M., Dismukes, W. E., Avent, C. K., & Cobbs, C. G., (1981). Clinical evaluation of Moxalactam. *Antimicrobial Agents and Chemotherapy, 20*(1), 88–97. doi: 10.1128/AAC.20.1.88.
233. McCracken, G. H. J., Threlkeld, N., Mize, S., Baker, C. J., Kaplan, S. L., Faingezicht, I., Feldman, W. E., & Schaad, U., (1984). Moxalactam therapy for neonatal meningitis due

to gram-negative enteric bacilli. A prospective controlled evaluation. *JAMA, 252*(11), 1427–1432. doi: 10.1001/jama.1984.03350110027024.
234. Ke, W., Bethel, C. R., Papp-Wallace, K. M., Pagadala, S. R. R., Nottingham, M., Fernandez, D., Buynak, J. D., et al., (2012). Crystal structures of KPC-2 β-lactamase in complex with 3-nitrophenyl boronic acid and the penam sulfone PSR-3-226. *Antimicrobial Agents and Chemotherapy, 56*(5), 2713–2718. doi: 10.1128/AAC.06099-11.
235. Rodkey, E. A., Winkler, M. L., Bethel, C. R., Pagadala, S. R. R., Buynak, J. D., Bonomo, R. A., & Van, D. A. F., (2014). Penam sulfones and β-lactamase inhibition: SA2-13 and the importance of the C2 side chain length and composition. *PLoS One, 9*(1), e85892/1–e85892/11. doi: 10.1371/journal.pone.0085892.
236. Tzouvelekis, L. S., Gazouli, M., Prinarakis, E. E., Tzelepi, E., & Legakis, N. J., (1997). Comparative evaluation of the inhibitory activities of the novel penicillanic acid sulfone Ro 48-1220 against β-lactamases that belong to groups 1, 2b, and 2be. *Antimicrobial Agents and Chemotherapy, 41*(2), 475–477. doi: 10.1128/AAC.41.2.475.
237. Jones, R. N., & Johnson, D. M., (1995). Comparative *in vitro* activity of apalcillin alone and combined with Ro 48-1220, a novel penam β-lactamase inhibitor. *Clinical Microbiology and Infection, 1*(2), 86–100. doi: 10.1111/j.1469-0691.1995.tb00451.x.
238. Drawz, S. M., Bethel, C. R., Doppalapudi, V. R., Sheri, A., Pagadala, S. R. R., Hujer, A. M., Skalweit, M. J., et al., (2010). Penicillin sulfone inhibitors of class D β-lactamases. *Antimicrobial Agents and Chemotherapy, 54*(4), 1414–1424. doi: 10.1128/AAC.00743-09.
239. Kalp, M., Sheri, A., Buynak, J. D., Bethel, C. R., Bonomo, R. A., & Carey, P. R., (2007). Efficient inhibition of class a and class D β-lactamases by Michaelis complexes. *Journal of Biological Chemistry, 282*(30), 21588–21591. doi: 10.1074/jbc.C700080200.
240. Bou, G., Santillana, E., Sheri, A., Beceiro, A., Sampson, J. M., Kalp, M., Bethel, C. R., et al., (2010). Design, synthesis, and crystal structures of 6-alkylidene-2'-substituted penicillanic acid sulfones as potent inhibitors of *Acinetobacter Baumannii* OXA-24 carbapenemase. *Journal of the American Chemical Society, 132*(38), 13320–13331. doi: 10.1021/ja104092z.
241. Ke, W., Pattanaik, P., Bethel, C. R., Sheri, A., Buynak, J. D., Bonomo, R. A., & Van, D. A. F., (2012). Structures of SHV-1 β-lactamase with penem and penam sulfone inhibitors that form cyclic intermediates stabilized by carbonyl conjugation. *PLoS One, 7*(11), e49035/1–e49035/10.
242. Kousovista, R., Athanasiou, C., Liaskonis, K., Ivopoulou, O., Ismailos, G., & Karalis, V., (2021). Correlation between *Acinetobacter Baumannii* resistance and hospital use of meropenem, cefepime, and ciprofloxacin: Time series analysis and dynamic regression models. *Pathogens, 10*(4), 480/1–480/13. doi: 10.3390/pathogens10040480.
243. Higashitani, F., Hyodo, A., Ishida, N., Inoue, M., & Mitsuhashi, S., (1990). Inhibition of β-lactamases by tazobactam and *in vitro* antibacterial activity of tazobactam combined with piperacillin. *Journal of Antimicrobial Chemotherapy, 25*(4), 567–574. doi: 10.1093/jac/25.4.567.
244. Jacqueline, C., Howland, K., & Chesnel, L., (2017). *In vitro* activity of ceftolozane/tazobactam in combination with other classes of antibacterial agents. *Journal of Global Antimicrobial Resistance, 10*, 326–329. doi: 10.1016/j.jgar.2017.04.003.

CHAPTER 2

Antimicrobial Activity of β-Lactams and Bacterial Resistance

2.1 BRIEF INTRODUCTION OF BACTERIA

Bacteria are microscopic, single-celled organisms that contain a cell wall, plasma membrane, cytoplasm, DNA, ribosomes, and pilli. Some bacteria also contain a capsule on the outside of the cell wall (outer lipid membrane), flagellum for the purpose of moving, or a plasmid of extra circular genetic material which renders the bacterium some advantage over other bacteria. Among these constitutional components, the outer lipid membrane and cell wall are very important, as the cell wall gives bacteria its shape [e.g., spherical (cocci), rod (bacilli), spiral (spirilla), comma (vibrios) or corkscrew (spirochaetes)], and the outer lipid membrane provides bacteria additional protection and somewhat identity. The plasma membrane transports chemicals. Particularly, bacteria are often classified into Gram-positive and Gram-negative groups, according to their response in Gram staining experiment, on the basis of their difference in the thickness of cell wall and the presence or absence of the outer lipid membrane. The Gram staining technique, which was developed in 1884 by the Danish bacteriologist Hans Christian Gram, can quickly narrow down the list of potential bacteria for further tests if necessary [1].

During the Gram staining experiment, a liquid bacterial culture is directly dropped onto a clean glass microscope slide by a sterile loop and gently smeared into an area of approximately 1 cm diameter in a circular motion. Very dense bacterial cultures should be diluted before being loaded to the glass slide to ensure that individual bacterial cells can be seen under a microscope after staining, whereas a loop of colony material from a bacterial

The Chemistry and Biology of Beta-Lactams. Zerong Wang, PhD

plate should be resuspended in a sterile phosphate-buffered saline (PBS) and transferred onto a glass slide in a similar manner. After air dried, the smeared slide is passed through a flame two or three times to kill the microbes and fix the sample to the slide. The glass slide is stained with a saturated solution of crystal violet for 1 minute, rinsed with tap water or distilled water; then the slide is gently flooded with Gram's iodine solution ($KI + I_2$) for 1 minute, and rinsed with tap water or distilled water again. Up to this point, the smear of Gram-positive bacteria on the glass slide will appear purple. In order to visualize the Gram-negative bacteria, the smear can be decolorized with 95% ethanol or acetone, by tilting the slide slightly and dropping the alcohol or acetone over the smear until the solvent becomes clear (about 5–10 seconds) and rinsed immediately with water to avoid over-decolorizing. The smeared area on the slide was covered with safranin counterstain for 45 seconds and rinsed with tap water or distilled water. Safranin will stain bacterial cells in a color of light red, allowing visualization of Gram-negative cells without interfering with the purple color of Gram-positive cells. After being dried with filter paper, the smeared area can be visualized with a light microscope under oil immersion.

Gram-positive bacteria have a thick layer of peptidoglycan without an additional outer lipid membrane; whilst Gram-negative bacteria have a thin layer of peptidoglycan and an outer lipid membrane. Therefore, during the Gram staining experiment, a water-soluble dye of crystal violet (i.e., *N*-(4-(bis(4-(dimethyl-amino)phenyl)methylene)-cyclohexane-2,5-dien-1-ylidene)-*N*-methyl-methanaminium chloride, **1**, Figure 2.1) directly enters the peptidoglycan layer in the bacterial cell wall, which has a maximum absorption wavelength at 579 nm [2], 590 nm [3], 600 nm [4], or 610 nm (depending on the condition) [5], and forms complex with the incoming iodine. After rinsing the smear with 95% ethanol or acetone, the layer of peptidoglycan is dehydrated and shrunk, and the encapsulated large complex of crystal violet-iodine is unable to escape the thick layer of peptidoglycan, resulting in purple stained cells. In the case of Gram-negative bacteria, the outer lipid membrane is degraded during decolorizing stage, and the thin layer of peptidoglycan is unable to retain the complex of crystal violet and iodine so that the purple color fades and disappears during the decolorizing. In contrast, weakly water-soluble safranin (i.e., 3,7-diamino-2,8-dimethyl-5-phenylphenazin-5-ium chloride, **2**, Figure 2.1) has a maximum absorption wavelength at 520 nm [6], 525 nm [7], or 530 nm (depending on the associated ion) [8], can code bacteria in pale reddish color when the smear is observed under a light microscope.

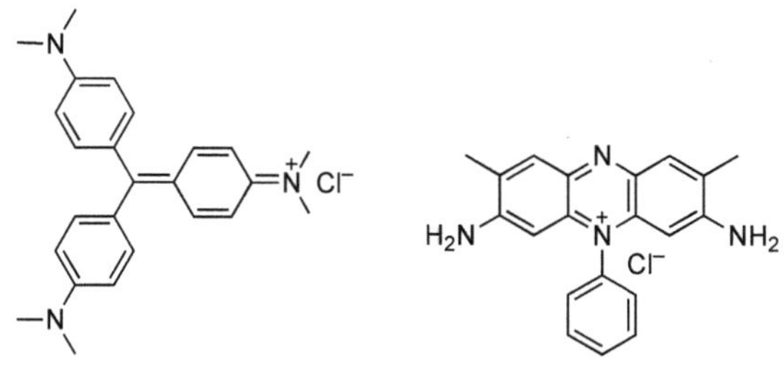

1
(Crystal Violet)

2
(Safranin)

FIGURE 2.1 The structure of crystal violet and safranin used in the Gram stain procedure

2.2 NOMENCLATURE AND CLASSIFICATION OF BACTERIA

The nomenclature of bacteria follows a binomial system introduced by Carl Linnaeus, a Swedish botanist, zoologist, and physician, known as the "father of modern taxonomy" [9]. This binomial nomenclature also applies to animals and plants. However, the nomenclature of bacteria is regulated by the International Code of Nomenclature of Prokaryotes, but the classification of actual species is not. All names are organized into a hierarchical taxonomic system, for the purpose of providing biological insights based on phylogeny. The hierarchical system of the official nomenclature is as follows: Phylum (or Division), Class (named after the type genus of the type order of the class); Order (named after the type genus of the order), Family (named after the type genus of the family), Genus, Species (and if necessary, Subspecies). According to the binomial nomenclature system, a species name should have two parts: a generic name and a specific epithet. The approved list of bacterial names before 1980 has been compiled by Skerman, McGowan, and Sneath [10]. Any subsequent changes of names and addition of new names for bacteria are published by the International Journal of Systematic and Evolutionary Microbiology (IJSEM) (formerly International Journal of Systematic Bacteriology (IJSB)). Any taxonomic proposals for the changes of bacterial names or the addition of new names to bacteria published in this journal will be reviewed and officially validated by the "Notification List" if such changes are valid. In addition, bacterial names published in journals other than IJSEM can be officially "validated" by sending the publication to IJSEM with proof that (for species and subspecies) the designated type strain is available without restrictions from at least two public culture collections located in different countries, and will be inventoried in the "Validation List."

The latest validation is published in 2019, which contains conserved and rejected family names, conserved names of genera of bacteria, conserved specific epithets in names of species of prokaryotes, rejected names of genera and subgenera of prokaryotes, and rejected specific and subspecific epithets in names of species and subspecies of prokaryotes, as well as many rules and principles regarding the nomenclature [11].

2.3 ANTIBACTERIAL MECHANISM

Bacteria have existed on the earth as early as the Devonian Period (419.2 million to 358.9 million years ago), according to Encyclopedia Britannica (https://www.britannica.com/science/bacteria/Evolution-of-bacteria). Bacteria possibly have even been present since early Precambrian time, about 3.5 billion years ago. Thus, bacteria have had plenty of time to adapt themselves to their environments and evolve to numerous descendant forms. Bacteria grow and multiply when suitable condition allows, i.e., enough nutrients, proper temperature, adequate moisture, etc. They take in nutrients and reproduce by simply dividing. While each type of bacteria has its own preferred conditions for growth, under ideal conditions, many types of bacteria can double their numbers every 20 minutes. Certain bacterial genera can develop spores to survive under chemical and physical inactivation conditions, such as to form an endospore, a dormant, tough, and non-reproductive structure.

The cell wall of bacteria is a biopolymer of glycan strands cross-linked with oligopeptides, where the glycan strands consist of repeating disaccharides of *N*-acetylglucosamine (GlcNAc) β-1,4-tethered to *N*-acetylmuramic acid (MurNAc), the latter is an ether of *L*-lactic acid attaching to the 3-hydroxy group of *N*-acetylglucosamine. The cross-linked oligopeptides (*D*-Ala-*D*-Ala-*L*-Lys-γ-*D*-Glu-*L*-Ala) are attached to the carboxyl end of the *L*-lactic acid residue of MurNAc. This layer of biopolymer structure is known as peptidoglycan, as illustrated in Figure 2.2 [12]. The rigid cell wall resists the turgor pressure within the cell and is responsible for support and maintenance of bacterial morphology, as well as protection of cellular contents. This layer of peptidoglycan on the cell wall within the endospores produced by species of Bacillus, Clostridium, and several other bacterial genera, is essential for the maintenance of the dehydrated state, allowing bacteria to survive for long periods under extreme conditions, which is the major determinant of spore heat resistance [13]. The peptidoglycans in bacterial spores form two distinct though contiguous layers: the thin inner layer (i.e., the germ cell wall) which serves as the initial cell wall of the germinated spore, and the thicker outer layer (i.e., the spore cortex), which can rapidly degrade during spore germination. The most dramatic modification of the cortex structure is the removal of most peptide side chains from the glycan strands, resulting in a decrease of the peptide to disaccharide ratio from one peptide per 2.3 ~ 2.9 disaccharides to one peptide per 35 disaccharides. The loss of these peptides limits the cross-linking potential of the peptidoglycan and ensures the dehydration of the spore core [14].

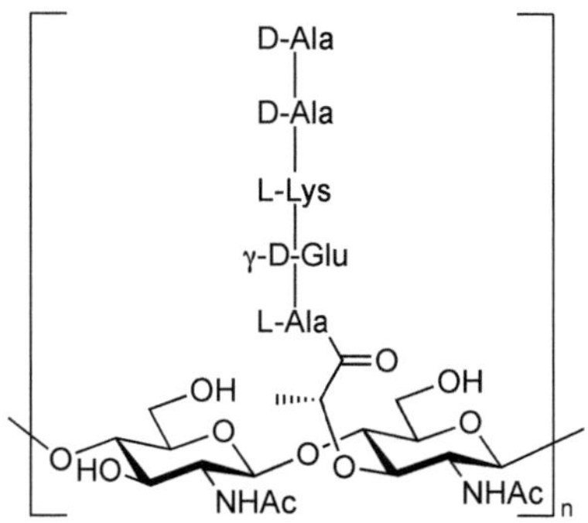

FIGURE 2.2 The structure of glycan on the bacterial cell wall

A family of enzymes collectively called penicillin-binding proteins (PBPs) is involved in the assembly, maintenance, and regulation of such peptidoglycan structure in both Gram-positive and Gram-negative bacteria [15]. This name just reflects the fact that the first protein of its kind was discovered, and also indicates its characteristic affinity for penicillins. PBPs are members of a subgroup of enzymes knowns as transpeptidases, specifically, the *DD*-transpeptidases. The *DD*-transpeptidase (https://enzyme.expasy.org/EC/3.4.16.4), is alternatively known as *DD*-peptidase, *DD*-carboxypeptidase, δ-alanyl-*D*-alanine carboxypeptidase, δ-alanyl-*D*-alanine-cleaving-peptidase, δ-alanine carboxypeptidase, δ-alanyl carboxypeptidase, and serine-type δ-Ala-*D*-Ala carboxypeptidase. These proteins are located in the inner membrane of bacteria with their active sites present in periplasmic space, and catalyze the transfer of the R-*L*-aca-*D*-alanyl moiety of R-*L*-aca-*D*-alanyl-*D*-alanine carbonyl donors to the γ-OH of their active-site serine and from there to a final acceptor, i.e., the peptidoglycan on the bacterial cell wall as mentioned above [16]. In many Gram-negative bacteria, the cross-linking occurs primarily between meso-diaminopimelate (meso-Dap) and *D*-Ala residues, which is catalyzed by the *D,D*-transpeptidase activity of the PBPs [17]. Also, some bacteria (e.g., *Mycobacterium tuberculosis* in stationary phase) contain peptidoglycan of a substantial amount of cross-links between meso-Dap residues, catalyzed by

L,D-transpeptidases employing a nucleophilic cysteine rather than a serine in the *D,D*-transpeptidases [18]. The transpeptidation also transfers the peptidyl-alanyl moieties containing *N*-acyl substituents on the δ-alanine. In addition, some PBPs have exhibited enzymatic activity of carboxypeptidase, endopeptidase, and even transglycosylase [19]. In all bacteria that have been studied, PBP has a penicillin-insensitive transglycosylase *N*-terminal domain (involved in the formation of linear glycan strands) and a penicillin-sensitive transpeptidase *C*-terminal domain (involved in cross-linking of the peptide subunits) and the serine at the active site is conserved in all members of the PBP family [20].

There are two types of PBPs, according to their difference in molecular weights, i.e., low molecular weight and high molecular weight PBPs. The high-molecular-weight PBPs (generally $\geq$ 60 kDa) carry the transglycosylase and transpeptidase activities involved in polymerization (to form glycan chain) and cross-linking of the glycan strands (with peptide side chains). In contrast, the low-molecular-weight PBPs have commonly been found to possess *D,D*-carboxypeptidase activity, which removes the terminal δ-alanine of the peptide side chains and thereby prevents the side chain from serving as a donor for peptide cross-linking [14]. Analysis of the *Bacillus subtilis* genome reveals six genes that encode low molecular-weight PBPs, i.e., dacA, dacB, dacC, dacF, pbpE, and pbpX, where the products of the four dac genes exhibit very high similarity in sequence to *D,D*-carboxypeptidases, as shown by PBP5 and PBP5*, the corresponding products of dacA and dacB, respectively. Similarly, genomic analysis of PBPs involved in the synthesis of the peptidoglycan layer of the cell wall that is encoded in 12 cyanobacterial genomes suggests that a minimum set of PBPs are needed to assemble the peptidoglycan layer in cyanobacteria, including one bifunctional transpeptidase-transglycosylase Class A high-molecular-weight PBP, two Class B high-molecular-weight PBPs, and one low-molecular-weight PBP, where one Class B PBP is involved in cellular elongation and the other is responsible for septum formation. In contrast, the low-molecular-weight PBPs of all cyanobacteria analyzed are putative endopeptidases [21].

Anyway, the synthesis of the bacterial cell wall is essential to the growth, cell division (thus reproduction), and maintenance of the cellular structure of bacteria. Obviously, inhibition of PBPs would affect both the formation of glycan strands and the cross-linking of peptide side chains, leading to defects in the cell wall structure and irregularities in cell shape, such as filamentation, *pseudo*-multicellular forms, and lesions. In the end, the inhibition of PBPs can lead to the formation of spheroplast, and eventual cell death and lysis. β-Lactam antibiotics bind to PBPs due to their similar chemical structures to the modular pieces that form the peptidoglycan [22], and inhibit PBPs by covalently modifying the nucleophilic serine (the 62nd serine residue of the active site) [15, 17, 23], forming a long-lasting acyl-enzyme intermediate via nucleophilic attack of the four-membered β-lactam ring by the serine residue on PBPs, as illustrated in Scheme 2.1 [24]. As a result, β-lactam drugs have become the most widely used class of antibiotics [25]. In addition to reacting with β-lactam antibiotics via the polar side chain of *L*-serine, PBPs may react with β-lactam antibiotics via other amino acid residues of a polar side chain, such as Thr and Asp [15]. Among the clinically available β-lactam antibiotics, carbapenems exhibit the greatest potency and broadest spectrum of activity against both Gram-negative and Gram-positive bacteria, which have once been reserved for the most serious infections and have now been widely used in hospitals and even outpatient settings [26].

It is commonly accepted that cyclic molecules except for the six-membered molecules all have ring strains. The smaller the size of the ring, the higher the ring strain is. The four-membered ring of the β-lactam antibiotic has been highly regarded for its effectiveness in binding with PBPs, primarily because of the ring strain. While the stability of a β-lactam is based on various factors, such as ring fusion, ring strain, etc., the C=O bond length in β-lactam correlates with its hydrolytic stability. Recent computation indicates that the ring strain energies of azetidin-2-one and penicillin G are comparable, within the range of 27.7–28.4 Kcal/mol [25]. However,

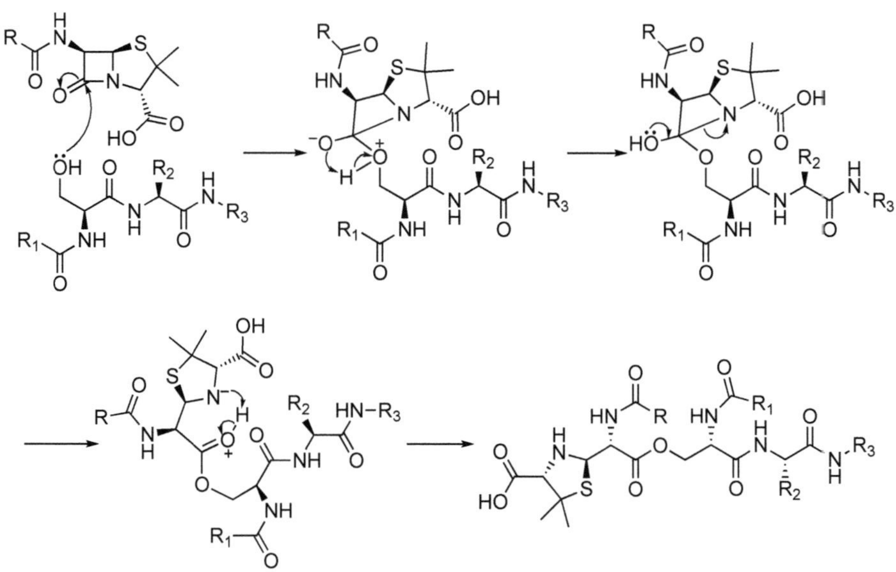

SCHEME 2.1 The mechanism for the interaction between β-lactam and PBPs

computational results also suggest that the effectiveness of β-lactam derivatives as of the strain in the four-membered ring is wrong because the overall ring strain is greatly reduced and substantially compensated by the stabilizing effect of amide resonance. It is found that the energy barrier for the rotation around the C-N bond in amides is about 20–30 Kcal/mol, indicating a significant angular dependence for resonance [19]. In addition, amides are less reactive than ketones toward the nucleophiles because of the resonance stabilization energy (ca. 18 Kcal/mol) resulting from the delocalization of the lone pair electrons on nitrogen over the carbonyl group [19]. The assumed β-lactam ring strain does not exert influence on the rate of hydrolysis because the transition state of the rate-determining step contains an unbroken β-lactam ring, indicating that the electronic factors (delocalization directly related to the amide force constant) have a larger effect on the kinetic variables [25]. While the β-lactam itself is not a necessary requirement for antibacterial activity, the transport, the binding of the antibiotic to its receptor site, and inhibition of the relevant enzyme are even more important. β-Lactams of weakly basic amines are about 500 times more reactive than an acyclic amide containing the same amine, and the geometrical constraints brought about by the bicyclic systems could increase their reactivity with respect to monocyclic β-lactams, resulting in ca. 100-fold increase of activity for the bicyclic antibiotics over monocyclic β-lactams with amines of similar basicity [19]. This trend is believed to be caused by the deviation of planarity of the amide functional group within the β-lactam. The nitrogen atom of an ideal amide is in sp^2-hybridization due to resonance, and sp^2-hybridized atoms have trigonal planar bond geometry. The four-membered ring forces a pyramidal bond geometry for the nitrogen atom in β-lactam, resulting in a reduced extent of resonance for the amide bond so that the carbonyl group becomes more ketone-like. Nobel laureate Robert Burns Woodward has proposed a parameter "h" as a measure of the height of the trigonal pyramid defined by the nitrogen (as the apex) and its three adjacent atoms [27]. This parameter corresponds to the strength of the β-lactam bond, where the lower the numbers, the more planar trigonal pyramid is, and the closer to the geometry of an ideal amide so that the corresponding β-lactam ring is stronger and less reactive. Following this suggestion, it has been found that the "h" values for monobactams, cephems, penams, carbapenems, and clavams are in the range of 0.05–0.10 (Å), 0.20–0.25 Å, 0.40–050 Å, and 0.50–0.60 Å, respectively, indicating that carbapenems and clavams are the most reactive β-lactams toward hydrolysis [28]. However, monocyclic β-lactams with suitable electron-withdrawing substituents may be as reactive as the bicyclic β-lactams [19].

In addition to presenting PBPs that catalyze the synthesis of the bacterial cell wall, bacteria also secrete another type of enzymes, generally known as β-lactamases, which inactivate β-lactam antibiotics in a manner like the reaction between PBPs and β-lactams for the serine-β-lactamases, whilst metallo-β-lactamases (MBLs) will react with β-lactam antibiotics in a different mechanism. Due to the presence of β-lactamases, as well as natural selection pressure, horizontal gene transfer among bacteria switches to *L*-form and formation of biofilm, as detailed in the following section, so that bacteria can quickly develop resistance against the applied antibiotics, including the β-lactam antibiotics. Consequently, antibiotic resistance is always a concern when selecting a feasible β-lactam antibiotic. For example, the threat posed by MBLs to public health is growing, and finding effective inhibitors of these MBLs is undoubtedly a race against time [29].

In order to counter the clinical problems posed by the β-lactamase-mediated resistance, two strategies have been employed: (a) development of new β-lactam drugs that are less favorable substrates for many common β-lactamases, so-called extended-spectrum β-lactams, such as the third-generation cephalosporins cefotaxime, ceftazidime, and ceftriaxone, the monobactam aztreonam, and the carbapenem imipenem; (b) administration of a mechanism-based β-lactamase inactivator (or inhibitor), such as clavulanate, sulbactam, or tazobactam, together with penicillin, to preserve the antibacterial activity of penicillin and to prolong the effective life of antibiotic drugs [30]. The administered β-lactamase inactivator should have broad-spectrum inhibitory activity and good synergistic effects with the antibiotics [29]. For example, clavulanic acid, produced by *Streptomyces clavuligerus* ATCC 27064, is a medicinally important inhibitor of class A β-lactamases, which play an important role in bacterial resistance to penicillins [31, 32].

On the other hand, the effectiveness of killing bacteria is related to the length of time that bacteria are exposed to and the concentration of antibiotic that exceeds the minimum inhibitory concentration (MIC) [33]. Maintaining the concentrations of antibiotics over MIC is not only important for the efficiency of clinical therapeutics but also important to suppress the generation of antibiotic resistance. However, β-lactam antibiotics with a very unstable and reactive four-membered ring are readily degradable in aqueous solution during clinical administrations due to a nucleophilic or electrophilic attack, resulting in the loss of antimicrobial activity and increased risk of therapeutic failure. For this reason, many physicochemical strategies have been implemented to increase the stability of β-lactam antibiotics, such as the addition of buffer salts to keep them in the pH range of maximum stability.

Overall, treatment of bacterial infections and eradicating those health-threatening bacteria is a never-ending battle, and it is expected that more broad-spectrum antibiotics will be created in the future.

2.4 RESISTANCE OF ANTIBACTERIAL ACTIVITY

Bacteria survive and adapt themselves to their living environments and conditions according to the same rules of evolution that apply to other living things on earth. Bacteria reproduce themselves asexually, by directly copying their genes to their offspring. During the process of reproduction, gene mutations occasionally occur either naturally or due to a sudden change of environmental factors, such as salinity and radiation [34], while the genetic code is being copied or transferred. The resulting mutation can be passed on to some or all of the organism's offspring, and they can then pass it on again. Most mutations are either detrimental or benign. If the mutation is deleterious for bacteria to survive, the descendants who receive this mutated gene will have fewer descendants themselves over time, so the mutated gene will eventually disappear. On the other hand, organisms with benign mutations will leave more surviving offspring, and such mutations will become widespread. The genetic material of bacteria in the chromosome is passed on from one bacterium to its daughter cells when it divides. In addition to the chromosome DNA, bacteria also contain genetic material on plasmids that reside outside the chromosome, which can be passed from one bacterium to another when they come into contact (either direct cell-to-cell contact [35] or by a bridge-like connection between two cells [36]), a process known as bacterial conjugation. In addition, bacteria may take up foreign genetic material (naked DNA) from their living environment, a process known as bacterial transformation for horizontal gene transfer [37]. Also, a host cell of bacteria may incorporate into the genome of a bacterial virus (bacteriophage), a process known as transduction, and carry to another host cell when the bacteriophage initiates another cycle of infection [38]. Anyway, genes on plasmids can spread much more quickly than those in the chromosome, and can even spread to unrelated types of bacteria. The transferred genetic information is often beneficial to the recipient. Benefits may include antibiotic resistance, xenobiotic tolerance, and/or the ability to use new metabolites [39].

Most antibiotic resistance is transmitted by means of plasmids. Once treated with antibiotics, the bacteria with only a weak defense against a particular antibiotic will die and leave no offspring, whereas the bacteria with a better defense due to a gene mutation for a trait, such as a stronger membrane that keeps antibiotics out, will likely survive and divide to their offspring. Over time, selective pressure leads to the greater and greater resistance of the antibiotic as the antibiotic-resistant bacteria increase in abundance. As a result, the applied antibiotic eventually no longer works for the infection caused by the particular strain of bacteria [40].

Another aspect of resistance to antibacterial activity is that the vast majority of commercial antibiotics are not synthesized from scratches, but are either the natural products of bactericides, such as penicillin, that may have already existed in nature over millions of years, or semi-synthetic products of bactericides with similar or superior antibacterial activities. For example, penicillin is produced by certain types of fungi to protect themselves from bacteria, and some bacteria have gradually evolved to counter this defense by producing penicillinases to degrade penicillin, prior to Fleming's discovery of penicillin. This natural arms race between fungi and bacteria are spreading much more rapidly nowadays because of the creation of greater selective pressure resulting from the popular use of antibiotics. By the mid-1950s, penicillin resistance has been commonly notified as penicillinase-producing strains of bacteria had become universally present in hospitals [41], so that chemists have to make new antibiotics, including methicillin. However, resistance to methicillin has developed almost immediately upon or even prior to the introduction of this even stronger antibiotic [40]. Unfortunately, methicillin-resistant *Staphylococcus aureus* (MRSA) has become one of the most well-known and deadliest types of resistant superbugs in the 1990s and early 2000s [42].

2.5 MECHANISMS FOR ANTIBIOTICS RESISTANCE

By which way and how do bacteria inactivate the applied antibiotics? The primary mechanism of bacterial resistance to β-lactam antibiotics is the secretion of β-lactamases, and the alternative mechanism may involve the formation of biofilm or switch to *L*-form.

Many bacteria under certain conditions retain the ability to switch into a cell wall deficient state known as the *L*-form [43], which is completely resistant to antibiotics (including β-lactam antibiotics) that specifically block the cell wall synthesis [44]. Such unusual pleomorphic variants (growing into spheroplasts or protoplasts) have been initially named the "*L*-form" in honor of the "Lister Institute in London," and have been provided with alternative names over the years, including *L*-phase bacteria, *L*-variants, *L*-organisms, and cell wall-deficient bacteria [43]. These *L*-forms can be differentiated into four types, depending on their ability to revert to the parental, cell-walled form and to the extent of their cell-wall modification. In *Bacillus subtilis*, the switch from normal cell shape to irregular *L*-form generally requires two genetic changes, i.e., mutation to favor the synthesis of membrane rather than cell wall, and mutation to relieve oxidative stress that *L*-forms may incur due to their unbalanced metabolism [45]. The promoted synthesis of membrane involves a range of relatively disorganized membrane blebbing or vesiculation events due to an unusual form of proliferation used by *L*-forms. Repression or inhibition of synthesis for cell wall precursors can stimulate the *L*-form transition in a wide range of both Gram-positive and Gram-negative bacteria. In addition, *L*-forms are potentially important for the survival of bacteria as the lack of a cell wall can be advantageous in a range of production or strain improvement applications. On the other hand, the mutation to relieve oxidative stress occurs either by downregulating respiratory chain activity, or upregulating oxidative stress response genes, to support the growth of *L*-form by counteracting an increase in the cellular levels of reactive oxygen species (ROS) originating from the respiratory chain [46]. Also, *L*-form proliferation may involve a simple biophysical effect based on an increased rate of surface area synthesis with respect to volume synthesis, causing deformations of cell shapes that lead to spontaneous scission [44]. Another common property of *L*-forms is that *L*-forms are able to rebuild a peptidoglycan cell wall *de novo*, which may contribute to chronic or recurrent infections [45]. While *L*-forms are completely resistant to most antibiotics (e.g., penicillins, and cephalosporins) working specifically on the cell wall synthesis, the switch to the *L*-form state to avoid the killing by β-lactam antibiotics depends on two key factors: escape from the cell wall sacculus by exogenous lytic enzymes and the avoidance of ROS toxicity via reduced glycolytic flux [44]. It has been suggested that the treatment with β-lactams may also effectively facilitate the generation of *L*-form within the human body due to the wide presence of hydrolases of peptidoglycan and gluconeogenic substrates, such as lactate, glycerol, and amino acids.

Another mechanism for the avoidance of β-lactam treatment is to form bacterial biofilms. Generally, a biofilm comprises any syntrophic consortium of microorganisms where cells stick to each other and become embedded within a slimy extracellular matrix that is composed of extracellular polymeric substances [47]. The extracellular polymeric substances are typically a polymeric conglomeration of extracellular polysaccharides, proteins, lipids, and DNA [48]. Development of biofilm starts with an initial attachment of planktonic cells (i.e., floating as single cells in water) to a surface, which then grow into clusters of multicellular colonies. Bacterial biofilms are three-dimensional communities of bacteria adhered to a surface. Since bacteria are enclosed in an extracellular matrix once bacterial biofilm is created, bacteria in biofilms are up to 1000 times more resistant to antibiotics than their planktonic counterparts and provide a reservoir for chronic infection. Currently, there has not been an FDA-approved antibiotic to prevent or disperse pathogenic bacterial biofilms [49]. Nearly 12 million Americans are estimated to be affected by biofilm-related infections each year, resulting in an expense of about $6 billion for the treatment of such type of infection, by means of prolonged, high-dose antibiotic therapy on biofilm-infected indwelling medical devices (IMDs) [50]. For severe cases, the IMDs are discarded. Particularly, about 80% of IMD infections are caused by one or more *Staphylococcus* species, especially *Staphylococcus aureus* and *Staphylococcus epidermidis,* which account for 66% of all IMD infections [51], and *Staphylococcus epidermidis* is the most common healthcare- and medical device-related biofilm infections [49], whereas *Staphylococcus aureus* secretes a large number of toxins and virulence factors that create serious and pervasive infections. It is the formation of biofilm that will reduce antibiotic efficacy so that the concentrations of antibiotics required to eradicate biofilms are ten to a thousand-fold higher than the concentrations required to kill planktonic bacteria, creating a burden on both public health and the economy from increased medical costs.

Compared to the switch to *L*-form and the creation of bacterial biofilm, the use of naturally existing β-lactamase to inactivate β-lactam antibiotics is the primary and most effective mechanism for bacterial resistance. β-Lactamases are a large family of hydrolases that catalyze the hydrolysis of the amide bond of the β-lactam antibiotics, in a manner similar to the interaction between PBPs and β-lactam antibiotics. Due to the large number of β-lactamases reported (e.g., ca 850 β-lactamases in 2009 [52], 1,300 natural β-lactamases [53], over 1,800 β-lactamases [54], or 4935 enzymes [55]), the β-lactamases have been divided into different groups according to one of the two characteristics: i.e., molecular, and functional characteristics. The classification of β-lactamases based on molecular characteristics has been conducted according to the similarity of a full nucleotide or amino acid sequence, whereas the classification based on functional characteristics compares the substrate and inhibition profiles. In addition, numerous classification schemes have been proposed, where the two most cited schemes are the ones developed by Ambler and Bush, respectively [56]. In the Ambler classification system, β-lactamases are divided into four classes (A, B, C, and D) based on their molecular structures (i.e., amino acid sequences) [57]. Initially, there are only two classes, which are class A (i.e., β-lactamases with serine at the active site) and class B (the MBLs that require a bivalent metal ion, usually Zn^{2+}). New β-lactamases discovered after the initial classification, which also have a serine at the active site but are distinctively different in structures from the class A β-lactamases, are known as class C and D β-lactamases. Whilst A, C, and D β-lactamase classes contain sufficient structural similarity and are homologous with evidence showing that they are descended from a common ancestor, they are also sufficiently different from each other that the alignment programs such as BLAST (basic local alignment search tool) fail to find detectable sequence similarity [58]. These enzymes require a serine to acylate the β-lactam substrate during the hydrolysis reaction, in which class C β-lactamases tend to have higher molecular weights than the other two classes [52]. Although there is only one class of MBLs, these enzymes cannot be aligned by either DNA or protein sequence, and are further divided into three subgroups, i.e., B1, B2, and B3 [59]. Class B β-lactamases contain at least one functional Zn^{2+} atom at the active site that participates in the hydrolytic process. In this classification system, class A enzymes include penicillinases that generally hydrolyze penicillins and some early-generation cephalosporins only; extended-spectrum beta-lactamases (ESBL) which hydrolyze late-generation cephalosporins (such as CTX-M-type); and class A carbapenemases which hydrolyze penicillins, cephalosporins, and carbapenems. These enzymes are inhibited or partially inhibited by class A inhibitors such as clavulanate or tazobactam. Class B enzymes hydrolyze all β-lactams except aztreonam and are not inhibited by clavulanate/tazobactam. Among these enzymes, three types including NDM (New Delhi MBL), VIM (Verona integron-encoded MBL), and IMP (imipenemase) are widely distributed. Class C enzymes are AmpC or cephalosporinases which more efficiently hydrolyze cephalosporins in comparison to benzylpenicillin, including the CMY enzyme family (CMY-1 to CMY-50). Class D enzymes are oxacillinases that hydrolyze cloxacillin or oxacillin, as well as carbapenem. Classes A and D lactamases are plasmid-mediated enzymes, whilst classes B and C are chromosomal. Currently, β-lactamase-mediated resistance does not spare even the newest and most powerful β-lactams (i.e., carbapenems), whose activity is challenged by the MBLs (e.g., VIM, IMP, and NDM) as well as classes A and D serine-carbapenemases [e.g., Klebsiella pneumoniae carbapenemase (KPC), IMIpenem-hydrolyzing β-lactamase (IMI), Guiana-extended-spectrum β-lactamase (GES), oxacillinases (e.g., OXA-48, OXA-23, OXA-40,...). According to the similarities and differences among classes A, C, and D as well as the similarities and differences among the subgroups of B1, B2, and B3, the β-lactamases have been suggested to be classified into two main groups: i.e., the serine β-lactamases and MBLs, which are further divided into subgroups of classes SA, SC, and SD, corresponding to the present classes A, C, and D; and classes MB and ME, where classes MB includes the current subgroups of B1 and B2, and ME is equivalent to subgroup B3 [60].

In contrast to the Ambler classification, the Bush-Jacoby-Medeiros classification combines the functional properties with the known molecular sequences of β-lactamases from both Gram-positive and Gram-negative bacteria [61], which has been performed through assessing the ability of β-lactamase to hydrolyze specific β-lactams and its inactivation by inhibitors [62]. The functional groups are identified according to their inhibitory properties primarily based on clavulanic acid (an inactivator of β-lactamases that contain an active site serine) and ethylenediamine-tetraacetic acid (EDTA, a chelating agent to identify the zinc-containing MBLs). This classification system takes into account substrate and inhibitor profiles to group enzymes in ways that can be correlated with their phenotype in clinical isolates while still retaining a good correlation with the Ambler classification. As a result, this system can better help clinicians and laboratory microbiologists to correlate the properties of a specific enzyme with the observed microbiological resistance profile for a clinical isolate. In this system, β-lactamases are divided into three groups, where group 1 cephalosporinases correspond to class C; group 2 enzymes are broad-spectrum, inhibitor-resistant, and extended-spectrum β-lactamases and serine carbapenemases that belong to classes A and D; and group 3 enzymes are MBLs. Each main group is further divided into several new subgroups based on specific attributes of individual enzymes [63]. For example, group 1 cephalosporinases are more active on cephalosporins than benzylpenicillin and are usually resistant to inhibition by clavulanic acid. When produced in large amounts, particularly in a host with reduced β-lactam accumulation, group 1 enzymes can provide resistance to carbapenems, especially ertapenem [64]. There are more than 50 enzymes in group 1 and subgroup 1e (e.g., CMY-1 to CMY-50) already [63]. Group 2 enzymes include A and D β-lactamase classes that are plasmid-mediated enzymes, reflecting the original TEM (contraction of Temoniera, the name of a patient from whom resistant bacteria were isolated) and SHV (contraction of sulfhydryl variable) genes. Examples include penicillinases and cephalosporinases that are inhibited by clavulanic acid. Enzymes of this group are further divided into many subgroups, where group 2a enzymes are penicillinase, belonging to molecular class A; group 2b enzymes while also belonging to molecular class A, are broad-spectrum β-lactamases to inactivate penicillins and cephalosporins at the same rate. Group 2c enzymes are carbenicillinases, also belonging to molecular class A, which inactivate carbenicillin more than benzylpenicillin, with some effect on cloxacillin; group 2d enzymes are cloxacilanases, belonging to either molecular class A or D, which inactivate cloxacillin more than benzylpenicillin, but are less effectively inhibited by clavulanic acid; group 2e enzymes are cephalosporinases, also belonging to molecular class A, which deactivate cephalosporins and monobactams, but are inhibited by clavulanic acid; group 2f enzymes are serine-based carbapenemases, also belonging to molecular class A, but are different from the zinc-based carbapenemases that belong to group 3. Group 2 enzymes also contain subgroups 2be, 2br, 2ber, etc., where "e" stands for "extended spectrum of activity," and "r" denotes the reduced binding to clavulanic acid and sulbactam. Subgroup 2be enzymes, belonging to molecular class A enzymes, are ESBL that are capable of inactivating the third-generation cephalosporins as well as monobactams. Subgroup 2br enzymes are inhibitor-resistant TEM-derivative enzymes. Subgroup 2d enzymes may be further subdivided according to their ability to hydrolyze extended-spectrum cephalosporins (group 2de) or carbapenems (group 2df). Group 3 enzymes are zinc-based or MBLs, corresponding to molecular class B enzymes, which can inactivate penicillins, cephalosporins, and carbapenems, but are not inhibited by clavulanic acid. Finally, group 4 enzymes are penicillinases that are not inhibited by clavulanic acid yet have not been assigned a corresponding molecular class. More details for functional property-based classification have recently been updated [63].

The major families of β-lactamases of clinical importance are summarized in Table 2.1.

TABLE 2.1 The Major Families of β-Lactamases [63]

Ambler Class	Enzyme Family	Full Enzyme Name	Functional Group or Subgroup	Representative Enzymes
Class A	BEL	Beta-Lactamase	–	BEL-1 to BEL-4
	BIC	Bicêtre Hospital in France	–	BIC-1
	CARB	Carbenicillin-hydrolyzing beta-lactamase	–	CARB-1 to CARB-49
	CTX-M	Cefotaxime-M (CTX-M) β-lactamases	2be	CTX-M-1, CTX-M-44 (Toho-1) to CTX-M-92
	FRI	may be "Fribourg," a university name	–	FRI-1
	GES	Guiana extended-spectrum beta-lactamase	2f	GES-2 to GES-7 (IBC-1) to GES-15
	IMI	Imipenem-hydrolyzing beta-lactamase	–	IMI-1 to IMI-16
	KPC	Klebsiella pneumoniae carbapenamase	2f	KPC-2 to KPC-10
	LEN	Strain name LEN-1	–	LEN-1 to LEN-31
	PER	Pseudomonas extended-resistant	2be	PER-1 to PER-5
	SFC	Serratia fonticola UTAD54	–	SFC-1
	SHV	Sulfydryl variable beta-lactamase	2b	SHV-1, SHV-11, SHV-89
			2be	SHV-2, SHV-3, SHV-115
			2br	SHV-10, SHV-72
	SME	Serratia marcescens extended serine beta-lactamase	2f	SME-1 to SME-3
	TEM	Temoneira (Greek patient's name)	2b	TEM-1, TEM-2, TEM-13
			2be	TEM-3, TEM-10, TEM-26

TABLE 2.1 *(Continued)*

Ambler Class	Enzyme Family	Full Enzyme Name	Functional Group or Subgroup	Representative Enzymes
			2br	TEM-30 (IRT-2), TEM-31 (IRT-1), TEM-163
			2ber	TEM-50 (CMT-1), TEM-158 (CMT-9)
	VCC	Vibrio cholerae strain	–	VCC-1
	VEB	Vietnam extended-spectrum beta-lactamase	2be	VEB-1 to VEB-7
Class B (metallo)	GIM	German imipenemase	–	GIM-1, GIM-2
	IMP	Imipenem-resistant *Pseudomonas aeruginosa*	3a	IMP-1 to IMP-26
	IND	*Chryseobacterium indologenes*	3a	IND-1, IND-2, IND-2a, IND-3 to IND-7
	NDM	New Delhi metallo-beta-lactamase	–	NDM-1 to NDM-19
	VIM	Verona integron-encoded metallo-beta-lactamase	3a	VIM-1 to VIM-23
Class C (AmpC type)	ACC	Ambler class C	–	ACC-1 to ACC-5
	ACT	AmpC type	–	ACT-1 to ACT-52
	ADC	Acinetobacter derived cephalosporinase	–	ADC-1 to ADC-81
	CFE	Citrobacter freundii extended serine beta-lactamase	–	CFE-1
	CMY	Cephamycin	1, 1e	CMY-1 to CMY-155
	DHA	Dhahran Hospital in Saudi Arabia	–	DHA-1 to DHA-25
	FOX	Cefoxitin	–	FOX-1 to FOX-14
	LAT	Latamoxef	–	LAT-1, LAT-2

TABLE 2.1 *(Continued)*

Ambler Class	Enzyme Family	Full Enzyme Name	Functional Group or Subgroup	Representative Enzymes
	MIR	Miriam Hospital	–	MIR-1 to MIR-21
	MOX	Moxalactam (Latamoxef)	–	MOX-1 to MOX-12
	PDC	Pseudomonas derived cephalosporinase	–	PDC-1 to PDC-10
Class D	OXA	Oxacillinase	2d	OXA-1, OXA-2, OXA-10
			2de	OXA-11, OXA-14, OXA-15
			2df	OXA-23 (ARI-1), OXA-51, OXA-58

2.6 β-LACTAMASES DATABASES

So far, as many as 4,935 β-lactamases have been known already [55]. It is expected that more and more β-lactamases will evolve along with the applications of new variants of antibiotics due to the strong selection pressure mentioned above, so it will be very difficult to judge whether a clinically isolated enzyme is a new or existing enzyme, and which group it belongs to. If the new isolate is a new enzyme, it must be classified like other β-lactamases so that it can be well characterized. In fact, newer enzyme families within each class of β-lactamases still arise due to the strong selection pressure. In order to help organize and classify these potentially new β-lactamases, it is of high necessity to create databases to archive the information of known β-lactamases, which allows easy access and comparison as well as the capability of expansion. To correspond to this need, different β-lactamases databases have been created, and the most influential databases are summarized here in alphabetical order.

2.6.1 ANTIBIOTIC RESISTANCE GENES DATABASE (ARDB)

Bacteria develop resistance to antibiotics through mutations or through the acquisition of resistance genes. A manually curated database, i.e., the antibiotic resistance genes database (ARDB) (http://ardb.cbcb.umd.edu/), has been created to facilitate the identification and characterization of these genes. Each gene and its resistance type have been annotated with rich information, including resistance profile, mechanism of action, ontology, COG (clusters of orthologous groups of proteins), and CDD (conserved domains database) annotations, as well as external links to sequence and protein databases. This database also supports searches for sequence similarity and implements an initial version of a tool for characterizing common mutations that confer antibiotic resistance. Currently, ARDB contains resistance information for 13,293 genes, 377 types, 257 antibiotics, 632 genomes, 933 species, and 124 genera [65]. However, as of July 3, 2009, the site (http://ardb.cbcb.umd.edu/) was no longer maintained, and the contents have been actively maintained at: https://card.mcmaster.ca/, a new database known as the comprehensive antibiotic resistance database (CARD), vide infra 2.6.7.

2.6.2 ANTIBIOTIC RESISTANCE GENES ONLINE (ARGO)

This database (www.argodb.org) collects gene sequences conferring resistance to vancomycin and β-lactams antibiotics. However, the URL address is not working now [66].

2.6.3 BACMET: ANTIBACTERIAL BIOCIDE AND METAL RESISTANCE GENES DATABASE

BacMet (http://bacmet.biomedicine.gu.se/) is a manually curated database of genes with experimentally confirmed resistance function against metals and/or antibacterial biocides, collecting bioinformatics resources of antibacterial biocide- and metal-resistance genes. The database of predicted genes is generated by sequence similarity searches in public databases, using a uniform cut-off for genes found on plasmids, and individually set cut-offs for chromosomal genes. The genes in the databases can be accessed either through the browsing option, where one can browse genes by compounds to which they confer resistance to, or by their name. Alternatively, one may search for any term in the database, including gene name, name of biocide or metal and chemical class. The entire database can also be downloaded for offline analysis of larger datasets [67]. As of March 11, 2018, this database has collected 155,512 predicted resistance genes, with 550 chromosomal-borne and 203 plasmid/transposon-borne resistance genes, as well as 268 biocide resistance genes, 420 metal resistance genes, and 65 genes with both biocide- and metal-resistance potential.

2.6.4 BLAD β-LACTAMASE DATABASE

This database (www.blad.co.in) categorizes all β-lactamases (non-MBLs such as TEM, CTXM, SHV, OXA, KPC, and CARB and MBLs, such as IMP, VIM, and NDM) and their associated variants along with the information on their epidemiology and resistance pattern that are very helpful to the researchers working on multidrug-resistant bacteria. In addition, the β-lactamases gene sequences are linked with structural data, phenotypic data (i.e., antibiotic resistance), and literature references to experimental studies [68]. It also provides the connections among the existing databases (NCBI) and can be used directly through a web-based interface. Information about crystal structures of ESBLs and MBLs is also collected in this database.

2.6.5 BETA-LACTAMASE DATABASE (BLDB)

Beta-lactamase database (BLDB, http://bldb.eu), updated on a weekly basis, is a comprehensive, manually curated public resource, providing up-to-date structural and functional information (e.g., hydrolytic profiles) focused on β-lactamase superfamily based on the Ambler classification, with a great impact on antibiotic resistance [55]. All three-dimensional structures of β-lactamases present in the protein data bank (PDB) are also shown. BLDB also provides extra tools to analyze β-lactamases and important links to the related web resources (NCBI, PDB, etc.), and can be queried using different protein- and nucleotide-based BLAST searches. As of February 13, 2020, this database has collected 4935 enzymes, with 1575 for class A, 707 for class B (508 subclass B1, 22 subclass B2, and 177 subclass B3), 1475 for class C, and 996 for class D β-lactamases. It has collected 1204 structural information, with 474 for class A, 354 for class B (275, 14, and 65 for subclasses B1, B2, and B3, respectively), 190 for class C, and 182 for class D. In addition, it has collected a total of 167 mutants and 47 kinetic data, according to the database.

2.6.6 BACTERIAL ANTIMICROBIAL RESISTANCE REFERENCE GENE DATABASE (BIOPROJECT ACCESSION NUMBER PRJNA313047)

This database (https://www.ncbi.nlm.nih.gov/bioproject/313047) contains annotated sequence records for representative DNA sequences that encode proteins conferring or contributing to resistance to various antibiotics, with a primary focus on acquired resistance. This collection includes contributions from the Lahey Clinic Database, Resfinder, the Center for Veterinary Medicine at FDA, the comprehensive antimicrobial resistance database (CARD), the Institut Pasteur, Dr. Marilyn Roberts, and Dr. Derrick Crook. Curation has been aided by the use of hidden Markov models (HMMs) from ResFams, Pfam, and TIGRFAMs collections. A total of 6260 Nucleotide (Genomic DNA) have been constructed along with the corresponding 5334 protein sequences, where nucleotide records also include flanking regions of 100 base pairs on each side of the coding region if such regions are available in the original sequence reports.

2.6.7 THE COMPREHENSIVE ANTIBIOTIC RESISTANCE DATABASE (CARD)

This database (https://card.mcmaster.ca/) integrates disparate molecular and sequence data, provides a unique organizing principle in the form of antibiotic resistance Ontology (ARO), and can quickly identify putative antibiotic resistance genes in new unannotated genome sequences. This unique platform provides an informatics tool that bridges antibiotic resistance concerns in health care, agriculture, and the environment. This monthly updated database has collected 6627 Ontology Terms, 5010 Reference Sequences, 1933 SNPs, 3004 Publications, and 5057 AMR Detection Models and allows resistome predictions for 377 pathogens, 21079 chromosomes, 2662 genomic islands, 41828 plasmids, 155606 WGS assemblies, and 322710 alleles. It helps researchers to study the genetics and genomics of antibiotic resistance and how it impacts bacterial populations, ecology, and the clinic [69].

2.6.8 COMPREHENSIVE Β-LACTAMASE MOLECULAR ANNOTATION RESOURCE (CBMAR)

This database facilitates comprehensive molecular annotation and discovery of novel β-lactamases by providing information useful for molecular and biochemical characterization of each β-lactamase family [70]. The basic architecture of CBMAR is based on the Ambler classification, where each class (A-D) is further divided into several families on the basis of their hydrolytic character. In this database, each enzyme family is annotated with: i) sequence variability, ii) antibiotic resistance profile, iii) inhibitor susceptibility, iv) active site, v) family fingerprints, vi) mutational profile, vii) variants, viii) gene location, ix) phylogenetic of tree, and several other features. In addition, each entry has external links to the relevant protein/nucleotide sequence and structure databases. The database supports the comparison of sequence similarity using BLAST and assigns a new β-lactamase protein to its respective family on the basis of the family-specific fingerprint. The database is built on an Apache HTTP Server 2.2.17 with MySQL Ver 14.14 and hosted on Ubuntu 11.04 Linux platform, taking the primary sources from GenBank and Uniprot. Unfortunately, the URL address (http://14.139.227.92/mkumar/lactamasedb) provided in the cited paper cannot be accessed, possibly due to an updated URL address or refusal of connection by the time of preparation for this chapter.

2.6.9 COMPREHENSIVE ONLINE DATABASE OF Β-LACTAMASE ENZYMES (CODLE)

This database (http://ifr48.timone.univ-mrs.fr/beta-lactamase/public/) is manually curated and incorporates the primary amino acid sequences from the PDB as well as the functional profiles and phylogenetic trees of four molecular classes (A, B, C, and D) of β-lactamases. The functional profiles are presented according to the MICs and kinetic parameters that make them more useful for the investigators. The database, collecting a total of 1,147 β-lactam resistance genes, is implemented in MySQL and the related website is developed with Zend Framework 2 on an Apache server, supporting all major web browsers. Users can easily retrieve and visualize biologically important information using a set of efficient queries from a graphical interface [53].

2.6.10 EUROPEAN ANTIMICROBIAL RESISTANCE SURVEILLANCE NETWORK (EARS-NET)

This database is a central and comprehensive database as of the collaborative effort of 29 European Union countries except for Poland, Norway, and Iceland, which focuses on seven different bacterial pathogens in the European Union. The seven bacterial pathogens include *Acinetobacter species*, *Enterococcus faecalis* (or *Enterococcus faecium*), *Escherichia coli*, *Klebsiella pneumoniae*, *Pseudomonas aeruginosa*, *Staphylococcus* aureus, and *Streptococcus pneumoniae*. This database was established in 1998 as EARSS, funded by the European Commission's Directorate-General for Health and Consumer Affairs and the Dutch Ministry of Health, Welfare, and Sports. It was transferred in January 2010 to the European Center for Disease Prevention and Control (ECDC) where it was renamed as EARS-Net. It can be accessed at: https://www.ecdc.europa.eu/en/about-us/partnerships-and-networks/disease-and-laboratory-networks/ears-net.

2.6.11 THE FUNCTIONAL ANTIBIOTIC-RESISTANT METAGENOMIC ELEMENT (FARME)

This is a database of DNA and protein sequences derived exclusively from environment sequences showing antibiotic resistance in laboratory experiments. It has been compiled based on publicly available DNA sequences, predicted protein sequences conferring antibiotic resistance and additional regulatory and mobile genetic elements, and predicted proteins flanking the antibiotic-resistant genes. FARME is the first database to focus on functional metagenomic antibiotic resistance gene elements and provides a resource to better understand antibiotic resistance in the 99% of bacteria which cannot be cultured, as well as to know the relationship between environmental antibiotic resistance sequences and antibiotic genes derived from clinical antibiotic-resistant isolates. It can be accessed at: http://staff.washington.edu/jwallace/farme/.

2.6.12 INTEGRALL

Integrall (http://integrall.bio.ua.pt/) is a database to document and annotate integrons and all other transposable elements that confer resistance to antibiotics in bacteria. Integrall, built on PHP5 using MySQL 5.0, seeks to be a comprehensive and unified database that facilitates the understanding and usage of integron information [72]. This is because that transposable elements and integrons in bacteria, which allow bacteria to develop resistances through interbacterial interactions, are a major threat in the field of antimicrobial drug research. It has already collected 11957 entries, with 1509 integrase genes, 8562 gene cassettes, 218 genera, and 482 species.

2.6.13 INTERPRO

InterPro (https://www.ebi.ac.uk/interpro/) provides functional analysis of proteins by classifying them into families and predicting domains and important sites. For the classification of protein families, InterPro uses predictive models, known as signatures, provided by several different databases (referred to as member databases) that make up the InterPro consortium. InterPro capitalizes on the strengths of individual member databases to produce a powerful integrated database and diagnostic tool. The entry of beta-lactamase-related (IPR001466), represents the serine beta-lactamase-like superfamily. This diverse group of sequences includes δ-alanyl-*D*-alanine carboxypeptidase B, aminopeptidase (DmpB), alkaline δ-peptidase, animal δ-Ala-*D*-Ala carboxypeptidase homologs, and the class A and C beta-lactamases and eukaryotic beta-lactamase homologs which are variously described as transesterases, non-ribosomal peptide synthetases, and hypothetical proteins. The beta-lactamases are classified as both S11 (*D*-Ala-*D*-Ala carboxypeptidase A family) and S12 (*D*-Ala-*D*-Ala carboxypeptidase B family, clan SE) non-peptidase homologs. These enzymes have been found experimentally to lack either peptidase activity or amino acid residues that are believed to be essential for catalytic activity. The three classes of serine beta-lactamase are evolutionarily related and belong to a superfamily that also includes *DD*-peptidases and other penicillin-binding proteins [73]. All these proteins contain an S-x-x-K motif, the Ser being the active site residue, but vary considerably outside the active site.

2.6.14 KYOTO ENCYCLOPEDIA OF GENES AND GENOMES (KEGG)

Kyoto Encyclopedia of Genes and Genomes (KEGG) is a database resource for understanding high-level functions and utilities of the biological system, such as the cell, the organism, and the ecosystem, from molecular-level information, especially large-scale molecular datasets generated by genome sequencing and other high-throughput experimental technologies. It is a part of the Japan genome project. In the particular section for antimicrobial resistance: beta-lactamase genes (https://www.genome.jp/kegg/annotation/

br01553.html), each entry contains the entry code, gene name, definition KO, taxonomy lineage, organism, SSDB, motif, amino sequence, and related reference information. In addition, the KEGG entry code has been tied to PubMed ID, GenBank ID, KO ID, and species. It has collected 1939 entries [227 TEM (from TEM-1 to TEM-277), 200 SHV (SHV-1 to SHV-199 and SHV-2a), 31 LEN (LEN-1 to LEN-31), 2 LAP, 204 CTX-M (CTX-M-1 to CTX-M-204), 9 PER (PER-1 to PER-9), 19 VEB (VEB-1 to VEB-19), 4 BEL (BEL-1 to BEL-4), 32 KPC (KPC-1 to KPC-32), 33 GES (GES-1 to GES-33), 16 IMI (IMI-1 to IMI-16), 5 SME (SME-1 to SME-5), 49 CARB (CARB-1 to CARB-49), 566 OXA (OXA-1 to OXA-566), 156 CMY (CMY-1 to CMY-155 and CMY-8b), 1 CFE (CFE-1), 2 LAT (LAT-1, LAT-2), 12 MOX (MOX-1 to MOX-12), 25 DHA (DHA-1 to DHA-25), 14 FOX (FOX-1 to FOX-14), 9 ACC (ACC-1 to ACC-5, plus ACC-1a to ACC-1d), 52 ACT (ACT-1 to ACT-52), 21 MIR (MIR-1 to MIR-21), 81 ADC (ADC-1 to ADC-81), 10 PDC (PDC-1 to PDC-10), 67 IMP (IMP-1 to IMP-67), 54 VIM (VIM-1 to VIM-54), 19 NDM (NDM-1 to NDM-19), 17 IND (IND-1 to IND-16 and IND-2a), and 2 GIM (GIM-1, GIM-2)].

2.6.15 LACTAMASE ENGINEERING DATABASE (LACED)

The lactamase engineering database (LacED: http://www.laced.uni-stuttgart.de/) integrates information on mutations, sequences, and structures of TEM and SHV β-lactamases, in order to help assign new β-lactamases, identify inconsistencies in public databases, and facilitate protein engineering. Databases of other lactamase classes will be added later. As of 2016, the TEM component of this database has already collected 474 TEM variants, where 259 variants form a large scale-free network of highly connected point mutants. The network has been divided into three subnetworks that are enriched by single phenotypes: one network with predominantly 2be and two networks with 2br phenotypes. Almost 50% of the variants are part of a quartet. Thus, two single mutations that result in functional enzymes can be combined into a functional protein [74]. Likewise, the SHV component of this database, i.e., the SHV β-lactamase engineering database (SHVED), has been developed to collect the SHV β-lactamase sequences from the NCBI protein database and the SHV mutation table (containing 132 SHV variants and 22 unassigned SHVs). It serves as a tool for the detection and reconciliation of inconsistencies, and for the identification of new SHV variants and amino acid substitutions. As of 2010, 452 protein sequence entries from NCBI protein database and

132 protein sequences from the SHV mutation table have been collected and parsed into the SHVED, resulting in 200 distinct protein entries, with distinct sequences and 20 crystal structures. 83 Protein sequences have been included in both the SHV mutation table and NCBI protein database, while 35 and 82 protein sequences are only available in the SHV mutation table and the NCBI protein database, respectively. The SHVED is updated monthly and supplies all data for download [75].

2.6.16 THE LAHEY CLINIC DATABASE

This database contains information on β-lactamase classification and amino acid sequences for TEM, SHV, and OXA extended-spectrum and inhibitor-resistant enzymes. It has been mainly developed by Drs. Karen Bush, George Jacoby, and Timothy Palzkill, and as of July 1, 2015, the original site was no long functional, but the database can still be accessed at: https://externalwebapps.lahey.org/studies/. In addition, the contents have been merged into an extended database now managed at NCBI (https://www.ncbi.nlm.nih.gov/pathogens/submit-beta-lactamase/). The new database contains information for β-lactamase, which confer resistance to β-lactam antibiotics; and mobile quinolone resistance genes (Qnr), which confer resistance to quinolone antibiotics, the second largest group of antibiotics. The resources collected in this database include: (a) the storage of current β-lactamase, Qnr, and MCR alleles along with links to any publication when available; (b) the curation (assignment of) new β-lactamase, Qnr, and MCR alleles; and (c) capability of retrieving β-lactamase, Qnr, and MCR nucleotide and protein accessions along with the antibiotic susceptibility profile of a transconjugant/ transformant bearing the β-lactamase gene(s).

2.6.17 MEGARES: AN ANTIMICROBIAL DATABASE FOR HIGH-THROUGHPUT SEQUENCING

The MEGARes database (http://megares.meglab.org) is an antimicrobial resistance database made for high throughput sequencing. It contains sequence data for approximately 8,000 hand-curated antimicrobial resistance genes accompanied by an annotation structure that is optimized for use with high throughput sequencing. The acyclical annotation graph of MEGARes allows for accurate, count-based, hierarchical statistical analysis of resistance at the population level, much like microbiome analysis. MEGARes has been designed for use in the computational analysis of large-scale sequencing data (on the order of terabytes) in a way that is fast and statistically accurate for count-based data and the construction of sequence classifiers. The latest update in MEGARes 2.0 incorporates previously published resistance sequences for antimicrobial drugs, while also including published sequences for metal and biocide resistance determinants. MEGARes annotation consists of three hierarchical levels when looking at AMR genes: drug class, mechanism, and group [76].

2.6.18 MUSTARD: ANTIBIOTIC RESISTANCE DETERMINANTS (ARDS) DATABASE

This database (http://mgps.eu/Mustard/) contains antibiotic resistance determinants (ARDs) and curated antibiotic resistance genes identified from metagenomics. This database is the product of MetaGenoPolis, which has developed a new 3-dimensional modeling-based approach called pairwise comparative modeling (PCM) to accurately predict functions of proteins that are distantly related to proteins with known functions. This project aims to determine the census (i.e., the resistome) of gut microbiota which harbors an allegedly vast diversity of ARDs. This website has been designed to provide easy access to the PCM predictions in various metagenomic datasets. So far, 6,095 ARDs from 20 families in the human gut microbiota have been predicted. As of September 30, 2017, this site has already collected information for 1109 β-lactamase class A, 292 β-lactamase class B, 329 β-lactamase class C, and 375 β-lactamase class D.

2.6.19 NATIONAL DATABASE OF ANTIBIOTIC-RESISTANT ORGANISMS (NDARO)

This is a collaborative, cross-agency, centralized hub for researchers to access antimicrobial resistance (AMR) data to facilitate real-time surveillance of pathogenic organisms. It is the product of the National Action Plan for Combating Antibiotic-Resistant Bacteria developed by the White House in 2015, in response to the rising threat of AMR. The agencies that participated in this National Action Plan include: National Center for Biotechnology Information (NCBI), U.S. Food and Drug Administration (FDA), U.S. Centers for Disease Control and Prevention (CDC), United

States Department of Agriculture (USDA), World Health Organization (WHO), Public Health Emergency (PHE), and others. The objectives of this database include: (a) to increase standardization, by developing and maintaining a curated database of AMR genes; (b) to make AMR-related data more widely available; (c) to make more effective use of bacterial genomic data, using AMRFinderPlus to identify AMR genes in bacterial genomes; and (d) to assist researchers and public health officials to identify bacterial genomes with AMR genes. This database allows users to browse Pathogen Detection Reference Gene Catalog, all AMR genes from all pathogens, and genomes with AMR genotypes or phenotypes. In addition, it also allows users to download AMRFinderPlus to identify AMR genes and submit sequence and phenotype data related to AMR. The URL address is: https://www.ncbi.nlm.nih.gov/pathogens/antimicrobial-resistance/.

2.6.20 PATHOSYSTEMS RESOURCE INTEGRATION CENTER (PATRIC)

The PATRIC (https://www.patricbrc.org/), a project of Virginia Tech's Cyberinfrastructure Division, is funded by the NIH's National Institutes of Allergy and Infectious Diseases. It is the Bacterial Bioinformatics Resource Center, an information system designed to support the biomedical research community's work on bacterial infectious diseases via the integration of vital pathogen information with rich data and analysis tools. PATRIC sharpens and hones the scope of available bacterial phylogenomic data from numerous sources specifically for the bacterial research community, in order to save biologists time and effort when conducting comparative analyzes. The freely available PATRIC platform, using RAST (rapid annotation using subsystem technology) annotation service to annotate bacterial genomes for free in less than 24 hours, provides an interface for biologists to discover data and information and conduct comprehensive comparative genomics and other analyzes in a one-stop-shop. PATRIC can conduct advanced searches based on Taxonomy, Gene Name, Locus Tag, Protein Function/Families, Pathways, EC Numbers, GO Terms, Host-Pathogen Interactions, and more. Also, PATRIC performs a BLAST search against plasmid-specific BLAST databases containing genomic sequences or proteins in PATRIC. Starting September 14, 2022, the PATRIC website will automatically redirect to the new BV-BRC website at: https://www.bv-brc.org/.

2.6.21 REPOSITORY OF ANTIBIOTIC RESISTANCE CASSETTES (RAC)

This database (http://www2.chi.unsw.edu.au/rac) uses the automatic Attacca annotation system to comprehensively annotate gene cassettes and transposable elements in a streamlined manner in order to discover novel gene cassettes. Many resistance genes in Gram-negative bacteria are found as part of small mobile genetic elements called gene cassettes, which are generally integrated into larger elements called integrons so that the integrons associated with mobile elements are designated "mobile resistance integrons" (MRIs). More than one cassette can be inserted in the same integron to create arrays that contribute to the spread of multi-resistance. RAC provides an archive of gene cassettes that include alternative gene names from multiple nomenclature systems and allows the community to contribute new cassettes. RAC also allows users to submit sequences containing cassettes or arrays for annotation using the automatic annotation system Attacca. Attacca recognizes features (gene cassettes, integron regions) and identifies cassette arrays as patterns of features and can also distinguish minor cassette variants that may encode different resistance phenotypes (aacA4 cassettes and bla cassettes-encoding β-lactamases). However, the provided URL address is not accessible [77].

2.7 OTHER BIOLOGICAL ACTIVITIES OF β-LACTAMS

Based on many references cited above and possibly even more uncited literature that are related to the antimicrobial activity of β-lactams, it could be concluded that β-lactams are only associated with antimicrobial activity (or antibacterial activity). In fact, β-lactams have also demonstrated other pharmacological features, such as antidiabetic, anti-HIV, anti-inflammatory, antimalarial, antiparasitic, antiparkinsonian, antithrombotic, anti-tuberculosis, antitumor, antiviral activities, as well as inhibitors for APAF-1 (apoptotic protease activating factor 1), cholesterol absorption, chymase, α-glucosidase, human tryptase, and antagonist of vasopressin V1a, and others. For example, ceftezole [CAS 26973-24-0], as one of the first-generation cephalosporins, has been identified as an α-glucosidase inhibitor, which demonstrates *in vivo* antidiabetic activity [78]. Also, monocyclic β-lactams have been reported to have antidiabetic activity, because they can control hypercholesterolemia [79]. Oligopeptides prepared from 7β-amino-desacetoxycephalosporanic acid and hydrophobic amino acids have been identified as weak inhibitors against human immunodeficiency virus type 1 proteinase (HIV-1 PR) by both spectroscopic assays and HPLC assay, with IC_{50} values in the millimolar range [80]. When a small β-lactam library containing 126 individual β-lactams is screened by colorimetric microtiter plate method against HIV-1 protease, several β-lactams demonstrate greater than 60% inhibition of the enzyme [81]. Regarding the anti-inflammatory activity, it has been shown that chromeno β-lactam hybrid, particularly 2-amino-4-(4-(((2*S*,3*R*)-2-(4-chlorophenyl)-4-oxo-1-(*p*-tolyl)azetidin-3-yl)oxy)phenyl)-7,7-dimethyl-5-oxo-5,6,7,8-tetrahydro-4*H*-chromene-3-carbonitrile (**3**, Figure 2.3) and its precursor 4-(((2*S*,3*R*)-2-(4-chlorophenyl)-4-oxo-1-(*p*-tolyl)azetidin-3-yl)oxy)benzaldehyde (**4**, Figure 2.3) showed an anti-inflammatory ratio of 19.8 and 53.4, respectively, and the aldehyde appears more active than the well-known dexamethasone corticosteroid used for the treatment of rheumatoid and skin inflammation [82]. Monamphilectine A, i.e., *N*-(1-((1*S*,3*S*,3a*R*,3a^1*S*,6*S*,6a*S*,9a*S*)-6a-isocyano-3,6-dimethyl-9-methylenedodecahydro-1*H*-phenalen-1-yl)-2-methyl-propan-2-yl)-2-(2-oxoazetidin-1-yl)acetamide (**5**, Figure 2.3), an isolated diterpenoid β-lactam alkaloid in milligram quantities following bioassay-directed extraction of a Puerto Rican marine sponge *Hymeniacidon sp.* has been identified to possess potent antimalarial activity [83]. Also, some novel benzothiazole-substituted β-lactam hybrids of high diastereoselectivity have been synthesized starting from (benzo[d]thiazol-2-yl)phenoxyacetic acid as the ketene precursor. Antimalarial activity screening indicates that adding methoxyphenyl or ethoxyphenyl group to the β-lactam ring makes the β-lactams more potent, as represented by (3*R*,4*S*)-3-(4-(benzo[*d*]thiazol-2-yl)phenoxy)-1-(4-ethoxyphenyl)-4-(4-nitrophenyl)azetidin-2-one (**6**, Figure 2.3) [84]. In addition, several nanoparticles that are synthesized in water by emulsion polymerization of an acrylate β-lactam (e.g., 4-((2*R*,3*S*)-2-(4-methoxyphenyl)-4-oxo-3-phenoxyazetidin-1-yl)phenyl acrylate (**7**, Figure 2.3) and 4-((2*R*,3*S*)-2-(4-methoxyphenyl)-4-oxo-3-phenoxyazetidin-1-yl)benzyl acrylate (**8**, Figure 2.3)) pre-dissolved in a mixture of comonomers in the presence of sodium dodecyl sulfate (SDS) as a surfactant and potassium persulfate as a radical initiator have shown antimalarial activities as screened against chloroquine-resistant *Plasmodium faliparum* K1 strain [85].

Moreover, diastereoselectively synthesized *cis*-β-lactams such as (3*S*,4*R*)-4-(3,4-dimethoxyphenyl)-3-(((1*R*,2*S*,5*R*)-2-isopropyl-5-methylcyclohexyl)oxy)-1-((*S*)-1-phenylethyl)azetidin-2-one (**9**, Figure 2.3) and

(3*S*,4*R*)-3-(((1*R*,2*S*,5*R*)-2-isopropyl-5-methylcyclohexyl)oxy)-4-(3-methoxyphenyl)-1-((*S*)-1-phenylethyl)-azetidin-2-one (**10**, Figure 2.3) obtained through a [2+2] cycloaddition of chiral imines and a chiral ketene also demonstrate excellent antimalarial activity [86]. Similarly, 2-((2*R*,3*S*)-1-(4-(dimethylamino)phenyl)-4-oxo-3-phenoxyazetidin-2-yl) anthracene-9,10-dione (**11**, Figure 2.3) synthesized from imine derived from anthraquinone-2-carbaldehyde and ketene shows excellent antimalarial activity against Plasmodium falciparum K14 resistant strain [87].

While not a specific β-lactam has been reported for antiparasitic activity, it has been reported that certain β-lactams do show antiparasitic activity [88]. In searching β-lactams for antiparkinsonian, ethyl naphthalen-1-ylacetate was condensed with an equimolar quantity of hydrazine hydrate in methanol to afford 2-(naphthalen-1-yl)acetohydrazide, which was then condensed with different aromatic aldehydes, affording the respective Schiff bases. The Schiff bases then underwent cycloaddition with ketene formed *in situ* from chloroacetyl chloride in triethylamine to yield naphthalene-bearing β-lactams. Biological screening of antiparkinson's activity according to 6-hydroxydopamine lesioned rat's model (6-OHDA) indicates that β-lactam with nitrophenyl group at position-2 has the highest antiparkinson activity [89]. In another practice, 1-((3*S*,5*S*,7*S*)-adamantan-1-yl)-3-chloro-4-(2-(2,4-dihydroxyphenyl)-5-methoxy-1*H*-indol-3-yl)azetidin-2-one (**12**, Figure 2.3) and 1-((3*S*,5*S*,7*S*)-adamantan-1-yl)-3-chloro-4-(2-(2,4-dihydroxyphenyl)-5-ethoxy-1*H*-indol-3-yl)azetidin-2-one (**13**, Figure 2.3) have been identified as the most effective antiparkinsonian β-lactams among 1-adamantyl-3-chloro-4-(2-aryl-1*H*-indol-3-yl) β-lactams prepared from the cycloaddition between 5-substituted 3-adamantylimino-2-aryl-1*H*-indoles and chloroacetyl chloride in the presence of Et_3N [90].

As for the antithrombotic activity, it has been tested that the culture extract of *Penicillium chrysogenum* (PCL 501) significantly lengthened thrombin and blood clotting time, and inhibited the formation of blood clot, where the delay in thrombin and blood clotting time was proportionate to the concentration of extract. Particularly, the commercial benzyl penicillin also exhibits a similar inhibitory pattern [91].

It is so obvious for β-lactams of anti-tubercular activity due to the general anti-bacterial activity of β-lactams as outlined above; this is because tuberculosis (TB) is an infectious disease usually caused by *Mycobacterium tuberculosis* bacteria which generally affects the lungs. Carbapenems (e.g., imipenem, meropenem, and ertapenem), targeting the *Mycobacterium tuberculosis* cell wall and β-lactamase, are effective against *Mycobacterium*

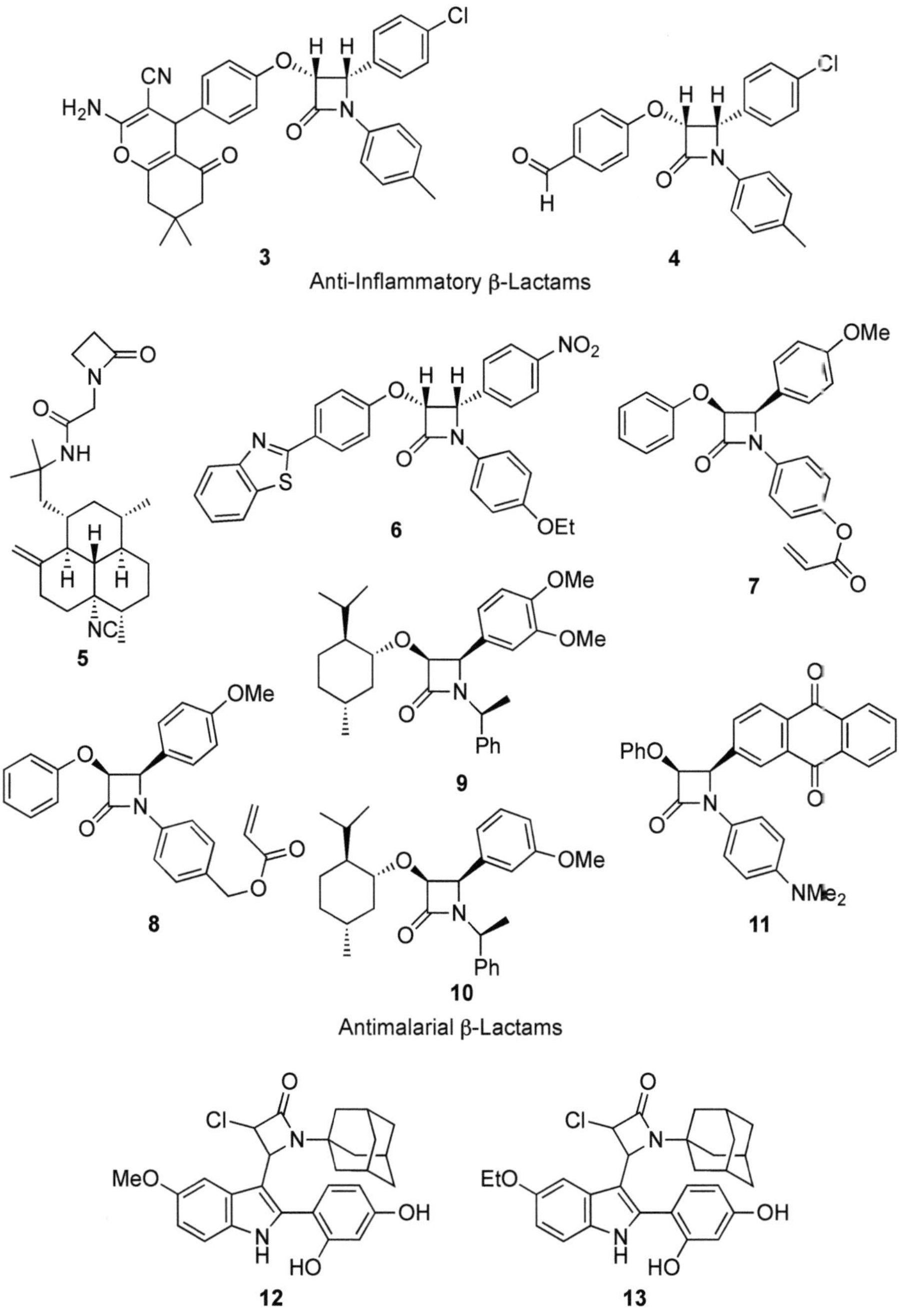

FIGURE 2.3 β-Lactams with anti-inflammatory, antimalarial and antiparkinsonian activities

tuberculosis, particularly for multidrug-resistant and extensively drug-resistant (XDR) tuberculosis [92]. The effectiveness of carbapenems against TB is because that *L,D*-transpeptidase LdtMt2 confers resistance to traditional β-lactams whereas carbapenems repress the activity of LdtMt2. The interaction mechanisms between LdtMt2 and carbapenems reveals that the presence of 1-β-methyl group in carbapenems is related to the ligand affinity of LdtMt2 and that the presence of Y308 and Y318 residues in LdtMt2 stabilizes the conformation of the LdtMt2-carbepenem adduct [93]. A particular experiment tested the antituberculosis activity of five β-lactams (meropenem, biapenem, tebipenem, ampicillin, and amoxicillin) alone or in combination with β-lactamase inhibitors (clavulante, avibactam) against 41 clinical isolates of *Mycobacterium tuberculosis*, including multidrug-resistant and XDR strains, and demonstrated oral carbapenem tebipenem as the most potent anti-TB agent [94]. Besides carbapenems, other types of β-lactams have shown anti-tubercular activity. For example, the quinolone-cephalosporin conjugates, particularly (6*R*,7*R*)-3-(((4-(3-carboxy-1-cyclopropyl-6-fluoro-4-oxo-1,4-dihydroquinolin-7-yl)piperazine-1-carbonyl) oxy)methyl)-8-oxo-7-(2-phenyl-acetamido)-5-thia-1-azabicyclo[4.2.0] oct-2-ene-2-carboxylic acid (**14**, Figure 2.4) has demonstrated antimycobacterial activity, including anti-TB activity [95]. Also, monobactams possessing hydroxamate functionality, such as 4-(1-(benzyloxy)-2-oxo-2-phenylethyl)-1-(*tert*-butyldimethylsilyl)azetidin-2-one (**15**, Figure 2.4) and *N*-(benzyloxy)-*N*-((2*S*,3*S*)-4-oxo-3-(2-phenoxyacetamido)azetidin-2-yl)benzamide (**16**, Figure 2.4) exhibited significant anti-TB (H37Rv) activity from microplate Alamar blue assay [96].

Several β-lactams have also demonstrated anticancer activity. For example, stilbenoid combretastatin α-4 (CA-4, i.e., 2-methoxy-5-[(*Z*)-2-(3,4,5-trimethoxy-phenyl)-vinyl]-phenol) [97], existing in either *cis*- or *trans*-form, is a well-known tubulin polymerization inhibitor. Given its role in cell division, tubulin becomes a potent target in cancer chemotherapy. CA-4 in *cis*-form binds much better to the "colchicine" site on tubulin to inhibit polymerization. For this reason, novel 1,4-diaryl-β-lactams of CA-4 analogs have been synthesized and evaluated *in vitro* for antiproliferative activity, antiapoptotic activity and inhibition of tubulin polymerization. The results indicate that among these analogs, (3*S*)-4-(3-fluoro-4-methoxyphenyl)-3-hydroxy-1-(3,4,5-trimethoxyphenyl)azetidin-2-one (**17**, Figure 2.4) and (3*S*)-4-(3-fluoro-4-methoxyphenyl)-3-phenyl-1-(3,4,5-trimethoxyphenyl)-azetidin-2-one (**18**, Figure 2.4) demonstrate improved activity over CA-4 in CA-4 resistant HT-29 colon cancer cells while retaining potency in MCF-7 breast cancer cells. The former also induces mitotic arrest at low concentration in both cell lines (MCF-7 and HT-29) together with down-regulation of expression of antiapoptotic proteins Mcl-1, Bcl-2, and survivin in MCF-7 cells [98]. While *N*-alkylthiolated β-lactams have been known to possess DNA-damaging and apoptosis-inducing activity in various tumor lines rather than non-transformed cells, careful examination on the structures of C3 and N1 substituents shows that 1-(methylthio)-2-oxo-4-phenylazetidin-3-yl 2-butyloctanoate (**19**, Figure 2.4) and 1-(*sec*-butylthio)-2-oxo-4-phenylazetidin-3-yl 2-butyloctanoate (**20**, Figure 2.4), both with a branched-chain moiety at C3 of the lactam ring, exhibit potent apoptosis-inducing activity. In addition, the branched β-lactams are able to inhibit the growth of mice bearing breast cancer xenografts, associated with induction of DNA damage and apoptosis in tumor tissues [99]. Similarly, another group of seven β-lactams have been evaluated for their apoptosis-inducing properties against human leukemia, breast, prostate, and head-and-neck cancer cells, showing that 4-(2-chlorophenyl)-3-methoxy-1-(methylthio)azetidin-2-one (**21**, Figure 2.4) is able to induce DNA damage and inhibit DNA replication in Jurkat T cells within a 2-h treatment [100]. Also, (3*R*,4*R*)-1-(chrysen-6-yl)-2-oxo-4-phenylazetidin-3-yl acetate (**22**, Figure 2.4) obtained from Staudinger reaction with chiral carbohydrate as the ketene component with an achiral imine has anticancer activity [79, 101].

Antivirals are drugs to treat people who have already been infected by a virus. They also can be used to prevent or limit infection within a certain time frame before or after exposure but prior to the event that illness occurs. While β-lactams have been widely applied as antibacterial agents, β-lactams being able to react with serine-based protease are also effective agents against certain viruses, such as human cytomegalovirus (HCMV), a member of the family *Herperviridae*. This virus is an opportunistic pathogen in immunocompromised individuals such as AIDS patients and organ transplant recipients, and can cause significant morbidity and mortality. HCMV encodes a protease that is essential for viral replication because it is expressed as part of a larger precursor together with the viral capsid assembly protein. Although many β-lactam derivatives are known inhibitors of classical serine proteases, HCMV protease has little similarity with other serine proteases [102], for which 4-thioalkyl-β-lactams have recently been identified as potent inhibitors [103–105]. Prompted by this result, a series of monobactams possessing a heterocyclic thiomethyl side chain at C-4 have been synthesized and subject to antiviral activity screening in a plaque reduction assay. The results indicate that (*R*)-*N*-methyl-2-(((1-methyl-1*H*-tetrazol-5-yl)thio) methyl)-4-oxo-*N*-(4-(trifluoromethyl)benzyl)azetidine-1-carboxamide (**23**, Figure 2.4) and (*R*)-*N*-methyl-2-(((1-methyl-1*H*-tetrazol-5-yl)thio)methyl)-*N*-(4-nitrobenzyl)-4-oxoazetidine-1-carboxamide (**24**, Figure 2.4) are most effective against the HCMV protease, which exhibit antiviral activity in a plaque reduction assay and cell-based assay [106].

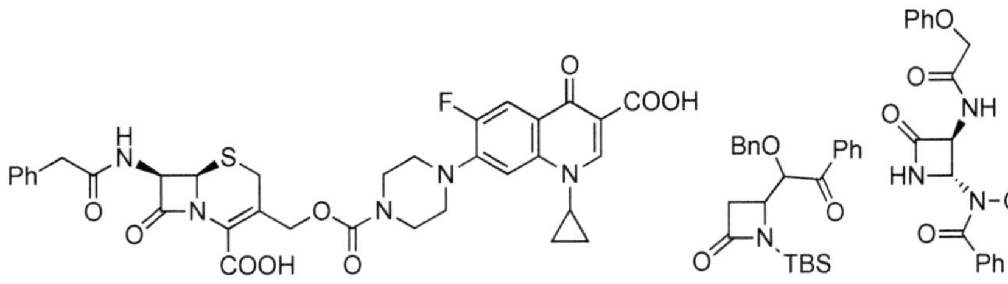

14 **15** **16**

Anti-Tuberculosis β-Lactams

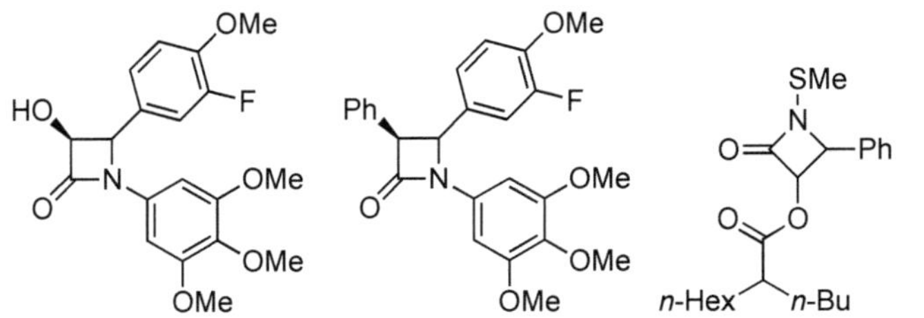

17 **18** **19**

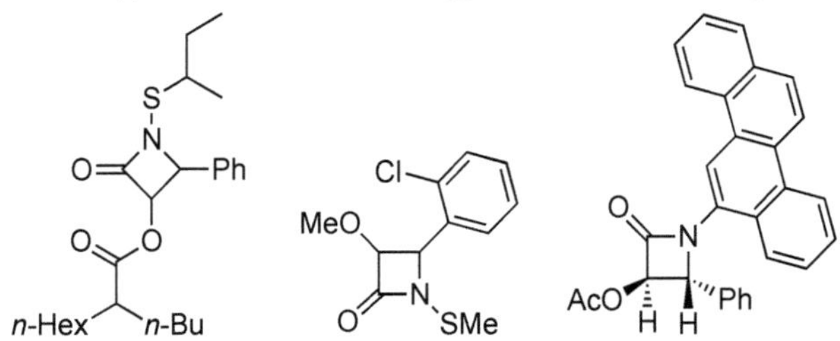

20 **21** **22**

Anticancer β-Lactams

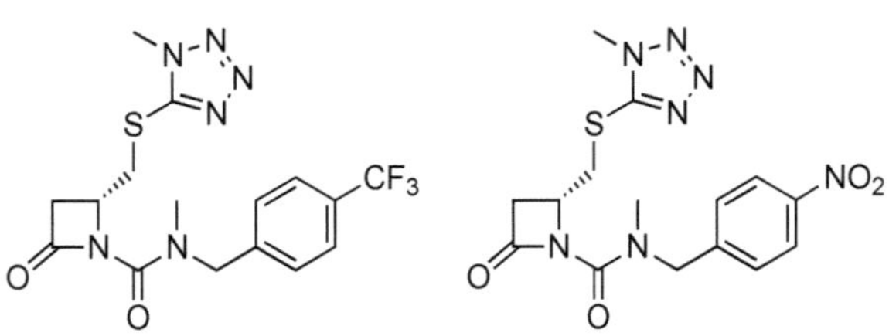

23 **24**

Antiviral β-Lactams

FIGURE 2.4 β-Lactams of anti-tuberculosis, anticancer and antiviral activities

In addition to the above potential medical usages, β-lactams have been identified as inhibitors of many other proteins and enzymes, in addition to the PBPs and β-lactamases (already described above). As one of these proteins, APAF-1, also known as APAF-1 or APAF1, plays a central role in apoptosis regulatory network. Due to the importance of this protein, a series of heterocycles have been prepared by means of Ugi's 4CC reaction, among which *N*-(2,4-dichlorophenethyl)-2-(2-(2,4-dichlorophenethyl)-2*H*-1,2,3-triazol-4-yl)-1-(3,3-diphenylpropyl)-4-oxoazetidine-2-carboxamide (**25**, Figure 2.5) has been identified as an effective inhibitor of APAF-1 [107]. Chymase (EC 3.4.21.39), also known as mast cell protease 1, skeletal muscle protease, skin chymotryptic proteinase, and mast cell serine proteinase, is the second major human mast cell protease, found at high levels in connective tissues such as the skin and submucosa of the gut and at low levels in the lung. An 1-oxacephem derivative, i.e., 2-(5-((((6*R*,7*R*)-7-methoxy-7-(2-methoxybenzamido)-2-(((3-methylbenzyl)oxy)carbonyl)-8-oxo-5-oxa-1-azabicyclo[4.2.0]oct-2-en-3-yl)-methyl)thio)-1*H*-tetrazol-1-yl)acetic acid (**26**, Figure 2.5) has been identified as highly selective human chymase inhibitor [108].

Monobactams of no antibiotic activity have been widely used as cholesterol absorption inhibitors, for the purpose of reducing the circulating low-density lipoproteins (LDL), and treatment of hypercholesterolemia [109]. Dietary cholesterol will be absorbed by intestinal enterocyte in a form of bile salt micelle and reassembled into intestinal lipoproteins called chylomicrons, which are then secreted into the lymphatics and circulated to the liver. These cholesterol particles are then secreted into blood as very low-density lipoproteins (VLDL) particles from the liver. Cholesterol absorption inhibitors block the uptake of micellar cholesterol, and reduce the incorporation of cholesterol esters into chylomicron particles that will be delivered back to the liver. Examples of β-lactams that are cholesterol absorption inhibitors are SCH 47947 (i.e., (3*R*,4*S*)-1,4-bis(4-methoxyphenyl)-3-(3-phenylpropyl)azetidin-2-one, **27** in Figure 2.5) [110, 111], Ezetimibe (i.e., (3*R*,4*S*)-1-(4-fluorophenyl)-3-((*S*)-3-(4-fluorophenyl)-3-hydroxypropyl)-4-(4-hydroxyphenyl)-azetidin-2-one, **28** in Figure 2.5) [112], Ezetimibe analogs (i.e., (3*R*,4*R*)-1-(4-fluorophenyl)-3-(((*R*)-2-(4-fluorophenyl)-2-hydroxyethyl)amino)-4-(4-hydroxy-phenyl)azetidin-2-one (**29**, Figure 2.5) and (3*R*,4*R*)-1-(4-fluorophenyl)-3-(((*S*)-2-(4-fluorophenyl)-2-hydroxyethyl)amino)-4-(4-hydroxyphenyl)azetidin-2-one (**30**,Figure2.5)[113],4-(4-((3*S*)-1-(4-bromophenyl)-2-(3,4-dimethoxyphenyl)-4-oxoazetidin-3-yl)-2-oxobutyl)benzyl nitrate (**31**, Figure 2.5) [114], (4*S*)-1-(4-fluorophenyl)-3-(2-(4-hydroxy-6-oxotetrahydro-2*H*-pyran-2-yl)ethyl)-4-(4-hydroxyphenyl)azetidin-2-one (**32**, Figure 2.5) [115], and many others (e.g., ezetimibe glycosides).

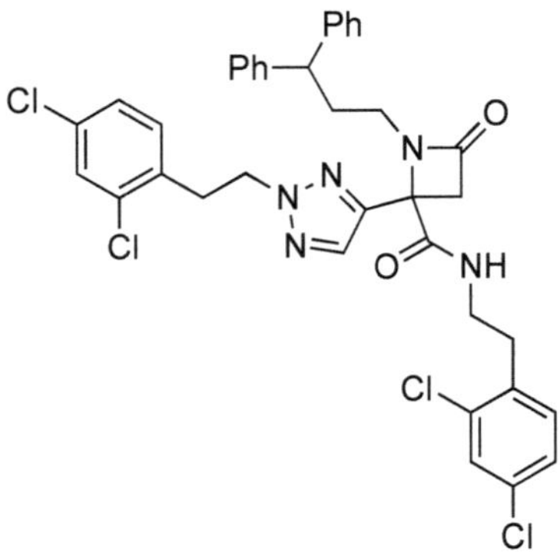

25
APAF-1 Inhibitor

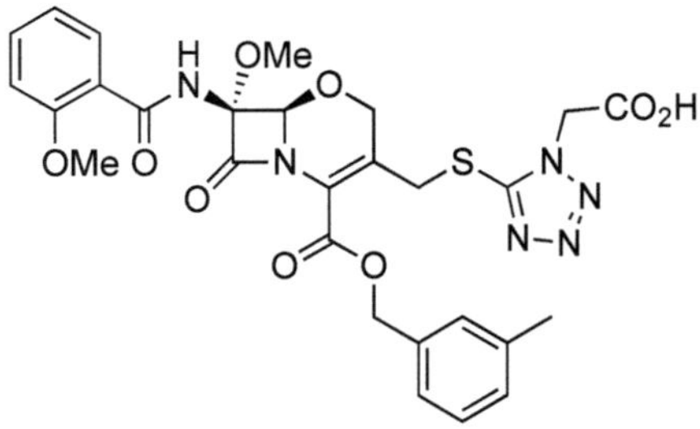

26
Human Chymase Inhibitor

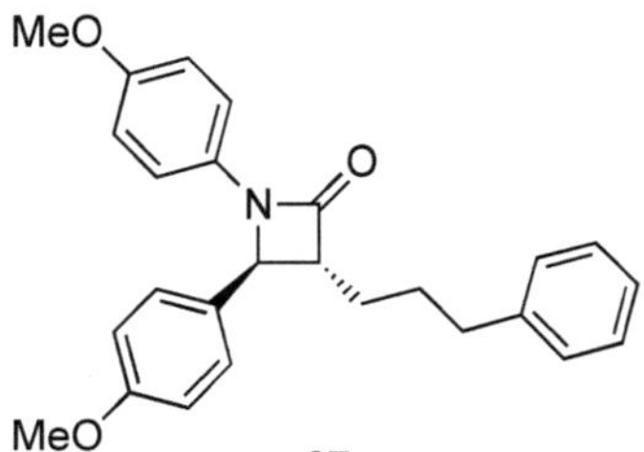

27
SCH 47947

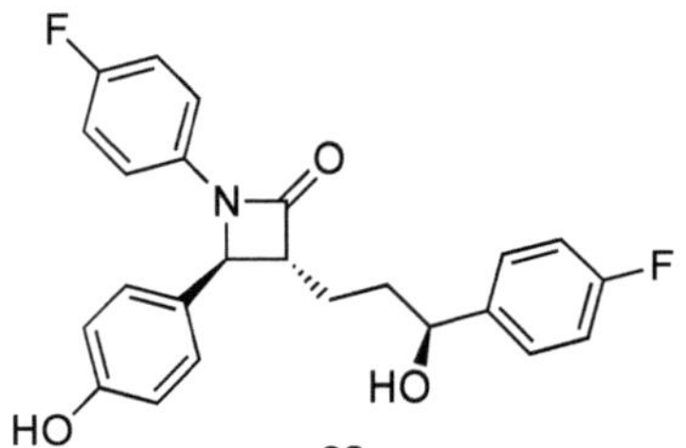

28
Ezetimibe

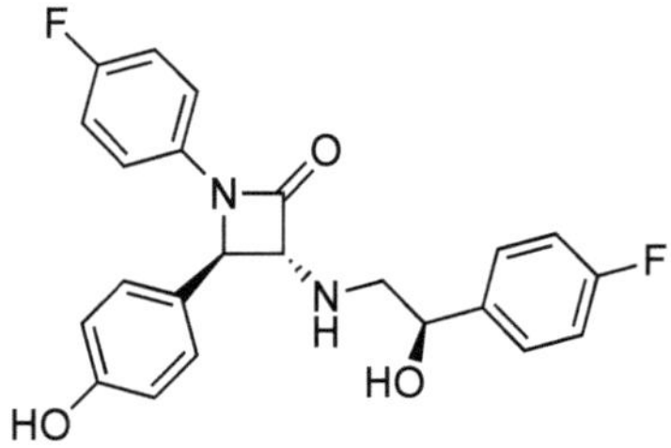

29

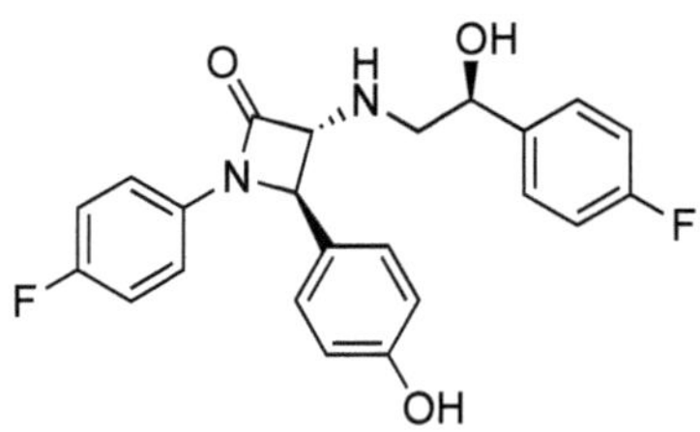

30

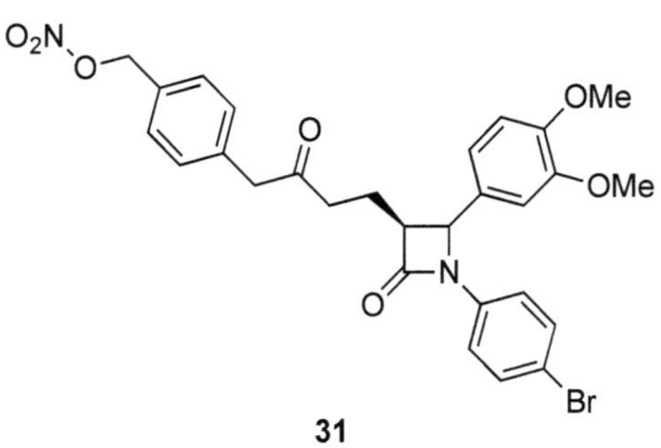

31

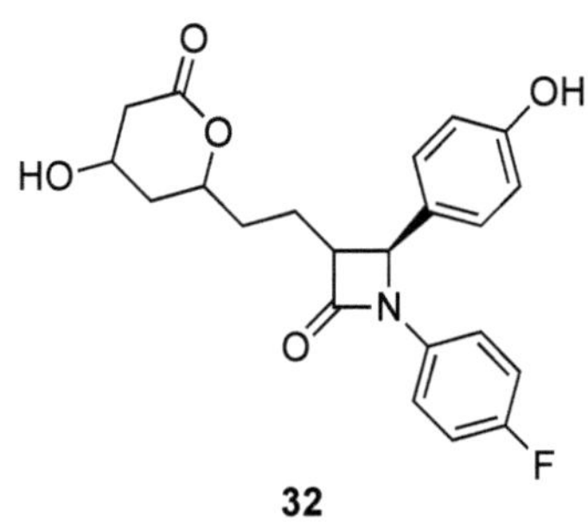

32

Cholesterol Absorption Inhibitors

FIGURE 2.5 β-Lactams as APAF-1, human chymase, and chlolesterol absorption inhibitors

Some other β-lactams have been identified as galactosidase inhibitors, such as (4*R*,5*S*,6*R*)-3,4,5-trihydroxy-1-azabicyclo[4.2.0]octan-8-one (**33**, Figure 2.6) and (4*R*,5*S*,6*R*)-3,4,5-trihydroxy-6-methyl-1-azabicyclo[4.2.0]octan-8-one (**34**, Figure 2.6) [116], and *N*-acylethanolamine acid amidase (NAAA) inhibitors that suppresses inflammation and pruritus, such as (2,2-difluorobenzo[*d*][1,3]dioxol-5-yl)methyl (2-oxoazetidin-3-yl) carbamate (**35**, Figure 2.6), 4-cyclohexylbutyl ((2*S*)-2-methyl-1-(4-(methylsulfonyl)phenoxy)-4-oxoazetidin-3-yl)carbamate (**36**, Figure 2.6) and 4-cyclohexylbutyl (2-oxoazetidin-3-yl)carbamate (**37**, Figure 2.6) [117, 118]. More functional enzyme inhibitors, including the inhibitors of α-glucosidase, human leukocyte elastase, porcine pancreatic elastase, tryptase, fatty acid amide hydrolase, protein phosphatase methylesterase-1, serine protease responsible for tumor proliferation, cysteine proteases, matrix metalloproteinases, human 20s proteasome, human fatty acid synthase, bacterial RNase A and Leishmania δ-mannosyl phosphate transferase and antagonist of vasopressin V1a can be found in several review articles [71, 119, 120].

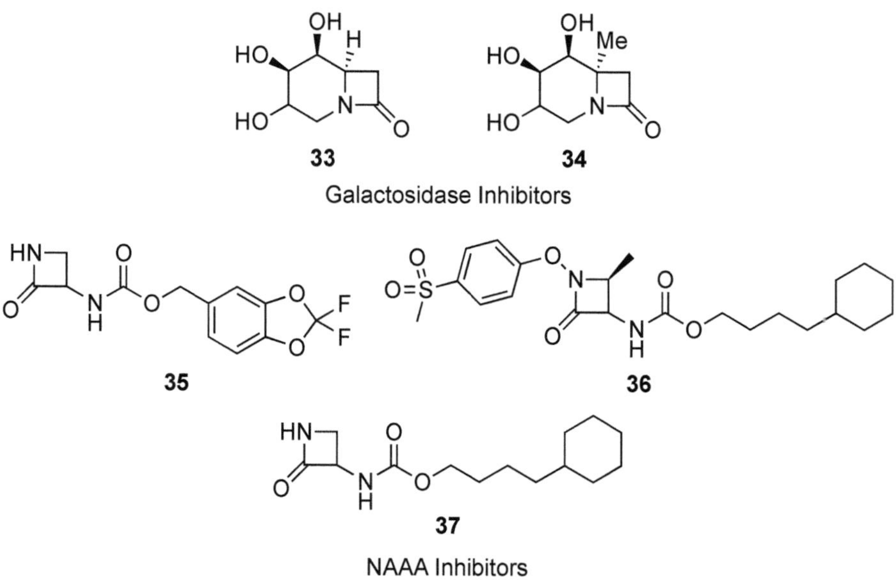

FIGURE 2.6 β-Lactams as galactosidase and *N*-acylethanolamine acid amidase inhibitors

KEYWORDS

- **extended-spectrum beta-lactamases**
- **metallo-β-lactamases**
- **methicillin-resistant Staphylococcus aureus**
- **penicillin-binding proteins**
- **β-lactamase database**
- **inhibitor**
- **Ambler classification**
- **Bush-Jacoby-Medeiros classification**
- **BLAST**

REFERENCES

1. Peck, M., & Badrick, T., (2017). A review of contemporary practice and proficiency with gram staining in anatomical pathology laboratories. *Journal of Histotechnology, 40*(2), 54–61. doi: 10.1080/01478885.2017.1327474.
2. Lu, M., Zhan, H., & Li, Q., (2015). Microcrystalline anthracene enrichment-spectrophotometry determination of trace crystal violet. *Asian Journal of Chemistry, 27*(2), 507–510. doi: 10.14233/ajchem.2015.16998.
3. Masadome, T., (2001). Flow injection spectrophotometric determination of anionic polyelectrolytes using the cationic dyes. *Analytical Letters, 34*(15), 2711–2719. doi: 10.1081/AL-100108417.
4. Zhang, Z. Q., Yan, H. T., & Yue, X. F., (2004). Catalytic determination of trace formaldehyde with a flow injection system using the indicator reaction between crystal violet and bromate. *Microchimica Acta, 146*(3, 4), 259–263. doi: 10.1007/s00604-003-0178-y.
5. Yue, X. F., Zhang, Z. Q., & Yan, H. T., (2004). Flow injection catalytic spectrophotometric simultaneous determination of nitrite and nitrate. *Talanta, 62*(1), 97–101. doi: 10.1016/S0039-9140(03)00421-1.
6. Yin, F., & Wu, Y., (2001). A study on the determination of trace nitrite-N by spectrophotometry with safranin-T. *Huaihai Gongxueyuan Xuebao, 10*(2), 42, 43.
7. Geokchyan, N. O., (2013). Diazin raw safranin "T" as a new reagent for determination osmium (IV) by extractions-spectrophotometric method. *Zekuytsner – Hayastani Gitut'yunneri Azgayin Akademia, 113*(2), 174–179.
8. Naboikin, Y. V., Ogurtsova, L. A., Podgornyi, A. P., & Pokrovskaya, F. S., (1969). Generation of light in allowed transitions of organic molecules. *Kvantovaya Elektronika (Kiev),* (4), 229–250.
9. Calisher, C. H., (2007). Taxonomy: What's in a name? Doesn't a rose by any other name smell as sweet? *Croatian Medical Journal, 48*(2), 268–270.

10. Skerman, V. B. D., McGowan, V., & Sneath, P. H. A., (1980). Approved lists of bacterial names free. *International Journal of Systematic and Evolutionary Microbiology, 30*(1), 225–420. doi: 10.1099/00207713-30-1-225.
11. Parker, C. T., Tindall, B. J., & Garrity, G. M., (2019). International code of nomenclature of prokaryotes: Prokaryotic code (2008 revision). *International Journal of Systematic and Evolutionary Microbiology, 69*(1A), S1–S111. doi: 10.1099/ijsem.0.000778.
12. Dik, D. A., Madukoma, C. S., Tomoshige, S., Kim, C., Lastochkin, E., Boggess, W. C., Fisher, J. F., et al., (2019). Slt, MltD, and MltG of *Pseudomonas aeruginosa* as targets of bulgecin A in potentiation of β-lactam antibiotics. *ACS Chemical Biology, 14*(2), 296–303. doi: 10.1021/acschembio.8b01025.
13. Nakashio, S., & Gerhardt, P., (1985). Protoplast dehydration correlated with heat resistance of bacterial spores. *Journal of Bacteriology, 162*(2), 571–578. doi: 10.1128/jb.162.2.571-578.1985.
14. Popham, D. L., Gilmore, M. E., & Setlow, P., (1999). Roles of low-molecular-weight penicillin-binding proteins in *Bacillus subtilis* spore peptidoglycan synthesis and spore properties. *Journal of Bacteriology, 181*(1), 126–132. doi: 10.1128/jb.181.1.126-132.1999.
15. Verma, P., Upadhyay, A., Tiwari, M., & Tiwari, V., (2016). In-silico approach explains evolution of beta-lactamases from penicillin binding proteins. *Journal of Proteomics & Bioinformatics, 9*(10), 248–254. doi: 10.4172/jpb.1000413.
16. Grandchamps, J., Nguyen-Distèche, M., Damblon, C., Frère, J. M., & Ghuysen, J. M., (1995). *Streptomyces* K15 active-site serine *DD*-transpeptidase: Specificity profile for peptide, thiol ester and ester carbonyl donors and pathways of the transfer reactions. *Biochemical Journal, 307*(2), 335–339. doi: 10.1042/bj3070335.
17. Lohans, C. T., Chan, H. T. H., Malla, T. R., Kumar, K., Kamps, J. J. A. G., McArdle, D. J. B., Van, G. E., et al., (2019). Non-hydrolytic β-lactam antibiotic fragmentation by *L,D*-transpeptidases and serine β-lactamase cysteine variants. *Angewandte Chemie, International Edition, 58*(7), 1990–1994. doi: 10.1002/anie.201809424.
18. Gupta, R., Lavollay, M., Mainardi, J. L., Arthur, M., Bishai, W. R., & Lamichhane, G., (2010). The mycobacterium tuberculosis protein LdtMt2 is a nonclassical transpeptidase required for virulence and resistance to amoxicillin. *Nature Medicine (New York, NY, United States), 16*(4), 466–469. doi: 10.1038/nm.2120.
19. Page, M. I., (1984). The mechanisms of reactions of β-lactam antibiotics. *Accounts of Chemical Research, 17*(4), 144–151. doi: 10.1021/ar00100a005.
20. Basu, J., Chattopadhyay, R., Kundu, M., & Chakrabarti, P., (1992). Purification and partial characterization of a penicillin-binding protein from mycobacterium smegmatis. *Journal of Bacteriology, 174*(14), 4829–4832. doi: 10.1128/jb.174.14.4829-4832.1992.
21. Leganes, F., Blanco-Rivero, A., Fernandez-Pinas, F., Redondo, M., Fernandez-Valiente, E., Fan, Q., Lechno-Yossef, S., & Wolk, C. P., (2005). Wide variation in the cyanobacterial complement of presumptive penicillin-binding proteins. *Archives of Microbiology, 184*(4), 234–248. doi: 10.1007/s00203-005-0046-8.
22. Nguyen-Distèche, M., Leyh-Bouille, M., & Ghuysen, J. M., (1982). Isolation of the membrane-bound 26000-Mr penicillin-binding protein of *Streptomyces* strain K15 in the form of a penicillin-sensitive δ-alanyl-D-alanine-cleaving transpeptidase. *Biochemical Journal, 207*(1), 109–115. doi: 10.1042/bj2070109.
23. Palacios, A. R., Mojica, M. F., Giannini, E., Taracila, M. A., Bethel, C. R., Alzari, P. M., Otero, L. H., et al., (2019). The reaction mechanism of metallo-β-lactamases is tuned by

the conformation of an active-site mobile loop. *Antimicrobial Agents and Chemotherapy, 63*(1), e01754-18/1–e01754-18/15. doi: 10.1128/AAC.01754-18.
24. Hargis, J. C., Vankayala, S. L., White, J. K., & Woodcock, H. L., (2014). Identification and characterization of noncovalent interactions that drive binding and specificity in *DD*-peptidases and β-lactamases. *Journal of Chemical Theory and Computation, 10*(2), 855–864. doi: 10.1021/ct400968v.
25. Ibeji, C. U., Lawal, M. M., Tolufashe, G. F., Govender, T., Naicker, T., Maguire, G. E. M., Lamichhane, G., et al., (2019). The driving force for the acylation of β-lactam antibiotics by *L,D*-transpeptidase 2: Quantum mechanics/molecular mechanics (QM/MM) study. *ChemPhysChem, 20*(9), 1126–1134. doi: 10.1002/cphc.201900173.
26. Mangat, C. S., Vadlamani, G., Holicek, V., Chu, M., Larmour, V. L. C., Vocadlo, D. J., Mulvey, M. R., & Marka, B. L., (2019). Molecular basis for the potent inhibition of the emerging carbapenemase VCC-1 by avibactam. *Antimicrobial Agents and Chemotherapy, 63*(4), e02112-18/1–e02112-18/13. doi: 10.1128/AAC.02112-18.
27. Woodward, R. B., (1980). Penems and related substances. *Philosophical Transactions of the Royal Society of London, Series B: Biological Sciences, 289*(1036), 239–250. doi: 10.1098/rstb.1980.0042.
28. Nangia, A., Biradha, K., & Desiraju, G. R., (1996). Correlation of biological activity in β-lactam antibiotics with Woodward and Cohen structural parameters – a Cambridge database study. *Journal of the Chemical Society, Perkin Transactions 2: Physical Organic Chemistry,* (5), 943–953. doi: 10.1039/P29960000943.
29. Shi, C., Chen, J., Kang, X., Shen, X., Lao, X., & Zheng, H., (2019). Approaches for the discovery of metallo-β-lactamase inhibitors: A review. *Chemical Biology & Drug Design, 94*(2), 1427–1440. doi: 10.1111/cbdd.13526.
30. Massova, I., & Mobashery, S., (1997). Molecular bases for interactions between β-lactam antibiotics and β-lactamases. *Accounts of Chemical Research, 30*(4), 162–168. doi: 10.1021/ar960007e.
31. Lloyd, M. D., Merritt, K. D., Lee, V., Sewell, T. J., Wha-Son, B., Baldwin, J. E., Schofield, C. J., et al., (1999). Product-substrate engineering by bacteria: Studies on clavaminate synthase, a trifunctional dioxygenase. *Tetrahedron, 55*(33), 10201–10220. doi: 10.1016/S0040-4020(99)00547-5.
32. Baldwin, J. E., Merritt, K. D., Schofield, C. J., Elson, S. W., & Baggaley, K. H., (1993). Studies on the stereospecificity of the clavaminic acid synthase-catalyzed hydroxylation reaction. *Journal of the Chemical Society, Chemical Communications,* (16), 1301, 1302. doi: 10.1039/c39930001301.
33. Olivares, B., Martínez, F., Rivas, L., Calderón, C., Munita, J. M., & Campodonico, P. R., (2018). A natural deep eutectic solvent formulated to stabilize β-lactam antibiotics. *Scientific Reports, 8*(1), 14900/1–14900/12. doi: 10.1038/s41598-018-33148-w.
34. Jozic, S., & Solic, M., (2017). Effect of environmental conditions on *Escherichia coli* survival in seawater. In: Samie, A., (ed.), *Escherichia Coli: Recent Advances on Physiology, Pathogenesis and Biotechnological Applications* (pp. 169–186).
35. Koraimann, G., (2004). Bacterial conjugation: Cell-cell contact-dependent horizontal gene spread. In: Miller, R. V., & Day, M. J., (eds.), *Microbial Evolution* (pp. 111–124).
36. Ionescu, M. D., Balazs, D., Draghici, D., Dobrovolski, D., Sasarman, E., & Mocanu, E., (1988). Thin bridgelike connections between cells in *Mycobacterium bovis* BCG and in *Corynebacterium glutamicum. Archives Roumaines de Pathologie Experimentales et de Microbiologie, 47*(1), 5–10.

37. Inniss, N. L., Prehna, G., & Morrison, D. A., (2019). The pneumococcal σ^X activator, ComW, is a DNA-binding protein critical for natural transformation. *Journal of Biological Chemistry, 294*(29), 11101–11118. doi: 10.1074/jbc.ra119.007571.
38. Watson, B. N. J., Staals, R. H. J., & Fineran, P. C., (2018). CRISPR-Cas-mediated phage resistance enhances horizontal gene transfer by transduction. *mBio, 9*(1), e02406-17/1–e02406-17/14. doi: 10.1128/mBio.02406-17.
39. Mittal, P., Prasoodanan, P. K. V., Dhakan, D. B., Kumar, S., & Sharma, V. K., (2019). Metagenome of a polluted river reveals a reservoir of metabolic and antibiotic resistance genes. *Environmental Microbiome, 14*(1), 1–12. doi: 10.1186/s40793-019-0345-3.
40. Harkins, C. P., Pichon, B., Doumith, M., Parkhill, J., Westh, H., Tomasz, A., De Lencastre, H., et al., (2017). Methicillin-resistant *Staphylococcus aureus* emerged long before the introduction of methicillin into clinical practice. *Genome Biology, 18*, 130/1–130/11. doi: 10.1186/s13059-017-1252-9.
41. Robinson, D. A., Kearns, A. M., Holmes, A., Morrison, D., Grundmann, H., Edwards, G., O'Brien, F. G., et al., (2005). Re-emergence of early pandemic *Staphylococcus aureus* as a community-acquired meticillin-resistant clone. *Lancet (London, England), 365*(9466), 1256–1258. doi: 10.1016/S0140-6736(05)74814-5.
42. Washer, P., & Joffe, H., (2006). The "hospital superbug": Social representations of MRSA. *Social Science & Medicine, 63*(8), 2141–2152. doi: 10.1016/j.socscimed.2006.05.018.
43. Allan, E. J., Hoischen, C., & Gumpert, J., (2009). Bacterial L-forms. *Advances in Applied Microbiology, 68*, 1–39. doi: 10.1016/S0065-2164(09)01201-5.
44. Kawai, Y., Mercier, R., Mickiewicz, K., Serafini, A., Sorio De, C. L. P., & Errington, J., (2019). Crucial role for central carbon metabolism in the bacterial L-form switch and killing by β-lactam antibiotics. *Nature Microbiology, 4*(10), 1716–1726. doi: 10.1038/s41564-019-0497-3.
45. Errington, J., Mickiewicz, K., Kawai, Y., & Wu, L. J., (2016). L-form bacteria, chronic diseases and the origins of life. *Philosophical Transactions of the Royal Society of London. Series B, Biological Sciences, 371*(1707), 20150494/1–20150494/8. doi: 10.1098/rstb.2015.0494.
46. Kawai, Y., Mercier, R., Wu, L. J., Domínguez-Cuevas, P., Oshima, T., & Errington, J., (2015). Cell growth of wall-free L-form bacteria is limited by oxidative damage. *Current Biology, 25*(12), 1613–1618. doi: 10.1016/j.cub.2015.04.031.
47. López, D., Vlamakis, H., & Kolter, R., (2010). Biofilms. *Cold Spring Harbor Perspectives in Biology, 2*(7), a000398/1–a000398/11. doi: 10.1101/cshperspect.a000398.
48. Aggarwal, S., Stewart, P. S., & Hozalski, R. M., (2015). Biofilm cohesive strength as a basis for biofilm recalcitrance: Are bacterial biofilms overdesigned? *Microbiology Insights, 8*(Suppl 2), 29–32. doi: 10.4137/MBI.S31444.
49. Cutrona, N., Gillard, K., Ulrich, R., Seemann, M., Miller, H. B., & Blackledge, M. S., (2019). From antihistamine to anti-infective: Loratadine inhibition of regulatory PASTA kinases in staphylococci reduces biofilm formation and potentiates β-lactam antibiotics and vancomycin in resistant strains of *Staphylococcus aureus*. *ACS Infectious Diseases, 5*(8), 1397–1410. doi: 10.1021/acsinfecdis.9b00096.
50. Wolcott, R. D., & Ehrlich, G. D., (2008). Biofilms and chronic infections. *Journal of the American Medical Association, 299*(22), 2682–2684. doi: 10.1001/jama.299.22.2682.
51. Oliveira, W. F., Silva, P. M. S., Coelho, L. C. B. B., Silva, R. C. S., Silva, G. M. M., Machado, G., & Correia, M. T. S., (2018). *Staphylococcus aureus* and *Staphylococcus*

epidermidis infections on implants. *The Journal of Hospital Infection, 98*(2), 111–117. doi: 10.1016/j.jhin.2017.11.008.

52. Bush, K., (2010). Chapter 15. Other β-lactam antibiotics. In: Finch, R. G., Greenwood, D., Norrby, S. R., & Whitley, R. J., (eds.), *Antibiotic and Chemotherapy: Anti-Infective Agents and Their Use in Therapy* (9th edn., pp. 226–244). Saunders/Elsevier: Edinburgh.
53. Keshri, V., Diene, S. M., Estienne, A., Dardaillon, J., Chabrol, O., Tichit, L., Rolain, J. M., et al., (2019). An integrative database of β-lactamase enzymes: Sequences, structures, functions, and phylogenetic trees. *Antimicrobial Agents and Chemotherapy, 63*(5), e02319-18/1–e02319-18/8. doi: 10.1128/AAC.02319-18.
54. Chen, Y., Xu, M., Xu, W., Song, H., Hu, L., Xue, S., Zhang, S., et al., (2019). Highly selective and wash-free visualization of resistant bacteria with a relebactam-derived fluorogenic probe. *Chemical Communications (Cambridge, United Kingdom), 55*(67), 9919–9922. doi: 10.1039/c9cc04533c.
55. Naas, T., Oueslati, S., Bonnin, R. A., Dabos, M. L., Zavala, A., Dortet, L., Retailleau, P., & Iorga, B. I., (2017). Beta-lactamase database (BLDB) – structure and function. *Journal of Enzyme Inhibition and Medicinal Chemistry, 32*(1), 917–919. doi: 10.1080/14756366.2017.1344235.
56. Li, J., Li, Z., & Liu, L., (2007). Research advances in PER-type β-lactamases. *Guoji Jianyan Yixue Zazhi, 28*(12), 1104–1106.
57. Das, S., Li, J., Riccobene, T., Carrothers, T. J., Newell, P., Melnick, D., Critchley, I. A., et al., (2019). Dose selection and validation for ceftazidime-avibactam in adults with complicated intra-abdominal infections, complicated urinary tract infections, and nosocomial pneumonia. *Antimicrobial Agents and Chemotherapy, 63*(4), e02187-18/1–e02187-18/17. doi: 10.1128/AAC.02187-18.
58. Hall, B. G., & Barlow, M., (2003). Structure-based phylogenies of the serine β-lactamases. *Journal of Molecular Evolution, 57*(3), 255–260. doi: 10.1007/s00239-003-2473-y.
59. Galleni, M., Lamotte-Brasseur, J., Rossolini, G. M., Spencer, J., Dideberg, O., Frère, J. M., Amicosante, G., et al., (2001). Standard numbering scheme for class B β-lactamases. *Antimicrobial Agents and Chemotherapy, 45*(3), 660–663. doi: 10.1128/AAC.45.3.660-663.2001.
60. Hall, B. G., & Barlow, M., (2005). Revised ambler classification of β-lactamases. *Journal of Antimicrobial Chemotherapy, 55*(6), 1050, 1051. doi: 10.1093/jac/dki130.
61. Yoo, J. H., Huh, D. H., Choi, J. H., Kim, Y. R., Shin, W. S., & Kang, M. W., (1997). Characterization of a constitutive β-lactamase from a *Pseudomonas aeruginosa*: A preliminary report. *Bulletin of the Catholic Research Institutes of Medical Science, Catholic University of Korea, 25*(1), 8–16.
62. Furlan, J. P. R., & Stehling, E. G., (2017). Presence of β-lactamases encoding genes in soil samples from different origins. *Water, Air, & Soil Pollution, 228*(4), 1–6. doi: 10.1007/s11270-017-3318-4.
63. Bush, K., & Jacoby, G. A., (2010). Updated functional classification of β-lactamases. *Antimicrobial Agents and Chemotherapy, 54*(3), 969–976. doi: 10.1128/AAC.01009-09.
64. Jacoby, G. A., Mills, D. M., & Chow, N., (2004). Role of β-lactamases and porins in resistance to ertapenem and other β-lactams in *Klebsiella pneumoniae*. *Antimicrobial Agents and Chemotherapy, 48*(8), 3203–3206. doi: 10.1128/AAC.48.8.3203-3206.2004.
65. Liu, B., & Pop, M., (2009). ARDB-antibiotic resistance genes database. *Nucleic Acids Research, 37*(Database Issue), D443–D447. doi: 10.1093/nar/gkn656.

66. Scaria, J., Chandramouli, U., & Verma, S. K., (2005). Antibiotic resistance genes online (ARGO): A database on vancomycin and beta-lactam resistance genes. *Bioinformation, 1*(1), 5–7. doi: 10.6026/97320630001005.
67. Pal, C., Bengtsson-Palme, J., Rensing, C., Kristiansson, E., & Larsson, D. G. J., (2014). BacMet: Antibacterial biocide and metal resistance genes database. *Nucleic Acids Research, 42*, D737–D743. doi: 10.1093/nar/gkt1252.
68. Danishuddin, M., Hassan, B. M., Kaushal, L., & Khan, A. U., (2013). BLAD: A comprehensive database of widely circulated beta-lactamases. *Bioinformatics, 29*(19), 2515, 2516. doi: 10.1093/bioinformatics/btt417.
69. McArthur, A. G., Waglechner, N., Nizam, F., Yan, A., Azad, M. A., Baylay, A. J., Bhullar, K., et al., (2013). The comprehensive antibiotic resistance database. *Antimicrobial Agents and Chemotherapy, 57*(7), 3348–3357. doi: 10.1128/AAC.00419-13.
70. Srivastava, A., Singhal, N., Goel, M., Virdi, J. S., & Kumar, M., (2014). CBMAR: A comprehensive β-lactamase molecular annotation resource. *Database: The Journal of Biological Databases and Curation,* bau111/1–bau111/8. doi: 10.1093/database/bau111.
71. Galletti, P., & Giacomini, D., (2011). Monocyclic β-lactams: New structures for new biological activities. *Current Medicinal Chemistry, 18*, 4265–4283.
72. Moura, A., Soares, M., Pereira, C., Leitão, N., Henriques, I., & Correia, A., (2009). INTEGRALL: A database and search engine for integrons, integrases and gene cassettes. *Bioinformatics, 25*(8), 1096–1098. doi: 10.1093/bioinformatics/btp105.
73. Joris, B., Ghuysen, J. M., Dive, G., Renard, A., Dideberg, O., Charlier, P., Frere, J. M., et al., (1988). The active-site-serine penicillin-recognizing enzymes as members of the *Streptomyces* R61 *DD*-peptidase family. *Biochemical Journal, 250*(2), 313–324. doi: 10.1042/bj2500313.
74. Zeil, C., Widmann, M., Fademrecht, S., Vogel, C., & Pleiss, J., (2016). Network analysis of sequence-function relationships and exploration of sequence space of TEM β-lactamases. *Antimicrobial Agents and Chemotherapy, 60*(5), 2709–2717. doi: 10.1128/AAC.02930-15.
75. Thai, Q. K., & Pleiss, J., (2010). SHV lactamase engineering database: A reconciliation tool for SHV β-lactamases in public databases. *BMC Genomics, 11*, 563/1–563/8. doi: 10.1186/1471-2164-11-563.
76. Lakin, S. M., Dean, C., Noyes, N. R., Dettenwanger, A., Ross, A. S., Doster, E., Rovira, P., et al., (2017). MEGARes: An antimicrobial resistance database for high throughput sequencing. *Nucleic Acids Research, 45*, D574–D580. doi: 10.1093/nar/gkw1009.
77. Cassettes, Tsafnat, G., Copty, J., & Partridge, S. R., (2011). RAC: Repository of antibiotic resistance. *Database: The Journal of Biological Databases and Curation,* bar054/1–bar054/7. doi: 10.1093/database/bar054.
78. Lee, D. S., Lee, J. M., Kim, S. U., Chang, K. T., & Lee, S. H., (2007). Ceftezole, a cephem antibiotic, is an α-glucosidase inhibitor with *in vivo* anti-diabetic activity. *International Journal of Molecular Medicine, 20*(3), 379–383. doi: 10.3892/ijmm.20.3.379.
79. Banik, B. K., (2013). Synthesis and biological studies of novel β-lactams. In: Brahmachari, G., (ed.), *Chemistry and Pharmacology of Naturally Occurring Bioactive Compounds* (pp. 31–72).
80. Pitlik, J., Bagossi, P., Jeko, J., & Tözsér, J., (1996). Synthesis of cephalosporin oligopeptides as potential proteinase inhibitors. *Pharmazie, 51*(10), 700–704.
81. Sperka, T., Pitlik, J., Bagossi, P., & Tözsér, J., (2005). Beta-lactam compounds as apparently uncompetitive inhibitors of HIV-1 protease. *Bioorganic & Medicinal Chemistry Letters, 15*(12), 3086–3090. doi: 10.1016/j.bmcl.2005.04.020.

82. Borazjani, N., Sepehri, S., Behzadi, M., Jarrahpour, A., Rad, J. A., Sasanipour, M., Mohkam, M., et al., (2019). Three-component synthesis of chromeno β-lactam hybrids for inflammation and cancer screening. *European Journal of Medicinal Chemistry, 179*, 389–403. doi: 10.1016/j.ejmech.2019.06.036.
83. Aviles, E., & Rodriguez, A. D., (2010). Monamphilectine A, a potent antimalarial β-lactam from marine sponge *Hymeniacidon sp*: Isolation, structure, semisynthesis, and bioactivity. *Organic Letters, 12*(22), 5290–5293. doi: 10.1021/ol102351z.
84. Alborz, M., Jarrahpour, A., Pournejati, R., Karbalaei-Heidari, H. R., Sinou, V., Latour, C., Brunel, J. M., et al., (2018). Synthesis and biological evaluation of some novel diastereoselective benzothiazole β-lactam conjugates. *European Journal of Medicinal Chemistry, 143*, 283–291. doi: 10.1016/j.ejmech.2017.11.053.
85. Ebrahimi, E., Jarrahpour, A., Heidari, N., Sinou, V., Latour, C., Brunel, J. M., Zolghadr, A. R., & Turos, E., (2016). Synthesis and antimalarial activity of new nanocopolymer β-lactams and molecular docking study of their monomers. *Medicinal Chemistry Research, 25*(2), 247–262. doi: 10.1007/s00044-015-1477-7.
86. Jarrahpour, A., Ebrahimi, E., Sinou, V., Latour, C., & Brunel, J. M., (2014). Diastereoselective synthesis of potent antimalarial Cis-β-lactam agents through a [2 + 2] cycloaddition of chiral imines with a chiral ketene. *European Journal of Medicinal Chemistry, 87*, 364–371. doi: 10.1016/j.ejmech.2014.09.077.
87. Jarrahpour, A., Ebrahimi, E., Khalifeh, R., Sharghi, H., Sahraei, M., Sinou, V., Latour, C., & Brunel, J. M., (2012). Synthesis of novel β-lactams bearing an anthraquinone moiety, and evaluation of their antimalarial activities. *Tetrahedron, 68*(24), 4740–4744. doi: 10.1016/j.tet.2012.04.011.
88. Wu, J. P., (2015). Progress of bioactivities other than antibacterial of β-lactam derivatives. *Zhongguo Shenghua Yaowu Zazhi, 35*(10), 161–167.
89. Gomathy, S., Gowramma, B., Yadav, S., Antony, A. S., Jubie, S., Kalirajan, R., & Elango, K., (2012). Synthesis and biological evaluation of some novel N-(3-chloro-2-oxo-4-substituted azetidin-1-yl)-2-(naphthalen-1-yl) acetamide derivatives as antiparkinson's agent. *Journal of Pharmacy Research, 5*(11), 5120–5122.
90. Kumar, S., Kaur, H., Saxena, K. K., Sharma, M., Vishwakarma, P., & Kumar, A., (2010). Synthesis and antiparkinsonian activity of some new adamantyl thiazolidinonyl/azetidinonyl indole derivatives. *Indian Journal of Chemistry, Section B: Organic Chemistry Including Medicinal Chemistry, 49B* (10), 1398–1405.
91. Onyegeme-Okerenta, B. M., Okochi, V. I., & Chinedu, S. N., (2014). Antithrombotic and anticoagulant properties of *Penicillium chrysogenum* (PCL 50) culture extracts. *Journal of Pharmaceutical and Scientific Innovation, 3*(1), 20–24. doi: 10.7897/2277-4572.03199.
92. Jaganath, D., Lamichhane, G., & Shah, M., (2016). Carbapenems against mycobacterium tuberculosis: A review of the evidence. *The International Journal of Tuberculosis and Lung Disease: The Official Journal of the International Union against Tuberculosis and Lung Disease, 20*(11), 1436–1447. doi: 10.5588/ijtld.16.0498.
93. Wang, X., Gu, X., Zhang, C., Zhao, F., & Deng, K., (2020). Structural and biochemical analyses of the LdtMt2-panipenem adduct provide new insights into the effect of the 1-β-methyl group on carbapenems. *Biochemical and Biophysical Research Communications, 523*(1), 6–9. doi: 10.1016/j.bbrc.2019.12.020.
94. Horita, Y., Maeda, S., Kazumi, Y., & Doi, N., (2014). *In vitro* susceptibility of mycobacterium tuberculosis isolates to an oral carbapenem alone or in combination with β-lactamase inhibitors. *Antimicrobial Agents and Chemotherapy, 58*(11), 7010–7014. doi: 10.1128/AAC.03539-14.

95. Zhao, G., Miller, M. J., Franzblau, S., Wan, B., & Moellmann, U., (2006). Syntheses and studies of quinolone-cephalosporins as potential anti-tuberculosis agents. *Bioorganic & Medicinal Chemistry Letters, 16*(21), 5534–5537. doi: 10.1016/j.bmcl.2006.08.045.
96. Majewski, M. W., Watson, K. D., Cho, S., Miller, P. A., Franzblau, S. G., & Miller, M. J., (2016). Syntheses and biological evaluations of highly functionalized hydroxamate containing and *N*-methylthio monobactams as anti-tuberculosis and β-lactamase inhibitory agents. *MedChemComm, 7*(1), 141–147. doi: 10.1039/C5MD00340G.
97. Pettit, G. R., Singh, S. B., Boyd, M. R., Hamel, E., Pettit, R. K., Schmidt, J. M., & Hogan, F., (1995). Antineoplastic agents. 291. Isolation and synthesis of combretastatins α-4, α-5, and α-6. *Journal of Medicinal Chemistry, 38*(10), 1666–1672. doi: 10.1021/jm00010a011.
98. Malebari, A. M., Fayne, D., Nathwani, S. M., O'Connell, F., Noorani, S., Twamley, B., O'Boyle, N. M., et al., (2020). β-Lactams with antiproliferative and antiapoptotic activity in breast and chemoresistant colon cancer cells. *European Journal of Medicinal Chemistry, 189*, 112050/1–112050/25. doi: 10.1016/j.ejmech.2020.112050.
99. Frezza, M., Garay, J., Chen, D., Cui, C., Turos, E., & Dou, Q. P., (2008). Induction of tumor cell apoptosis by a novel class of *N*-thiolated β-lactam antibiotics with structural modifications at N1 and C3 of the lactam ring. *International Journal of Molecular Medicine, 21*(6), 689–695. doi: 10.3892/ijmm.21.6.689.
100. Smith, D. M., Kazi, A., Smith, L., Long, T. E., Heldreth, B., Turos, E., & Dou, Q. P., (2002). A novel β-lactam antibiotic activates tumor cell apoptotic program by inducing DNA damage. *Molecular Pharmacology, 61*(6), 1348–1358. doi: 10.1124/mol.61.6.1348.
101. Banik, B. K., Banik, I., & Becker, F. F., (2010). Asymmetric synthesis of anticancer β-lactams via Staudinger reaction: Utilization of chiral ketene from carbohydrate. *European Journal of Medicinal Chemistry, 45*(2), 846–848. doi: 10.1016/j.ejmech.2009.11.024.
102. Rawlings, N. D., & Barrett, A. J., (1994). Families of serine peptidases. *Methods in Enzymology, 244(Proteolytic Enzymes: Serine and Cysteine Peptidases),* 19–61.
103. Borthwick, A. D., Weingarten, G., Haley, T. M., Tomaszewski, M., Wang, W., Hu, Z., Bedard, J., et al., (1998). Design and synthesis of monocyclic β-lactams as mechanism-based inhibitors of human cytomegalovirus protease. *Bioorganic & Medicinal Chemistry Letters, 8*(4), 365–370. doi: 10.1016/S0960-894X(98)00032-8.
104. Déziel, R., & Malenfant, E., (1998). Inhibition of human cytomegalovirus protease no with monocyclic β-lactams. *Bioorganic & Medicinal Chemistry Letters, 8*(11), 1437–1442. doi: 10.1016/S0960-894X(98)00240-6.
105. Yoakim, C., Ogilvie, W. W., Cameron, D. R., Chabot, C., Guse, I., Hache, B., Naud, J., et al., (1998). β-Lactam derivatives as inhibitors of human cytomegalovirus protease. *Journal of Medicinal Chemistry, 41*(15), 2882–2891. doi: 10.1021/jm980131z.
106. Yoakim, C., Ogilvie, W. W., Cameron, D. R., Chabot, C., Grand-Maître, C., Guse, I., Haché, B., et al., (1998). Potent β-lactam inhibitors of human cytomegalovirus protease. *Antiviral Chemistry & Chemotherapy, 9*(5), 379–387. doi: 10.1177/095632029800900502.
107. Corredor, M., Bujons, J., Orzaez, M., Sancho, M., Perez-Paya, E., Alfonso, I., & Messeguer, A., (2013). Optimizing the control of apoptosis by amide/triazole isosteric substitution in a constrained peptoid. *European Journal of Medicinal Chemistry, 63*, 892–896. doi: 10.1016/j.ejmech.2013.03.004.
108. Aoyama, Y., Uenaka, M., Konoike, T., Iso, Y., Nishitani, Y., Kanda, A., Naya, N., & Nakajima, M., (2000). Synthesis and structure-activity relationships of a new class of

1-oxacephem-based human chymase inhibitors. *Bioorganic & Medicinal Chemistry Letters, 10*(21), 2397–2401. doi: 10.1016/S0960-894X(00)00488-1.

109. Burnett, D. A., (2004). β-Lactam cholesterol absorption inhibitors. *Current Medicinal Chemistry, 11*(14), 1873–1887. doi: 10.2174/0929867043364865.
110. Takabatake, T., Yoneda, T., Otsuka, J., Kagawa, N., & Toyota, M., (2020). Artificial intelligence-designed stereoselective one-pot synthesis of trans-β-lactams and its application to cholesterol absorption inhibitor SCH 47949 synthesis. *Heterocycles, 100*(1), 60–84. doi: 10.3987/com-19-14130.
111. Takabatake, T., Yoneda, T., Otsuka, J., Kagawa, N., & Toyota, M., (2019). Artificial intelligence designed drug synthesis: One-pot preparation of trans β-lactams and application to cholesterol absorption inhibitor SCH 47949 synthesis. *Tetrahedron Letters, 60*(34), 150942/1–150942/4. doi: 10.1016/j.tetlet.2019.07.033.
112. Isoda, M., Sato, K., Kunugi, Y., Tokonishi, S., Tarui, A., Omote, M., Minami, H., & Ando, A., (2016). Rh-catalyzed reductive mannich-type reaction and its application towards the synthesis of (±)-Ezetimibe. *Beilstein Journal of Organic Chemistry, 12*, 1608–1615. doi: 10.3762/bjoc.12.157.
113. Drazic, T., Molcanov, K., Sachdev, V., Malnar, M., Hecimovic, S., Patankar, J. V., Obrowsky, S., et al., (2014). Novel amino-β-lactam derivatives as potent cholesterol absorption inhibitors. *European Journal of Medicinal Chemistry, 87*, 722–734. doi: 10.1016/j.ejmech.2014.10.014.
114. Wang, Y., Zhang, H., Qian, H., & Huang, W., (2011). Synthesis and biological evaluation of 2-azetidinone derivatives as cholesterol absorption inhibitors. *Zhongguo Yaoke Daxue Xuebao, 42*(3), 213–219.
115. He, X. L., Xu, X. X., Zheng, L. Y., & Bai, X., (2011). Merger of Ezetimibe and Statin: A potential dual inhibitor. *Chemical Research in Chinese Universities, 27*(2), 232–236.
116. Pandey, G., Dumbre, S. G., Khan, M. I., Shabab, M., & Puranik, V. G., (2006). A β-lactam-azasugar hybrid as a competitive potent galactosidase inhibitor. *Tetrahedron Letters, 47*(45), 7923–7926. doi: 10.1016/j.tetlet.2006.09.005.
117. Nuzzi, A., Fiasella, A., Ortega, J. A., Pagliuca, C., Ponzano, S., Pizzirani, D., Bertozzi, S. M., et al., (2016). Potent α-amino-β-lactam carbamic acid ester as NAAA inhibitors. Synthesis and structure-activity relationship (SAR) studies. *European Journal of Medicinal Chemistry, 111*, 138–159. doi: 10.1016/j.ejmech.2016.01.046.
118. Petracca, R., Ponzano, S., Bertozzi, S. M., Sasso, O., Piomelli, D., Bandiera, T., & Bertozzi, F., (2017). Progress in the development of β-lactams as *N*-acylethanolamine acid amidase (NAAA) inhibitors: Synthesis and SAR study of new, potent N-O-substituted derivatives. *European Journal of Medicinal Chemistry, 126*, 561–575. doi: 10.1016/j.ejmech.2016.11.039.
119. Anusha, K., Kumar, Y. P., Prasad, M. V., Raju, V. B. M., & Gopinath, C., (2015). A review on 2-azetedinones. *Journal of Global Trends in Pharmaceutical Sciences, 6*(1), 2388–2402.
120. Veinberg, G., Potorocina, I., & Vorona, M., (2014). Recent trends in the design, synthesis and biological exploration of β-lactams. *Current Medicinal Chemistry, 21*(4), 393–416. doi: 10.2174/09298673113206660268.

CHAPTER 3

Synthetic Methods of β-Lactams

3.1 BRIEF INTRODUCTION

Due to the importance of β-lactams, particularly for their applications as antibiotics, many synthetic strategies and methods have been developed and reviewed already [1–20]. From the retrosynthetic point of view, β-lactams can be formed via two main strategies, i.e., cycloaddition and cyclization. The primary approach of cycloaddition is [2+2] cycloaddition, which can be further classified into Staudinger's ketene-imine cycloadditions, Wolff rearrangement and subsequent reaction with imines, ester enolate-imine cycloadditions, alkene-isocyanate cycloadditions, Kinugasa reaction, and carbonylative cycloaddition. Regarding the cyclization approach, the monocyclic azetidin-2-ones can be accomplished via the formation of N_1-C_2, N_1-C_4, C_2-C_3, and C_3-C_4 bonds. The cyclization method involving the formation of the N_1-C_2 bond employs *N*-protected or unprotected β-amino esters, as well as aziridines as substrates. The cyclization method involving the formation of the C_3-C_4 bond is represented by intramolecular Michael addition, photolysis of α,β-unsaturated amides, and *N,N*-disubstituted α-halo amides. Whereas the cyclization methods involving the formation of the N_1-C_4 bond include intramolecular amidation, carbonylation of allylamines, and cyclization of hydroxamate esters. The representative approaches for making azetidin-2-ones are summarized in Scheme 3.1, where the cycloaddition approaches are highlighted in blue color. It should be pointed out that many other reaction approaches have not been included in Scheme 3.1 yet.

3.2 FORMATION OF Β-LACTAMS THROUGH CYCLOADDITIONS

According to the Woodward-Hoffman orbital-symmetry rules [21], Fukui's frontier molecular orbital theory [22] as well as Dewar-Zimmerman

The Chemistry and Biology of Beta-Lactams. Zerong Wang, PhD

rule [23], regular [2+2] cycloaddition is thermally forbidden but photochemically allowed. However, under special conditions, the thermal [2+2] cycloadditions also take place readily, such as the cycloaddition involving chemical species with consecutive double bond moiety, like ketenes (e.g., Staudinger's [2+2] cycloaddition) and isocyanates, and the [2+2] cycloaddition involving heavy atoms, such as the Wittig reaction (phosphorus involved) [24], Peterson olefination (silicon involved) and Tebbe reaction (titanium involved) [25].

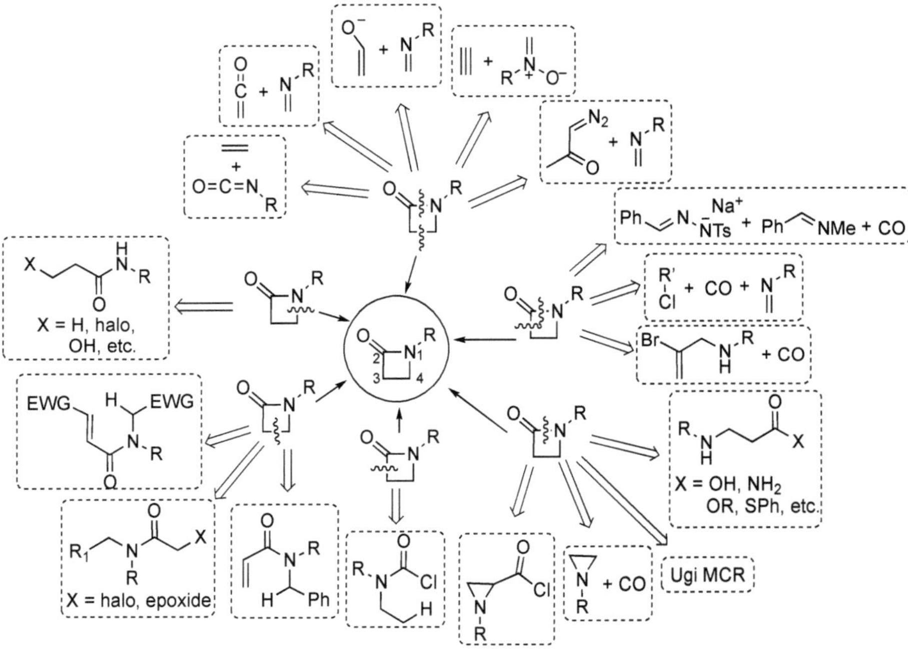

SCHEME 3.1 Potential synthetic routes to form β-lactams

3.2.1 STAUDINGER'S [2+2] CYCLOADDITION

In fact, Staudinger's [2+2] cycloaddition is the first example to give β-lactams, reported by Hermann Staudinger in 1907 for the reaction of the Schiff base of aniline and benzaldehyde with diphenylketene (Scheme 3.2) [26, 27], which is much earlier than the discovery of penicillin by Fleming. Since then, this reaction has been greatly explored and extended to a variety of imines and ketenes. Imines are the condensation products of primary amines and aldehydes (and/or ketones), where the imines are generally known as the

Schiff bases if an alkyl or aryl group rather than hydrogen is attached to the nitrogen atom of the imines [28].

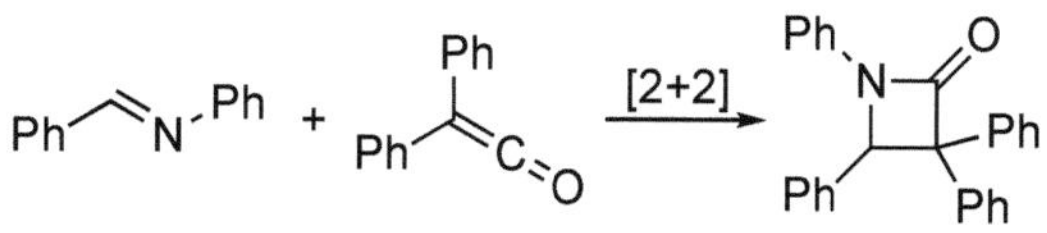

SCHEME 3.2 The first [2+2] cycloaddition to yield β-lactam

For the general [2+2] cycloaddition, such as the one between ketene and alkene, more and more experimental evidence supports that this reaction occurs via a stepwise mechanism rather than a concerted mechanism. These pieces of evidence include the formation of both *cis*- and *trans*-cycloadducts from *trans*-alkenes, a very large polar effect, and a kinetic isotopic effect [29–31]. In contrast, the experimental proofs in favor of a concerted mechanism for the [2+2] cycloaddition are: (a) the higher reactivity of *cis*-alkenes than the *trans*-alkenes in the cycloaddition; and (b) the stereoselectivity for the addition of ketene to *cis*-alkenes [32, 33]. However, for the Staudinger reaction, i.e., the [2+2] cycloaddition between ketene and imine, the case might be different as many imines are in *trans*-configuration. In this case, the Staudinger reaction may be regarded as a cascade reaction involving the nucleophilic addition of imine to the ketene and subsequent Mannich reaction (i.e., the intramolecular addition of the enolate moiety to iminium component to form β-lactam), as shown in Scheme 3.3 [34, 35].

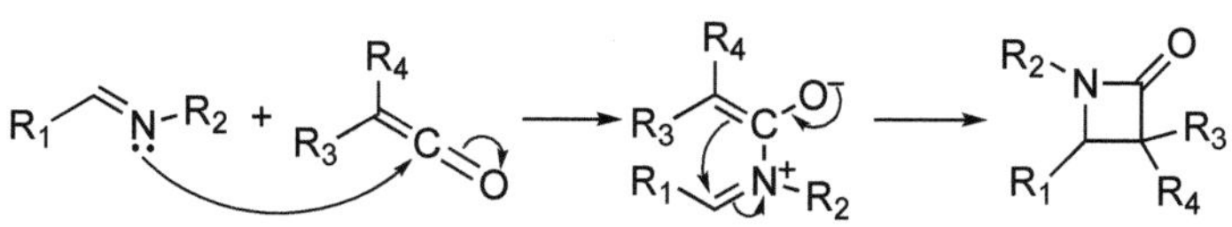

SCHEME 3.3 The plausible mechanism for the Staudinger reaction

In order to form β-lactams of good stereoselectivity, chiral amines have been applied as a catalyst for the reaction between ketene and imine. For example, the reaction between 1-ethyl-1-phenyl-ketene (or 1-isobutyl-1-phenyl-ketene) and (*Z*)-*N*-alkylidene-4-methylbenzenesulfonamide (or (*Z*)-*N*-arylidene-4-methylbenzenesulfonamide) in toluene at room temperature in the presence of 0.1 equivalent chiral ferrocene catalyst formed from 1,2,3,4,5-pentamethyl-cyclopenta-2,4-dien-1-ide and 4-(pyrrolidin-1-yl)-7*H*-cyclopenta[*b*]pyridin-7-ide affords *cis*-β-lactam in yield of 88–98%, with 89–98% ee and > 8:1 diastereoselectivity, as shown in Scheme 3.4 [36]. Quite interestingly, reaction of the same ketene with (*Z*)-*N*-alkylidene-1,1,1-trifluoromethanesulfonamide or (*Z*)-*N*-arylidene-1,1,1-trifluoromethanesulfonamide in the presence of the same catalyst leads to the formation of β-lactams in favor of *trans*-selectivity [37].

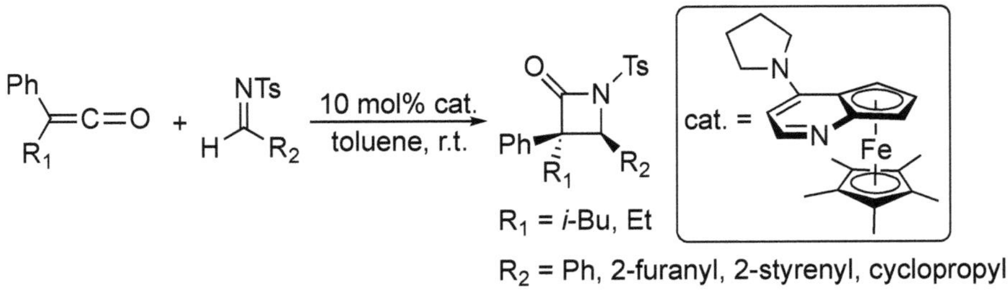

SCHEME 3.4 A chiral ferrocene catalyzed Staudinger reaction

In addition to the chiral ferrocene-based pyridine in the promotion of the reaction between ketenes and imines, asymmetric *N*-heterocyclic carbenes (NHC) have also been developed for the purpose of enhancing the stereoselectivity of β-lactams. For example, the reaction of 1,1-diphenylketene and (*E*)-*N*-arylidene-4-methylbenzenesulfonamide in the presence of 0.1 equivalent of either (3a*R*, 7a*R*)-1,3-dibenzyl-3a,4,5,6,7,7a-hexahydro-1*H*-benzo[*d*]imidazol-3-ium tetra-fluoroborate or (5a*R*,10b*S*)-2-phenyl-5a,10b-dihydro-4*H*,6*H*-indeno[2,1-*b*][1,2,4]-triazolo[4,3-*d*][1,4]oxazin-2-ium tetrafluoroborate and strong base KHMDS in ether at room temperature affords β-lactams in favor of the *S*-configuration except for Ar = 2-furanyl, with good to excellent yield (79–96%) and moderate stereoselectivity (Scheme 3.5) [38]. Concurrently, a different chiral NHC has been developed for the reaction of 1-ethyl-1-phenyl-ketene and *tert*-butyl (*E*)-arylidenecarbamate, in which the NHC is generated by means of the treatment of (*S*)-5-(((*tert*-butyldimethylsilyl)oxy)diphenylmethyl)-2-phenyl-6,7-dihydro-5*H*-pyrrolo[2,1-*c*][1,2,4]triazol-2-ium tetrafluoroborate with cesium carbonate in THF at room temperature. The reaction under this condition gives *cis*-β-lactams in preference, with 90–99% ee, as shown in Scheme 3.6 [39].

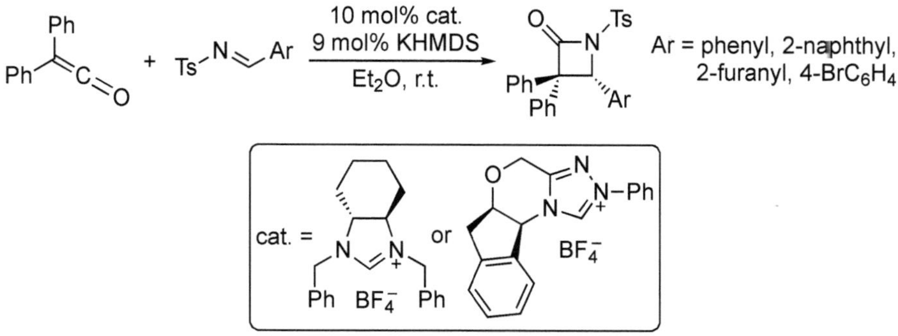

SCHEME 3.5 A chiral NHC catalyzed Staudinger reaction involving diphenylketene

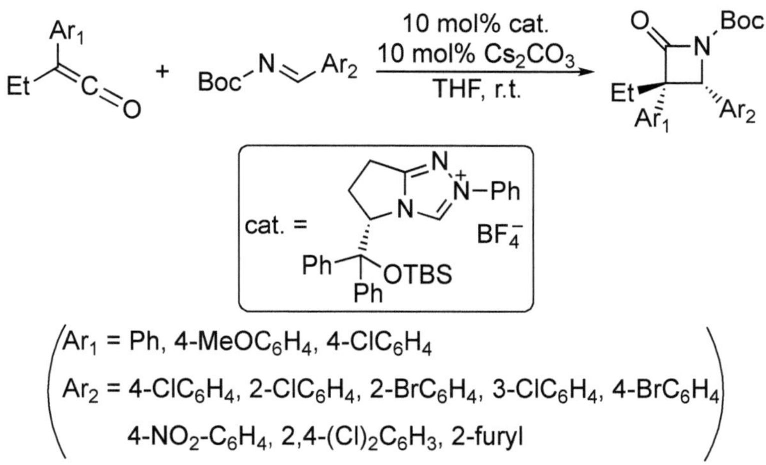

SCHEME 3.6 A chiral NHC catalyzed Staudinger reaction involving 1-aryl-1-ethyl-ketene

However, due to the high reactivity and instability of ketenes, in most cases, ketenes used in β-lactam preparation via the Staudinger reaction are generated in situ or prepared freshly before the reaction. Normally, ketenes are prepared by the weak base (e.g., Et_3N) treatment of acyl chlorides due to the relative acidity of the α-hydrogen in the acyl chlorides as of the synergistic electron-withdrawing effect of both acyl and chloro groups [40]. For example, the first synthetic penicillin (i.e., methyl 6-(2,5-dioxo-pyrrolidin-1-yl)-3,3-dimethyl-7-oxo-5-phenyl-4-thia-1-aza-bicyclo[3.2.0] heptane-2-carboxylate, classified as a penam) was prepared by a [2+2] cycloaddition of 1-(2-oxovinyl)pyrrolidin-2,5-dione, *in situ* formed from 2-(2,5-dioxopyrrolidin-1-yl)acetyl chloride (by treatment with Et_3N), and methyl 5,5-dimethyl-2-phenyl-4,5-dihydrothiazole-4-carboxylate in 1950.

This compound was selectively hydrolyzed in the presence of base followed by esterification with diazomethane to afford methyl 5-phenyl-(2-carbomethoxy-ethyl)-penicillinate, i.e., methyl 6-(4-methoxy-4-oxobutanamido)-3,3-dimethyl-7-oxo-5-phenyl-4-thia-1-azabicyclo[3.2.0]heptane-2-carboxylate (Scheme 3.7) [41].

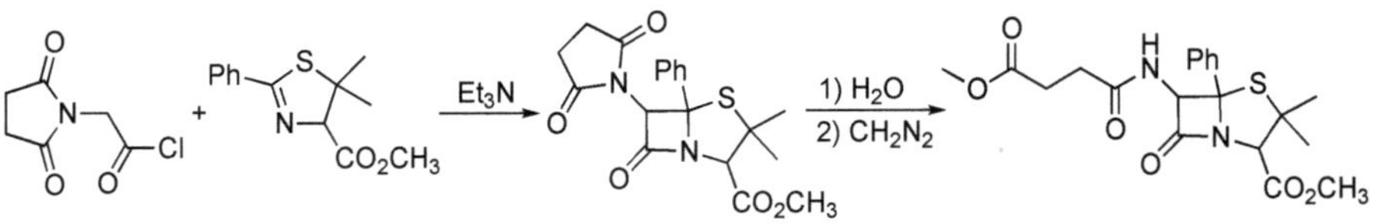

SCHEME 3.7 The first synthetic penicillin via the Staudinger reaction

Similarly, 3-(3-*t*-butoxyl)succinimidyl acetal chloride, prepared from malic acid through a series of transformations, reacted with Schiff base to afford eight 3-(3-*t*-butoxyl)succinimidyl monocyclic β-lactam derivatives, as illustrated in Scheme 3.8 [42]. Extension of this particular reaction series indicates that the configurations at C_3 and C_4 are in *trans*-relationship, presumably involving a stepwise mechanism where the nitrogen of imine adds to the *in situ* generated ketene first, and isomerization occurs prior to the cyclization to yield the β-lactams [43].

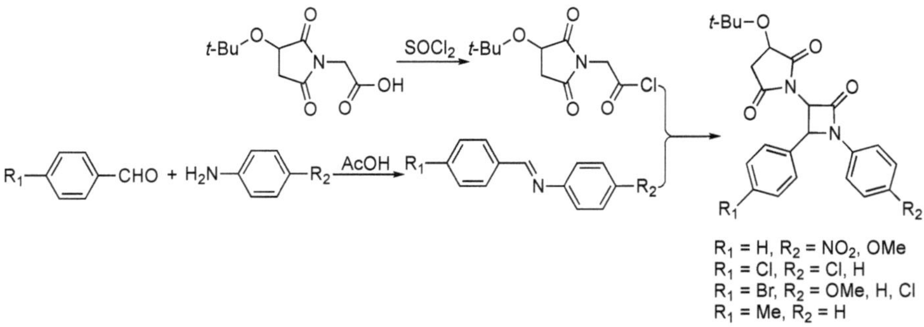

SCHEME 3.8 Synthesis of pyrrolidin-2,5-diones containing the β-lactam moiety

According to the molecular orbital theory, if the [2+2] cycloaddition is a concerted process, the reaction should be stereospecific; however, if this reaction involves a stepwise mechanism, then the stereospecificity is not fully guaranteed. While earlier work suggests that the diastereoselectivity of β-lactams obtained from the chiral Schiff bases deriving from optically active amines and achiral aldehydes varies, the configuration of β-lactams obtained from chiral Schiff bases deriving from achiral amines and optically active aldehydes is totally stereospecific. This has been demonstrated by the [2+2] cycloaddition of phenoxyacetyl chloride with a series of chiral Schiff bases derived from esters of *L*-valine and *L*-alanine and achiral aldehydes (e.g., 3,4-dimethoxybenzaldehyde) in the presence of Et_3N, where the obtained *cis*-β-lactams are of poor diastereomeric excess (de 6–10%) [44]. Evans, Moore, and Sheehan have studied the substituent effects on the stereoselectivity of β-lactams resulting from ketenes of different groups. The "Bose-Evans ketenes" possessing strong electron-donating substituents R (e.g., *O*-alkyl, *O*-aryl, and *N*-alkylaryl) favor the formation of *cis*-β-lactams due to the large rate constant of the direct ring closure. The "Moore ketenes" having very weak electron-donating substituents R (such as *S*-alkyl, *S*-aryl, alkyl, and aryl) primarily lead to the formation of *trans*-β-lactams due to the slow reaction of direct ring closure, rendering the opportunity of isomerization prior to the ring closure. For comparison, the "Sheehan ketenes" lead to β-lactams of complex stereochemical outcomes due to a moderate rate constant of the direct ring closure [45]. In addition, the Staudinger [2+2] cycloaddition under either microwave or photo-irradiation seems to afford products, of which the stereoselectivity is similar to the ones obtained under regular thermal conditions. This is because the influence of the photoirradiation (under a medium-pressure or high-pressure mercury lamp) on photochemical Staudinger reactions is limited, whereas the stereoselectivity is predominately controlled by structural features of ketenes and imines, not the reaction conditions [45].

Recently, a solventless preparation of imine has been carried out by grinding a mixture of mono-methoxy-benzaldehyde (up to tri-methoxy-benzaldehyde) and 1-aminoindane or 1,2,3,4-tetrahydro-1-naphthylamine in a mortar without further purification. The resulting imines are then treated with 3-phenylpropionyl chloride as a ketene precursor and triethylamine as a base at high temperature in order to accelerate the isomerization of the imine moiety in the zwitterionic intermediate and to override the direct ring-closure leading to *cis*-isomer, to diastereoselectively afford β-lactams of desired *trans*-C_3,C_4-configuration. Analysis of the product configuration indicates that the reaction selectivity is dependent on the chiral auxiliary moiety and the number and position of methoxy groups in benzaldehyde component, for which 1-aminoindane is a more effective auxiliary group than 1,2,3,4-tetrahydro-1-naphthylamine except in the case of imine formed from 3,4-dimethoxy-benzaldehyde. Also, imine prepared from either *p*-methoxy- or *m*-methoxy-benzaldehyde leads to the β-lactam of reasonable selectivity, whereas the imine formed from *o*-methoxy-benzaldehyde yields β-lactam of low selectivity. The best selectivity (52%) was achieved with imine derived from 3,4,5-trimethoxy-benzaldehyde and 1,2,3,4-tetrahydro-1-naphthylamine [46].

When 2-(benzyloxy)acetyl chloride or 2-phenoxyacetyl chloride was treated with Et_3N at room temperature, the resulting ketene was allowed to undergo the [2+2] cycloaddition with (*S*,*E*)-1-(2,2-dimethyl-1,3-dioxolan-4-yl)-*N*-(prop-2-yn-1-yl)methanimine, (*S*,*E*)-*N*-(but-3-yn-1-yl)-1-(2,2-dimethyl-1,3-dioxolan-4-yl)methanimine, (*S*,*E*)-1-(2,2-dimethyl-1,3-dioxolan-4-yl)-*N*-(pent-4-yn-1-yl)methanimine or (*S*,*E*)-*N*-allyl-1-(2,2-dimethyl-1,3-dioxolan-4-yl)methanimine in benzene at room temperature to afford the corresponding β-lactam. After subsequent several transformations, β-lactam-fused tricyclic molecules can be obtained involving a Pauson-Khand reaction with $Co_2(CO)_8$. For example, (3*R*,4*S*)-3-(benzyloxy)-1-(but-3-yn-1-yl)-4-((*S*)-2,2-dimethyl-1,3-dioxolan-4-yl) azetidin-2-one from the [2+2] cycloaddition of (*S*,*E*)-*N*-(but-3-yn-1-yl)-1-(2,2-dimethyl-1,3-dioxolan-4-yl)methanimine was hydrolyzed with *p*-toluenesulfonic acid in aqueous THF to give (3*R*,4*S*)-3-(benzyloxy)-1-(but-3-yn-1-yl)-4-((*S*)-1,2-dihydroxyethyl)azetidin-2-one, which was then treated with di(1*H*-imidazol-1-yl)methanethione in THF under refluxing followed by thermal treatment with triethyl phosphite, resulting (3*R*,4*S*)-3-(benzyloxy)-1-(but-3-yn-1-yl)-4-vinylazetidin-2-one. Upon treatment with $Co_2(CO)_8$ in toluene, and further heated in toluene, the β-lactam-fused tricyclic molecule, i.e., (1*R*,8a*S*,8b*S*)-1-(benzyloxy)-1,4,5,8,8a,8b-hexahydroazeto[1,2-*a*]-cyclopenta[*c*]pyridine-2,7-dione was obtained in high yield, as illustrated in Scheme 3.9 [47]. A series of penam and penem analogs have been prepared involving [2+2] cycloaddition of the *in situ* generated ketenes and imines and halogen-induced sulfide cyclization, as displayed in Scheme 3.10. In this practice, 2-(benzylthio)acetyl chloride (or 2-diethylthio)acetyl chloride (not shown) is allowed to react with (*E*)-*N*-benzyl-3-phenylprop-2-yn-1-imine or (*E*)-*N*-benzyl-3-methyl-but-2-en-1-imine (or other imine derivatives) in the presence of Et_3N in CH_2Cl_2 to create the corresponding β-lactams, such as 1-benzyl-3-(benzylthio)-4-(phenyl-ethynyl)azetidin-2-one and 1-benzyl-3-(benzylthio)-4-(2-methylprop-1-en-1-yl)-azetidin-2-one. The former is then treated with iodine in CH_2Cl_2 to yield (1*S*,5*S*)-6-benzyl-4-iodo-3-phenyl-2-thia-6-azabicyclo[3.2.0]hept-3-en-7-one, and the latter is converted into (1*S*,5*S*)-6-benzyl-4-bromo-3,3-dimethyl-2-thia-6-azabi-cyclo[3.2.0]heptan-7-one with Br_2 in CH_2Cl_2, as shown in Scheme 3.10, respectively [48].

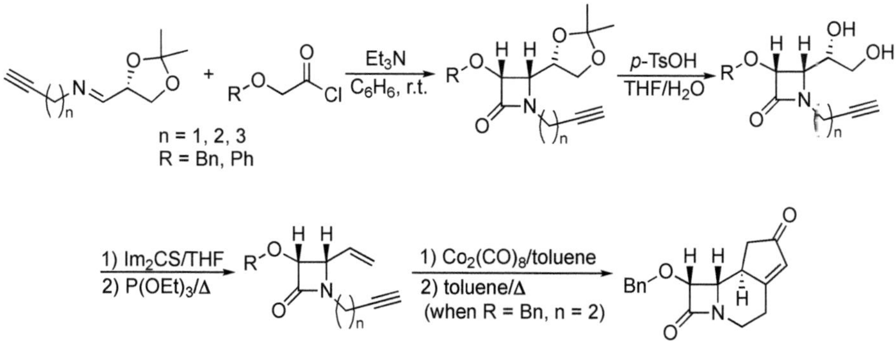

SCHEME 3.9 The synthesis of (1*R*,8a*S*,8b*S*)-1-(benzyloxy)-1,4,5,8,8a,8b-hexahydroazeto[1,2-a]cyclopenta[c]pyridin-2,7-dione

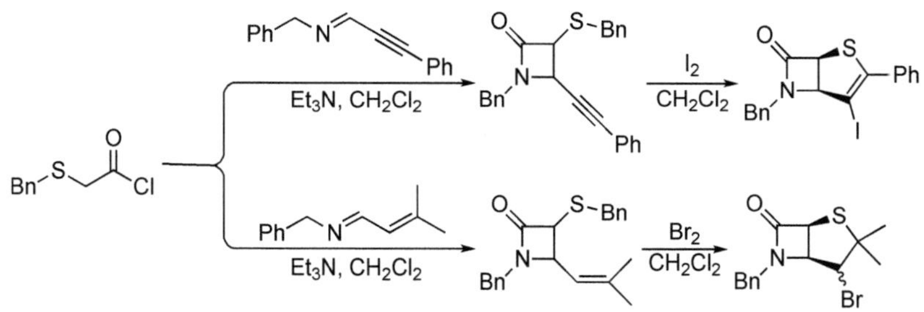

SCHEME 3.10 Preparation of (1*S*,5*S*)-6-benzyl-4-iodo-3-phenyl-2-thia-6-azabicyclo[3.2.0]hept-3-en-7-one and (1*S*,5*S*)-6-benzyl-4-bromo-3,3-dimethyl-2-thia-6-azabicyclo[3.2.0]heptan-7-one from 2-(benzylthio)acetyl chloride

The Staudinger reaction has been extended to a multi-component reaction (MCR) of dialdehyde (e.g., pyridine-2,6-dicarbaldehyde and terephthalaldehyde) and diamine (e.g., 2,2'-(ethane-1,2-diylbis(oxy))bis(ethan-1-amine) and 1,3-phenylenedimethanamine) in CH_2Cl_2 followed by addition of Et_3N and methoxyacetyl chloride at -50°C to form macrocyclic compounds with simultaneous creation of four β-lactam units [49].

This similar approach has been applied to simultaneously synthesize amino-protected and carboxy-activated forms (e.g., by means of NCA functionality) of polyhydroxylated α-amino acids without the necessity to prepare each individual α-amino acid prior to connecting them. For example, the diimine, i.e., (1*E*,1'*E*)-1,1'-((4*R*,5*R*)-2,2-dimethyl-1,3-dioxolane-4,5-diyl)bis(*N*-benzylmethanimine) was allowed to react with 1-benzoxyketene formed from 2-(benzyloxy)-acetyl chloride and triethylamine, followed by Pd/C catalyzed hydrogenation to afford 40% of (3*R*,3'*R*,4*R*,4'*R*)-4,4'-((4*R*,5*R*)-2,2-dimethyl-1,3-dioxolane-4,5-diyl)-

bis(1-benzyl-3-hydroxyazetidin-2-one). Upon oxidation with sodium hypochlorite in the presence of KBr and catalytic amount of TEMPO (2,2,6,6-tetramethylpiperidine 1-oxyl radical), the corresponding *N*-carboxyanhydride (NCA), i.e., (4*S*,4'*S*)-4,4'-((4*R*,5*R*)-2,2-dimethyl-1,3-dioxolane-4,5-diyl) bis(3-benzyloxazolidin-2,5-dione) was obtained in 98% yield. The NCA is then coupled with methyl *L*-phenylalaninate in dichloromethane to give 80% of dimethyl 2,2'-(((2*S*,2'*S*)-2,2'-((4*R*,5*R*)-2,2-dimethyl-1,3-dioxolane-4,5-diyl)bis(2-(benzylamino) acetyl))bis-(azanediyl))-(2*S*,2'*S*)-bis(3-phenylpropanoate), as illustrated in Scheme 3.11 [50]. Several other derivatives have been prepared in a similar manner. In addition, the conversion of α-benzyloxy-β-lactam into α-amino acid based NCA with NaOCl and TEMPO has been extended to the preparation of peptides containing unusual β,γ-dihydroxyl-α-amino acid [51].

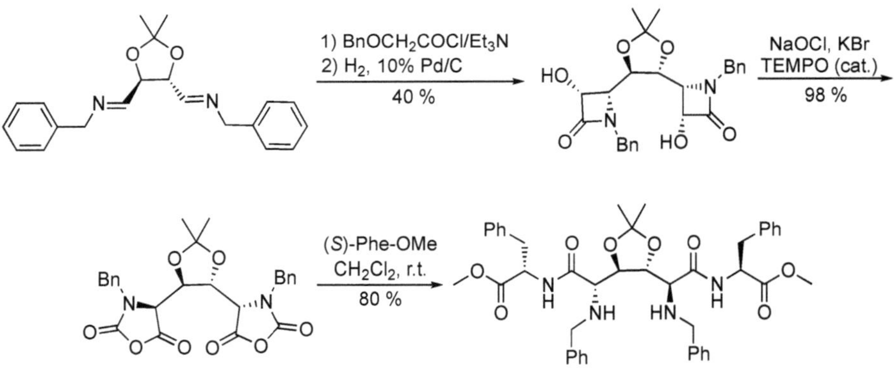

SCHEME 3.11 Synthesis of dimethyl 2,2'-(((2*S*,2'*S*)-2,2'-((4*R*,5*R*)-2,2-dimethyl-1,3-dioxolane-4,5-diyl)bis(2-(benzylamino)acetyl))bis(azanediyl))(2*S*,2'*S*)-bis(3-phenylpropanoate) via β-lactam intermediate

The treatment of acyl chloride with triethylamine to form the corresponding ketene has been applied in the synthesis of cephalosporin derivatives, i.e., *DL*-3-phenyl-, 3-*p*-carbomethoxyphenyl-, and 3-(4-thiazolyl)-7β-(2-thienyl)acetamido-decephalosporanic acids, as demonstrated in Scheme 3.12 for the preparation of 3-*p*-carbomethoxyphenyl-7β-(2-thienyl) acetamidodecephalosporanic acid. In this preparation, 4-methoxybenzyl 2-(diethoxyphosphoryl)-2-methanethioamidoacetate was deprotonated with K_2CO_3 in acetone, and the resulting carbanion undergoes regular S_N2 reaction with 2-chloro-1-phenylethan-1-one to afford 4-methoxybenzyl (*Z*)-2-(diethoxyphosphoryl)-2-((((2-oxo-2-phenylethyl)thio)methylene) amino)acetate, which was then deprotonated with sodium hydride in glyme to generate 4-methoxybenzyl 5-phenyl-6*H*-1,3-thiazine-4-carboxylate via an intramolecular Horner-Wadsworth-Emmons (HWE) reaction. Then, this compound was treated with 2-azidoacetyl chloride in the presence of Et_3N to afford 4-methoxybenzyl (6*R*,7*S*)-7-azido-8-oxo-3-phenyl-5-thia-1-azabicyclo-[4.2.0]oct-2-ene-2-carboxylate via [2+2] cycloaddition. After several consecutive transformations, (6*R*,7*S*)-8-oxo-3-phenyl-7-(2-(thiophen-2-yl) acetamido)-5-thia-1-azabicyclo[4.2.0]oct-2-ene-2-carboxylic acid was obtained [52].

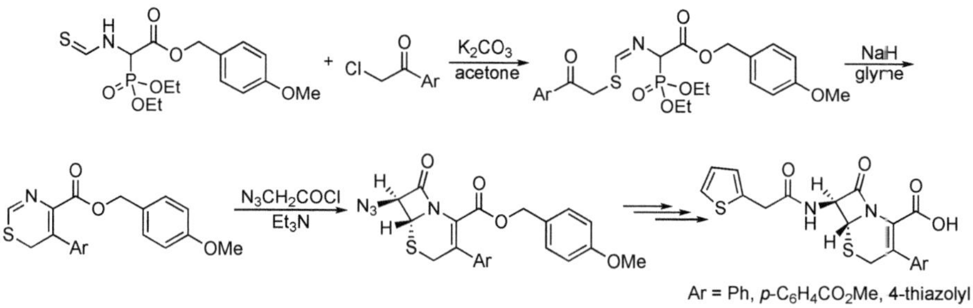

SCHEME 3.12 Synthesis of (6*R*,7*S*)-8-oxo-3-phenyl-7-(2-(thiophen-2-yl)acetamido)-5-thia-1-azabicyclo[4.2.0]oct-2-ene-2-carboxylic acid and analogs

In another practice, a series of *cis*-3-hydroxy-β-lactams have been efficiently prepared in three steps by means of the formation of imine in THF or CH_2Cl_2 with benzylamine or isopropylamine and benzaldehyde, butyraldehyde or isobutyraldehyde in the presence of two equivalences of $MgSO_4$, followed by the treatment of the resulting imine with 1.1–1.5 equivalent of 2-chloro-2-oxoethyl acetate and 3.0 equivalents of Et_3N in THF or CH_2Cl_2, and final removal of the 3-acetyl protecting group with 2.0 equivalents of K_2CO_3 in aqueous MeOH. The prepared *cis*-3-hydroxy-β-lactams are then implemented for the selective formation of the corresponding 3-*O*-α- and β-glucosides, using sucrose as a glucosyl donor in the presence of two carbohydrate-active enzymes. The 3-*O*-α-glucosides are obtained in the presence of a mutant glycoside phosphorylase from *Thermoanaerobacterium thermosaccharolyticum* (TtSPP_R134A), whereas 3-*O*-β-glucosides are formed in the presence of glycosyltransferase (UGT-76G1Sr) from *Stevia rebaudiana*, respectively [53].

The Staudinger [2+2] cycloaddition has been performed under microwave irradiation, using a series of 2,4-diaryl-2,3-dihydrobenzo[*b*][1,4] thiazepines such as 2,4-diphenyl-2,3-dihydrobenzo[*b*][1,4]thiazepine and

2-(2,5-dioxopyrrolidin-1-yl)acetyl chloride or 2-bromo(chloro) acetyl chloride in the presence of Et_3N (Scheme 3.13). This particular reaction affords 1-(1-oxo-2a,4-diaryl-2,2a,3,4-tetrahydro-1*H*-azeto[1,2-*d*]benzo[*b*][1,4] thiazepin-2-yl)pyrrolidin-2,5-diones, 2-bromo-2a,4-diaryl-2,2a,3,4-tetrahydro-1*H*-azeto[1,2-*d*]benzo[*b*][1,4]thiazepin-1-ones or 2-chloro-2a,4-diphenyl-2,2a,3,4-tetrahydro-1*H*-azeto[1,2-*d*]benzo[*b*][1,4]-thiazepin-1-ones, depending on the acetyl chloride used [54].

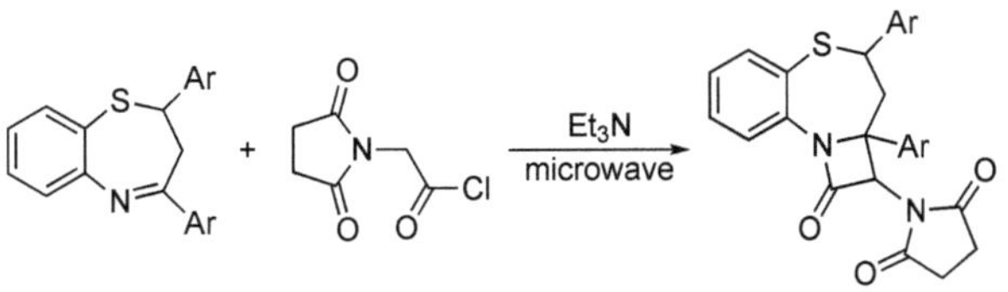

SCHEME 3.13 The Staudinger reaction to prepare 1-(1-oxo-2a,4-diaryl-2,2a,3,4-tetrahydro-1*H*-azeto[1,2-*d*]benzo[*b*][1,4]thiazepin-2-yl)pyrrolidin-2,5-diones and analogs

A very interesting Staudinger [2+2] cycloaddition has been reported recently, affording β-lactams of different stereospecificity depending on the base used to generate the ketene. For example, when the imines prepared from the condensation between five different benzaldehydes (H, 2-MeO, 4-Cl, 2,4-di-Cl, and 4-*i*-Pr) and 1 equivalent of 2-bromoethan-1-amine hydrogen bromide or 3-bromopropan-1-amine hydrogen bromide in the presence of 3 equivalents of Et_3N, and 2 equivalents of $MgSO_4$ in CH_2Cl_2 at room temperature are allowed to react with 2-methoxy-acetyl chloride, 2-benzyloxy-acetyl chloride or 2-phenoxy-acetyl chloride and 3 equivalents of Et_3N, 60–73% of β-lactams of *cis*-configuration for $C_3(S)$ and $C_4(R)$ substituents. For comparison, when the imine formed from benzaldehyde or 2-fluorobenzaldehyde, with one equivalent of isopropylamine, *n*-propylamine or benzylamine in the presence of 2 equivalents of $MgSO_4$ in CH_2Cl_2 is subject to react with 5-chloropentanoyl chloride or 4-chlorobutanoyl chloride in the presence of 3.0 equivalents of 2,6-lutidine in benzene, 41–55% of expected β-lactams of *trans*-configuration (C_3-*R* and C_4-*S*) was obtained, as illustrated in Scheme 3.14. The resulting β-lactams from both protocols are used to synthesize purine-β-lactam hybrids through *N*-alkylation of the purines with ω-haloalkyl-substituted β-lactams, and the β-lactams hybrids have been screened for their antiviral activities and cytotoxicity [55].

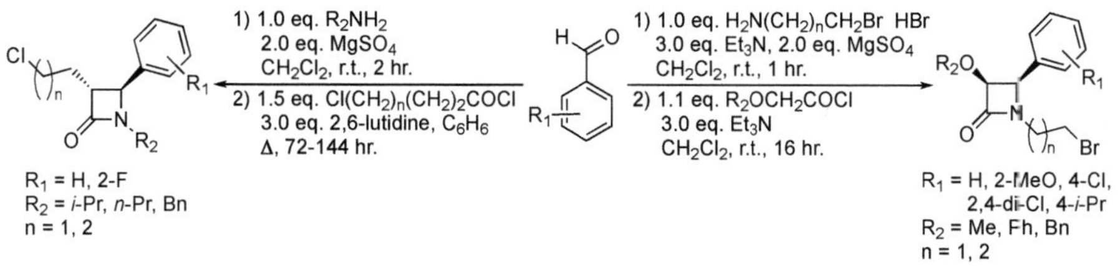

SCHEME 3.14 The impact of base on the stereospecificity of β-lactams in the Staudinger reaction

Other examples of using Et_3N in the preparation of β-lactam include the reaction between 2-ethylbutanoyl chloride and (*Z*)-2-((4-methoxyphenyl) imino)-1-phenylethan-1-one, yielding 4-benzoyl-3,3-diethyl-1-(4-methoxyphenyl)azetidin-2-one, of which the 4-methoxyphenyl group has been removed by means of oxidation with CAN [56]; and the reaction of (*E*)-2-(benzylideneamino)-4-methylthiazole-5-carboxylic acid with chloroacetyl chloride in refluxing benzene to form 2-(3-chloro-2-oxo-4-phenylazetidin-1-yl)-4-methylthiazole-5-carboxylic acid derivatives [57].

In addition to Et_3N, other tertiary amines have also been applied to the *in-situ* generation of ketenes from acyl halides, such as 2,6-lutidine displayed in Scheme 3.14. Also, the tertiary amine of *N*-methyl morpholine has been used in the [2+2] cycloaddition under microwave irradiation. In this practice, *L*-glyceraldehyde acetonide was treated with a few primary aromatic amines (aniline, *p*-anisidine, and *p*-methylaniline) in CH_2Cl_2 to form the corresponding imines, which were dried thoroughly with anhydrous sodium sulfate. The individual imine was then mixed with acetoxyacetyl chloride, methoxyacetyl chloride, or phenoxyacetyl chloride in dichloroethane in the presence of *N*-methyl morpholine and irradiated for 2 minutes. The expected β-lactams of single enantiomer (3*R*,4*S*) were obtained in 80–90% yields [58]. Tertiary amine of diisopropylethylamine (DIPEA) has been applied to generate imine *in situ*, e.g., methyl (*Z*)-3-((2-methoxy-2-oxoethylidene) amino)propanoate) from methyl 3-(chloro(2-methoxy-2-oxoethyl)amino) propanoate in CH_2Cl_2 and chloroketene from dichloroacetyl chloride. The subsequent reaction affords a mixture of methyl 3,3-dichloro-1-(3-methoxy-3-oxopropyl)-4-oxoazetidine-2-carboxylate, methyl 3-((1,2-dichloro-2-oxoethyl)amino)propanoate and (*E*)-1,1,1-trichloro-4-(diisopropyl-amino) but-3-en-2-one, as shown in Scheme 3.15 [59].

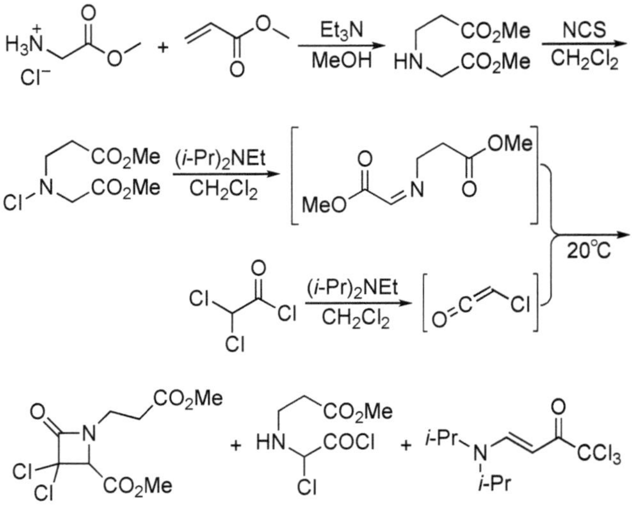

SCHEME 3.15 The preparation of methyl 3,3-dichloro-1-(3-methoxy-3-oxopropyl)-4-oxoazetidine-2-carboxylate from methyl glycinate, methyl acrylate and 2,2-dichloroacetyl chloride

The Staudinger [2+2] cycloaddition is very easy to proceed with so it is not necessary to start from acyl halide, instead, it can be carried out in such a way that a mixture of the preformed imine and carboxylic acid is treated with Et_3N and an acylating agent, such as $POCl_3$. However, the reaction should not occur in the absence of the acylating agent. Unfortunately, it has not been clarified in one piece of literature [60]. Besides the acylating agent $POCl_3$, phosgene fits in the reaction perfectly. As an example, the carboxylic acid and phosgene have been refluxed in CH_2Cl_2 for a while (e.g., 2 hours), then the preformed imine is added to the refluxing solution of acid, followed by Et_3N to obtain the targeted β-lactams [61]. For three special β-lactams in this study, they have been alternatively prepared from three different imines and the ketene generated *in situ* by the treatment of ethyl 2-bromo-2-phenylacetate with zinc and trimethylsilyl chloride in benzene under microwave irradiation, as illustrated in Scheme 3.16 [61].

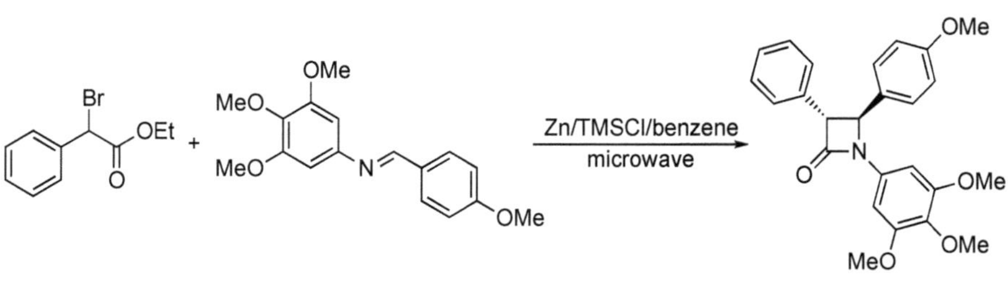

SCHEME 3.16 The preparation of (3*R*,4*S*)-4-(4-methoxyphenyl)-3-phenyl-1-(3,4,5-trimethoxyphenyl)azetidin-2-one

It should be pointed out that the imines for the Staudinger reaction is usually prepared by the reaction of a primary amine and an aldehyde in solution under either acid or base catalysis, and such preparation is often carried out under thermal condition, although it is also feasible under microwave irradiation. The reaction between the primary amine and ketone is not welcome due to potential steric hindrance of the resulting imine in the subsequent [2+2] cycloaddition, and even the severe challenge of regioselectivity should asymmetric ketone be used. This protocol of making imines often suffers from unacceptably low yields of imines and the challenges in obtaining imines of high purity. Recently, a fast solventless preparation of imine has been reported by a solid melt process where the mixture of aromatic aldehyde and primary amine (either aromatic or aliphatic) is heated until all reactants melt without any solvent. The resulting imine once cooled to room temperature is then dissolved in anhydrous CH_2Cl_2, and subject to the Staudinger reaction with phenoxyacetic acid and *p*-toluenesulfonyl chloride (*p*-TsCl) in the presence of Et_3N, in a molar ratio of 1:1.5:1.5:7.5 at room temperature. This multi-component method affords good to an excellent yield of the desired β-lactam derivatives, exclusively in *cis*-stereoselectivity. Overall, this simple approach demonstrates several features such as shorter reaction times, operational simplicity, easy workup procedure, and good yields. One typical preparation of β-lactam is illustrated by fusing 4-bromoaniline (m.p., 60–64°C) and 2-chloroquinoline-3-carbaldehyde (m.p., 148–150°C) and subsequent reaction with phenoxyacetic acid (1.5 eq.), *p*-TsCl (1.5 eq.) and 7.5 equivalents of Et_3N in CH_2Cl_2 at room temperature to afford 93% of (3*S*,4*R*)-1-(4-bromophenyl)-4-(2-chloroquinolin-3-yl)-3-phenoxyazetidin-2-one, as displayed in Scheme 3.17. In contrast, the same reaction only leads to 30% of the β-lactam of pure *trans*-selectivity when refluxed in toluene [62]. A similar Staudinger reaction between (1*E*,2*E*)-*N*-(4-fluorophenyl)-3-phenylprop-2-en-1-imine and 2-(1*H*-indol-1-yl)acetic acid in CH_2Cl_2 in the presence of *p*-TsCl and Et_3N also takes an *exo* approach to afford (3*S*,4*R*)-1-(4-fluorophenyl)-3-(1*H*-indol-1-yl)-4-((*E*)-styryl)azetidin-2-one, whereas (3*R*,4*R*)-1-(4-fluorophenyl)-3-(1*H*-indol-1-yl)-4-((*E*)-styryl)azetidin-2-one via an *endo* approach has not been formed [63]. The *cis*-stereoselectivity has been validated by both NMR investigation and X-ray crystallography.

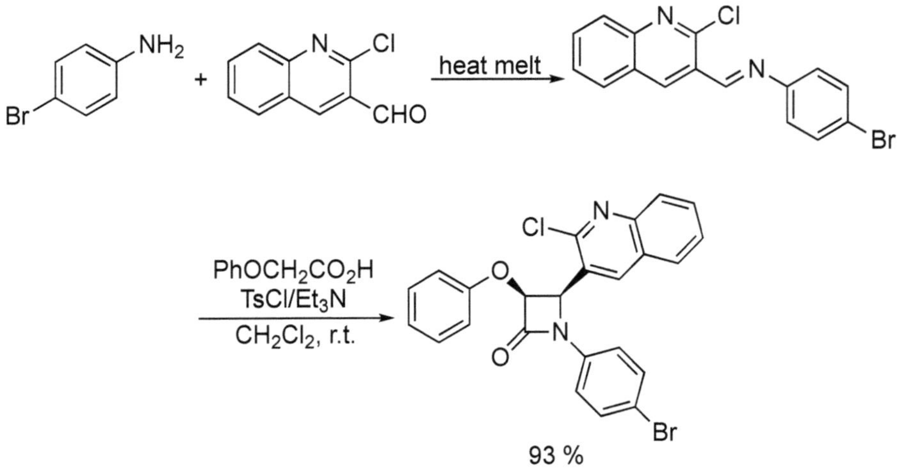

SCHEME 3.17 Simple preparation of (3*S*,4*R*)-1-(4-bromophenyl)-4-(2-chloroquinolin-3-yl)-3-phenoxyazetidin-2-one

Besides the conversion of carboxylic acids into the corresponding acyl chloride with $POCl_3$, *p*-TsCl, or TMSCl, pivaloyl chloride (PivCl) has also been applied for this purpose, as demonstrated in the preparation of a series of spirooxindole β-lactams, with potential antibacterial, antiviral, and antifungal activities, such as the spirocyclic oxindolo-β-lactams, commercially known as charteline A, B, and C which are isolated from the marine bryozoan *Chartella papyracea* [64, 65]. In this protocol, the mixture of substituted phenylacetic acid and half equivalence of *tert*-butyl (*Z*)-(1-benzyl-2-oxoindolin-3-ylidene) carbamate was first treated with 1 equivalent of PivCl and Et_3N at 0°C, respectively, and the *in-situ* generated ketene was then treated with additional 0.75 equivalent Et_3N and 20 mol% of (2*S*,3*R*)-3-isopropyl-2-phenyl-3,4-dihydro-2*H*-benzo[4,5]thiazolo[3,2-*a*]pyrimidine (HBTM) at -78°C to afford *tert*-butyl (2*R*,3*S*)-1'-benzyl-2',4-dioxo-3-aryl-spiro[azetidine-2,3'-indoline]-1-carboxylate (Scheme 3.18) [35]. In this series of reactions, it is found that substituted phenylacetic acids with electron-donating groups (Me, MeO) at the *para*-position afford the corresponding spirooxindole β-lactams in high yields (up to 97%) with excellent stereoselectivities (dr up to 94:6; ee up to >99%). For comparison, phenylacetic acids with electron-withdrawing groups (F, Cl, Br) at the *para*-position lead to the corresponding β-lactams in high yields (up to 97%) with excellent enantioselectivities (up to >99%) but lower diastereoselectivities (dr = 70:30 to 83:17), due to the existence of α-H on the amide group, causing epimerization via keto-enol tautomerization in the presence of a base [35]. Similar preparations of the spirooxindole β-lactams have been reported by means of the reaction of ketene (e.g., 1-phenyl-1-ethyl ketene) and isatin-derived ketimines such as *tert*-butyl (*E*)-(1-benzyl-2-oxoindolin-3-ylidene)carbamate in the presence of a catalytic amount of NHC in THF, as displayed in Scheme 3.19. In this reaction, a series of NHCs have been tested, and the NHC generated from the reaction between (*S*)-2-benzyl-5-(bis(3,5-bis(trifluoromethyl)phenyl)(hydroxy)-methyl)-6,7-dihydro-5*H*-pyrrolo[2,1-*c*][1,2,4]triazol-2-ium tetrafluoroborate and cesium carbonate works the best, affording *tert*-butyl (2*S*,3*S*)-1'-benzyl-3-ethyl-2',4-dioxo-3-phenylspiro[azetidine-2,3'-indoline]-1-carboxylate in 89% yield, with 94% of ee% and > 95% dr [66].

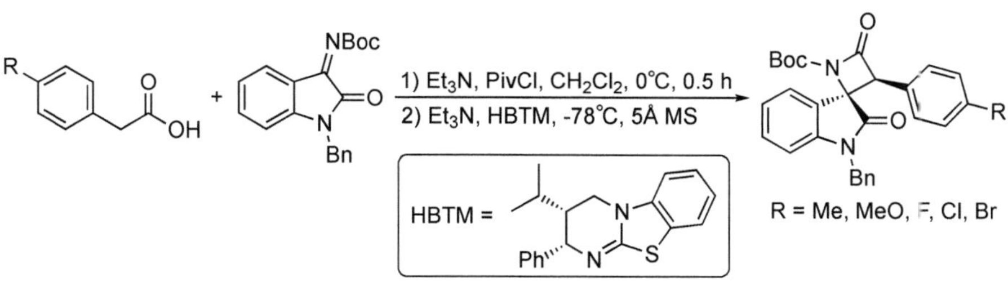

SCHEME 3.18 Preparation of *tert*-butyl (2*R*,3*S*)-1'-benzyl-2',4-dioxo-3-(*p*-tolyl)spiro[azetidine-2,3'-indoline]-1-carboxylate and analogs

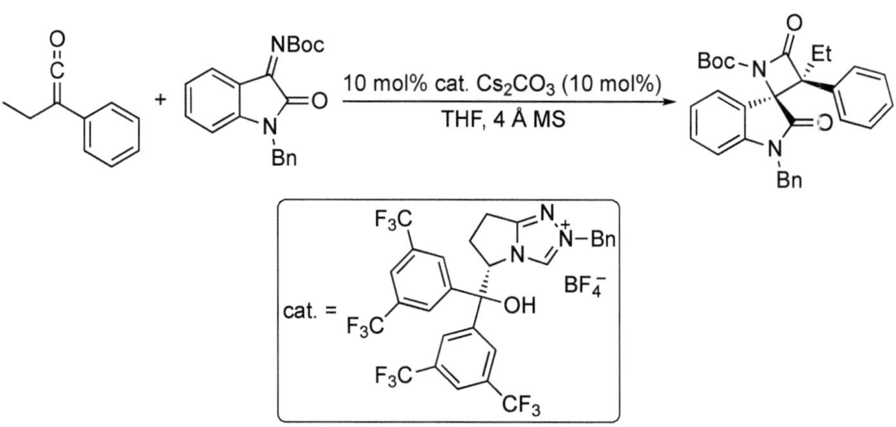

SCHEME 3.19 The Staudinger reaction to form *tert*-butyl (2*S*,3*S*)-1'-benzyl-3-ethyl-2',4-dioxo-3-phenylspiro[azetidine-2,3'-indoline]-1-carboxylate

It has been suspected that the use of tertiary amines for the dehydrohalogenation of acid halides to form ketenes for the Staudinger reaction is complicated by the fact that they are usually too nucleophilic, as often seen for the immediately dark color of acetyl chloride after Et_3N is added. To overcome this issue, a new strategy of catalytic, asymmetric synthesis of β-lactams has been developed, involving the use of a thermodynamically strong but kinetically non-nucleophilic base. This protocol, termed "shuttle deprotonation," utilizes a chiral nucleophile in catalytic amount to dehydrohalogenate the acid chloride at the first step, which is then regenerated by means of hexahydroimidazo base in stoichiometric amount, such as K_2CO_3, NaH in combination with 15-crown-5 or proton sponge (PS). The catalytic chiral nucleophilic base is (*R*)-(6-methoxyquinolin-4-yl)((1*S*,2*S*,4*S*,5*R*)-5-vinylquinuclidin-2-yl)methyl benzoate, abbreviated as BQ, and the PS is naphthalene-1,8-diamine. The concept of this strategy is outlined in Scheme 3.20, in which the base BQ either functions the role like Et_3N in other similar treatments of acid chlorides to directly form ketenes, and the protonated BQ is then deprotonated by the PS. Alternatively, the base BQ may undergo scenario 2 where BQ nucleophilically substitutes chloride in acid chloride, and the resulting acyl ammonium is deprotonated by PS to form enolate which then reacts with imine, yielding the β-lactam of high stereoselectivity due to the chiral BQ component. The role of PS can be replaced by K_2CO_3 or NaH in combination with crown ether (15-crown-5) in a similar manner so these two bases are not included in this mechanism [34]. A related strategy has utilized 4-chloro-*N*-(2-(diethylamino)ethyl)benzamide, 2-(diethylamino)ethyl 4-chlorobenzoate, or 1-isopropyl-1*H*-benzo[*d*]imidazole as a nucleophilic catalyst in the reaction between methyl phenylketene and ethyl (*E*)-2-(tosylimino)acetate in THF [67].

In addition to *in situ* generating an acyl chloride with $POCl_3$, *p*-TsCl, PivCl, or TMSCl, for the purpose of creating ketene in the preparation of β-lactams, Mukaiyama reagent has been commonly used in β-lactam formation. The Mukaiyama reagent, i.e., 2-chloro-1-methylpyridinium iodide (CMPI) was introduced by Teruaki Mukaiyama in 1975 [68], as a useful condensation reagent for the generation of esters and amides. It can be easily prepared from the reaction of 2-chloropyridine with methyl iodide [69]. The features of the Mukaiyama reagent in the synthesis of β-lactam can be demonstrated in either direct conversion of *N*-substituted β-amino acids (i.e., $RNHCHR_1CHR_2CO_2H$, R = CH_2Ph, H; R_1 = Me, Pr, H, Ph; R_2 = H, Me) into β-lactams in 83–95% yields [70], or more often in the Staudinger reaction between carboxylic acids and imines, particularly by means of solid-state synthesis. In one of these syntheses, pent-4-enoic acid was mounted to Wang resin with DIC/DMAP coupling, affording the resin immobilized pent-4-enoate, which was converted into (*E*)-5-(4-(chloromethyl)phenyl)pent-4-enoate by means of olefin metathesis with 1-(chloromethyl)-4-vinylbenzene in the presence of

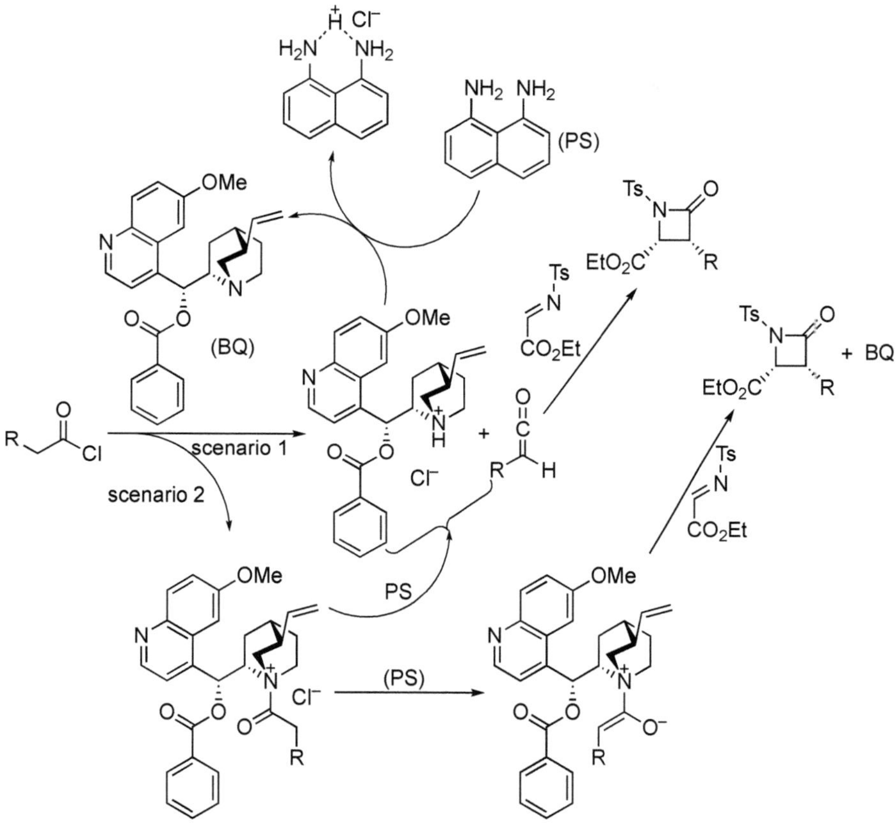

SCHEME 3.20 Application of BQ and proton sponge in the preparation of chiral β-lactams

Grubbs or Hoveyda-Grubbs catalyst. Mild hydrolysis of the chloromethyl group followed by oxidation with DMSO led to the formation of resin immobilized (*E*)-5-(4-formylphenyl)pent-4-enoate, which was then condensed with *para*-substituted anilines to form (*E*)-5-(4-((*E*)-(*p*-substituted-phenylimino) methyl)phenyl)pent-4-enoate on the resin. This resin mounted imine was then treated with 2.5 equivalents of carboxylic acid, 6 equivalents of Et3N, and 3.0 equivalents of Mukaiyama reagent in chloroform under refluxing, yielding the resin immobilized β-lactam in *trans*-selectivity. Two approaches have been applied to cleave the β-lactam from the Wang resin, i.e., treatment of the β-lactam with 10% TFA in CH_2Cl_2 followed by diazomethane and metathesis of the β-lactam with another alkene to afford β-lactam with alternative side chain, as displayed in Scheme 3.21 [71].

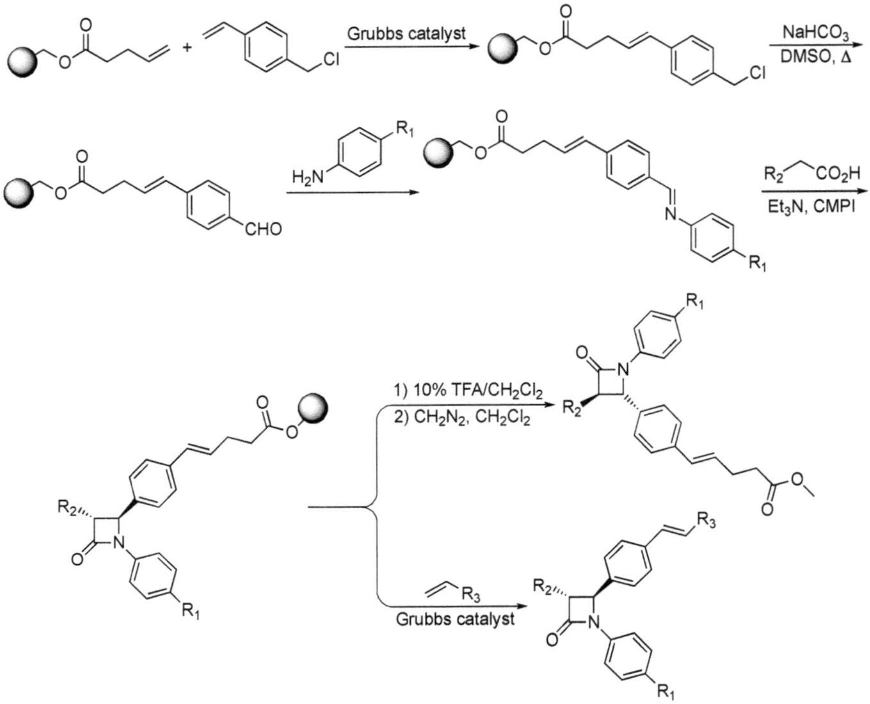

SCHEME 3.21 Resin-supported synthesis of chiral β-lactams

Another solid phase supported synthesis (SPSS) of β-lactams has been performed based on (4-diethylsilylbutyl)polystyrene resin which was activated by chlorination with *N*-chlorosuccinimide (NCS). Acetic acid was then mounted to the polystyrene resin terminated with diethylsilyl chloride by means of coupling with methyl 2-(4-hydroxyphenoxy)acetate in the presence of imidazole in CH_2Cl_2, where the methyl ester was hydrolyzed with trimethyltin hydroxide. This substituted acetic acid was then treated with Et_3N (6 eq.), (*E*)-*N*-benzyl-1-(4-methoxyphenyl)methanimine (5 eq.), and Mukaiyama reagent (6 eq.) in chloroform under refluxing to give (3*S*,4*R*)-1-benzyl-3-(4-hydroxyphenoxy)-4-(4-methoxyphenyl)azetidin-2-one, which was cleaved from the polymer support by tetrabutylammonium fluoride (TBAF) or hydrogen fluoride in the presence of pyridine, as displayed in Scheme 3.22. Unfortunately, this approach only produced β-lactam in 12–14% yield. In order to improve the yield, methyl 2-(4-hydroxyphenoxy)acetate was mounted to the Wang resin under modified Mitsunobu reaction conditions (TMAD, Bu_3P), and saponified with potassium trimethylsilanoate to give the immobilized carboxylic acid. After that, the solid-phase version of the Staudinger reaction was carried out using the Mukaiyama reagent as the activating agent [72]. The latter strategy for the Wang resin supported carboxylic acid has been successfully applied to the preparation of multicyclic β-lactams in reaction with cyclic imines such as phenanthridine or 6,7-dimethoxy-1-phenyl-3,4-dihydroisoquinoline, as illustrated in Scheme 3.23.

The resulting β-lactams, i.e., (1*S*,11b*S*)-1-(4-hydroxyphenoxy)-1,11b-dihydro-2*H*-azeto[1,2-*f*]-phenanthridin-2-one and (1*S*,9b*S*)-1-(4-hydroxyphenoxy)-7,8-dimethoxy-9b-phenyl-1,4,5,9b-tetrahydro-2*H*-azeto[2,1-*a*]isoquinolin-2-one were cleaved from the resin with 10% TFA in CH_2Cl_2, in 57% and 77% yields, respectively [73]. Some of these multicyclic β-lactams have emerged as a new generation of antibiotics known as trinems, with good stability against β-lactamase and dehydropeptidase deactivation, whereas other β-lactams of this kind have potent activity against class A and class C β-lactamases, as well as anti-cancer activity.

In addition to the application of the Mukaiyama reagent in the formation of β-lactams, cyanuric chloride, i.e., 2,4,6-trichloro-1,3,5-triazine (TCT), in complexation with DMF works just fine for the generation of β-lactams from the direct reaction of carboxylic acid and imine, as illustrated in Scheme 3.24 [74]. In the presence of TCT, two related but different chiral imines all arising from δ-threonine have been allowed to react with carboxylate, affording *cis*-β-lactams. For example, the reaction between methyl (2*S*,3*R*)-3-hydroxy-2-(((1*Z*,2*E*)-3-phenylallylidene)amino)butanoate (i.e., the condensation product from methyl threoninate and cinnamaldehyde) and potassium

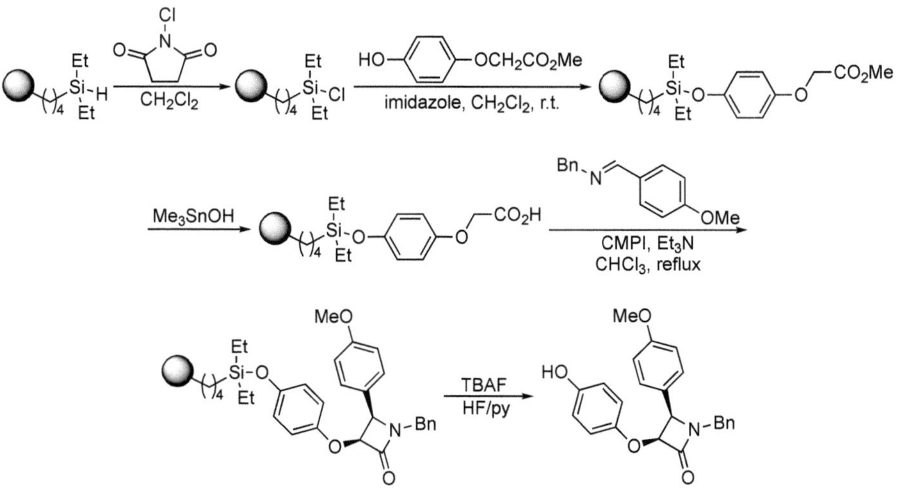

SCHEME 3.22 (4-Diethylsilylbutyl)polystyrene resin supported the synthesis of chiral β-lactams

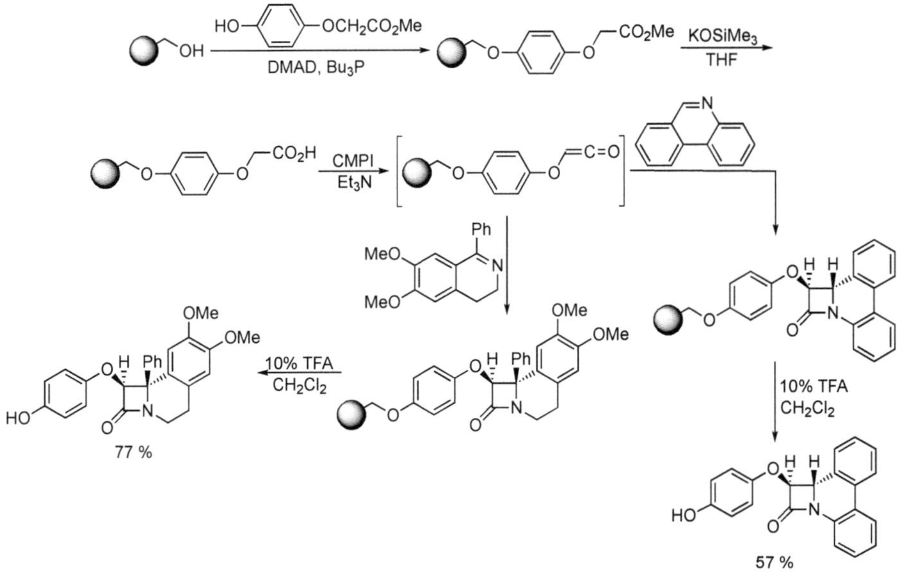

SCHEME 3.23 Solid-phase supported the synthesis of (1*S*,9b*S*)-1-(4-hydroxyphenoxy)-7,8-dimethoxy-9b-phenyl-1,4,5,9b-tetrahydro-2*H*-azeto[2,1-*a*]isoquinolin-2-one and (1*S*,11b*S*)-1-(4-hydroxyphenoxy)-1,11b-dihydro-2*H*-azeto[1,2-*f*]phenanthridin-2-one

(*Z*)-(4-methoxy-4-oxobut-2-en-2-yl)glycinate in the presence of TCT and Et_3N gives methyl (*Z*)-3-(((3*S*,4*R*)-1-((2*S*,3*R*)-3-hydroxy-1-methoxy-1-oxobutan-2-yl)-2-oxo-4-((*E*)-styryl)-azetidin-3-yl)amino)but-2-enoate and methyl (*Z*)-3-(((3*R*,4*S*)-1-((2*S*,3*R*)-3-hydroxy-1-methoxy-1-oxobutan-2-yl)-2-oxo-4-((*E*)-styryl)azetidin-3-yl)amino)but-2-enoate in nearly equal amount, as shown in Scheme 3.25. In this case, the nearly complete lack of diastereoselectivity during the cycloaddition could be ascribed to an almost planar structure for the Schiff base caused by strong hydrogen bonding between the β-hydroxyl and the α-ester carbonyl groups. This planar structure would permit nearly equal ease of approach by a reactant from either the *si* or the *re* face of the Schiff base. When phenylserine is used rather than threonine, the two faces were no longer equally accessible because of the bulk phenyl group, and two diastereomeric *cis*-β-lactams are formed in 80:20 proportion [75]. When methyl *L*-threoninate was sequentially converted into methyl (2*R*,3*R*)-2,3-dihydroxybutanoate, (4*R*,5*R*)-2,2,5-trimethyl-1,3-dioxolane-4-carbaldehyde, and the resulting imine (*Z*)-*N*-(4-methoxyphenyl)-1-((4*S*,5*R*)-2,2,5-trimethyl-1,3-dioxolan-4-yl)methanimine reacted with phenoxyacetic acid under the similar condition to yield (3*R*,4*S*)-1-(4-methoxyphenyl)-3-phenoxy-4-((4*S*,5*R*)-2,2,5-trimethyl-1,3-dioxolan-4-yl) azetidin-2-one. Similarly, 2-chloro-1-methylpyridin-1-ium iodide has been applied for the *in situ* generation of ketene for the synthesis of β-lactams [76].

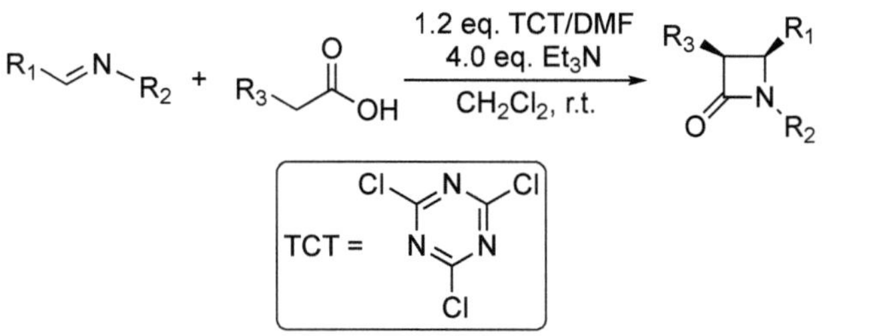

R_1 = 4-MeC_6H_4, 3,4-$(MeO)_2C_6H_3$, 4-$NO_2C_6H_4$, 4-ClC_6H_4, Ph, CH=CHPh

R_2 = 4-$EtOC_6H_4$, 4-$MeOC_6H_4$, Bn, Me, PhN=NC_6H_4

R_3 = 4-ClC_6H_4O, PhO, PhthN, MeO, 2,4-$(Cl)_2C_6H_3O$, Cl, N_3, 2-NaphthO

SCHEME 3.24 TCT/DMF promoted the formation of β-lactams from imines and carboxylic acids

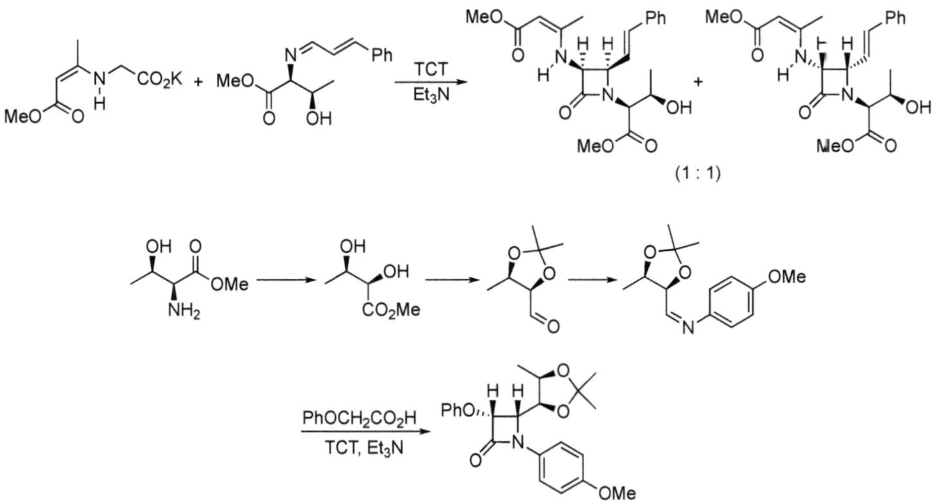

SCHEME 3.25 TCT promoted the synthesis of β-lactams starting with *D*-threonine

A special case of the [2+2]-cycloaddition involving imine to afford β-lactam is to treat lactone with a strong base (e.g., LDA), as shown in Scheme 3.26 for the synthesis of (3*S*,4*S*)-3-((*S*)-2-(4-fluorophenoxy)-1-hydroxyethyl)-1-(4-fluoro-phenyl)-4-(4-hydroxyphenyl)azetidin-2-one, which is a cholesterol absorption inhibitor. In this synthesis, (*S*)-4-hydroxydihydrofuran-2(3*H*)-one was treated with 2.0 equivalents of LDA in DMF in the presence of *N,N'*-dimethylpropyleneurea (DMPU), and the *in situ* generated ketene then reacted with 1-(4-(benzyloxy)-phenyl)-*N*-(4-fluorophenyl)methanimine to give (3*S*,4*S*)-4-(4-(benzyloxy)phenyl)-3-((*S*)-1,2-dihydroxyethyl)-1-(4-fluorophenyl)azetidin-2-one in 64% yield. The terminal hydroxyl group was then protected with tosyl chloride in the presence of pyridine and then treated with (4-fluorophenoxy)zinc(II) bromide to afford the final product due to the push-pull effect [77].

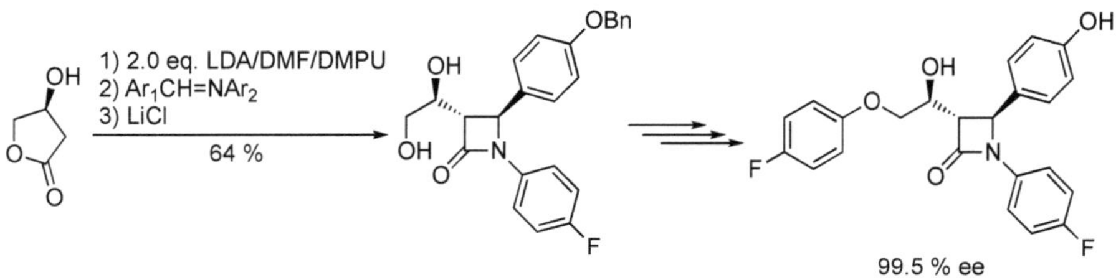

SCHEME 3.26 The synthesis of (3*S*,4*S*)-3-((*S*)-2-(4-fluorophenoxy)-1-hydroxyethyl)-1-(4-fluorophenyl)-4-(4-hydroxyphenyl)azetidin-2-one

In contrast to the treatment of lactone with a strong base as illustrated in Scheme 3.26, there are in fact many examples of making β-lactams via the [2+2]-cycloadditions between imines and enolate intermediates directly generated from esters, pioneered by Gilman and Speeter [78]. In this preliminary practice, 0.2 mol of *N*,1-diphenylmethanimine, zinc foil, and ethyl 2-bromoacetate in dry toluene were refluxed in the presence of some iodine to form 56% of 1,4-diphenylazetidin-2-one, with or without hydrolysis in the presence of NH_4OH. Similarly, the reaction between *N*,1-diphenylmethanimine and ethyl 2-bromopropanoate yielded 84% of 3-methyl-1,4-diphenylazetidin-2-one. Due to its versatility, this one-pot condensation between ester and imine to afford β-lactam has been applied to prepare tri- and tetra-substituted β-lactams, which is generally known as the Gilman-Speeter reaction [79–82]. The latest application of this approach appeared in the preparation of the next-generation taxoids, such as SB-T-1214, a highly potent cytotoxic agent with remarkable efficacy against drug-resistant tumors *in vivo*, including those that overexpress the P-glycoprotein efflux pump. In this practice, both the Staudinger reaction and the Gilman-Speeter reaction have been applied. For example, when (*E*)-*N*-(4-methoxyphenyl)but-2-yn-1-imine was subject to the traditional

Staudinger reaction with 1-acetoxyketene generated from the treatment of 2-acetoxyacetyl chloride with Et_3N, (3*R*,4*S*)-1-(4-methoxyphenyl)-2-oxo-4-(prop-1-yn-1-yl)azetidin-3-yl acetate and its enantiomer were obtained in 63% yield. Deprotection of the acetyl group and re-protection of the hydroxyl with triisopropylsilyl group led to 78% of (3*R*,4*S*)-1-(4-methoxyphenyl)-4-(prop-1-yn-1-yl)-3-((triisopropylsilyl)oxy)azetidin-2-one. For comparison, under the Gilman-Speeter reaction condition, (2*R*)-2-phenylcyclohexyl 2-((triisopropylsilyl)oxy)acetate was treated with LDA in THF at -40 to 0°C, and the resulting lithium enolate was allowed to react with (*E*)-*N*-(4-methoxyphenyl)but-2-yn-1-imine to give 57% of nearly enantiomerically pure (3*R*,4*S*)-1-(4-methoxyphenyl)-4-(prop-1-yn-1-yl)-3-((triisopropylsilyl)oxy)azetidin-2-one.

The control of stereochemistry is attributed to the chair transition state. Subsequent deprotection of the *p*-anisyl group with CAN oxidation and protection of the free amino group with *t*-Boc, tributyltin was attached to the side chain of the β-lactam, allowing this moiety to be connected with (2a*R*,4*S*,4a*S*,6*R*,9*S*,11*S*,12*S*, 12a*R*,12b*S*)-12b-acetoxy-6-((cyclopropanecarbonyl)oxy)-9,11-dihydroxy-4a,8,13,13-tetramethyl-5-oxo -4-((triethylsilyl)oxy)-2a,3,4,4a,5,6,9,10,11,12,12a, 12b-dodecahydro-1*H*-7, 11-methanocyclodeca[3,4]benzo[1,2-*b*]oxet-12-yl benzoate. In the end, the target taxoid molecule SB-T-1214, i.e., (2a*R*,4*S*,4a*S*,6*R*,9*S*, 11*S*,12*S*,12a*R*,12b*S*)-12b-acetoxy-9-(((2*R*,3*S*)-3-((*tert*-butoxycarbonyl) amino)-2-hydroxy-5-methylhex-4-enoyl)oxy)-6-((cyclopropanecarbonyl)oxy)-4,11-dihydroxy-4a,8,13,13-tetramethyl-5-oxo-2a,3,4,4a, 5,6,9,10,11,12,12a,12b-dodecahydro-1*H*-7,11-methanocyclodeca[3,4] benzo[1,2-*b*]oxet-12-yl benzoate was obtained, in which the moiety from the β-lactam component is highlighted in red color, as demonstrated in Scheme 3.27 [83].

Similarly, the one-pot reaction between (*E*)-1-(benzo[*d*][1,3]dioxol-5-yl)-*N*-(4-fluorophenyl)methanimine and methyl 2-(2,2,5,5-tetramethyl-1,2,5-azadisilo-lidin-1-yl)acetate in THF at -78°C overnight gave 69% of 3-amino-4-(benzo[*d*] [1,3]dioxol-5-yl)-1-(4-fluorophenyl)azetidin-2-one, with a *cis*/*trans*-β-lactam ratio of 88:12, whereas in low enantioselectivity, with a (3*R*,4*S*)-/(3*S*,4*R*)-enantiomer ratio of 46:54. Likewise, the reaction between (*E*)-*N*-(4-fluorophenyl)-1-(4-methoxyphenyl)methanimine and methyl 2-(2,2,5,5-tetramethyl-1,2,5-azadisilolidin-1-yl)acetate under the same condition yield 60% of 3-amino-1-(4-fluorophenyl)-4-(4-methoxyphenyl)azetidin-2-one, with a *cis*/*trans*-β-lactam ratio of 95:5, but in low enantioselectivity, with a (3*R*,4*S*)-/(3*S*,4*R*)-enantiomer ratio of 52:48

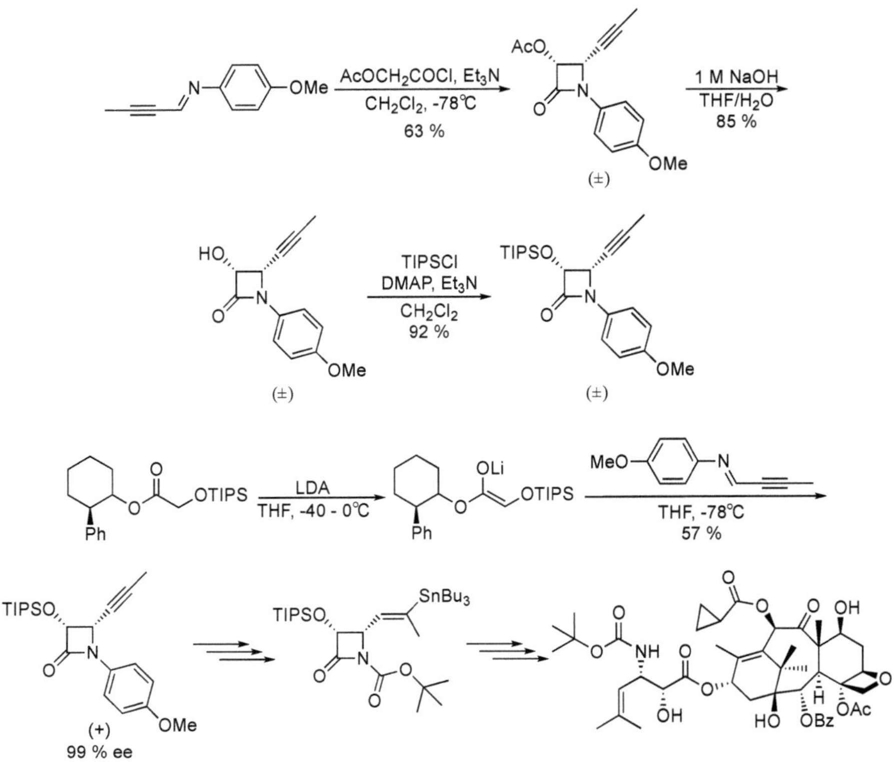

SCHEME 3.27 The preparation of β-lactam and its application in the synthesis of (2a*R*,4*S*,4a*S*,6*R*,9*S*,11*S*,12*S*,12a*R*,12b*S*)-12b-acetoxy-9-(((2*R*,3*S*)-3-((*tert*-butoxycarbonyl)amino)-2-hydroxy-5-methylhex-4-enoyl)oxy)-6-((cyclopropanecarbonyl)oxy)-4,11-dihydroxy-4a,8,13,13-tetramethyl-5-oxo-2a,3,4,4a,5,6,9,10,11,12,12a,12b-dodecahydro-1*H*-7,11-methanocyclodeca[3,4]benzo[1,2-*b*]oxet-12-yl benzoate

(Scheme 3.28). The preference in the formation of *cis*-β-lactam is due to the very large axial repulsion in the chair transition state which leads to the formation of *trans*-β-lactam [84]. While the *trans*-β-lactams are not favored under this condition, they are still achievable under this reaction condition using the menthyl ester group as the chiral auxiliary. For example, the reaction of (*E*)-1-(4-fluorophenyl)-*N*-(4-methoxyphenyl)methanimine and (1*R*,2*S*,5*R*)-2-isopropyl-5-methylcyclohexyl 2-(2,2,5,5-tetramethyl-1,2,5-azadisilolidin-1-yl)acetate under this condition afforded 55% of (3*R*,4*R*)-3-amino-4-(4-fluorophenyl)-1-(4-methoxyphenyl)azetidin-2-one, in all *trans*-configuration, with > 99% ee. The possible mismatched reaction pair between (*E*)-*N*-(4-methoxyphenyl)-1-phenyl-methanimine and (1*S*,2*R*,5*S*)-2-isopropyl-5-methylcyclohexyl 2-(2,2,5,5-tetramethyl-1,2,5-azadisilolidin-1-yl)acetate led to 65% of isolated β-lactams, consisting of 26% of (3*S*,4*S*)-3-amino-1-(4-methoxyphenyl)-4-phenylazetidin-2-one (54% ee) and 74% of (3*S*,4*R*)-3-amino-1-(4-methoxyphenyl)-4-phenylazetidin-2-one (21% ee) (Scheme 3.28) [85]. Obviously, condensation of a chiral lithium enolate of ester with an imine renders much better control of stereochemistry, as shown in the additional examples of chiral enolates from (1*R*,2*S*)-2-phenylcyclohexyl 2-((triisopropylsilyl)oxy)acetate [86], (1*S*,2*S*,3*S*,4*S*)-3-(dimethyl-amino)-1,7,7-trimethylbicyclo[2.2.1]heptan-2-ylisobutyrate,(1*S*,2*R*,3*R*,4*S*)-3-(dimethylamino)-1,7,7-trimethylbicyclo[2.2.1]-heptan-2-yl isobutyrate, (1*S*,2*S*,3*R*, 4*S*)-3-(dimethylamino)-1,7,7-trimethylbicyclo[2.2.1]-heptan-2-yl isobutyrate, (1*S*, 2*R*,3*S*,4*S*)-3-(dimethylamino)-1,7,7-trimethylbicyclo[2.2.1]heptan-2-yl isobutyrate,and(1*S*,2*S*,3*R*,4*S*)-1,7,7-trimethyl-3-(pyrrolidin-1-yl)bicyclo[2.2.1] heptan-2-yl isobutyrate [87]. Likewise, the examples of condensation between lithium enolate of esters and chiral imines have been reported, as illustrated in the coupling reaction of chiral imine acetal (*E*)-1-((4*S*,5*S*)-4,5-bis(methoxymethyl)-2-(3-methoxyphenyl)-1,3-dioxolan-2-yl)-*N*-(4-methoxyphenyl)methanimine and LiOC(=CH_2)OMe to give (*S*)-4-((4*S*,5*S*)-4,5-bis(methoxymethyl)-2-(3-methoxyphenyl)-1,3-dioxolan-2-yl)-1-(4-methoxyphenyl)azetidin-2-one in a highly diastereoselective manner [88].

In addition, the asymmetric reaction of achiral lithium enolates of esters with prochiral imines can also be achieved in the presence of an external chiral ligand for the lithium cation. For example, the reaction of lithium 2-methyl-1-(pentan-3-yloxy)prop-1-en-1-olate with (*E*)-*N*-(4-methoxyphenyl)-1-phenylmethanimine, (*E*)-*N*,1-bis(4-methoxyphenyl) methanimine, (*E*)-*N*-(4-methoxyphenyl)-1-(naphthalen-1-yl)methanimine, (*E*)-*N*-(4-methoxyphenyl)-1-(naphthalen-2-yl)methanimine, (1*E*)-*N*-(4-methoxyphenyl)-2-methyl-3-phenylprop-2-en-1-imine, or (*E*)-*N*-(4-methoxyphenyl)-3-phenylpropan-1-imine in the presence of 20 mol% of (1*R*,2*R*)-2-(2-methoxyethoxy)-*N*,*N*-dimethyl-1,2-diphenylethan-1-amine at -20°C, all giving the corresponding β-lactams in excellent yields (90% yield or above) with 65 to 90% ee, respectively [89]. The same reaction has been performed in the presence of 2.6 equivalents of (1*R*,2*R*)-1,2-dimethoxy-1,2-diphenylethane in toluene at -20°C, demonstrating control in stereochemistry [90, 91]. In addition, (4*S*,4'*S*)-4,4'-diisopropyl-4,4',5,5'-tetrahydro-2,2'-bioxazole, (4*S*,4'*S*)-2,2'-(propane-2,2-diyl) bis(4-isopropyl-4,5-dihydrooxazole), (4*S*,4'*S*)-2,2'-(propane-2,2-diyl)bis(4-(*tert*-butyl)-4,5-dihydrooxazole) and (4*S*,4'*S*)-2,2'-(propane-2,2-diyl)bis-(4-phenyl-4,5-dihydrooxazole) have been applied as external chiral ligands to improve the control in stereochemistry [91]. Specifically, the reaction between lithium 2-methyl-1-(pentan-3-yloxy)prop-1-en-1-olate and (*E*)-*N*-(4-methoxyphenyl)-1-phenylmethanimine to give (*S*)-1-(4-methoxyphenyl)-3,3-dimethyl-4-phenylazetidin-2-one has been performed in the presence of seven different chiral amines such as *N*-(2-((1*R*,2*R*)-2-(dimethylamino)-1,2-diphenylethoxy)-ethyl)propan-2-amine as the external controller for the stereochemistry [92].

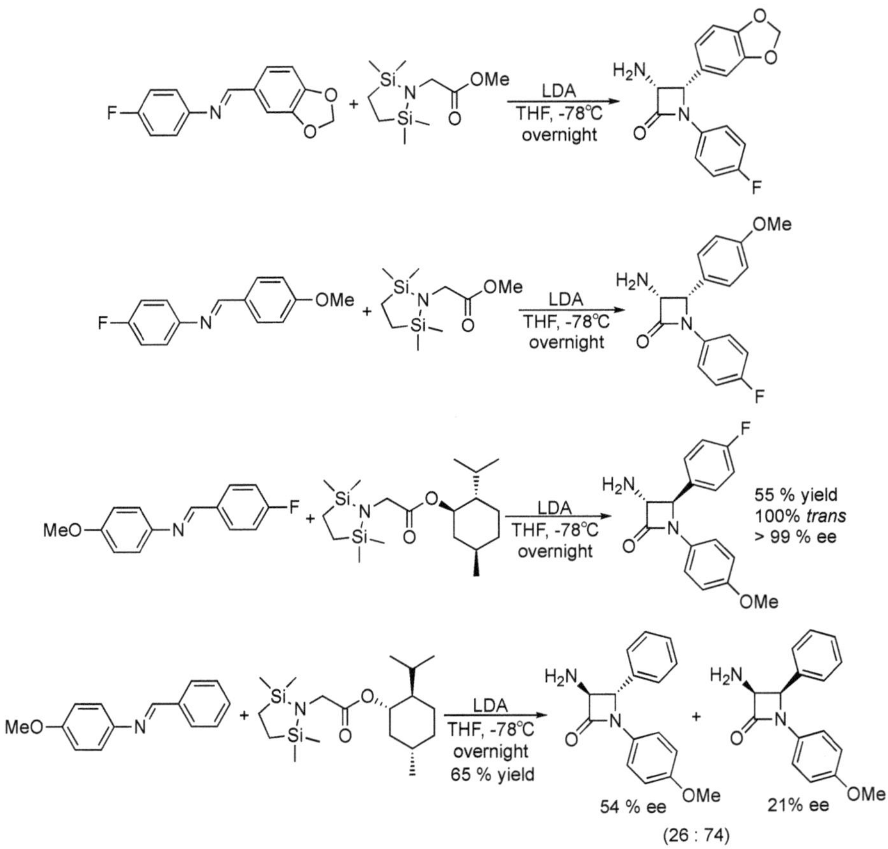

SCHEME 3.28 The [2+2] cycloaddition between imines and ester enolates to afford β-lactams

The intramolecular version for the ester enolate-imine cyclization can be applied to make polycyclic β-lactams and cyclic β-amino acid derivatives upon opening of the β-lactam ring. For example, ethyl (*S*,*E*)-3-(3-fluoro-2-(((1-(4-methoxyphenyl)ethyl)imino)methyl)phenyl)propanoate upon treatment with 1.1 equivalent of NaHMDS undergoes intramolecular cyclization reactions in the presence of 1.1 equivalent of 15-crown-5 in THF to afford (2a*R*,7b*R*)-7-fluoro-1-((*S*)-1-(4-methoxyphenyl)ethyl)-1,2a,3,7b-tetrahydro-2*H*-indeno[1,2-*b*]azet-2-one in 79% yield, with a diastereoselectivity of 95:5 as shown in Scheme 3.29 [93].

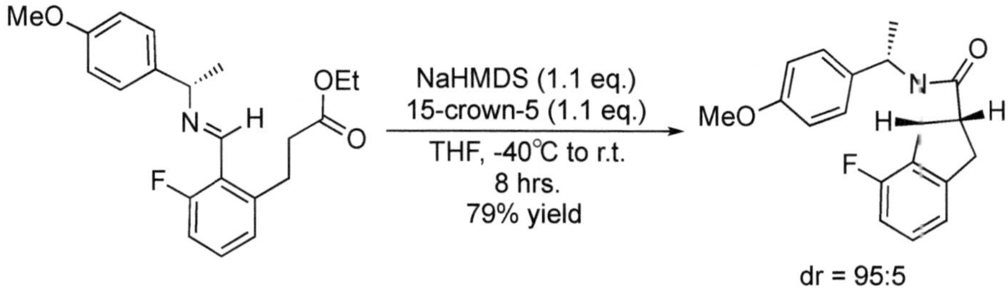

SCHEME 3.29 The preparation of (2a*R*,7b*R*)-7-fluoro-1-((*S*)-1-(4-methoxyphenyl)ethyl)-1,2a,3,7b-tetrahydro-2*H*-indeno[1,2-b]azet-2-one via intramolecular [2+2] cycloaddition

It is interesting that the unstable epoxide ring can sustain under the Gilman-Speeter reaction condition, as all the tested oxiranecarbaldimines react with lithium enolates derived from aliphatic esters to give exclusively β-lactams bearing the strained oxirane moiety. For example, when the mixture of methyl cyclohexanecarboxylate and (1*E*,1'*E*)-*N*,*N*'-(naphthalene-1,5-diyl)bis(1-(3-methyl-oxiran-2-yl)methanimine) was treated with LDA at –78°C, 57% of tetrasubstituted β-lactam, i.e., 2,2'-(naphthalene-1,5-diyl) bis(3-(3-methyloxiran-2-yl)-2-azaspiro-[3.5]nonan-1-one) was obtained, still containing the two oxirane moieties (Scheme 3.30). However, it should be pointed out that in most of these reactions with imines containing unstable oxirane moieties, the chemical yields are very low [94].

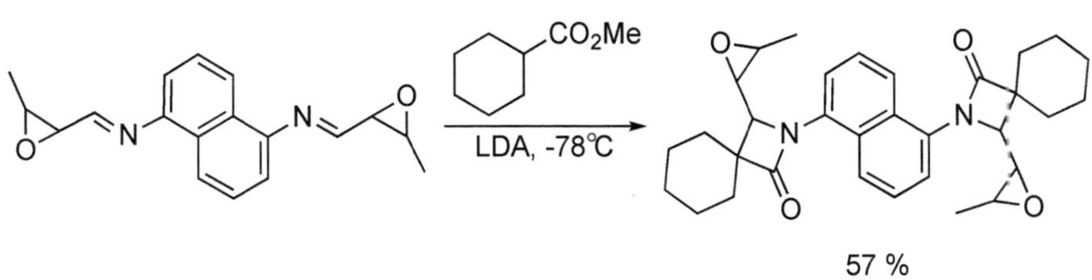

SCHEME 3.30 The preparation of 2,2'-(naphthalene-1,5-diyl)bis(3-(3-methyloxiran-2-yl)-2-azaspiro[3.5]nonan-1-one)

In addition, diazo and triazene functional groups can also tolerate this reaction condition, which renders the possibility of making β-lactams by means of solid-phase synthesis with a triazene linker. For example, in one model experiment, methyl (*E*)-(4-(3,3-dibenzyltriaz-1-en-1-yl)benzoyl) alaninate was treated with 2.2 equivalents of LiHMDS in THF at –78°C for 1 hour, then 1–3 equivalent of (*E*)-*N*,1-diphenylmethanimine in THF was added at –78°C, and the reaction mixture was allowed to warm to room temperature in 23 hours to afford 71% of 4-((*E*)-3,3-dibenzyltriaz-1-en-1-yl)-*N*-((3*R*,4*R*)-3-methyl-2-oxo-1,4-diphenylazetidin-3-yl)benzamide, with greater than 96% of diastereoselectivity. Similarly, methyl (*E*)-2-(4-(3,3-dibenzyltriaz-1-en-1-yl)phenyl)propanoate was subject to the same reaction condition to yield 69% of (3*R*,4*R*)-3-(4-((*E*)-3,3-dibenzyltriaz-1-en-1-yl)phenyl)-3-methyl-1,4-diphenylazetidin-2-one, with 46% of diastereoselectivity. Based on these model reactions, the diazotized esters were loaded onto the surface of benzylamine resin. For the particular case of methyl benzoylalaninate, it was mounted to the resin up to 84% of the loading capacity, and treated with 2.2 equivalents of LiHMDS for 1.5 hours in THF at –78°C, then allowed to react with 3.0 equivalents of (*E*)-1-(4-methoxyphenyl)-*N*-phenyl-methanimine for 23 hours in THF at a temperature raising from –78°C to room temperature, then washed with water. Treatment of the resin loaded β-lactam with 5% trifluoroacetic acid in CH_2Cl_2 cleaved the β-lactam from the resin, yielding 4-(((2*R*,3*R*)-2-(4-methoxyphenyl)-3-methyl-4-oxo-1-phenylazetidin-3-yl)-carbamoyl) benzenediazonium trifluoroacetate, which was then heated in a mixture of THF/DMF (5:2) at 60°C for 15 minutes to afford 69% of *N*-((2*R*,3*R*)-2-(4-methoxyphenyl)-3-methyl-4-oxo-1-phenylazetidin-3-yl)benzamide, with 96% of diastereoselectivity (Scheme 3.31) [95, 96].

In addition to the solid-phase support, carbosilane dendrimers as soluble supports have been developed for the synthesis of β-lactams. In this practice, hydroxy-functionalized dendritic carbosilanes $Si[(CH_2)_3SiMe_2(C_6H_4CH(R)OH)]_4$ (G_0-OH, R = H or (S)-Me) and $Si[(CH_2)_3Si[(CH_2)_3SiMe_2(C_6H_4CH(R)OH)]_3]_4$ (G_1-OH, R = H or (S)-Me) were prepared and the hydroxyl groups were subsequently protected with either 2-arylacetyl or pivaloyl groups. The resulting carbosilane dendrimers supported esters were treated with LDA in THF at –78°C, followed by $ZnCl_2$ to form zinc enolates, which were allowed to react with imines, such as (*E*)-*N*-*tert*-butyl-1-(pyridin-2-yl) methanimine, (*E*)-*N*-(1-phenylethyl)-1-(pyridin-2-yl)-methanimine or (*E*)-1-(pyridin-2-yl)-*N*-(trimethylsilyl)methanimine. For example, when the carbosilane dendrimer protected with both phenylacetyl and pivaloyl group was treated with LDA and $ZnCl_2$, then subject to react with a mixture of imines containing (*E*)-*N*-*tert*-butyl-1-(pyridin-2-yl)methanimine, and (*E*)-*N*-(1-phenylethyl)-1-(pyridin-2-yl)methanimine, all four possible β-lactams,

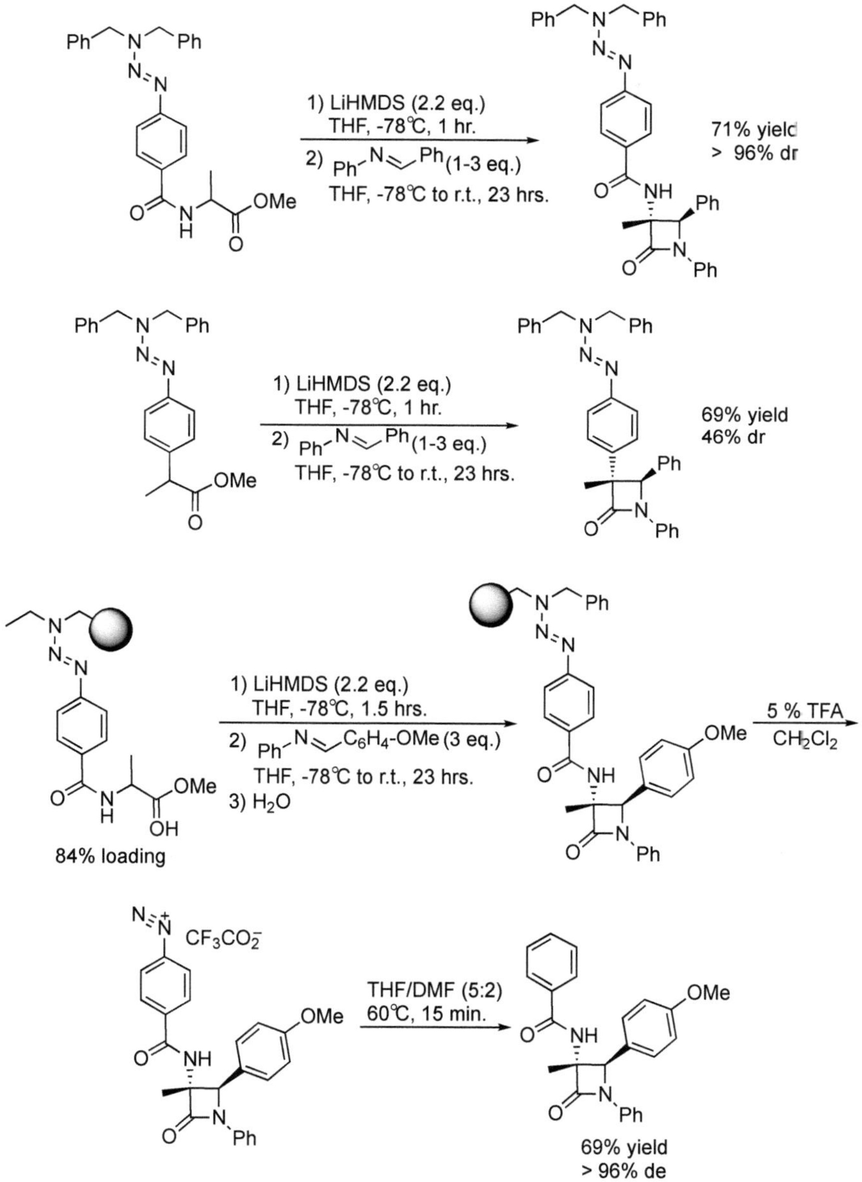

SCHEME 3.31 Conversion of triazene-containing esters into β-lactams

i.e., 1-(*tert*-butyl)-3-phenyl-4-(pyridin-2-yl)azetidin-2-one, 3-phenyl-1-(1-phenylethyl)-4-(pyridin-2-yl)azetidin-2-one, 1,3-di-*tert*-butyl-4-(pyridin-2-yl)azetidin-2-one, and 3-(*tert*-butyl)-1-(1-phenylethyl)-4-(pyridin-2-yl) azetidin-2-one, have been obtained, in equal amount with an overall yield of 85%. However, the zinc-enolate derived from pivalate is not reactive toward (*E*)-1-(pyridin-2-yl)-*N*-(trimethylsilyl)-methanimine, resulting in only 3,4-diphenylazetidin-2-one from the reaction of 2-phenylacetyl group. This result indicates the importance of the match of enolates and imines, as shown in Scheme 3.32. The different reactivity between imine and ester groups on the periphery of carbosilane dendrimers implies the potential for combinatorial synthesis of β-lactams by selectively removing a particular ester group [97].

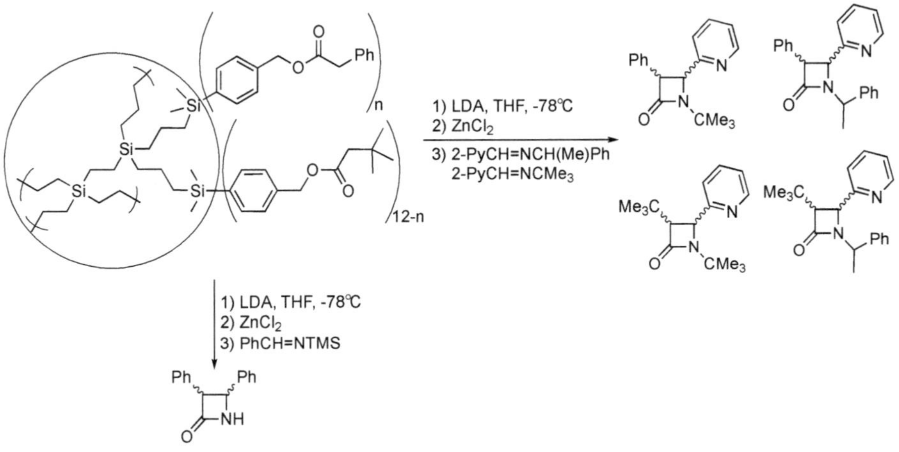

SCHEME 3.32 Carbosilane dendrimers supported the synthesis of β-lactams

In addition to the cycloaddition between enolate and imine, a catalytic version of cycloaddition between imine and vinyl ether in the presence of a chiral catalyst has been established, as represented in the reaction between (*E*)-(1,2-diphenoxyvinyl)trimethylsilane and ethyl (*E*)-2-((4-methoxyphenyl)imino)acetate in the presence of 0.1 equivalent of 6,6,19,19-tetra-(*p*-tolyl)-5,6,19,20-tetrahydrobenzo[*h*]dinaphtho[2,1-*b*:1',2'-*d*][1,6] diphosphecine-6,19-diium difluoride, which afforded 88% of ethyl (2*R*,3*S*)-1-(4-methoxyphenyl)-4-oxo-3-phenoxyazetidine-2-carboxylate with a diastereoselectivity of 88:1 [98].

For the reaction between the enolate of methyl acetate and formaldimine to form β-lactam, *ab initio* calculation using MP2-FC at 6-31+G* basis set taking into account the electrostatic effect of solvent by means of a self-consistent reaction field continuum model indicates that the reaction involves three main steps: the formation of C_3-C_4 bond, the closure of β-lactam ring, and the elimination of methoxide ion. The formation of the C_3-C_4 bond is the rate-determining step which is irreversible [99]. More examples of the approach of ester enolate and imine condensation to make β-lactams can be found in the literature reviews [100–104].

Additional examples for the preparation of β-lactams involving the Staudinger [2+2]-cycloadditions can be found in the individual reports [105–120] and review [121].

3.2.2 WOLFF REARRANGEMENT PATHWAY TO Β-LACTAMS

In addition to the generation of ketene from the treatment of acid chloride with base, Wolff rearrangement [122–125], discovered by Ludwig Wolff in 1902 [126], provides another approach to forming ketenes from the photochemical or thermal decomposition of α-diazocarbonyl compounds by extrusion of nitrogen gas accompanied with 1,2-rearrangement, where the resulting ketenes are normally not isolated but directly subject to the subsequent reactions. Also, the decomposition of α-diazocarbonyl compound can be initiated by a transition metal, such as silver (I) oxide or other silver salts. Like conjugated dienes, α-diazocarbonyl compounds also exist in s-*cis* and s-*trans* conformations, where the distribution of such conformation may influence the mechanism of the rearrangement. In general, the α-diazocarbonyl compounds in s-*cis* conformation undergo a concerted mechanism under photolysis due to the antiperiplanar relationship between the leaving and migrating groups, whereas the compounds in s-*trans* conformation take a stepwise approach through a carbene intermediate or do not rearrange. Also, the α-diazocarbonyl compounds with better migratory groups prefer a concerted mechanism, whereas the transition metal-mediated Wolff rearrangements generally involve the formation of metal carbene intermediates. Regarding the ability to migrate, the hydrogen atom migrates the best, whereas alkyl and aryl groups migrate at different rates according to the reaction condition, where the migration of the alkyl group is favored under photolysis, whereas the migration of the aryl group is preferred under thermolysis. Electron deficient alkyl and aryl groups cannot compete with the electron enriched groups, whereas heteroatoms, in general, are poor migratory groups, due to their ability to donate electron density from their p orbitals into the π^* C=O bond to decrease the migratory ability. Except for NO_2, the effect of other substituents on aryl groups is negligible, but NO_2 reduces the attendance to migrate. Overall, the general migratory trend under photochemical conditions is H > alkyl ≥ aryl >> SR > OR ≥ NR_2, and the trend under thermal conditions is: H > aryl ≥

alkyl, where heteroatoms do not migrate [125]. An illustrative mechanism for the Wolff rearrangement is provided in Scheme 3.33.

As an early example of synthesizing β-lactam involving the Wolff rearrangement, *N*-(*t*-butoxycarbonylacetyl)-*DL*-alanine ethyl ester prepared by coupling *DL*-alanine ethyl ester and *t*-butyl hydrogen malonate in the presence of DCC, was cyclized with potassium *t*-butoxide in benzene. The resulting 5-methylpyrrolidin-2,4-dione was then treated with methanesulfonyl azide in the presence of Et_3N for diazo transfer to afford 3-diazo-5-methylpyrrolidin-2,4-dione in 95% yield. Photolysis of this diazo compound in benzene, in the presence of 1.1 eq. *t*-butyl carbazate with a medium-pressure mercury lamp in a Pyrex vessel for 1 hour at room temperature yields 36% of *tert*-butyl 2-((2*R*,3*S*)-2-methyl-4-oxoazetidine-3-carbonyl)hydrazine-1-carboxylate and

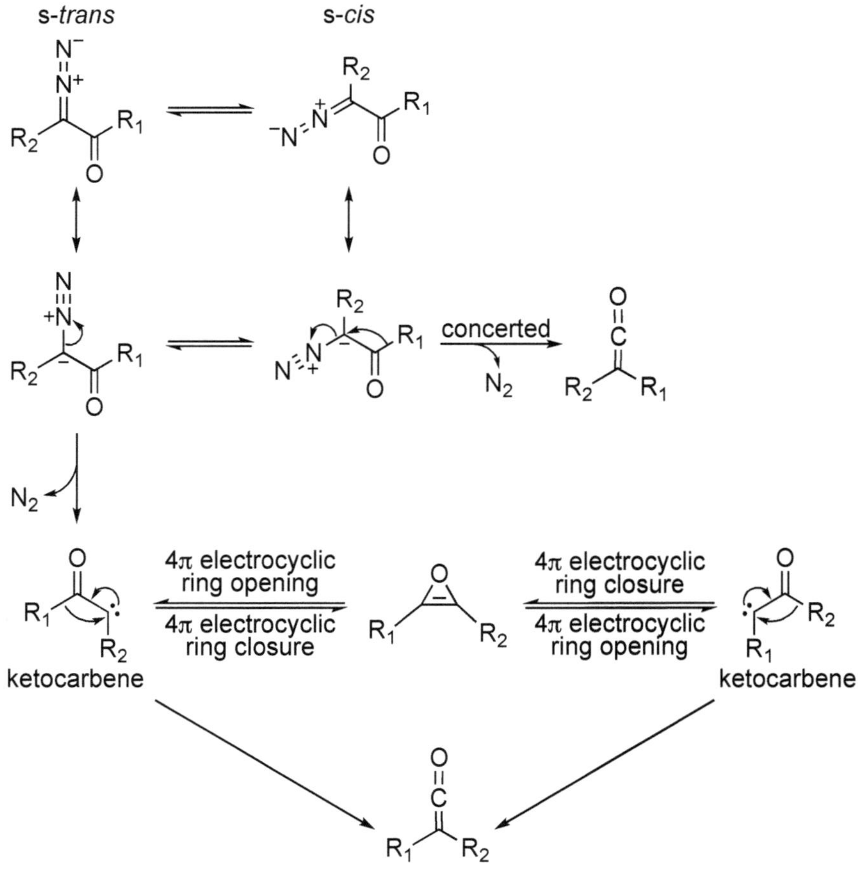

SCHEME 3.33 The mechanism for the decomposition of α-diazocarbonyl compounds into ketenes

55% of *tert*-butyl 2-((2*R*,3*R*)-2-methyl-4-oxoazetidine-3-carbonyl)hydrazine-1-carboxylate, as displayed in Scheme 3.34 [127]. Under a similar condition, although the irradiation of benzyl (3*R*,7a*R*)-6-diazo-5,7-dioxohexahydro-1*H*-pyrrolizine-3-carboxylate at room temperature in the presence of 1.0 equivalent of β-methylphenethyl carbazate gave no detectable β-lactam, the photoreaction performed at –70°C in ether afforded mainly benzyl (2*R*,5*R*,6*R*)-7-oxo-6-(2-(((2-phenylpropan-2-yl)oxy)carbonyl)hydrazine-1-carbonyl)-1-azabicyclo[3.2.0]heptane-2-carboxylate as shown by the IR and NMR of the reaction mixture. However, purification of this product by TLC led to rapid decomposition [128]. Likewise, irradiation of a solution of 0.43 g (*S*)-1-benzyl-5-((benzylthio)methyl)-3-diazopyrrolidin-2,4-dione and 0.19 g of *t*-butyl carbazate in 175 mL dry benzene under nitrogen for 1 hour with a Hanovia 450 W medium-pressure mercury lamp, followed by preparative liquid chromatography (acetone/light petroleum = 7/13) afford 0.42 g of *tert*-butyl 2-((2*S*,3*S*)-1-benzyl-2-((benzylthio)methyl)-4-oxoazetidine-3-carbonyl) hydrazine-1-carboxylate and *tert*-butyl 2-((2*S*,3*R*)-1-benzyl-2-((benzylthio) methyl)-4-oxoazetidine-3-carbonyl)hydrazine-1-carboxylate, in a ratio of 2:3 [129].

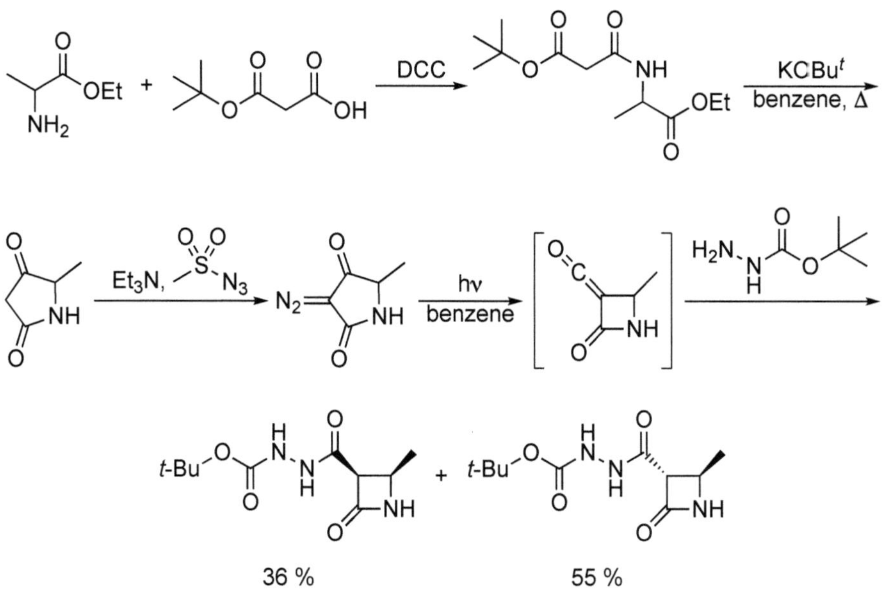

SCHEME 3.34 The preparation of 3-diazo-5-methylpyrrolidin-2,4-dione and its conversion into β-lactam

In contrast, the photolysis of acyclic α-diazocarbonyl compounds, such as 2-diazo-*N,N*-dimethylacetamide, in methanol gave OH insertion product (i.e., 2-methoxy-*N,N*-dimethylacetamide), β-lactam (i.e., 1-methylazetidin-2-one) and the Wolff rearrangement product (i.e., methyl dimethylglycinate). Similarly, direct photolysis of 2-diazo-*N,N*-diethyl-3-oxobutanamide in methanol afforded 16% of 3-acetyl-1-ethyl-4-methylazetidin-2-one and 30% of methyl 3-(diethylamino)-2-methyl-3-oxopropanoate. Quite interestingly, sensitized photolysis of 2-diazo-*N,N*-diethyl-3-oxobutanamide in the presence of benzophenone afforded 3-acetyl-1-ethyl-4-methylazetidin-2-one exclusively, in 25% yield. A mechanism has been proposed to justify the experimental phenomena, as shown in Scheme 3.35. Among these four possible conformers, *E,Z*-conformer and *Z,Z*-conformer are favored as a result of electrostatic attraction whereas *E,E*-conformer is less stable due to the repulsion between two C=O groups, as is predicted from the canonical form of *Z'*,*Z'*-conformer. On the other hand, the *E,E*-conformer would suffer from steric repulsion in the Me-N_2-NEt_2 group, and the nonbonding repulsion between Me and NEt_2 would be severe in the *Z,Z*-conformer. Therefore, it is arguable that the *E,Z*-conformer and *Z,E*-conformer would be more stable than the other two conformers. The formation of the single product under sensitized photolysis is difficult to understand but has been suggested that 3-acetyl-1-ethyl-4-methylazetidin-2-one is derived from a triplet carbene, which is stabilized by the resonance interaction with neighboring carbonyl groups. Such stabilization allows the triplet species to undergo intramolecular H-abstraction from the weaker methylene C-H bonds, followed by recombination [130].

Besides the intramolecular rearrangement of the α-diazocarbonyl compounds, the combination of the generation of ketene and subsequent Staudinger reaction also leads to the formation of β-lactams. For example, a series of α-diazocarbonyl compounds originating from α-amino acids (e.g., alanine, valine, leucine, isoleucine, and phenylalanine) have been irradiated in Et_2O at –30°C in the presence of (*E*)-*N*-benzyl-1-phenylmethanimine, forming diastereomeric mixture of β-lactams, as shown in Scheme 3.36 [131]. This experimental condition has been further extended and optimized [132]. While photolysis of 1,1'-(1,3-phenylene)bis(2-diazoethan-1-one) in toluene leads to the formation of 1,3-bisketenylbenzene (i.e., 2,2'-(1,3-phenylene) bis(ethen-1-one)), which then reacts selectively with (*E*)-*N*,1-diphenylmethanimine to give 58% of 3,3'-(1,3-phenylene)bis(1,4-diphenylazetidin-2-one) as a 1:1 diastereomeric mixture (i.e., (3*S*,4*R*)-3-(3-((3*R*,4*S*)-2-oxo-1,4-diphenylazetidin-3-yl)phenyl)-1,4-diphenylazetidin-2-one and its diastereomer); the reaction between (*E*)-*N*,1-diphenylmethanimine and 2,2'-(1,2-phenylene)

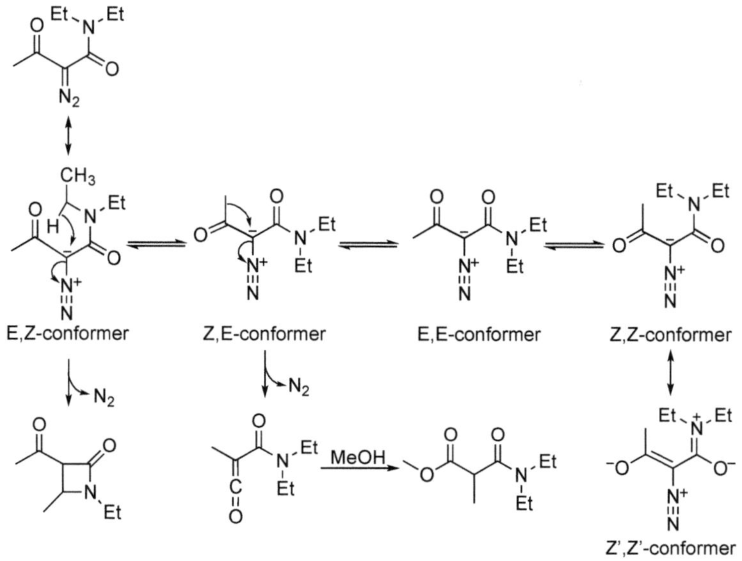

SCHEME 3.35 The possible products from the photolysis of acyclic α-diazocarbonyl compounds in methanol

bis(ethen-1-one) arising from the treatment of 2,2'-(1,2-phenylene)diacetyl chloride with N^1,N^1,N^8,N^8-tetramethylnaphthalene-1,8-diamine afford a mixture of *trans,trans*-3,3'-(1,2-phenylene)bis(1,4-diphenylazetidin-2-one) (chiral), *trans,trans*-3,3'-(1,2-phenylene)bis(1,4-diphenylazetidin-2-one) (achiral), *trans,cis*-3,3'-(1,2-phenylene)bis(1,4-diphenylazetidin-2-one) and 2-(2-((3*S*,4*R*)-2-oxo-1,4-diphenylazetidin-3-yl)phenyl)-*N*-phenylacetamide, in a ratio of 73:12:4:11, according to NMR analysis [133]. The formation of 2-(2-((3*S*,4*R*)-2-oxo-1,4-diphenylazetidin-3-yl)phenyl)-*N*-phenylacetamide

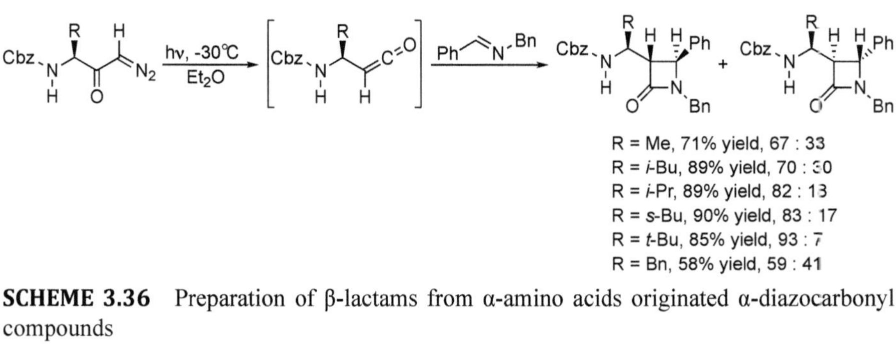

SCHEME 3.36 Preparation of β-lactams from α-amino acids originated α-diazocarbonyl compounds

in the case of 2,2'-(1,2-phenylene)bis(ethen-1-one) can be attributed to the steric hindrance.

The photolysis of α-diazocarbonyl compounds has been recently adopted to form enantiomerically pure *trans*-β-lactams from α-amino acids, as illustrated in Scheme 3.37 starting from *L*-aspartic acid. In this synthesis, *L*-aspartic acid was esterified to 72% of (*S*)-2-amino-4-methoxy-4-oxobutanoic acid according to the literature procedure, which was then sequentially treated with trimethylsilyl chloride, trityl chloride, methanol, and 1,1'-carbonyldiimidazole to afford 93% of methyl (*S*)-4-(1*H*-imidazol-1-yl)-4-oxo-3-(tritylamino)butanoate. Further treatment of this compound with the Weinreb amide and LHMDS led to 72% of methyl (*S*)-6-(methoxy(methyl)amino)-4,6-dioxo-3-(tritylamino) hexanoate. Subsequent diazo transfer reaction with methanesulfonyl azide gave 94% of methyl (*S*)-5-diazo-6-(methoxy(methyl)amino)-4,6-dioxo-3-(tritylamino)hexanoate. Photolysis of this α-diazocarbonyl compounds in the presence of DBU afforded methyl 2-((2*S*,3*S*)-3-(methoxy(methyl)carbamoyl)-4-oxo-1-tritylazetidin-2-yl)acetate. After that, the Weinreb amide was converted

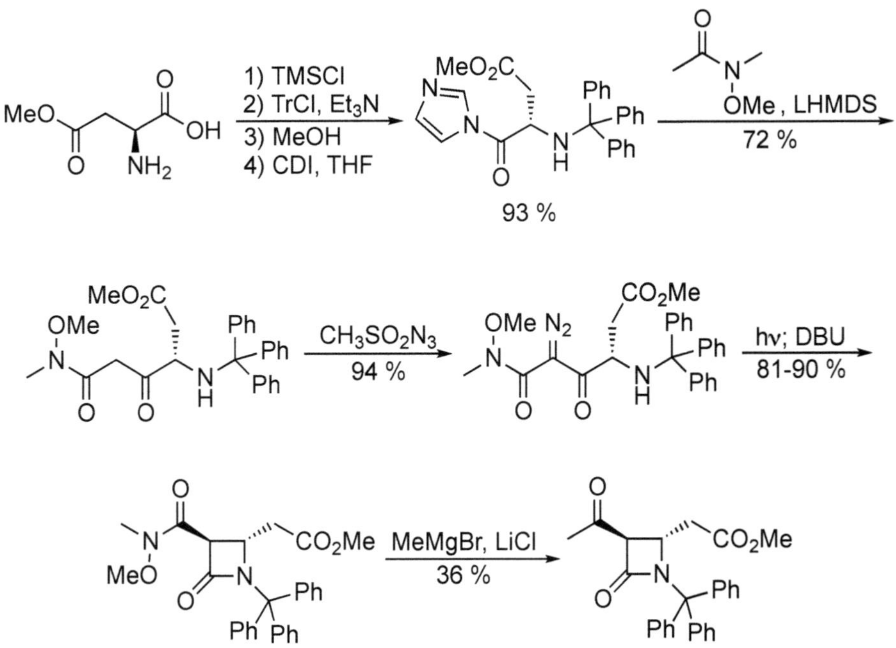

SCHEME 3.37 *L*-Aspartic acid based synthesis of methyl 2-((2*S*,3*R*)-3-acetyl-4-oxo-1-tritylazetidin-2-yl)acetate

into acetyl group with MeMgBr, yielding 36% of methyl 2-((2*S*,3*R*)-3-acetyl-4-oxo-1-tritylazetidin-2-yl)acetate [134]. The photolytic Wolff rearrangement has been extended to the reaction of α-amino acid or dipeptide-derived diazo-ketones in the presence of imine (e.g., *N*-benzylbenzaldimine) or α-amino acid or tripeptide derived imines in ether at -20°C, leading to the formation of peptide-containing β-lactams in *trans*-configuration for the substituents at positions C-3 and C-4. Nearly all suitably protected amino acids can be employed in this reaction [135].

For the example of transition metal-catalyzed conversion of α-diazocarbonyl compounds into ketenes and subsequent Staudinger reaction, $Rh_2(OAc)_4$ has been applied to the reaction between *S*-phenyl 2-diazo-ethanethioate and *N*,1-diphenylmethanimine in CH_2Cl_2 at 40°C for 3 hours, yielding 85% of (3*S*,4*S*)-1,4-diphenyl-3-(phenylthio)azetidin-2-one [136]. Similarly, a novel three-component reaction involving *N*-hydroxyanilines, enynones, and α-diazocarbonyl compounds has been developed by means of relay catalysis of $Rh_2(OAc)_4$ and chiral amine, affording highly functionalized β-lactams with two quaternary carbon centers in good yields and excellent diastereoselectivities. This protocol features a sequential reaction of Rh-catalyzed imine formation, Wolff rearrangement, and benzoylquinine-catalyzed Staudinger cycloaddition, as displayed in Scheme 3.38. In this protocol, $Rh_2(OAc)_4$ catalyzed reaction between *N*-phenylhydroxylamine and enynones, such as 3-(3-phenylprop-2-yn-1-ylidene)pentan-2,4-dione in CH_2Cl_2 at room temperature, leads to the formation of (*E*)-1-(2-methyl-5-(phenyl(phenylimino)methyl)furan-3-yl)ethan-1-one, in 97% yield. This imine is then allowed to react with ethyl 2-diazo-3-oxo-3-phenylpropanoate in the presence of 1 mol% $Rh_2(OAc)_4$ and 3 mol% of (*R*)-(6-methoxyquinolin-4-yl)((1*S*,2*S*,4*S*,5*R*)-5-vinylquinuclidin-2-yl)methyl benzoate in CH_2Cl_2 at 80°C, affording 90% of ethyl (2*S*,3*S*)-2-(4-acetyl-5-methylfuran-2-yl)-4-oxo-1,2,3-triphenylazetidine-3-carboxylate. In contrast, no β-lactam is formed in the absence of $Rh_2(OAc)_4$. These two-step reactions can be carried out in a one-pot manner, giving β-lactams of high diastereoselectivity [137]. In a very similar fashion, a reaction between *N*-hydroxyaniline and methyl 2-diazo-2-phenylacetate has been screened for rhodium catalysts, for the purpose of making methyl (*Z*)-2-phenyl-2-(phenylimino)acetate efficiently. This reaction can be carried out in a one-pot manner in the presence of methyl (*E*)-2-diazo-3-oxo-5-phenylpent-4-enoate to form β-lactam directly. Among the tested rhodium catalysts, $Rh_2(OAc)_4$ is the most effective one [138].

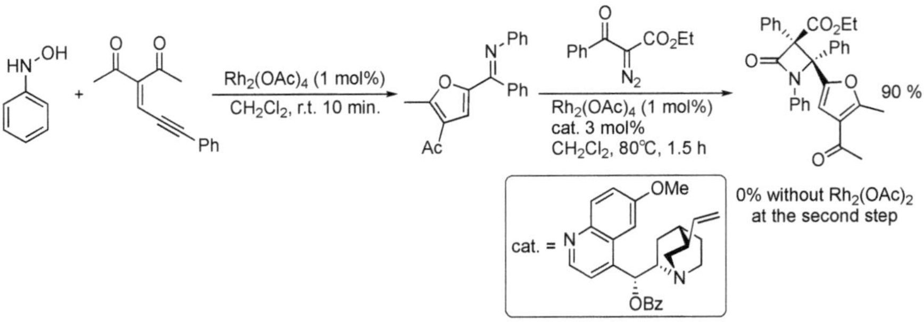

SCHEME 3.38 $Rh_2(OAc)_4$ catalyzed preparation of ethyl (2*S*,3*S*)-2-(4-acetyl-5-methylfuran-2-yl)-4-oxo-1,2,3-triphenylazetidine-3-carboxylate

The example of thermal decomposition of α-diazocarbonyl compounds in the formation of β-lactams in conjunction with imines can be demonstrated in the reaction of 1-((3*R*,5*R*,7*R*)-adamantan-1-yl)-2-diazoethan-1-one with a series of imines, affording a mixture of substituted (3*R*,4*S*)-3-((3*R*,5*R*,7*R*)-adamantan-1-yl)-1,4-diphenylazetidin-2-one and (3*R*,4*R*)-3-((3*R*,5*R*,7*R*)-adamantan-1-yl)-1,4-di-phenylazetidin-2-one (major product), as outlined in Scheme 3.39. This reaction has been performed by initial heating of 0.2 mmol imine in 10 mL of dry xylene up to reflux, followed by the addition of 0.26 mmol of 1-((3*R*,5*R*,7*R*)-adamantan-1-yl)-2-diazoethan-1-one in 5 mL xylene within 10 minutes. The reaction mixture was refluxed for 12 hours and the residue after removal of solvent was purified by flash chromatography to remove the highly polar impurities, and the crude mixture was further purified by silica gel column chromatography [139]. Recently, a single-mode bench-top resonator was applied for a Wolff-Staudinger cascade reaction involving the microwave-assisted flow generation of primary ketenes by thermal decomposition of α-diazocarbonyl compounds at high temperature (165°C) and high pressure (20 bars) and subsequent reaction with imines, affording β-lactams of preferential *trans*-configuration in most cases. In this study, the α-diazocarbonyl compounds are: 1-diazopropan-2-one, 1-diazo-3,3-dimethylbutan-2-one, 2-diazo-1-(4-methoxyphenyl)ethan-1-one, 1-(benzofuran-2-yl)-2-diazo-ethan-1-one and 2-diazo-1-(pyridin-3-yl)ethan-1-one, which have been prepared as 0.06 M solution in acetonitrile and allowed to pass through the resonator at a speed of 0.5 mL/min. Meanwhile, acetonitrile solution of 13 different imines each at 0.125 M is passed through the resonator at the same speed. The mechanism of the reaction has been discussed as well [140]. In addition, the Wolff-Staudinger cascade reaction between 2-diazo-1-(ferrocenyl) ethan-1-one and *N*,1-diphenylmethanimine at various temperatures in toluene (80°C, 110°C) or xylene (140°C) reveals the preference at higher temperature with a higher yield. Extension of this reaction to 2-diazo-1-(4-methoxyphenyl)ethan-1-one, 2-diazo-1-(4-nitrophenyl)ethan-1-one and 2-diazo-1-phenylethan-1-one with different imines at 140°C in xylene, all lead to the formation of β-lactams in favored *trans*-configurations [141]. In addition, 20 α-diazocarbonyl compounds containing a trifluoromethane-sulfonyl group have been applied to a one-pot two steps synthesis of the corresponding β-lactams, involving the initial thermal decomposition of the diazo-compounds in toluene at 100°C and subsequent addition of the imine derivatives. In this case, the substituent effect on the reactivity of the diazo-compounds has been clearly displayed. For example, the reactivity decreases as the electron-withdrawing effects of the aryl functional groups increase (CF_3 > Cl > OMe) due to the lower migration abilities of the electron-deficient aryl groups to generate the ketene intermediates under the Wolff rearrangement [142].

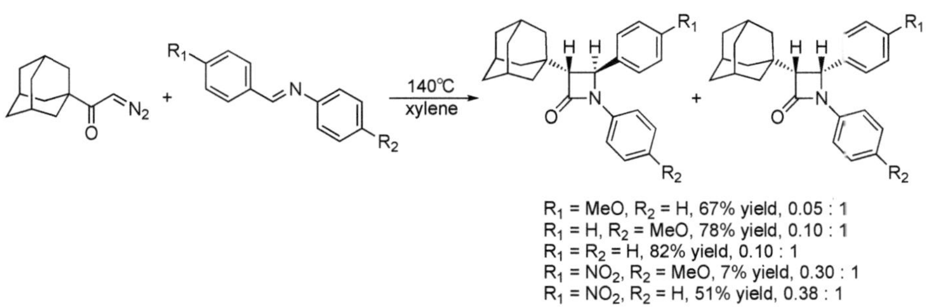

SCHEME 3.39 The thermal reaction of 1-((3*R*,5*R*,7*R*)-adamantan-1-yl)-2-diazoethan-1-one with imines to afford β-lactams

It should be pointed out that even though a series of β-lactams can be prepared by the addition of the α-diazocarbonyl compounds into the refluxing solution of imines generated from the condensation between aldehydes and primary amines, as illustrated in the reaction between methyl 2-diazo-3-oxobutanoate and *N*-benzyl-1-(*p*-tolyl)methanimine that affords 86% of methyl (3*R*,4*S*)-1-benzyl-3-methyl-2-oxo-4-(*p*-tolyl) azetidine-3-carboxylate, the reaction of methyl 2-diazo-4,4-dimethyl-3-oxopentanoate with the same imine does not proceed effectively, giving 0% of the expected β-lactam. This might be caused by the failure of this compound to undergo the regular Wolff rearrangement, as no ethyl (S)-3,3-dimethyl-2-(*p*-tolylcarbamoyl)butanoate has been obtained from the reaction between ethyl 2-diazo-4,4-dimethyl-3-oxopentanoate and *p*-toluidine after being refluxed in toluene for 20–24 hours. Similarly, the reaction between *tert*-butyl 2-diazo-3-oxobutanoate and *N*-benzyl-1-cyclohexylmethanimine, and the reaction between methyl 2-diazo-3-oxobutanoate and *N*-(furan-2-ylmethyl)-2-methylpropan-1-imine fail to generate the expected β-lactam, instead, enamides and other side products are formed. The formation of enamides has been attributed to the formation of α-C-H imine that potentially undergoes the proton transfer as displayed in Scheme 3.40 [143].

More recent examples for the synthesis of β-lactams involving Wolff rearrangement can be found in the individual reports [144–147].

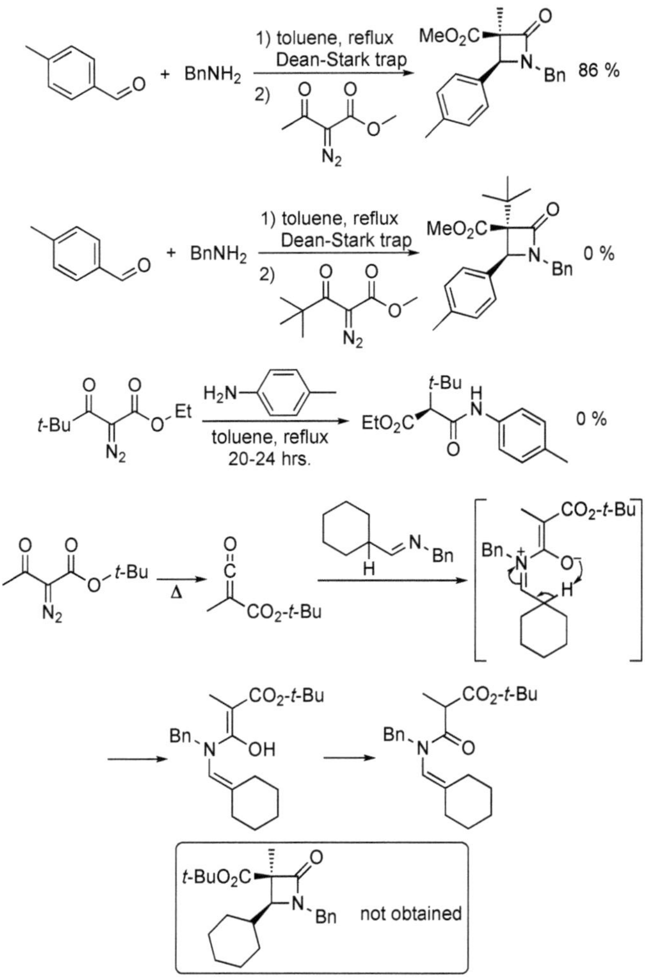

SCHEME 3.40 The potential failure of applying the Wolff rearrangement in the preparation of β-lactams

3.2.3 [2+2]-CYCLOADDITION BETWEEN ALKENE AND ISOCYANATE

In addition to the Staudinger reaction ([2+2] cycloaddition between ketene and imine) and the Wolff rearrangement to *in situ* generate ketene that then reacts with imine, another method to create β-lactam involving [2+2]-cycloaddition is the reaction between alkene and isocyanate. In many of these types of cycloadditions, chlorosulfonyl isocyanate (CSI) has often been applied to react with a variety of olefins. CSI reacts with alkenes to give *N*-chlorosulfonyl β-lactams that can be readily reduced to the corresponding β-lactams. However, CSI is unreactive with electron-deficient alkenes [148]. For example, treatment of the [2+2]-cycloaddition product between CSI and (*S,E*)-*tert*-butyldimethyl((1-(phenylthio)pent-1-en-3-yl)oxy)silane in diisopropyl ether with ethanethioic *S*-acid (i.e., $CH_3C(O)SH$) leads to the formation of (3*R*,4*S*)-3-((*S*)-1-((*tert*-butyldimethylsilyl)oxy)propyl)-4-(phenylthio)azetidin-2-one, which is then converted into sodium (5*R*,6*R*)-6-((*S*)-1-hydroxypropyl)-7-oxo-3-((*R*)-tetrahydrofuran-2-yl)-4-thia-1-azabicyclo[3.2.0]hept-2-ene-2-carboxylate after sequential transformations, for the purpose of making broad-spectrum anti-methicillin-resistant *Staphylococcus aureus* β-lactam antibiotics, as shown in Scheme 3.41 [149].

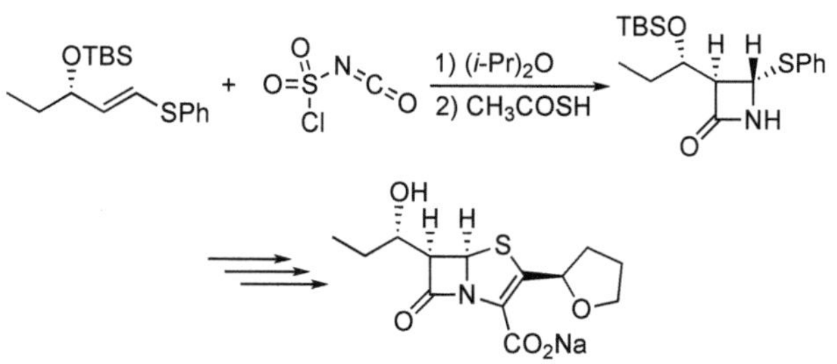

SCHEME 3.41 Application of CSI in preparation of (3*R*,4*S*)-3-((*S*)-1-((*tert*-butyldimethylsilyl)oxy)propyl)-4-(phenylthio)azetidin-2-one

Investigation of the reaction rates for the cycloadditions between CSI and alkenes at various temperatures indicates that CSI reacts readily with electron-enriched olefins, whereas it reacts sluggishly with electron-deficient alkenes as well as many monofluoroalkenes. In one kinetic study, the specific reaction rates for the cycloaddition of CSI or fluorosulfonyl isocyanate to olefins have been determined by infrared spectroscopy, where the effect of chlorine or fluorine, the nature of the olefin, and the polarity of solvent on the reaction rates have been examined [150]. When the reaction is performed above room temperature, a bimolecular reaction profile predominates, whereas when the reaction is carried out below room temperature, the reaction between CSI and alkene takes a two-step approach to initially form a CSI/alkene complex, which then undergoes a slow unimolecular reaction in second-order to form the corresponding *N*-chlorosulfonyl β-lactam in a concerted mechanism [151], or involves a single electron transfer (SET) process to give the SET intermediate, which then slowly transforms into chlorosulfonyl β-lactam. In this case, electron-deficient alkenes, typically with an ionization potential of 8.9 eVs or above react with CSI via the concerted mechanism, whereas olefins with an ionization potential of 8.5 eVs or less react with CSI by the SET pathway, as outlined in Scheme 3.42 [148, 152]. Therefore, the efficiency of the reaction between CSI and alkene can be improved at temperatures between –15°C and 25°C because CSI and the alkene are in equilibrium with an intermediate at these temperatures, and the resulting unimolecular reaction of the intermediate is more efficient than the bimolecular reaction of the dissociated reagents above the room temperature. Hexahydroimidazo advantage of such cycloaddition at low temperature is the less susceptible rearrangement and decomposition of the *N*-chlorosulfonyl β-lactam, as indicated in the decomposition of many *N*-chlorosulfonyl β-lactams in nitromethane above room temperature. For example, a series of alkenes (e.g., cyclopentene, cyclohexene, cycloheptene, (*Z*)-cyclooctene, cyclohex-1,3-diene, styrene, indene, norbornene, and benzonorborndiene) dissolved in 20 mL of CH_2Cl_2 was allowed to react with 1.1 equivalent of CSI for 2 hours at 0°C, and the reaction mixture was added to methanol and stirred for 1 hour to afford the corresponding sulfamide ester in good yield, as illustrated by the reaction of cycloheptene to yield 80% of methyl (1*R*,2*S*)-2-((methoxysulfonyl)amino)cycloheptane-1-carboxylate, instead of forming (1*R*,7*S*)-9-oxo-8-azabicyclo[5.2.0]nonane-8-sulfonyl chloride or methyl (1*R*,7*S*)-9-oxo-8-azabicyclo[5.2.0]nonane-8-sulfonate (Scheme 3.43) [153]. Similarly, a series of cyclic β-*N*-3,5-dinitrobenzoylamido esters have been prepared, taking the advantage of the instability of *N*-chlorosulfonyl β-lactams, which were further purified on chiral HPLC. For a particular case, ring opening of (1*R*,6*S*)-7-azabicyclo[4.2.0]octan-8-one obtained from the treatment of (1*R*,6*S*)-8-oxo-7-azabicyclo[4.2.0]octane-7-sulfonyl chloride from the reaction of cyclohexene and CSI with sodium sulfite and esterification of the resulting β-amino acid (i.e., (1*R*,2*S*)-2-aminocyclohexane-1-carboxylic acid) was accomplished by heating a 1:1.2:1.2 molar ratio of the β-lactam, undecenyl alcohol, and methanesulfonic acid, respectively, in benzene by means of azeotropic removal of water overnight. Subsequent acylation of the crude amino ester undec-10-en-1-yl (1*R*,2*S*)-2-aminocyclohexane-1-carboxylate with 3,5-dinitrobenzoyl chloride led to the final product undec-10-en-1-yl

(1*R*,2*S*)-2-(3,5-dinitrobenzamido)cyclohexane-1-carboxylate, as shown in Scheme 3.44 where n = 1 [154]. In addition, β-lactams formed from cycloaddition of CSI with alkenes also undergo thermal rearrangement to yield γ-lactams [155]. In a series of reactions of CSI with 4-methyloct-4-ene, (4*E*,8*E*)-4,8-dimethyldodeca-4,8-diene and 3-ethyl-2-methylpent-1-ene to mimic each type of the microstructures of polyisoprene, free lactams and amides have been generated by basic hydrolysis, and amino acid hydrochlorides, as well as unsaturated nitriles, have also been formed, as illustrated in Scheme 3.45 for the reaction between CSI and 3-ethyl-2-methylpent-1-ene. In this reaction, the initial reaction intermediate may cyclize to afford the expected β-lactam, i.e., 2-methyl-4-oxo-2-(pentan-3-yl)azetidine-1-sulfonyl chloride, or undergo proton abstraction by the nitrogen atom to afford either (4-ethyl-3-methylhex-3-enoyl)sulfamoyl chloride or (4-ethyl-3-methylene-hexanoyl)sulfamoyl chloride. Alternatively, the initial reaction intermediate may undergo [1,2]-H shift to give a new carbocation species, which cyclizes to afford γ-lactam, i.e., 2,2-diethyl-3-methyl-5-oxopyrrolidine-1-sulfonyl chloride, or γ-lactone, i.e., 5,5-diethyl-4-methyldihydrofuran-2(3*H*)-one, by means of oxygen attacking of the carbocation center followed by hydrolysis. This example clearly demonstrates the potential complexity of the reaction of CSI with alkenes [156].

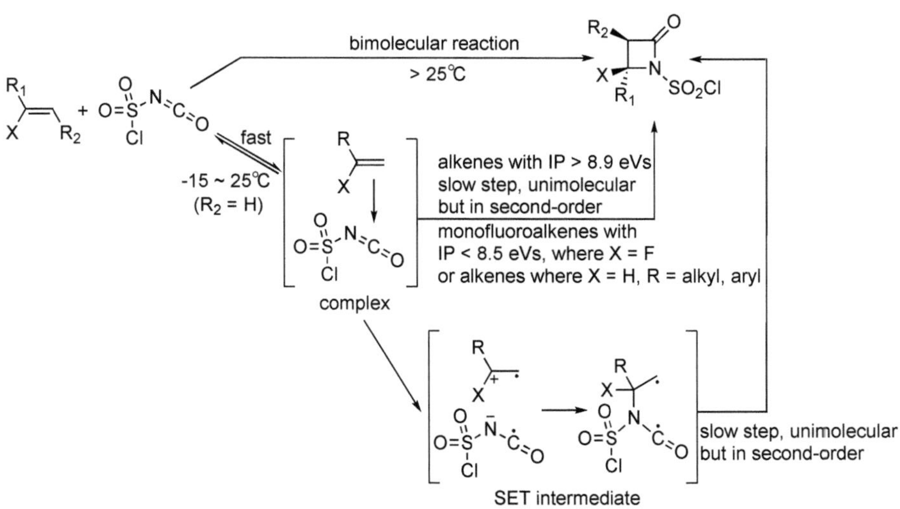

SCHEME 3.42 The reaction mechanisms between CSI and alkene at different temperatures

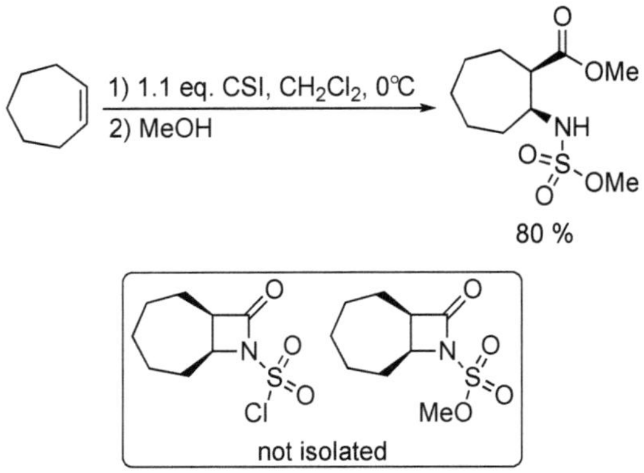

SCHEME 3.43 The reaction between cycloheptene and CSI that affording methyl (1*R*,2*S*)-2-((methoxysulfonyl)amino)cycloheptane-1-carboxylate

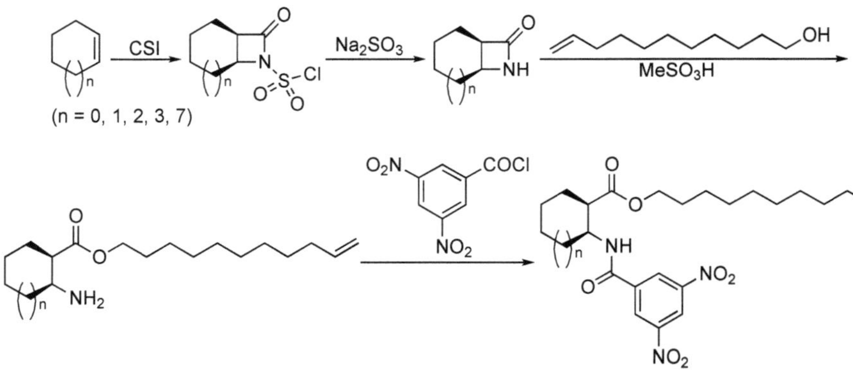

SCHEME 3.44 The preparation of bicyclic β-lactams from cycloalkenes and CSI and subsequent reaction with alcohol

The complexity of the reaction involving CSI also appears in the reaction with conjugated dienes, such as buta-1,3-diene, isoprene, 2,3-dimethylbuta-1,3-diene, (*E*)-penta-1,3-diene, (*Z*)-penta-1,3-diene, (*E*)-hexa-1,3-diene and (2*E*,4*E*)-hexa-2,4-diene. In these cases, CSI plays an antarafacial role as the π^2a component in concerted reactions with π^2s systems at the low reaction temperature. The resulting β-lactams are Markovnikov-oriented 1,2-cycloadducts in which CSI has added to the terminal double bond. Interestingly, no symmetry-allowed π^4s + π^2s cycloadducts have been observed. However, the initially formed *N*-chlorosulfonyl-β-lactams thermally rearrange to a variety of products, as demonstrated in Scheme 3.46 for the reaction of 2,3-dimethylbuta-1,3-diene with CSI. In this particular reaction, the initially

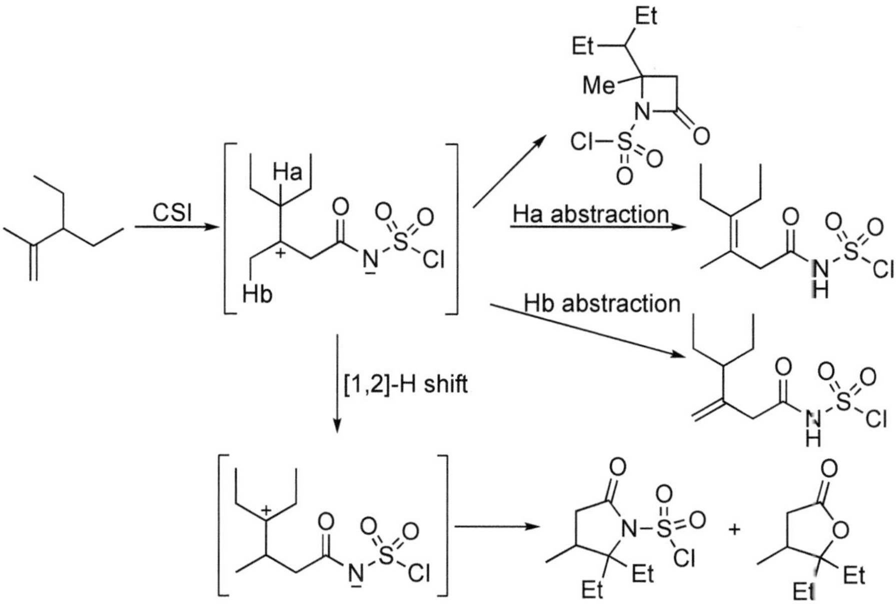

SCHEME 3.45 The possible products from the reaction between CSI and 3-ethyl-2-methylpent-1-ene

formed 2-methyl-4-oxo-2-vinyl-azetidine-1-sulfonyl chloride from isoprene thermally rearranges to (*E*)-(4-methyl-3,6-dihydro-2*H*-pyran-2-ylidene)sulfamoyl chloride, 4-methyl-2-oxo-3,6-dihydropyridine-1(2*H*)-sulfonyl chloride, (*E*)-(3-methylpenta-2,4-dienoyl)sulfamoyl chloride and (*E*)-3-methylpenta-2,4-dienamide. In addition, (*E*)-(4-methyl-3,6-dihydro-2*H*-pyran-2-ylidene) sulfamoyl chloride further rearranges to (*E*)-(4-methyl-5,6-dihydro-2*H*-pyran-2-ylidene)sulfamoyl chloride or decomposes to 4-methyl-3,6-dihydro-2*H*-pyran-2-one, where (*E*)-(4-methyl-5,6-dihydro-2*H*-pyran-2-ylidene) sulfamoyl chloride then decomposes to 4-methyl-5,6-dihydro-2*H*-pyran-2-one and (*Z*)-3-methylpenta-2,4-dienoic acid [157]. A similar reaction has been reported elsewhere [158]. In comparison, when the reaction of CSI and isoprene in equimolar amount was performed in ether at -65°C, followed by a slow temperature rise to -10°C and cooling again to -65°C, 2-methyl-4-oxo-2-vinylazetidine-1-sulfonyl chloride can be obtained in 60–80% yield. This unstable compound was then hydrolyzed with saturated NaOH-MeOH in Et_2O to give 50% of 4-methyl-4-vinylazetidin-2-one [159].

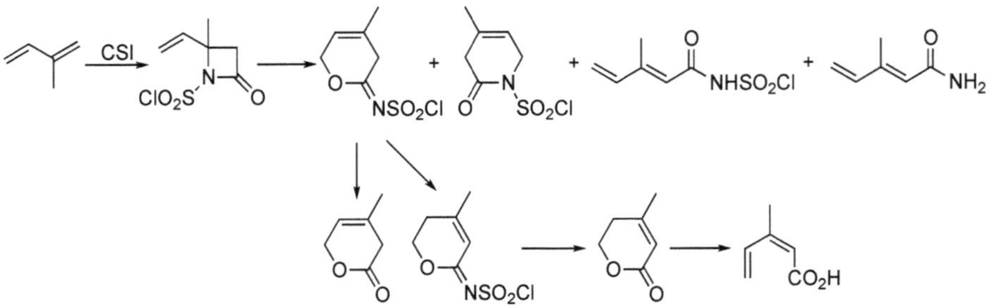

SCHEME 3.46 The products from the reaction between CSI and isoprene

The reaction between CSI and alkenes might be stereoselective, as indicated in the reactions between CSI and stilbenes where the reaction of *cis*-stilbene yields > 90% of *cis*-2-oxo-3,4-diphenylazetidine-1-sulfonyl chloride whereas the reaction of *trans*-stilbene affords exclusively *trans*-2-oxo-3,4-diphenylazetidine-1-sulfonyl chloride [160]. Also, the reaction of CSI with ε- or *Z*-prop-1-en-1-ylbenzene stereospecifically affords *N*-chlorosulfonyl-*trans*- and *cis*-3-methyl-4-phenyl-2-azetidinone accordingly. Similarly, the reaction of CSI with ε- or *Z*-hex-3-ene generates the stereospecific cycloadduct of *N*-chlorosulfonyl-*trans*- and *cis*-3,4-diethyl-2-azetidinone [161]. It should be pointed out that the pyrolysis of β-lactams is also stereospecific [162].

Using CSI, the β-lactam of methyl (1*R*,2*R*,5*S*,7*R*,9*S*)-4-oxo-3-azatricyclo[5.3.1.02,5]undecane-9-carboxylate has been prepared from the reaction with methyl (1*S*,3*S*,5*R*)-bicyclo[3.3.1]non-6-ene-3-carboxylate and subsequent reduction of methyl (1*R*,2*R*,5*S*,7*R*,9*S*)-3-(chlorosulfonyl)-4-oxo-3-azatricyclo-[5.3.1.02,5]undecane-9-carboxylate with Na_2SO_3 [163]. Similarly, several Fmoc-protected non-proteogenic amino acids, including (1*S*,2*R*)-2-((((9*H*-fluoren-9-yl)methoxy)carbonyl)amino)cyclopentane-1-carboxylic acid, (1*S*,2*R*)-2-((((9*H*-fluoren-9-yl)methoxy)carbonyl)amino) cyclohexane-1-carboxylic acid and (1*R*,2*S*, 3*R*,4*S*)-3-((((9*H*-fluoren-9-yl) methoxy)carbonyl)amino)bicyclo[2.2.1]-heptane-2-carboxylic acid have been prepared in large scale (up to 1 kg) by means of the reaction of CSI with cyclopentene, cyclohexene, and norbornene in CH_2Cl_2, followed by the treatment of *N*-chlorosulfonyl β-lactams with sodium sulfite in aqueous NaOH solution (pH = 5–7). Acidic hydrolysis of the β-lactam leads to the generation of non-proteogenic β-amino acid, which is then protected with Fmoc. It is found that during the acidic hydrolysis of β-lactam in the presence of concentrated HCl, dipeptide may form as a diastereomeric mixture that contaminates the desired β-amino acid, as indicated in the formation of (1*R*,2*S*)-2-((1*S*,2*R*)-2-aminocyclopentane-1-carboxamido) cyclopentane-1-carboxylic acid hydrochloride and (1*S*,2*R*)-2-((1*S*,2*R*)-2-aminocyclopentane-1-carboxamido)-cyclopentane-1-carboxylic acid hydrochloride during the hydrolysis of (1*S*,5*R*)-6-azabicyclo[3.2.0]heptan-7-one [164]. For volatile olefins, such as buta-1,3-diene with its boiling point at –5°C, the corresponding β-lactams are usually obtained in poor to moderate yields by the normal procedure. However, when the reaction is carried out in a pressure bottle at low temperature, and the intermediate chlorosulfonyl β-lactams are immediately reduced with bisulfite, the yields of β-lactams can be dramatically enhanced, as demonstrated in the reaction of CSI with buta-1,3-diene, isoprene, isobutene, and vinyl acetate [165].

The versatility for the generation of β-lactams with CSI can be clearly demonstrated in the reaction between CSI and cyclohexa-1,4-diene, for which the racemic mixture of 8-oxo-7-azabicyclo[4.2.0]oct-3-ene-7-sulfonyl chloride intermediate is treated with 5 *N* NaOH to afford 7-azabicyclo[4.2.0] oct-3-en-8-one. From this key β-lactam, several different β-lactams as well as β-amino acids can be prepared via the manipulation on the existing olefinic functional group, including (1*S*,9*R*)-5,5-dimethyl-4,6-dioxa-10-azabicyclo[7.2.0]undecan-11-one, (3*S*,4*R*)-1-(*tert*-butyldimethylsilyl)-3,4-bis(2-iodoethyl)azetidin-2-one, *tert*-butyl (2-((2*R*,3*S*)-2-ethyl-4-oxoazetidin-3-yl) ethyl)carbamate, *tert*-butyl ((1*S*,3*S*,6*R*)-7-(*tert*-butyldimethylsilyl)-8-oxo-7-azabicyclo[4.2.0]octan-3-yl)carbamate and their enantiomers as well as (1*S*,6*R*)-6-((S)-3,3,3-trifluoro-2-methoxy-2-phenylpropanamido)cyclohex-3-ene-1-carboxylic acid, along with many intermediates prepared during the sequential transformations [166]. Moreover, the application of CSI in the formation of β-lactam has been applied to prepare a series of highly substituted cyclobutane fused ring systems through a one-pot domino process. In this protocol, the fused aminocyclobutanes are derived from a domino [3,3]-rearrangement/6π-electrocyclization process in a highly diastereoselective fashion. This approach works well even if adjacent quaternary centers are present, and has been applied to make systems containing up to four fused rings, as illustrated in the treatment of 1,2-dimethylenecyclohexane with CSI in aqueous Na_2SO_3, leading to the formation of 5-methylene-1-azaspiro[3.5] nonan-2-one. This β-lactam is then treated with 1-iodocyclohept-1-ene in THF in the presence of catalytic amount of CuI and Cs_2CO_3, and base N^1,N^2-dimethylethane-1,2-diamine under microwave irradiation, affording 81% of (7a*R*,8a*S*,13a*R*)-4,5,6,7,8a,9,10,11,12,13-decahydro-8*H*-cyclohepta[2,3] cyclo-buta[1,2-*j*]isoquinolin-2(1*H*)-one in one-pot fashion. In the latter domino process, CuI catalyzes the coupling of 1-iodocyclohept-1-ene to the nitrogen atom, leading to the formation of 1-(cyclohept-1-en-1-yl)-5-methylene-1-azaspiro[3.5]nonan-2-one, which undergoes [3,3]-sigmatropic rearrangement to yield (*Z*)-1,2,3,4,5,8,9,10,11,12,13,13a-dodecahydro-7*H*-benzo[*e*]cyclohepta[*b*]-azocin-7-one. In addition, it is believed that CuI functions as a Lewis acid to complex with the carbonyl group (and possibly the nitrogen atom as well) of (*Z*)-1,2,3,4,5,8,9,10,11,12,13,13a-dodecahydro-7*H*-benzo[*e*]cyclohepta[*b*]azocin-7-one, which activates the α-hydrogen and facilitates the deprotonation by N^1,N^2-dimethylethane-1,2-diamine. The *in situ* generated enolate, i.e., (5a*Z*,7*E*)-2,3,4,5,9,10,11,12,13,13a-decahydro-1*H*-benzo[*e*]cyclohepta[*b*]azocin-7-olate easily undergoes the electrocyclic ring closure reaction and subsequent tautomerization to afford the final product in diastereoselective fashion. This process is displayed in Scheme 3.47 [167].

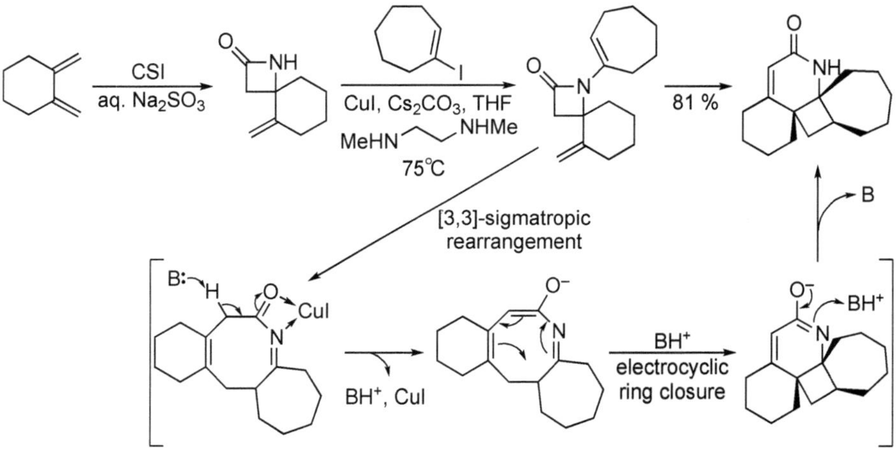

SCHEME 3.47 The preparation of (7a*R*,8a*S*,13a*R*)-4,5,6,7,8a,9,10,11,12,13-decahydro-8*H*-cyclohepta[2,3]cyclobuta[1,2-*j*]isoquinolin-2(1*H*)-one from the reaction of 1,2-dimethylenecyclohexane with CSI

In addition to the commonly used CSI, some other substituted sulfonyl isocyanates, including fluorosulfonyl isocyanate [150], *N*-(2,2,2-trichloroethoxy)sulfonyl isocyanate, 2,2,2-trichloroethanesulfonyl isocyanate, and trifluoroacetyl isocyanate [168], have been shown to be applicable in preparation of β-lactams. Like the treatment in the case of CSI, the corresponding β-lactams can be obtained by reductive desulfonylation of the expected sulfonyl β-lactams, whereas the β-lactams formed from the reaction of trifluoroacetyl isocyanate with alkenes, e.g., methylenecyclohexane, 5-methyl-3,4-dihydro-2*H*-pyran, and 5-benzyl-3,4-dihydro-2*H*-pyran can be obtained after the detrifluoroacetylation on Florisil [168].

In addition to so many experimental examples for the reaction of isocyanate and alkenes, theoretical treatment of this [2+2]-cycloaddition has been carried out at different levels. For example, *ab initio* calculations based on 3-21G, 6-31G, and 6-311G basis sets predict incorrect linear geometries of isocyanates, whereas semiempirical calculations using MNDO and AM1 give significantly different geometries of transition states from the ones predicted with *ab initio* methods, and PM3 is the most accurate semiempirical method in prediction of the heat of formation for HNCO [169]. *Ab initio* calculations, such as MP2/6-31+G*, MP3/6-31+G*, and MP4SDQ/6-31+G*, indicate that the [2+2]-cycloaddition takes place via concerted transition structures with a zwitterionic character that retains the configuration of the starting olefins. When olefins bearing π-donating groups and/or isocyanates carrying electron-withdrawing groups, the synchronicity as well as the activation energy of the reaction diminish, leading to exclusively the 4-substituted regio-isomers. In addition, solvent also diminishes the synchronicity of the reaction, resulting in the transition of a concerted mechanism to a two-step process, in line with the loss of stereospecificity observed in the reaction between sulfonyl isocyanates and vinyl ethers [170]. *Ab initio* calculations (HF/6-31G* and MP2/6-31G**) as well as density functional theory calculation (pBP86/6-31G*) yield similar results, i.e., the reaction proceeds via a concerted one-step mechanism, and olefins with an electron-donating group or isocyanate with an electron-withdrawing group facilitate the cycloaddition; although the transition state is slightly asynchronous, where C_2-C_3 bond is shorter than the N_1-C_4 bond [171]. Different *ab initio* calculations also indicate a concerted suprafacial mechanism for the reaction between isocyanates (HNCO or ClNCO) with ethene, propene, and vinyl alcohol [172]. Density functional theory calculation (B3LYP/6-31G*) indicates that the intramolecular [2+2]-cycloaddition of 4-methylpent-4-ene-1-sulfonyl isocyanate or 3-methylbut-3-ene-1-sulfonyl isocyanate to form 6-methyl-2-thia-1-azabicyclo[4.2.0]octan-8-one 2,2-dioxide or 5-methyl-2-thia-1-azabi-cyclo[3.2.0]heptan-7-one 2,2-dioxide is an endergonic process. This result is consistent with the experimental fact that the above isocyanates once formed from the treatment of the corresponding sulfonylcarbamates with BCl_3 and Et_3N in refluxing benzene, were immediately transformed into sulfonylureas in the presence of amine, whereas the expected cycloadducts could not be achieved under any of the reaction conditions applied [173].

More examples of the [2+2]-cycloaddition involving CSI and alkenes can be found in several reviews [174, 175], and the computational studies on the synthesis of β-lactams via [2+2]-cycloaddition has been collectively summarized recently [176].

Finally, it should be pointed out that cationic iron(II) vinylidenes behave as ketene surrogates in their reactions with imines and 2-thiazolines. Subsequent oxidation of the resultant mono- and bicyclic carbene complexes lead to the formation of mono- and bicyclic β-lactams. However, several reagents including iodosobenzene, aqueous bromine, pyridine *N*-oxide, and oxygen have been proved unsuitable for oxidation, as very low yields of β-lactams are obtained during decomposition of the carbene complexes with the above oxidizing agents. In addition, the low diastereoselectivity for monocyclic carbene complexes might be caused by the intermediacy of vinyl iron intermediates and/or either (*E*/*Z*) isomerization prior to or low facial selectivity on ring closure [177]. Similarly, the iron vinylidene complex $[((MeO)_3P)Cp(CO)Fe{=}C{=}CMe_2]^+CF_3SO_3^-$ where Cp = η^5-cyclopentadienyl reacted in a stepwise manner with PhCH=NMe, 2-thiazoline, ethyl 2-thiazoline-4-carboxylate and methyl 5,5-dimethyl-2-thiazoline-4-carboxylate to afford the corresponding [2+2] azetidylidene cycloadducts in 51–82% yields. Subsequent oxidation with iodosobenzene in ethanol solution gives the β-lactams [178].

More examples of the synthesis of β-lactams from the [2+2] cycloaddition involving alkenes and isocyanates can be found in additional individual reports [179–187].

3.2.4 KINUGASA REACTION

In addition to the above three common methods for the generation of β-lactams, Kinugasa reaction has gradually become another popular approach. The Kinugasa reaction was initially reported in 1972 by Kinugasa and Hashimoto for the treatment of a copper acetylide with a nitrone to afford β-lactam [188]. Due to its several obvious features, including its optimal atom economy, employment of readily accessible starting materials, a convergent approach, and potentially asymmetrical synthesis of β-lactams, the Kinugasa reaction has drawn tremendous attention recently, as indicated in the number of recent reviews [189–196]. However, this reaction is not perfect though; two of its limitations are the formation of multiple byproducts throughout the course of the reaction and the modest yields of the desired β-lactams. As a result, this reaction has been regarded as "an 'ugly duckling' of β-lactam chemistry" in one of the above reviews [189]. For this particular reaction, several mechanisms have been proposed [196–201]. Deuteriation and isomerization studies indicate that the *trans*-β-lactams are produced from the initially formed *cis*-isomers and that the amount of *cis*-products may be increased by using non-basic solvents [200]. Also, the presence of sterically encumbered secondary amines is beneficial for the superior yield of β-lactam [199]. Further investigation of the Kinugasa reaction, using tetrakis(acetonitrile)copper(I) tetrafluoroborate as the Cu(I) source, tris((1-cyclopentyl-1*H*-1,2,3-triazol-4-yl)methyl)amine as the ligand, and diisopropylamine as the base, indicates that the *cis*-diastereomer of β-lactams is favored, but in low yield as about 60% of the initial starting materials have been converted into the corresponding imine and amide in nearly 1:1 ratio. The formation of imines has been attributed to the Cu-catalyzed deoxygenation of nitrone, and the decomposition of the Cu-dihydroisoxazolide may justify the 1:1 ratio of amine and amide. In addition, the substitution of diisopropylamine with a stronger, non-nucleophilic amine base, such as DBU (1,8-diazabicyclo-(5.4.0)undec-7-ene) led to an increased yield of β-lactam (from 17% to 78%), with a concomitant reduction of imine formation, where the β-lactam was predominantly in *trans*-configuration [201]. In addition, the acceptable yield of β-lactams can be obtained in the presence of only a catalytic amount of CuI without a decrease in diastereoselectivity [202]. Based on the latest experimental results, an updated mechanism has been proposed for the Kinugasa reaction, as shown in Scheme 3.48, in which B stands for a base, L represents a ligand, and BH^+ is the conjugated acid of the base.

A recent application of the Kinugasa reaction in preparation of β-lactam has been demonstrated in a reaction between copper(I) acetylides and cyclic nitrones derived from chiral amino alcohols and glyoxylic acid, in which the alkyne molecule approaches exclusively *anti* to the large substituent next to the nitrogen atom of the nitrone component to provide preferentially *cis*-substituted β-lactam. As an example, (*S*)-6-oxo-3-phenyl-3,6-dihydro-2*H*-1,4-oxazine 4-oxide reacts with 3,3-diethoxyprop-1-yne in acetonitrile at –35°C in the presence of 2.0 equivalents of cuprous chloride and triethylamine, affording 51% of (2*S*,6*S*,7*R*)-7-(diethoxymethyl)-2-phenyl-4-oxa-1-azabicyclo[4.2.0]octane-5,8-dione and (2*S*,6*S*, 7*S*)-7-(diethoxymethyl)-2-phenyl-4-oxa-1-azabicyclo[4.2.0]octane-5,8-dione, in a ratio of 10:3. Similarly, the reaction of this nitrone with 2-ethynylisoindoline-1,3-dione under a similar condition but at –30°C, yielded 48% of (2*S*,6*S*,7*S*)-7-(1,3-dioxoisoindolin-2-yl)-2-phenyl-4-oxa-1-azabicyclo[4.2.0]-octane-5,8-dione and (2*S*,6*S*,7*R*)-7-(1,3-dioxoisoindolin-2-yl)-2-phenyl-4-oxa-1-azabicyclo[4.2.0]-octane-5,8-dione, in a ratio of 10:6. These two reactions are displayed in Scheme 3.49 [203].

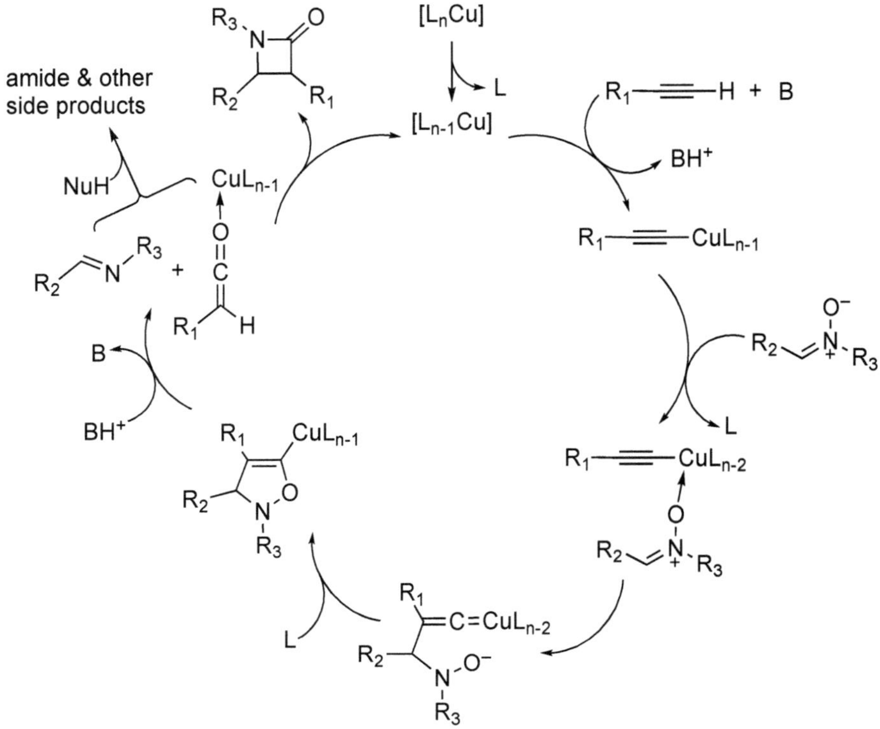

SCHEME 3.48 The mechanism of the Kinugasa reaction in the formation of β-lactam

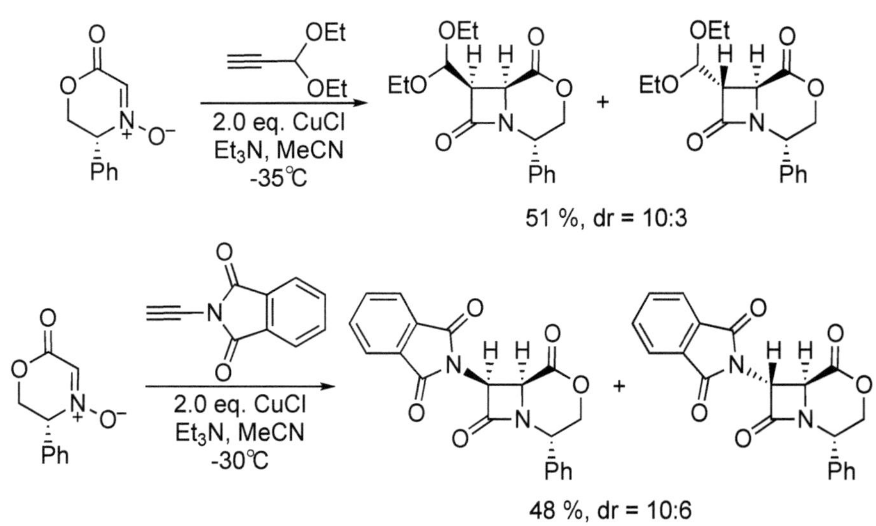

SCHEME 3.49 The Kinugasa reaction of (*R*)-6-oxo-3-phenyl-3,6-dihydro-2*H*-1,4-oxazine 4-oxide with different alkynes

While it has been found that the stereochemical outcome of the Kinugasa reaction involving cyclic nitrones is primarily controlled by the cyclic nitrones, regardless of reaction partner as an achiral or optically active alkyne, the Kinugasa reaction involving acyclic nitrones has demonstrated a different trend, as shown in a series of reactions between acyclic nitrones with or without a chiral center (e.g., (*Z*)-*N*-benzyl-1-(5,5-dimethyl-1,3-dioxan-2-yl) methanimine oxide, (*S*,*Z*)-*N*-benzyl-1-(2,2-dimethyl-1,3-dioxan-4-yl)methanimine oxide, (*R*,*Z*)-*N*-benzyl-1-(2,2-dimethyl-1,3-dioxan-4-yl)methanimine oxide, (*S*,*Z*)-*N*-benzyl-1-(2,2-dimethyl-1,3-dioxolan-4-yl)methanimine oxide and (*R*,*Z*)-*N*-benzyl-1-(2,2-dimethyl-1,3-dioxolan-4-yl)methanimine oxide) and acetylenes. The acetylenes explored in this study include (*R*)-4-ethynyl-2,2-dimethyl-1,3-dioxolane, (*S*)-4-ethynyl-2,2-dimethyl-1,3-dioxolane, (*S*)-4-ethynyl-2,2-dimethyl-1,3-dioxane, (*R*)-4-ethynyl-2,2-dimethyl-1,3-dioxane and 3,3-diethoxyprop-1-yne. Among these reaction pairs, it is shown that the stereogenic center present in the sidechains of chiral acyclic nitrones does not provide sufficient stereo-differentiation of the faces of the corresponding nitrones in reaction with 3,3-diethoxyprop-1-yne, leading to a low asymmetric induction and poor stereoselectivity of β-lactams. In contrast, the reaction between (*Z*)-*N*-benzyl-1-(5,5-dimethyl-1,3-dioxan-2-yl)methanimine oxide and chiral acetylenes proceeds with high stereoselectivity. When both reaction partners possess a dioxane moiety, the Kinugasa reaction proceeds with high stereoselectivity and yields predominantly the *cis*-products. Therefore, chiral acetylene appears to play a decisive role in the context of asymmetric induction, where the configuration of the stereogenic center next to the triple bond fully controls the configuration at C-4 of the azetidin-2-one ring, whereas the configuration of the nitrone stereogenic center can be neglected, due to the free rotation of the nitrone substituents. In addition, owing to the linear symmetry of the triple bond, the nitrone approaches the acetylene component from the place between the small and medium substituents on the stereogenic center of acetylene [204]. Similar results for stereoselectivity have been observed in a series of reactions between (*R*)-4-ethynyl-2,2-dimethyl-1,3-dioxolane (or (*R*)-(but-3-yn-2-yloxy)(*tert*-butyl)dimethylsilane) and one of the following nitrones, i.e., (*Z*)-1-(furan-2-yl)-*N*-(4-methoxyphenyl)methanimine oxide, (*Z*)-*N*-(2-ethoxy-2-oxoethyl)-1-(4-methoxyphenyl)methanimine oxide and (*Z*)-*N*-(2-ethoxy-2-oxoethyl)-1-(furan-2-yl)methanimine oxide, in the presence of different base, such as Et_3N, tetramethylguanidine, DIPEA, DBU, and 2,6-lutidine [205].

Moreover, the effect of solvent and temperature on the catalytic version of Kinugasa reaction, in the presence of ligand 4-(2-(diphenylphosphaneyl)-

naphthalen-1-yl)-*N*-((*R*)-1-phenylethyl)phthalazin-1-amine, has been studied using the reaction pair of 3,3-diethoxyprop-1-yne and (*Z*)-*N*,1-diphenyl-meth-animine oxide. At room temperature, acetonitrile works the best among the tested solvents including EtOH, CH_2Cl_2, THF, DMF, toluene, water, EtOAc, and CH_3CN. When the reaction is performed in CH_3CN, there is nearly no temperature effect on the *cis*/*trans* selectivity. On the other hand, the same reaction in CH_3CN in the presence of a different copper catalyst (e.g., CuI, CuBr, CuCl, CuCN, CuOAc, $Cu(OAc)_2$, CuOTf·toluene (2:1), $Cu(OTf)_2$, CuSPh) does not show a dramatic variation on the *cis*/*trans* selectivity, where CuI appears to be superior to other copper catalysts [206]. In another study on the solvent effect, the reaction between propargyl alcohol and (1*E*,2*E*)-*N*,3-diphenylprop-2-en-1-imine oxide in the presence of Et_3N and CuI clearly indicates that DMF outperforms other solvents tested, including DMSO, CH_3CN, and THF. In addition, the same reaction in the presence of DIPEA or K_2CO_3 nearly fails to afford the expected product. According to the optimized reaction condition, propargyl alcohol or propargyl bromide has been subject to react with different α,β-unsaturated nitrones, affording the expected *cis*-β-lactams in very good yields [207].

Also, two fluorine-containing nitrones, i.e., (*Z*)-*N*-benzyl-2,2-difluoroethan-1-imine oxide and (*Z*)-*N*-benzyl-2,2,2-trifluoroethan-1-imine oxide, particularly the latter, have been applied to react with a series of acetylenes in the presence of Et_3N and CuI in acetonitrile at room temperature, affording fluorine-containing β-lactams where the fluorine-containing group is always located at C_4 of the β-lactams. When a catalytical amount of chiral ligand was added, as shown in the reaction between (*Z*)-*N*-benzyl-2,2,2-trifluoroethan-1-imine oxide and ethynyl-benzene, the β-lactam was obtained in high diastereoselectivity (Scheme 3.50). In the presence of any ligand from L1 to L4, i.e., 2-((*R*)-4-phenyl-4,5-dihydrooxazol-2-yl)-6-((*S*)-4-phenyl-4,5-dihydrooxazol-2-yl) pyridine, bis((3a*S*,8a*R*)-3a,8a-dihydro-8*H*-indeno[1,2-*d*]oxazol-2-yl)methane, (3a*S*,3a'*S*,8a*R*,8a'*R*)-2,2'-(propane-2,2-diyl)bis(3a,8a-dihydro-8*H*-indeno[1,2-*d*]oxazole), and (*S*)-[1,1'-binaphthalene]-2,2'-diol), only one diastereomer was observed, whereas in the presence of ligand L5 (i.e., 4-(2-(diphenylphosphaneyl)naphthalen-1-yl)-*N*-((*R*)-1-phenylethyl)phthalazin-1-amine), a mixture of *cis*/*trans* (6:4) β-lactams was obtained. Interestingly, in the presence of ligand L1, *cis*-β-lactam, i.e., (3*S*,4*S*)-1-benzyl-3-phenyl-4-(trifluoromethyl) azetidin-2-one was obtained exclusively, whereas in the presence of ligand L2 to L4, the *trans*-β-lactam, i.e., (3*R*,4*S*)-1-benzyl-3-phenyl-4-(trifluoromethyl) azetidin-2-one was obtained. In addition, the presence of chiral ligand also enhanced the enantioselectivity, with *ee* up to 26% [208].

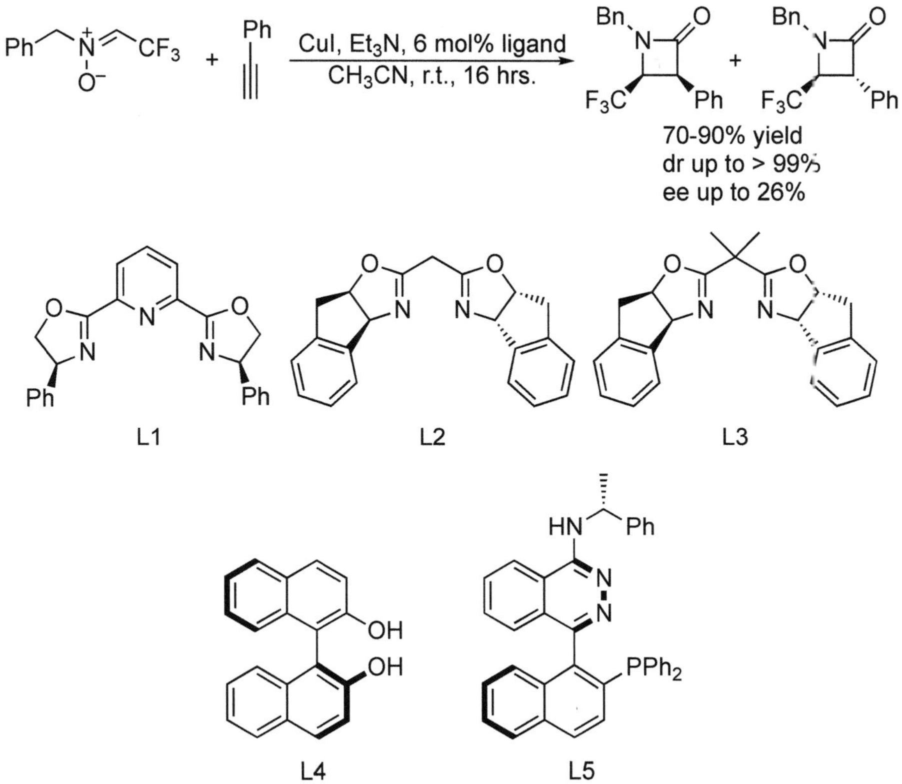

SCHEME 3.50 The Kinugasa reaction between (*Z*)-*N*-benzyl-2,2,2-trifluoroethan-1-imine oxide and ethynylbenzene in the presence of chiral ligand

In addition to the ligands displayed in Scheme 3.50, another series of eight chiral prolinol-phosphine ligands have been explored for the Kinugasa reaction between (*Z*)-1-cyclohexyl-*N*-phenylmethanimine oxide and ethynylbenzene in the presence of diethylamine and a catalytical amount of $Cu(OTf)_2$, of which (*S*)-1-((*S*)-1-(2-(bis(4-methoxyphenyl)phosphaneyl) benzyl)pyrrolidin-2-yl)-2-(tripropyl-silyl)-ethan-1-ol outperforms other similar ligands in terms of chemical yield, diastereoselectivity, and enantioselectivity of β-lactams. Based on these preliminary results, this ligand has been further extended to a series of Kinugasa reactions by changing the reaction partners of either nitrones or acetylenes. For a particular reaction between 1-ethynyl-4-methoxybenzene and (*Z*)-1-cyclohexyl-*N*-phenylmethanimine oxide in the presence of diethylamine and a catalytic amount of $Cu(OTf)_2$ and the ideal ligand, 95% of β-lactam was obtained, with a 96:4 ratio for the *cis*- and *trans*-diastereomers, as shown in Scheme 3.51. The enantiomeric excesses for (3*R*,4*S*)-4-cyclohexyl-3-(4-methoxyphenyl)-1-phenylazetidin-2-one and (3*S*,4*S*)-4-cyclohexyl-3-(4-methoxyphenyl)-1-phenylazetidin-2-one have been measured at 92% and 36%, respectively. For comparison, the reaction between 2-ethynyl-5-methylthiophene and the same nitrone under this condition afforded 71% of β-lactams, in a ratio of 83:17 for the *cis*- and *trans*-diastereomers, each with an ee% greater than 98%. In most cases, the tested reaction pairs favor the *cis*-β-lactams [209]. In addition, eight different chiral prolinol-phosphine ligands have been screened for the reaction between nitrones and propargyl alcohol or its derivatives such as *tert*-butyl prop-2-yn-1-yl carbonate, 3-(methoxymethoxy)prop-1-yne, prop-2-yn-1-yl benzoate, diethyl prop-2-yn-1-yl phosphate, for the purpose of making chiral α-alkylidene-β-lactams. In this reaction protocol, (*S*)-1-((*S*)-1-(2-(bis(4-(trifluoromethyl)phenyl)phosphaneyl)benzyl)pyrrolidin-2-yl)-2-(trimethylsilyl)-ethan-1-ol outperforms other seven prolinol-phosphine ligands [210].

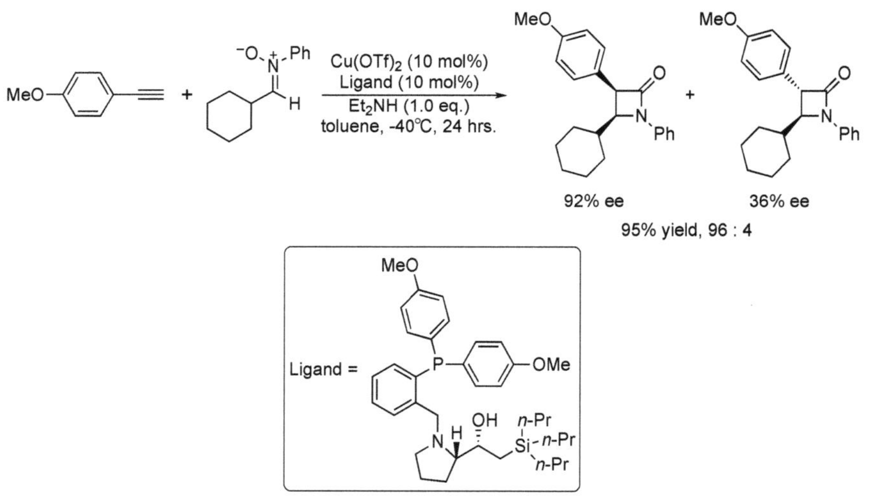

SCHEME 3.51 The Kinugasa reaction between (*Z*)-1-cyclohexyl-*N*-phenylmethanimine oxide and 1-ethynyl-4-methoxybenzene

Interestingly, the reaction between diethyl ethynylphosphonate and (*E*)-*N*-benzyl-1-phenylmethanimine oxide, or other nitrones such as (*E*)-*N*-benzyl-1-(4-methoxyphenyl)methanimine oxide, (*E*)-*N*-benzyl-1-(4-(trifluoromethyl)phenyl)-methanimine oxide, (*E*)-*N*-benzyl-1-(4-bromophenyl)methanimine oxide, (*E*)-*N*-benzyl-1-(furan-2-yl)methanimine oxide, (*E*)-*N*-methyl-1-phenylmethanimine oxide, (*E*)-*N*-methyl-1-(*p*-tolyl) methanimine oxide or (*E*)-*N*,1-diphenyl-methanimine oxide, in the presence of Et_3N and CuI in acetonitrile, all lead to the formation of *trans*-β-lactams exclusively [211].

Very recently, the intramolecular version of the Kinugasa reaction on *in situ* generated carbohydrate-derived alkynylnitrones has been reported, in order to find out the effects of chain length, their mutual configuration, and the influence of experimental conditions on product distribution and feasibility of the β-lactam ring construction. In contrast to the intermolecular version of the Kinugasa reaction, the intramolecular Kinugasa reaction proceeds with high stereoselectivity and provides only one product in each case. In this protocol, δ-glucose has been acetalized with acetone and converted into (*R*)-1-((3a*R*,5*R*,6*S*,6a*R*)-6-(benzyloxy)-2,2-dimethyltetrahydrofuro[2,3-d][1,3]dioxol-5-yl)ethane-1,2-diol, of which the primary hydroxyl group is then protected with trityl chloride in pyridine, yielding (*R*)-1-((3a*R*,5*R*,6*S*,6a*R*)-6-(benzyloxy)-2,2-dimethyltetrahydrofuro[2,3-*d*]-[1,3]dioxol-5-yl)-2-(trityloxy)ethan-1-ol. The remaining secondary hydroxyl group is allowed to react with propargyl bromide in the presence of phase transfer catalyst tetrabutyl ammonium bromide (TBAB), giving 57.9% of (3a*R*,5*R*,6*S*,6a*R*)-6-(benzyloxy)-2,2-dimethyl-5-((*R*)-1-(prop-2-yn-1-yloxy)-2-(trityloxy)ethyl)tetra-hydrofuro[2,3-*d*][1,3]dioxole. Subsequent removal of the trityl protecting group in the presence of toluenesulfonic acid yields 85.7% of (*R*)-2-((3a*R*,5*R*,6*S*,6a*R*)-6-(benzyloxy)-2,2-dimethyltetra-hydrofuro[2,3-*d*][1,3]dioxol-5-yl)-2-(prop-2-yn-1-yloxy)ethan-1-ol, which is then converted into ethyl 2-((*R*)-2-((3a*R*,5*R*,6*S*,6a*R*)-6-(benzyloxy)-2,2-dimethyltetra-hydrofuro[2,3-*d*][1,3]dioxol-5-yl)-2-(prop-2-yn-1-yloxy)ethoxy)acetate by means of ethyl bromoacetate. Mild reduction of the ester group with DIBAL-H affords 94% of 2-((*R*)-2-((3a*R*,5*R*,6*S*,6a*R*)-6-(benzyloxy)-2,2-dimethyltetrahydrofuro[2,3-d][1,3]dioxol-5-yl)-2-(prop-2-yn-1-yloxy)ethoxy)-acetaldehyde, which is then converted into the nitrone functionality with *N*-benzylhydroxylamine. Subsequent addition of cuprous bromide and Et_3N prompts the intramolecular Kinugasa reaction, affording a single product of (1*S*,4*R*,8*R*)-9-benzyl-4-((3a*R*,5*R*,6*S*,6a*R*)-6-(benzyloxy)-2,2-dimethyltetrahydrofuro[2,3-*d*][1,3]-dioxol-5-yl)-3,6-dioxa-9-azabi-cyclo[6.2.0]decan-10-one, as shown in Scheme 3.52 [212].

Another intramolecular Kinugasa reaction has been applied to efficiently prepare fused tricyclic ring systems with very good levels of enantioselectivity in the presence of a planar-chiral Cu/phosphaferrocene-oxazoline catalyst, as displayed in Scheme 3.53. In one of these intramolecular Kinugasa reactions, (*E*)-1-(2-(but-3-yn-1-yl)phenyl)-*N*-(4-(ethoxycarbonyl) phenyl)methanimine oxide, in the presence of 0.5 equivalent of *N*-cyclohexyl-*N*-methylcyclohexanamine and a catalytic amount of CuBr and phosphaferrocene-oxazoline ligand, has been converted into 74% of ethyl 4-((2a*S*,8b*S*)-2-oxo-2a,3,4,8b-tetrahydronaphtho[1,2-*b*]azet-1(2*H*)-yl) benzoate, with 88% of ee [197].

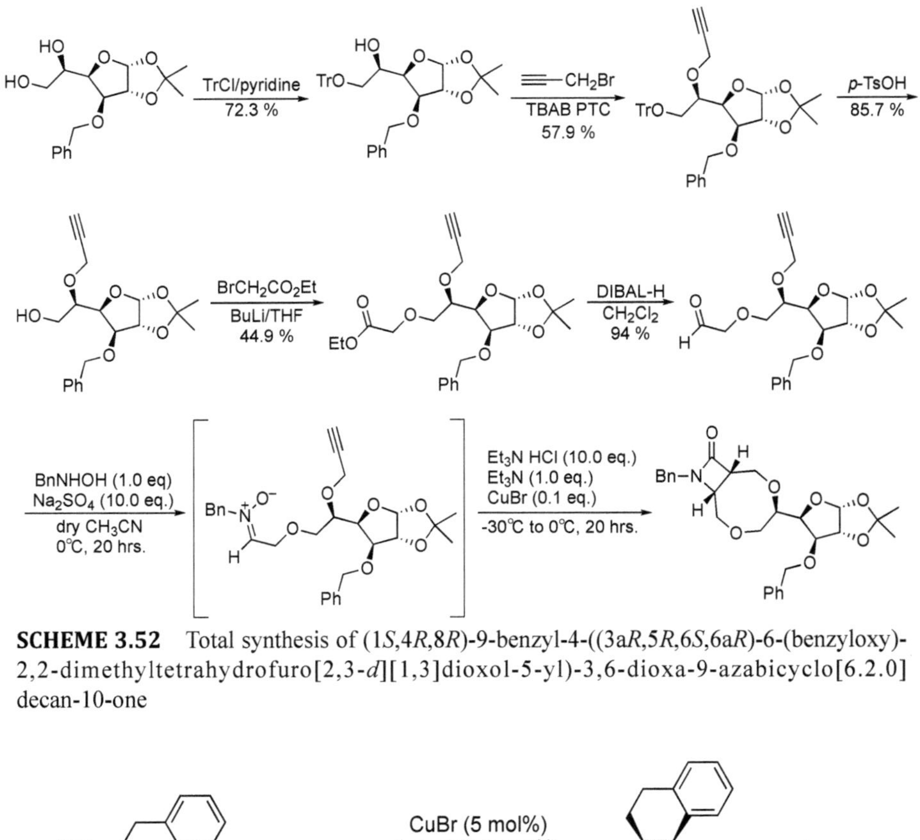

SCHEME 3.52 Total synthesis of (1*S*,4*R*,8*R*)-9-benzyl-4-((3a*R*,5*R*,6*S*,6a*R*)-6-(benzyloxy)-2,2-dimethyltetrahydrofuro[2,3-*d*][1,3]dioxol-5-yl)-3,6-dioxa-9-azabicyclo[6.2.0]decan-10-one

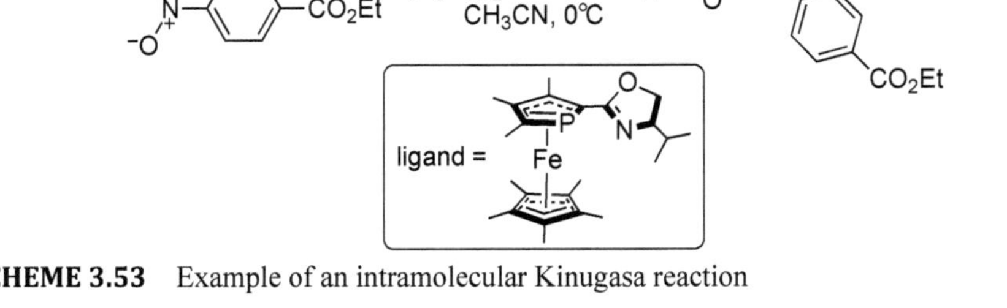

SCHEME 3.53 Example of an intramolecular Kinugasa reaction

Recently, a scalable and highly chemo-, regio-, diastereo-, and enantioselective cascade reaction involving the Kinugasa reaction and Michael addition has been reported for the preparation of spirocyclic β-lactam bearing a fused bicyclic and spiro-fused bicyclic framework with four contiguous stereocenters. To optimize the reaction condition, seven different chiral ligands have been tested for the model reaction between 1-(prop-2-yn-1-yloxy)-[1,1'-biphenyl]-4(1*H*)-one and (*Z*)-*N*,1-diphenylmethanimine oxide in the presence of DIPEA and a catalytic amount of $Cu(OTf)_2$, where the ligand (3a*S*,3a'*S*,8a*R*,8a'*R*)-2,2'-(1,3-bis(4-(*tert*-butyl)phenyl)propane-2,2-diyl)bis(3a,8a-dihydro-8*H*-indeno[1,2-*d*]oxazole) stands out to be the superior ligand. Then, this reaction has been further optimized for a suitable base, among DIPEA, dibutylamine, dicyclohexylamine, and diisobutylamine. The ideal reaction condition has been illustrated in Scheme 3.54, for the generation of (3*R*,3a'*R*,4*R*,7a'*R*)-1,3a',4-triphenyl-3',3a',7',7a'-tetrahydrospiro[azetidine-3,1'-indene]-2,6'(2'*H*)-dione with 83% yield and 94% ee. This ideal reaction condition has been extended to different acetylene and nitrone combinations, all demonstrating excellent diastereoselectivity as well as enantioselectivity, although with moderate to excellent chemical yields. This cascade reaction protocol has overcome several potential challenges, including: (a) chemoselectivity as the enone moiety of cyclohexadienone could also undergo a [3+2] cycloaddition with the nitrone component; (b) regioselectivity, as the Kinugasa intermediate could lead to spirocyclic β-lactams through a Kinugasa/Michael domino reaction, whereas the Baldwin intermediate could lead to spirocyclic aziridines through a Baldwin/Michael domino reaction; and (c) identification of a suitable catalytical system for alkylalkyne-tethered cyclohexadienones [213].

It should be emphasized that all the above examples of Kinugasa reactions involve the application of substituted acetylenes. An approach of using unsubstituted acetylene directly generated from cheap calcium carbide (CaC_2) has recently been reported, which bears several advantages over acetylene gas, such as low cost, convenience to handle, and perceivable safety. As shown in the reaction mechanism in Scheme 3.48, a potential challenge in the Kinugasa reaction is to avoid the formation of an imine side product. In the case of using unsubstituted acetylene, the ketene formed *in situ* is prone to undergo dimerization. In order to overcome these challenges, $TBAF\cdot 3H_2O$ in the presence of CuCl and *N*-methylimidazole (NMI) is used to activate calcium carbide, and the *in situ* generated acetylene is trapped as acetylide "ate" complex. The optimal reaction condition requires 2.4 equivalents of calcium carbide, 3.0 equivalents of NMI, 0.34 equivalent of TBAF, 0.76 equivalent of CuCl, and

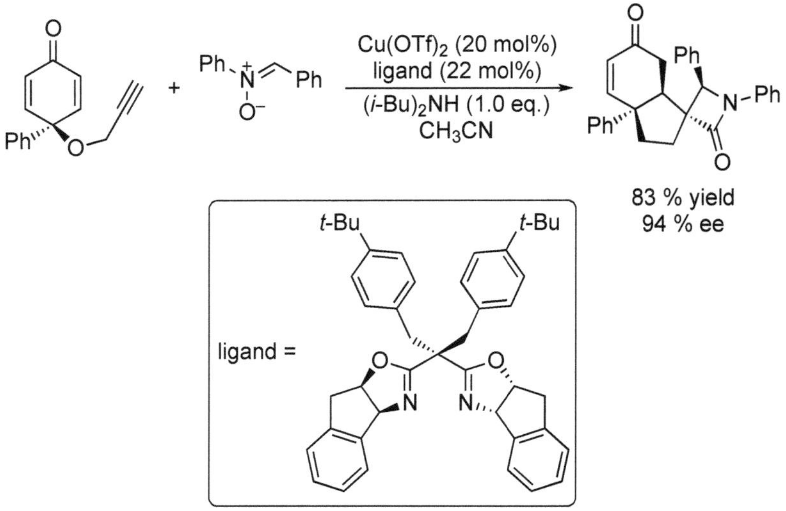

SCHEME 3.54 The Kinugasa reaction between 1-(prop-2-yn-1-yloxy)-[1,1'-biphenyl]-4(1*H*)-one and (*Z*)-*N*,1-diphenylmethanimine oxide in the presence of chiral ligand

5.6 equivalents of water in THF. The 3-unsubstituted β-lactams can be obtained in moderate to very good yields, depending on the nitrones used [214].

A new approach related to the Kinugasa reaction takes a two-step approach involving a regioselective [3+2]-cycloaddition between nitrone and ethynyl trimethylsilane to give substituted 5-(trimethylsilyl)-2,3-dihydroisoxazole and subsequent desilylation with TBAF in THF at 0°C, as illustrated in Scheme 3.55. In this protocol, the cycloaddition between (*E*)-*N*-benzyl-1-phenylmethanimine oxide and ethynyltrimethylsilane yields 91% of 2-benzyl-3-phenyl-5-(trimethylsilyl)-2,3-dihydroisoxazole, which upon the treatment with TBAF in THF, leads to 37% of 1-benzyl-4-phenylazetidin-2-one. Likewise, the 1,3-dipolar cycloaddition between (*E*)-1-((2*R*,4*R*,5*R*)-5-(2,5-bis(benzyloxy)phenyl)-5-methyl-2-phenyl-1,3-dioxolan-4-yl)-*N*-methylmethanimine oxide and ethynyltrimethylsilane affords 82% of 3-((2*R*,4*R*,5*R*)-5-(2,5-bis(benzyloxy)phenyl)-5-methyl-2-phenyl-1,3-dioxolan-4-yl)-2-methyl-5-(trimethylsilyl)-2,3-dihydroisoxazole. Similar treatment of this intermediate yields 32% of 4-((2*R*,4*R*,5*R*)-5-(2,5-bis(benzyloxy)phenyl)-5-methyl-2-phenyl-1,3-dioxolan-4-yl)-1-methylazetidin-2-one. As illustrated in Scheme 3.55, upon the desilylation with TBAF, the substituted 5-(trimethylsilyl)-2,3-dihydroisoxazole such as 2-benzyl-3-phenyl-5-(trimethylsilyl)-2,3-dihydroisoxazole decomposes to release trimethylsilyl fluoride and an amide intermediate with a ketene functionality, which intramolecularly cyclizes to afford the β-lactam [215].

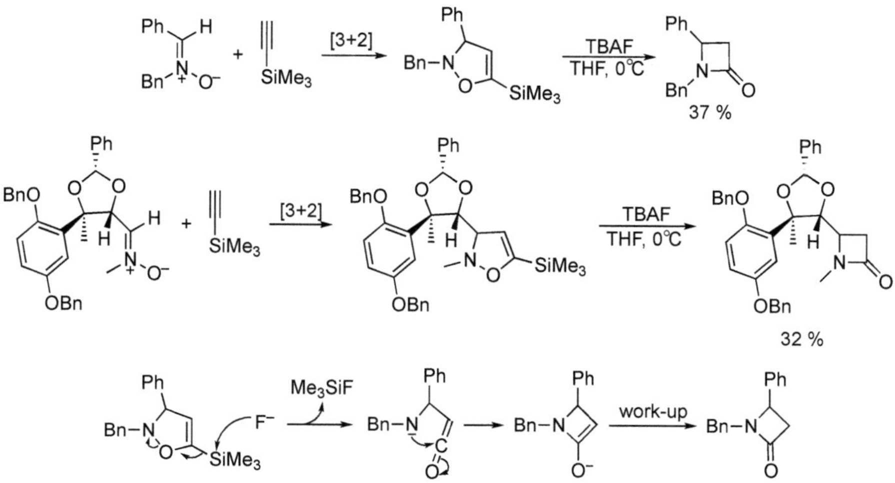

SCHEME 3.55 The Kinugasa reaction involving ethynyltrimethylsilane and the mechanism of desilylation

Recently, a new approach that might be considered as one variant of the Kinugasa reaction involves a ruthenium-catalyzed oxidative transformation of a terminal alkyne into ketene using the intramolecularly tethered sulfoxide as the oxidant, and subsequent Staudinger reaction between the *in situ* generated ketene and imine to afford the β-lactam. However, the alkynes with intramolecularly tethered sulfone or thioether rather than sulfoxide functionality do not work under this condition, even in the presence of an excess amount of sulfoxide. The resulting β-lactams can be obtained in excellent yields, with exceptionally high diastereoselectivity. For example, the reaction of (*E*)-*N*,1-diphenylmethanimine and 1.2 equivalent of 1-ethynyl-2-((phenylsulfinyl)methyl)-benzene in the presence of 5 mol% of ruthenium catalyst, 10 mol% of sodium tetrakis(3,5-bis(trifluoromethyl) phenyl)borate ($NaBAr^F_4$), 10 mol% of 2-(diphenylphosphaneyl)-6-(2,4,6-triisopropylphenyl)pyridine (functions as the ligand) and 4 Å molecular sieves in dichloroethane at 60°C, affords 95% of (3*S*,4*R*)-1,4-diphenyl-3-(2-((phenylthio)methyl)phenyl)azetidin-2-one with greater than 100:1 of diastereoselectivity, as displayed in Scheme 3.56 [216]. A little excess for the sulfoxide component is to ensure the oxidation of the alkyne functional group to activate the ketene species. For this particular reaction condition, a plausible mechanism has been proposed to rationalize intramolecular oxidation to form the ketene species.

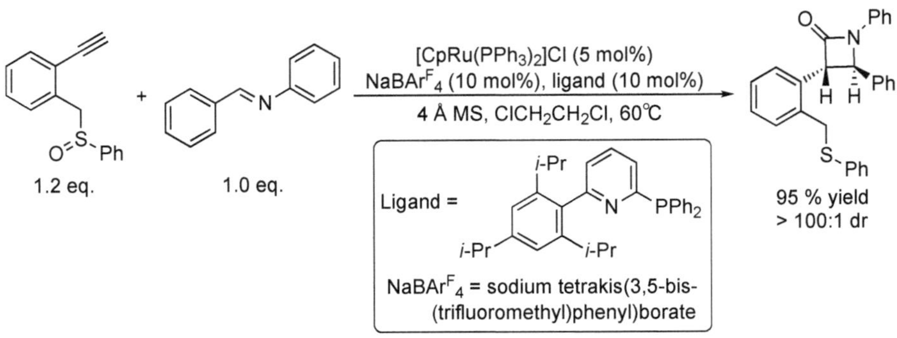

SCHEME 3.56 Ruthenium catalyzed Kinugasa reaction between 1-ethynyl-2-((phenylsulfinyl)methyl)benzene and (*E*)-*N*,1-diphenylmethanimine

Similar to the oxidative transformation of a terminal alkyne into active ketene species with intramolecularly tethered sulfoxide moiety, a similar intermolecular approach using 4-picoline *N*-oxide as the oxidizing agent in the presence of Wilkinson's catalyst also provides β-lactam in reaction with imine, in favor of *trans*-configuration. For example, the reaction of 1-ethynyl-4-methoxybenzene with (*E*)-1-(4-methoxyphenyl)-*N*-methylmethanimine in the presence of 5 mol% of the Wilkinson catalyst in acetonitrile at 50°C yields 80% of (3*R*,4*S*)-3,4-bis(4-methoxyphenyl)-1-methylazetidin-2-one. However, the outcome of this approach is somewhat different from the Kinugasa reaction and Staudinger reaction, as demonstrated in Scheme 3.57. Under the Kinugasa reaction condition, the reaction between 1-ethynyl-4-methoxybenzene and (*Z*)-*N*-methyl-1-phenylmethanimine oxide in the presence of 10 mol% CuCl and 2,2'-bipyridine leads to 20% of (3*R*,4*S*)-3-(4-methoxyphenyl)-1-methyl-4-phenylazetidin-2-one, whereas no such product has been formed in the presence of the Wilkinson catalyst rather than CuCl. For comparison, the regular Staudinger reaction between 2-(4-methoxyphenyl)acetyl chloride and (*E*)-*N*-methyl-1-(4-nitrophenyl)methanimine in the presence of Et_3N yields (3*R*,4*R*)-3-(4-methoxyphenyl)-1-methyl-4-(4-nitrophenyl)azetidin-2-one as the major product, but (3*R*,4*S*)-3-(4-methoxyphenyl)-1-methyl-4-(4-nitrophenyl)azetidin-2-one becomes the major product once 1.0 equivalent of the Wilkinson catalyst was added in this reaction. Very likely, the mechanism involved in this reaction in the presence of the Wilkinson catalyst is similar to the oxidative transformation of the terminal alkyne into ketene species with intramolecularly tethered sulfoxide [217].

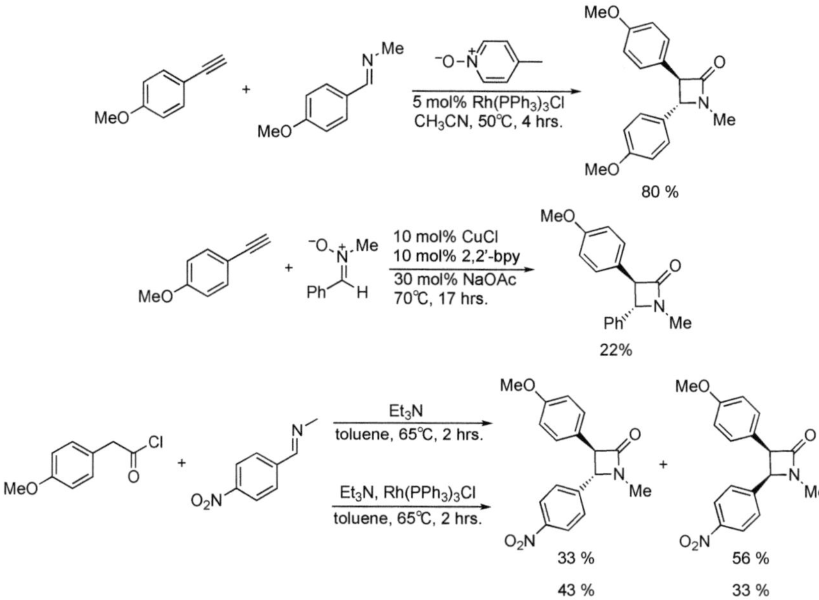

SCHEME 3.57 The formation of β-lactams in the presence of the Wilkinson catalyst

In addition to the above examples of Kinugasa reaction for the preparation of β-lactams, there are still many reports to cover the different scopes of this reaction, such as the preparation of enantiopure *cis*- and *trans*-β-lactams from chiral oxazolidinyl propynes and phenylnitrones [218], synthesis of *cis*- and *trans*-β-lactam nucleoside chimeras [219], intramolecular Kinugasa reaction for β-lactam-fused enediynes [220, 221], reaction between terminal copper acetylides and nonracemic cyclic nitrones [222], Kinugasa reaction between terminal alkynes and nitrones [223], reaction between chiral acetylenes and five-membered nitrones [224], Kinugasa reactions between terminal acetylenes and six-membered ring nitrones [225], synthesis of chiral α-amino-β-lactams involving ynamide [226], synthesis of 4-phosphonylated β-lactams from terminal alkynes and *N*-methyl-*C*-(diethoxyphosphonyl)nitrone [227], synthesis of α-methylene- or α-alkylidene-β-lactams from alkynes bearing a nucleofuge in the propargylic position [228], synthesis of exo-alkylidene β-lactams with a fluorine in the vinylic position from propargylic gem-difluorides [229], cascade Kinugasa reaction followed by rearrangement involving chiral propargyl alcohols [230], and synthesis of Ezetimibe [231].

Also, the Kinugasa reaction has been performed in the presence of different catalysts/ligands, such as the application of a C_2-symmetric planar-chiral bis(azaferrocene) ligand [232], or a range of HETPHOX ligands

[233], and Kinugasa reaction between terminal alkynes substituted by metal fragments, including sandwich, half-sandwich, and arene-tethered metal-carbene (Fischer) complexes [234]. There are many Cu(I) and Cu(II) mediated Kinugasa reactions as well, including the Cu(I)-mediated reaction [235, 236], Cu(I)-mediated reaction in the presence of chiral tris(oxazoline) ligand [237], Cu(I)-mediated reaction between five-membered cyclic nitrones and terminal acetylenes [238], Cu(I)-mediated Kinugasa reaction/rearrangement cascade reaction between terminal acetylene derived from δ-lactic acid and five-membered cyclic nitrone [239], Cu(I)-catalyzed asymmetric three-component interrupted Kinugasa reaction [240, 241], Cu(II)-mediated reaction in the presence of 20 mol.% of IndaBox and di-*sec*-butylamine [242], or chiral *N,N,N* imine-containing ligands derived from TsDPEN (*N*-(*p*-tosyl)-1,2-diphenylethylene-1,2-diamine) [243], Cu(II)-nanoparticle as heterogeneous catalyst (e.g., $CuFe_2O_4$) [244], micelle-promoted, copper-catalyzed multicomponent Kinugasa reactions [245]. Recently, the mechanism of the Kinugasa reaction has been revisited [246].

3.3 INTRAMOLECULAR SUBSTITUTIVE CYCLIZATION

The formation of β-lactams by means of intramolecular nucleophilic substitution can be proceeded using either carbanion as the nucleophile or nitrogen atom as the nucleophile. In addition, β-lactam scaffolds can be generated by intramolecular amide formation via a typical condensation reaction between amines and carboxylic acid derivatives, generally known as acyl substitution. These concepts are generally summarized in Scheme 3.58.

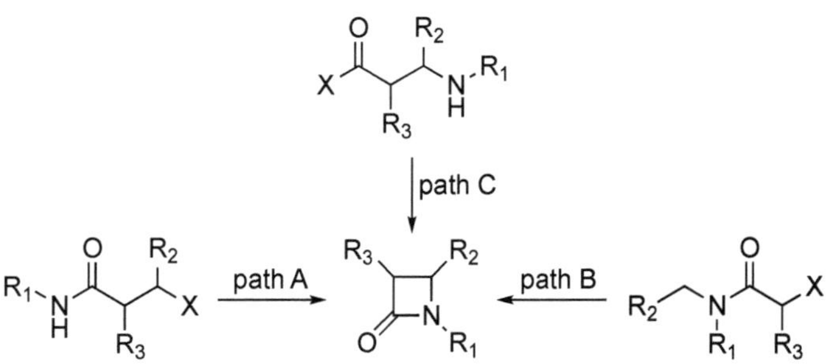

SCHEME 3.58 Cyclization approaches to form β-lactams

3.3.1 INTRAMOLECULAR MICHAEL ADDITION AND ELECTROCYCLIC RING CLOSURE

The Michael addition is a nucleophilic addition of a carbanion or more generally a nucleophile to an α,β-unsaturated compound containing a strong electron-withdrawing group directly attached to the double bond, which belongs to a larger class of conjugate additions. This reaction is one of the most useful methods for the mild formation of C–C bonds. Regarding the preparation of β-lactams, a less-extended approach has been developed recently based on the 4-*exo-trig* intramolecular cyclization of fumaramides that can be prepared from enantiomerically pure α-amino acid derivatives and activated fumaric acid. For example, ethyl *L*-phenylalaninate hydrochloride once treated with *p*-methoxybenzyl chloride in the presence of DIPEA, the resulting ethyl (4-methoxybenzyl)-*L*-phenylalaninate was acylated with *tert*-butyl (*E*)-4-chloro-4-oxobut-2-enoate also in the presence of DIPEA to give *tert*-butyl (*S*,*E*)-4-((1-ethoxy-1-oxo-3-phenylpropan-2-yl)(4-methoxybenzyl)-amino)-4-oxobut-2-enoate. Under the ideal reaction condition using Cs_2CO_3 as the base in ethanol at 20°C, the carbanion after deprotonation undergoes intramolecular conjugate addition to afford 80% of β-lactam upon protonation of the enolate side chain. The diastereoselectivity was measured at 91:9, where the major product of ethyl (2*S*,3*S*)-2-benzyl-3-(2-(*tert*-butoxy)-2-oxoethyl)-1-(4-methoxybenzyl)-4-oxoazetidine-2-carboxylate displayed a 91% ee, and the minor product showed an 85% ee, as shown in Scheme 3.59 [247]. Likewise, the 4-*exo-trig* cyclization of axially chiral enolates prepared from *L*-serine and *L*-cysteine derivatives, using Cs_2CO_3 as a base and CF_3CH_2OH as a proton source, leads to the formation of chiral β-lactams bearing contiguous tri- and tetrasubstituted carbon centers, with enantioselectivity up to 96% ee. The high enantioselectivity is attributed to the presence of strong electron-withdrawing groups [248]. For comparison, when the benzyl moiety on the nitrogen atom of the amide was lithiated with LDA, the benzyllithium thus formed undergoes 5-*endo-trig* anionic cyclization (i.e., intramolecular conjugate addition) to yield 5-member lactam, i.e., γ-lactam. For example, (*E*)-*N*-benzyl-*N*-(*tert*-butyl)-3-(dimethyl(phenyl)silyl)-acrylamide without active hydrogen, was treated with either LDA or LiTMP, and the resulting benzyllithium underwent the conjugate addition to afford lithium (4*S*,5*R*)-1-(*tert*-butyl)-4-(dimethyl(phenyl)silyl)-5-phenyl-4,5-dihydro-1*H*-pyrrol-2-olate, which was then treated with benzaldehyde (Aldol reaction) to give 45% of (3*R*,4*S*,5*R*)-1-(*tert*-butyl)-4-(dimethyl(phenyl)silyl)-3-((*S*)-hydroxy(phenyl)methyl)-5-phenylpyrrolidin-2-one [249].

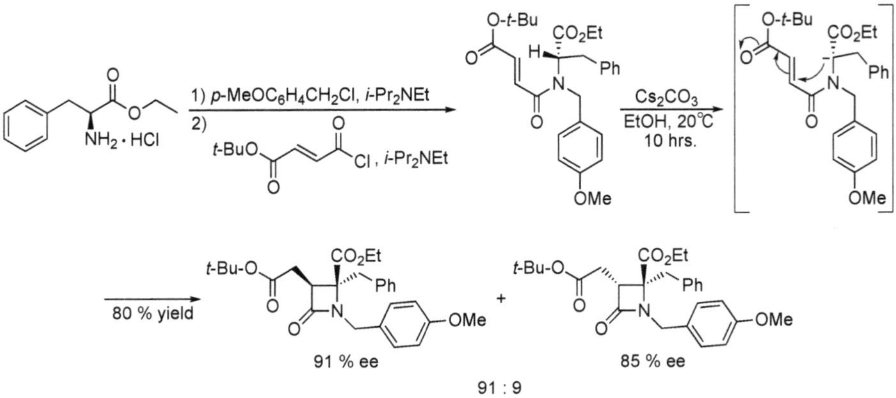

SCHEME 3.59 Formation of β-lactams via intramolecular Michael addition

Based on these precedent explorations, [2]-rotaxane interlocked *N*-benzylfumaramide has been prepared recently which was allowed to undergo a one-pot three-step transformation (i.e., deprotonation, an intramolecular ring closure followed by a thermally induced dethreading step) to yield *trans*-β-lactam in high yield, and in regio- and diastereoselective manner. In this protocol, the activation effect of mechanical bonds markedly differs from the more common shielding protection of threaded functions by the macrocycle, promoting an unusual and disfavored 4-*exo-trig* ring closure. An illustrative reaction is provided in Scheme 3.60, where the cyclization between isophthaloyl dichloride and *p*-xylylenediamine in the presence of Et_3N in DMF leads to the formation of a [2]-rotaxane, i.e., 3,7,11,15-tetraaza-1,9(1,3),5,13(1,4)-tetrabenzenacyclohexa decaphane-2,8,10,16-tetraone, within which the N^1,N^1,N^4,N^4-tetrabenzyl fumaramide is fixed by hydrogen bonds. Treatment of the fumaramide with Cs_2CO_3 in DMF at 60°C affords the, [2]-rotaxane interlocked *N,N*-dibenzyl-2-((3*S*,4*R*)-1-benzyl-2-oxo-4-phenylazetidin-3-yl) acetamide in nearly quantitative yield [250, 251]. Further study indicates that only the highly polar solvents such as DMF or DMSO were suitable for this transformation. While Cs_2CO_3 is ideal for this transformation, other alkali carbonates, such as Na_2CO_3 or K_2CO_3 do not work under identical conditions, however, stronger bases such as NaOH, KOH, and CsOH do work, even at room temperature. For comparison, in the absence of the [2]-rotaxane, treatment of N^1,N^1,N^4,N^4-tetrabenzylfumaramide with 10 eq. Cs_2CO_3 in DMF at 120°C for 7 days, yielded only 45% of β-lactam with diastereoselectivity of 2:1 for the *trans*- and *cis*-β-lactam. Similarly, treatment of this fumaramide with 1.0 eq. of CsOH at room temperature afforded 51% of the β-lactam with a diastereoselectivity of 2.5:1 in favor of *trans*-β-lactam. Therefore, the presence of the mechanical bond with rotaxane is the origin of enhanced reactivity of *N*[1],*N*[1],*N*[4],*N*[4]-tetrabenzylfumaramide, leading to accelerated cyclization process, full control of the regio- and diastereoselectivity of β-lactam, and protection of the β-lactam moiety from decomposition under basic condition [252].

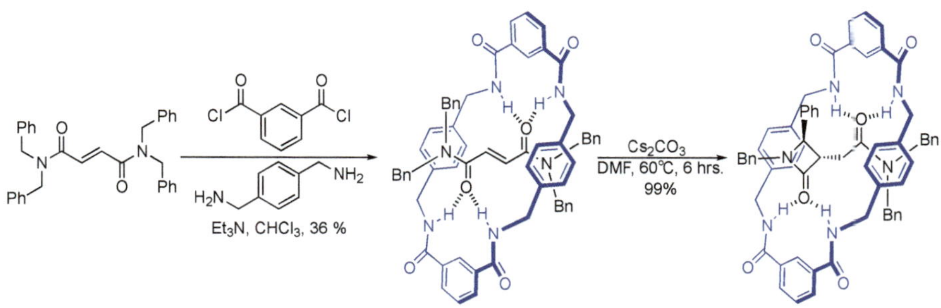

SCHEME 3.60 Formation of the [2]-rotaxane interlocked *N,N*-dibenzyl-2-((3*S*,4*R*)-1-benzyl-2-oxo-4-phenylazetidin-3-yl)acetamide

Alternatively, a variety of reaction conditions have been examined for the conversion of ethyl (*E*)-4-(benzylamino)-4-oxobut-2-enoate into ethyl 2-((3*R*,4*S*)-2-oxo-4-phenylazetidin-3-yl)acetate, where several organic acids (TsOH, CSA), organic bases (Et_3N, LDA, DBN, and DBU), as well as inorganic acid (HCl, H_2SO_4) and inorganic bases (KOH, K_2CO_3, Cs_2CO_3) are not suitable for this conversion. However, the presence of 2 equivalents of *N,O*-bis(trimethylsilyl)acetimidate (i.e., trimethylsilyl (*E*)-*N*-(trimethylsilyl) acetimidate, BSA), particularly in the presence of Et_3N, DABCO, or DMAP makes such transformation feasible. While the conjugate addition mechanism as mentioned in the previous examples cannot be completely ruled out, an alternative mechanism involving electrocyclic ring closure makes more sense, as outlined in Scheme 3.61, for the conversion of ethyl (*E*)-4-(benzylamino)-4-oxobut-2-enoate [253]. In this novel approach, BSA functions as both base as well as trimethylsilylation agent. The coherent interaction of base (e.g., Et_3N) and BSA converts ethyl (*E*)-4-(benzylamino)-4-oxobut-2-enoate into ethyl (2*E*,4*Z*)-4-(benzylimino)-4-((trimethylsilyl)oxy)but-2-enoate and then (*E*)-*N*-((4*Z*,6*Z*)-7-ethoxy-2,2,9,9-tetramethyl-3,8-dioxa-2,9-disiladeca-4,6-dien-4-yl)-1-phenylmethanimine. The rotation around the C-N bond leads to (*E*)-*N*-((4*Z*,6*Z*)-7-ethoxy-2,2,9,9-tetramethyl-3,8-dioxa-2,9-disiladeca-4,6-dien-4-yl)-1-phenyl-methanimine, which undergoes electrocyclic ring closure in a conrotatory manner to afford (2*S*,3*R*)-3-((*E*)-2-ethoxy-2-((trimethylsilyl) oxy)vinyl)-2-phenyl-4-((trimethylsilyl)oxy)-2,3-dihydroazete. Desilylation under acidic condition provides the final product of ethyl 2-((3*R*,4*S*)-2-oxo-4-phenylazetidin-3-yl)acetate. It should be pointed out that the electrocyclic reaction is highly stereospecific that guarantees the *trans*-configuration.

Simply for the pure electrocyclization, 2-(4-methoxyphenyl)acetyl chloride and (*E*)-1-phenyl-*N*-(trimethylsilyl)methanimine were first treated with trimethylsilyl chloride, followed by triethylamine to give (*E*)-*N*-((*Z*)-2-(4-methoxyphenyl)-1-((trimethylsilyl)oxy)vinyl)-1-phenylmethanimine, which was subject to microwave irradiation in chlorobenzene at 135°C for 40 minutes in the presence of a catalytic amount of europium(III) tris(1,1,2,2,3,3-heptafluoro-7,7-dimethyl-4,6-octanedionate) (EuFod) to give 55% of (3*S*,4*R*)-3-(4-methoxyphenyl)-4-phenylazetidin-2-one [254]. A similar reaction has been performed and reported elsewhere [255].

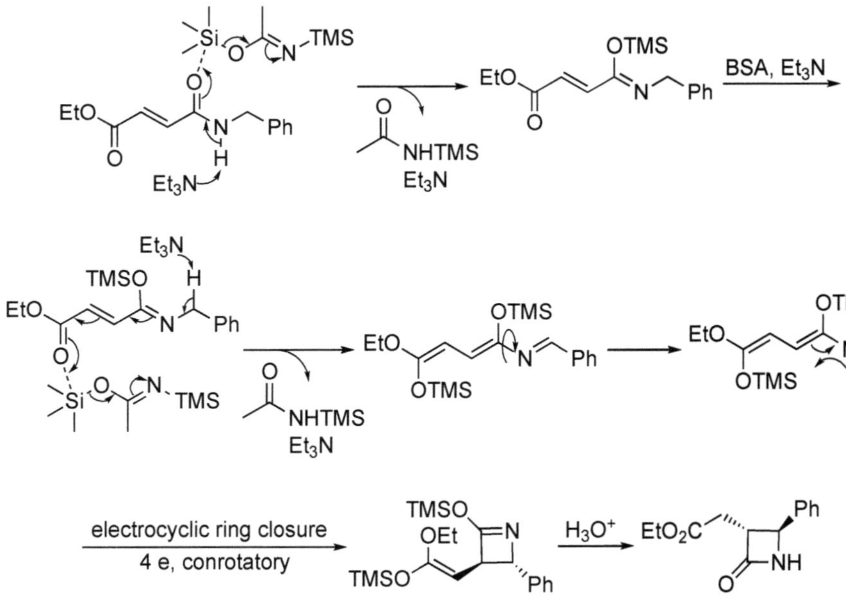

SCHEME 3.61 The mechanism for the formation of ethyl 2-((3*R*,4*S*)-2-oxo-4-phenylazetidin-3-yl)acetate

The 4π-electrocyclic ring closure reaction to give β-lactams have been studied theoretically [256–259]. Also, additional experimental practice for the electrocyclic ring closure to afford β-lactams include the Lewis acid-catalyzed electrocyclization of 2-aza-1,3-butadienes to NH-β-lactams [260], N-silylimines [261], and others [262, 263].

3.3.2 PHOTOLYSIS OF α,β-UNSATURATED AMIDES

In addition to the two possible mechanisms (conjugate addition of carbanion and electrocyclic ring closure), α,β-unsaturated amides may take a different approach involving radical intermediate under photochemical conditions. For example, the stable atropisomeric enone carboxamides with high racemization barriers at room temperature undergo photochemical hydrogen abstraction leading to the formation of spiro-β-lactam, under either direct irradiation (ca. 300 or 350 nm) or sensitized irradiation (using visible light with thioxanthone as a triplet sensitizer/catalyst). It was found that direct irradiation of *N*-benzyl-*N*-(2-(*tert*-butyl) phenyl)-2,2-dimethyl-4-oxo-3,4-dihydro-2*H*-pyran-6-carboxamide in MeCN resulted in moderate to high conversions (62–89%) of this enone carboxamide and led to moderate yield of (3*R*,4*S*)-2-(2-(*tert*-butyl)phenyl)-6,6-dimethyl-3-phenyl-5-oxa-2-azaspiro[3.5]nonane-1,8-dione, as shown in Scheme 3.62. Decreasing the temperature to -20°C diminished the conversion. In comparison, the photo-transformation proceeded efficiently under visible light-sensitized irradiation with thioxanthone as a sensitizer except in methanol and tetrahydrofuran, giving moderate conversions. In contrast, photochemical irradiation of *N*-benzyl-2,2-dimethyl-4-oxo-*N*-phenyl-3,4-dihydro-2*H*-pyran-6-carboxamide yields a mixture of (4a*R*,10b*R*)-6-benzyl-3,3-dimethyl-2,3,6,10b-tetrahydro-1*H*-pyrano[2,3-*c*]quinoline-1,5(4a*H*)-dione and (4a*S*,10b*R*)-6-benzyl-3,3-dimethyl-2,3,6,10b-tetrahydro-1*H*-pyrano[2,3-*c*]quinoline-1,5(4a*H*)-dione in 2:1 ratio, with a total yield of 75%, as shown in Scheme 3.62 [264]. For the first case, due to the presence of atropisomeric *t*-butyl group, the rotation around the C(O)-N bond is not feasible. Photochemical irradiation leads to triplet biradical, which undergoes benzylic hydrogen abstraction. After intersystem crossing (ISC), the combination of singlet biradical gives the end product of β-lactam. For comparison, in the case of *N*-benzyl-2,2-dimethyl-4-oxo-*N*-phenyl-3,4-dihydro-2*H*-pyran-6-carboxamide, the photo-irradiation generated triplet biradical adds to the phenyl ring, and 6-benzyl-3,3-dimethyl-1-oxo-2,3,10a,10b-tetrahydro-1*H*-pyrano[2,3-*c*]quinolin-6-ium-5-olate forms after ISC, which tautomerizes to the end products.

Similarly, the photo-irradiation of an inclusion crystal of ((2*S*,3*S*)-1,4-dioxaspiro[4.5]decane-2,3-diyl)bis(diphenylmethanol) with *N*-benzyl-*N*-isopropyl-3-oxocyclohex-1-ene-1-carboxamide (2:1 ratio) in a water suspension for 4 hours yielded an optically pure (-)-2-benzyl-3,3-dimethyl-2-azaspiro[3.5]nonane-1,6-dione as an oil in 53% yield. Likewise, the photo-irradiation of an inclusion crystal of *N*-benzyl-*N*-isopropyl-3-oxocyclohex-1-ene-1-carboxamide with ((2*S*,3*S*)-1,4-dioxaspiro[4.4]nonane-2,3-diyl)bis (diphenylmethanol) (1:2) gave the same product in 51% yield. However, photo-irradiation of *N*-benzyl-*N*-isopropyl-3-oxocyclohex-1-ene-1-carboxamide in CH_3CN for 1.5 hours afforded 14% of 2-isopropyl-3-phenyl-2-azaspiro[3.5]nonane-1,6-dione (Scheme 3.63) [265]. Similarly, photo-irradiation of *N*-methyl-3-oxo-*N*-(1-phenylethyl)cyclohex-1-ene-1-carboxamide in CH_3CN at 313 nm for 45 minutes gave 56% of 1:1 ratio of (3*R*)- and (3*S*)-2,3-dimethyl-3-phenyl-2-azaspiro[3.5]nonane-1,6-diones. However, photo-irradiation of *N*-benzyl-*N*-(3-oxocyclohex-1-ene-1-carbonyl)benzamide in CH_3CN for 75 minutes yielded 70% of (3*S*)-2-benzoyl-3-phenyl-2-azaspiro[3.5]nonane-1,6-dione. Likewise, the photo-irradiation of *N*-benzoyl-*N*-benzyl-2,2-dimethyl-4-oxo-3,4-dihydro-2*H*-pyran-6-carboxamide under the same condition for 1 hour yielded 67% of (3*S*)-2-benzoyl-6,6-dimethyl-3-phenyl-5-oxa-2-azaspiro[3.5]-nonane-1,8-dione [266]. Similarly, irradiation of 1:1 inclusion crystal of *N*-isopropyl-*N*-(6-oxocyclohex-1-en-1-yl)cyclohexanecarboxamide with (*R*,*R*)-(-)-*trans*-4,5-bis-(hydroxydiphenylmethyl)-2,2-dimethyl-1,3-dioxacyclopentane or (*R*,*R*)-(-)-*trans*-4,5-bis(hydroxydiphenylmethyl)-1,4-dioxaspiro[4.4] nonane afforded 37% and 61% of highly optically pure 13-isopropyl-13-aza-dispiro[$5.0.5^7.2^6$]tetradecane-8,14-dione, with > 99.9% ee [267].

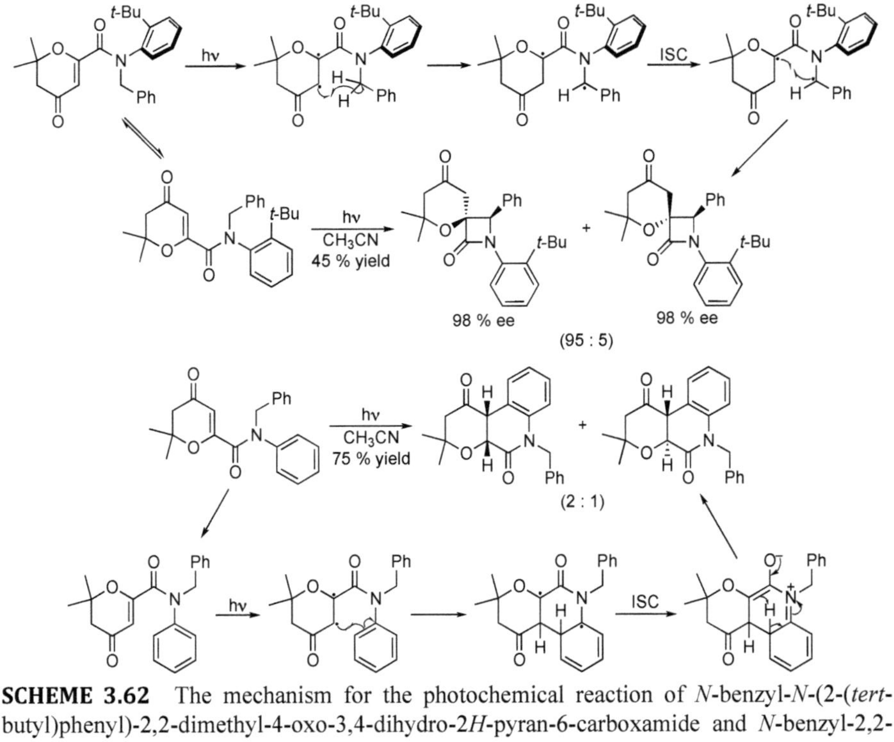

SCHEME 3.62 The mechanism for the photochemical reaction of *N*-benzyl-*N*-(2-(*tert*-butyl)phenyl)-2,2-dimethyl-4-oxo-3,4-dihydro-2*H*-pyran-6-carboxamide and *N*-benzyl-2,2-dimethyl-4-oxo-*N*-phenyl-3,4-dihydro-2*H*-pyran-6-carboxamide

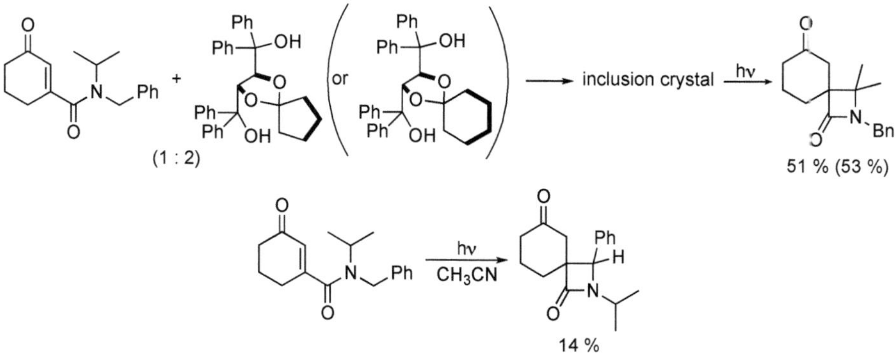

SCHEME 3.63 The photochemical reaction of *N*-benzyl-*N*-isopropyl-3-oxocyclohex-1-ene-1-carboxamide under different conditions

Regarding the photolysis of α,β-unsaturated amides, a series of *N*-substituted pyridin-2(1*H*)-ones, including 1-methylpyridin-2(1*H*)-one, 1-ethylpyridin-2(1*H*)-one, 1-propylpyridin-2(1*H*)-one, 1-benzylpyridin-2(1*H*)-one, 1-phenethylpyridin-2(1*H*)-one and 1-(3-phenylpropyl)pyridin-2(1*H*)-one undergo electrocyclic reaction (ring closure) under photochemical condition, in disrotatory manner, to afford 2-substituted (1*R*,4*S*)- and (1*S*,4*R*)-2-azabicyclo[2.2.0]hex-5-en-3-ones [268]. The enantiomeric excess of the β-lactams can be improved by excitation of *N*-alkyl pyridones adsorbed within the modified chiral zeolites, where the chiral environment of commercially available achiral Y zeolite was created by the inclusion of chiral inductor molecules (e.g., (-)-norephedrine, (-)-ephedrine, (+)-pseudoephedrine, (-)-camphor quinone-3-oxime) into the super-cages of MY zeolite. In this approach, a higher ratio of the chiral inductor to the reactant (10: 1) was applied for the purpose to maximize the chances of the reactant being closer to the chiral inductor molecule. The presence of a higher number of cations inside the zeolite is critical to enhancing the enantioselectivity, as indicated in the irradiation of 1-(3-phenylpropyl)pyridin-2(1*H*)-one in NaY zeolite with a Si/Al = 40, affording 2-(3-phenylpropyl)-2-azabicyclo[2.2.0]hex-5-en-3-one of only 2% enantioselectivity, whereas irradiation of the same substrate in NaY with a Si/Al = 2.4 of a higher number of cation gave the β-lactam of 50% enantioselectivity. In addition, the enantioselectivity is also dependent on the nature of the alkali ion present in a zeolite. For example, in the case of 1-phenethylpyridin-2(1*H*)-one, irradiation in (-)-ephedrine-modified alkali ion exchanged Y zeolites led to enantioselectivity of 23% in LiY, 5% in NaY, 53% in KY, 40% in RbY, and 12% in CsY. Furthermore, decreasing the temperature from 25°C to -55°C increased the enantioselectivity [268]. A representative reaction is illustrated in Scheme

3.64 for the conversion of 1-phenethylpyridin-2(1*H*)-one into 2-phenethyl-2-azabicyclo[2.2.0]hex-5-en-3-one, with 53% ee. Additional examples for the light-promoted electrocyclic ring closure reaction of 2-pyridones to give cyclobutene-fused β-lactams can be found in alternative references [269–273]. For the same principle, photo-irradiation of the 1:1 inclusion complex in solid state between ((2*R*,3*R*,9*R*,10*R*)-1,4,8,11-tetraoxadispiro[4.1.47.35]tetradecane-2,3,9,10-tetrayl)tetrakis(diphenylmethanol) and 1-ethylpyridin-2(1*H*)-one, 1-propylpyridin-2(1*H*)-one, 1-isopropylpyridin-2(1*H*)-one, 1-butylpyridin-2(1*H*)-one or 1-isobutylpyridin-2(1*H*)-one leads to the formation of optically active β-lactams in good yields, with 91–99.5% ee [274]. Likewise, photo-irradiation of solvent-free inclusion complexes prepared by mechanically mixing *N*-methyl pyridone or *N*-ethyl pyridone with β-cyclodextrin yields the corresponding chiral 2-azabicyclo[2.2.0]-hex-5-en-3-ones, in 60% enantiomeric excess [275]. Moreover, irradiation of a series of 1-(methoxymethyl)-3-methylpyridin-2(1*H*)-one, 1-(methoxymethyl)-4-methylpyridin-2(1*H*)-one, 1-(methoxymethyl)-5-methylpyridin-2(1*H*)-one and 1-(methoxymethyl)-6-methylpyridin-2(1*H*)-one yields 2-(methoxymethyl)-4-methyl-2-azabicyclo[2.2.0]hex-5-en-3-one, 2-(methoxymethyl)-5-methyl-2-azabicyclo[2.2.0]hex-5-en-3-one, 2-(methoxymethyl)-6-methyl-2-azabicyclo[2.2.0]hex-5-en-3-one and 2-(methoxymethyl)-1-methyl-2-azabicyclo[2.2.0]hex-5-en-3-one, respectively. Epoxidation of the existing double bond and subsequent ring opening lead to the formation of 4-(methoxymethyl)-1,4-oxazepin-5(4*H*)-ones [276]. In another example, (1*R*,6*R*)-1,6-bis(2-chlorophenyl)-1,6-diphenylhexa-2,4-diyne-1,6-diol or (2,2-dimethyl-1,3-dioxolane-4,5-diyl)bis(diphenylmethanol) has been applied to form the inclusion complex with substituted pyridin-2(1*H*)-ones, in most cases in 1:1 ratio, except for pyridin-2(1*H*)-one and 1,4-dimethylpyridin-2(1*H*)-one which form 1:2 ratio of complex. Irradiation of the inclusion complexes of pyridin-2(1*H*)-ones and these two optically active host compounds in the solid state gave optically active β-lactams [277]. The remaining double bond can be converted into other functional groups, such as epoxidation, dihydroxylation, halogenation, ozonolysis, etc.

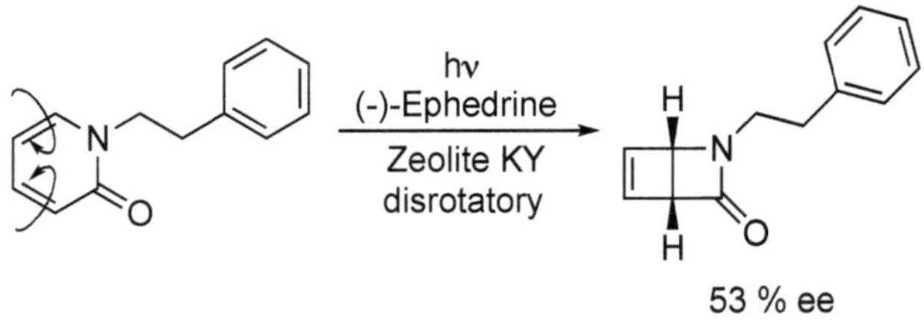

SCHEME 3.64 The photochemical reaction of 1-phenethylpyridin-2(1*H*)-one into (1*S*,4*R*)-2-phenethyl-2-azabicyclo[2.2.0]hex-5-en-3-one

3.3.3 INTRAMOLECULAR CYCLIZATION WITH NITROGENOUS NUCLEOPHILES

For the first type of intramolecular substitutive cyclization that gives β-lactam, a full synthetic route taking the path A approach in Scheme 3.58 has been displayed in Scheme 3.65 for the transformation of *L*-phenylalanine into (*S*)-3-benzylazetidin-2-one. In this approach, *L*-phenylalanine was first reduced with $LiAlH_4$, and the resulting (*S*)-2-amino-3-phenylpropan-1-ol was then acylated with acetic formic anhydride to generate (*S*)-2-formamido-3-phenylpropyl formate. Upon treatment with $POCl_3$, it was converted into (*S*)-2-isocyano-3-phenylpropyl formate, which underwent pyrolysis at 585°C at a very high vacuum (10^{-4} torr) to yield (*R*)-2-cyano-3-phenylpropyl formate. Subsequent alcoholysis in methanol in the presence of HCl led to methyl (*S*)-2-benzyl-3-hydroxypropanoate, which was converted into methyl (*R*)-2-benzyl-3-chloropropanoate with PCl_5. Then, the ester group was catalytically hydrolyzed in water with Pork Liver Esterase to give 85% of (*R*)-2-benzyl-3-chloropropanoic acid. This acid was then condensed with 1.3 equivalent of benzyloxyamine hydrochloride in the presence of 3-(((ethylimino)methylene)amino)-*N*,*N*-dimethylpropan-1-amine in water/THF (1:1) to yield 91% of (*R*)-2-benzyl-*N*-(benzyloxy)-3-chloropropanamide. Basic treatment with DBU resulted in intramolecular substitutive cyclization, affording 53% of (*S*)-3-benzyl-1-(benzyloxy)azetidin-2-one. Finally, catalytical hydrogenation using Pd-C removed the benzyl group, and subsequent treatment with $TiCl_3$ completed this synthetic approach [278].

In addition to β-chloro as the leaving group, the sulfenyl group works just fine in a special case. When *N*-benzylmethacrylamide was treated with 1.2 equivalents of phenylsulfenyl chloride in CH_2Cl_2 at room temperature for 2 hours, a mixture of *N*-benzyl-3-chloro-2-methyl-2-(phenylthio)propenamide and *N*-benzyl-2-chloro-2-methyl-3-(phenylthio)propenamide was obtained, in a total yield of 88%. The major product is in three times the minor one; and being the kinetic product, it gradually converts into a more stable product at room temperature or by means of treatment with silica gel. The kinetic product once treated with 1.0 equivalent of KOH in aqueous benzene at 40–50°C, in the presence of a small amount of phase transfer catalyst such as tetrabutylammonium bromide (TBAB), 94% of (*R*)-1-benzyl-3-methyl-3-(phenylthio) azetidin-2-one was obtained. Similarly, when the thermodynamic product, i.e., *N*-benzyl-2-chloro-2-methyl-3-(phenylthio)-propenamide, was treated under the same condition, a nearly equal amount of (*R*)-1-benzyl-3-methyl-3-(phenylthio)azetidin-2-one was obtained. Therefore, it is assumed that either addition product between *N*-benzylmethacrylamide and phenylsulfenyl chloride likely undergoes intramolecular substitution to form transient (1*R*,2*S*)-2-(benzylcarbamoyl)-2-methyl-1-phenylthiiran-1-ium, which upon deprotonation with K_2CO_3, proceeds the intramolecular substitution to afford the β-lactam, as outlined in Scheme 3.66 [279].

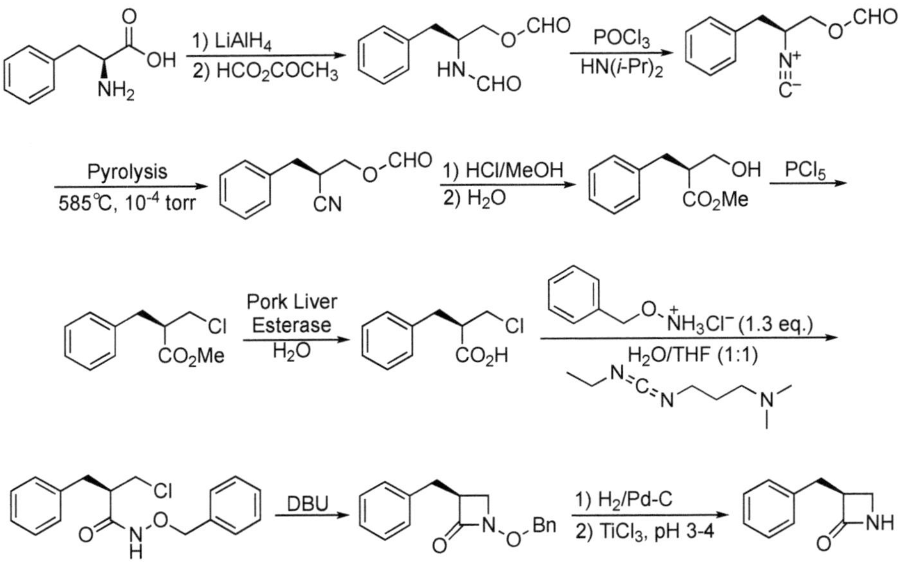

SCHEME 3.65 The transformation of *L*-phenylalanine into (*S*)-3-benzylazetidin-2-one

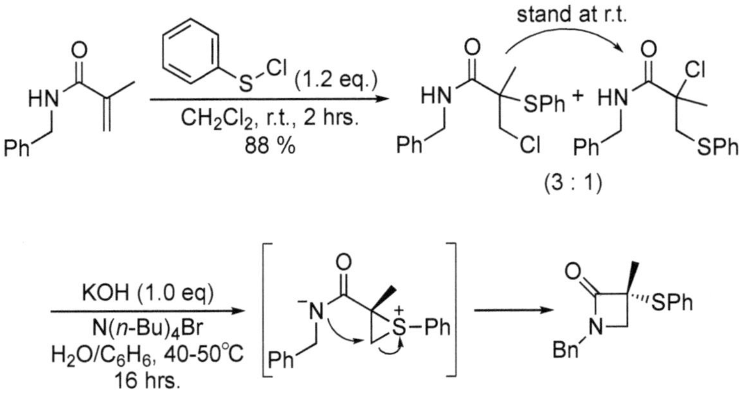

SCHEME 3.66 Transformation of *N*-benzylmethacrylamide into (*R*)-1-benzyl-3-methyl-3-(phenylthio)azetidin-2-one

Clearly, this strategy is very attractive because of its biosynthetic analogy and potential ability to use chiral amino acid derivatives or other modified amides with β-leaving groups as the starting materials [280]. Following this strategy, one-pot synthesis of β-lactams has been successfully achieved by the reaction of α,β-dibromo-α-methylpropionyl chloride with *L*-phenylalanine or (*R*)-2-amino-2-phenylacetic acid in a mixture of aqueous 30% NaOH solution and CH_2Cl_2 in the presence of a phase transfer catalyst, such as benzyltriethylammonium chloride, affording (2*R*)-2-(3-bromo-3-methyl-2-oxoazetidin-1-yl)-3-phenylpropanoic acid and (2*R*)-2-(3-bromo-3-methyl-2-oxoazetidin-1-yl)-2-phenylacetic acid in 45–65% yield. Further optimization of the reaction condition revealed that 5% NaOH gave a much higher yield of β-lactam than the 30% NaOH. Even the simple α-amino acid can function as a catalyst and facilitate the transformation of *N*-benzyl-3-bromo-2,2-dimethylpropanamide, *N*-benzyl-2,3-dibromo-2-methylpropanamide, or 3-chloro-2,2-dimethyl-*N*-phenylpropanamide into the corresponding β-lactams in good to excellent yields [281]. Similarly, bromination of methyl 3-bromo-2-methoxy-2-((*R*)-2-(4-methoxyphenyl)-4-methyl-4,5-dihydrothiazole-4-carboxamido)-3-methyl-butanoate with NBS leads to methyl 3-bromo-2-((4*R*)-5-bromo-2-(4-methoxyphenyl)-4-methyl-4,5-dihydrothiazole-4-carboxamido)-2-methoxy-3-methylbutanoate, which is then treated with KH-$LiClO_4$ in THF to give methyl 3-bromo-2-methoxy-2-((1*R*)-3-(4-methoxyphenyl)-1-methyl-7-oxo-4-thia-2,6-diaza-bicyclo[3.2.0]hept-2-en-6-yl)-3-methylbutanoate in an overall yield of 22% [282].

Following this strategy, β,γ-unsaturated amidosulfamoyl esters and related structures can be directly converted into 4-halomethyl *N*-sulfonylated β-lactams in a manner of one-pot synthesis when they are treated with bromine or iodine under basic conditions. The halo-β-lactams can be dehalogenated with tributyltin hydride. For example, when a 0.5 *M* bromine solution in CH_2Cl_2 was added dropwise at room temperature to *N*-tosylbut-3-enamide in an aqueous $NaHCO_3$ solution, 67% of 4-(bromomethyl)-1-tosylazetidin-2-one was obtained, which was reduced to 92% of 4-methyl-1-tosylazetidin-2-one with Bu_3SnH. Other analogs, including methyl but-3-enoylsulfamate, ethyl but-3-enoylsulfamate, 2,2,2-trichloroethyl but-3-enoylsulfamate, and 2,2,2-trichloroethyl (3-methylbut-3-enoyl)sulfamate have been successfully transformed into the corresponding halo-β-lactams. When *N*-tosylcyclohexa-2,5-diene-1-carboxamide was brominated in aqueous $NaHCO_3$, 60% of (1*R*,5*R*,6*R*)-5-bromo-7-tosyl-7-azabicyclo[4.2.0]oct-2-en-8-one was obtained. However, when 2,2,2-trichloroethyl (*E*)-pent-3-enoylsulfamate was treated with I_2 in aqueous $NaHCO_3$, 90% of 4-iodo-5-methyldihydrofuran-2(3*H*)-one was obtained [283].

Besides halogen, particularly bromine as the leaving group in this type of nucleophilic substitution, methanesulfonate has also been applied as the leaving group. For example, methylsulfonation of 3-hydroxy-*N*-(4-methoxyphenyl)-2-methylenepentanamide with methylsulfonyl chloride in the presence of Et_3N in CH_2Cl_2 at 0°C gave 57% of 2-((4-methoxyphenyl) carbamoyl)pent-1-en-3-yl methanesulfonate, which was then treated with potassium *tert*-butoxide in THF under argon atmosphere for 2 hours to afford 79% of 4-ethyl-1-(4-methoxyphenyl)-3-methyleneazetidin-2-one after cyclization. Oxidation with ceric ammonium nitrate (CAN) in aqueous CH_3CN at -15°C for an hour gave 57% of 4-ethyl-3-methyleneazetidin-2-one, as displayed in Scheme 3.67 [284].

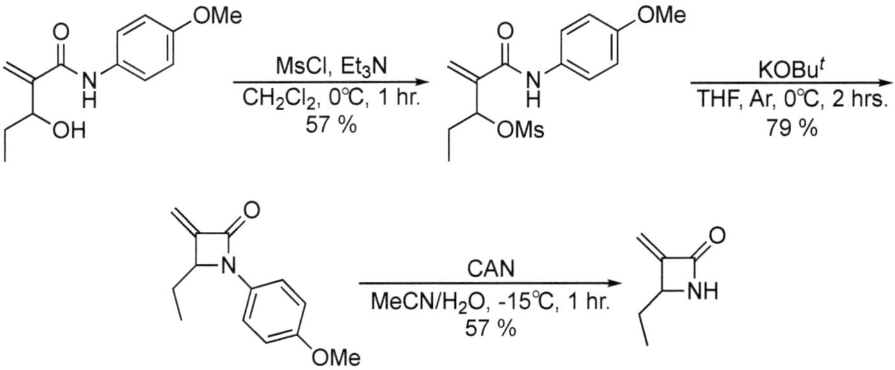

SCHEME 3.67 The procedure to convert 3-hydroxy-*N*-(4-methoxyphenyl)-2-methylenepentanamide into 4-ethyl-3-methyleneazetidin-2-one

In another example, *tert*-butyl ((2*S*,3*R*)-1-amino-3-hydroxy-1-oxobutan-2-yl)carbamate was prepared from *L*-threonine, which was then methanesulfonated with methylsulfonyl chloride and Et_3N to yield (2*R*,3*S*)-4-amino-3-((*tert*-butoxycarbonyl)amino)-4-oxobutan-2-yl methanesulfonate. Further treatment of this compound with chlorosulfonic acid in the presence of 2-methylpyridine in CH_2Cl_2 at -7°C, mixed with Na_2HPO_4 aqueous solution, and extracted with tetrabutylammonium bisulfate to afford tetrabutylammonium (*N*-(*tert*-butoxycarbonyl)-*O*-(methylsulfonyl)-*L*-threonyl)sulfuramidite. This compound in CH_2Cl_2 was then treated with $KHCO_3$ at 30°C, subsequently with formic acid to give (2*S*,3*S*)-3-amino-2-methyl-4-oxoazetidine-1-sulfinic acid, the precursor for aztreonam [285]. Similarly, starting from (2*R*,3*S*)-2,3,4-trihydroxybutanoic acid, (2*R*,3*R*)-4-amino-3-(((benzyloxy)carbonyl)amino)-2-((methylsulfonyl)oxy)-4-oxobutyl 2-chloroacetate was obtained after several sequential transformations. This compound was then treated with SO_3, 2-picoline in CH_2Cl_2, then $KHSO_4$, tetrabutylammonium bisulfate, and potassium bicarbonate in aqueous dichloroethane to afford 92% of λ^1-azaneyl *N*-((2*S*,3*S*)-3-(((benzyloxy) carbonyl)amino)-2-(hydroxymethyl)-4-oxoazetidin-1-yl)tetrabutyl-λ^5-oxidanesulfonimidate [286]. Similarly, β-lactam-containing surfactants have been prepared by treatment of 3-hydroxy-2-(hydroxymethyl)-2-methyl-*N*-octylpropanamide or other analogous with *N,N,N',N',N",N"*-hexamethylphosphanetriamine and carbon tetrachloride, and subsequent KPF_6 to afford tris(dimethylamino)(2-(hydroxymethyl)-2-methyl-3-(octylamino)-3-oxopropoxy)phosphonium hexafluorophosphate in good yields. Alkaline treatment of the phosphonium salt with K_2CO_3 gave 50–55% β-lactam-containing surfactants (e.g., 3-(hydroxymethyl)-3-methyl-1-octylazetidin-2-one). Likewise, phenylalanine has been subject to similar treatment to yield a different type of β-lactam-containing surfactants, as shown in Scheme 3.68 [287].

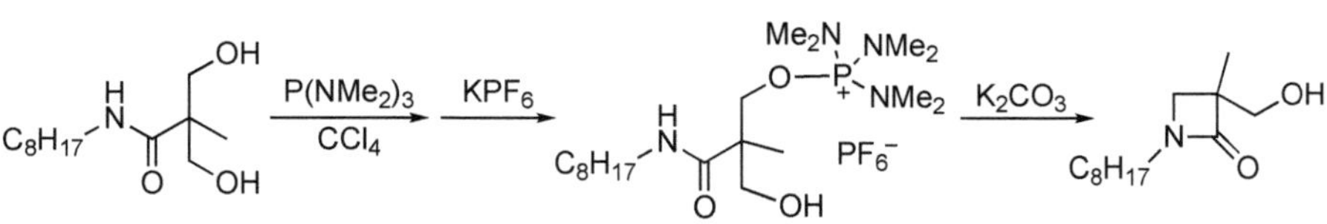

SCHEME 3.68 Preparation of 3-(hydroxymethyl)-3-methyl-1-octylazetidin-2-one from 3-hydroxy-2-(hydroxymethyl)-2-methyl-*N*-octylpropanamide

It should be emphasized that a large portion of the synthetic practices to form β-lactams using the approach of route A in Scheme 3.58 takes advantage of the Mitsunobu reaction, which connects alcohol (R-OH) and acidic compound (H-Nu) to a new compound (R-Nu) using triphenylphosphine and an azodicarboxylate such as diethyl azodicarboxylate (DEAD) or diisopropyl azodicarboxylate (DIAD), where triphenylphosphine is oxidized to triphenylphosphine oxide and the azodicarboxylate is reduced to the hydrazine. This reaction is generally limited to primary and secondary alcohols, and clean inversion of stereochemistry is always observed in the case of secondary alcohols. Examples of acidic compounds generally have an aqueous pKa < 15, with exceptions for intramolecular versions, such as carboxylic acids and phenols; nitrogen nucleophiles such as imides, hydroxamates, and heterocycles; sulfur nucleophiles such as thiols and thioamides; and carbon nucleophiles such as β-ketoesters [288]. Therefore, the retention of configuration at C_3 and clean inversion at C_4 during the intramolecular Mitsunobu reaction implies that essentially any chiral β-lactams could be made by simply choosing an appropriate chiral starting β-hydroxy acid [289]. This approach has been initially adopted by Miller, as illustrated in Scheme 3.69 for the synthesis of methyl 2-(2-(((2*R*,3*S*)-1-(benzyloxy)-3-ethyl-4-oxoazetidin-2-yl)methyl)-1,3-dioxolan-2-yl)acetate starting from *L*-cysteine [280].

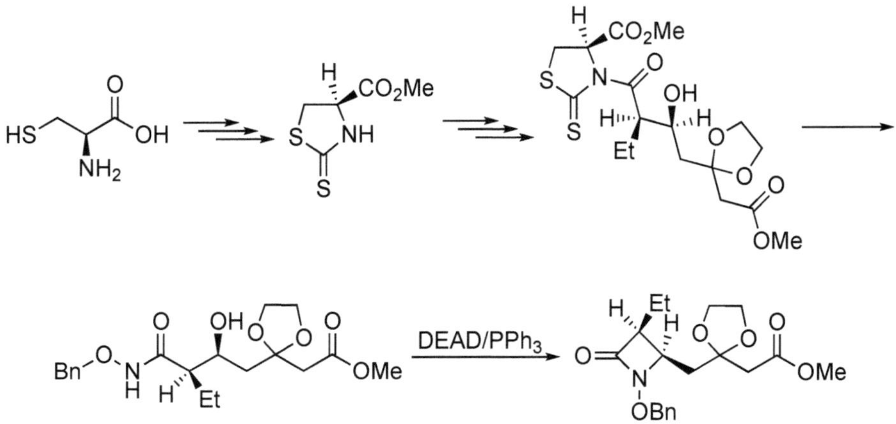

SCHEME 3.69 Application of the Mitsunobu reaction in the preparation of methyl 2-(2-(((2*R*,3*S*)-1-(benzyloxy)-3-ethyl-4-oxoazetidin-2-yl)methyl)-1,3-dioxolan-2-yl)acetate from *L*-cysteine

Using the same strategy, Miller has converted δ-(+)-3,6-glucuronolactone (i.e., (3*S*,3a*R*,6*R*,6a*R*)-3,5,6-trihydroxytetrahydrofuro[3,2-*b*]furan-2(3*H*)-one) into (1*S*,3*S*,6*R*)-3,7-bis(benzyloxy)-2-oxa-7-azabicyclo[4.2.0]oct-4-en-8-one, which was then selectively oxidized to 70% of (1*S*,6*R*)-7-(benzyloxy)-2-oxa-7-azabicyclo-[4.2.0]oct-4-ene-3,8-dione in CH_3CN using CrO_3 and acetic acid, as shown in Scheme 3.70. In this total synthesis, the starting material was first treated with methanol in the presence of a catalytic amount of NaOH to form methyl (2*S*)-2-hydroxy-2-((2*S*,3*R*,4*R*)-3,4,5-trihydroxytetrahydrofuran-2-yl)acetate, then the remaining hydroxyl groups were protected with acetyl groups. After two more steps of transformations, (2*S*,3*S*,4*R*)-2-(methoxycarbonyl)-3,4-dihydro-2*H*-pyran-3,4-diyl diacetate was obtained. The subsequent Ferrier rearrangement in the presence of $BF_3 \cdot OEt_2$ and benzyl alcohol in CH_2Cl_2 at -10°C for 4 days afforded 53% of methyl (2*S*,3*S*,6*S*)-3-acetoxy-6-(benzyloxy)-3,6-dihydro-2*H*-pyran-2-carboxylate, with > 99% selectivity. Cleavage of the methyl ester was achieved in THF in the presence of potassium trimethylsilanolate (KOTMS). After removal of the solvent and dissolution of the salt in dry DMF, simultaneous addition of *O*-benzylhydroxylamine hydrochloride (OBHA·HCl) and 1-ethyl-3-(3-dimethyl-aminopropyl) carbodiimide hydrochloride (EDAC·HCl) resulted in the formation of (2*S*,3*S*,6*S*)-6-(benzyloxy)-2-((benzyloxy)carbamoyl)-3,6-dihydro-2*H*-pyran-3-yl acetate in 75% yield. Then, treatment of this substrate with DEAD and triphenylphosphine in CH_3CN yielded 70% pure (1*S*,3*S*,6*R*)-3,7-bis(benzyloxy)-2-oxa-7-azabicyclo[4.2.0]oct-4-en-8-one after crystallization [290].

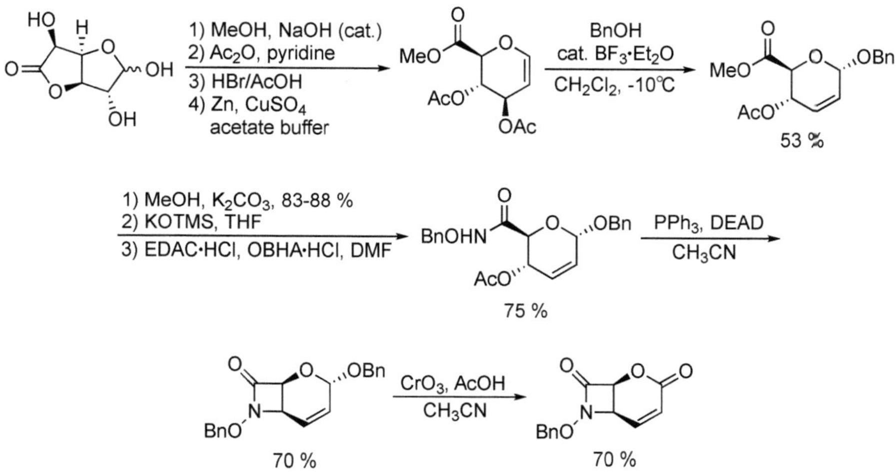

SCHEME 3.70 Synthesis of (1*S*,6*R*)-7-(benzyloxy)-2-oxa-7-azabicyclo[4.2.0]oct-4-ene-3,8-dione starting from *D*-(+)-3,6-glucuronolactone

In a similar example, (2*S*,3*S*)-2-amino-3-hydroxyhept-6-enoic acid was sequentially transformed into 4-nitrophenyl ((2*S*,3*S*)-2-(1,3-dioxoisoindolin-2-yl)-3-hydroxyhept-6-enoyl)glycinate, which was then treated with a combination of phosphite and azodicarboxylate to yield the expected 4-nitrophenyl 2-((2*R*,3*S*)-2-(but-3-en-1-yl)-3-(1,3-dioxoisoindolin-2-yl)-4-oxoazetidin-1-yl) acetate with inversion of stereochemistry at the secondary hydroxyl group. Specifically, when trimethyl phosphite and DEAD were applied for this reaction, only 50% of *cis*-β-lactam was obtained, in addition to 50% of 4-nitrophenyl ((2*S*,3*S*)-3-((dimethoxy-phosphoryl)oxy)-2-(1,3-dioxoisoindolin-2-yl) hept-6-enoyl)glycinate as the reaction intermediate. In comparison, when the substrate was treated with triphenylphosphine and DIAD in THF at 25°C, 80% of the *cis*-β-lactam was yielded. Increasing the reaction temperature may lead to the generation of a diastereomeric mixture, as indicated in the cases when the substrate was treated with triethylphosphite and DIAD in toluene at either 70°C or 80°C, afford 84% and 87% of the *cis*-β-lactam, along with 2% and 0.6% of the *trans*-β-lactam, respectively (Scheme 3.71) [291].

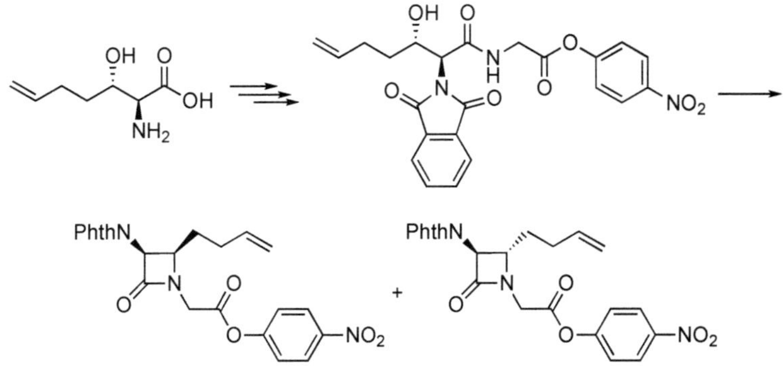

$(MeO)_3P$, DEAD	50 %	
PPh_3, DIAD, THF, 25℃	80 %	
$(EtO)_3P$, DIAD, toluene, 70℃	84 %	2 %
$(EtO)_3P$, DIAD, toluene, 80℃	87 %	0.6 %

SCHEME 3.71 The temperature effect on the diastereoselectivity of cyclization to form β-lactams under the Mitsunobu reaction condition

In addition to the above examples, several representative transformations of β-hydroxyl amides into the corresponding β-lactams by means of the Mistunobu reaction are: (a) treatment of (3*S*,5*S*)-*N*-(benzyloxy)-5-(2-((*R*,*E*)-4-bromo-2-((triisopropylsilyl)oxy)pent-3-en-1-yl)thiazol-4-yl)-3-hydroxyhexanamide with triphenylphosphine and DIAD to give (*R*)-1-(benzyloxy)-4-((*S*)-2-(2-((*R*,*E*)-4-bromo-2-((triisopropylsilyl)oxy) pent-3-en-1-yl)thiazol-4-yl)propyl)azetidin-2-one in excellent yield, an important intermediate in the total synthesis of immunosuppressive agent (-)-pateamine A [292]; (b) conversion of methyl (4*S*,5*S*)-6-((benzyloxy) amino)-4-hydroxy-6-oxo-5-(2-oxo-4,5-diphenyloxazol-3(2*H*)-yl) hexanoate in THF into methyl 3-((2*R*,3*S*)-1-(benzyloxy)-4-oxo-3-(2-oxo-4,5-diphenyloxazol-3(2*H*)-yl)azetidin-2-yl)propanoate with PPh_3 and DIAD in 57% of yield [293]; (c) synthesis of (3*S*,4*S*)-1-(allyloxy)-3-(1-hydroxyethyl)-4-phenylazetidin-2-one and its isomers [294]; (d) treatment of *tert*-butyl ((2*S*)-1-((benzyloxy)amino)-4-(carbamoyloxy)-3-hydroxy-3-methyl-1-oxobutan-2-yl)carbamate with DEAD and PPh_3 to give 40% of *tert*-butyl ((3*R*)-1-(benzyloxy)-2-((carbamoyloxy)methyl)-2-methyl-4-oxoazetidin-3-yl)carbamate, or alternative conversion of the same substrate into β-lactam using the modified Mitsunobu reaction using PPh_3 and CCl_4 in the presence of Et_3N, with similar yield of β-lactam [295]; (e) treatment of solid-phase supported benzyl ((2*S*,3*R*)-3-hydroxy-1-(alkoxyamino)-1-oxobutan-2-yl)carbamate with 5.0 eq. of DEAD and 10.0 eq. of PPh_3 in THF to yield benzyl ((2*S*,3*S*)-1-alkoxy-2-methyl-4-oxoazetidin-3-yl) carbamate, where the β-lactam can be cleaved from the solid-phase using SmI_2 to give benzyl ((2*S*,3*S*)-2-methyl-4-oxoazetidin-3-yl)carbamate or 5% TFA in CH_2Cl_2 to afford benzyl ((2*S*,3*S*)-1-hydroxy-2-methyl-4-oxoazetidin-3-yl)carbamate [296]; (f) synthesis of methyl 3-(4-(benzyloxy)phenyl)-2-((*S*)-3-(dibenzylamino)-2-oxoazetidin-1-yl)-propanoate via the treatment of methyl 3-(4-(benzyloxy)phenyl)-2-((*S*)-2-(dibenzylamino)-3-hydroxypropanamido)propanoate with PPh_3 and DIAD in THF, giving nearly quantitative yield of the β-lactam with a diastereoselectivity of 22:1 [297]. These applications of Mitsunobu reaction in formation of β-lactams are summarized in Scheme 3.72.

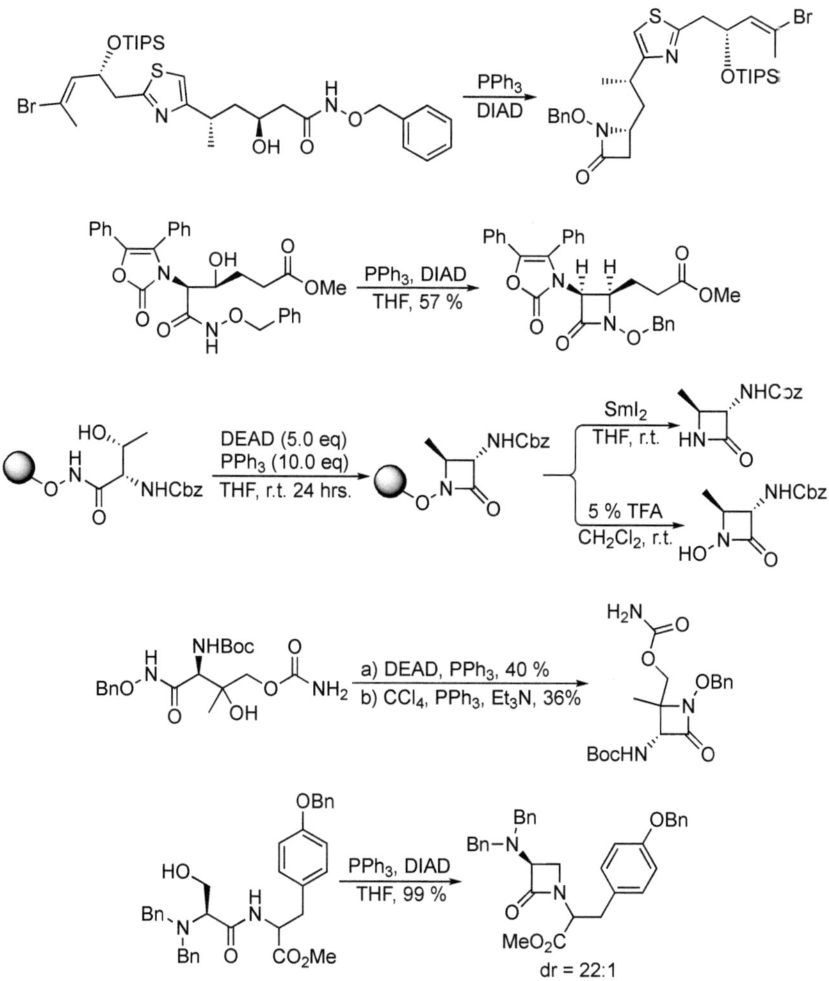

SCHEME 3.72 Examples to form β-lactams under the Mitsunobu reaction conditions

While most synthetic designs using path A in Scheme 3.58 involve a β-hydroxyl amide where the β-hydroxyl group is a secondary alcohol, when the β-hydroxyl group is a tertiary alcohol, a different approach can be applied, although the commonly used Mitsunobu reaction also works as shown in Scheme 3.73 [295]. For example, treatment of *tert*-butyl ((2*S*,3*R*)-1-((benzyloxy)amino)-3-hydroxy-3-methyl-1-oxopentan-2-yl)carbamate with a tertiary hydroxyl group with pyridine sulfur trioxide at 65°C allows the generation of (2*S*,3*R*)-1-((benzyloxy)amino)-2-((*tert*-butoxycarbonyl) amino)-3-methyl-1-oxopentan-3-yl hydrogen sulfate, which is then treated with K_2CO_3 in refluxing aqueous EtOAc to give 83% of *tert*-butyl ((2*S*,3*S*)-1-(benzyloxy)-2-ethyl-2-methyl-4-oxoazetidin-3-yl)carbamate. In comparison, transformation of its diastereomer, i.e., *tert*-butyl ((2*S*,3*S*)-1-((benzyloxy) amino)-3-hydroxy-3-methyl-1-oxopentan-2-yl)carbamate, under the same condition, only affords 65% of *tert*-butyl ((2*R*,3*S*)-1-(benzyloxy)-2-ethyl-2-methyl-4-oxoazetidin-3-yl)carbamate, as shown in Scheme 3.73 [295].

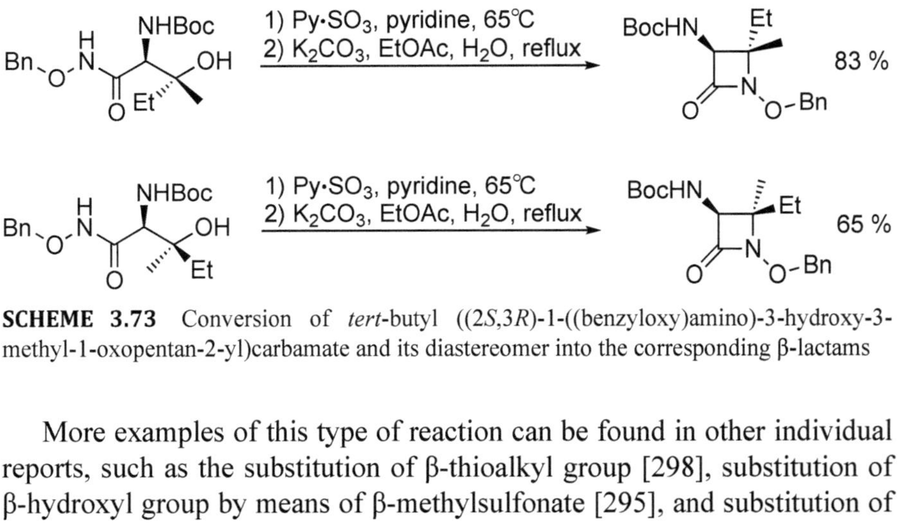

SCHEME 3.73 Conversion of *tert*-butyl ((2*S*,3*R*)-1-((benzyloxy)amino)-3-hydroxy-3-methyl-1-oxopentan-2-yl)carbamate and its diastereomer into the corresponding β-lactams

More examples of this type of reaction can be found in other individual reports, such as the substitution of β-thioalkyl group [298], substitution of β-hydroxyl group by means of β-methylsulfonate [295], and substitution of β-hydroxyl group involving the Mitsunobu reaction [299, 300].

3.3.4 INTRAMOLECULAR CYCLIZATION WITH CARBON NUCLEOPHILES

In path B synthetic approach outlined in Scheme 3.58, a carbon-based nucleophile needs to be generated, so that often a strong electron-withdrawing group at this carbon atom is essentially critical because the amido group also attaching to this carbon atom would make the creation of carbanion difficult due to its electron-donating nature. On the other hand, the α-halo is a very good leaving group due to the activating feature of the carbonyl group. As an example, a series of methyl (*S*)-2-(2-chloro-*N*-(4-methoxybenzyl)acetamido)-2-phenylacetate, methyl *N*-(2-chloroacetyl)-*N*-(4-methoxybenzyl)-*L*-phenylalaninate, methyl (*S*)-2-(2-chloro-*N*-(4-methoxybenzyl)acetamido)-4-phenylbutanoate when treated with a base, i.e., *tert*-butylimino-tri(pyrrolidino)phosphorane (BTPP) with or without the presence of chiral additive such as ((4*R*,5*R*)-2,2-dimethyl-1,3-dioxolane-4,5-diyl)bis(diphenylmethanol) (i.e., (-)-TADDOL), have been converted into the corresponding β-lactams, i.e., methyl (*S*)-1-(4-methoxybenzyl)-4-oxo-2-phenylazetidine-2-carboxylate, methyl (*R*)-2-benzyl-1-(4-methoxybenzyl)-4-oxoazetidine-2-carboxylate, methyl (*R*)-1-(4-methoxybenzyl)-4-oxo-2-phenethylazetidine-2-carboxylate with certain degree of memory for their stereochemistry, with methyl (*S*)-2-(2-chloro-*N*-(4-methoxybenzyl)acetamido)-2-phenylacetate being most effective in retaining the stereochemistry, resulting in the highest enantioselectivity (74% ee) of the corresponding β-lactam. The enantiomers of the above three amino acid derivatives give similar results in these transformations. All these α-amino acid derivatives carry an ester group as the electron-withdrawing group to activate the carbon atom so that it can be readily converted into carbanion or an enolate type of nucleophile. The addition of a chiral additive seems to increase the chemical yield of the β-lactams but does not improve the memory of stereochemistry [301]. When *L*-phenylalanine was converted to (2*R*)-1-((*N*,*N*-dicyclohexylsulfamoyl)methyl)-7,7-dimethylbicyclo[2.2.1]heptan-2-yl *N*-(2-chloroacetyl)-*N*-(4-methoxybenzyl)-*L*-phenylalaninate, and (1*R*,2*R*,4*R*)-1-((*N*,*N*-dicyclohexylsulfamoyl)methyl)-7,7-dimethylbicyclo[2.2.1]heptan-2-yl *N*-(2-chloroacetyl)-*N*-(4-methoxybenzyl)-*L*-phenylalaninate, the stereochemistry of the corresponding β-lactams would be affected largely from the chirality of the ester group, yielding (2*R*)-1-((*N*,*N*-dicyclohexylsulfamoyl)methyl)-7,7-dimethyl-bicyclo[2.2.1]heptan-2-yl (2*S*)-2-benzyl-1-(4-methoxybenzyl)-4-oxoazetidine-2-carboxylate and (1*R*,2*R*,4*R*)-1-((*N*,*N*-dicyclohexylsulfamoyl)methyl)-7,7-dimethyl-bicyclo[2.2.1]heptan-2-yl (*R*)-2-benzyl-1-(4-methoxybenzyl)-4-oxoazetidine-2-carboxylate, respectively [302]. Inspired by this result, methyl (4-methoxybenzyl)-*L*-phenylalaninate and methyl (4-methoxybenzyl)-*D*-phenylalaninate are condensed with (*S*)-2-chloropropanoic acid or (*R*)-2-chloropropanoic acid in the presence of peptide condensation reagent benzotriazol-1-yloxytris(dimethylamino)phosphonium hexafluorophosphate (BOP) in THF in the presence of Et_3N, affording four diastereomers, i.e., methyl

N-((*S*)-2-chloropropanoyl)-*N*-(4-methoxybenzyl)-*L*-phenylalaninate, methyl *N*-((*S*)-2-chloropropanoyl)-*N*-(4-methoxybenzyl)-*D*-phenylalaninate, methyl *N*-((*R*)-2-chloropropanoyl)-*N*-(4-methoxybenzyl)-*L*-phenylalaninate and methyl *N*-((*R*)-2-chloropropanoyl)-*N*-(4-methoxybenzyl)-*D*-phenylalaninate. When these four diastereomers were treated with base (either Cs_2CO_3 or BTPP in CH_3CN), β-lactams were obtained accordingly, showing that the stereochemical control of the cyclization to the β-lactam ring is fully dominated by the configuration of the 2-chloropropionyl group, which totally abolishes the asymmetric induction by the memory of chirality observed in the case of acetyl analogs, as shown in Scheme 3.74 [303].

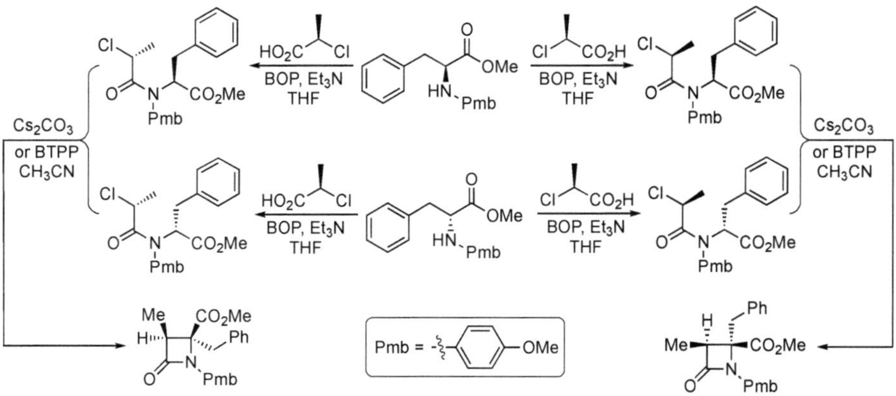

SCHEME 3.74 The examples for the stereochemical control of the cyclization to form β-lactam ring

In another example, the formation of β-lactam is a critical step in the total synthesis of BMS-262084 from δ-ornithine in an overall 30% yield, a sub-nanomolar inhibitor of tryptase which suppresses the induced inflammation in animal lungs. In this synthesis, δ-ornithine hydrochloride was converted into (*R*)-2-bromo-5-(2,2,2-trifluoroacetamido)pentanoic acid; meanwhile, dimethyl 2-((2,4-dimethoxy-benzyl)amino)malonate was prepared by reductive condensation between 2,4-dimethoxybenzaldehyde and dimethyl 2-aminomalonate hydrochloride. Combination of these two intermediates under basic condition transiently gave dimethyl (*R*)-2-(2-bromo-*N*-(2,4-dimethoxybenzyl)-5-(2,2,2-trifluoroacetamido)-pentanamido)malonate, which underwent intramolecular cyclization to afford dimethyl (*R*)-1-(2,4-dimethoxybenzyl)-4-oxo-3-(3-(2,2,2-trifluoroacetamido)-propyl)azetidine-2,2-dicarboxylate in 95% yield. Very interestingly, one of the methoxycarbonyl groups on the β-lactam ring can be selectively removed in aqueous DMF in the presence of LiCl, yielding 93% of methyl (2*S*,3*R*)-1-(2,4-dimethoxybenzyl)-4-oxo-3-(3-(2,2,2-trifluoroacetamido)-propyl)azetidine-2-carboxylate. This important intermediate was subsequently converted to the target molecule of BMS-262084, i.e., (2*S*,3*R*)-1-(4-(*tert*-butylcarbamoyl)piperazine-1-carbonyl)-3-(3-guanidinopropyl)-4-oxoazetidine-2-carboxylic acid, as shown in Scheme 3.75 [304]. In a very similar transformation, *L*-threonine was first converted into (2*R*,3*S*)-3-bromo-4-chloro-4-oxobutan-2-yl acetate, which is then condensed with dimethyl 2-((4-methoxyphenyl)-amino)malonate. The resulting dimethyl 2-((2*S*,3*R*)-3-acetoxy-2-bromo-*N*-(4-methoxyphenyl) butanamido)malonate was then treated with DBU in benzene at room temperature to give dimethyl (*R*)-3-((*R*)-1-acetoxyethyl)-1-(4-methoxyphenyl)-4-oxoazetidine-2,2-dicarboxylate. Treatment of this compound in pyridine with 1 *N* NaOH at 5°C removed one methoxycarbonyl group on the β-lactam ring, but not stereoselectively, affording methyl (3*R*)-3-((*R*)-1-acetoxyethyl)-1-(4-methoxyphenyl)-4-oxoazetidine-2-carboxylate [305]. These two syntheses using two ester groups (from malonate) to activate the carbonucleophile have been adopted to make β-lactams as early as 1950 [306, 307].

When there is no obvious electron-withdrawing group to activate the carbon, it is very difficult to proceed with this route of synthesis for β-lactams. Luckily, a recent report using palladium-catalyzed activation of C-H bond nicely solved this problem, as demonstrated in the transformation of *N*,*N*-dibenzyl-2-chloroacetamide into (*S*)-1-benzyl-4-phenylazetidin-2-one in the presence of catalyst Pd(dba)$_2$, cocatalyst (3*R*,5*R*,7*R*)-adamantane-1-carboxylic acid (AdCO$_2$H), Cs$_2$CO$_3$, and ligand 1-((3a*R*,8a*R*)-4,4,8,8-tetrakis(3,5-di-*tert*-butyl-4-methoxyphenyl)-2,2-dimethyltetrahydro-[1,3]dioxolo[4,5-*e*][1,3,2] dioxaphosphepin-6-yl)pyrrolidine in toluene at 100°C (Scheme 3.76) [308]. In this approach, excellent control of stereochemistry from the ligand has been observed, as shown in the reaction of (*R*)-2-chloro-*N*-(4-nitrobenzyl)-*N*-(1-phenylethyl)-acetamide and (*S*)-2-chloro-*N*-(4-nitrobenzyl)-*N*-(1-phenylethyl) acetamide under the same condition, where the former substrate leads to 70% of β-lactams, with greater than 95:5 diastereomeric ratio of (*S*)-4-(4-nitrophenyl)-1-((*R*)-1-phenyl-ethyl)azetidin-2-one over (*R*)-4-(4-nitrophenyl)-1-((*R*)-1-phenylethyl)-azetidin-2-one; and the latter due to its mismatched enantiomeric nature, leads to reduced yield of 50% for the two diastereomers of (*S*)-4-(4-nitrophenyl)-1-((*S*)-1-phenylethyl)azetidin-2-one and (*R*)-4-(4-nitrophenyl)-1-((*S*)-1-phenylethyl)-azetidin-2-one, with the diastereomeric ratio of 94: 6. In addition, the effect of electron-withdrawing group has been observed, as illustrated in a 4.8:1 preference for the reaction occurring on 4-cyanobenzyl moiety rather than 4-methoxybenzyl moiety when 2-chloro-*N*-(4-cyanobenzyl)-*N*-(4-methoxybenzyl)-acetamide was subject to this reaction condition.

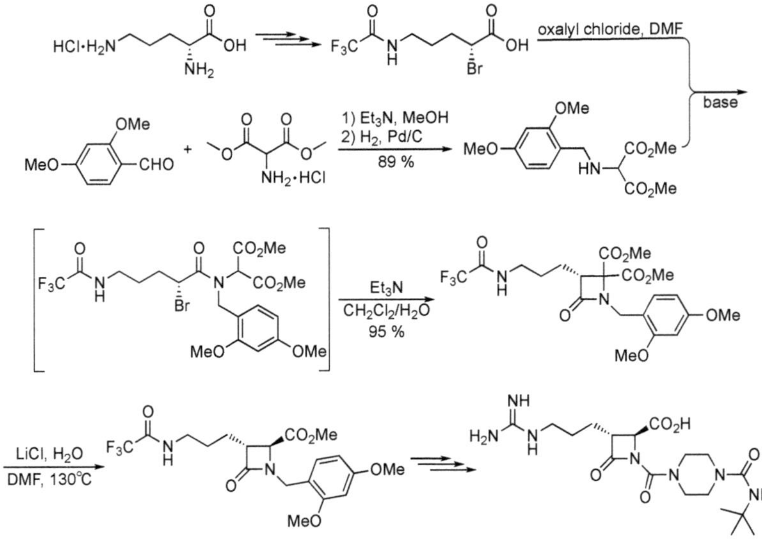

SCHEME 3.75 The key steps in the preparation of (2*S*,3*R*)-1-(4-(*tert*-butylcarbamoyl) piperazine-1-carbonyl)-3-(3-guanidinopropyl)-4-oxoazetidine-2-carboxylic acid

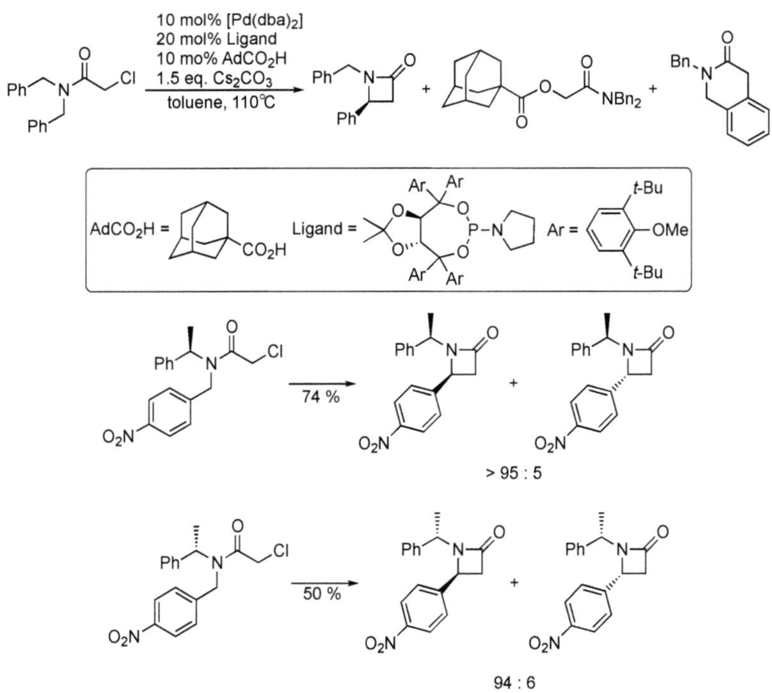

SCHEME 3.76 A palladium-catalyzed *C-H* activation for an intramolecular S_N2 reaction to form β-lactams

Besides the α-chloro as the leaving group, due to the high strain of the epoxide ring, the α-epoxide has been explored following path B in Scheme 3.58. For example, when *L*-threonine was treated with sodium nitrite under acidic conditions, the resulting diazonium was replaced with bromide, affording (2*S*,3*R*)-2-bromo-3-hydroxybutanoic acid in 94% yield. Upon treatment with KOH in ethanol, the epoxide structure formed. Then the carboxyl group was treated with thionyl chloride in THF in the presence of a small amount of pyridine, the resulting (2*R*,3*R*)-3-methyloxirane-2-carbonyl chloride was condensed with 1-(4-chlorophenyl)-2-((4-methoxyphenyl)amino) ethan-1-one to afford (2*R*,3*R*)-*N*-(2-(4-chlorophenyl)-2-oxoethyl)-*N*-(4-methoxyphenyl)-3-methyloxirane-2-carboxamide. This epoxide intermediate was then treated with a base to form the β-lactam. However, when this epoxide intermediate was treated with a strong base like lithium bis(trimethylsilyl) amide, 74% of (3*S*,4*S*)-4-(4-chlorobenzoyl)-3-((*R*)-1-hydroxyethyl)-1-(4-methoxyphenyl)azetidin-2-one was obtained, along with 26% of unexpected (1*S*,2*R*,4*R*,5*R*)-4-(4-chlorophenyl)-4-hydroxy-6-(4-methoxyphenyl)-2-methyl-3-oxa-6-azabicyclo[3.2.0]heptan-7-one. Similarly, when this epoxide intermediate was treated with K_2CO_3 in DMF for 4–24 hours, 81% of the expected β-lactam was obtained, as well as 13% of the bicyclic hemiacetal side product (Scheme 3.77) [309]. This bicyclic hemiacetal product was believed to form via an intramolecular cyclization between the hydroxyl group and carbonyl group on the same side of the β-lactam ring, i.e., via (3*S*,4*R*)-4-(4-chlorobenzoyl)-3-((*R*)-1-hydroxyethyl)-1-(4-methoxyphenyl)azetidin-2-one. This has been clearly demonstrated when (2*R*,3*R*)-*N*-(4-methoxyphenyl)-3-methyl-*N*-(2-oxo-2-phenylethyl)oxirane-2-carboxamide was subject to the same reaction condition with K_2CO_3 as the base, affording 74% of (3*S*,4*S*)-4-benzoyl-3-((*R*)-1-hydroxyethyl)-1-(4-methoxyphenyl)azetidin-2-one, 7% of (1*S*,2*R*,4*R*,5*R*)-4-hydroxy-6-(4-methoxyphenyl)-2-methyl-4-phenyl-3-oxa-6-azabicyclo[3.2.0]-heptan-7-one and 4% of (3*S*,4*R*)-4-benzoyl-3-((*R*)-1-hydroxyethyl)-1-(4-methoxyphenyl)azetidin-2-one.

Similarly, *L*-threonine was converted into (2*R*,3*R*)-*N*-(4-methoxyphenyl)-3-methyl-*N*-(2-oxo-2-phenylethyl)oxirane-2-carboxamide, which was then treated with K_2CO_3 in DMF at 60°C for 3–4 hours to give 75% of (3*S*,4*S*)-4-benzoyl-3-((*R*)-1-hydroxyethyl)-1-(4-methoxyphenyl)azetidin-2-one. Then the hydroxyl group was protected with *t*-butyldimethylsilyl group, and the *p*-methoxyphenyl group on the β-lactam ring was removed by CAN oxidation in CH_3CN at -10°C, yielding (3*S*,4*S*)-4-benzoyl-3-((*R*)-1-((*tert*-butyldimethylsilyl)-oxy)ethyl)azetidin-2-one. This key intermediate was sequentially transformed to the target molecule of sodium

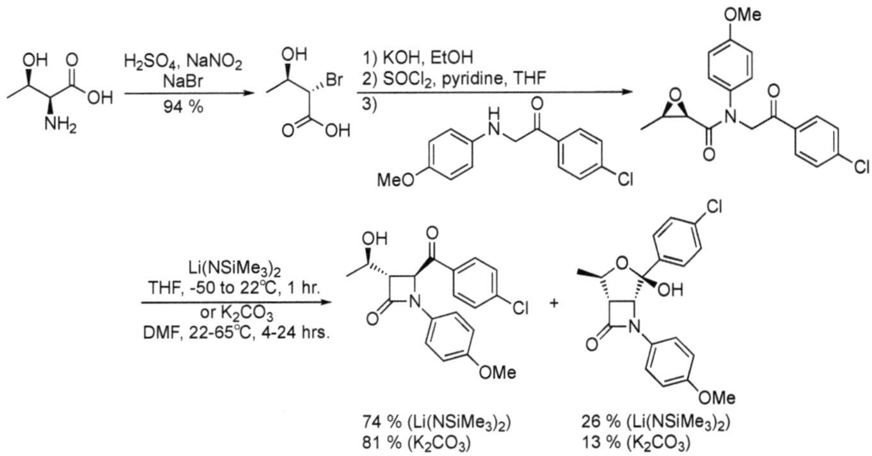

SCHEME 3.77 Synthesis of β-lactam from *L*-threonine invovling an epoxide intermeidate

(5*R*,6*S*)-3-((carbamoyloxy)methyl)-6-((*R*)-1-hydroxyethyl)-7-oxo-4-thia-1-azabicyclo[3.2.0]hept-2-ene-2-carboxylate [310].

3.3.5 INTRAMOLECULAR AMIDE FORMATION

Intramolecular amide formation can be completed by two approaches, either via the condensation between an amino group and a carboxyl group in the presence of an amide condensation agent, such as DCC, or by means of an acyl substitution reaction between the amino group and carboxylic derivative (e.g., ester group). For the example of direct condensation between free amino group and carboxyl group, 0.25 mmol of (2*R*,3*S*)-2-((*R*)-1-(benzyloxy) ethyl)-3-((*tert*-butoxycarbonyl)amino)-4-((*tert*-butyldiphenylsilyl)oxy) butanoic acid in 35 mL CH_3CN was converted into (3*S*,4*R*)-3-((*R*)-1-(benzyloxy)ethyl)-4-(2-((*tert*-butyldiphenylsilyl)oxy)ethyl)azetidin-2-one in 56% yield when it was treated with 5 mL trifluoroacetic acid, 0.11 mL of Et_3N and 0.55 mmol of DCC at 65–68°C for 10 hours, as shown in Scheme 3.78. It is assumed that the *t*-Boc protecting group was initially deprotected by TFA, yielding the intermediate of (2*R*,3*S*)-3-amino-2-((*R*)-1-(benzyloxy) ethyl)-4-((*tert*-butyldiphenylsilyl)oxy)butanoic acid, within which the free amino group then was condensed intramolecularly with the carboxyl group to give the final product. Under the same condition, (2*R*,3*S*)-3-((*tert*-butoxycarbonyl)amino)-4-((*tert*-butyldiphenylsilyl)oxy)-2-ethylbutanoic acid was converted into (3*R*,4*R*)-4-(2-((*tert*-butyldiphenylsilyl)oxy) ethyl)-3-ethylazetidin-2-one in 91% yield. Similarly, (2*R*,3*R*)-3-amino-2-ethylbutanoic acid, (2*R*,3*R*)-3-amino-2-isopropylbutanoic acid and (2*S*,3*R*)-3-amino-2-((*S*)-1-(benzyloxy)ethyl)butanoic acid have been transformed into the corresponding β-lactams, i.e., (3*R*,4*R*)-3-ethyl-4-methylazetidin-2-one, (3*R*,4*R*)-3-isopropyl-4-methylazetidin-2-one and (3*S*,4*R*)-3-((*S*)-1-(benzyloxy)-ethyl)-4-methylazetidin-2-one in 71%, 82% and 76% yields, respectively [311]. The intramolecular amide formation has also been performed for the conversion of 3-amino-2-methylpropanoic acid into 3-methylazetidin-2-one, and 3-aminobutanoic acid into 4-methylazetidin-2-one by treatment of 2-chloro-1-methylpyridin-1-ium iodide and Et_3N in CH_3CN in 57% and 65% yields, respectively [312]. With the same strategy, allyl group in allyl (*S*)-3-((2-methoxy-2-oxoethyl)amino)-2-(tritylamino) propanoate was deprotected with $Pd(Ph)_3Cl$, HOBt, and Bu_3SnH in CH_2Cl_2 at room temperature for a half-hour, then peptide condensation reagent EDC was added in combination with DIPEA in CH_2Cl_2 to give overall 85% of methyl (*S*)-2-(2-oxo-3-(tritylamino)azetidin-1-yl)acetate for the two consecutive steps [313].

In another example, the free amino group has been condensed with amido group to form the β-lactam, as demonstrated in the treatment of (*R*,*E*)-8-(((*S*)-1-phenylethyl)amino)cyclooct-1-ene-1-carboxamide and (*S*,*E*)-8-(((*S*)-1-phenyl-ethyl)amino)cyclooct-1-ene-1-carboxamide in CH_2Cl_2 with DCC and *tert*-butanol at room temperature for 8 hours, affording 78% and 62% of (*R*,*E*)-9-((*S*)-1-phenylethyl)-9-azabicyclo[6.2.0]dec-1-en-10-one and (*S*,*E*)-9-((*S*)-1-phenylethyl)-9-azabicyclo[6.2.0]dec-1-en-10-one, respectively. However, Pd/C catalyzed hydrogenation of these two β-lactams in methanol appears to be different, as the former leads to 78% of (1*S*,8*R*)-9-((*S*)-1-phenylethyl)-9-azabicyclo[6.2.0]decan-10-one, whereas the latter affords a mixture of (8*R*)-9-((*S*)-1-phenylethyl)-9-azabicyclo[6.2.0]decan-10-one, as displayed in Scheme 3.79 [314]. Very similarly, the tosyl protected amino group is still able to react with the amide in the presence of di-*tert*-butyl dicarbonate $(t\text{-Boc})_2O$ and DMAP to form β-lactam, as demonstrated in conversion of (2*S*,3*R*,4*R*)-5-((*tert*-butyldimethylsilyl)oxy)-2-((*R*)-1-((*tert*-butyl dimethylsilyl)oxy)ethyl)-4-methyl-3-((4-methylphenyl)sulfonamido) pentanamide into (3*S*,4*R*)-3-((*R*)-1-((*tert*-butyldimethylsilyl)oxy)ethyl)-4-((*R*)-1-((*tert*-butyldimethylsilyl)oxy)propan-2-yl)-1-tosylazetidin-2-one via intermediate of *tert*-butyl ((2*S*,3*R*,4*R*)-5-((*tert*-butyldimethylsilyl)oxy)-2-((*R*)-1-((*tert*-butyldimethylsilyl)oxy) ethyl)-4-methyl-3-((4-methylphenyl) sulfonamido)pentanoyl)carbamate [315].

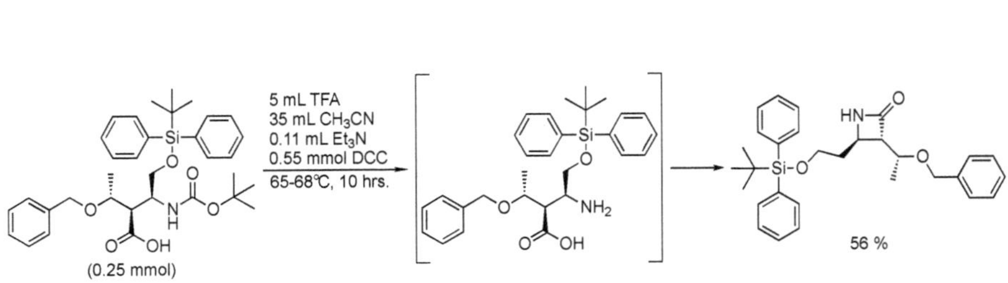

SCHEME 3.78 Formation of β-lactam via intramolecular amidization of β-amino acid

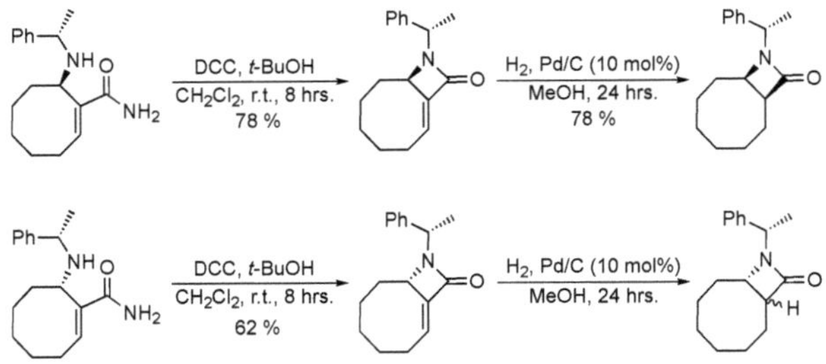

SCHEME 3.79 Formation of β-lactam via intramolecular amidization of β-amino amide

In the second approach to form β-lactam by means of the nucleophilic acyl substitution with an amino group, a series of α,β-diamino esters in dry THF at 0°C were treated with LiHMDS solution in THF for 12 hours and quenched with $NaHCO_3$ solution. Extraction with CH_2Cl_2 and purification by silica gel column chromatography (hexanes/EtOAc = 4:1) afforded the β-lactams in high yield. For example, when methyl (*R*)-2-benzyl-3-(((*R*)-2-((*tert*-butyldimethylsilyl) oxy)-1-phenylethyl)amino)-2-((2-nitrophenyl)sulfonamido)propanoate was treated with LiHMDS, 93% of *N*-((*R*)-3-benzyl-1-((*R*)-2-((*tert*-butyldimethylsilyl)oxy)-1-phenyl-ethyl)-2-oxoazetidin-3-yl)-2-nitrobenzenesulfonamide was obtained as oil, as illustrated in Scheme 3.80 [316].

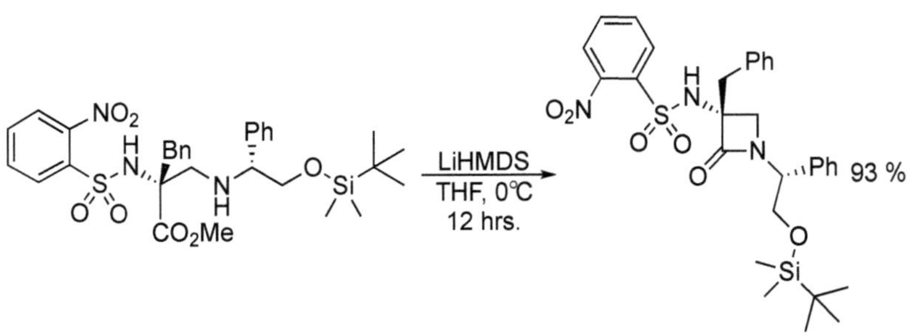

SCHEME 3.80 Formation of *N*-((*R*)-3-benzyl-1-((*R*)-2-((*tert*-butyldimethylsilyl)oxy)-1-phenylethyl)-2-oxoazetidin-3-yl)-2-nitrobenzenesulfonamide via intramolecular amidization of β-amino ester

In another practice, ethyl 2-(acetoxy(phenyl)methyl)acrylate and its analogs have been coupled with substituted anilines in the presence of $[Pd(allyl)Cl]_2$ and 2.5 mol% ligand in CH_2Cl_2 in the presence of aqueous K_2CO_3 to afford ethyl (*S*)-2-(phenyl(phenylamino)methyl)acrylate in 89% yield with 96% ee, and the minor product of ethyl (*E*)-3-phenyl-2-((phenylamino)methyl)acrylate arising from the S_N2' substitution, where the ligand is (5a*R*,8a*R*)-1,13-bis(bis(3,5-dimethylphenyl)phosphaneyl)-5a,6,7,8,8a,9-hexahydro-5*H*-chromeno[3,2-*d*] xanthene. Subsequently, ethyl (*S*)-2-(phenyl(phenylamino)methyl)acrylate was allowed to undergo intramolecular acyl substitution in refluxing toluene in the presence of *N*,*N*,*N'*,*N'*-tetrakis(trimethylsilyl)-λ^2-stannanediamine ($Sn[N(TMS)_2]_2$) to afford 85% of (*S*)-3-methylene-1,4-diphenylazetidin-2-one with 96% ee, as shown in Scheme 3.81. In this practice, it is interesting to note that a diastereomeric mixture (2:1) of ethyl 2-((*S*)-1-(4-(benzyloxy) phenyl)-2-(4-fluorophenyl)ethyl)-5-(4-fluorophenyl)-5-oxopentanoate was converted into a single isomer of (3*R*,4*S*)-4-(4-(benzyloxy)phenyl)-1-(4-fluorophenyl)-3-(3-(4-fluorophenyl)-3-oxopropyl)-azetidin-2-one in 77% yield with 95% ee, when it was treated with LiHMDS in THF at -20°C for 40 minutes (Scheme 3.81) [317]. In addition, $Sn[N(TMS)_2]_2$ has recently been applied in the formation of (*R*)-4-(3-((*tert*-butyldimethylsilyl)oxy)-4-methoxyphenyl)-3-methylene-1-(3,4,5-trimethoxyphenyl)azetidin-2-one from ethyl (*R*)-2-((3-((*tert*-butyldimethylsilyl)oxy)-4-methoxyphenyl) ((3,4,5-trimethoxyphenyl)amino)methyl)acrylate in 86% yield, and its enantiomer from ethyl (*S*)-2-((3-((*tert*-butyldimethylsilyl)oxy)-4-methoxyphenyl) ((3,4,5-trimethoxyphenyl)-amino)methyl)acrylate in 92% yield [318].

An interesting phenomenon has been noticed during the addition of ketene acetal such as ((1-ethoxy-2-methylprop-1-en-1-yl)oxy) trimethylsilane or *tert*-butyl-((1-ethoxy-2-methylprop-1-en-1-yl)oxy) dimethylsilane to chiral imines such as (*R*,*E*)-2-(benzylideneamino)-1-morpholino-2-phenylethan-1-one, the diastereo selectivity can be switched over in the presence of different titanium(IV) halide. For example, in the presence of 2.0 equivalents of TiF_4, the reaction of ((1-ethoxy-2-methylprop-1-en-1-yl)oxy)trimethylsilane yielded 67% of ethyl (*S*)-2,2-dimethyl-3-(((*R*)-2-morpholino-2-oxo-1-phenylethyl)amino)-3-phenylpropanoate and ethyl (*R*)-2,2-dimethyl-3-(((*R*)-2-morpholino-2-oxo-1-phenylethyl) amino)-3-phenylpropanoate, in a ratio of 85:15. In contrast, in the presence of 2.0 equivalents of $TiCl_4$ or $TiBr_4$, only 34% or 33% of the addition products were obtained, in a ratio of 9:91 or 13:87, respectively. Subsequent treatment of ethyl (*S*)-2,2-dimethyl-3-(((*R*)-2-morpholino-2-oxo-1-phenylethyl)amino)-3-phenylpropanoate with trimethylaluminum in

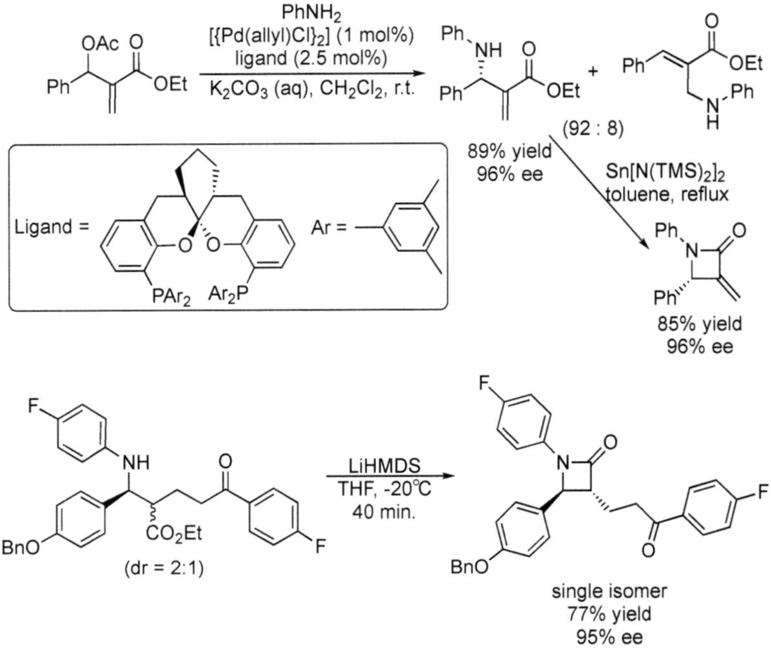

SCHEME 3.81 Strong base promoted intramolecular amidization of β-amino ester to yield β-lactams

toluene afforded 69% of (*S*)-3,3-dimethyl-1-((*R*)-2-oxo-1-phenylpropyl)-4-phenylazetidin-2-one. In order to confirm the stereochemistry of this compound, ethyl (*S*)-3-azido-2,2-dimethyl-3-phenylpropanoate was catalytically hydrogenated in the presence of Pd-C in methanol, and the resulting ethyl (*S*)-3-amino-2,2-dimethyl-3-phenylpropanoate was triggered to undergo intramolecular acyl substitution with MeMgBr, yielding 74% of (*S*)-3,3-dimethyl-4-phenylazetidin-2-one with 65% ee (Scheme 3.82). This compound was identical to the one after removal of the amino protecting group from (*S*)-3,3-dimethyl-1-((*R*)-2-oxo-1-phenylpropyl)-4-phenylazetidin-2-one [319]. It should be emphasized that the Grignard reagent should react with the ester group to afford a tertiary alcohol, but it does not react with the ester group in this case, instead it simply functions as a base to deprotonate the amino group. Another example of using the Grignard reagent to trigger the intramolecular acyl substitution to afford β-lactam is the two-step transformation of (2*S*,3*S*,4*R*)-2-amino-4-hydroxy-3-(methoxycarbonyl)pentanoic acid into (2*S*,3*S*)-4-oxo-1-(trimethylsilyl)-3-((*R*)-1-((trimethylsilyl)oxy)ethyl)azetidine-2-carboxylic acid in 88% yield, where the hydroxyl group on the starting material was first protected with trimethylsilyl after the treatment of substrate in toluene at 95°C with TMSCl and HMDS for 12 hours, and the resulting intermediate was then treated with *tert*-butyl magnesium chloride in THF at room temperature for 12 hours. In this case, even the free carboxyl group is not protected during the acyl substitution [320].

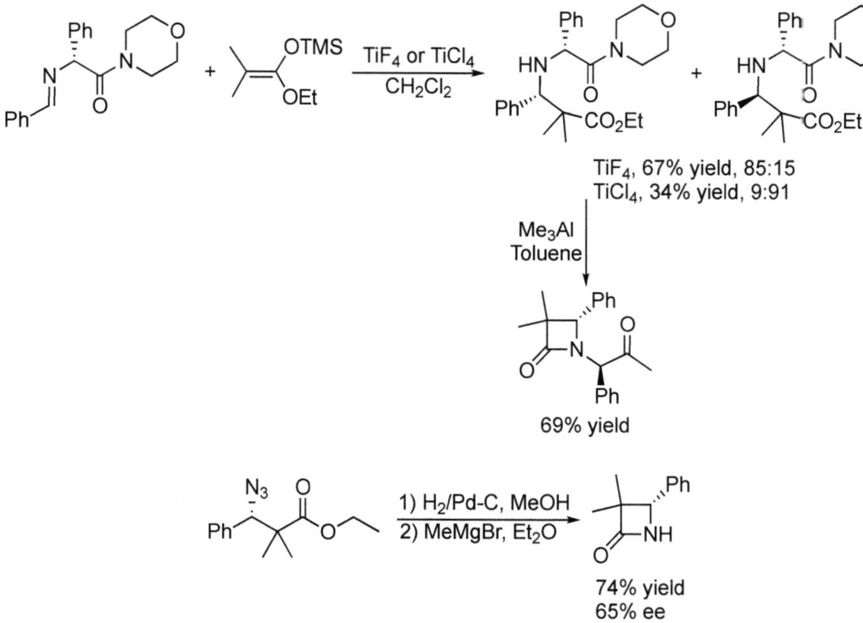

SCHEME 3.82 Example of Me_3Al and Grignard reagent promoted intramolecular amidization to afford β-lactams

Furthermore, β-lactam has been applied as an important intermediate in the total synthesis of (-)-haouamine B pentaacetate starting from commercially available 3,4-dimethoxyphenol in 40 steps in 0.055% overall yield with an average 83% chemical yield for each step [321]. Haouamines exhibit highly specific and strong cytotoxicity against the HT-29 human colon carcinoma cell line. In this total synthesis, the β-amino ester of methyl (2*R*,3*S*)-3-amino-4-(2-(benzyloxy)-4,5-dimethoxyphenyl)-2-((triethylsilyl)oxy)butanoate was treated with *t*-butyl magnesium chloride in THF at 0°C to give (3*R*,4*S*)-4-(2-(benzyloxy)-4,5-dimethoxybenzyl)-3-((triethylsilyl)oxy)azetidin-2-one, of which the free amino group was subsequently protected with benzyl group via KHMDS/BnBr treatment in the presence of phase transfer agent tetrabutylammonium iodide, and triethylsilyl group was removed by TBAF to afford (3*R*,4*S*)-1-benzyl-4-(2-(benzyloxy)-4,5-dimethoxybenzyl)-3-hydroxyazetidin-2-one in 61% of overall yield for these three steps of transformations. Following operations led to the important intermediate of (2a*R*,8a*S*)-1-benzyl-5,6-dimethoxy-2a-(3-methoxyphenyl)-8,8a-dihydro-1*H*-chromeno[3,2-*b*]azet-2(2a*H*)-one and final (-)-haouamine B pentaacetate, i.e., (11a*S*,16b*S*)-16b-(3-acetoxyphenyl)-9,11a,12,16b,17,18-hexahydro-10*H*-5,8-etheno-11,18-methanobenzo[*f*]indeno[2,1-*b*][1]-azacyclotridecine-2,15,16,21-tetrayl tetraacetate, as displayed in Scheme 3.83 [321]. Also, *tert*-butyl magnesium chloride has been applied in the conversion of methyl 3-hydroxy-2-((((*S*)-1-phenylethyl)amino)methyl)butanoate into 3-(1-hydroxyethyl)-1-((*S*)-1-phenylethyl)azetidin-2-one in THF in 61% yield [322]. In another example, menthyl 4-methyl (2*S*,3*S*)-2-amino-3-((*R*)-1-(((2,3-dimethylbutan-2-yl)dimethylsilyl)oxy)ethyl)succinate (11.4 mmol) in 25 mL of Et_2O was treated with equivalent amount of trimethylsilyl chloride and Et_3N at room temperature for 2 hours, followed by four equivalents of isopropyl magnesium bromide for 18 hours to yield 91% of menthyl (2*S*,3*S*)-3-((*R*)-1-(((2,3-dimethylbutan-2-yl)dimethylsilyl)oxy)ethyl)-4-oxoazetidine-2-carboxylate [323].

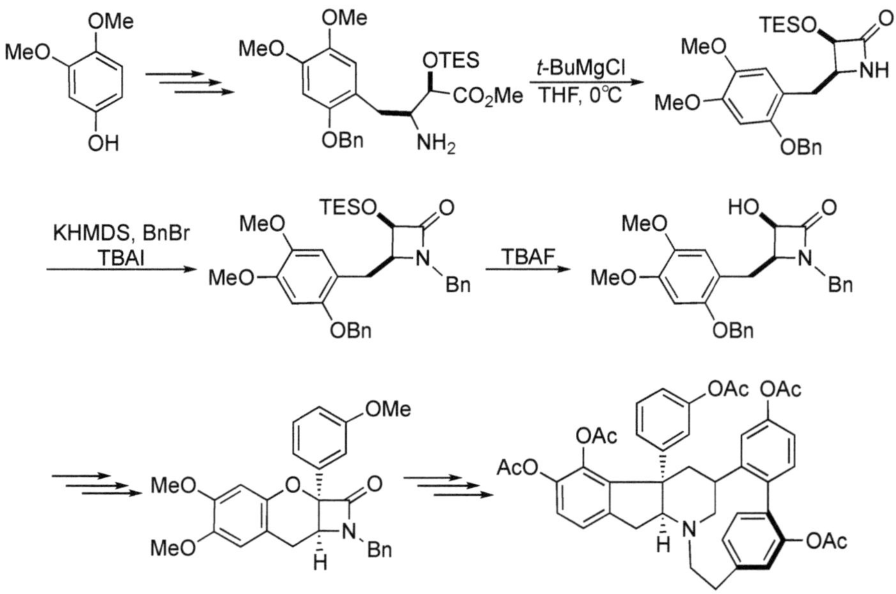

SCHEME 3.83 The key steps in the total synthesis of (11a*S*,16b*S*)-16b-(3-acetoxyphenyl)-9,11a,12,16b,17,18-hexahydro-10*H*-5,8-etheno-11,18-methanobenzo[*f*]indeno[2,1-*b*][1]azacyclotridecine-2,15,16,21-tetrayl tetraacetate involving the β-lactamization

A neutral condition to directly convert the β-amino acids into β-lactams takes advantage of Mukaiyama's reagent (Ph_3P-$(PyS)_2$), which is very useful for peptide synthesis through the oxidation-reduction condensation first introduced in 1970 [324]. For example, when 0.01 M of 3-(benzylamino) butanoic acid was refluxed in CH_3CN for 4.5 hours, 96% of 1-benzyl-4-methylazetidin-2-one was obtained (Scheme 3.84). In this practice, CH_3CN has been proven to be the ideal solvent, which remarkably increases the yields of the corresponding β-lactams. The features of this approach include the general applicability of this system to a variety of β-amino acids under neutral conditions, the preference for high dilution (0.01–0.05 M) and higher temperatures (reflux), tolerance of other functional groups such as amino, hydroxyl, and ester groups [325]. In another example, 300 mg of (2*S*,3*R*)-2-(((*tert*-butoxycarbonyl)amino)methyl)-3-((*tert*-butyldimethylsilyl)*oxy*) butanoic acid in 2 mL of CH_2Cl_2 at 0°C was added 0.5 mL of 30% TFA in CH_2Cl_2 to remove the *t*-Boc protecting group. When no starting material could be detected by TLC, the reaction mixture was concentrated *in vacuo* and dissolved in 17.3 mL of CH_3CN. To this solution was added with 0.12 mL of Et_3N, 272 mg of PPh_3 and 22.82 mg of 2-aldrithiol (i.e., 1,2-di(pyridin-2-yl)disulfane). The resulting solution was heated at 60°C for 12 hours and worked up to give 71% of (*S*)-3-((*R*)-1-((*tert*-butyldimethylsilyl)oxy)ethyl) azetidin-2-one as a white solid [326].

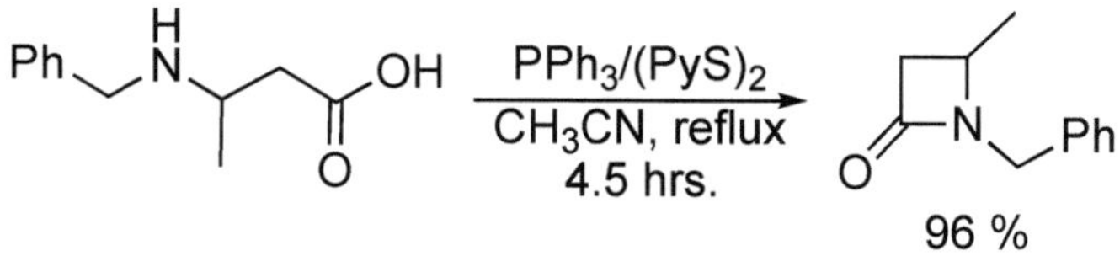

SCHEME 3.84 The application of Mukaiyama's reagent in the direct conversion of β-amino acid into β-lactam

Another useful reagent for the creation of β-lactam is cuprous trifluoromethanesulfonate (Cu(I)OTf) which converts the β-amino thioester into β-lactam. For example, treatment of butyl (2*S*,3*R*)-2-((4-methoxyphenyl) amino)-3-methyl-4-oxo-4-(phenylthio)butanoate in refluxing toluene with 1.2 equivalents of CuOTf in combination with 2 equivalents of calcium carbonate yielded 82% of butyl (2*S*,3*R*)-1-(4-methoxyphenyl)-3-methyl-4-oxoazetidine-2-carboxylate. Interestingly, when *S*-phenyl (2*R*,3*S*)-2-ethyl-3-(((*R*)-1-phenylethyl)amino)-5-(trimethylsilyl)pent-4-ynethioate was treated under the same condition, 87% of (3*R*,4*S*)-3-ethyl-4-ethynyl-1-((*R*)-1-phenylethyl) azetidin-2-one was obtained along with the loss of trimethylsilyl group. Surprisingly, when *S*-phenyl (2*R*,3*S*)-3-(benzylamino)-2-methyl-5-(trimethylsilyl) pent-4-ynethioate was treated with the same reagent but in dioxane, the phenylsulfenyl group migrated to the end of the triple bond, yielding 55% of (3*R*,4*S*)-1-benzyl-3-methyl-4-((phenylthio)ethynyl)azetidin-2-one. In addition, CuOTf even works for the conversion of *S*-phenyl (2*R*,3*R*)-3-(benzyl(9-borabicyclo[3.3.1]nonan-9-yl)amino)-2-methylhexanethioate into the corresponding β-lactam, i.e., (3*R*,4*R*)-1-benzyl-3-methyl-4-propylazetidin-2-one, in yield of 85% (Scheme 3.85). The *S*-phenyl (2*R*,3*R*)-3-(benzyl(9-borabicyclo[3.3.1]nonan-9-yl)amino)-2-methylhexanethioate was formed by the boron enolate-imine condensation reaction, of which the hydrolysis to β-amino thiol ester occasionally requires a rather drastic reaction condition, resulting in the formation of epimerized β-amino thiol esters. However, the transformation of *S*-phenyl (2*R*,3*R*)-3-(benzyl(9-borabicyclo[3.3.1]nonan-9-yl)amino)-2-methylhexanethioate into β-lactam in refluxing toluene in the presence of CuOTf proceeds smoothly without any epimerization [327].

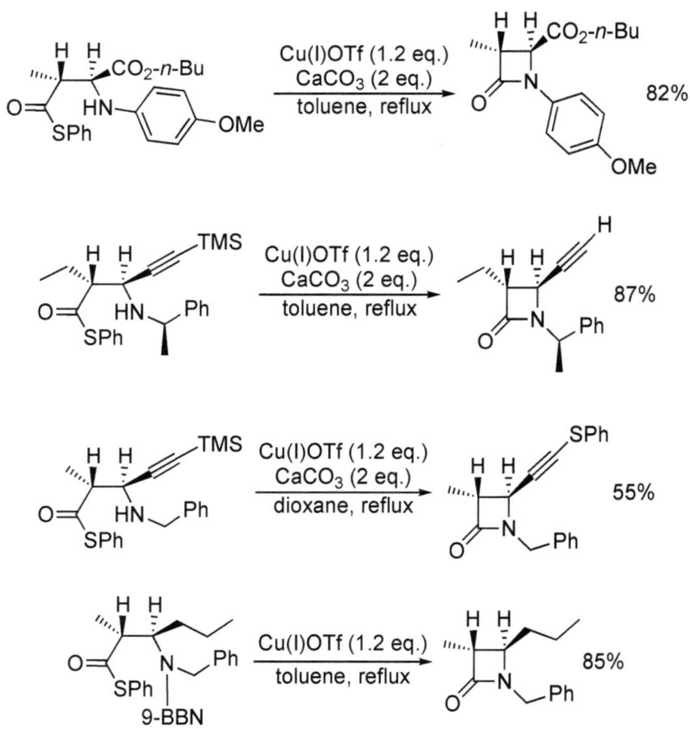

SCHEME 3.85 The application of Cu(I)OTf in the conversion of β-amino thioesters into the corresponding β-lactams

Also, phenyl phosphorodichloridate in combination with Et_3N is very effective in promoting the formation of β-lactams from β-amino acids in CH_3CN at room temperature. The nine tested reactions all give the corresponding β-lactams in very good to excellent yields. In addition, other phosphorylating reagents, such as diphenyl phosphorochloridate and dimethylphosphoramidic dichloride, also provide good to excellent yields of the expected β-lactams, as demonstrated in the conversion of (2*R*,3*S*,*E*)-2-phenoxy-5-phenyl-3-(*p*-tolylamino)pent-4-enoic acid into (3*R*,4*S*)-3-phenoxy-4-((*E*)-styryl)-1-(*p*-tolyl)azetidin-2-one (Scheme 3.86) [328].

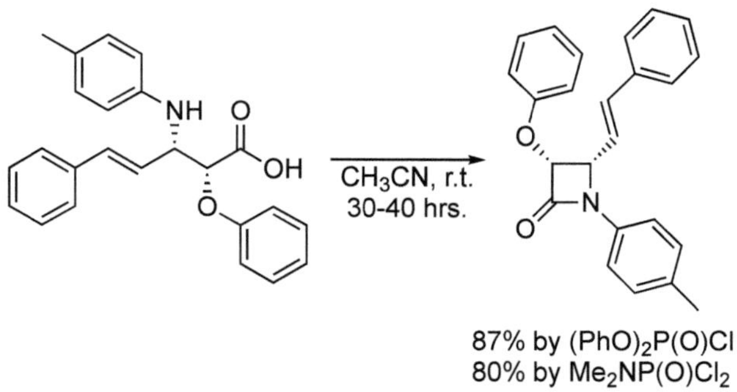

SCHEME 3.86 diphenyl phosphorochloridate and dimethylphosphoramidic dichloride promoted the formation of β-lactam from β-amino acid

A recent one-pot approach involving $Fe(acac)_3$ catalyzed reaction between substituted benzaldehydes and ethyl 2-cyanoacetate and subsequent reduction with $NaBH_4$ to give substituted 3-benzylazetidin-2-ones in good yields is suspicious as both double bond in ethyl (*E*)-2-cyano-3-arylacrylate and cyano group have been reduced by $NaBH_4$ [329].

3.3.6 FORMATION OF Β*-LACTAMS VIA AMIDATION*

By definition, amidation is a method for the synthesis of amide. More specifically in this chapter, the amidation methods for the generation of β-lactams refer to the activation of the C-H bond and subsequent formation of the C-N bond or the activation of the N-H bond and upcoming formation of the C-N bond in β-lactams. The activation of the C-H bond can be conducted either by transition metal catalysts or enzymes. For example, when carbamoyl chlorides in xylene or mesitylene are treated with a catalytic amount of palladium catalyst (10 mol%), in the presence of 20 mol% ligand, 30 mol% pivalic acid, 1.5 or 3.0 equivalents of Cs_2CO_3 and source of carbon monoxide (either balloon or precursor), the corresponding β-lactams can be obtained up to excellent yields. The advantages of this approach are: (a) easily accessible starting material of carbamoyl chloride from the treatment of secondary amines with triphosgene, in many cases without further purification; and (b) enantioselectivity induced by using either a chiral ancillary ligand or a chiral base. In addition, β-lactams with mono- or disubstituents at the C_β position can be easily prepared. In one typical reaction, isopropyl(2,4,6-trimethylbenzyl)carbamic chloride was enantioselectively converted into (*R*)-4-methyl-1-(2,4,6-trimethylbenzyl) azetidin-2-one in the presence of $PdCl_2$ (10 mol%), 20 mol% of ligand (5*S*,6*S*)-4,4,7,7-tetrakis(3,5-bis(trifluoromethyl)phenyl)-2-(3,5-di-*tert*-butyl-4-methoxyphenyl)-5,6-dimethoxy-1,3,2-dioxaphosphepane, 0.3 equivalent of pivalic acid, 1.5 equivalent of Cs_2CO_3 in mesitylene at 120°C under two different conditions, where the condition A was provided with 3.0 equivalents of 9-methylfluorene-9-carbonyl chloride (Cogen) as the source of CO, and condition B was provided with CO in CO balloon (Scheme 3.87) [330]. The 2,4,6-trimethylbenzyl group can be oxidatively removed in aqueous CH_3CN with $K_2S_2O_8$, affording 76% of (*R*)-4-methylazetidin-2-one. Based on many tested reaction conditions, it is observed that in most cases, the conditions with Cogen are superior or equal to the CO-balloon conditions, as excellent yields are obtained with reactants bearing a trisubstituted β-position, including enantiopure reactants, whereas markedly reduced yields are noticed for the reactions from reactants with disubstituted β-positions. In addition to the primary C-H bonds, activated tertiary cyclopropyl C-H bonds and benzylic secondary C-H bonds are also reactive, leading to spirocyclic and fused β-lactams in high yields. It should be pointed out that it is critical to run this reaction under a source of CO, according to the proposed reaction mechanism as displayed in Scheme 3.88. Under this reaction condition, the palladium catalyst (e.g., $Pd(PPh_3)_4$, $Pd(OAc)_2$, $PdCl_2$, etc.), if not at 0 valence, will be reduced by amines or amide to 0 valence, which then undergoes oxidative addition to the carbamoyl chloride. In this case, decarbonylation is quite possible under heating, and the presence of CO atmosphere would suppress the decarbonylation process, and facilitates the abstraction of β-hydrogen either by cesium pivalate or simply Cs_2CO_3; after that, reductive elimination regenerates the palladium(0) catalyst along with the formation of β-lactam [330].

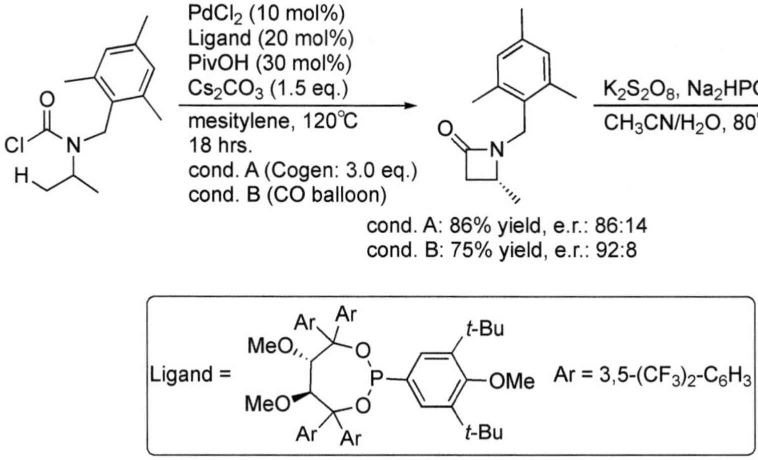

SCHEME 3.87 Palladium-catalyzed conversion of isopropyl(2,4,6-trimethylbenzyl) carbamic chloride into (*R*)-4-methyl-1-(2,4,6-trimethylbenzyl)azetidin-2-one in the presence of ligand and CO

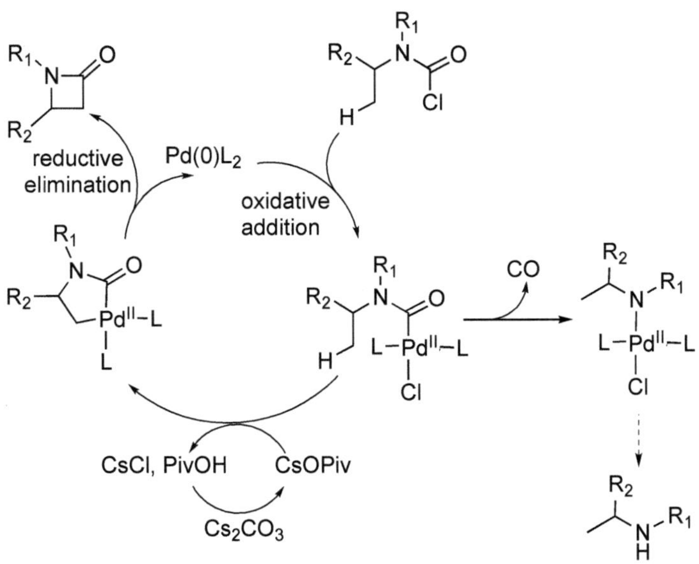

SCHEME 3.88 A possible mechanism for the palladium-catalyzed amidation

Also with palladium catalyst, secondary amides with β-H on the carboxyl component can be enantioselectively transformed into the corresponding β-lactams, as shown in the case for the preparation of

(*R*)-4-(3-methoxyphenyl)-1-(2-(pyridin-2-yl)propan-2-yl)azetidin-2-one from 3-(3-methoxyphenyl)-*N*-(2-(pyridin-2-yl)propan-2-yl)propenamide in *tert*-butanol at 115°C, in the presence of Pd(hfac)$_2$, additive 2-fluoro-1-iodo-4-nitrobenzene, ligand 3,3'-dichloro-[1,1'-binaphthalene]-2,2'-diol and base K_2CO_3, in 58% chemical yield and 90% ee (Scheme 3.89). The original ligand with Pd, i.e., hfac, is 1,1,1,5,5,5-hexafluoropentane-2,4-dione or hexafluoroacetylacetone [331]. More exciting method is a streamlined asymmetric synthesis of β-aryl β-lactams from propionamide and aryl iodides via Pd-catalyzed sequential C(sp^3)-H functionalization. In this approach, *N*-(quinolin-8-yl)propionamide undergoes sequential β-arylation with aryl iodides and cyclization to generate β-aryl β-lactams. One typical example is to make α-amino-β-lactams from materials that can be conveniently prepared from alanine through palladium(II)-catalyzed sequential C(sp^3)-H monoarylation/ amidation, as shown in Scheme 3.90. In this protocol, the coupling of 1-(3-iodo-1*H*-indol-1-yl)ethan-1-one to the β-position of (*S*)-2-(1,3-dioxoisoindolin-2-yl)-*N*-(2-(pyridin-2-yl)propan-2-yl)propanamide was achieved in the presence of 10 mol% Pd(OAc)$_2$, 1.5 equivalents of CuF_2, 5.0 equivalent of DMPU under nitrogen atmosphere, yielding 70% of (*S*)-3-(1-acetyl-1*H*-indol-3-yl)-2-(1,3-dioxoisoindolin-2-yl)-*N*-(2-(pyridin-2-yl)propan-2-yl)propenamide, with greater than 30:1 of monoarylation over diarylation. It is found that the *N*,*N*-bidentate 2-pyridyl-isopropyl amide moiety is an effective directing group both in controlling the selectivity in the arylation step for monoarylation and in enhancing the reactivity in subsequent amidation. This amidation step was performed in the

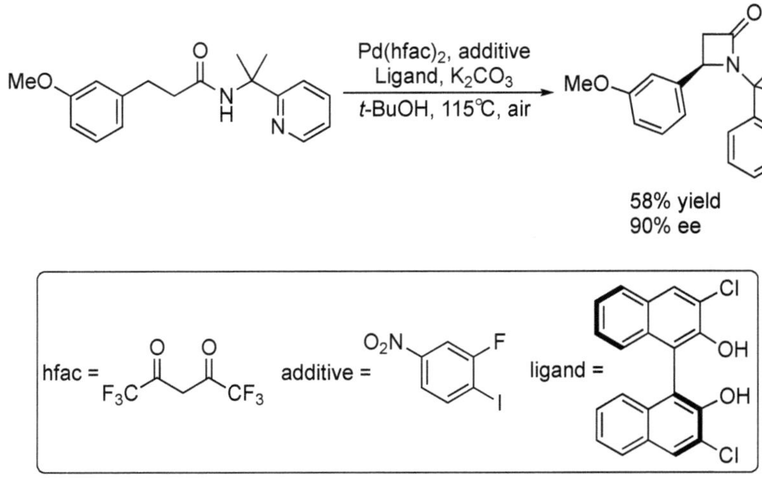

SCHEME 3.89 Pd(hfac)$_2$ catalyzed direct *C*-*N* bond formation to yield β-lactam

presence of 10 mol% $Pd(OAc)_2$, 2.0 equivalents of $NaIO_3$, 10.0 equivalents of Ac_2O in CH_3CN at 70°C under inert atmosphere for 48 hours, yielding 46% of 2-((2*S*,3*S*)-2-(1-acetyl-1*H*-indol-3-yl)-4-oxo-1-(2-(pyridin-2-yl)propan-2-yl)azetidin-3-yl)isoindoline-1,3-dione along with tiny quantity of (1*S*,2*S*)-1-(1-acetyl-1*H*-indol-3-yl)-2-(1,3-dioxoisoindolin-2-yl)-3-oxo-3-((2-(pyridin-2-yl)propan-2-yl)amino)propyl acetate [332].

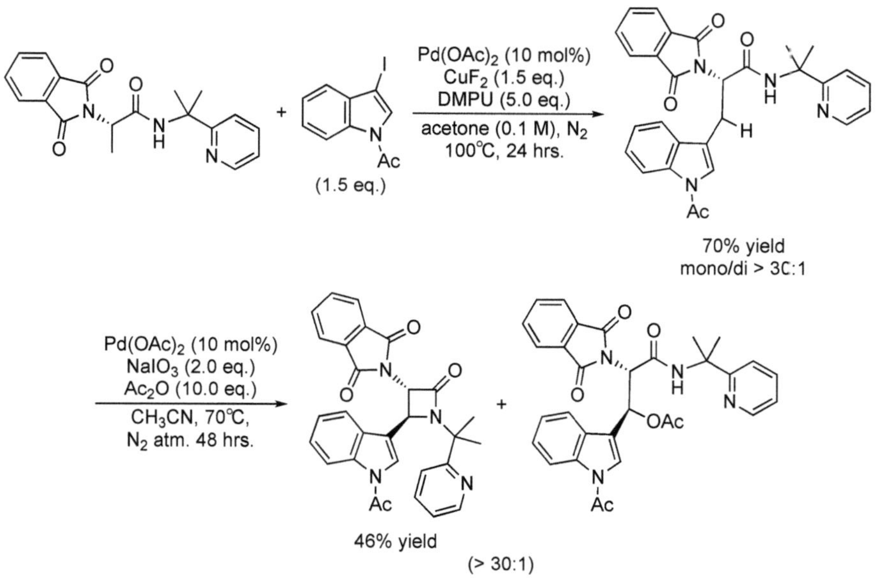

SCHEME 3.90 $Pd(OAc)_2$ catalyzed preparation of β-lactam via *C-C* and *C-N* bond formation

In addition, the lactam-forming reaction provides an example of Pd(II)-catalyzed enantioselective intramolecular $C(sp^3)$-H amidation reaction which proceeds up to 94% ee. The use of 2-methoxy-5-chlorophenyl iodide as an oxidant is critical in controlling the competing C-N versus C-C reductive elimination pathways of the Pd(IV) intermediate to achieve the desired chemoselectivity. For a particular reaction, in the presence of $PdCl_2(PhCN)_2$, ligand 3,3'-difluoro-[1,1'-binaphthalene]-2,2'-diol, 1.5 equivalents of Cs_2CO_3, and 4.0 equivalents of oxidant, in a 3:2 mixed solvent of 1,3-ditrifluoromethylbenzene and *tert*-amyl alcohol, 3-phenyl-*N*-(quinolin-8-yl)propenamide was converted into 90% of 4-phenyl-1-(quinolin-8-yl)azetidin-2-one with 89% ee, although the actual configuration of the chiral center has not been specified. Meanwhile, 6% of 3-(5-chloro-2-methoxyphenyl)-3-phenyl-*N*-(quinolin-8-yl)propanamide was obtained as well (Scheme 3.91)

[333]. In addition, the amides formed from (*S*)-2-(1,3-dioxoisoindolin-2-yl) propanoic acid and 5-methoxyquinolin-8-amine have been subject to similar monoarylation and subsequent amidation to afford various α-amino-β-lactams. These two reports demonstrate the potential for the stereoselective synthesis of orthogonally protected diamino-arylpropanoates [334].

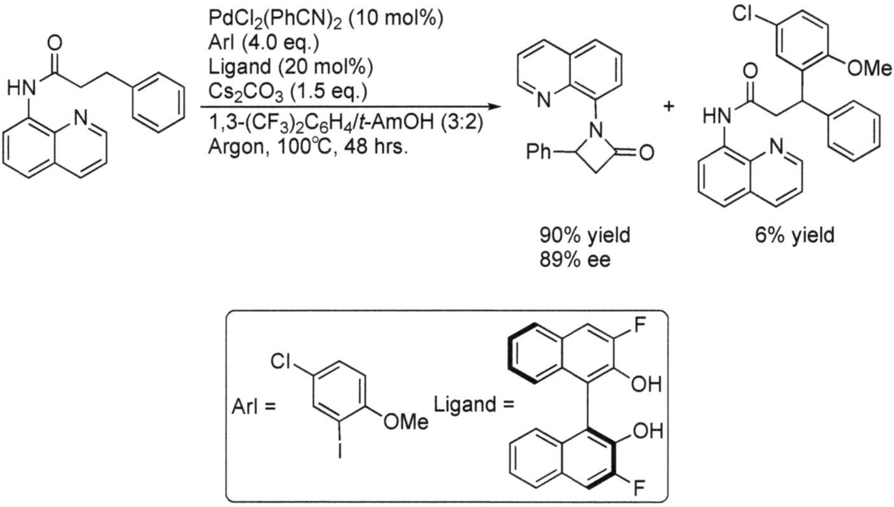

SCHEME 3.91 Palladium-catalyzed preparation of β-lactam via *C-N* bond formation

The palladium-catalyzed intramolecular amidation of unactivated β-C(sp^3)-H bond of amides has been successfully extended to form diazabicyclic β-lactams stereoselectively. For example, ((benzyloxy)carbonyl)-*L*-proline was allowed to react with the easily removable 5-methoxyquinolin-8-amine under peptide forming condition, i.e., in the presence of 1.5 equivalents of water-soluble 1-ethyl-3-(3-dimethylaminopropyl)carbodiimide (EDCI), 0.1 equivalent of DMAP in CH_2Cl_2 to afford benzyl (*R*)-2-((5-methoxyquinolin-8-yl)carbamoyl)pyrrolidine-1-carboxylate in nearly quantitative yield. Then, treatment of this substrate with a catalytic amount of $Pd(OAc)_2$ in the presence of 1.2 equivalents of AgOAc and 6.8 equivalents of fluorobenzene under microwave irradiation at 160°C for 1.5 hours led to the isolation of 86% of benzyl (1*S*,5*R*)-6-(5-methoxyquinolin-8-yl)-7-oxo-2,6-diazabicyclo[3.2.0] heptane-2-carboxylate with excellent control in stereochemistry. After subsequent removal of the 5-methoxyquinolin-8-amine component via CAN oxidation and (benzyloxy)carbonyl group by means of hydrogenation, the diazabicyclic β-lactam, i.e., (1*S*,5*R*)-2,6-diazabicyclo[3.2.0]-heptan-7-one was obtained, as shown in Scheme 3.92 [335, 336]. In addition to this particular diazabicyclic β-lactam, some other diazabicyclic β-lactams have been prepared successfully, such as the syntheses of (2a*S*,3a*S*,7a*R*,7b*R*)-decahydro-2*H*-azeto[3,2-*b*]indol-2-one from (2*S*,3a*S*,7a*S*)-octahydro-1*H*-indole-2-carboxylic acid; (1*S*,6*R*)-2,7-diazabicyclo[4.2.0]octan-8-one from (*S*)-1-((benzyloxy)-carbonyl)piperidine-2-carboxylic acid; and (2a*R*,8b*R*)-2a,3,4,8b-tetrahydroazeto-[3,2-*c*]isoquinolin-2(1*H*)-one from (*R*)-1,2,3,4-tetrahydroisoquinoline-3-carboxylic acid. However, it is unclear why indoline-2-carboxylic acid and (2*S*,4*R*)-4-hydroxypyrrolidine-2-carboxylic acid are not suitable for this protocol, even with the hydroxyl group protected in the latter one. An extensive search of substrates has revealed that many amides actually do not work in this approach either [335].

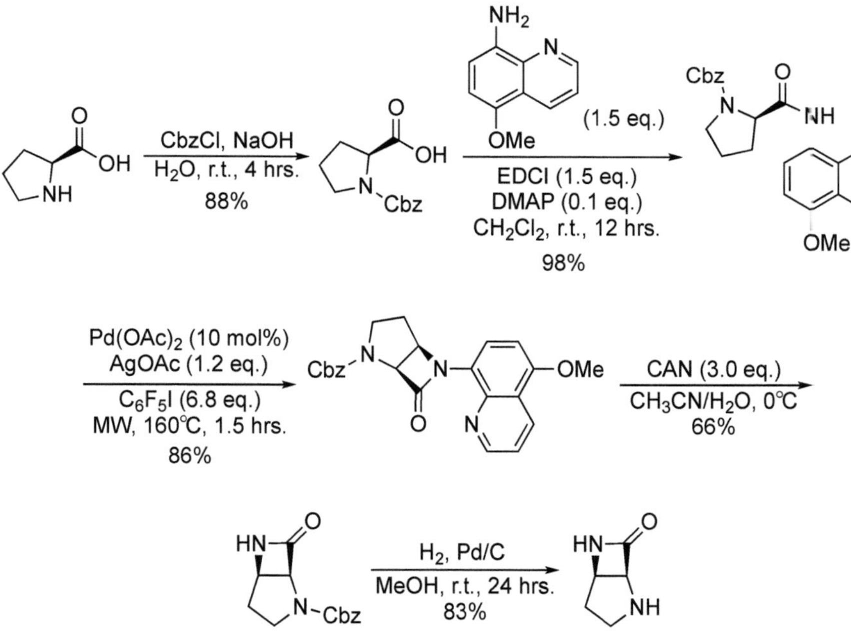

SCHEME 3.92 The synthetic route for conversion of *L*-proline into (1*S*,5*R*)-2,6-diazabicyclo[3.2.0]heptan-7-one involving palladium-catalyzed *C-N* bond formation

In addition to using an expensive palladium catalyst, some other transition-metal elements have been found to be effective for the activation of the C(sp^3)-H bond for the purpose of making β-lactams. For example, a Kharasch-Sosnovsky amidation-based approach with excellent tolerance of functional group using $CuCl_2$ as the catalyst (5 mol%) in combination with DMAP (15 mol%) in the presence of 3.0 equivalents of *tert*-butyl peroxide $(t\text{-BuO})_2$ can easily convert 3,*N*-bisaryl propanamides into the corresponding β-lactams in moderate yields, without pre-functionalization of C(sp^3)-H bonds or the installation of a directing group. In this protocol, it is critical to use the peroxide as the oxidizing agent, as the reaction vanishes in the presence of radical scavengers such as (2,2,6,6-tetramethylpiperidin-1-yl)oxyl (TEMPO) or butylated hydroxytoluene (BHT). Evidence for the intramolecular version of this reaction is supported by the isotope labeling experiment, and the kinetic isotope effect has been found at $k_H/k_D = 8.1$. Interestingly, the yield of the model reaction decreased by nearly 10% when the amount of $CuCl_2$ and DMAP are doubled [337]. When copper catalysts (e.g., $Cu(OAc)_2$, $CuBr_2$, $CuCl_2$, CuOAc, CuI, CuBr, CuCl) were screened for a model reaction of *N*-(quinolin-8-yl)pivalamide with a bidentate ligand to afford 3,3-dimethyl-1-(quinolin-8-yl)azetidin-2-one, cuprous chloride (CuCl) worked the best in the presence of sodium benzoate as base and 2,3,5,6-tetramethyl-2,5-cyclohexadiene-1,4-dione (duroquinone) as the oxidizing agent when heated in *o*-xylene at 160°C (Scheme 3.93). This ideal condition has been extended to many analogs of *N*-(quinolin-8-yl) amides, where the activation of the C(sp^3)-H bond of amides with β-methyl groups proceeds well, involving a five-membered ring intermediate in the cyclometalation step. In contrast, the aromatic C(sp^2)-H bonds are less reactive as the reaction proceeds via a six-membered ring intermediate. Moreover, this reaction condition is suitable for amides with a quaternary α-carbon atom, whereas amides such as *N*-(quinolin-8-yl)propionamide, *N*-(quinolin-8-yl) isobutyramide, *N*-(quinolin-8-yl)cyclohexanecarboxamide, and *N*-(quinolin-8-yl)cyclopentane carboxamide, with either a secondary or tertiary α-carbon atom all failed to afford the expected β-lactams under this reaction condition [338]. Similar results have been obtained using cupric acetate $Cu(OAc)_2$ as a catalyst and silver carbonate (Ag_2CO_3) as the oxidizing agent in dichloroethane, affording β-lactams up to gram scale, as shown in the conversion of 2-methyl-2-phenyl-*N*-(quinolin-8-yl)propenamide into 65% of 3-methyl-3-phenyl-1-(quinolin-8-yl)azetidin-2-one, along with 31% of 3,3-dimethyl-1-(quinolin-8-yl)indolin-2-one. In contrast, when the same substrate is treated with cupric chloride ($CuCl_2$) and Ag_2CO_3 under an O_2 atmosphere in DMSO, the chemoselectivity is inversed, affording only 4% of the expected β-lactam, but 89% of 3,3-dimethyl-1-(quinolin-8-yl)indolin-2-one, arising from the C(sp^2)-H activation, as shown in Scheme 3.94. However, it is not sure whether this inversion of chemoselectivity is caused by different copper catalyst or solvent. A study on the kinetic isotope effect indicates that the activation of the carbon-hydrogen bond is the rate-determining step. Also, this reaction proceeds at a terminal methyl group, as well as at the internal benzylic position of an alkyl chain [339].

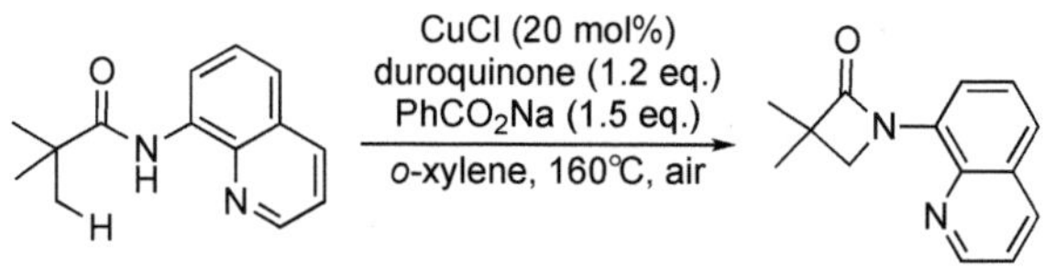

87%

SCHEME 3.93 CuCl-catalyzed preparation of β-lactam via *C-N* bond formation

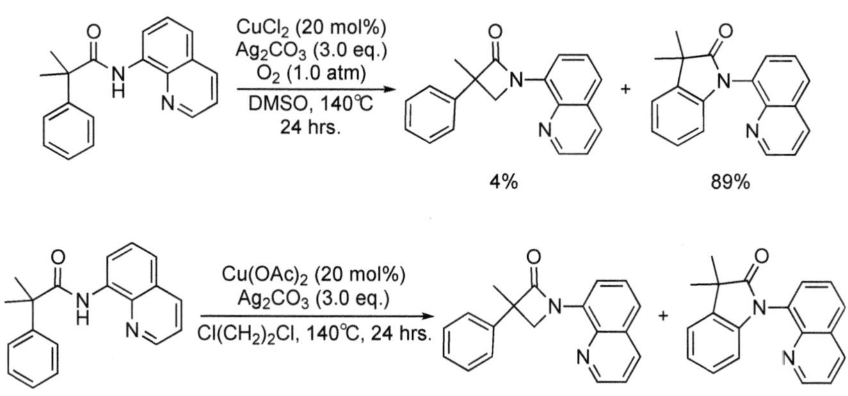

4% 89%

65% 31%

SCHEME 3.94 Cu(II)-catalyzed preparation of β-lactam via *C-N* bond formation

Besides palladium and copper, nickel compounds are also effective in catalytical amidation of amide to form β-lactams. For example, when 2-iodination of *N*-(5-chloroquinolin-8-yl)benzamide in the presence of $Ni(OTf)_2$, additive, 2.0 equivalents of base and I_2 was applied to *N*-(5-chloroquinolin-8-yl)pivalamide, various amounts of 1-(5-chloroquinolin-8-yl)-3,3-dimethylazetidin-2-one (0 ~ 91%) were obtained depending on the base (Na_2CO_3, Cs_2CO_3, K_2CO_3, NaOAc), additive (e.g., Ag_2CO_3), and solvent (e.g., toluene, 1,2-dichloroethane, DMSO, DMF) used. Extension of the ideal reaction condition to other α,α,α-trisubstituted aliphatic amides containing an *N*-(5-chloroquinolin-8-yl) moiety affords the corresponding β-lactams in good yields. Analysis of the structures of the obtained β-lactams indicates that the reactions proceed exclusively at the methyl group in a highly regioselective manner. For examples, the β-methylene C−H bonds in *N*-(5-chloroquinolin-8-yl)-2,2,3-trimethylbutanamide and *N*-(5-chloroquinolin-8-yl)-2-ethyl-2-methylbutanamide, cyclic β-methylene C−H bonds in*N*-(5-chloroquinolin-8-yl)-1-methylcyclohexane-1-carboxamideand*N*-(5-chloroquinolin-8-yl)-1-methylcyclopentane-1-carboxamide, and γ-aromatic C−H bonds in *N*-(5-chloroquinolin-8-yl)-2,2-dimethyl-3-phenylpropanamide and *N*-(5-chloroquinolin-8-yl)-3-(3-methoxyphenyl)-2,2-dimethylpropanamide do not react with I_2 under this reaction condition. In addition, the cleavage of C−H bonds is irreversible and no H/D exchange occurs in the deuterium labeling experiments at 160°C in the presence of I_2. However, the addition of radical scavengers, such as TEMPO and BHT drastically suppresses the reaction [340]. Comparably, the reaction catalyzed by [Ni(dme)$_2$I$_2$] (dme = 1,2-dimethoxyethane (DME)) in the presence of 2.0 equivalents of base (K_2HPO_4), 0.1 equivalent of tetrabutylammonium iodide and 3.0 equivalents of TEMPO in a mixed solvent of nitrile (butyronitrile/benzonitrile 3:2) offers improved yield of the corresponding β-lactams from the α,α,α-trisubstituted aliphatic amides containing the *N*-quinolin-8-yl moiety. This reaction condition favors the C-H bonds of β-Me groups over the γ-Me or β-methylene groups. Also, there appears a predominant preference for the β-Me C-H bonds over the aromatic sp^2 C-H bonds. A representative reaction is provided in Scheme 3.95 [341].

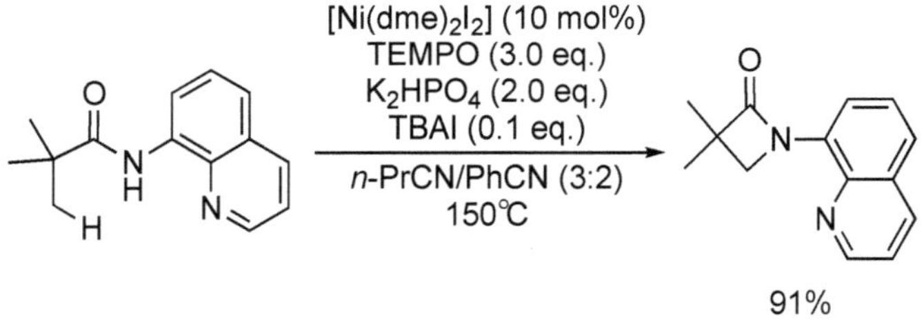

SCHEME 3.95 Ni(II)-catalyzed preparation of β-lactam involving *C-N* formation

Besides Pd, Cu, and Ni, the cobalt compound is another effective catalyst for the generation of β-lactams. For example, in the presence of Co(OAc)$_2$, 2.5 equivalents of Ag_2CO_3, 0.5 equivalent of base sodium benzoate, 2-ethyl-2-methyl-*N*-(quinolin-8-yl)pentanamide was converted into 3-ethyl-3-propyl-1-(quinolin-8-yl)azetidin-2-one in 91% yield via intramolecular amidation, as shown in Scheme 3.96. Extension of this optimal reaction condition to other propionamide and butyramide derivatives with an 8-aminoquinolinyl moiety as the bidentate directing group reveals that the reaction favors the C–H bonds of β-methyl groups over those of the β-methylene and γ- or δ-methyl groups, providing the β-lactams in a high site- and diastereoselective manner.

Also, a predominant preference for the functionalization of γ-benzylic over β-methyl C–H bonds was observed, producing γ-lactams as the major products. Very interestingly, treatment of 2-ethyl-2-methyl-*N*-(quinolin-8-yl) pentanamide with $Co(acac)_3$, Ag_2CO_3, K_2HPO_4, $B(OH)_3$ in the presence of 2,2,3,3,4,4,4-heptafluorobutanamide and 3Å molecular sieves in (trifluoromethyl)benzene at 160°C led to 50% of 2-ethyl-2-((2,2,3,3,4,4,4-heptafluorobutanamido)methyl)-*N*-(quinolin-8-yl)pentanamide, arising from the intermolecular amidation [342].

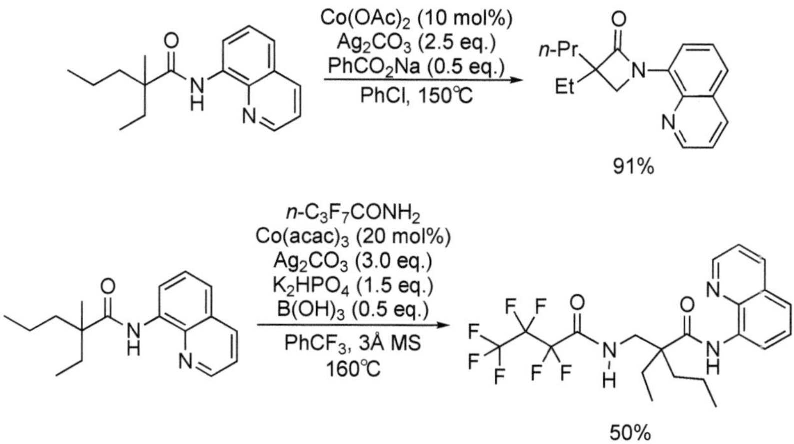

SCHEME 3.96 Cobalt-catalyzed *C-N* bond formation

In addition to the above transition metal complexes-catalyzed amidation to yield β-lactams, recently engineered cytochrome P450 enzymes also accomplish the C-N bond formation via reactive iron-bound carbonyl nitrenes generated from nature-inspired acyl-protected hydroxamate precursors, such as 3-substituted-*N*-(pivaloyloxy)propanamides and its analogs. This innovative method has been inspired by the transformation of 3-phenyl-*N*-(pivaloyloxy)propenamide into 1.5% of (*S*)-4-phenylazetidin-2-one with 90% ee and 1.0% of 3,4-dihydroquinolin-2(1*H*)-one in the presence of cytochrome P411 variant E10-V78F S438A (E10FA), which was cytochrome P450BM3 variant having the axial heme-ligating cysteine residue substituted with serine and originally engineered to catalyze carbene transfer to alkynes. The ideal enzyme was obtained by substituting the amino acid residues near the heme iron in a closely related crystal structure of P411 E10 for sequential rounds of site-saturation mutagenesis and subsequent screening. During this process, two rounds of site saturation mutagenesis and screening introduced the mutations F78L and A264G, leading to a

28-fold improvement in the total turnover number (TTN) of the enzyme to 52,000 for creation of (*S*)-4-phenylazetidin-2-one while greatly suppressing the formation of 3,4-dihydroquinolin-2(1*H*)-one. Another round of double site-saturation mutagenesis and screening subsequently added mutations T327R and V328M, which further increased the TTN of the new variant to 71,000 for (*S*)-4-phenylazetidin-2-one with 96% ee. Additional double site saturation led to the discovery of the final variant, LS_{sp3}, containing mutations L263N and A268F. This final enzyme is effective for a TTN of 223,000 in the generation of the desired (*S*)-4-phenylazetidin-2-one in 96% yield and 96% ee. Furthermore, the TTN of a reaction by addition of LS_{sp3} expressing *E. coli* and 3-phenyl-*N*-(pivaloyloxy)propenamide portion-wise to give (*S*)-4-phenylazetidin-2-one was boosted to 1,020,000. Under this condition, substrates with two methyl groups adjacent to the carbonyl group, with secondary or tertiary aliphatic C–H bonds, even adjacent to a trifluoromethyl group, or bearing heteroatom substituents, such as amide and silyl groups, all are converted into the expected β-lactams. In addition, olefin-containing substrates could also undergo enantioselective C–H amidation without the formation of aziridines, highlighting the complete chemoselectivity of this enzyme. Also, the α-C-H bond of the alkynyl group can be readily amidated. This reaction can be performed on a preparative scale without losing yield and enantioselectivity, as demonstrated in the synthesis of 1.62 g of (*S*)-4-(4-chlorophenyl)azetidin-2-one from 2.84 g of 3-(4-chlorophenyl)-*N*-(pivaloyloxy)propenamide, in 86% isolated yield by simple filtration of the aqueous reaction mixture, with 92% ee. It should be pointed out that this protocol can be applied to engineer enzymes for the synthesis of γ- or δ-lactams [343, 344].

Another type of amidation involves the activation of the N-H bond followed by the formation of the C-N bond, as shown by the Cu(I)-catalyzed intramolecular C-N coupling of amides with vinyl bromides. Particularly, the 4-*exo* ring closure is fundamentally preferred over other modes, such as 5-*exo*, 6-*exo*, and 6-*endo* under copper catalysis. For example, in the presence of 5 mol% of CuI, 10 mol% of dimethylglycine hydrochloride, 2.0 equivalents of K_2CO_3, (*E*)-3-bromo-2-methyl-*N*-phenyldec-3-enamide was transformed into (*E*)-4-heptylidene-3-methyl-1-phenylazetidin-2-one in 97% yield after refluxing in THF for 28 hours. The same yield of (*Z*)-4-heptylidene-3-methyl-1-phenylazetidin-2-one was obtained when (*Z*)-3-bromo-2-methyl-*N*-phenyldec-3-enamide was refluxed in THF for 32 hours in the presence of CuI, dimethylglycine, and K_2CO_3, as shown in Scheme 3.97. Under this condition, 2-bromo-1-methyl-*N*-phenylcyclohex-2-ene-1-carboxamide can be transformed into 1-methyl-7-phenyl-7-azabicyclo[4.2.0] oct-5-en-8-one in nearly quantitative yield [345]. The exclusive preference of 4-*exo* cyclization over 5-*exo*, 6-*exo* cyclization has been indicated in the nearly quantitative transformations of 4-bromo-2-(1-bromovinyl)-2-methyl-*N*-phenylpent-4-enamide and 5-bromo-2-(1-bromovinyl)-2-methyl-*N*-phenylhex-5-enamide into 3-(2-bromoallyl)-3-methyl-4-methylene-1-phenylazetidin-2-one and 3-(3-bromo-but-3-en-1-yl)-3-methyl-4-methylene-1-phenylazetidin-2-one after refluxing for 18 and 19 hours, respectively. Otherwise, 3-(1-bromovinyl)-3-methyl-5-methylene-1-phenylpyrrolidin-2-one and 3-(1-bromovinyl)-3-methyl-

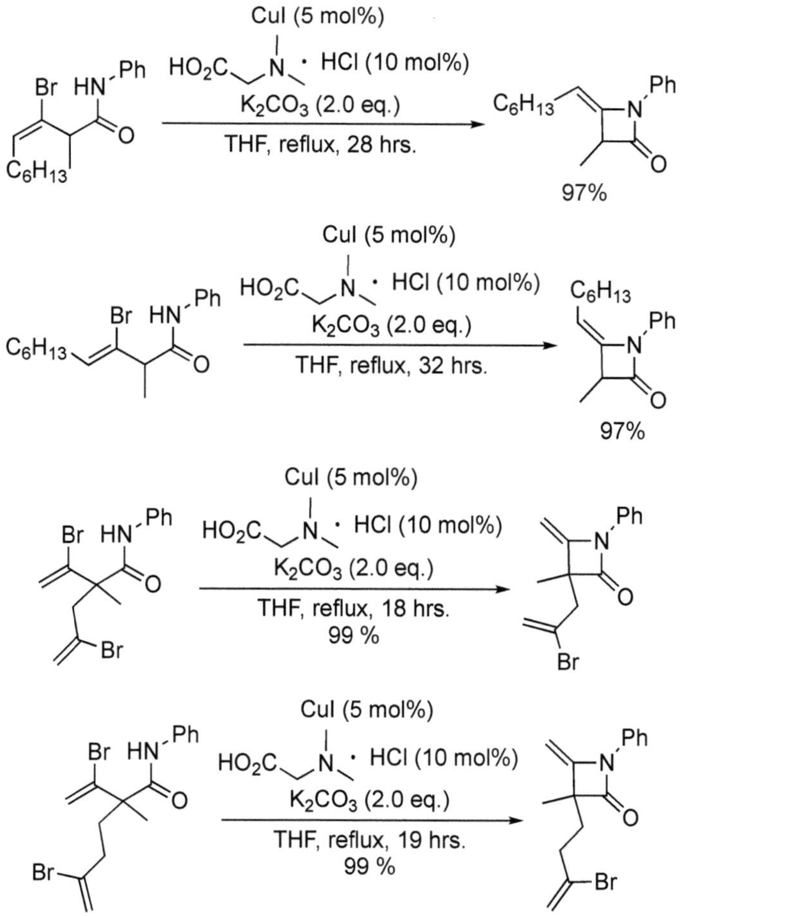

SCHEME 3.97 CuI catalyzed the preparation of β-lactam via intramolecular coupling between vinyl bromide and amino moieties

6-methylene-1-phenylpiperidin-2-one should be obtained instead [345]. More examples for the preparation of β-lactams via intramolecular amidation can be found in the reviews [2, 346].

3.3.7 SYNTHESES OF Β*-LACTAMS FROM AZIRIDINES*

Aziridines are organic compounds containing an aziridine functional group, a three-membered heterocycle with one carbon atom substituted with nitrogen [347]. Due to the high ring strain of the aziridine moiety, it often breaks up or undergoes rearrangement to other structural scaffolds. In fact, aziridines have been applied in the syntheses of natural products [348, 349]. In this part of the contents, only representative references pertinent to the formation of β-lactams from aziridines will be commented.

One of the early practices for the ring-expansion of aziridines to give β-lactams is the treatment of sodium 1-(*tert*-butyl)aziridine-2-carboxylate with thionyl chloride ($SOCl_2$) to afford 33% of 1-(*tert*-butyl)-3-chloroazetidin-2-one, instead of the expected 1-(*tert*-butyl) aziridine-2-carbonyl chloride. Similarly, treatment of sodium 1-(*tert*-butyl)aziridine-2-carboxylate in benzene with oxalyl chloride with or without the presence of Et_3N also leads to the formation of the same β-lactam in 29 and 26% yields, respectively [350]. Treatment of 1-(*tert*-butyl)-3-methylaziridine-2-carboxylic anhydride with Et_4NCl in CH_3CN yield 75% of 1-(*tert*-butyl)-3-chloro-4-methylazetidin-2-one [351]. This particular reaction is assumed to form the transient acyl chloride, followed by intramolecular acyl substitution to form an unstable intermediate of 1-(*tert*-butyl)-2-oxo-1-azabicyclo[1.1.0]butan-1-ium chloride, as shown in Scheme 3.98 [350–352]. However, different products might form under slightly varied conditions. For example, when sodium (2*R*,3*R*)-1-benzyl-3-phenylaziridine-2-carboxylate was treated with 1.2 equivalents of oxalyl chloride and Et_3N in benzene for 45 minutes, 62% of (3*S*,4*R*)-1-benzyl-3-chloro-4-phenylazetidin-2-one was obtained [352]. In contrast, when (2*R*,3*R*)-1-benzyl-3-phenylaziridine-2-carboxylic acid rather than the sodium salt was treated with 2.1 equivalents of oxalyl chloride in CH_2Cl_2 for 1 hour, 82% of (*S*)-4-benzyl-5-((*S*)-chloro(phenyl)methyl) morpholine-2,3,6-trione was obtained, as shown in Scheme 3.99 [353]. In addition, the reaction took a different turn with aziridine-2-carboxylic acids bearing additional substituent at position-2. For example, when (2*R*,3*R*)-1-benzyl-2-methyl-3-phenylaziridine-2-carboxylic acid was treated with 2.3 equivalents of oxalyl chloride in CH_2Cl_2 at 0°C for 5 minutes, then 25°C for 1 hour, 51% of (*S*)-3-benzyl-4-((*S*)-chloro(phenyl)methyl)-4-methyloxazolidine-2,5-dione was obtained. However, aziridine-2-carboxylic acids with an alkyl substituent at position-3 exclusively lead to β-lactams upon reaction with oxalyl chloride, without any detectable amount of *N*-carboxyanhydrides (NCAs) or morpholinetriones. The reactions of *cis*-substituted aziridines yield *cis*-substituted β-lactams. Compared to the reaction with oxalyl chloride, the reaction with oxalyl bromide is faster, resulting in a certain level of isomerization as a 1:2 mixture of (3*S*,4*R*)-1-benzyl-3-bromo-4-cyclohexylazetidin-2-one and (3*R*,4*R*)-1-benzyl-3-bromo-4-cyclohexylazetidin-2-one, when the reaction mixture of (2*R*,3*R*)-1-benzyl-3-cyclohexylaziridine-2-carboxylic acid and oxalyl bromide is simply concentrated at 25°C without an aqueous quench. The reaction is stereospecific with oxalyl chloride as the reaction of (2*S*,3*R*)-1-benzyl-3-cyclohexylaziridine-2-carboxylic acid with oxalyl chloride gives exclusively (3*R*,4*R*)-1-benzyl-3-chloro-4-cyclohexylazetidin-2-one, and the reaction of *cis*-starting material affords exclusively the *cis*-β-lactam.

Furthermore, addition of DMF also changes the reaction path, as indicated in the drastic decreasing of the yield of (*S*)-4-benzhydryl-5-((*S*)-chloro(phenyl)methyl)morpholine-2,3,6-trione from the reaction of (2*R*,3*R*)-1-benzhydryl-3-phenylaziridine-2-carboxylic acid with 2.0 equivalents of oxalyl chloride, from 92% (measured by NMR) in the absence of DMF to 6% with 4.0 equivalents of DMF added. When 1.0 equivalent of DMF was present, 28% of (*S*)-4-benzhydryl-5-((*S*)-chloro(phenyl)methyl)morpholine-2,3,6-trione was obtained, indicating that the acid can compete with DMF for oxalyl chloride. Presumably, the reaction between DMF and oxalyl chloride leads to the formation of the Vilsmeier reagent ($[Me_2N{=}CHCl]Cl$). When this reagent is separately prepared and mixed with (2*R*,3*R*)-1-benzhydryl-3-phenylaziridine-2-carboxylic acid, only (3*S*,4*R*)-1-benzhydryl-3-chloro-4-phenylazetidin-2-one was obtained in 83% yield. It is found that the ring expansion of aziridine-2-carboxylic acids is much more efficient with the Vilsmeier reagent than it is with thionyl chloride, as demonstrated in the yields of (3*S*,4*R*)-1-benzyl-3-chloro-4-phenylazetidin-2-one at 74% and 56%, and the yields of (3*S*,4*R*)-1-benzhydryl-3-chloro-4-cyclohexylazetidin-2-one at 100% and 62% when their corresponding aziridine-2-carboxylic acids are treated with the Vilsmeier reagent and $SOCl_2$, respectively. Considering these experimental results, a mechanism is outlined in Scheme 3.100 to illustrate the potential reaction routes for the ring expansion of aziridine-2-carboxylic acid with oxalyl chloride [353].

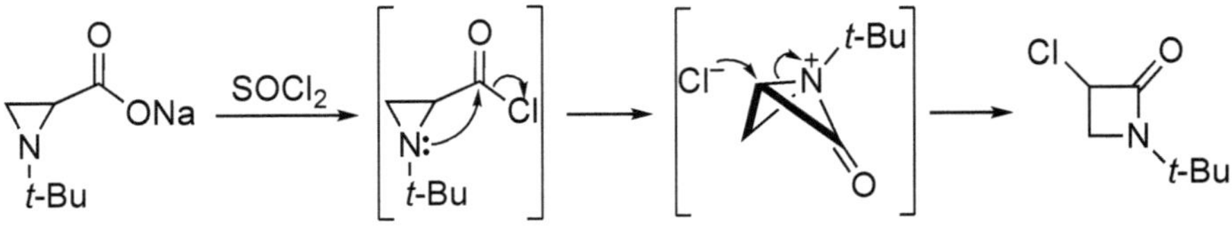

SCHEME 3.98 Possible reaction mechanism for the conversion of aziridine-2-carbonyl chloride into β-lactam

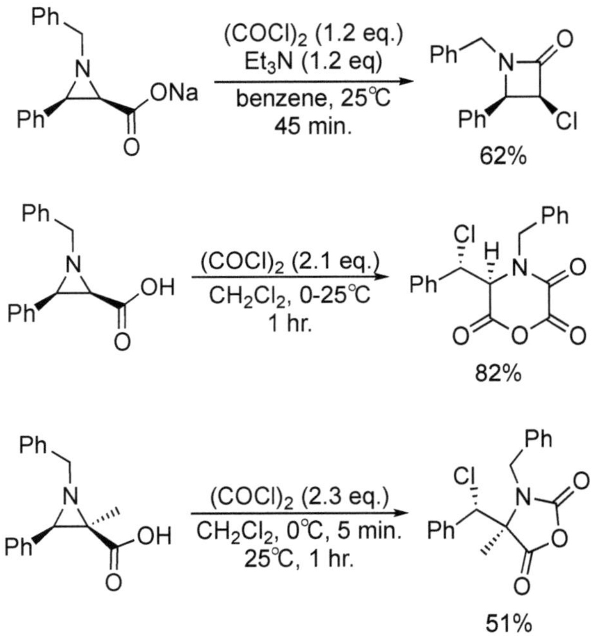

SCHEME 3.99 The potential reactivity of 1-benzyl-3-phenylaziridine-2-carboxylic acid with oxalyl chloride

Besides the above methods involving the treatment of aziridine carboxylic acids with acyl halides to form β-lactams, more common methods to form β-lactams from aziridines are organometallic complexes catalyzed carbonylation of aziridines initially reported by Howard Alper et al. by treatment of azirines with carbon monoxide (CO) in the presence of a catalytic amount of $Pd(PPh_3)_4$ at 40°C to affords bicyclic β-lactams [354]. Subsequently, this method has been optimized to convert aziridines to regiospecific monocyclic β-lactams in the presence of CO and a catalytic amount of $[Rh(CO)_2Cl]_2$. In this case, treatment of

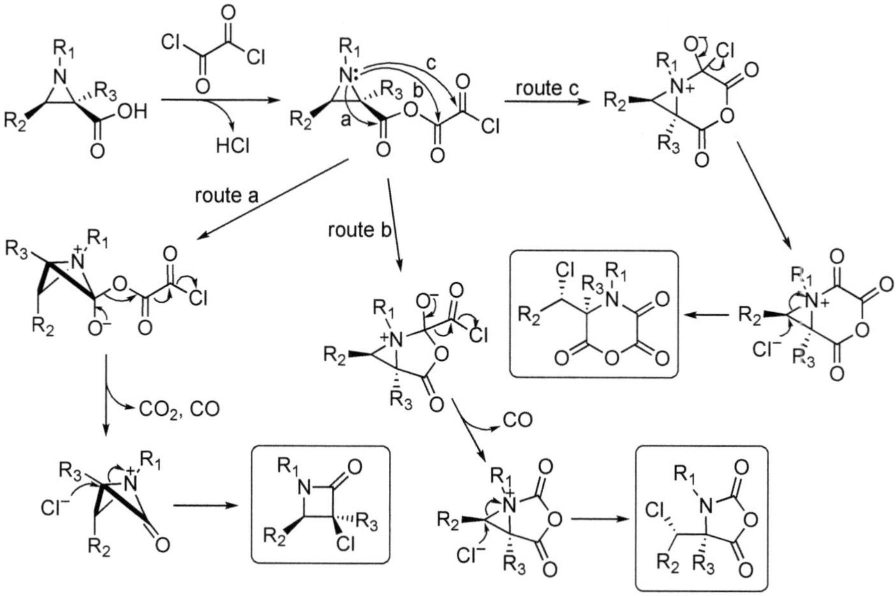

SCHEME 3.100 The reaction mechanism of aziridine acid with oxalyl chloride

(*S*)-1-(*tert*-butyl)-2-phenylaziridine with CO in benzene in the presence of 5% of $[Rh(CO)_2Cl]_2$ at 90°C and 20 atms afforded (*S*)-1-(*tert*-butyl)-3-phenylazetidin-2-one in quantitative yield. Extension of this condition to other aziridines, such as (*S*)-2-(4-bromophenyl)-1-(*tert*-butyl)aziridine, (*S*)-2-([1,1'-biphenyl]-4-yl)-1-(*tert*-butyl)aziridine, (*S*)-1-(1-adamantyl)-2-phenylaziridine, (*S*)-1-(1-adamantyl)-2-(4-bromophenyl)aziridine and (*S*)-1-(1-adamantyl)-2-([1,1'-biphenyl]-4-yl)aziridine, the corresponding β-lactams were obtained also in quantitative yields with complete regiospecificity, where the more substituted C-N bonds in the aziridines were inserted with the carbonyl group. A mechanism was proposed to rationalize this ring-expansion carbonylation, as shown in Scheme 3.101 [355]. Furthermore, the rhodium complex with CO and chloride has been crafted to the terminal of dendrimer resin prepared from 3,5-diaminobenzoic acid, lysine, and glycine with diphenylmethylene phosphine ligands. The resin-based rhodium complex has been screened for 1-(*tert*-butyl)-2-phenylaziridine, 1-(1-adamantyl)-2-phenylaziridine, 2-([1,1'-biphenyl]-4-yl)-1-(*tert*-butyl)aziridine, 2-(4-bromophenyl)-1-(*tert*-butyl)aziridine and 1-(1-adamantyl)-2-([1,1'-biphenyl]-4-yl)-aziridine, to test the effects of reaction temperature, solvent, time, and pressure of carbon monoxide on this transformation. The dendritic catalysts have demonstrated a comparable activity to the homogeneous analogs, with features of easy recovery by simple filtration and retainability of catalytic activity [356].

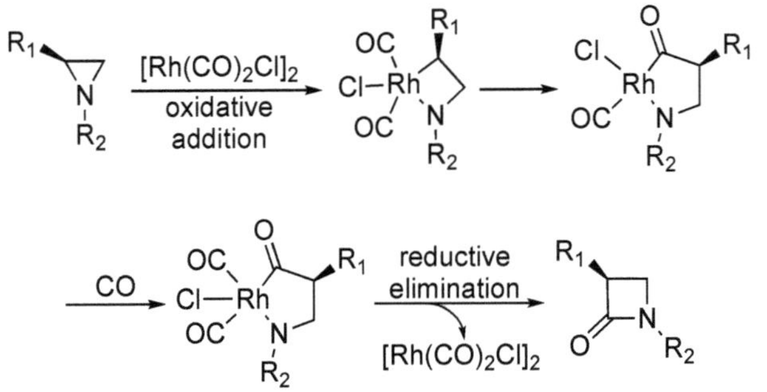

SCHEME 3.101 The mechanism for β-lactamization under the Alper condition

Computational study with DFT method (B3LYP/6-31G(d) (LANL2DZ for Rh) taking into account of solvent effects has been performed on the regioselectivity and enantiospecificity of the $[Rh(CO)_2Cl]_2$-catalyzed carbonylative ring expansions of *N-tert*-butyl-2-phenylaziridine to yield 1-(*tert*-butyl)-3-phenylazetidin-2-one and the inertness of *N-tert*-butyl-2-methylaziridine along this process. It is found that different degree of activation when breaking the C-N bond in the initial aziridine-$Rh(CO)_2Cl$ complex is related to its hyperconjugation interaction with the substituent on the carbon atom. For the case of *N-tert*-butyl-2-phenylaziridine, the hyperconjugation with the C_α-N bond facilitates the insertion of the metal atom into this bond, whereas for the case of *N-tert*-butyl-2-methylaziridine, larger stability of the initial complex along with a lower stabilization through hyperconjugation of the transition state for insertion of the Rh atom into the C_α-N bond makes the ring expansion of this aziridine unviable. In addition, the stereogenic center has not been perturbed to change its configuration so that the ring expansion is enantiospecific [357].

Additionally, this method has been extended to make α-methylene-β-lactams in good yields by the palladium $[Pd(PPh_3)_4]$ catalyzed regiospecific carbonylation of methyleneaziridines under exceptionally mild conditions [358]. As of these initiative explorations, the ring-expansion of aziridine by metal-catalyzed CO insertion is known as the Alper reaction [79].

A complementary condition for carbonylative ring expansion has been reported, using a stoichiometric amount of inexpensive nickel tetracarbonyl $[Ni(CO)_4]$ and an inert atmosphere, where the less substituted C-N bond of the aziridine will be carbonylated [359]. Under this condition,

1-benzyl-2-methylaziridine and 2-methyl-1-(methylsulfonyl)aziridine have been converted into 1-benzyl-4-methylazetidin-2-one and 4-methyl-1-(methylsulfonyl)azetidin-2-one, respectively. A similar yield of β-lactams can be obtained under one atmosphere of carbon monoxide or an argon atmosphere, with net retention of configuration between the aziridines and the β-lactams [360]. The first step of this approach is an S_N2 ring opening of the aziridine by lithium iodide in refluxing tetrahydrofuran. If this step is not performed, and aziridine is allowed to reflux in THF for 3 hours with only $Ni(CO)_4$, or if the lithium iodide, $Ni(CO)_4$ and aziridine are combined in one step, no β-lactam is observed. Extension of the reaction time to more than 3 hours results in a reduced yield of product [361].

Following this trend, several carbonylative ring-expansion conditions for aziridines have been reported. For example, (2*R*,3*R*)-2-(2-((*tert*-butyldiphenylsilyl)oxy)ethyl)-1-tosyl-3-vinylaziridine was transformed into 46% of (3*R*,4*R*)-4-(2-((*tert*-butyldiphenylsilyl)oxy)ethyl)-1-tosyl-3-vinylazetidin-2-one in the presence of excess amount of PPh_3, CO, and 15 mol% of tri(dibenzylideneacetone)dipalladium(0)-chloroform complex in benzene at 50°C, with complete retention of the initial configuration, as shown in Scheme 3.102 [362]. For this reaction, it is assumed that the π-allyl complex was formed initially by the attack of Pd(0) on the vinyl aziridine with inversion at the reaction center. Then the isomerization occurs to allow the internal coordination between nitrogen and palladium. Insertion of CO with retention of stereochemistry is followed by reductive elimination and regeneration of the catalyst. The entire sequence thus becomes regio- and stereoselective.

Diiron nonacarbonyl [$Fe_2(CO)_9$], which is a more reactive source of Fe than $Fe(CO)_5$ and less dangerous to handle, has been applied to form a stable π-allyltricarbonyliron complex with (*E*)-4-((2*S*,3*R*)-1-benzyl-3-methylaziridin-2-yl)but-3-en-2-one and its analogs. Additional functionalization of the side chain by addition of a variety of nucleophiles into the carbonyl group on the side chain proceeds in good to excellent diastereoselectivity, and subsequent treatment with trimethylamine *N*-oxide decomposes the moiety of π-allyltricarbonyliron complex and forms the corresponding β-lactams. For example, when (*E*)-4-((2*S*,3*R*)-1-benzyl-3-methylaziridin-2-yl)but-3-en-2-one was sonicated with $Fe_2(CO)_9$ in benzene at 30°C for 3 hours, 77% of the π-allyltricarbonyliron complex was obtained, with 10:1 of diastereoselectivity. Treatment of this complex with allyltributyltin in CH_2Cl_2 in the presence of $BF_3{\cdot}OEt_2$, the π-allyltricarbonyliron complex with tertiary allyl alcohol moiety was obtained in 89% yield. Final decomposition

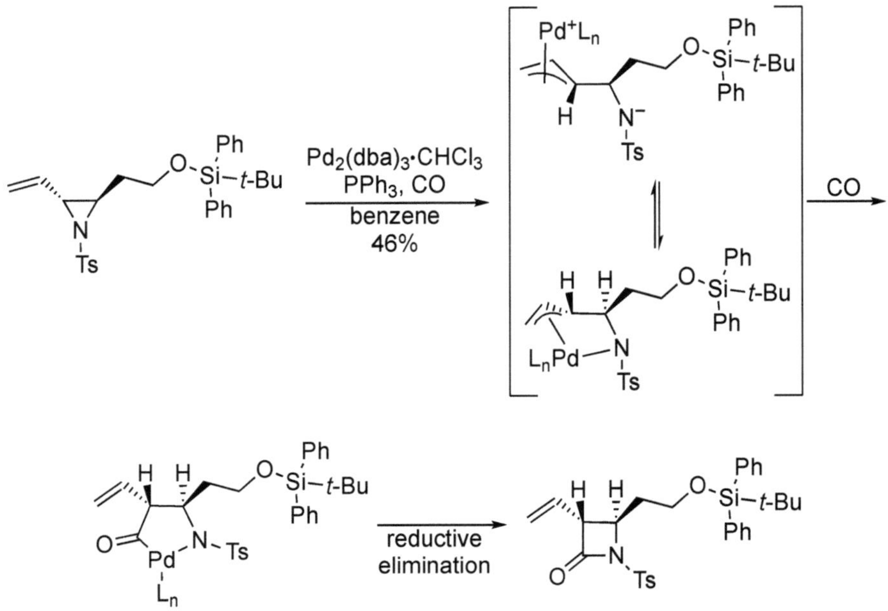

SCHEME 3.102 Conversion of 3-vinyl aziridine derivative into β-lactam under the Alper condition

of the π-allyltricarbonyliron complex in THF at 0°C with trimethylamine *N*-oxide for 2 hours led to the formation of (3*S*,4*R*)-1-benzyl-3-((*R*,*E*)-3-hydroxy-3-methylhexa-1,5-dien-1-yl)-4-methylazetidin-2-one, in 69% yield (Scheme 3.103) [363].

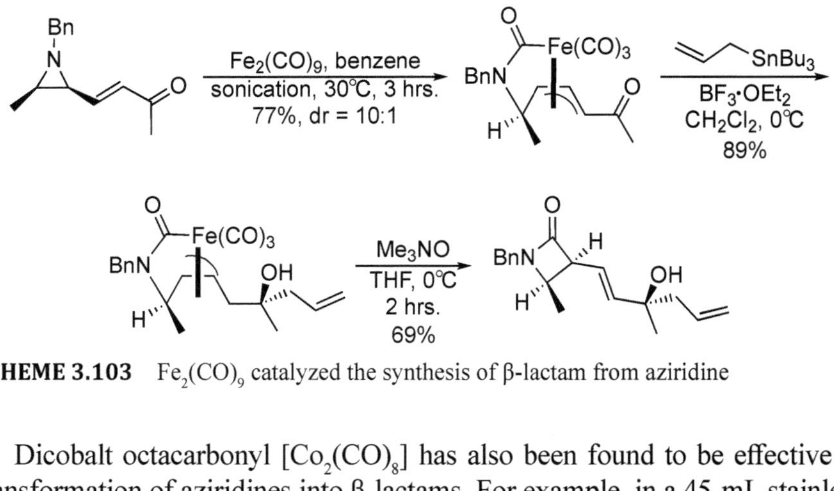

SCHEME 3.103 Fe$_2$(CO)$_9$ catalyzed the synthesis of β-lactam from aziridine

Dicobalt octacarbonyl [Co$_2$(CO)$_8$] has also been found to be effective in transformation of aziridines into β-lactams. For example, in a 45-mL stainless steel autoclave equipped with a glass liner and a stirring bar was added 1.82

mmol of (2*S*,3*S*)-1-benzyl-2-(((*tert*-butyldimethylsilyl)oxy)methyl)-3-methylaziridine, 0.15 mmol of $Co_2(CO)_8$ and 10 mL of freshly distilled anhydrous and O_2-free DME. After being purged with 300 psi CO 4 times, the autoclave was charged with 500 psi CO and heated in an oil bath at 95°C for 16 hours. After releasing the remaining CO, the reaction mixture was worked up and purified to give 531.9 mg of (3*R*,4*S*)-1-benzyl-4-(((*tert*-butyldimethylsilyl)oxy) methyl)-3-methylazetidin-2-one and 49.1 mg of (3*R*,4*S*)-1-benzyl-3-(((*tert*-butyldimethylsilyl)oxy)methyl)-4-methylazetidin-2-one, for a ratio of 92:8, in 99.8% of total yield, as displayed in Scheme 3.104 [364]. Similarly, when (2*R*,3*R*)-1-benzyl-2-(((*tert*-butyldimethylsilyl)oxy)methyl)-3-ethylaziridine was treated with 8.3 mol% of $Co_2(CO)_8$ in DME in the presence of CO at 500 psi at 105°C for 14 hours, a 83:17 mixture of (3*S*,4*R*)-1-benzyl-4-(((*tert*-butyldimethylsilyl)oxy)methyl)-3-ethylazetidin-2-one (92% ee) and (3*S*,4*R*)-1-benzyl-3-(((*tert*-butyldimethylsilyl)oxy)methyl)-4-ethylazetidin-2-one (92% ee) was obtained, in a 93% total isolated yield [365].

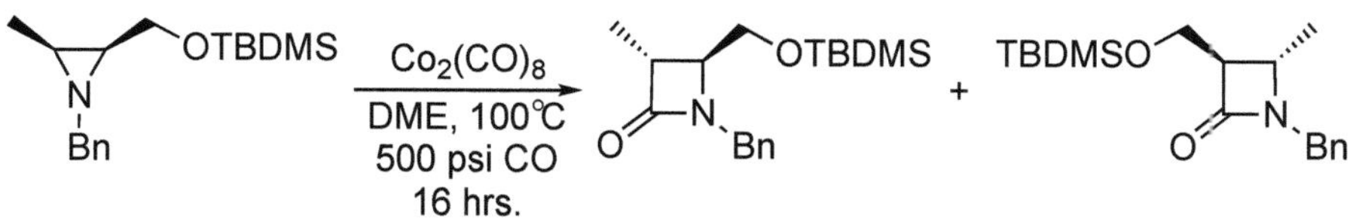

SCHEME 3.104 $Co_2(CO)_8$ catalyzed the conversion of aziridine into β-lactam

Further explorations of $Co_2(CO)_8$ in the preparation of β-lactams reveal the range of aziridines that can be used as substrates for the $Co_2(CO)_8$-catalyzed carbonylative ring-expansion of aziridines. For example, the aziridines containing electron-withdrawing substituents on ring carbon atoms, such as a trfluoromethyl or an alkoxycarbonyl group, are incompatible with carbonylation, whereas aziridines with a hydroxymethyl (protected or unprotected), an aminomethyl group as ring carbon substituents, as well as an alkoxycarbonylmethyl group on the ring nitrogen are suitable for this type of transformation [366].

Computational study with DFT (B3LYP) in conjunction with the conductor polarizable continuum model/united atom Kohm-Sham method to take into account solvent effects for the $Co(CO)_4$-catalyzed carbonylative ring expansion of *N*-benzoyl-2-methylaziridine to afford 1-benzoyl-4-methylazetidin-2-one and 1-benzoyl-3-methylazetidin-2-one indicate that except for the rate-determining nucleophilic ring-opening step, the most favorable reaction mechanism differs from the experimental results [367].

More examples of the transition metal complexes catalyzed carbonylative ring-expansion of aziridines to give β-lactams can be found in the related reviews [368, 369], as well as individual reports, such as the cobalt carbonyl [370–373], palladium [374], and rhodium [375] catalyzed carbonylative expansion of aziridines. Other methods of converting aziridines into β-lactams include the treatment of aziridine with a halogenating agent ($SOCl_2$, oxalyl chloride, PPh_3/Br_2) [376], reduction of 2-hydroxymethyl-*N*-tosyl aziridine with Red-Al, followed by oxidation of hydroxyl group to carboxyl and subsequent cyclization [377], and treatment of aziridine with SbF_3 to give imines which then undergoes cycloaddition with ketene to afford β-lactams [378].

3.3.8 CARBONYLATION OF ALLYLAMINES

In addition to the carbonylative ring expansion of aziridines for the generation of β-lactams, carbon monoxide has been applied to convert other types of substrates into β-lactams. For example, in the presence of palladium catalyst, 2-bromoallylamine can be converted into 3-methyleneazetidin-2-one in the presence of CO atmosphere, as indicated in the treatment of benzyl 2-(4-(benzyloxy)phenyl)-2-((2-bromoallyl)amino)acetate in the presence of tributylamine, a catalytic amount of $Pd(OAc)_2$ and ligand PPh_3. In this reaction, 36.5% of benzyl 2-(4-(benzyloxy)phenyl)-2-(3-methylene-2-oxoazetidin-1-yl)acetate was obtained after 3 hours of reaction at 1 atm of CO at 100°C, whereas 62.7% of β-lactam was obtained at 80°C at 4 atms of CO. Lowering reaction temperature to room temperature vanishes the reaction, without any β-lactam formed. The starting material of this step was prepared from the S_N2 reaction of 2-(benzyloxy)-1-(4-(benzyloxy)phenyl)-2-oxoethan-1-aminium chloride with 2,3-dibromoprop-1-ene under basic condition, as shown in Scheme 3.105 [379]. The resulting β-lactam was an important intermediate in the total synthesis of Nocardicin A.

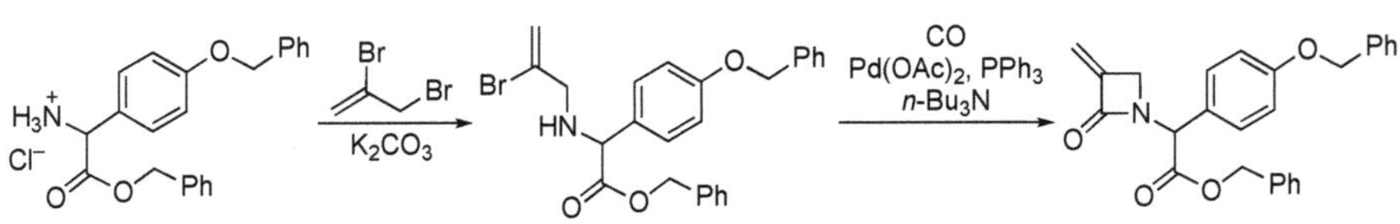

SCHEME 3.105 $Pd(OAc)_2$ catalyzed the synthesis of β-lactam from allylamine

For the case of allylamine in the absence of halogen, it is difficult to connect the allyl group to the organometallic complex. However, in the presence of oxidizing agent and CO atmosphere, the allylamine can be transformed into β-lactam. As a model reaction, when *N*-allyl-4-methylaniline in DMF at 1 atm of CO at 100°C was treated with 10 mol% $PdCl_2$, tricyclohexylphosphine (20 mol%), 2.0 equivalents of $Cu(OAc)_2$ and pivalic acid, 71% of 3-methylene-1-(*p*-tolyl)azetidin-2-one was obtained, along with the *N*-acetylation product presumably arising from $Cu(OAc)_2$. Replacement of $Cu(OAc)_2$ with $Cu(OPiv)_2$ as the oxidant completely suppresses the acylation side product and increases the yield of 3-methylene-1-(*p*-tolyl)azetidin-2-one to 74%. Other palladium catalysts, such as $[PdCl_2(PPh_3)_2]$ and $[Pd(dba)_2]$, are less effective in this oxidative carbonylation, leading to lower yields of 62% and 12%, respectively. Also, lowering the reaction temperature by 20°C decreased the yield of β-lactam to 54%. When acetic acid was added as the additive, the yield of β-lactam decreased as well. This approach works for various *N*-allylamines such as *N*-allylanilines, *N*-allylbenzylamines, *N*-allylalkylamines, and branched *N*-allylamines, except for electron-deficient *N*-allylanilines, e.g., *N*-allyl-4-(trifluoromethyl)aniline, *N*-allyl benzylamines, *N*-allyl phenethylamine, and *N*-(3-phenylpropyl)prop-2-en-1-amine as well as long-chain *N*-allylamine. *N*-Allylamines with a cyclopropyl group all exhibit good reactivities. In addition, enantioenriched *N*-allylamines could be readily transformed into their corresponding β-lactam derivatives without loss of enantioselectivity under the standard reaction conditions. A feasible mechanism is illustrated in Scheme 3.106 [380].

In addition to convert the *N*-allylamines into β-lactams in the presence of CO with transition metal complex, cobalt(II)tetramethyltetraaza[14]

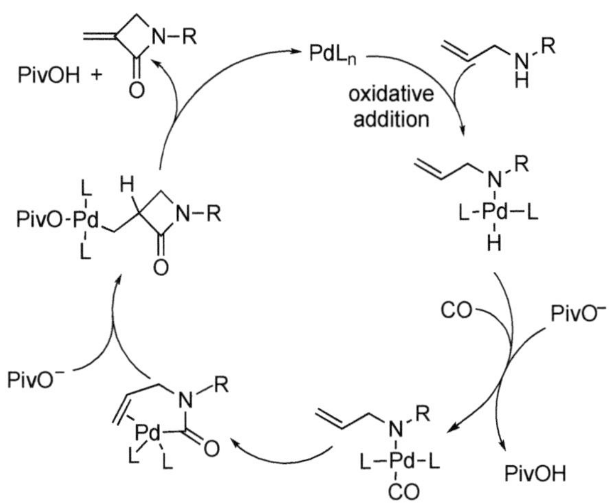

SCHEME 3.106 The mechanism for converting allylamines into β-lactams

annulene complex [Co(MeTAA)] formed from a template condensation between 2,4-pentandione and 1,2-diaminobenzene, yielding (6*Z*,8*E*,15*Z*,17*E*)-6,8,15,17-tetramethyl-5,14-dihydrodibenzo[*b*,*i*][1,4,8,11]tetraazacyclotetradecine, in the presence of cobalt(II) acetate tetrahydrate is an effective catalyst for the reaction between *N*-tosylhydrazone and imine to form β-lactam under CO atmosphere [381]. In order to investigate the effects of temperature, solvent, additive, and CO pressure on this reaction, the [Co(MeTAA)] catalyzed model reaction between sodium (*E*)-2-benzylidene-1-tosylhydrazin-1-ide and *N*-methyl-1-phenylmethanimine (1.2 eq.) to give (3*R*,4*S*)-1-methyl-3,4-diphenylazetidin-2-one was performed in toluene at 60°C with 20 bars of CO as the initial condition, as shown in Scheme 3.107.

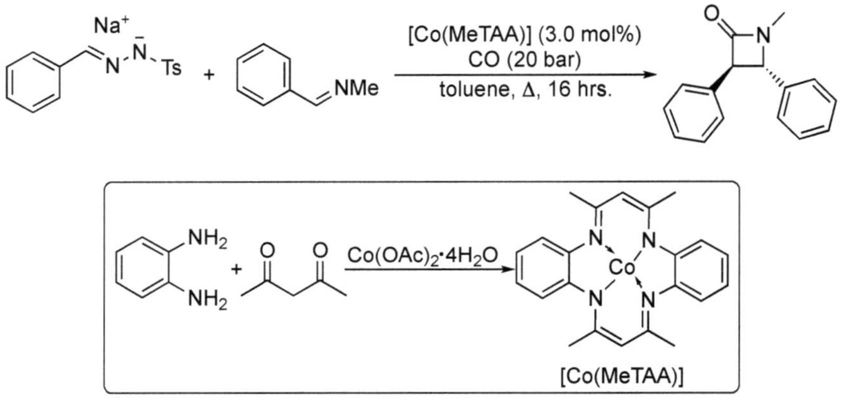

SCHEME 3.107 The preparation of (3*R*,4*S*)-1-methyl-3,4-diphenylazetidin-2-one from sodium (*E*)-2-benzylidene-1-tosylhydrazin-1-ide and (*Z*)-*N*-methyl-1-phenylmethanimine

Screening of solvents among toluene, THF, chlorobenzene, 1,2-dichloroethane, acetonitrile, 1,4-dioxane, and trifluoromethyl benzene indicates that toluene is the best solvent for this reaction, and polar solvents are not suitable for this reaction as the conversion drops significantly. The addition of pyridine, 4-dimethylaminopyridine, or 1-methylimidazole as an additive led to a lower yield of β-lactam than those reactions without these additives. Variation of the CO pressure indicates that β-lactam can be formed at 2 bars of CO, but higher yields are obtained at higher CO pressures. Also, temperature plays an important role, as higher temperature helps the formation of carbene radical intermediate from the benzaldehyde tosylhydrazone sodium salt. Further examination of the substituent effect on this reaction revealed that non-substituted *N*-methyl-1-phenylmethanimine gave the highest yield of β-lactam, followed by the *N*-methyl-1-phenylmethanimines with an electron-donating group at the *para*-position, and the ones with an electron-withdrawing group at the *para*-positions led to β-lactams of substantially lower yields. On the other hand, the hydrazone component with either electron-donating or electron-withdrawing group at the *para*-position worked under this condition, and the one with a strong electron-withdrawing group, such as nitro, nearly did not produce any β-lactam. While this reaction is assumed to proceed via a [2+2] reaction between ketene and imine, it occurs in pressurized systems at elevated temperatures of 140°C, which is different from the normal Staudinger cycloadditions that are usually performed below 0°C. In all cases, the reactions proved to be highly *trans*-selective, producing the *trans*-β-lactams in excellent diastereoselectivity. Utilization of chiral cobalt(II) catalysts has no influence on the control of stereochemistry in this reaction [381]. It should be pointed out that the reaction between sodium (*E*)-2-benzylidene-1-tosylhydrazin-1-ide and imine can also be performed at atmospheric pressure in the presence of $Pd_2(dba)_3$ in dichloroethane at 60°C, affording *trans*-β-lactams in very good diastereoselectivity (~9:1) [382].

Another carbonylative cycloaddition to afford β-lactam occurs between benzyl halides and imines using NHC as ligands in the presence of CO. The reaction between benzyl chloride and (*E*)-*N*-benzyl-1-phenylmethanimine has been selected as the model reaction to optimize the conditions, including the variation of temperature, the pressure of CO, ligand, solvent, base, etc. [383]. Screening of the synthesized carbene-palladium complexes, including bis(1-butyl-3-methyl-2,3-dihydro-1*H*-imidazol-2-yl)palladium(IV) chloride [$(Bmim)_2PdCl_2$], bis(1-butyl-3-methyl-2,3-dihydro-1*H*-imidazol-2-yl) palladium(IV) bromide [$(Bmim)_2PdBr_2$], bis(1-butyl-3-methyl-2,3-dihydro-1*H*-imidazol-2-yl)-palladium(IV) iodide [$(Bmim)_2PdI_2$] and [$(Bmim)PdI_2]_2$ revealed [$(Bmim)PdI_2]_2$ to be the most efficient catalyst, while other complexes led to lower yields of β-lactams. Also, the reaction is temperature-dependent as the reaction nearly vanished when the temperature was lowered from 100°C to 80°C. Likewise, when the CO pressure was reduced from 30 atms to 20 atms, the yield of β-lactam decreased from 96% to 62%. The reaction was found to proceed more efficiently in polar solvents rather than in non-polar solvents, and CH_3CN was superior to other tested solvents. Regarding the bases as additives, inorganic bases are completely inefficient, so for strong organic bases such as 1,4-diazabicyclo[2.2.2]octane (DABCO) and tetramethylethylenediamine (TMEDA), as well as acetate, sulfonate; whereas nitrogen-based weaker bases, such as Et_3N, *n*-Pr_3N, and *i*-Pr_2NEt, are crucial to the success of the reaction. As far as the substrate is concerned, both benzyl halides and phosphate undergo the cycloaddition reaction to give β-lactams in good yields and with excellent regioselectivities, and a variety of aldimines are suitable for this reaction to afford *trans*-β-lactams in high yields. Besides the imines derived from benzylamine, the imines derived from aniline and *n*-propylamine also undergo the carbonylative cycloaddition reaction. A slight steric hindrance effect on the reactivity has been observed, as demonstrated in the reaction of 1-(chloromethyl)-4-methoxybenzene and 1-(chloromethyl)-2-methoxybenzene. A representative reaction of this kind is illustrated in Scheme 3.108 between 1-chloro-4-(chloromethyl) benzene and (*R*,*E*)-1-phenyl-*N*-(1-phenylethyl)methanimine to indicate the diastereoselectivity, yielding (3*R*,4*S*)-3-(4-chlorophenyl)-4-phenyl-1-((*R*)-1-phenylethyl)-azetidin-2-one and (3*S*,4*R*)-3-(4-chlorophenyl)-4-phenyl-1-((*R*)-1-phenylethyl)-azetidin-2-one in 7:3 ratio [383]. Another reaction of this type utilizes a palladium catalyst and chiral Lewis base to convert benzyl bromides and *N*-tosylimines to β-lactams of very high enantioselectivity and good yield in the presence of CO [384].

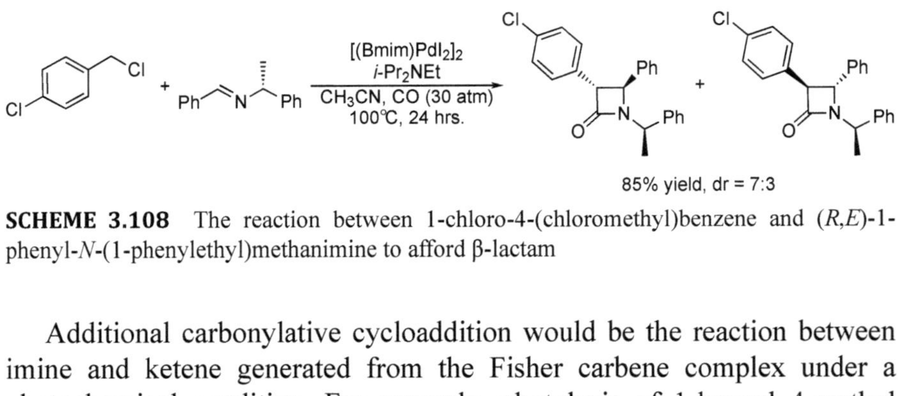

SCHEME 3.108 The reaction between 1-chloro-4-(chloromethyl)benzene and (*R*,*E*)-1-phenyl-*N*-(1-phenylethyl)methanimine to afford β-lactam

Additional carbonylative cycloaddition would be the reaction between imine and ketene generated from the Fisher carbene complex under a photochemical condition. For example, photolysis of 1-benzyl 4-methyl (*S*)-4-methyl-4,5-dihydro-1*H*-imidazole-1,4-dicarboxylate with the (methoxymethylcarbene) chromium complex in Pyrex tube with a 500 W Hanovia medium pressure Hg lamp at 30°C afforded methyl (2*S*,5*R*,6*S*)-6-methoxy-2,6-dimethyl-7-oxo-1,4-diazabicyclo-[3.2.0]heptane-2-carboxylate in low yield, with chromium pentacarbonyl imidazoline complex as the major byproduct. However, under the optimal photochemical condition with four 500 W quartz/halogen lamps as the light source at 70°C, and subsequent deprotection of benzyloxycarbonyl via Pd/C catalyzed hydrogenation, methyl (2*S*,5*R*,6*S*)-6-methoxy-2,6-dimethyl-7-oxo-1,4-diazabicyclo[3.2.0] heptane-2-carboxylate can be obtained on a multigram scale in good yield. Extension of this reaction to a bis-Fisher complex allows the preparation of bis-dioxocyclam, i.e., dimethyl 6,6'-(propane-1,3-diylbis(oxy))-(2*S*,2'*S*,5*R*,5'*R*,6*S*,6'*S*)-bis(2,6-dimethyl-7-oxo-1,4-diazabicyclo[3.2.0] heptane-2-carboxylate) in 71% yield, as shown in Scheme 3.109 [385].

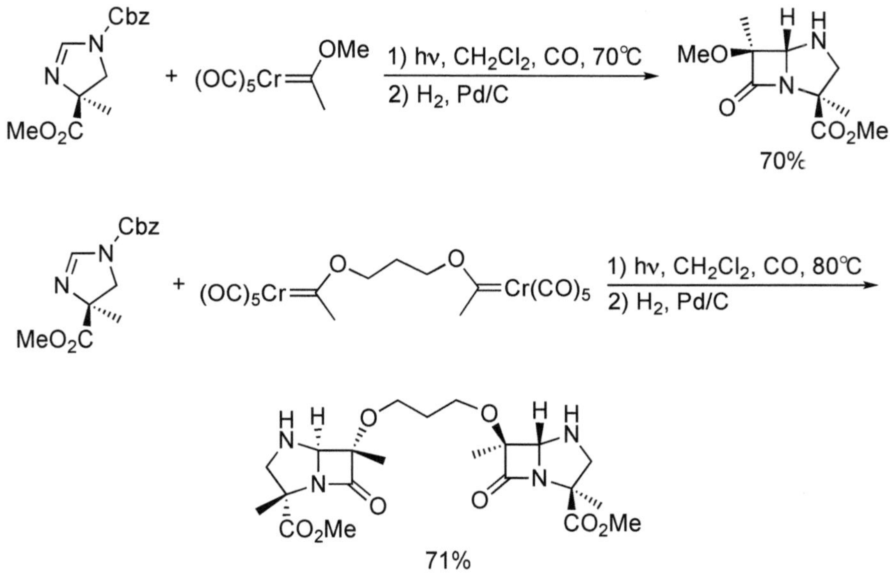

SCHEME 3.109 The conversion of 1-benzyl 4-methyl (*S*)-4-methyl-4,5-dihydro-1*H*-imidazole-1,4-dicarboxylate into β-lactam in the presence of Fischer carbene complex

3.3.9 UGI MULTI-COMPONENT REACTION (MCR)

The Ugi multi-component reaction (MCR), initially reported by Ivar Karl Ugi in 1960 [386], is a one-pot synthesis of bis-amide involving four components, including an aldehyde (or ketone), an amine, a carboxylic acid, and an isocyanide. It has wide applications in peptide chemistry [387, 388], and heterocyclic chemistry [389, 390] For the synthesis of β-lactams, only aldehydes, β-amino acids and isonitriles are needed, as illustrated in some early practices, such as: (a) a smooth reaction of 3-aminopropanoic acid with isobutyraldehyde and 2-isocyano-2-methylpropane in methanol to afford *N*-(*tert*-butyl)-3-methyl-2-(2-oxoazetidin-1-yl)butanamide [391]; (b) reaction of (*S*)-3-amino-2-hydroxypropanoic acid with 4-(benzyloxy)benzaldehyde and (isocyanomethylene) dibenzene to form *N*-benzhydryl-2-(4-(benzyloxy)phenyl)-2-((*S*)-3-hydroxy-2-oxoazetidin-1-yl)acetamide, which was applied to the synthesis of Nocardicin A and Nocardicin D [392]; (c) reaction of 3-aminopropanoic acid with (isocyanomethylene)dibenzene and a series of aldehydes to afford β-lactams which were further treated with N_2O_4 and NaOAc in $CHCl_3$ to yield diphenylmethyl esters of 2-oxo-1-azetidineacetic acids [393]; and (d) reaction of (*S*)-3-amino-4-(benzyloxy)-4-oxobutanoic acid (i.e., α-benzyl *L*-aspartate) with eight combinations of different aldehydes and isonitriles [394].

As of the convenience and diversity of this approach, the Ugi four-center three-component reaction (U-4C-3CR) has been widely applied to the preparation of β-lactams. For example, a series of alicyclic β-amino acids, including (1*R*,2*S*)-2-aminocyclohexane-1-carboxylic acid, (1*R*,2*R*)-2-amino cyclohexane-1-carboxylic acid, (1*R*,2*S*)-2-aminocyclopentane-1-carboxylic acid, (1*R*,2*R*)-2-aminocyclopentane-1-carboxylic acid, (1*R*,6*S*)-6-amino-cyclohex-3-ene-1-carboxylic acid, (1*S*,2*R*,3*S*,4*R*)-3-aminobicyclo[2.2.1]heptane-2-carboxylic acid and (1*R*,2*R*,3*S*,4*S*)-3-aminobicyclo[2.2.1]hept-5-ene-2-carboxylic acid, have been converted into respective β-lactams in parallel liquid-phase reaction in combination with substituted benzaldehydes (e.g., 4-NO2, 3-Cl, 3-MeO-benzaldehydes) and isonitriles in methanol in one to 4 days in moderate to good yields [395]. For the Ugi three-component reaction involving isatin, isocyanide, and β-amino acid, polar solvent such as trifluoroethanol is superior to the commonly used methanol, where non-polar solvents such as toluene or less polar solvents such as CH_2Cl_2 are invalid for this reaction. Also, the reaction at 0.5 *M* concentration seems to be the ideal concentration, too high or too low concentration leads to a lower yield of β-lactam-containing 3,3-disubstituted oxindoles [396]. In order to identify a lead compound as a potential human leukocyte elastase inhibitor, a 126-member library of monocyclic β-lactams has been generated in a parallel fashion by solution-phase Ugi four-component condensation reaction between β-amino acids, aldehydes, and isocyanides [397].

Recently, a modified Ugi condition adopted an aqueous media to construct β-lactam libraries, where precipitation occurred when less water-soluble β-amino acids and an appropriate amount of water were used. In this way, the reactions were complete in 1 day at room temperature, whereas the reaction in methanol ended in 3 days. In these reactions, noteworthy increases in diastereoselectivity occurred when bicyclic β-amino acids (e.g., with a norbornane or norbornene skeleton) and bulky *tert*-butyl isocyanide were used as the starting materials. The reaction in water demonstrated several advantages, such as environmentally benign and mild conditions, shorter reaction time, and simple purification. A representative reaction of (1*R*,2*R*,3*S*,4*S*)-3-aminobicyclo-[2.2.1]hept-5-ene-2-carboxylic acid with 4-methoxybenzaldehyde and 2-isocyano-2-methylpropane (i.e., *t*-butyl isonitrile) at room temperature is displayed in Scheme 3.110, which afforded 57% of *N*-(*tert*-butyl)-2-(4-methoxyphenyl)-2-((1*S*,2*S*,5*R*,6*R*)-4-oxo-3-azatricyclo[4.2.1.0^{2,5}]non-7-en-3-yl)acetamide with a diastereomeric ratio of 83:17 [398]. Following this approach, 2-, 3- or 4-formylbenzoic acid was immobilized on the super acid sensitive resin (Sasrin) through its carboxy group, and the remaining aldehyde group is allowed to react with cyclic β-amino acid and isonitrile in a mixed solvent of CH_2Cl_2/MeOH (1:1) to afford similar β-lactams [399]. However, the reaction of (1*S*,2*R*,3*S*,4*R*)-3-amino-7-oxabicyclo[2.2.1]heptane-2-carboxylic acid with an aldehyde (e.g., benzaldehyde, isobutyraldehyde, and 2-phenylacetaldehyde) and *t*-butyl isonitrile (or *n*-butyl isonitrile) in methanol, although give tricyclic β-lactams in very good to excellent yields, the corresponding diastereoselectivity is not very high [400]. Interestingly, a comparable study for the reaction of (1*R*,2*S*,3*R*,4*S*)-3-amino-7-oxabicyclo[2.2.1]hept-5-ene-2-carboxylic acid with aldehydes (e.g., propionaldehyde, pivalaldehyde, 4-chlorobenzaldehyde, 4-methylbenzaldehyde and 4-methoxybenz-aldehyde) and *tert*-butyl isonitrile or cyclohexyl isonitrile in either water or methanol reveals that the expected β-lactams can be obtained only in moderate to good yields, only in a few cases the diastereoselectivities of β-lactams have been improved from the reactions in water [401]. The factors that can accelerate the reaction rate may include but not limited to the hydrophobic effect, enhanced hydrogen bonding in the transition state, the high cohesive energy density of water, and salting out effect caused either by an ionic solute (e.g., LiCl) or non-ionic solute (e.g., glucose) [402]. The modification of traditional Ugi reaction to include primary amine, isonitrile, and stable β-keto acid in 1.0 *M* aqueous glucose solution affords β-lactams in good yields within 3 days. Extension of reaction time to 6 days results in a dramatic decrease in yields of β-lactams, possibly due to the instability of β-lactams in an aqueous solution. A representative reaction using β-keto acid is illustrated in Scheme 3.111, where the mixture of lithium 2-oxocyclohexane-1-carboxylate and furan-2-ylmethanamine in 1.0 *M* aqueous glucose solution was stirred for nearly 1 hour then *t*-butyl isonitrile was added. After 6 days, 49% of *N*-(*tert*-butyl)-7-(furan-2-ylmethyl)-8-oxo-7-azabicyclo-[4.2.0]octane-6-carboxamide was obtained [402].

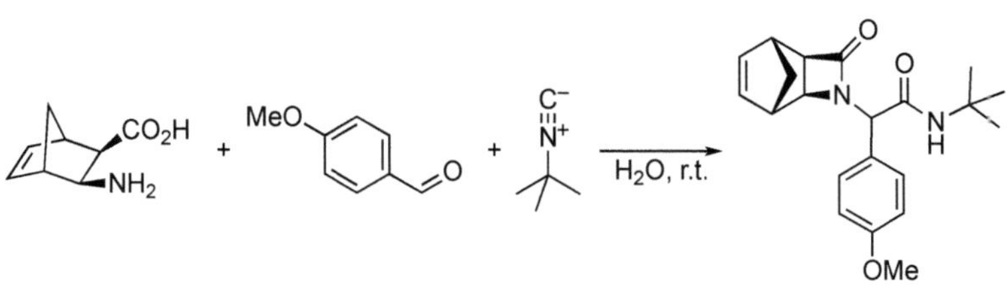

SCHEME 3.110 The three-component reaction to afford β-lactam

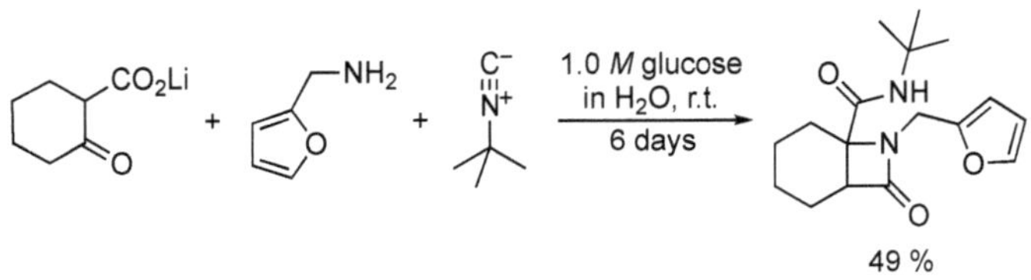

SCHEME 3.111 The three-component reaction involving ketone, amine, and isonitrile to afford β-lactam

Excitingly, very good to excellent diastereoselectivity has been observed in the Ugi reaction between β-amino acid, aldehyde, and N$^{\beta}$-Fmoc-amino alkyl isonitrile, as illustrated specifically in Scheme 3.112. In this practice, (*R*)-3-((((9*H*-fluoren-9-yl)methoxy)carbonyl)amino)-5-(benzyloxy)-5-oxo-pentanoic acid was treated with 4-methylmorpholine (NMM) and isobutyl carbonochloridate (IBC-Cl), then sodium azide. After being heated in toluene at 65°C, the product was treated with 98% formic acid and *N*,*N*-dimethylpyr-idin-4-amine in CH_2Cl_2 at -15°C to afford benzyl (*S*)-3-((((9*H*-fluoren-9-yl) methoxy)carbonyl)amino)-4-formamidobutanoate. Subsequent treatment with the Burgess reagent, i.e., (methyl *N*-(triethylammoniumsulfonyl)carba-mate), the benzyl (*S*)-3-((((9*H*-fluoren-9-yl)-methoxy)carbonyl)amino)-4-formamidobutanoate was converted into benzyl (*S*)-3-((((9*H*-fluoren-9-yl) methoxy)carbonyl)amino)-4-isocyanobutanoate, which was then allowed to react with (*S*)-3-amino-4-methoxy-4-oxobutanoic acid and 3-methylbutanal in methanol at room temperature to afforded 63% of methyl (2*R*)-1-(1-(((*S*)-2-((((9*H*-fluoren-9-yl)methoxy)carbonyl)amino)-4-(benzyloxy)-4-oxobutyl) amino)-4-methyl-1-oxopentan-2-yl)-4-oxoazetidine-2-carboxylate, with a diastereoselectivity of 100:0 [403].

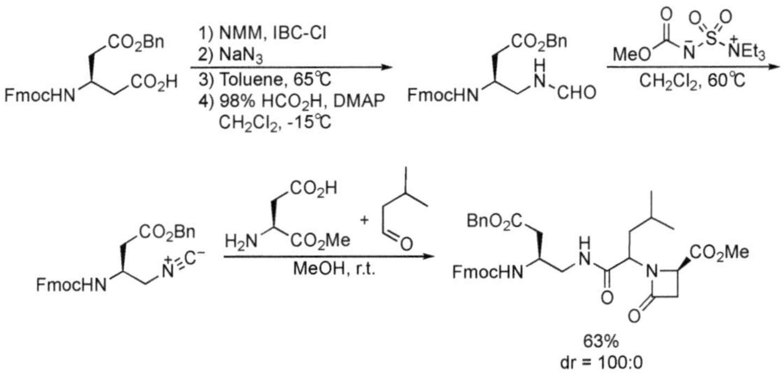

SCHEME 3.112 The synthesis of methyl (2*R*)-1-(1-(((*S*)-2-((((9*H*-fluoren-9-yl)methoxy) carbonyl)amino)-4-(benzyloxy)-4-oxobutyl)amino)-4-methyl-1-oxopentan-2-yl)-4-oxoazetidine-2-carboxylate from (*R*)-3-((((9*H*-fluoren-9-yl)methoxy)carbonyl)amino)-5-(benzyloxy)-5-oxopentanoic acid

Similarly, chiral β-amino acids, such as (*S*)-3-amino-2-hydroxypropanoic acid and (*S*)-3-amino-2-methylpropanoic acid have been recently applied to react with formaldehyde and terpene-based isonitrile in order to confirm the structures of monamphilectines B and C, which are amphilectane-type diterpenes containing isocyanide group isolated from Caribbean sponge *Svenzea flava* collected near Mona Island, off the west coast of Puerto Rico via bioassay-guided fractionation [404]. In these syntheses, the slightly less sterically hindered isonitrile group of the two tertiary isocyanide moieties selectively reacts with the β-amino acids and formaldehyde, as shown in Scheme 3.113. In this report, one-pot reaction of (1*S*,3*S*,3a*S*,3a¹*S*,6a*S*,7*S*,9a*R*)-6a-isocyano-3-(2-isocyano-2-methylpropyl)-1,7-dimethyl-4-methylenedodecahydro-1*H*-phenalene with formaldehyde and (S)-3-amino-2-methylpropanoic acid in ethanol led to 73% of *N*-(1-((1*S*,3*S*,3a*R*,3a¹*S*,6*S*,6a*S*,9a*S*)-6a-isocyano-3,6-dimethyl-9-methylenedodecahydro-1*H*-phenalen-1-yl)-2-methylpropan-2-yl)-2-((*S*)-3-methyl-2-oxoazetidin-1-yl)acetamide. Comparably, the reaction of the same substrate with (*S*)-3-amino-2-hydroxypropanoic acid under the same condition gave 70% of 2-((*S*)-3-hydroxy-2-oxoazetidin-1-yl)-*N*-(1-((1*S*,3*S*,3a*R*,3a¹*S*,6*S*,6a*S*,9a*S*)-6a-isocyano-3,6-dimethyl-9-methylenedodecahydro-1*H*-phenalen-1-yl)-2-methylpropan-2-yl)-acetamide. Treatment of this compound in Et_2O with diazomethane gave 72% of *N*-(1-((1*S*,3*S*,3a*R*,3a¹*S*,6*S*,6a*S*,9a*S*)-6a-isocyano-3,6-dimethyl-9-methylenedodecahydro-1*H*-phenalen-1-yl)-2-methylpropan-2-yl)-2-((*S*)-3-methoxy-2-oxoazetidin-1-yl)acetamide. The structures of these two β-lactams are identical to the structures of Monamphilectine B and Monamphilectine C, respectively [404]. In addition, Monamphilectine A has been semi-synthesized by the Ugi reaction involving (1*S*,3*S*,3a*S*,3a¹*S*,6a*S*,7*S*,9a*R*)-6a-isocyano-3-(2-isocyano-2-methylpropyl)-1,7-dimethyl-4-methylene-dodecahydro-1*H*-phenalene, formaldehyde, and 3-aminopropanoic acid in MeOH at room temperature for 12–24 hours [405]. Besides these diterpene-related β-lactams, a monoterpene-based β-amino acid has been applied to make six conformationally constrained tricyclic β-lactam enantiomers via Ugi four-center three-component reaction. In this practice, (1*R*,2*R*,3*S*,5*R*)-2-amino-6,6-dimethylbicyclo[3.1.1]heptane-3-carboxylic acid was allowed to react with propionaldehyde (or pivalaldehyde, or benzaldehyde) and *t*-butyl isonitrile (or (isocyanomethyl)benzene) under three conditions (i.e., in methanol or water or neat) to afford tricyclic β-lactams in good to excellent diastereoselectivity. As an example, the reaction of this particular β-amino acid with *t*-butyl isonitrile and pivalaldehyde affords 75%, 51% and 53% of (*S*)-*N*-(*tert*-butyl)-2-((1*R*,2*R*,5*S*,7*R*)-8,8-dimethyl-4-oxo-3-azatricyclo[5.1.1.$0^{2,5}$]

nonan-3-yl)-3,3-dimethylbutanamide in methanol (1 day), water (12 hours) or neat (6 hours), all with 100:0 diastereoselectivity, respectively [406].

According to the nature of the Ugi reaction, it is a perfect reaction to explore the concept of combinatorial synthesis, efficiently creating a large library of β-lactams with wide structural diversity. For example, a combination of 3 isonitriles (i.e., *t*-butyl isonitrile, cyclohexyl isonitrile, benzyl isonitrile), 3 chiral β-amino acids (i.e., (1*R*,2*S*)-2-aminocyclohexane-1-carboxylic acid, (1*R*,6*S*)-6-aminocyclohex-3-ene-1-carboxylic acid and (1*R*,2*S*)-2-aminocyclopentane-1-carboxylic acid), and 15 different aldehydes in methanol at room temperature for 24 hours allows the creation of a small library of 135 β-lactams [407].

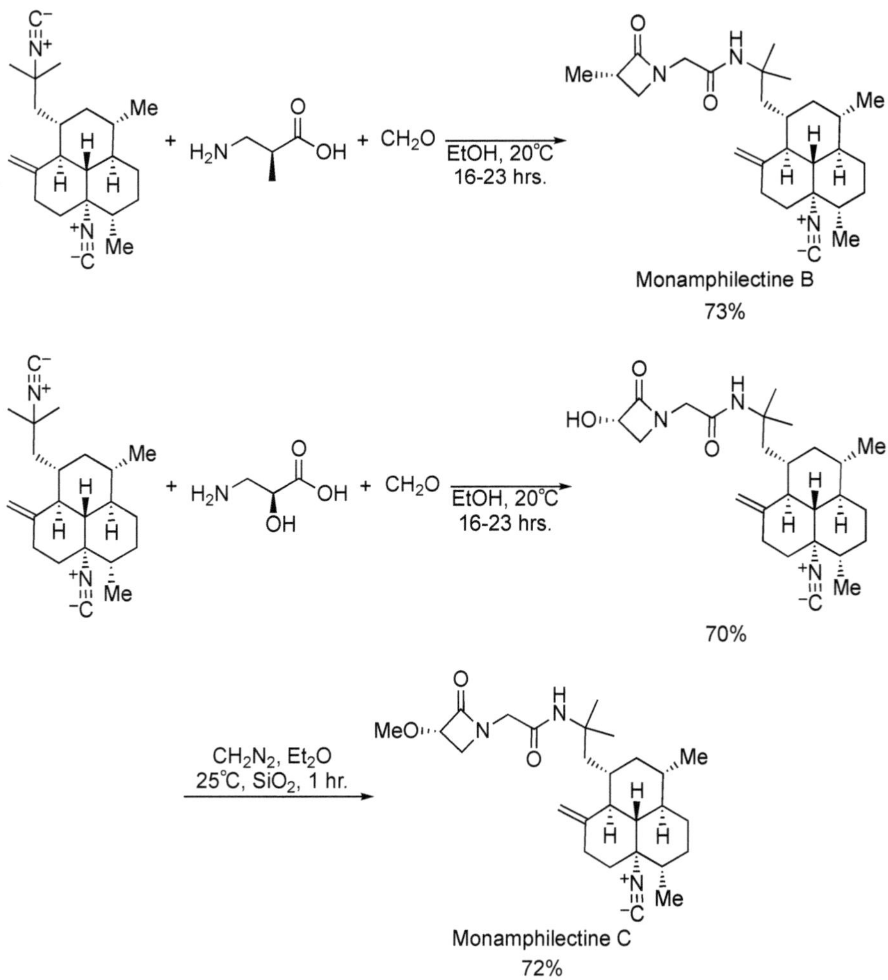

SCHEME 3.113 Application of the Ugi reaction to synthesize Monamphilectines

Recently, an interesting approach has been reported to form bis-β-lactams involving two consecutive creations of β-lactams which combines the Ugi three-component reaction and robustness of the CH_2I_2-based β-lactam formation. For example, the Ugi three-component reaction of 2-naphthaldehyde with isocyanocyclohexane and 3-aminopropanoic acid in MeOH under microwave irradiation at 120°C for 1 hour gave 69% of *N*-cyclohexyl-2-(naphthalen-2-yl)-2-(2-oxoazetidin-1-yl)acetamide. This mono-β-lactam after being treated with 2.5 equivalents of NaH in DMSO followed by 1.5 equivalents of CH_2I_2 at room temperature for 2–7 hours, 54% of 1'-cyclohexyl-3'-(naphthalen-2-yl)-[1,3'-biazetidine]-2,2'-dione was obtained, as outlined in Scheme 3.114. Alternatively, this two-step approach has been performed in a one-pot manner, in which MeOH was directly evaporated and NaH was added after the residue was dissolved in DMSO, followed by CH_2I_2, yielding a relatively higher yield of the bis-β-lactam [408].

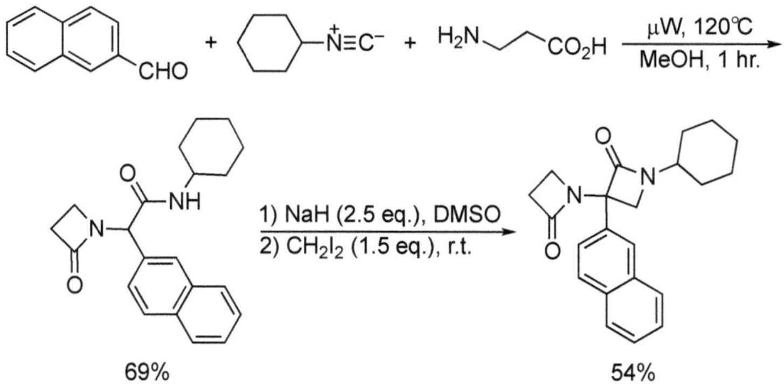

SCHEME 3.114 The preparation of bis-β-lactam

In addition to the three-component reaction involving β-amino acid, aldehyde, and isonitrile to yield β-lactam, the four-component one-pot reaction of primary amine, arylglyoxal, isonitrile, and α-haloacetic acid, also leads to the generation of β-lactam, as illustrated in Scheme 3.115. In this reaction, the four reactants of 2-bromoacetic acid, 2-(4-chlorophenyl)-2-oxoacetaldehyde, *t*-butylamine, and *t*-butylisonitrile in methanol in the presence of Cs_2CO_3 afforded 93% of *N*,1-di-*tert*-butyl-2-(4-chlorobenzoyl)-4-oxoazetidine-2-carboxamide. This reaction is believed to involve a tandem Ugi reaction and subsequent intramolecular C-alkylation at room temperature in the presence of Cs_2CO_3 [409].

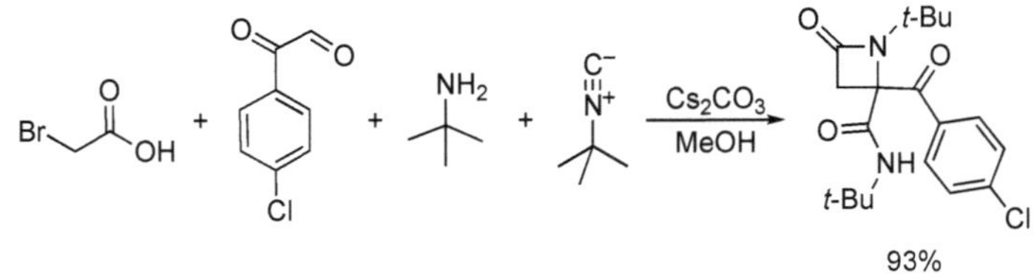

SCHEME 3.115 The one-pot four-component reaction to afford β-lactam

In another variant of the Ugi MCR, the combination of aldehyde, primary amine, isonitrile, and chloroacetic acid in methanol in the presence of a base (DBU or KOH) leads to the formation of both β-lactam and 2,5-diketopiperazine at different ratio. For an interesting comparison, the reaction of 2-(2,4-dichlorophenethyl)-2*H*-1,2,3-triazole-4-carbaldehyde with 3,3-diphenylpropan-1-amine, 2-isocyano-1,3-dimethylbenzene and chloroacetic acid in methanol in the presence of KOH gives 21% of 3-(2-(2,4-dichlorophenethyl)-2*H*-1,2,3-triazol-4-yl)-1-(2,6-dimethylphenyl)-4-(3,3-diphenylpropyl) piperazine-2,5-dione, but only 2.1% of 2-(2-(2,4-dichlorophenethyl)-2*H*-1,2,3-triazol-4-yl)-*N*-(2,6-dimethylphenyl)-1-(3,3-diphenylpropyl)-4-oxoazetidine-3-carboxamide. In contrast, the same reaction with isonitrile changed to 1-isocyano-2-methylbenzene in the presence of DBU rather than KOH affords 73% of 2-(2-(2,4-dichlorophenethyl)-2*H*-1,2,3-triazol-4-yl)-1-(3,3-diphenylpropyl)-4-oxo-*N*-(*o*-tolyl)-azetidine-3-carboxamide as a single product, without detection of the DKP component, as illustrated in Scheme 3.116 [410]. In most of these performed reactions, β-lactams are the major products, but the reason to affect the reaction is not clear yet.

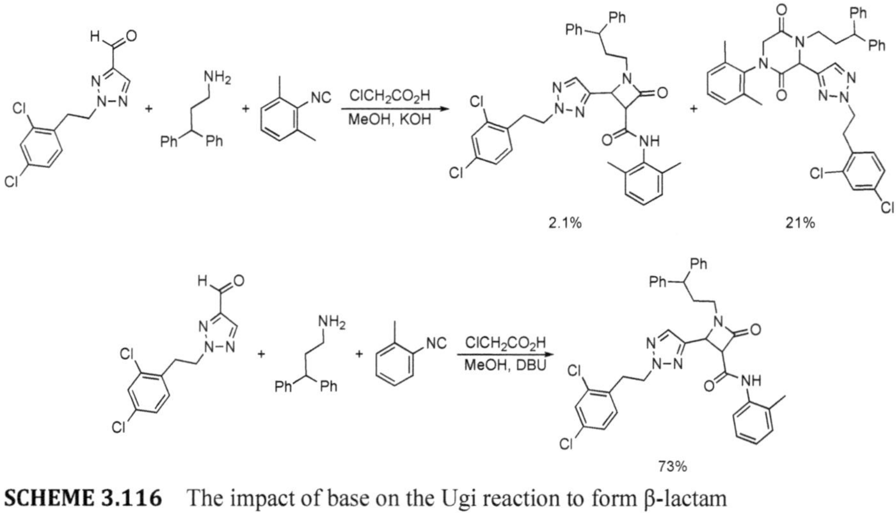

SCHEME 3.116 The impact of base on the Ugi reaction to form β-lactam

Different from the Ugi three-component reaction to give β-lactams, the Ugi four-component reaction involving aldehyde, primary amine, acid, and isonitrile seems to proceed via an intermediate of diamide, which is then converted into β-lactam in the presence of a base (K_2CO_3) in acetonitrile. However, when this diamide is treated with K_2CO_3 in refluxing methanol, pyrrolidine-2,5-dione derivative is formed instead. As a model reaction, the reaction of benzaldehyde and aniline in methanol at room temperature for 1 hour afforded the imine, to which 3-phenylpropiolic acid was added, and 15 minutes later, *t*-butyl isonitrile was added. The resulting mixture was stirred at room temperature for 24 hours to give *N*-(2-(*tert*-butylamino)-2-oxo-1-phenylethyl)-*N*,3-diphenylpropiolamide as a precipitate. When the isolated precipitate was added to CH_3CN in the presence of 1.0 equivalent of K_2CO_3 and heated at 80°C for 2 hours, 95% of (*S*,*E*)-3-benzylidene-*N*-(*tert*-butyl)-4-oxo-1,2-diphenylazetidine-2-carboxamide was obtained. In comparison, when this precipitate was refluxed in MeOH in the presence of K_2CO_3, 94% of (*R*,*E*)-4-benzylidene-1-(*tert*-butyl)-3-phenyl-3-(phenylamino)pyrrolidine-2,5-dione was formed, as displayed in Scheme 3.117 [411]. Likewise, the four-component reaction of 2-(2,4-dichlorophenethyl)-2*H*-1,2,3-triazole-4-carbaldehyde, 3,3-diphenylpropan-1-amine, ethyl 2-isocyanoacetate and chloroacetic acid in methanol at room temperature for 48 hours gave 70% of ethyl (2-(2-chloro-*N*-(3,3-diphenylpropyl)acetamido)-2-(2-(2,4-dichlorophenethyl)-2*H*-1,2,3-triazol-4-yl)acetyl)glycinate. This isolated intermediate can be easily converted into β-lactam, i.e., ethyl (2-(2-(2,4-dichlorophenethyl)-2*H*-1,2,3-triazol-4-yl)-1-(3,3-diphenylpropyl)-4-oxoazetidine-2-carbonyl)glycinate when treated with DBU in THF at room temperature, instead of heating in CH_3CN [412]. In comparison, optimization of a series of Ugi four-component reactions involving 2-aminoethan-1-ol, benzaldehydes, 3-phenylpropiolic acid, and *tert*-butyl isonitrile indicates that treatment of the amide intermediate with a stronger base, such as NaH, leads to the formation of functionalized pyrrolidine-2,5-diones in good to excellent yields. A mechanistic study reveals that the reaction proceeds through an intermediate of β-lactam formed by energetically favored intramolecular cyclization in a 4-*exo-dig* fashion, which then rearranges to give the pyrrolidine-2,5-dione with the help of the hydroxyethyl appendage [413]. Similarly, the Ugi four-component reaction product arising from aldehyde, acid, isonitrile, and primary amine in MeOH, once isolated or separated, can be converted into a different type of β-lactam when treated with 2.5 equivalents of NaH in DMSO in the presence of CH_2I_2 [414], an approach similar to the one outlined in Scheme 3.114.

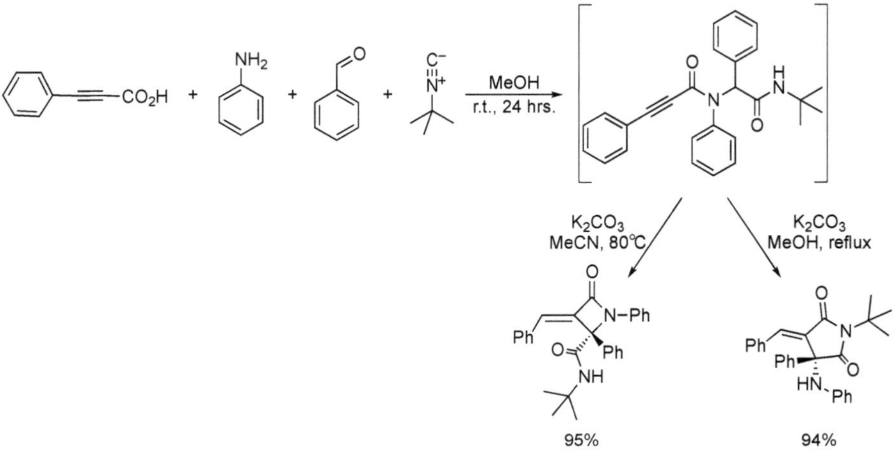

SCHEME 3.117 The solvent effect on the Ugi four-component reaction to yield β-lactam

Different from the post-treatment of the Ugi four-component reaction product as displayed in Scheme 3.117, the post-Ugi treatment of the reaction product with different Lewis acids in different solvents also leads to different structural scaffolds, as illustrated in Scheme 3.118. In this approach, the model Ugi four-component reaction of propiolic acid with imidazo[1,2-*a*]pyridine-2-carbaldehyde, (4-methoxyphenyl) methanamine, and isocyanocyclohexane in methanol at room temperature led to 84% of *N*-(2-(cyclohexylamino)-1-(imidazo[1,2-*a*]pyridin-2-yl)-2-oxoethyl)-*N*-(4-methoxybenzyl)propiolamide. This isolated Ugi reaction product was then treated with a Lewis acid that could be AgOTf, $AgSbF_6$, AuCl, $InCl_3$, $In(OTf)_3$, $AlCl_3$, $ZnCl_2$, or CuI in a different amount, in various solvents (e.g., dichloroethane, toluene, *o*-xylene, CH_3CN) to identify an ideal condition to form the β-lactam. It is found that in the presence of 30 mol% of $InCl_3$ in toluene at 120°C for 12 hours, the Ugi product was 100% converted into a mixture of two products, i.e., *N*-cyclohexyl-2-(imidazo[1,2-*a*]pyridin-2-yl)-1-(4-methoxybenzyl)-3-methylene-4-oxoazetidine-2-carboxamide and *N*-cyclohexyl-2-(imidazo[1,2-*a*] pyridin-2-yl)-1-(4-methoxybenzyl)-5-oxo-2,5-dihydro-1*H*-pyrrole-2-carboxamide, with isolated yield of 88% and 4%, respectively. In contrast, when this Ugi product was treated with 10 mol% of $AlCl_3$ in dichloroethane at 120°C for 12 hours, 75% of the second product was isolated, with 100%

of conversion. The experiments also indicated that in the presence of less or more $InCl_3$, the chemical yield of the β-lactam decreased slightly, whereas the treatment in a shorter duration (e.g., 6 hours) or lower temperature (e.g., 100°C) also resulted in a lower yield of the β-lactam. Further study on the post-Ugi treatment revealed that treatment of the Ugi product with 10 mol% of $InCl_3$ in dichloroethane also led to the formation of the second type products in very good to excellent yields [415]. Different from the treatment with K_2CO_3 in CH_3CN in Scheme 3.117, treatment of this model Ugi product with K_2CO_3 in dichloroethane at 120°C for 12 hours did not yield any products.

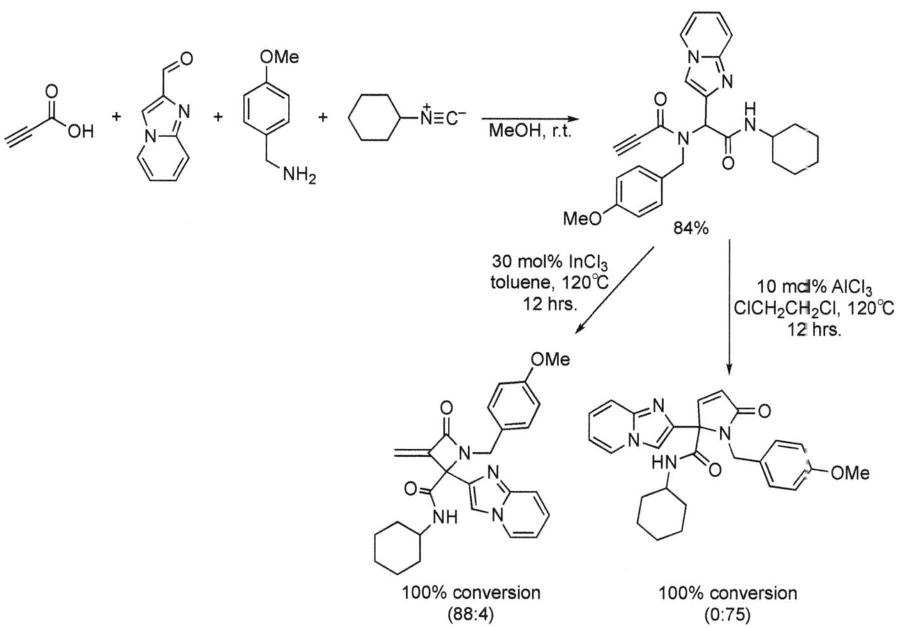

SCHEME 118 The impact of the post-Ugi treatment on the four-component reaction

Recently, it is reported that the Ugi four-component reaction involving 3-phenylpropiolic acid, benzaldehyde, 4-methoxyaniline, and isonitriles can form four different structural scaffolds depending on the reaction conditions. For example, when the four components are allowed to react in CH_3CN at room temperature for 12 hours, then treated with NaH at 0°C for 0.5 hour, (*E*)-3-benzylidene-*N*-(*tert*-butyl)-1-(4-methoxyphenyl)-4-oxo-2-phenylazetidine-2-carboxamide was obtained. For comparison, when the mixture was allowed to react in MeOH at room temperature for

12 hours, and the precipitate was then refluxed in MeOH in the presence of K_2CO_3, (*E*)-4-benzylidene-1-(*tert*-butyl)-3-((4-methoxyphenyl)amino)-3-phenylpyrrolidine-2,5-dione was obtained. Similarly, when the reaction mixture in MeOH was treated with iodine in the presence of sodium bicarbonate, *N*-(*tert*-butyl)-2-(3-iodo-2,8-dioxo-4-phenyl-1-azaspiro-[4.5]deca-3,6,9-trien-1-yl)-2-phenylacetamide was obtained. In contrast, when the mixture was allowed to react in dichloroethane at room temperature for 12 hours, and the precipitate was then treated with sulfuric acid for 2 hours, (5*R*,7a*R*,11a*R*)-1,5-diphenyl-7a,8-dihydro-3*H*-pyrrolo[1,2-*d*]quinoxaline-3,6,9(5*H*,7*H*)-trione was obtained. These four reaction conditions are displayed in Scheme 3.119 [416].

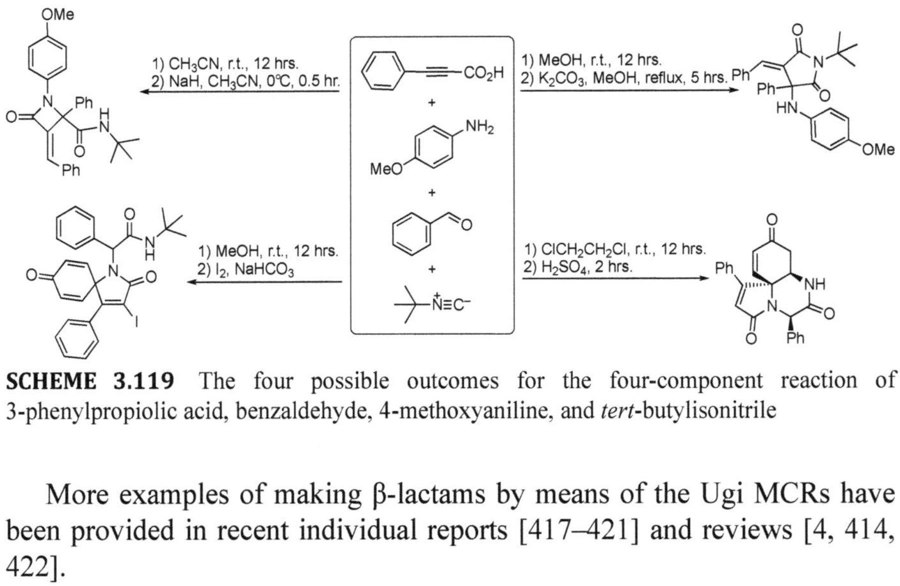

SCHEME 3.119 The four possible outcomes for the four-component reaction of 3-phenylpropiolic acid, benzaldehyde, 4-methoxyaniline, and *tert*-butylisonitrile

More examples of making β-lactams by means of the Ugi MCRs have been provided in recent individual reports [417–421] and reviews [4, 414, 422].

3.3.10 RING CONTRACTIONS

Ring contraction refers to a set of reactions that involve the conversion of larger rings to smaller, more strained rings. The scopes of ring contractions to form β-lactams can be demonstrated in the following reactions.

One of the early practices on ring contraction to form β-lactams is the consecutive treatment of 1-benzoyl-5,5-dimethylpyrazolidin-3-one in DME (glyme) by deprotonation with base, oxidation with 3.0 equivalents of HgO, and addition of 1.0 equivalent of *O*-(mesitylsulfonyl)-hydroxylamine in CH_2Cl_2 at room temperature, yielding 72% of

1-benzoyl-4,4-dimethylazetidin-2-one. Similarly, treatment of (3a*R*,7a*S*)-1-benzoyl-3a-methyloctahydro-3*H*-indazol-3-one under similar condition, about 30–50% of (1*R*,6*S*)-7-benzoyl-1-methyl-7-azabicyclo[4.2.0]octan-8-one was isolated (Scheme 3.120) [423]. However, in the absence of HgO, direct photolysis of 1-acetyl-5,5-dimethylpyrazolidin-3-one in MeOH gave 60% of *N*-(2,2-dimethyl-4-oxoazetidin-1-yl)acetamide (Scheme 3.120). Several other analogs undergo similar ring contractions to afford *N*-acylamino-4,4-dimethylazetidin-2-ones in various yields [424]. Likewise, the spiropyrazolidin-3-ones and bicyclic pyrazolidin-3-one have been found to undergo similar reactions under photochemical irradiation [425]. In this case, the remaining amino group upon deprotection of the acyl group can be removed by means of nitrosative deamination with *N*,*N*-diphenylnitrous amide in refluxing benzene, yielding *N*-unsubstituted β-lactams [426].

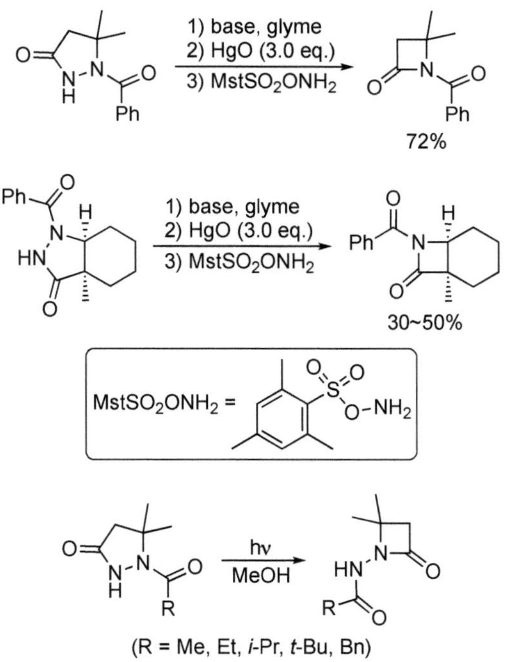

SCHEME 3.120 The oxidative contraction of 1-acylpyrazolidin-3-ones into β-lactams

Photochemical ring contraction of 3-diazopyrrolidin-2,4-diones is another early approach to form β-lactams, such as 2-azetidinone-3-carboxylic acids, although it faces difficulty in achieving steric control. A typical reaction is photolysis of 1.0 mmol of 3-diazo-1,5,5-trimethylpyrrolidin-2,4-dione in MeOH at 0°C through a Pyrex filter for 2.5 hours using a Hanovia 450 W medium pressure mercury arc. Subsequent filtration of solution through glass wool and evaporation afforded 93% of methyl 1,2,2-trimethyl-4-oxoazetidine-3-carboxylate (Scheme 3.121). A more general method is to photolyze 0.5 to 2.0 mmol of 3-diazopyrrolidin-2,4-dione in 100 mL of ether in the presence of a ketene trapping agent, such as 5 mL of MeOH or CF_3CH_2OH, 1.1 to 1.2 equivalent of hydrazine, or ether saturated with water or ammonia [427]. Extension of this approach to the photochemical decomposition of 4-diazopyrazolidin-3,5-diones in the presence of nucleophiles (e.g., EtOH, *t*-BuOH, Et_2NH, H_2O) afforded the *aza*-β-lactams in 4%–56% yield. However, the substituent on the nitrogen atom appears to play a certain role in controlling the reaction outcome. For example, photolysis of 1,2-dibenzyl-4-diazopyrazolidin-3,5-dione in the ether in the presence of EtOH resulted in 45% of ethyl 1,2-dibenzyl-4-oxo-1,2-diazetidine-3-carboxylate, whereas the photolysis of 4-diazo-1,2-diphenylpyrazolidine-3,5-dione under this condition did not afford the expected *aza*-β-lactam, but rather 39% of ethyl (*E*)-3-oxo-3-(phenylamino)-2-(phenylimino)propanoate and 2% of azobenzene. The corresponding 1,2-dibenzyl-4-oxo-1,2-diazetidine-3-carboxylic acid underwent quantitative decarboxylation in refluxing benzene to give 1,2-dibenzyl-1,2-diazetidin-3-one, which when reduced with diborane/THF, N^1,N^2-dibenzylethane-1,2-diamine was obtained. In comparison, when 1,2-dibenzyl-1,2-diazetidin-3-one was treated with LDA at -78°C followed by methyl iodide, 1-benzyl-2-phenylimidazolidin-4-one was obtained (Scheme 3.121) [428]. For the case of 3-diazopyrrolidin-2,4-diones, the reluctance of N-C bonds to migrate in the photochemical Wolff rearrangement is generally attributed to the participation of the nitrogen lone pair in amide resonance, especially in photo-excited states. However, the relative migratory aptitude of the *N*-substituents in the Wolff rearrangement can be explored in the case of asymmetrically substituted 4-diazopyrazolidin-3,5-diones, affording different *aza*-β-lactams. Experimental details reveal that although *N*-methyl, *N*-benzyl, and *N*-benzhydryl groups have similar migratory aptitudes leading to a 1:1 mixture of both possible *aza*-β-lactams from 1-benzyl-4-diazo-2-methylpyrazolidin-3,5-dione and 1-benzhydryl-4-diazo-2-methylpyrazolidin-3,5-dione, an *N*-phenyl group shows a slight preference to migrate over *N*-benzyl in the photolysis of 1-benzyl-4-diazo-2-phenylpyrazolidin-3,5-dione. Greater regioselectivity was observed in the photolysis of ethyl 2-(2-benzyl-4-diazo-3,5-dioxopyrazolidin-1-yl)acetate, where the *N*-benzyl group migrated more readily than the *N*-alkyl group bearing an electron-withdrawing ester substituent. This reluctance of an *N*-alkyl group bearing an ester to migrate is more marked in the ring contraction of *tert*-butyl 2-diazo-1,3-dioxohexahydro-1*H*-pyrazolo[1,2-*a*]pyridazine-5-carboxylate, affording highly regio- and stereoselective product of 2-(*tert*-butyl) 7-ethyl 8-oxo-1,6-diazabicyclo[4.2.0]octane-2,7-dicarboxylate or 2-(*tert*-butoxycarbonyl)-8-oxo-1,6-diazabicyclo[4.2.0]octane-7-carboxylic acid when photolyzed in the presence of EtOH or water, yet with low yields. These results exhibit a relative migratory aptitude of the nitrogen groups in the following order: *N*-Ph > *N*-$CHPh_2$ ~ *N*-CH_2Ph > *N*-Me > *N*-CH_2CO_2R [429].

Different from the photolytic ring contraction of 3-diazopyrrolidin-2,4-diones as displayed in the first part of Scheme 3.121, the photolytic ring contraction of 4-diazopyrrolidin-2,3-diones yields 3-carboxy-2-azetidinones. For example, photo-irradiation of 1-cyclohexyl-4-diazo-5-methoxy-5-(phenylethynyl)pyrrolidin-2,3-dione in anhydrous MeOH at -78°C with light greater than 300 nm (Hanovia 450-W lamp, Pyrex filter) under argon atmosphere for 1 hour led to more than 80% of methyl 1-cyclohexyl-2-methoxy-4-oxo-2-(phenylethynyl)azetidine-3-carboxylate, with 74% of isolated yield with a diastereomeric ratio of 5:1. Similarly, photolysis of 5-(3-(benzyloxy)prop-1-yn-1-yl)-1-cyclohexyl-4-diazo-5-methoxypyrrolidin-2,3-dione and 1-cyclohexyl-4-diazo-5-methoxy-5-(5-((tetrahydro-2*H*-pyran-2-yl)-oxy)pent-1-yn-1-yl)pyrrolidin-2,3-dione under the same condition afforded 57% and 63% of methyl 2-(3-(benzyloxy)prop-1-yn-1-yl)-1-cyclohexyl-2-methoxy-4-oxoazetidine-3-carboxylate and methyl 1-cyclohexyl-2-methoxy-4-oxo-2-(5-((tetrahydro-2*H*-pyran-2-yl)oxy)pent-1-yn-1-yl)azetidine-3-carboxylate, respectively, with the corresponding diastereomeric ratio of 6:1 and > 10:1 as shown in Scheme 3.122 [430]. In addition, this approach appears to be more advantageous in the synthesis of strained bicyclic β-lactams than other analogies, plus the easy synthesis of 4-diazopyrrolidin-2,3-diones from readily available precursors. For example, photolysis of methyl (3*S*,7a*R*)-7-diazo-2,2-dimethyl-5,6-dioxohexahydropyrrolo[2,1-*b*]thiazole-3-carboxylate in THF at -78°C in the presence of 1.0 equivalent of diisopropylamine yielded 72% of methyl (2*S*,5*R*,6*S*)-6-(diisopropylcarbamoyl)-3,3-dimethyl-7-oxo-4-thia-1-azabicyclo[3.2.0]heptane-2-carboxylate [430].

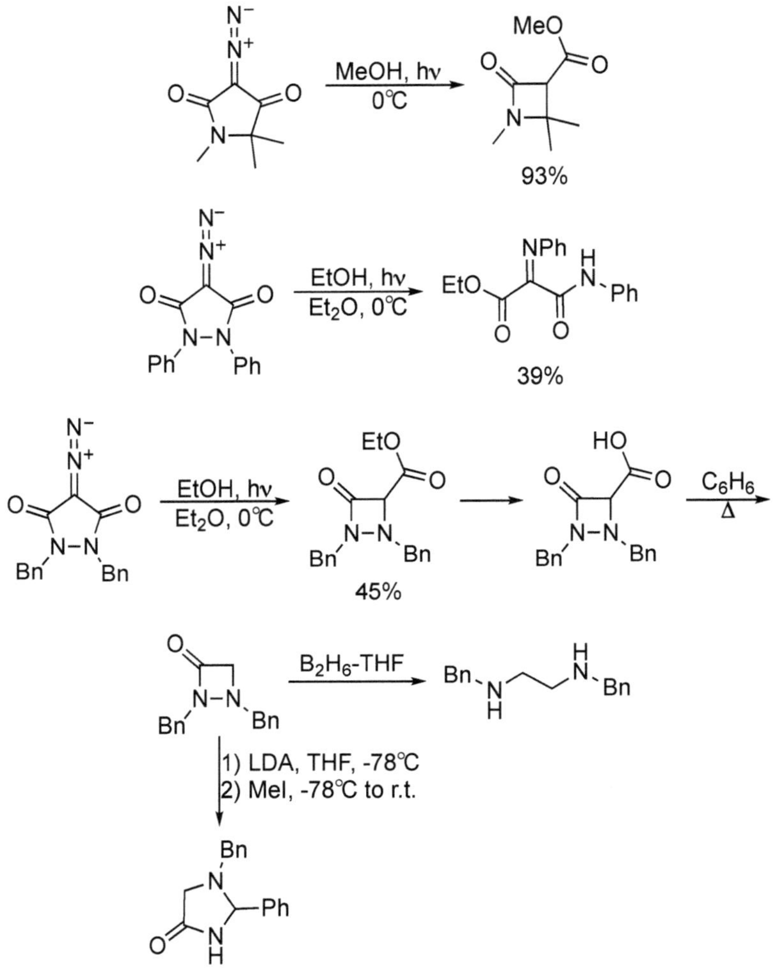

SCHEME 3.121 The photochemical reaction of 3-diazo-1,5,5-trimethylpyrrolidin-2,4-dione and 4-diazo-1,2-disubstituted pyrazolidin-3,5-diones

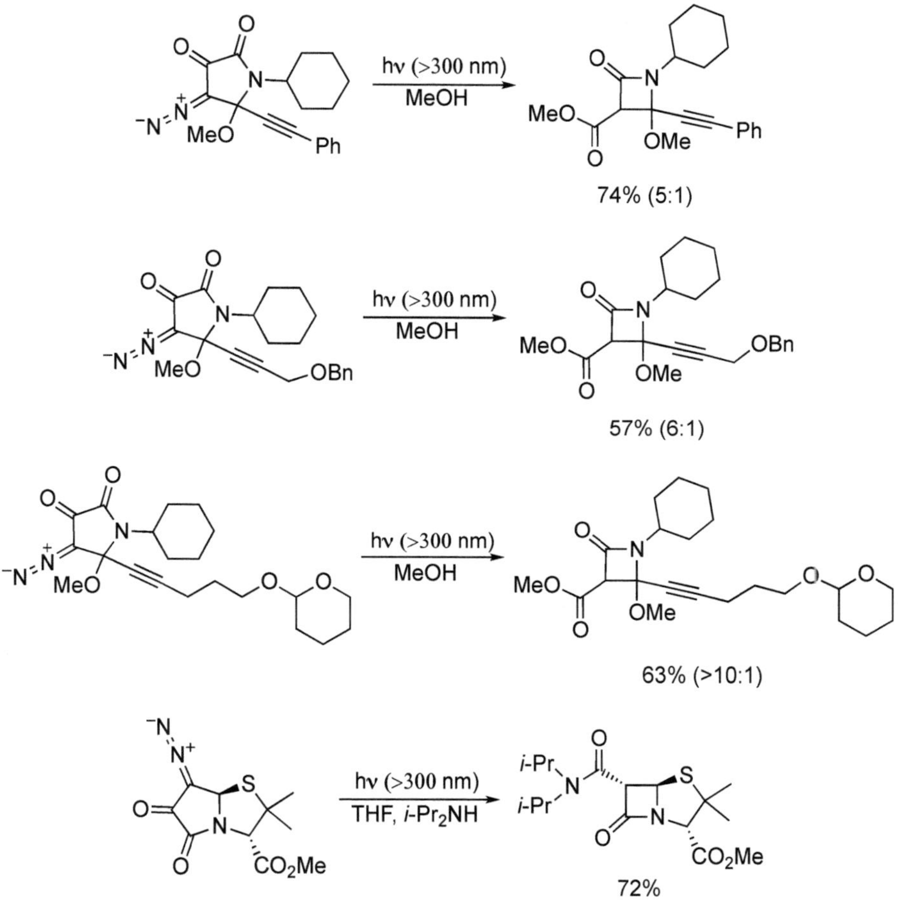

SCHEME 3.122 The photochemcial contraction of 4-diazopyrrolidin-2,3-diones to β-lactams

Another photochemical ring contraction to afford β-lactams has been demonstrated in the photochemical irradiation of 2,2-dimethyl-4-oxo-5-phenyl-3,4-dihydro-2*H*-pyrrole 1-oxide in benzene to yield both 1-benzoyl-4,4-dimethylazetidin-2-one and 4,4-dimethyl-2-phenyl-4,5-dihydro-6*H*-1,3-oxazin-6-one, in a ratio of 58% versus 42%, whereas the actual β-lactam was isolated in 42% yield. Similarly, treatment of 2,2,5-trimethyl-4-oxo-3,4-dihydro-2*H*-pyrrole 1-oxide under this condition afforded 50% of 1-acetyl-4,4-dimethylazetidin-2-one, as well as 2,4,4-trimethyl-4,5-dihydro-6*H*-1,3-oxazin-6-one, in a ratio of 64:36 by NMR analysis. However, photochemical irradiation of 5-(*tert*-butyl)-2,2-dimethyl-4-oxo-3,4-dihydro-2*H*-pyrrole 1-oxide in benzene led to a single product of 2-(*tert*-butyl)-4,4-dimethyl-4,5-dihydro-6*H*-1,3-oxazin-6-one, with an isolated yield of 47%. The formation of these two types of products has been rationalized in Scheme 3.123 [431].

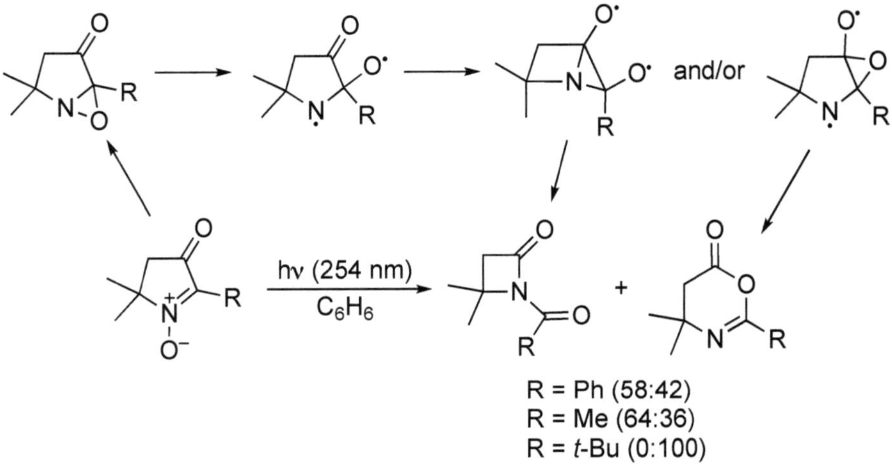

SCHEME 3.123 The mechanism for the photochemical reaction of 2,2-dimethyl-4-oxo-5-substituted-3,4-dihydro-2*H*-pyrrole 1-oxide

An additional photochemical ring contraction to yield β-lactams is the photo-irradiation of the derivatives of 5-(aminomethyl)-2,2,6-trimethyl-4*H*-1,3-dioxin-4-one in acetonitrile at 254 nm. For example, irradiation of 5-((benzylamino)methyl-2,2,6-trimethyl-4*H*-1,3-dioxin-4-one afforded 40% of (*S*)-3-acetyl-1-benzylazetidin-2-one. In contrast, when this starting material was refluxed in xylene, it was recovered unchanged; when this compound was heated in *o*-dichlorobenzene at 170°C, the starting material disappeared rapidly but no expected β-lactam was obtained. This reaction clearly demonstrates the applicability of photo-irradiation rather than thermal cyclo-reversion. Similarly, 5-(1-(benzylamino)ethyl)-2,2,6-trimethyl-4*H*-1,3-dioxin-4-one was converted into (3*S*,4*R*)-3-acetyl-1-benzyl-4-methylazetidin-2-one, and 3-(benzylamino)-3-(2,2,6-trimethyl-4-oxo-4*H*-1,3-dioxin-5-yl)propyl acetate was transformed into 40% of 2-((2*R*,3*S*)-3-acetyl-1-benzyl-4-oxoazetidin-2-yl)ethyl acetate. Even the presence of free amino group at the α-position of 2,2,6-trimethyl-4*H*-1,3-dioxin-4-one would not affect the progress of this photochemical ring contraction, as illustration in the conversion of 3-amino-3-(2,2,6-trimethyl-4-oxo-4*H*-1,3-dioxin-5-yl)propyl benzoate into 40% of 2-((2*R*,3*S*)-3-acetyl-4-oxoazetidin-2-yl)ethyl benzoate (Scheme

3.124). Stereochemistry favors the *trans*-relationship for the two side chains on the β-lactams [432].

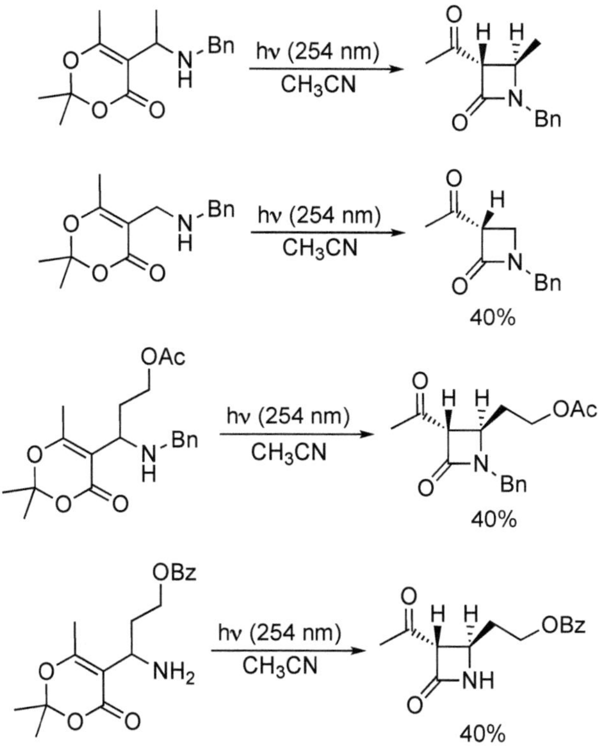

SCHEME 3.124 The photochemical contraction of 5-(aminomethyl)-2,2,6-trimethyl-4*H*-1,3-dioxin-4-ones to yield β-lactams

Another photochemical ring contraction that leads to the formation of β-lactams involves the photo-irradiation of 2,3,6-substituted pyrimidin-4(3*H*)-ones, as shown in Scheme 3.125 [433]. From a simple pyrimidin-4(3*H*)-one, such as 2,3,6-trimethylpyrimidin-4(3*H*)-one, a mixture of (*E*)-3-(1-aminoethylidene)-4-methoxy-1,4-dimethylazetidin-2-one and (*Z*)-3-(1-aminoethylidene)-4-methoxy-1,4-dimethylazetidin-2-one was obtained in a ratio of 67/33 after photo-irradiation in methanol with a Nikko Seiki 200-W high-pressure mercury lamp equipped with a quartz tube immersed in the reaction cell at 30–35°C, where the progress of the reaction was monitored by TLC and the reaction was terminated when the spot of byproduct was clearly visible to avoid further photochemical decomposition of the products. After being cooled, the crude crystal of the product was filtered, and alcohol was removed under vacuum, then ether and a small amount of *n*-hexane were added to the mother liquid, and the resulting solution was stored in a freezer for 2 days to yield additional crude crystals. Under this condition, bicyclic pyrimidin-4(3*H*)-ones as represented by 2-methyl-6,7,8,9-tetrahydro-4*H*-pyrido[1,2-*a*]pyrimidin-4-one and 2-methyl-7,8,9,10-tetrahydropyrimido[1,2-*a*]azepin-4(6*H*)-one have been converted into 7-(1-aminoethylidene)-6-methoxy-1-azabicyclo[4.2.0]octan-8-ones (*E*/*Z* = 59/41) and 8-(1-aminoethylidene)-7-methoxy-1-azabicyclo[5.2.0]nonan-9-ones (*E*/*Z* = 66/34) when irradiated in methanol, respectively. Similar results were obtained when these compounds were irradiated in ethanol [433].

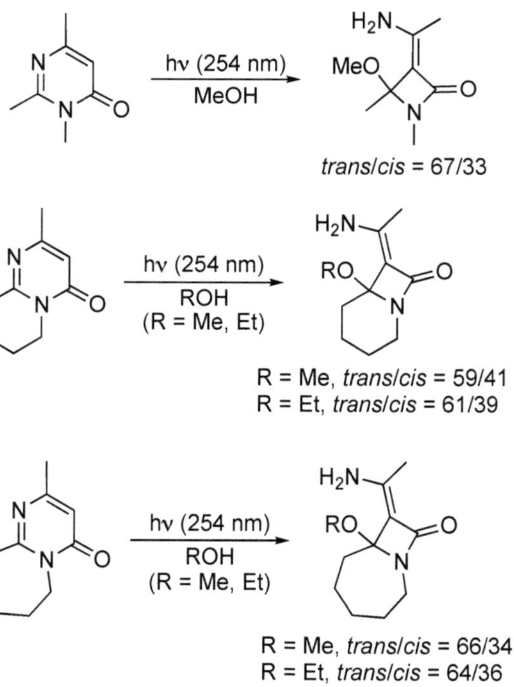

SCHEME 3.125 The photochemical contraction of 2,3-disubstituted 6-methylpyrimidin-4(3*H*)-ones into β-lactams

One more photochemical ring contraction to yield β-lactams is the photochemical extrusion of sulfur dioxide from 1,1-dioxo-4-thiazolidinones. For example, photolysis of 211 mg of (2*S*,5*R*)-3,5-dimethyl-2-phenylthiazolidin-4-one

1,1-dioxide in 175 mL of a 6:1 mixture of *tert*-butyl alcohol/acetonitrile purged with argon was irradiated with a 450-W medium-pressure Hanovia lamp through a Vycor filter for 40 minutes. Upon removal of the solvent under reduced pressure, the oily residue was taken up in CH_2Cl_2 and washed with water. After removal of solvent, NMR analysis of the residue with CH_2Cl_2 as an internal standard indicated 35% of (3*S*,4*S*)-1,3-dimethyl-4-phenylazetidin-2-one, and 9% of (3*S*,4*R*)-1,3-dimethyl-4-phenylazetidin-2-one, as well as 28% of the starting material. The actual yields of the *cis*- and *trans*-β-lactams based on the consumed starting material were 48% and 13%, respectively (Scheme 3.126). Similarly, photolysis of (2*S*,5*S*)-3,5-dimethyl-2-phenylthiazolidin-4-one 1,1-dioxide under the same condition yielded 7% and 14% of *cis*- and *trans*-β-lactams, respectively. The photochemical extrusion of sulfur dioxide is generally believed to proceed via radical intermediates, where bond formation by a possible diradical intermediate would have to be faster than bond rotation (with inversion) since the *cis*-1,1-dioxo-4-thiazolidinone gave mostly *cis*-β-lactam. It is possible but less likely, that bond formation is concerted with the extrusion of sulfur dioxide. In contrast, the thermolysis of 1,1-dioxo-4-thiazolidinone yielded products of different ratios. For example, a 25-mL round-bottomed flask containing 24 mg of (2*S*,5*S*)-3,5-dimethyl-2-phenylthiazolidin-4-one 1,1-dioxide was immersed in an oil bath maintained at 210°C for 12 minutes. The flask was allowed to cool, and NMR analysis of the residue indicated a 42% yield of (3*S*,4*R*)-1,3-dimethyl-4-phenylazetidin-2-one, 19% of recovered starting material, and 9% of isomerized sulfone, i.e., (2*S*,5*R*)-3,5-dimethyl-2-phenylthiazolidin-4-one 1,1-dioxide. For comparison, thermolysis of (2*S*,5*R*)-3,5-dimethyl-2-phenylthiazolidin-4-one 1,1-dioxide in an oil bath at 200°C for 5 minutes afforded 83% of (3*S*,4*R*)-1,3-dimethyl-4-phenylazetidin-2-one, with no starting material left [434].

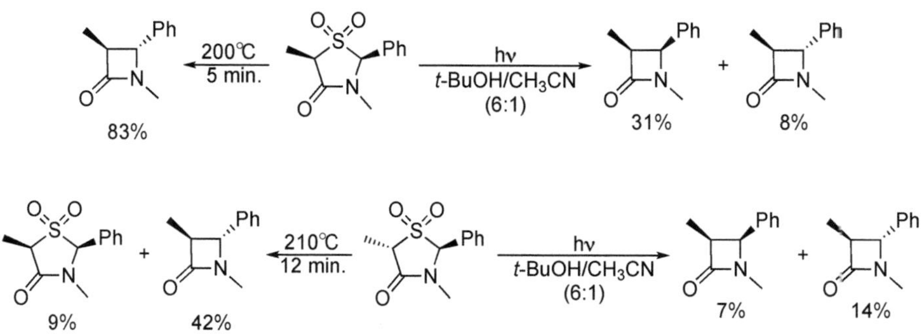

SCHEME 3.126 The extrusion of SO_2 from 3,5-dimethyl-2-phenylthiazolidin-4-one 1,1-dioxides to afford β-lactams

In addition, 2-phenyl-1,2-oxazinane-3,6-dione undergoes ring contraction to form β-lactam 1-phenylazetidin-2-one under either thermolysis or photolysis condition. When this compound was thermolyzed at 190°C, a complex mixture was obtained; however, 1-phenylazetidin-2-one was isolated as the only discrete product, in 16% yield. In comparison, photolysis of this compound also afforded 1-phenylazetidin-2-one as the only identifiable product in a slightly higher yield (20%). This compound is assumed to undergo homolytic bond cleavage to a form biradical intermediate, which immediately decarboxylates to give a new biradical, and intramolecular combination of radicals gives 1-phenylazetidin-2-one, as shown in Scheme 3.127. The deuterium labeling experiment indicates that the bond rotation is faster than the formation of the C-N bond. For comparison, the ring contraction of 2-(1-phenylethyl)-1,2-oxazinane-3,6-dione to form β-lactam 1-(1-phenylethyl)-azetidin-2-one only takes place under photochemical condition, possibly due to the active α-H on 1-phenylethyl group which promotes hydrogen abstraction from carboxyl radical to give (*E*)-4-oxo-4-((1-phenylethylidene)amino)butanoic acid [435]. More photochemical ring contractions to give β-lactams have been summarized in a short review [436].

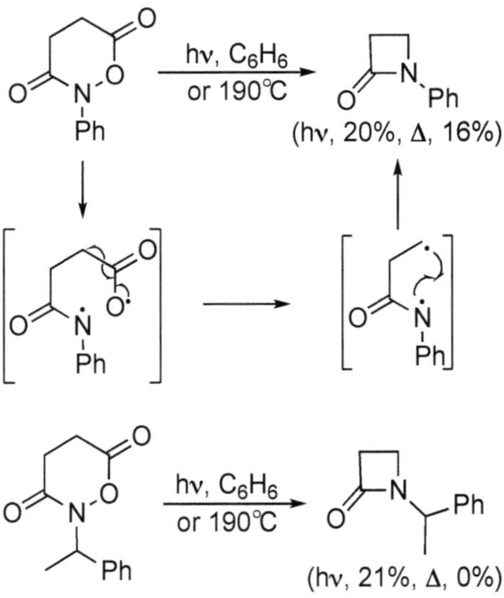

SCHEME 3.127 The mechanism for the photochemical contraction of 2-substituted-1,2-oxazinane-3,6-diones

Oxidative ring contraction of α-keto-γ-lactams with sodium periodate ($NaIO_4$) at room temperature and neutral pH is another approach to form β-lactams containing either mono- or difunctionality at the α-carbon, which is compatible with a variety of substituents [437]. Before the oxidation, a 0.2 *M* phosphate buffer was prepared by titrating phosphoric acid with aqueous LiOH, and the α-keto-γ-lactam was dissolved in the buffer at 0.02 *M*, then $NaIO_4$ was added and pH was adjusted to the desired value. After the oxidation, IO_4^- and IO_3^- were destroyed by the addition of an approximately stoichiometric amount of $NaHSO_3$ dissolved in a minimum of water while keeping the pH near 7 by the addition of 2 *M* NaOH. Following extraction of the neutral solution with CH_2Cl_2, the pH was adjusted to 2.0 with H_3PO_4, and the solution was continuously extracted with CH_2Cl_2. For the case of hexahydroindolizine-2,3-dione, 70% of (6*R*,7*R*)-8-oxo-1-azabicyclo-[4.2.0] octane-7-carboxylic acid was obtained. Similarly, 2-hydroxy-1-methyl-6,7,8,8a-tetrahydroindolizin-3(5*H*)-one was oxidized to 50% of (6*R*,7*R*)-7-methyl-8-oxo-1-azabicyclo[4.2.0]octane-7-carboxylic acid without other stereoisomers (Scheme 3.128) [437]. Additional study reveals the mechanism of this oxidative ring contraction [438, 439].

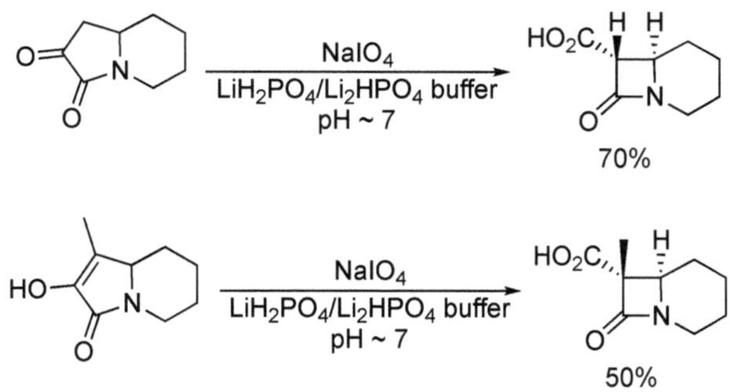

SCHEME 3.128 The oxidative contraction of hexahydroindolizine-2,3-diones to yield β-lactams

Raney nickel desulfurization of meso-ionic thiazol-4-ones provides another stereospecific formation of β-lactams, as illustrated in the treatment of 2,3,5-triphenylthiazol-3-ium-4-olate hydroxide with Raney nickel in MeOH, THF or acetone at room temperature to afford 85% of (3*S*,4*S*)-1,3,4-triphenylazetidin-2-one (Scheme 3.129). While several possible mechanisms have been proposed for this transformation, the most plausible one involves the conservation of dipolar character through the desulfurization and intermediacy of the dipole [440]. Another sulfur-containing cyclic system also undergoes ring-contraction to form β-lactam by means of Raney nickel desulfurization, as shown in Scheme 3.130. For example, reduction of *N*,2-dibutyl-3-oxo-4,5-diphenylisothiazolidine-5-carboxamide with Raney nickel in refluxing ethanol yielded 12% of *N*,1-dibutyl-4-oxo-2,3-diphenylazetidine-2-carboxamide, along with the major product of *N*¹,*N*⁴-dibutyl-2,3-diphenylsuccinamide. Similar treatment of 3-oxo-4,5-diphenyl-*N*,2-dipropylisothiazolidine-5-carboxamide afforded 16% of 4-oxo-2,3-diphenyl-*N*,1-dipropylazetidine-2-carboxamide and the major product of 2,3-diphenyl-*N*¹,*N*⁴-dipropylsuccinamide. In contrast, treatment of 3-oxo-*N*,2,4,5-tetraphenylisothiazolidine-5-carboxamide and *N*,2-di-*tert*-butyl-3-oxo-4,5-diphenylisothiazolidine-5-carboxamide with Raney nickel in refluxing ethanol led to the formation of only *N*¹,*N*⁴,2,3-tetraphenylsuccinamide and *N*¹,*N*⁴-di-*tert*-butyl-2,3-diphenylsuccinamide, possibly due to the steric hindrance during radical combination [441].

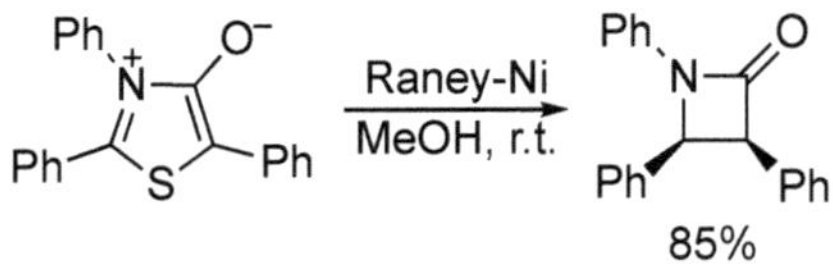

SCHEME 3.129 Raney nickel promoted desulfurization of 2,3,5-triphenylthiazol-3-ium-4-olate to afford β-lactam

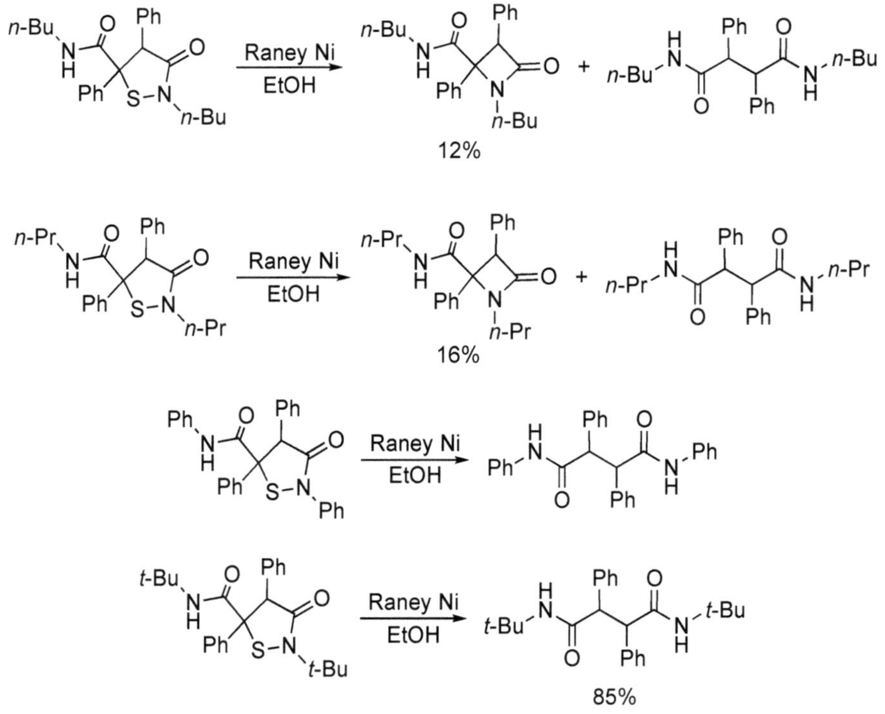

SCHEME 3.130 The desulfurization of 3-oxo-4,5-diphenylisothiazolidine-5-carboxamides

Besides these examples of ring contractions to form the β-lactam scaffold, the most commonly used method in this category is the formation of 4,6,7-trisubstituted-5-oxa-6-azaspiro[2.4]heptanes through 1,3-dipolar cycloaddition of nitrones and alkylidenecyclopropanes or arylidenecyclopropanes and their subsequent decomposition accompanied with extrusion of ethylene under acidic condition. It is reported that the 1,3-dipolar cycloaddition of (*Z*)-*N*-methyl-1-phenylmethanimine oxide and other nitrone derivatives with methyl (*E*)-2-benzylidenecyclopropane-1-carboxylate in toluene at 110°C proceeds in a highly regioselective fashion to give methyl 5-oxa-6-azaspiro[2.4]heptane-1-carboxylates as mixtures of two diastereomers, such as methyl (1*R*,3*S*,4*R*)-6-methyl-4,7-diphenyl-5-oxa-6-azaspiro[2.4]heptane-1-carboxylate and methyl (1*S*,3*S*,4*R*)-6-methyl-4,7-diphenyl-5-oxa-6-azaspiro[2.4]heptane-1-carboxylate, as shown in Scheme 3.131 [442]. Similarly, *N*-aryl *C*-(arylcarbamoyl) nitrones regioselectively add to methyl 2-(2-phenylcyclopropylidene)acetate and methyl 2-methylidene-3-phenylcyclopropanecarboxylate to give in each case two diastereoisomeric 5-oxa-6-azaspiro[2.4]heptane-4-carboxylates [443]. In addition to the intermolecular 1,3-dipolar cycloaddition, intramolecular 1,3-dipolar cycloaddition of nitrone and alkylidenecyclopropane moieties also occur, as represented in the preparation of (3a*S*,5*S*,6a*R*)-1-methylhexahydrospiro[cyclopenta[*c*]isoxazole-3,1'-cyclopropan]-5-yl acetate and (3a*S*,5*R*,6a*R*)-1-methylhexahydrospiro[cyclopenta[*c*]isoxazole-3,1'-cyclopropan]-5-yl acetate by treatment of 1-cyclopropylidene-5-oxopentan-3-yl acetate in Et_2O at 0°C with *N*-methylhydroxylamine hydrochloride and 1.0 equivalent of pyridine (Scheme 3.132). The total yield of the two diastereomers was 64%, with a diastereomeric ratio of 1:1. The subsequent decomposition of this compound takes two different approaches, depending on the reaction conditions. For example, (3a*S*,5*S*,6a*R*)-1-methylhexahydrospiro[cyclopenta[*c*]isoxazole-3,1'-cyclopropan]-5-yl acetate once heated in xylenes under refluxing for 6 hours, (4a*S*,6*S*,7a*R*)-1-methyl-4-oxooctahydro-1*H*-cyclopenta[*b*]pyridin-6-yl acetate was obtained in 75% via ring expansion. On the other hand, when the same compound was refluxed in toluene in the presence of 2.0 equivalents of trifluoroacetic acid (TFA) for 1 hour, it underwent ring contraction to provide (1*S*,3*S*,5*R*)-6-methyl-7-oxo-6-azabicyclo[3.2.0]heptan-3-yl acetate in 76% of yield, with the elimination of one equivalent of ethylene [444]. Both approaches are believed to involve radical intermediates [445], as illustrated in detail in Scheme 3.132.

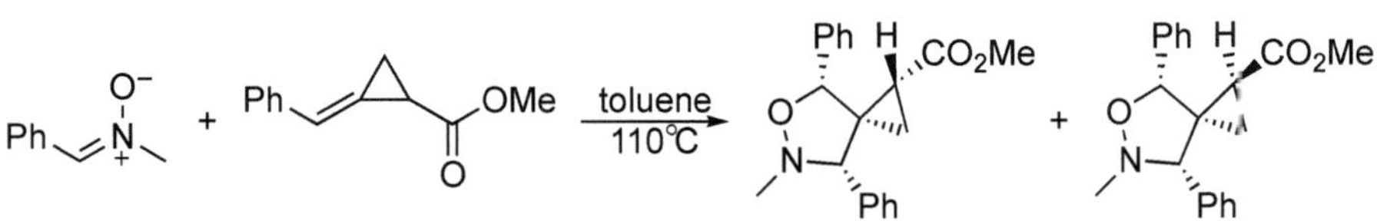

SCHEME 3.131 The 1,3-dipolar cycloaddition between (*Z*)-*N*-methyl-1-phenylmethanimine oxide and methyl (*E*)-2-benzylidenecyclopropane-1-carboxylate

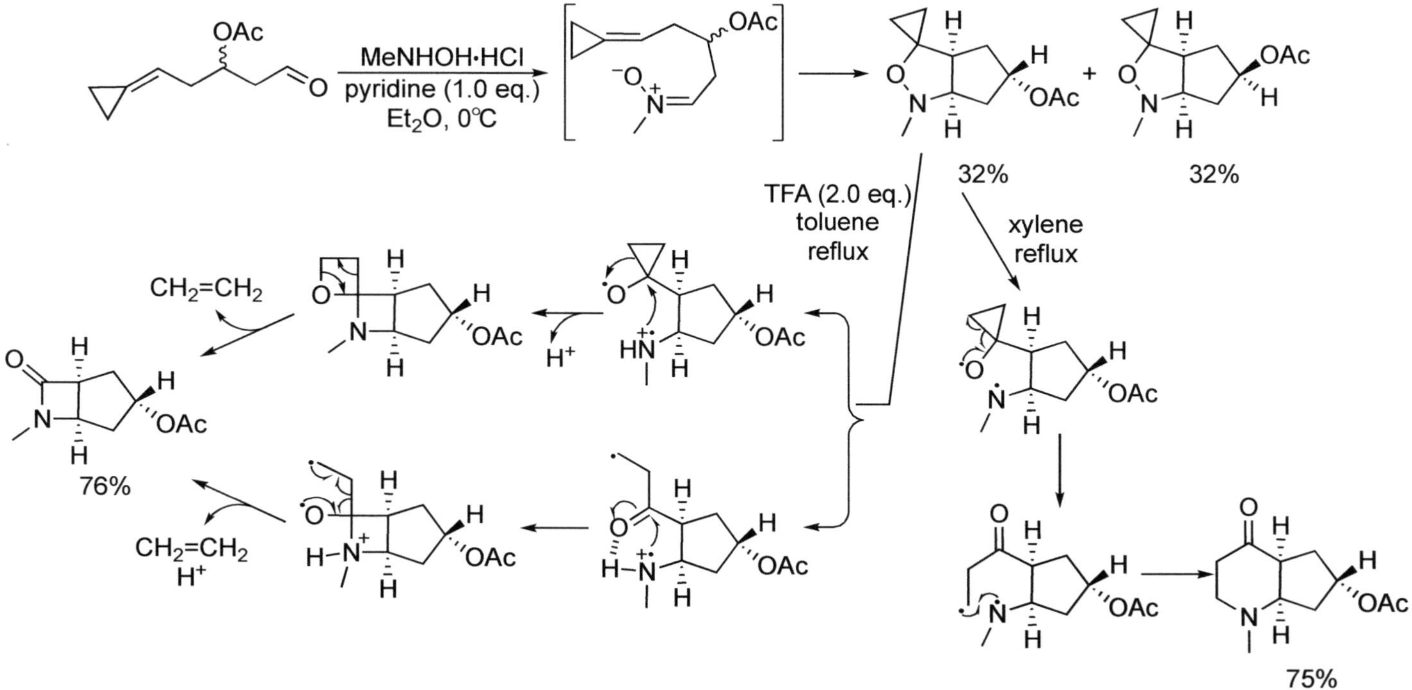

SCHEME 3.132 The mechanism for the intramolecular 1,3-dipolar cycloaddition of (*Z*)-3-acetoxy-5-cyclopropylidene-*N*-methylpentan-1-imine oxide

Similarly, treatment of *N*-(2-cyclopropylideneethyl)-*N*-(2-formylphenyl)-4-methylbenzenesulfonamide with *N*-methylhydroxylamine hydrochloride in refluxing toluene in the presence of Et_3N led to the formation of (3a'*R*,9b'*R*)-1'-methyl-5'-tosyl-3a',4',5',9b'-tetrahydro-1'*H*-spiro[cyclopropane-1,3'-isoxazolo[4,3-*c*]quinoline]. This compound when refluxed in toluene in the presence of a little excess of TFA, underwent ring contraction to afford 72% of (2a*R*,8b*R*)-1-methyl-4-tosyl-2a,3,4,8b-tetrahydroazeto[3,2-*c*]quinolin-2(1*H*)-one (Scheme 3.133). However, simple thermolysis of the 1,3-dipolar cycloaddition product failed to give 10-methyl-5-tosyl-6,6a,8,9,10,10a-hexahydrophenanthridin-7(5*H*)-one, as it was stable up to 120°C and decomposed at higher temperatures. Still, it is unclear whether the 1,3-dipolar cycloadduct decomposed at this temperature. For comparison, *N*-(2-cyclopropylideneethyl)-*N*-(2-formylphenyl)-4-nitrobenzenesulfonamide has been successfully converted into (3a'*R*,9b'*R*)-1'-methyl-5'-((4-nitrophenyl)sulfonyl)-3a',4',5',9b'-tetrahydro-1'*H*-spiro[cyclopropane-1,3'-isoxazolo[4,3-*c*]quinoline], which when heated in toluene in the presence of TFA, was transformed into (2a*R*,8b*R*)-1-methyl-4-((4-nitrophenyl)sulfonyl)-2a,3,4,8b-tetra-hydroazeto[3,2-*c*]quinolin-2(1*H*)-one in a similar yield. It should be pointed out that when 2-phenylhexahydr ospiro[cyclopropane-1,2'-isoxazolo[2,3-*a*]pyridine] was treated with an acid at room temperature, 1-azabicyclo[4.2.0]octan-8-one and styrene, as well as (*E*)-4-phenyl-1-(piperidin-2-yl)but-3-en-2-one and 4-phenyloctahydro-2*H*-quinolizin-2-one were obtained. For another spirotricyclic system, i.e., (3a'*R*,6'*S*,6a'*R*)-1',6'-dimethyl-5'-tosylhexahydrospiro[cyclopropane-1,3'-pyrrolo[3,4-*c*]isoxazole], when it was heated in refluxing toluene in the presence of TFA, 60% of (1*R*,4*S*,5*R*)-4,6-dimethyl-3-tosyl-3,6-diazabicyclo[3.2.0]heptan-7-one was obtained. Likewise, treatment of (3a'*R*,6'*S*,6a'*R*)-6'-isopropyl-1'-methyl-5'-tosylhexahydrospiro[cyclopropane-1,3'-pyrrolo[3,4-*c*]-isoxazole and (3a'*R*,6'*S*,6a'*R*)-6'-((1*H*-indol-3-yl)methyl)-1'-methyl-5'-tosylhexahydrospiro[cyclopropane-1,3'-pyrrolo[3,4-*c*]isoxazole under the same reaction condition led to 63% of (1*R*,4*S*,5*R*)-4-isopropyl-6-methyl-3-tosyl-3,6-diazabicyclo[3.2.0]heptan-7-one and 57% of (1*R*,4*S*,5*R*)-4-((1*H*-indol-3-yl)methyl)-6-methyl-3-tosyl-3,6-diazabicyclo[3.2.0]heptan-7-one, respectively (Scheme 3.133) [445, 446].

This strategy has been applied to the total synthesis of Gelsemium alkaloid (±)-gelsemoxonine through a spirocyclopropane isoxazolidine ring contraction, as shown in Scheme 3.134 [447]. In this total synthesis, the initial Henry reaction has been performed between nitromethane and 2-(2-cyclopropylideneethoxy)acetaldehyde to yield 70% of

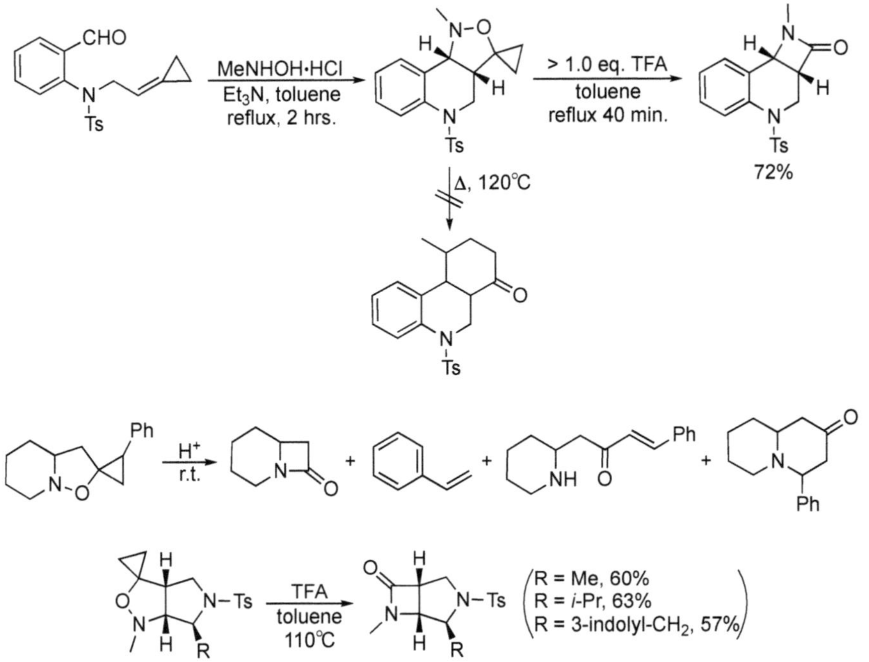

SCHEME 3.133 Acid promoted isomerization of (3a'*R*,9b'*R*)-1'-methyl-5'-tosyl-3a',4',5',9b'-tetrahydro-1'*H*-spiro[cyclopropane-1,3'-isoxazolo[4,3-*c*]quinoline], 2-phenylh exahydrospiro[cyclopropane-1,2'-isoxazolo[2,3-*a*]pyridine] and 6'-substituted-1'-methyl-5'-tosylhexahydrospiro[cyclopropane-1,3'-pyrrolo[3,4-*c*]isoxazole] into β-lactams

1-(2-cyclopropylideneethoxy)-3-nitropropan-2-ol, which was treated with di-*tert*-butyl dicarbonate and DMAP to undergo intramolecular 1,3-dipolar cycloaddition to afford 79% of 3a',4',6',7'-tetrahydrospiro[cyclopropane-1,3'-pyrano[4,3-*c*]isoxazole. Subsequent treatment with the Murry's reagent, i.e., dimethyldioxirane (DMDO), $InBr_3$ and ((1-ethoxyvinyl)oxy) trimethylsilane, 56% of ethyl 2-((3a'*R*,6'*S*,7'*R*)-7'-hydroxy-3a',4',6',7'-tetrahydrospiro[cyclopropane-1,3'-pyrano[4,3-*c*]isoxazol]-6'-yl) acetate was obtained. In order to achieve the correct stereochemistry during the introduction of the propynyl group, (3a'*R*,7'*R*)-3a',4',6',7'-tetrahydrospiro[cyclopropane-1,3'-pyrano[4,3-*c*]isoxazol]-7'-ol has been selected as the model compound to react with a variety of nucleophilic reagents, including but-1-en-2-yllithium, prop-1-en-2-yllithium, prop-1-yn-1-yllithium, vinyl magnesium bromide, triisopropoxy(prop-1-yn-1-yl) titanium, and prop-1-yn-1-ylcerium(III) chloride in the presence of different additive at -78°C [448]. The result indicates that prop-1-yn-1-ylcerium(III) chloride in the presence of $BF_3{\cdot}Et_2O$ is superior to other tested reagents or conditions. Therefore, ethyl 2-((3a'*R*,6'*S*,7'*R*)-7'-hydroxy-3a',4',6',7'-tetrahydrospiro[cyclopropane-1,3'-pyrano[4,3-*c*]isoxazol]-6'-yl)acetate was treated subsequently with 1-bromo-1-propene, *n*-BuLi, dry $CeCl_3$ and $BF_3{\cdot}Et_2O$ to afforded 78% of ethyl 2-((3a'*R*,6'*S*,7'*R*,7a'*R*)-7'-hydroxy-7a'-(prop-1-yn-1-yl)hexahydrospiro[cyclopropane-1,3'-pyrano[4,3-*c*] isoxazol]-6'-yl)acetate. The following treatment with TFA in CH_3CN led to the extrusion of ethylene and formation of β-lactam, i.e., ethyl 2-((1*R*,4*S*,5*R*,6*R*)-5-hydroxy-8-oxo-6-(prop-1-yn-1-yl)-3-oxa-7-azabicyclo[4.2.0]octan-4-yl)acetate, in 40–50% yield. Then, two important intermediates including *tert*-butyl(1*R*,4*S*,5*R*,6*R*,8*R*)-5-((*tert*-butoxycarbonyl) oxy)-4-(2-ethoxy-2-oxoethyl)-8-(hydroxymethyl)-6-(prop-1-yn-1-yl)-3-oxa-7-azabicyclo[4.2.0]octane-7-carboxylate, and *tert*-butyl (1*R*,3*S*,6*S*)-5-(2-(2-bromophenyl)-2-hydroxyacetyl)-10-((*tert*-butoxycarbonyl) oxy)-1-(prop-1-yn-1-yl)-7-oxa-2-azatricyclo[4.3.1.03,9]dec-4-ene-2-carboxylate, and the final product of (1'*S*,3'*S*,6'*S*)-10'-hydroxy-1-methoxy-1'-propionyl-7'-oxa-2'-azaspiro[indoline-3,5'-tricyclo[4.3.1.$0^{3,9}$]decan]-2-one, i.e., Gelsemoxonine, have been synthesized, as outlined in Scheme 3.134 [447].

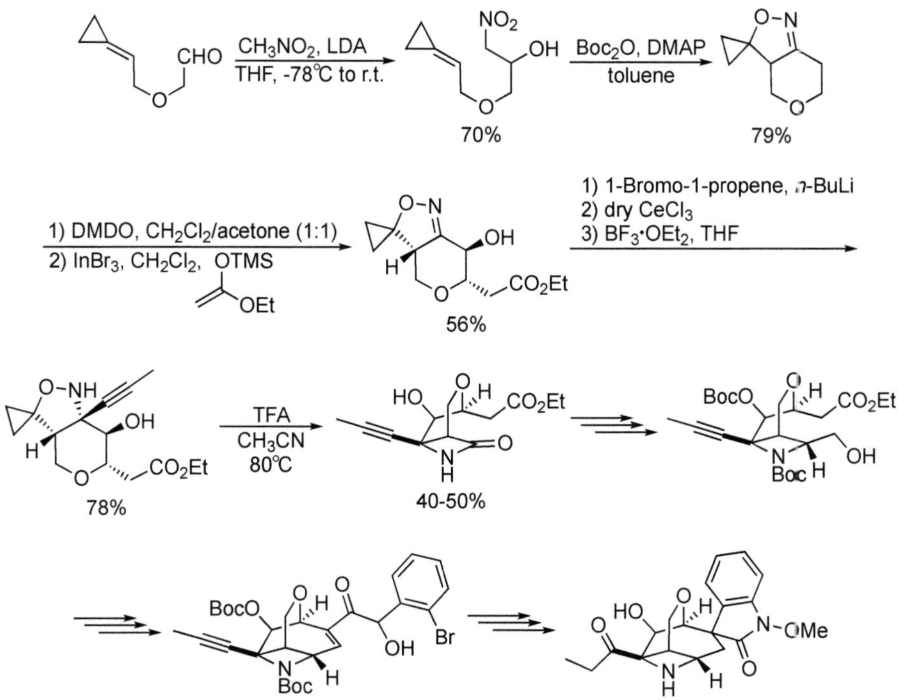

SCHEME 3.134 The key steps in the total synthesis of Gelsemoxonine

For this reaction, while possible mechanisms have been proposed as outlined in Scheme 3.132, further study in combination of experimental results and theoretical calculation (UB3LYP and UCCSD(T) level of theory) strongly suggests that the reaction involves a concerted rupture of the cyclopropane ring with retention of configuration and the labile N–O bond of the substrate. Experimentally, treatment of ((1*S*,2*S*)-5,6-dimethyl-4-oxa-5-azaspiro[2.4]heptane-1,2-diyl)bis(methylene) diacetate in CH_3CN with deuterated trifluoroacetic acid (CF_3CO_2D) at 80°C led to 77% of 1,4-dimethylazetidin-2-one, and (*E*)-but-2-ene-1,4-diyl diacetate in exclusive *trans*-configuration. Similarly, thermolysis of ((2'*R*,3*r*,3a*S*,3'*S*,7a*R*)-1-methylhexahydro-1*H*-spiro[benzo[*c*]isoxazole-3,1'-cyclopropane]-2',3'-diyl) bis(methylene) diacetate under this condition yielded 95% of (1*S*,6*R*)-7-methyl-7-azabicyclo[4.2.0]octan-8-one and exclusively (*Z*)-but-2-ene-1,4-diyl diacetate, as displayed in Scheme 3.135. These experimental results strongly suggest a concerted mechanism. Even though a stepwise mechanism cannot be ruled out, the involved intermediates would not undergo C–C bond rotation over the duration to form the product. Computation indicates the fragmentation of the cyclopropyl ring after protonation, leading to the extrusion of ethylene and the formation of carbonyl cation [449].

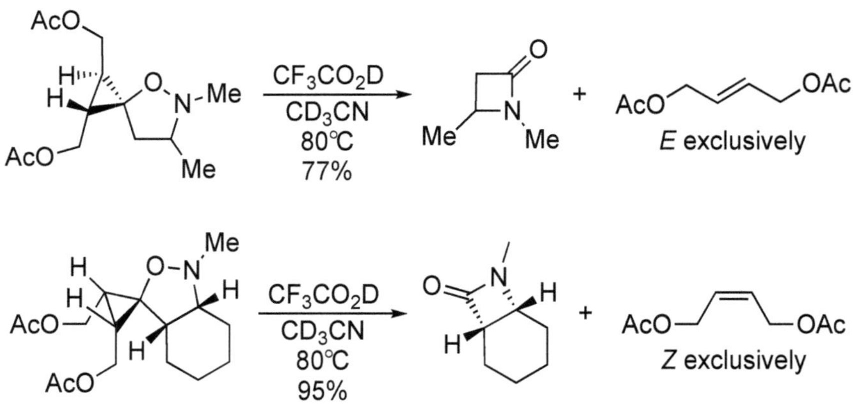

SCHEME 3.135 Experimental evidence supporting the concerted rupture of the cyclopropane ring during the formation of β-lactam

Additional experimental evidences for the retention of stereo-configuration have been demonstrated in the thermal treatment of methyl (6*R*,7*R*)-5-methyl-6-phenyl-4-oxa-5-azaspiro[2.4]heptane-7-carboxylate and methyl (6*R*,7*S*)-5-methyl-6-phenyl-4-oxa-5-azaspiro[2.4]heptane-7-carboxylate with *p*-toluenesulfonic acid at 80°C, affording 30% and

29% of methyl (3*R*,4*R*)-1-methyl-2-oxo-4-phenylazetidine-3-carboxylate and methyl (3*S*,4*R*)-1-methyl-2-oxo-4-phenylazetidine-3-carboxylate, respectively (Scheme 3.136). These two diastereomers have been prepared from 1,3-dipolar cycloaddition of (*Z*)-*N*-methyl-1-phenylmethanimine oxide and methyl 2-cyclopropylideneacetate at 110°C. Likewise, the cycloaddition of (*Z*)-*N*-benzyl-1-phenylmethanimine oxide with ethyl 2-cyclopropylideneacetate at 110°C afforded 28% of methyl (6*R*,7*R*)-5-benzyl-6-phenyl-4-oxa-5-azaspiro[2.4]heptane-7-carboxylate and 40% of methyl (6*R*,7*S*)-5-benzyl-6-phenyl-4-oxa-5-azaspiro[2.4]heptane-7-carboxylate, respectively. DIBAL reduction of the ester groups led to the formation of ((6*R*,7*S*)-5-benzyl-6-phenyl-4-oxa-5-azaspiro[2.4]heptan-7-yl) methanol in 73% of yield, and 78% of ((6*R*,7*R*)-5-benzyl-6-phenyl-4-oxa-5-azaspiro[2.4]heptan-7-yl)methanol, accordingly. Due to the presence of the hydroxyl group, treatment with *p*-toluenesulfonic acid at 50°C afforded a relatively higher yield of the corresponding β-lactams, i.e., 67% of (3*S*,4*R*)-3-(hydroxymethyl)-1-methyl-4-phenylazetidin-2-one and 78% of (3*S*,4*R*)-3-(hydroxymethyl)-1-methyl-4-phenylazetidin-2-one (Scheme 3.136) [450]. However, when the nitrogen atom of the β-lactam moiety is also a part of five-membered ring, the β-lactam moiety cannot sustain when it is treated with TFA at 110°C, affording *N*-trifluoroacetyl β-amino acid derivative, as shown in Scheme 3.137. In this reaction, 1,3-dipolar cycloaddition between (3*R*,4*R*)-3,4-di-*tert*-butoxy-3,4-dihydro-2*H*-pyrrole 1-oxide and methylene-cyclopropane at 42°C in 7 days yielded 64% of (3a'*R*,4'*R*,5'*R*)-4',5'-di-*tert*-butoxytetrahydro-3'*H*-spiro[cyclopropane-1,2'-pyrrolo[1,2-*b*]isoxazole and 8%of(3a'*S*,4'*R*,5'*R*)-4',5'-di-*tert*-butoxytetrahydro-3'*H*-spiro[cyclopropane-1,2'-pyrrolo[1,2-*b*]isoxazole, along with 4% of regio-isomer, i.e., (3a'*R*,4'*R*,5'*R*)-4',5'-di-*tert*-butoxytetrahydro-2'*H*-spiro[cyclopropane-1,3'-pyrrolo[1,2-*b*]isoxazole. After treatment of these two regular 1,3-cycloadducts with TFA in toluene at 110°C for just 2 minutes, 68% of 2-((2*R*,3*R*,4*R*)-3,4-di-*tert*-butoxy-1-(2,2,2-trifluoroacetyl)pyrrolidin-2-yl)acetic acid and 35% of 2-((2*S*,3*R*,4*R*)-3,4-di-*tert*-butoxy-1-(2,2,2-trifluoroacetyl)pyrrolidin-2-yl)acetic acid were obtained, respectively [450]. In a similar fashion, a variety of spirofluorenyl-β-lactams have been prepared in moderate yields over three steps involving 1,3-dipolar cycloaddition of *N*-aryl fluorenone nitrones and methylenecyclopropanes, reduction with $LiAlH_4$ and subsequent *p*-TsOH-catalyzed ring contraction under mild reaction conditions with retention of stereochemistry. This study also indicates that the nitrone preferably reacts with cycloproylidene moiety during the 1,3-dipolar cycloaddition when another C=C double bond is available [451].

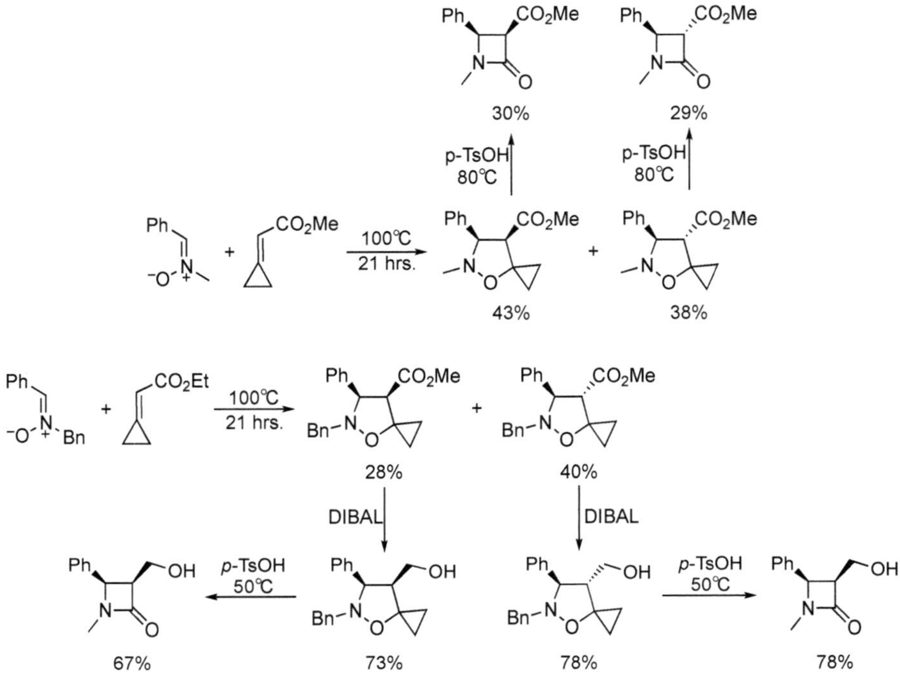

SCHEME 3.136 The 1,3-dipolar cycloaddition of methyl 2-cyclopropylideneacetate with *N*-substituted 1-phenylmethanimine oxide and subsequent ring-contraction to afford β-lactams

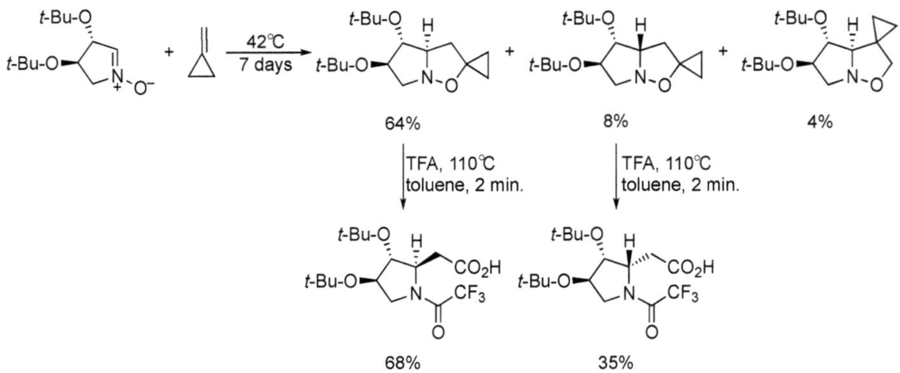

SCHEME 3.137 The 1,3-dipolar cycloaddition of (3*R*,4*R*)-3,4-di-*tert*-butoxy-3,4-dihydro-2*H*-pyrrole 1-oxide with methylenecyclopropane and subsequent treatment with TFA

In addition to the acid-promoted ring contraction of 5-substituted-4-oxa-5-azaspiro[2.4]heptanes that yield β-lactams, isoxazolidines can also be converted into β-lactams via ring-contraction. For example, 1,3-dipolar cycloaddition of (*Z*)-*N*-*tert*-butyl-1-phenylmethanimine oxide and (*E*)-3-nitroacrylonitrile yielded (3*R*,4*R*,5*S*)-2-(*tert*-butyl)-5-nitro-3-phenylisoxazolidine-4-carbonitrile and (3*S*,4*S*,5*S*)-2-(*tert*-butyl)-4-nitro-3-phenylisoxazolidine-5-carbonitrile. The former undergoes ring-contraction under either thermal or photochemical condition (254 nm), yielding thermodynamically more stable (3*R*,4*S*)-1-(*tert*-butyl)-2-oxo-4-phenylazetidine-3-carbonitrile as the final product when heated in MeOH, as shown in Scheme 3.138. Similarly, treatment of the 1,3-dipolar cycloaddition product from (*Z*)-*N*-methyl-1-phenylmethanimine oxide and (*E*)-3-nitroacrylonitrile afforded (3*S*,4*S*)-1-methyl-2-oxo-4-phenylazetidine-3-carbonitrile [452]. However, irradiation of (3*S*,4*R*,5*S*)-2-methyl-5-nitro-3-phenylisoxazolidine-4-carbonitrile at 310 nm rather than 254 nm leads to the generation of 2-methyl-3-phenyl-2,3-dihydroisoxazole-4-carbonitrile, instead of β-lactam. Similarly, treatment of methyl (3*S*,4*R*,5*S*)-2-(*tert*-butyl)-5-nitro-3-phenylisoxazolidine-4-carboxylate with DBN (1,5-diazabicyclo[4.3.0]non-5-ene) afforded 75% of methyl (3*R*,4*S*)-1-(*tert*-butyl)-2-oxo-4-phenylazetidine-3-carboxylate and 20% of methyl 2-(*tert*-butyl)-3-phenyl-2,3-dihydroisoxazole-4-carboxylate [453]. DFT calculations at various theory levels unequivocally show that this transformation involves a three-step, domino-type reaction rather than a ring opening step [454]. It is found that "the isoxazolidine to β-lactam transformation can be achieved with groups other than nitro in the 5-position of the heterocyclic ring" [455]. For example, the cyano group can be removed under basic conditions, as demonstrated in the treatment of (3*S*,3a*R*,6a*R*)-1-(4-(*tert*-butyl)phenyl) tetrahydro-1*H*,3*H*-thieno[3,4-*c*]isoxazole-3-carbonitrile in THF with LDA at -78°C for 2 hours, and subsequent work-up with aqueous NH_4Cl afforded 89% of (1*R*,5*R*)-7-(4-(*tert*-butyl)phenyl)-3-thiabicyclo[3.2.0]heptan-6-one, as shown in Scheme 3.139, along with possible reaction mechanism. Similarly, when ethyl (3*S*,3a*R*,6*S*,6a*R*)-1-(4-(*tert*-butyl)phenyl)-3-cyano-6-isopropyldihydro-1*H*,3*H*-thieno[3,4-*c*]isoxazole-3a(4*H*)-carboxylate was treated in a similar fashion, 65% of ethyl (1*R*,4*S*,5*R*)-6-(4-(*tert*-butyl) phenyl)-4-isopropyl-7-oxo-3-thiabicyclo[3.2.0]heptane-1-carboxylate was obtained [455].

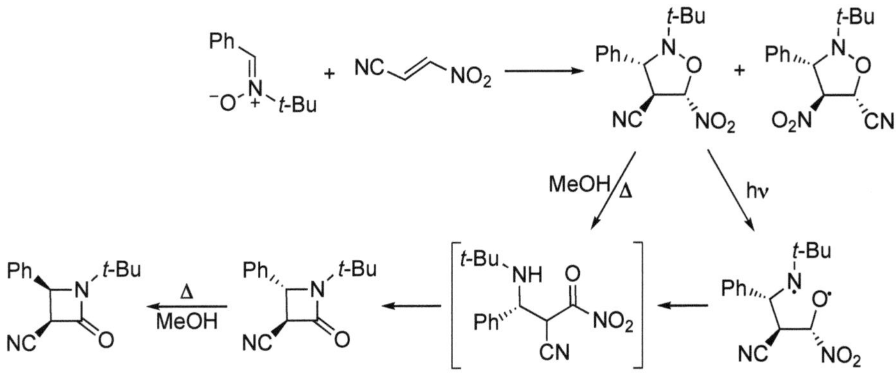

SCHEME 3.138 The mechanism of creation (3*R*,4*S*)-1-(*tert*-butyl)-2-oxo-4-phenylazetidine-3-carbonitrile from the reaction between (*Z*)-*N*-*tert*-butyl-1-phenylmethanimine oxide and (*E*)-3-nitroacrylonitrile

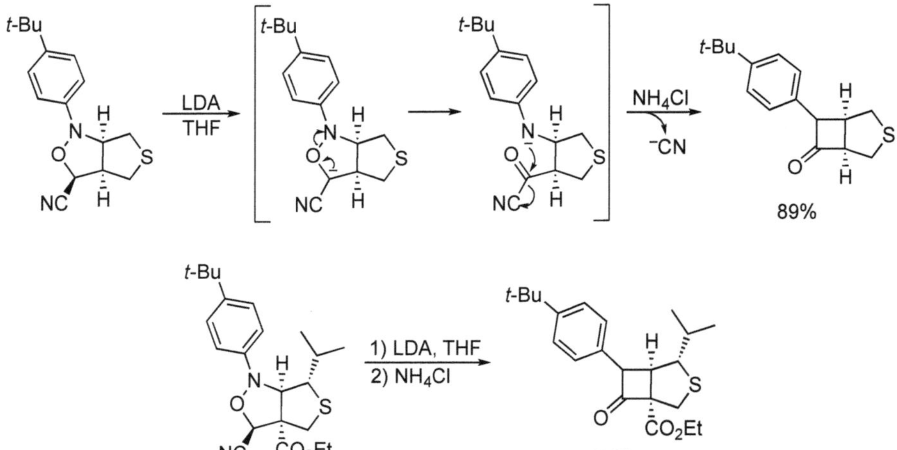

SCHEME 3.139 The mechanism for the conversion of isoxazolidine-5-carbonitriles into β-lactams with base treatment

In addition to the nitro and cyano groups at position 5 of isoxazolidines, thiophenyl group can be another candidate to be removed during ring contraction. In this case, 1,3-dipolar cycloaddition of phenyl vinyl sulfide to aryl nitrones regioselectively affords phenylthioisoxazolidines in primarily *cis*-configurations. Subsequent treatment of the 1,3-dipolar cycloadduct with BuLi or *tert*-BuLi at -78°C in THF gives β-phenylamino thioester and/or the corresponding β-lactam in good yield. Complete conversion of β-amino thioesters into β-lactams can in any case be accomplished by treating them with CH_3MgI/Et_2O at 0°C. For example, the cycloaddition between (*Z*)-1-(3-nitrophenyl)-*N*-phenylmethanimine oxide and vinyl phenyl sulfide in refluxing toluene yielded 78% of 3-(3-nitrophenyl)-2-phenyl-5-(phenylthio)isoxazolidine, with a *cis/trans* ratio of 9:1. Treatment of this cycloadduct in THF at -78°C with butyllithium afforded 7% of 4-(3-nitrophenyl)-1-phenylazetidin-2-one, whereas subsequent treatment of the reaction mixture with methyl magnesium iodide increased the overall yield of 4-(3-nitrophenyl)-1-phenylazetidin-2-one to 69% (Scheme 3.140). In contrast, direct treatment of 3-(4-methoxyphenyl)-2-phenyl-5-(phenylthio)isoxazolidine in THF at -78°C with butyllithium gave 60% of 4-(4-methoxyphenyl)-1-phenylazetidin-2-one [456].

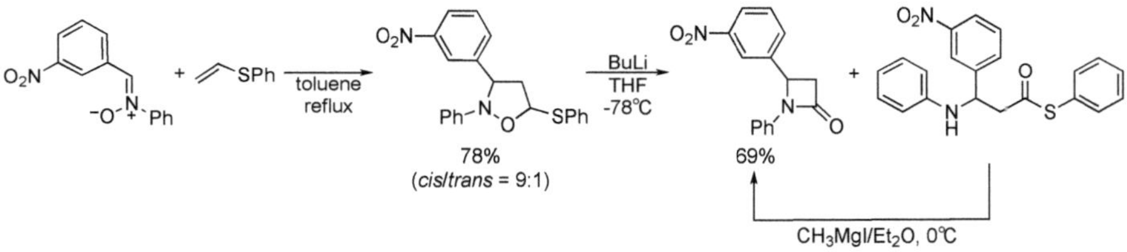

SCHEME 3.140 Formation of β-lactam from the 1,3-dipolar cycloaddition between (*Z*)-1-(3-nitrophenyl)-*N*-phenylmethanimine oxide and phenyl(vinyl)sulfane and subsequent base treatment

In a rare case, fluorine can be a leaving group at position 5 of isoxazolidines during ring contraction to form β-lactams. For example, when the mixture of 3,4-dihydroisoquinoline 2-oxide and (2,2-difluorovinyl)benzene in THF was refluxed for 3 hours, 64% of (1*R*,10b*S*)-2,2-difluoro-1-phenyl-1,5,6,10b-tetrahydro-2*H*-isoxazolo[3,2-*a*]isoquinoline was obtained. After the aqueous solution of Raney nickel in acetone was refluxed for 2 hours, the cycloadduct was added and refluxed for overnight to afford 46% of (1*R*,9b*S*)-1-phenyl-1,4,5,9b-tetrahydro-2*H*-azeto[2,1-*a*]isoquinolin-2-one (Scheme 3.141) [457].

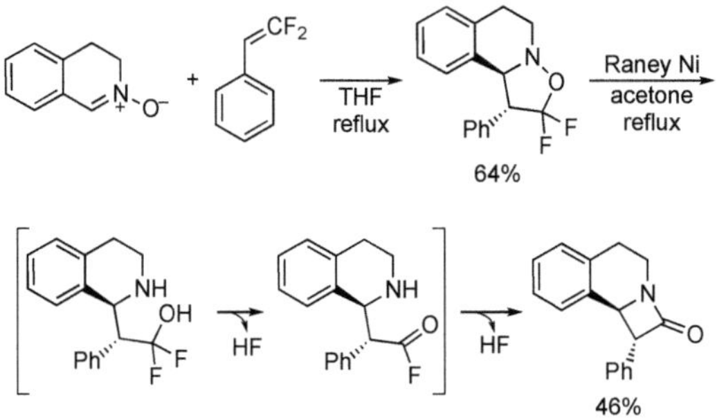

SCHEME 3.141 The preparation of β-lactam from the reaction of 3,4-dihydroisoquinoline 2-oxide and (2,2-difluorovinyl)benzene

Finally, the reaction between acrylates and substituted hydroxylamines affords 3-substituted isoxazolidinones in good yields, although this is not a 1,3-dipolar cycloaddition. In general, this reaction applies to a variety of β-substituted acrylates, with yields of isoxazolidinone generally being in the range of 70–95%. Of particular interest was the observation that the ratio of diastereoisomers produced in this reaction was usually more than 80:20 with a little variation for a series of β-substituents, except for the ratio of 68:32 for methyl cinnamate. In addition, the diastereoisomers can be separated either by simple chromatography or fractional crystallization. Subsequent hydrogenolysis of the labile N-O bond in isoxazolidinone with H_2/Pd-C followed by cyclization of the resultant β-amino acids cleanly affords β-lactams, where the α-methylbenzyl groups can be removed by brief exposure to Na/NH_3 to give simple 4-substituted β-lactams. For example, when a mixture of methyl (*E*)-but-2-enoate and (*R*)-*N*-(1-phenylethyl)-hydroxylamine was refluxed in benzene in the presence of K_2CO_3, 91% of (*R*)-3-methyl-2-((*R*)-1-phenylethyl) isoxazolidin-5-one and its diastereomer, i.e., (*S*)-3-methyl-2-((*R*)-1-phenylethyl)isoxazolidin-5-one were obtained, in a ratio of 8:2. Subsequent hydrogenation with H_2/Pd-C in EtOH, and treatment with tetrabutylammonium hydrosulfate and $KHCO_3$ in chloroform in the presence of methanesulfonyl chloride, (*R*)-4-methyl-1-((*R*)-1-phenylethyl)azetidin-2-one and its diastereomer (*S*)-4-methyl-1-((*R*)-1-phenylethyl)azetidin-2-one were obtained. Further treatment of these β-lactams in liquid ammonia with sodium gave (*R*)-4-methylazetidin-2-one and (*S*)-4-methylazetidin-2-one in 91% overall yield (Scheme 3.142). The stereochemistry at the 4-position of β-lactams is controlled by the stereochemistry of *N*-(1-phenylethyl)hydroxylamine. For comparison, the reaction of methyl (*E*)-but-2-enoate and (*S*)-*N*-(1-phenylethyl) hydroxylamine gave 81% of (*R*)-3-methyl-2-((*R*)-1-phenylethyl)isoxazolidin-5-one and its diastereomer, with a diastereomeric ratio of 81:19 [458].

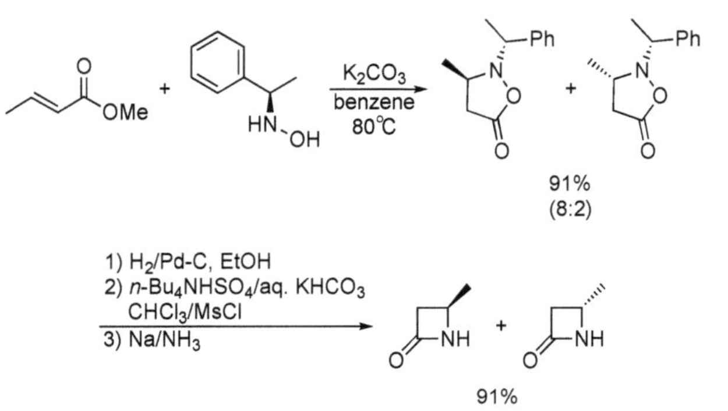

SCHEME 3.142 The formation of 3-methyl-2-(1-phenylethyl)isoxazolidin-5-ones from methyl (*E*)-but-2-enoate and (*R*)-*N*-(1-phenylethyl)hydroxylamine and their conversion into β-lactams

One more example of ring-contraction to yield β-lactams has been demonstrated in a bromine-induced rearrangement in the formation of a spiro-β-lactam moiety present in unique chartelline alkaloids. Chartellines, chartellamides, as well as securine, and securamines are members of a structurally unique class of natural products isolated from the bryozoa *Chartella papyracea* and *Securiflustra securifrons*, where both chartellines (A, B, and C) and chartellamides (A and B) contain a spiro-β-lactam scaffold. For the total synthesis of chartellines, a target molecule of methyl (*S*,8*Z*,14*Z*)-10,10-dimethyl-1-oxo-1,2,10,11-tetrahydroazeto[1',2':1,2]imidazo[4',5':7,8]azecino[3,2-*b*]indole-15-carboxylate has been selected as the key intermediate. This compound can be formed via a bromine-induced ring-contraction from the hypothesized molecule of methyl (4*Z*,8a*R*,13a*S*,14*Z*)-8a-bromo-16,16-dimethyl-7-oxo-1,7,8,8a,13,16-hexahydroimidazo[4,"5":6',7']azonino[1',2':1,2]pyrrolo[2,3-*b*]indole-5-carboxylate after cyclization and bromination from the molecule created by intramolecular Horner–Wadsworth–Emmons reaction of *tert*-butyl

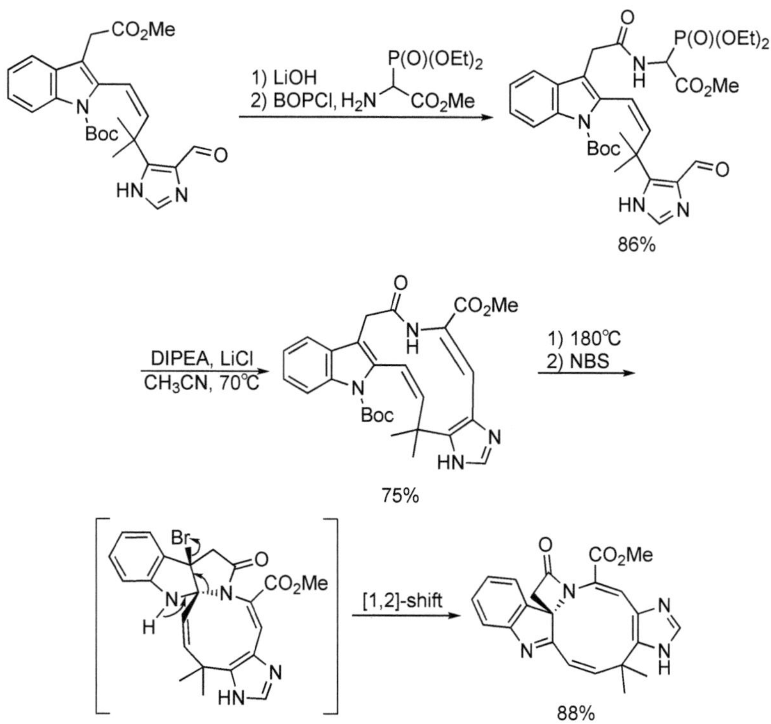

SCHEME 3.143 The key steps in the synthesis of methyl (*S*,8*Z*,14*Z*)-10,10-dimethyl-1-oxo-1,2,10,11-tetrahydroazeto[1',2':1,2]imidazo[4',5':7,8]azecino[3,2-*b*]indole-15-carboxylate

(*Z*)-3-(2-((1-(diethoxyphosphoryl)-2-methoxy-2-oxoethyl)amino)-2-oxoethyl)-2-(3-(4-formyl-1*H*-imidazol-5-yl)-3-methylbut-1-en-1-yl)-1*H*-indole-1-carboxylate, as displayed in Scheme 3.143. The total synthesis starts with methyl *O*-(*tert*-butyldimethylsilyl)serinate, and the key molecule of *tert*-butyl (*Z*)-2-(3-(4-formyl-1*H*-imidazol-5-yl)-3-methylbut-1-en-1-yl)-3-(2-methoxy-2-oxoethyl)-1*H*-indole-1-carboxylate (as the starting material in Scheme 3.143) was obtained after several transformations [459].

An additional example of ring-contraction to afford β-lactam can be found in the synthesis of β-lactam from isothiazolidinones [460].

3.4 ENZYMATIC SYNTHESIS OF Β-LACTAMS

This is another type of synthesis of β-lactam which can be performed in the presence of bacteria that secret the essential enzymes or the necessary enzymes. For the particular case of β-lactams, the enzymatic synthesis can go back to the initial discovery of penicillin by Fleming as mentioned early. Some features of enzymatic synthesis may include high efficiency, high specificity, and mild reaction conditions. The high efficiency can be demonstrated by the catalytic sense of enzymatic reaction where a tiny amount of enzyme is needed, and a high turnover number (e.g., up to 10^6 or even higher) can be achieved (times of reaction occurring on enzyme per second), and a minimum of waste is generated. Likewise, the high specificity of enzyme reaction is shown that for a given enzyme, a particular molecule of unique stereochemistry will be specifically created, whereas the total synthesis of this molecule in the laboratory is still challenging. In addition, the enzymatic reaction often occurs under mild conditions, such as neutral pH, in aqueous media, at room temperature, etc. In fact, many β-lactam antibiotics listed in Tables 1.1 to 1.4 of Chapter 1 have been initially isolated from the growth media of bacteria or fungi, definitely showing the power of enzymatic synthesis. However, the industrial implementation of the enzymatic synthesis of β-lactam antibiotics will face competition with well-established, fine-tuned chemical processes, as well as the challenges of undesired side reactions. For example, kinetically controlled synthesis using immobilized penicillin G acylase (PGA) in an aqueous environment demonstrates an advantage for simultaneous crystallization of the product, the enzyme may act either as a transferase or as a hydrolase, catalyzing the hydrolysis of the acyl side-chain in the precursor or the hydrolysis of the antibiotic itself [461]. Many enzymatic syntheses of β-lactams are actually semi-syntheses, by attaching an acyl group to the amino group. For example, the PGA isolated from *Achromobacter sp.* (NPGA) was significantly more efficient than the Penicillin acylase isolated from *Escherichia coli* (PGA), for the syntheses of ampicillin and amoxicillin (higher S/H ratio and product accumulation) in the whole range of substrate concentrations in kinetically controlled *N*-acylation. The degree of conversion of 6-aminopenicillanic acid (6-APA, i.e., (2*S*,5*R*,6*R*)-6-amino-3,3-dimethyl-7-oxo-4-thia-1-azabicyclo[3.2.0]heptane-2-carboxylic acid) to amoxicillin and ampicillin (160 mM 6-APA, 350 mM methyl ester, pH 6.3, 25°C, 200 minutes) with immobilized NPGA has reached 96.9% and 91.1%, respectively. This enzyme was highly thermostable with its maximum activity at 60°C (pH 8.0) and 65°C (pH 6.0) [462]. Recently, A newly isolated PGA from *Achromobacter xylosoxidans* PX02 has been engineered for site-directed mutagenesis at three important positions αR141, αF142, βF24 in order to improve the enzymatic synthesis of β-lactam antibiotics. The efficient mutant βF24A has reached a 99.1% and 98.7% conversion under optimal conditions for ampicillin and amoxicillin, respectively, from a high concentration (600 mM) of substrate 6-APA in the low acyl donor/nucleophile ratio (1.1:1). A large amount of ampicillin (561 mM) and amoxicillin (568 mM) have been precipitated from the aqueous reaction solution, corresponding to 93.5 and 94.6% of yields with high purity (99%), respectively. As a result, the downstream purification has been significantly simplified. This approach demonstrates an efficient synthesis process for β-lactam antibiotics with *in situ* product removal, which barely forms any byproducts [463].

For the synthesis of semisynthetic antibiotics such as amoxicillin and cephalexin in good yields, kinetically controlled methods are generally used with activated acyl-substrates, which are often unstable in aqueous media. Although co-solvents can be added to improve the yields, precipitation-driven, or "solid-to-solid" synthesis has been a proven method for high-yield enzymatic synthesis. After screening several counter-ions, Zn^{2+} was found to form a poorly soluble salt with amoxicillin anion. Despite the fact that degradation of β-lactam increased in the presence of Zn^{2+}, $ZnSO_4$ has been deliberately added to the synthetic media to reach a concentration of 0.1 *M* $ZnSO_4$, where the yield of amoxicillin can be enhanced at least 30-fold [464].

In another practice, penicillin V acylase from *Streptomyces mobaraensis* (Sm-PVA) with broad substrate specificity has shown a high acyl-transfer activity in reactions when methyl ester of the carboxylic acid is used as acyl donor during the formation of amides. For example, when methyl phenoxyacetate is applied as the acyl donor, and 6-APA and

7-aminodeacetoxycephalosporanic acid (7-ADCA, i.e., (6*R*,7*R*)-7-amino-3-methyl-8-oxo-5-thia-1-azabicyclo[4.2.0]-oct-2-ene-2-carboxylic acid) are used as nucleophiles, 66% and 69% of Penicillin V and deacetoxycephalosporin V are obtained, respectively. Likewise, when methyl *n*-hexanoate and methyl *n*-octanoate are applied as the acyl donors, in combination with 6-APA as the nucleophile, 90% and 91% of dihydropenicillin F and Penicillin K were obtained [465]. Interestingly, it is found that free penicillin-G acylase from *Escherichia coli* could perform transformations in frozen media, where the β-lactam antibiotics such as ampicillin and cephalexin have been obtained in 61% and 80%, respectively in ice at -20°C, when methyl (*R*)-2-amino-2-phenylacetate was applied as the acyl donor, and 6-APA and 7-ADCA were used as nucleophiles accordingly. For comparison, when this biocatalytic synthesis of β-lactam antibiotics is performed at 20°C, the yields of ampicillin and cephalexin were 16% and 22%, respectively. While this comparison might not be very valuable, as reaction time was much shorter at 20°C. However, when methyl (*R*)-2-amino-2-(4-hydroxyphenyl)acetate was applied as the acyl donor, 38% of amoxicillin (6 days) was obtained from 6-APA at -20°C versus 10% at 20°C (0.7 hour), no cefadroxil has been detected in the reaction with 7-ADCA at -20°C, whereas 22% of cefadroxil could be obtained after 0.7 hour. This result indicates that the temperature of the reaction media may affect the individual nucleophile/acyl-enzyme complex to a different extent [466].

Also, the technologies to make cephalexin and cefaclor using immobilized aminocephalosporin synthetase from *Xanthomonas sp.* as a biocatalyst [467], and to produce cefazolin by acyl transfer enzymatic synthesis with immobilized cefazolin synthetase from *Escherichia coli* as a biocatalyst and the following separation method [468], have been reported.

For a specific example, carbapenems contain at least three chiral centers, and all clinically used carbapenems have a (6*R*)-hydroxyethyl side-chain, and most of them are C-1 substituted, in order to increase their potency and avoid hydrolysis by dehydropeptidases. Three enzymes (CarA, B, and C) are known to catalyze the biosynthesis of (5*R*)-carbapen-2-em-3-carboxylate in *Pectobacterium carotovorum*, with multiple enzymes being involved in the biosynthesis of thienamycin in *Streptomyces cattleya.* CarB and ThnE are carboxymethylproline synthases of the crotonase superfamily, and are believed to catalyze a common step during the formation of (2*S*,5*S*)-carboxymethylproline (*t*-CMP) from malonyl-CoA and pyroline-5-carboxylate in *Pectobacterium carotovorum* and ThnE in *Streptomyces cattleya*, respectively. CarB/ThnE-catalyzed C–C bond formation is proposed to proceed via reaction of the enolate intermediate with the *Re* face of L-P5C to give a *t*-CMP-CoA thioester, which is hydrolyzed to afford *t*-CMP. In order to bio-catalytically synthesize carbapenem precursors functionalized at the C-1 and C-6-equivalent positions, engineered CMPSs have been applied solely, and/or in tandem with an alkylmalonyl-CoA forming enzyme, to catalyze the formation of 4,6-disubstituted-*t*-CMP stereoisomers, with three contiguous chiral centers. Some resulting molecules are converted by CarA to bicyclic β-lactams [469]. It should be pointed out that the biosynthetic study of penicillin N is by far the most advanced, where a tripeptide precursor δ-(*L*-α-amioadipyl)-*L*-cysteinyl-*D*-valine (i.e., (*S*)-2-amino-6-(((*R*)-1-(((*R*)-1-carboxy-2-methylpropyl)amino)-3-mercapto-1-oxopropan-2-yl)amino)-6-oxohexanoic acid), is cyclized into isopenicillin N (i.e., (2*S*,5*R*,6*R*)-6-((*S*)-5-amino-5-carboxypentanamido)-3,3-dimethyl-7-oxo-4-thia-1-azabicyclo[3.2.0]heptane-2-carboxylic acid) with impressive efficiency by isopenicillin N synthase in the presence of ferrous ion and molecular oxygen [470]. The synthesis of this tripeptide has been directed by the *acvA*, *hts1*, *esyn1*, and *cssA* genes in the fungal systems [471].

More references about the enzymatic synthesis of antibiotic β-lactams can be found in the following literature [9, 472–476].

3.5 MISCELLANEOUS METHODS

While the mainstream methods for the formation of β-lactams, such as Staudinger reaction, Wolff rearrangement and subsequent reaction with imine, [2+2]-cycloaddition between alkene and isocyanate, Kinugasa reaction, intramolecular Michael addition, photolysis of α,β-unsaturated amides, intramolecular substitutive cyclization using either nitrogen or carbon as the nucleophile, intramolecular amide condensation, intramolecular amidation, ring-expansion of aziridines, and carbonylative cycloaddition have been collected above, although not comprehensive, several methods work under specially occasions do appear in the literature, which cannot be classified into any specific methods outlined above, and are summarized in this section.

One occasional method involves the addition of a primary amine to a cationic iron complex containing cyclopentadienyl, CO, and olefin ligands (Fp(olefin)$^+$), and subsequent oxidative lactamization of an intermediate β-aminoalkyl Fp complex. This approach can be extended to the synthesis of bicyclic β-lactams using ω-amino olefins as the starting materials. As an example, the Fp complex with propene was treated with benzylamine and then oxidized to afford 1-benzyl-4-methylazetidin-2-one as shown in Scheme 3.144. Similarly, the reactions of pent-4-en-1-amine and hex-5-en-1-amine under this condition afford 1-azabicyclo[3.2.0]heptan-7-one and 1-azabicyclo[4.2.0]octan-8-one, respectively [477–479].

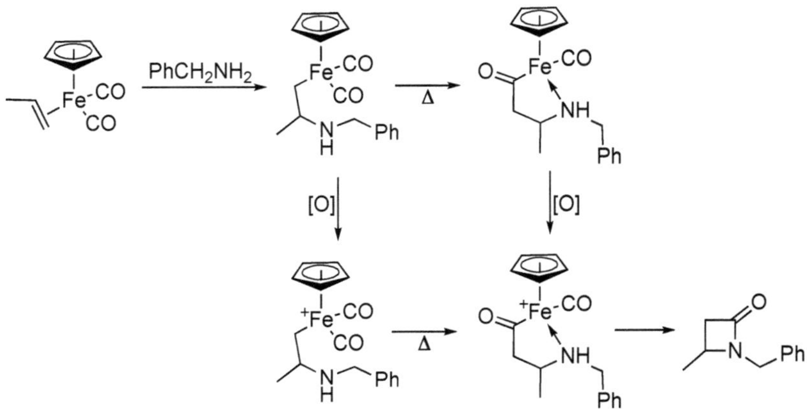

SCHEME 3.144 The mechanism for the reaction between benzylamine and cyclopentadienyl propene dicarbonyl complex that yields β-lactam

Another example of this kind involves an intramolecular lactamization occurring within the acyl ligand that is coordinated to iron(II). When the iron complex containing PPh_3, CO, cyclopentadienide, and 2-((2*R*,4*S*)-2,5,5-trimethyl-4-phenylpyrrolidin-2-yl)acetyl was treated with iodine/triethylamine in CH_2Cl_2, 62% of (3*S*,5*R*)-2,2,5-trimethyl-3-phenyl-1-azabicyclo[3.2.0] heptan-7-one was yielded, simply by means of cleaving the substituted acetyl ligand from the metal center along with the intramolecular lactamization, as displayed in Scheme 3.145 [480].

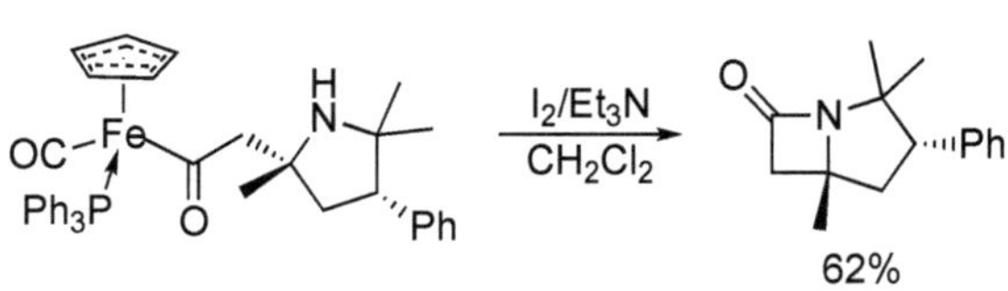

SCHEME 3.145 A preparation of (3*S*,5*R*)-2,2,5-trimethyl-3-phenyl-1-azabicyclo[3.2.0] heptan-7-one

A very recent report introduces a preparation of β-lactam from acyclic diaminocarbene and carbon monoxide. Acyclic diaminocarbenes (ADAC) exhibit a much higher electrophilicity and nucleophilicity than the NHCs and are even superior to cyclic (alkyl)(amino)carbenes (CAACs). In the presence of CO, the primary carbonylation product of an ADAC is the corresponding diaminoketene $(R_2N)_2C{=}C{=}O$, which is a transient species and undergoes rearrangement to afford β-lactam and other products. For example, *N*-isopropyl-*N*-((2,2,6,6-tetramethylpiperidin-1-yl)methylene)propan-2-aminium hexafluoro-phosphate prepared from diisopropylamine, triethylamine, 1-(chloromethylene)-2,2,6,6-tetramethylpiperidin-1-ium chloride in CH_2Cl_2 and ammonium hexafluorophosphate [481], was treated with sodium bis(trimethylsilyl)amide in THF at low temperature, and the resulting ADAC subsequently was exposed to an atmospheric pressure of carbon monoxide to afford 2-(diisopropylamino)-2-(2,2,6,6-tetramethylpiperidin-1-yl)ethen-1-one. This highly unstable ketene undergoes rearrangement to yield either 2-(diisopropylamino)-2-oxo-1-(2,2,6,6-tetramethylpiperidin-1-ium-1-ylidene)ethan-1-ide or *N*-isopropyl-*N*-(2-oxo-2-(2,2,6,6-tetramethylpiperidin-1-yl)ethan-1-id-1-ylidene)propan-2-aminium, a process similar to the retro-Wolff rearrangement. Intramolecular carbene C-H insertion of these two intermediates leads to the formation of final lactams, i.e., (*R*)-1-isopropyl-4,4-dimethyl-3-(2,2,6,6-tetramethylpiperidin-1-yl)azetidin-2-one and diastereomeric mixtures of (2*S*,8a*S*)-2-(diisopropylamino)-5,5,8a-trimethylhexahydroindolizin-3(2*H*)-one and (2*S*,8a*R*)-2-(diisopropylamino)-5,5,8a-trimethylhexahydroindolizin-3(2*H*)-one (Scheme 3.146) [482].

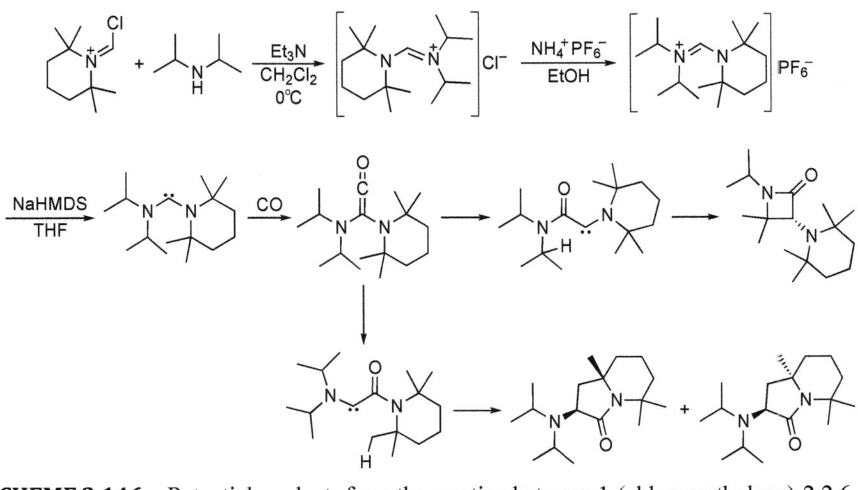

SCHEME 3.146 Potential products from the reaction between 1-(chloromethylene)-2,2,6,6-tetramethylpiperidin-1-ium and diisopropylamine

An interesting approach involving a one-pot process to afford either aziridine or β-lactam depending on the reaction condition was recently reported. In this approach, the intermediate arising from the addition of malononitrile with an appropriately tethered alkyne functionality to an α,β-unsaturated aldehyde condenses with *N*-aryl hydroxylamine, and the resulting product would undergo an intramolecular 1,3-dipolar cycloaddition to afford hexahydrobenzo[c] isoxazole. Owing to the inherent strain of the five-membered dihydroisoxazole ring, it further undergoes a Baldwin rearrangement to a densely functionalized aziridine carbonyl. Alternatively, if the cycloaddition step for the terminal alkyne functionalized intermediate is carried out in the presence of a CuI source and a suitable base, the Kinugasa reaction occurs to generate the β-lactam. It is found the stereocenter created in the initial addition step controls the formation of the other stereocenters in a highly selective manner, giving rise to single diastereoisomers of the optically active products. As an example, the reaction between (*E*)-hex-2-enal and 2-(prop-2-yn-1-yl)malononitrile in CH_2Cl_2 in the presence of 10 mol% of benzoic acid and (*S*)-2-(bis(3,5-bis(trifluoromethyl)phenyl)((trimethylsilyl)oxy)methyl)pyrrolidine (as a chiral base), afforded (*R*)-2-(1-oxohexan-3-yl)-2-(prop-2-yn-1-yl)malononitrile. Upon treatment with *N*-phenylhydroxylamine, (*R*,*E*)-4,4-dicyano-*N*-phenyl-3-propylhept-6-yn-1-imine oxide was resolved, which underwent intramolecular 1,3-dipolar cycloaddition to yield (6*R*,7a*S*)-1-phenyl-6-propyl-1,6,7,7a-tetrahydrobenzo[*c*]isoxazole-5,5(4*H*)-dicarbonitrile. Subsequent Baldwin rearrangement led to the final product of (1*R*,4*R*,6*S*)-1-formyl-7-phenyl-4-propyl-7-azabicyclo[4.1.0]heptane-3,3-dicarbonitrile, in 46% yield with 91% ee. On the other hand, when (*R*)-2-(1-oxohexan-3-yl)-2-(prop-2-yn-1-yl) malononitrile was treated with *N*-phenylhydroxylamine, then with 25 mol% cuprous iodide, 27.5 mol% of 1,10-phenanthroline, and 2.0 equivalents of Et_3N at room temperature for 24 hours, 58% of (1*R*,4*R*,6*S*)-1-methyl-8-oxo-7-phenyl-4-propyl-7-azabicyclo[4.2.0]octane-3,3-dicarbonitrile was obtained, with a diastereoselectivity greater than 20:1, and 88% ee (Scheme 3.147) [483].

A series of studies have been devoted to the manganese(III)-promoted free radical cyclization of enamides to form β-lactams [484, 485]. It is found that the alkyl substituents at the nitrogen atom of the substrates have a pronounced effect on the cyclization outcome, where the best yields of β-lactams are obtained when the alkyl group presenting a secondary or tertiary carbon close to the nitrogen. This might be attributed to the influence of the alkyl group on the geometry of the 4-*exo-trig* transition state. In order to enhance the diastereoselectivity of this approach, several chiral alkyl groups have been attached to the nitrogen atom of enamides, which can be easily prepared by condensation of readily available chiral amines with

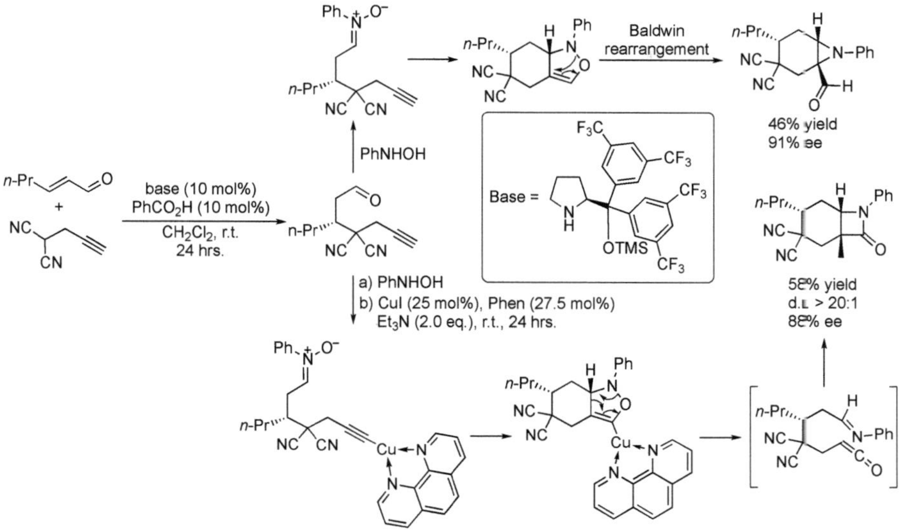

SCHEME 3.147 The different outcomes for the reaction between 2-(prop-2-yn-1-yl) malononitrile and (*E*)-hex-2-enal with and without CuI added

2,2-diphenylacetaldehyde, and subsequent reaction with methyl 3-chloro-3-oxopropanoate [486]. When the resulting enamides are treated with $Mn(OAc)_3 \cdot 2H_2O$ in glacial acetic acid at 70°C, the corresponding β-lactams are obtained in high yields. However, not much noticeable diastereoselectivity has been found from the enamides prepared from (*R*)-(+)- or (*S*)-(-)-phenylethylamine as well as methyl *L*-phenylalaninate, methyl *L*-alaninate, although the diastereoselectivity has been slightly enhanced in enamides prepared from (*S*)-1-cyclohexylethan-1-amine or (*R*)-1-(naphthalen-1-yl) ethan-1-amine. Fortunately, good diastereoselectivity has been observed in the enamides prepared from ethyl *L*-valinate, methyl (*S*)-2-amino-2-phenylacetate, and methyl (*S*)-2-amino-3,3-dimethylbutanoate. A representative example has been provided in Scheme 3.148 to illustrate this approach [486]. In addition to $Mn(OAc)_3$, ceric ammonium nitrate (CAN) is also effective in the promotion of free radical cyclization reactions of enamides to afford β-lactams [487].

Two approaches to making α-bromo-α-fluoro-β-lactams have been reported, all involving a Reformatsky type reaction of 2-bromo-2,2-difluoroacetic acid derivatives. Treatment of the mixture of ethyl 2-bromo-2,2-difluoroacetate and imines in Et_2O at -10°C for 1–2 hours with diethyl zinc (Et_2Zn) leads to good yields of *syn*-α-bromo-α-fluoro-β-lactams in

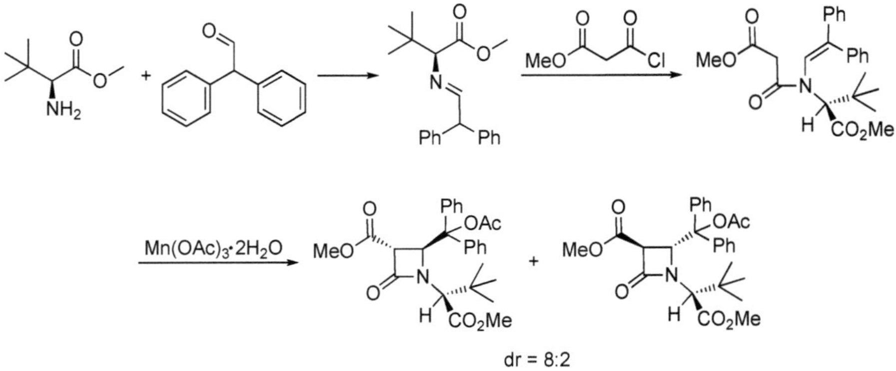

SCHEME 3.148 Preparation of β-lactam from the reaction of methyl (*S*)-2-amino-3,3-dimethylbutanoate, with 2,2-diphenylacetaldehyde, and methyl 3-chloro-3-oxopropanoate

excellent diastereoselectivity. Alternatively, treatment of the mixture of *N*-benzyl-2-bromo-2,2-difluoroacetamide and benzaldehyde with Et_2Zn in the presence of a catalytic amount of Wilkinson reagent [$Rh(PPh_3)_3Cl$] yielded 86% of *N*-benzyl-2-bromo-2-fluoro-3-hydroxy-3-phenylpropanamide, with very low diastereoselectivity. This amide can then be converted into α-bromo-α-fluoro-β-lactam by means of Mitsunobu reaction, with good diastereoselectivity (*syn*/*anti* = 7:1), as outlined in Scheme 3.149 [488]. A similar approach utilizes the Wilkinson reagent (2 mol%) and 1.5 equivalents of Et_2Zn to promote the reaction of α,β-unsaturated esters with imines to form β-lactams. It is believed

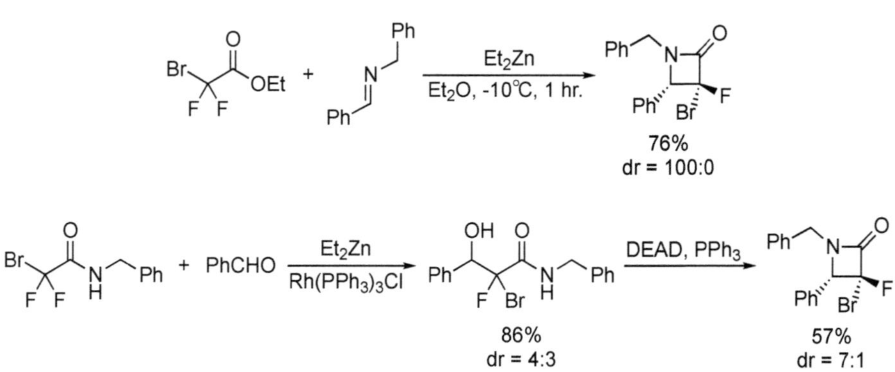

SCHEME 3.149 The approaches to make (3*R*,4*S*)-1-benzyl-3-bromo-3-fluoro-4-phenylazetidin-2-one

that a rhodium-hydride complex generated from Et_2Zn and the Wilkinson reagent catalyzes the 1,4-reduction of the α,β-unsaturated esters to form the rhodium enolate, which functions as a Reformatsky type reagent to react with various imines to afford *syn*-β-lactams in good to excellent yields with high diastereoselectivity. As an example, the reaction of methyl acrylate and (*E*)-1-(2-chlorophenyl)-*N*-(4-methoxyphenyl)methanimine in DMF at room temperature gave 93% of (3*R*,4*R*)-4-(2-chlorophenyl)-1-(4-methoxyphenyl)-3-methylazetidin-2-one, with a diastereomeric ratio of 100:0. For this particular approach, it is found that the substituent at the β-position of the α,β-unsaturated ester has very little impact on the outcome of the reaction. Methyl crotonate gave the β-lactam in low yield with reasonably high *syn*-selectivity, whereas methyl cinnamate failed to provide any of the desired products. However, α,β-unsaturated γ- and δ-lactones reacted smoothly to give the desired products, although with *anti*-configuration in the products [489].

A phenyl phosphorodichloridate promoted annulation of Schiff bases with trimethylsiloxyacetic acid provides a simple preparation of 3-hydroxy-β-lactams, in favor of the *cis*-configuration. For example, treatment of hydroxyacetic acid with trimethylsilyl chloride in benzene in the presence of Et_3N gives the initial product of trimethylsilyl 2-hydroxyacetate, which isomerizes to 2-((trimethylsilyl)oxy)acetic acid at room temperature. This acetic acid in CH_2Cl_2 undergoes annulation with (*E*)-*N*,1-bis(4-methoxyphenyl)methanimine in the presence of phenyl phosphorodichloridate and Et_3N, where the trimethylsilyl group cleaves during aqueous work-up, affording 65% of 3-hydroxy-1,4-bis(4-methoxyphenyl)azetidin-2-one as purely *cis*-isomer. Although the annulations with other Schiff bases, such as (*E*)-*N*,1-diphenylmethanimine, (*E*)-*N*-(4-methoxyphenyl)-1-phenylmethanimine, and (*E*)-1-(4-methoxyphenyl)-*N*-(*p*-tolyl)methanimine produce 3-hydroxy-β-lactam derivatives in both *cis*- and *trans*-configurations, the *cis*-isomer is always the predominant product. Interestingly, the annulation of 2-phenyl-2-((trimethylsilyl)oxy)acetic acid with Schiff base under the same condition affords the corresponding 3-trimethyloxy-β-lactam instead of the desilylated 3-hydroxy-β-lactam, where the trimethylsilyl group can be quantitatively removed via the hydrolysis in acetone in the presence of 1 *N* HCl at room temperature, or during the treatment with HF in MeOH/CH_2Cl_2 at room temperature, as shown in Scheme 3.150 [490].

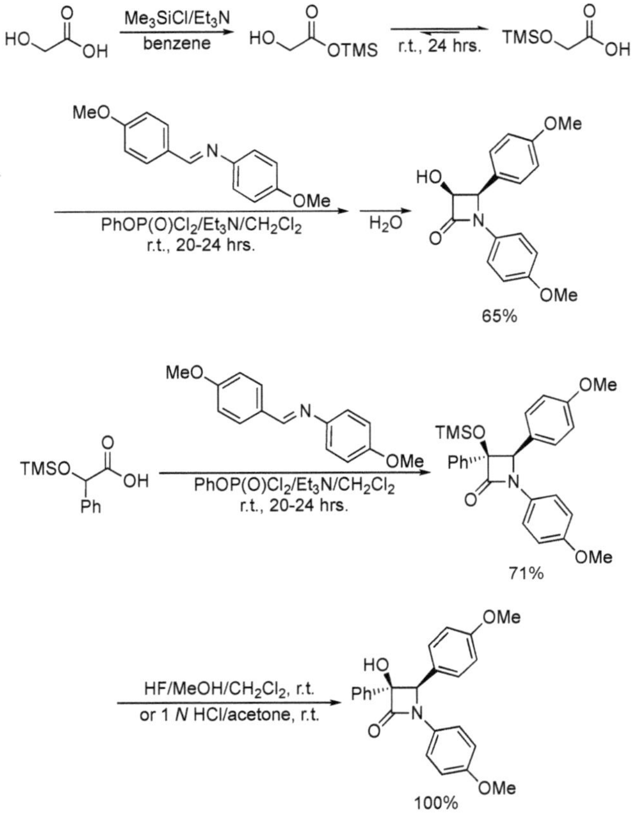

SCHEME 3.150 Formation of β-lactam from the reaction of 2-((trimethylsilyl)oxy)acetic acids with (*E*)-*N*,1-bis(4-methoxyphenyl)methanimine

Also, a unique and simple approach has been developed to form 4-unsubstituted β-lactams in one step from the reaction of lithium enolates of esters and secondary *N*-(cyanomethyl)amines, where the *N*-1 and *C*-3 substituents on β-lactams can be varied widely. In addition, β-lactams with nitrogen substituents at *C*-3 can be readily assembled in this way from amino acid precursors, and optically active β-lactams can be prepared from *N*-(cyanomethyl)amine derived from certain enantiomerically pure primary amines. The strong bases used include LDA and LHMDS. A few representative reactions are collected in Scheme 3.151 [491].

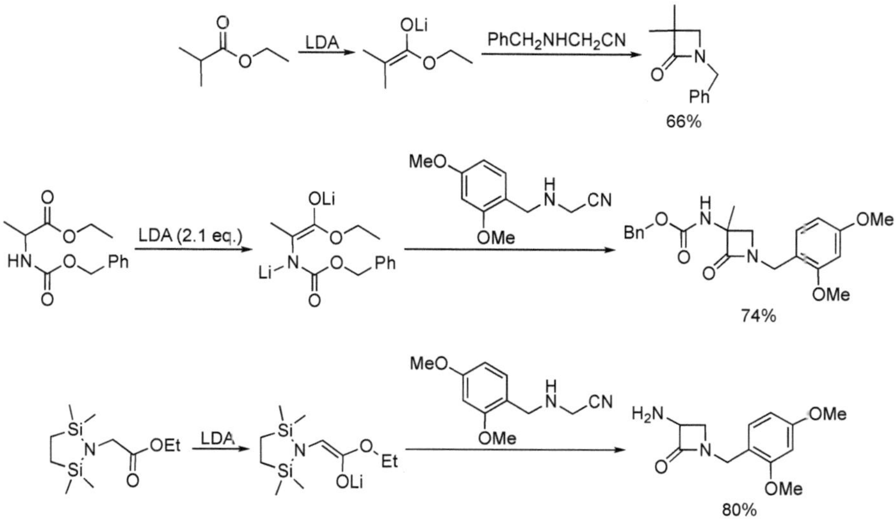

SCHEME 3.151 Examples of the β-lactam preparations from the reaction of ester enolates and α-aminonitriles

Another less commonly applied method to form β-lactams, particularly for the β-lactams with an *exo*-double bond at position 3 is the palladium-catalyzed intramolecular thiocarbamoylation or selenocarbamoylation of the propargyl moiety. For example, when *Se*-phenyl pent-4-yn-1-yl(prop-2-yn-1-yl)carbamoselenoate was refluxed in toluene in the presence of 1 mol% of $[Pd(PPh_3)_4]$, 50% of (*Z*)-1-(pent-4-yn-1-yl)-3-((phenylselanyl)methylene) azetidin-2-one was obtained, with *E*/*Z* ratio of 5:95, along with 10% of (*Z*)-3-((phenylselanyl)methylene)-1-(prop-2-yn-1-yl)piperidin-2-one, with an *E*/*Z* ratio of 11:89. Likewise, when *S*-phenyl methyl(prop-2-yn-1-yl)carbamothioate was refluxed in toluene in the presence of 5% $[Pd(PPh_3)_4]$, 72% of (*Z*)-1-methyl-3-((phenylthio)methylene)azetidin-2-one was obtained, as summarized in Scheme 3.152 [492].

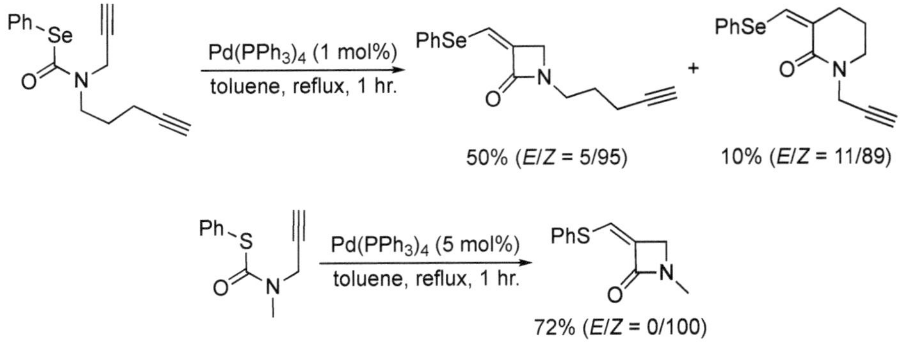

SCHEME 3.152 A palladium-catalyzed intramolecular reaction of *Se*-phenyl pent-4-yn-1-yl(prop-2-yn-1-yl)carbamoselenoate or *S*-phenyl methyl(prop-2-yn-1-yl)carbamothioate to yield the β-lactam

Similarly, a palladium catalyst-directed α-selective olefinic C-H activation using air as the sole oxidant allows the construction of 4-imino-β-lactam derivatives. For example, a mixture of *N*-methoxycinnamamide and 2.0 equivalents of *tert*-butyl isonitrile was converted into 68% of (*E*)-3-((*Z*)-benzylidene)-1-(*tert*-butyl)-4-(methoxyimino)azetidin-2-one when the mixture was heated in 1,4-dioxane at 100°C for 6 hours in the presence of 2 mol% of $Pd_2(dba)_3$. However, when there is not an α-olefinic hydrogen, a different scaffold of the heterocyclic molecule was obtained. For instance, when (*E*)-*N*-methoxy-2-methyl-3-phenylacrylamide and 2.0 equivalents of *tert*-butyl isonitrile were heated in 1,4-dioxane for 12 hours in the presence of 5 mol% of $Pd_2(dba)_3$, 87% of (*Z*)-1-(*tert*-butyl)-5-(methoxyimino)-3-methyl-4-phenyl-1,5-dihydro-2*H*-pyrrol-2-one was obtained, as displayed in Scheme 3.153 [493].

A very interesting reaction is the treatment of 6-methyl-7*H*-[1,2,4]triazolo[4,3-*b*][1,2,4]triazepin-8(9*H*)-one with acetic anhydride which affords 1-(6-methyl-1*H*-pyrazolo[5,1-*c*][1,2,4]triazol-1-yl)ethan-1-one and 1-acetyl-6-methyl-1*H*,5*H*-azeto[3',2':4,5]pyrazolo[5,1-*c*][1,2,4]triazol-7(8*H*)-one. It is assumed that the acetylation occurs first to yield 9-acetyl-6-methyl-7*H*-[1,2,4]triazolo[4,3-*b*][1,2,4]triazepin-8(9*H*)-one. Further acetylation generates 1,9-diacetyl-6-methyl-8-oxo-8,9-dihydro-7*H*-[1,2,4]triazolo[4,3-*b*][1,2,4]triazepin-1-ium. Deprotonation with acetate results in carbanion, which undergoes intramolecular addition to iminium, affording the key intermediate of 1,1'-(6-methyl-7-oxo-6a,7-dihydro-1*H*,8*H*-azeto[3',2':4,5]pyrazolo[5,1-*c*][1,2,4]triazole-1,8-diyl)bis(ethan-1-one). From this intermediate, *retro*-[2+2] cycloaddition of the β-lactam moiety leads to the formation of 1-(6-methyl-1*H*-pyrazolo[5,1-*c*][1,2,4]triazol-1-yl)ethan-1-one (path A in Scheme 3.154). In parallel, *retro*-Ene reaction of this intermediate yields

1-(7-hydroxy-6-methyl-1*H*,6a*H*-azeto[3',2':4,5]pyrazolo[5,1-*c*][1,2,4]triazol-1-yl)ethan-1-one (path B in Scheme 3.154), which undergoes tautomerization to regain the β-lactam moiety, affording 1-acetyl-6-methyl-1*H*,8*H*-azeto[3',2':4,5]-pyrazolo[5,1-*c*][1,2,4]triazol-7(6a*H*)-one. Subsequent 1,3-H shift leads to final β-lactam of 1-acetyl-6-methyl-1*H*,5*H*-azeto[3',2':4,5]pyrazolo[5,1-*c*][1,2,4]triazol-7(8*H*)-one [494].

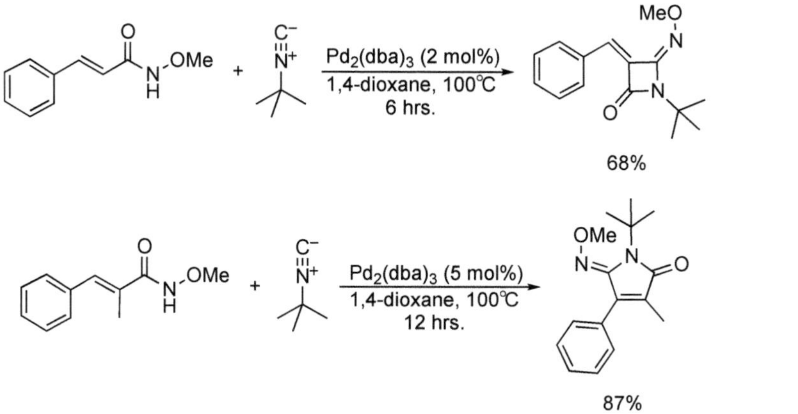

SCHEME 3.153 The preparation of β-lactam from the reaction of *N*-methoxycinnamamide and *t*-butyl isonitrile

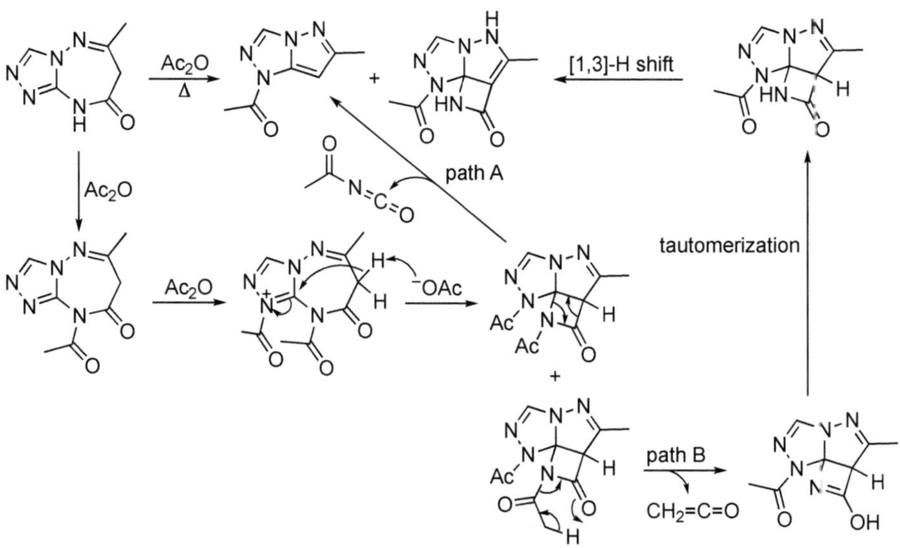

SCHEME 3.154 The reaction mechanism for the conversion of 6-methyl-7*H*-[1,2,4]triazolo[4,3-*b*][1,2,4]triazepin-8(9*H*)-one into 1-acetyl-6-methyl-1*H*,5*H*-azeto[3',2':4,5]pyrazolo[5,1-*c*][1,2,4]triazol-7(8*H*)-one with Ac_2O

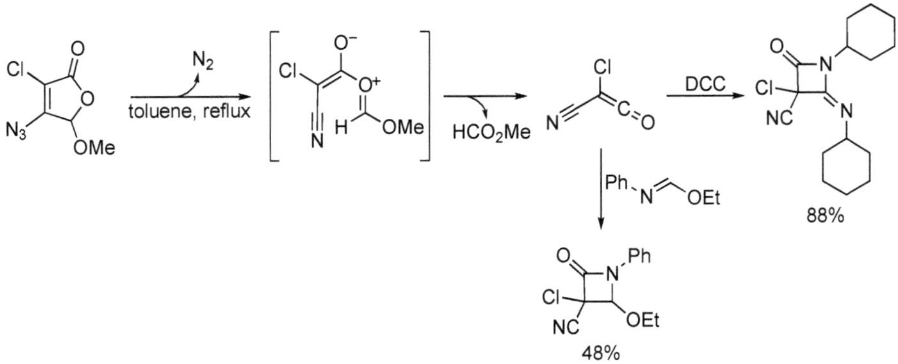

SCHEME 3.155 Conversion of 4-azido-3-chloro-5-methoxyfuran-2(5*H*)-one into β-lactams in the presence of different imines

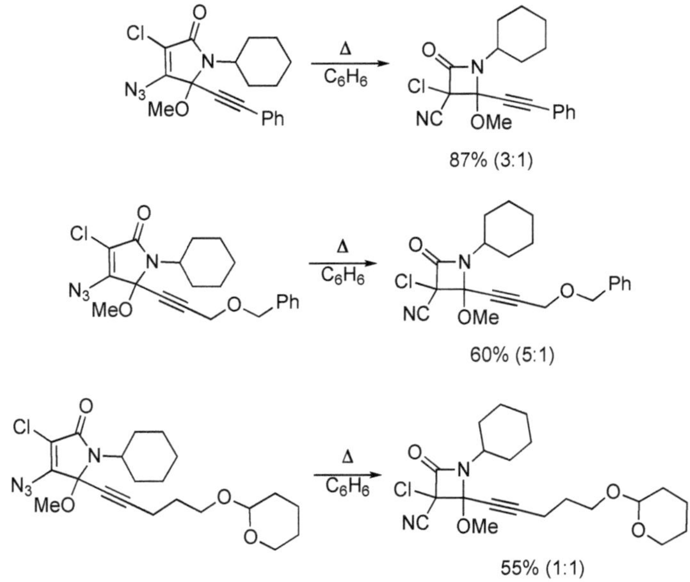

SCHEME 3.156 Thermolysis of 4-azido-3-chloro-1-cyclohexyl-5-(substituted)ethynyl-5-methoxy-1,5-dihydro-2*H*-pyrrol-2-ones into β-lactams

Moreover, thermolysis of 4-azido-3-chloro-1-methyl-5-methoxy-Δ^3-pyrrolinone leads to the formation of ketene, which immediately undergoes the Staudinger reaction with imines to form β-lactams, particularly, the highly substituted β-lactams. For example, when a benzene solution containing stoichiometric amounts of 4-azido-3-chloro-5-methoxyfuran-2(5*H*)-one and DCC was refluxed for 3 hours, 88% of (*E*)-3-chloro-1-cyclohexyl-2-(cyclohexylimino)-4-oxoazetidine-3-carbonitrile was isolated. Analogously, when the mixture of this compound and ethyl *N*-phenylformimidate was refluxed in benzene, 48% of 3-chloro-2-ethoxy-4-oxo-1-phenylazetidine-3-carbonitrile was isolated (Scheme 3.155). It should be noted that the cycloaddition between the *in situ* generated ketene and imine appears to be stereospecific, in that only one diastereomer was observed by NMR analysis of the crude reaction product, although the actual stereochemistry of the product has not been determined [495]. Further study elaborates the potential reaction mechanism and the scope of this approach, in addition to the preparation of 3-cyano-β-lactams [496]. For 4-azido-3-chloro-1-cyclohexyl-5-methoxy-5-(phenylethynyl)-1,5-dihydro-2*H*-pyrrol-2-one, it could thermolyze to form both ketene and imine moieties *in situ* which immediately undergo the Staudinger reaction to form β-lactam 3-chloro-1-cyclohexyl-2-methoxy-4-oxo-2-(phenylethynyl)azetidine-3-carbonitrile, as demonstrated in Scheme 3.156, in 87% of yield with 3:1 ratio of stereoisomers. Similarly, thermolysis of 4-azido-5-(3-(benzyloxy)prop-1-yn-1-yl)-3-chloro-1-cyclohexyl-5-methoxy-1,5-dihydro-2*H*-pyrrol-2-one and 4-azido-3-chloro-1-cyclohexyl-5-methoxy-5-(5-((tetrahydro-2*H*-pyran-2-yl)oxy)pent-1-yn-1-yl)-1,5-dihydro-2*H*-pyrrol-2-one in benzene yielded 60% and 55% of 2-(3-(benzyloxy)prop-1-yn-1-yl)-3-chloro-1-cyclohexyl-2-methoxy-4-oxoazetidine-3-carbonitrile and 3-chloro-1-cyclohexyl-2-methoxy-4-oxo-2-(5-((tetrahydro-2*H*-pyran-2-yl)oxy)pent-1-yn-1-yl) azetidine-3-carbonitrile, respectively [430].

A general method for the synthesis of 3-arylidene-4-acyl-β-lactams has recently been developed from the treatment of a mixture of propiolamide and α-bromo-ketone with 3.0 equivalents of K_2CO_3 in CH_3CN at 70°C in the presence of 20 mol% KI, as shown in Scheme 3.157. According to the mechanism outlined, it is possible that a little over 2.0 equivalents of K_2CO_3 might suffice the reaction. This approach has been screened for the effect of the base, solvent, reaction temperature, and various substituents. The results indicate that a base stronger than $NaHCO_3$ is necessary for this reaction, thus K_2CO_3 has been selected as the base. However, other bases such as DBU, K_3PO_4, and Cs_2CO_3 do not obviously improve the reaction compared to K_2CO_3. KI has been used as an additive. Among the tested solvents, including CH_3CN, DMSO, DMF, THF, toluene, and EtOH, CH_3CN appears to be superior to the rest solvents. In addition to α-bromo-ketone, other alkyl bromides with an electron-withdrawing group, such as ester, amidyl, or cyano, work comparably well in this reaction.

The deuterium-labeling experiment indicated that an anion intermediate was formed in the process. The reaction generally affords 3-arylidene-β-lactams in very good to excellent yields. As indicated in Scheme 3.157, the reaction of 2-bromo-1-phenylethan-1-one with *N*-phenyl-3-(*p*-tolyl)propiolamide, 3-(4-chlorophenyl)-*N*-phenylpropiolamide, *N*-phenyl-3-(*o*-tolyl)propiolamide, 3-(naphthalen-2-yl)-*N*-phenylpropiolamide, and *N*-phenyl-3-(thiophen-3-yl)propiolamide under this condition all afford the respective β-lactams in yields from 85% to 89% [497]. In addition to K_2CO_3, it is found that in the presence of a catalytic amount of triphenylphosphine (PPh_3), a number of 2-propiolamidoacetates, *N*-alkyl (or aryl)-*N*-(2-oxoalkyl)propiolamide and *N*-(cyanomethyl)-*N*-alkyl(or aryl) propiolamides can also be converted into 3-methylene-4-substituted β-lactams in refluxing ethanol [498].

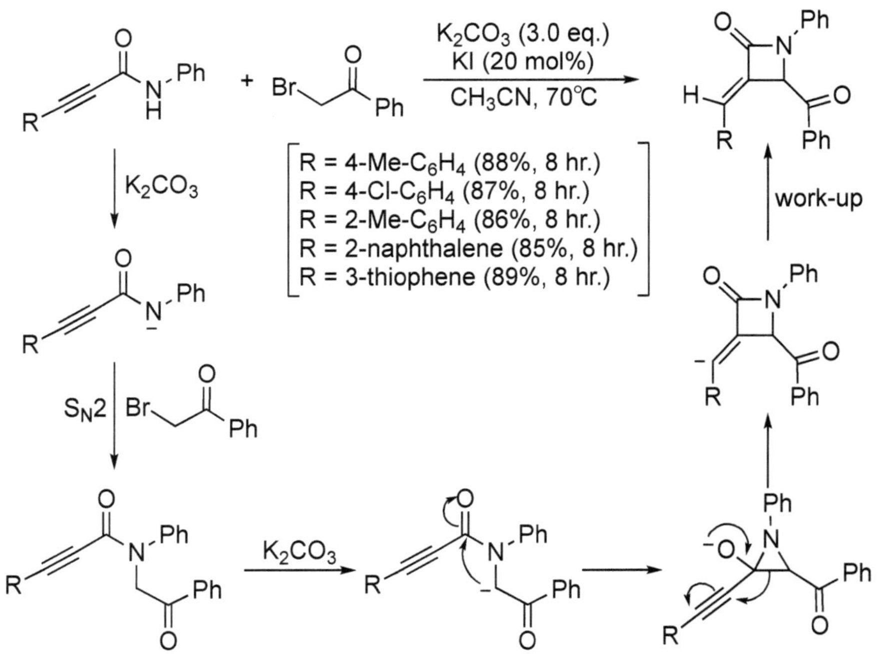

SCHEME 3.157 The mechanism of the reaction between 3-aryl,*N*-phenylpropiolamides and 2-bromoacetophenone that affords β-lactams

Similar to the approach outlined in Scheme 3.157 that potentially involves a ring-enlargement in the mechanism, another approach to form β-lactams also involving the ring enlargement is the treatment of cyclopropanone or a suitable cyclopropanone precursor such as 1-acetoxycyclopropanol in

CH_2Cl_2 with α-amino acid ester at -78°C. The resulting carbinolamine in CH_2Cl_2/CH_3CN was then treated with 1.0 equivalent of *tert*-butyl hypochlorite at about -10°C, followed by a threefold excess of silver nitrate. A representative reaction between cyclopropenone and methyl L-leucinate is illustrated in Scheme 158 [499]. In this particular reaction, treatment of ethyl (1-hydroxycyclopropyl)-*L*-leucinate with *tert*-butyl hypochlorite may lead to ethyl (1-(chlorooxy)cyclopropyl)-*L*-leucinate, from which the chlorine atom is removed by silver cation, generating unstable carbocation. Cyclization and workup complete the reaction process.

An elegant strategy to make cyclopentyl-fused β-lactams with four contiguous stereocenters has been established, which involves the NHC-catalyzed annulations of α,β-unsaturated aldehyde and chalcone-derived imines. This highly enantio- and diastereoselective approach is believed to take place via powerful *aza*-benzoin reaction and *oxy*-Cope rearrangement before the β-lactam moiety is formed. As an example, treatment of a mixture of 1.4 equivalents of (*E*)-but-2-enal and 1.0 equivalent of *N*-((1*Z*,2*E*)-1,3-diphenylallylidene)-4-methoxybenzenesulfonamide with a catalytic amount of chiral NHC, which is *in situ* generated from (5a*R*,10b*S*)-2-mesityl-5a,10b-dihydro-4*H*,6*H*-indeno[2,1-*b*][1,2,4]triazolo[4,3-*d*][1,4]oxazin-2-ium chloride and DBU in EtOAc at room temperature for 15 hours, 94% of (1*S*,2*R*,3*R*,5*R*)-6-((4-methoxyphenyl)sulfonyl)-2-methyl-3,5-diphenyl-6-abicyclo[3.2.0]heptan-7-one was obtained, with greater than 10:1 of diastereoselectivity and > 99% of enantioselectivity (Scheme 3.159) [500]. For this approach, a detailed mechanism has been proposed in the original report.

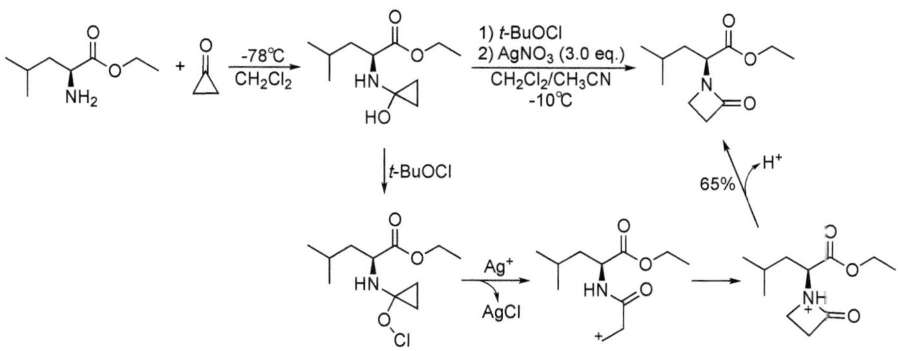

SCHEME 3.158 The mechanism for the creation of ethyl (*S*)-4-methyl-2-(2-oxoazetidin-1-yl)pentanoate from the reaction of cyclopropanone with ethyl *L*-leucinate

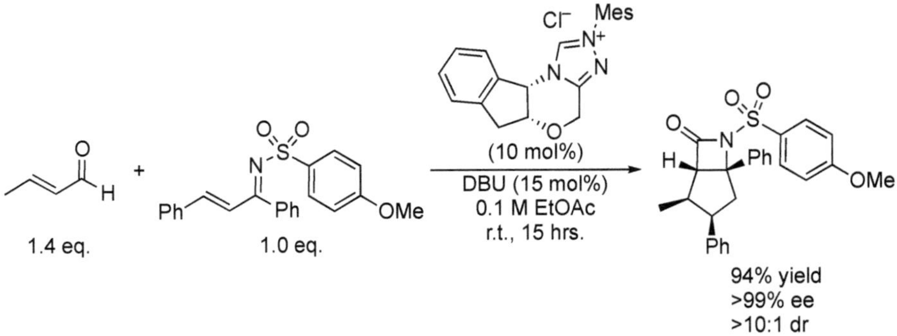

SCHEME 3.159 Stereoselective synthesis of (1*S*,2*R*,3*R*,5*R*)-6-((4-methoxyphenyl) sulfonyl)-2-methyl-3,5-diphenyl-6-azabicyclo[3.2.0]heptan-7-one

It is also worth mentioning a simple method to intramolecularly make 2-alkylideneazetidines from *N*-tosyl-3-halo-3-butenylamines under Ullmann-type coupling condition (0.1–0.2 equivalent of CuI, 0.2–0.4 equivalent of ligand N^1,N^2-dimethylethane-1,2-diamine, 2.0 equivalents of K_2CO_3 or Cs_2CO_3) in dioxane or THF. The resulting 2-alkylideneazetidines can be easily converted into the corresponding β-lactams by oxidation with ozone, as illustrated in Scheme 3.160. When *N*-(3-chloro-1-phenylbut-3-en-1-yl)-4-methylbenzenesulfonamide was selected as the model substrate, a suitable ligand has been screened among dimethylglycine hydrochloride, *L*-proline, 2-aminoethanol, 1,10-phenanthroline and N^1,N^2-dimethylethane-1,2-diamine. A nearly quantitative yield of 2-methylene-4-phenyl-1-tosylazetidine was obtained in the presence of the diamine ligand when the reaction was refluxed in dioxane for 2 hours in the presence of K_2CO_3 or Cs_2CO_3. However, the yield decreased to only 9% when the reaction was refluxed for extended hours (e.g., 15 hours). The study also indicates a reactivity order for halogen on the double bond as I > Br > Cl, as the vinyl chloride is refluxed at 100°C whereas vinyl iodide works at 40°C. Ozonolysis of the 2-alkylideneazetidines in the presence of PPh_3 in CH_2Cl_2 at room temperature gives the corresponding β-lactams in good to excellent yields [501].

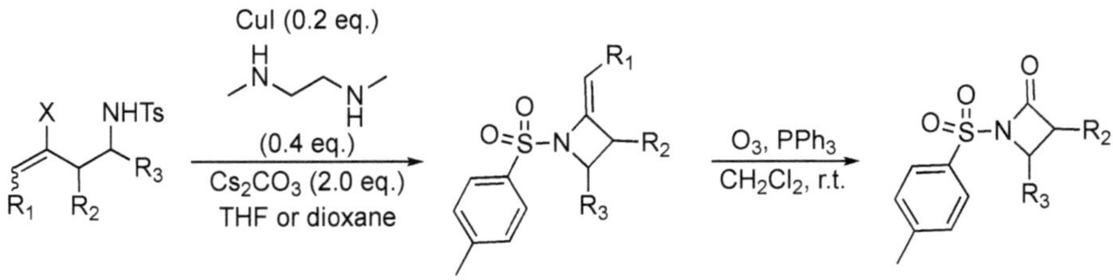

SCHEME 3.160 Formation of β-lactam via the ozonolysis of 2-alkylidene-azetidines

Different from the intramolecular Michael addition to form the β-lactam ring as demonstrated in Schemes 3.59 and 3.60, a palladium catalyst (Pd_2dba_3) catalyzed intramolecular carbon-carbon coupling at room temperature also generates β-lactams in good to excellent yield in most cases. In addition, a selection of ligands allows the possibility of controlling the diastereoselectivity for the resulting β-lactams. Specifically, when (*E*)-4-((2-(*tert*-butylamino)-2-oxo-1-(pyridin-2-yl)ethyl)(3,4-dimethoxyphenethyl) amino)-4-oxobut-2-en-1-yl ethyl carbonate was treated in CH_2Cl_2 in the presence of 1,2-bis(diphenylphosphino)ethane (dppe) ligand and Pd_2dba_3 for 24 hours, 75% of (2*S*,3*S*)-*N*-(*tert*-butyl)-1-(3,4-dimethoxyphenethyl)-4-oxo-2-(pyridin-2-yl)-3-vinylazetidine-2-carboxamide and (2*R*,3*S*)-*N*-(*tert*-butyl)-1-(3,4-dimethoxyphenethyl)-4-oxo-2-(pyridin-2-yl)-3-vinylazetidine-2-carboxamide were yielded, in a ratio of 31:69. For comparison, when the same reaction was performed in the presence of dicyclohexyl(2',6'-dimethoxy-[1,1'-biphenyl]-2-yl)phosphane (SPhos), these two diastereomers were yielded in nearly quantitative yield, in a ratio of 91:9 (Scheme 3.161) [502].

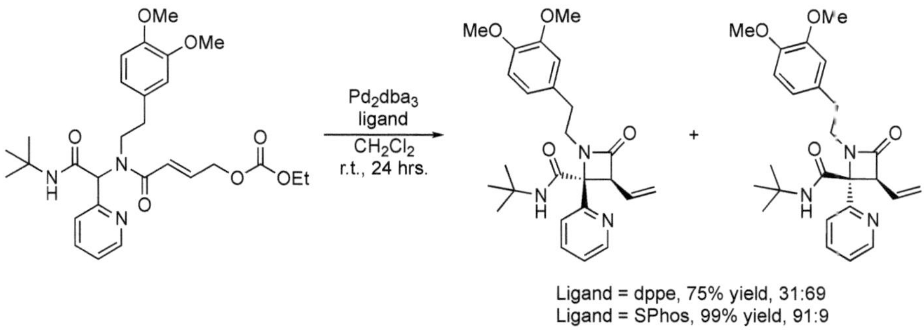

SCHEME 3.161 Formation of β-lactam via intramolecular *C-C* coupling

In contrast to the very popular Staudinger reaction involving the cycloaddition of ketene and imine to yield β-lactam as outlined above, a unique reaction involving a [2+2] cycloaddition between zinc enolate and imine to generate β-lactam was reported. As an example, when ethyl diethylglycinate was first treated with LDA, the resulting lithium enolate was *in situ* transformed into zinc enolate by the addition of $ZnCl_2$. This zinc enolate dimer was then refluxed with *N*-methyl-1-phenylmethanimine. Aqueous workup of reaction solution led to a mixture of (3*S*,4*S*)-3-(diethylamino)-1-methyl-4-phenylazetidin-2-one and its isomer. It is found that the ratio of these two isomers would be affected by the choice of solvent. For instance, when benzene was applied as the solvent, 95% of β-lactam was obtained, with a ratio of 23:77 in favor of the *trans*-isomer; whereas when THF was used as the solvent, 90% of the product was obtained, with a *cis/trans* ratio of 58:42, respectively (Scheme 3.162) [503]. Similarly, (1*R*,2*S*)-2-phenylcyclohexyl 2-((triisopropylsilyl)oxy)acetate was treated with LDA to form lithium (*E*)-1-(((1*R*,2*S*)-2-phenylcyclohexyl)oxy)-2-((triisopropylsilyl)oxy) ethen-1-olate, which underwent [2+2] cycloaddition with (*E*)-1-phenyl-*N*-(trimethylsilyl)methanimine to yield (3*R*,4*S*)-4-phenyl-3-((triisopropylsilyl) oxy)azetidin-2-one, as shown in Scheme 3.163 [504]. This compound was applied to prepare taxol analogs after the opening of the β-lactam ring.

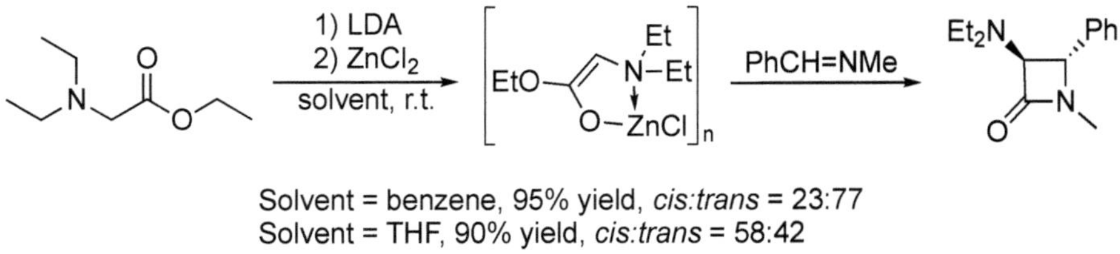

SCHEME 3.162 The solvent impact on the transformation of ethyl diethylglycinate into (3*S*,4*S*)-3-(diethylamino)-1-methyl-4-phenylazetidin-2-one

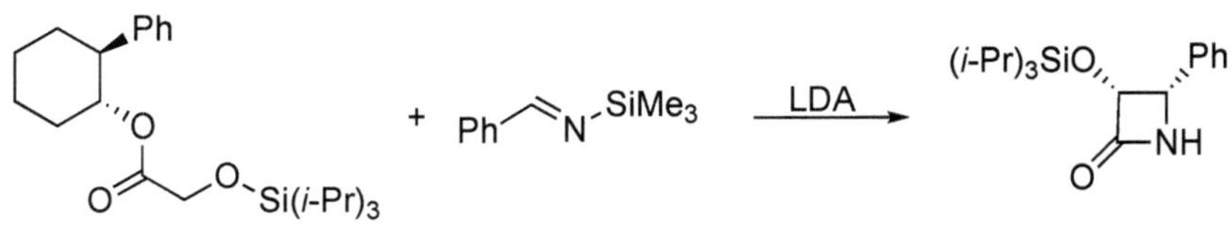

SCHEME 3.163 The preparation of (3*R*,4*S*)-4-phenyl-3-((triisopropylsilyl)oxy)azetidin-2-one

Another simple reaction to yield β-lactam is the photolysis of 2-oxo-amides, such as *N*,*N*-diisopropyl-2-oxo-2-phenylacetamide shown in Scheme 3.164. It is believed that upon photo-activation of the starting material,

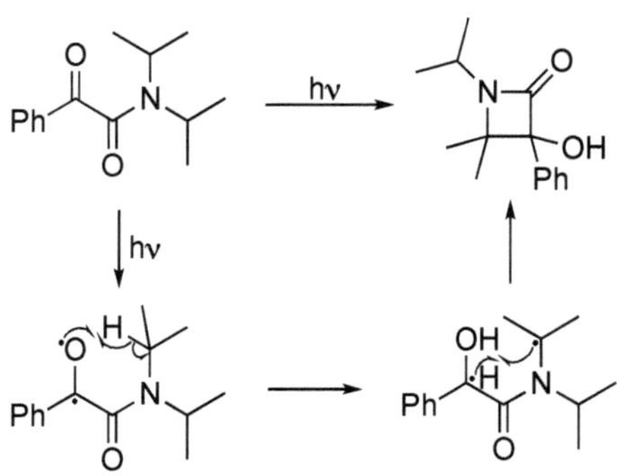

SCHEME 3.164 The mechanism for the photo-transformation of *N*,*N*-diisopropyl-2-oxo-2-phenylacetamide into 3-hydroxy-1-isopropyl-4,4-dimethyl-3-phenylazetidin-2-one

2-oxo functionality behaves like a biradial, and intramolecular abstraction of hydrogen from the isopropyl moiety by alkoxyl radical leads to the carbon radical, and subsequent combination of carbon radials yields 3-hydroxy-1-isopropyl-4,4-dimethyl-3-phenylazetidin-2-one [505]. However, this method would have limited application in the preparation of a variety of β-lactams.

More references on the preparation of β-lactams can be found in the reviews [18, 20, 186, 506].

KEYWORDS

- **Staudinger reaction**
- **Wolff rearrangement**
- **Kinugasa reaction**
- **Baldwin rearrangement**
- **Michael addition**
- **[2+2]-cycloaddition**
- **ring-expansion**
- **enzymatic synthesis**
- **lactamization**

REFERENCES

1. Aranda, M. T., Pérez-Faginas, P., & González-Muñiz, R., (2013). An update on the synthesis of β-lactams. In: *Advances in Organic Synthesis* (pp. 296–354).
2. Hosseyni, S., & Jarrahpour, A., (2018). Recent advances in β-lactam synthesis. *Organic & Biomolecular Chemistry, 16*(38), 6840–6852. doi: 10.1039/c8ob01833b.
3. Pitts, C. R., & Lectka, T., (2014). Chemical synthesis of β-lactams: Asymmetric catalysis and other recent advances. *Chemical Reviews (Washington, DC, United States), 114*(16), 7930–7953. doi: 10.1021/cr4005549.
4. Singh, G. S., & Sudheesh, S., (2014). Advances in synthesis of monocyclic β-lactams. *ARKIVOC (Gainesville, FL, United States),* (1), 337–385. doi: 10.3998/ark.5550190.p008.524.
5. Pilli, R. A., & Russowsky, D., (1997). The stereochemistry of the addition of carbon nucleophiles to imines and iminium ions. *Trends in Organic Chemistry, 6*, 101–123.
6. Southgate, R., (1994). The synthesis of natural β-lactam antibiotics. *Contemporary Organic Synthesis, 1*(6), 417–431. doi: 10.1039/CO9940100417.
7. Wegman, M. A., Janssen, M. H. A., Van, R. F., & Sheldon, R. A., (2001). Towards biocatalytic synthesis of β-lactam antibiotics. *Advanced Synthesis & Catalysis, 343*(6, 7), 559–576. doi: 10.1002/1615-4169(200108)343:6/7<559::AID-ADSC559>3.0.CO;2-Z.

8. Deaguero, A. L., Blum, J. K., & Bommarius, A. S., (2010). Biocatalytic synthesis of β-lactam antibiotics. In: Flickinger, M. C., (ed.), *Encyclopedia of Industrial Biotechnology* (Vol. 1, pp. 535–566).
9. Sklyarenko, A. V., El'darov, M. A., Kurochkina, V. B., & Yarotsky, S. V., (2015). Enzymatic synthesis of β-lactam acids (review). *Applied Biochemistry and Microbiology, 51*(6), 627–640. doi: 10.1134/S0003683815060150.
10. Banik, B. K., (2014). Novel synthesis of β-lactams and their biological evaluation. *Journal of the Indian Chemical Society, 91*(10), 1837–1860.
11. Tuba, R., (2013). Synthesis of β-lactams by transition metal promoted Staudinger reactions: Alternative synthetic approaches from transition metal enhanced organocatalysis to in situ, highly reactive intermediate synthesis and catalytic tandem reactions. *Organic & Biomolecular Chemistry, 11*(36), 5976–5988. doi: 10.1039/c3ob41048j.
12. Samanta, T. B., (2012). Enzymatic synthesis of β-lactams: Constraints and control. *Indian Journal of Biotechnology, 11*(1), 7–15.
13. Magriotis, P. A., (2001). Recent progress in the enantioselective synthesis of β-lactams: Development of the first catalytic approaches. *Angewandte Chemie, International Edition, 40*(23), 4377–4379. doi: 10.1002/1521-3773(20011203)40:23<4377::AID-ANIE4377>3.0.CO;2-J.
14. Kawabata, T., (2000). Asymmetric synthesis based on enolate chemistry: β-Lactam synthesis and memory of chirality. *Reviews on Heteroatom Chemistry, 22*, 33–58.
15. Palomo, C., & Aizpurua, J. M., (1998). Asymmetric synthesis of 3-amino-β-lactams via Staudinger ketene-imine cycloaddition reaction. *Chemistry of Heterocyclic Compounds, 34*(11), 1222–1236. doi: 10.1007/BF02256803.
16. Thomas, R. C., (1990). Synthetic aspects of monocyclic β-lactam antibiotics. In: Lukacs, G., & Ohno, M., (eds.), *Recent Progress in the Chemical Synthesis of Antibiotics* (pp. 533–564).
17. Koppel, G. A., (1983). The synthesis of the β-lactam function. In: *Chemistry of Heterocyclic Compounds (Chichester, United Kingdom)* (Vol. 42, pp. 219–441).
18. Nagpal, R., Bhalla, J., & Bari, S. S., (2019). A comprehensive review on C-3 functionalization of β-lactams. *Current Organic Synthesis, 16*(1), 3–16. doi: 10.2174/1570179415666181116103341.
19. Patil, C. J., & Shinde, A. H., (2018). β-Lactam: Part-I. An overview on synthesis and biological activities of some derivatives of β-lactam. *International Journal of Green and Herbal Chemistry, 7*(2), 201–221. doi: 10.24214/IJGHC/GC/7/2/20121.
20. Sierra, M. A., Casarrubios, L., & De La Torre, M. C., (2019). Bio-organometallic derivatives of antibacterial drugs. *Chemistry – A European Journal, 25*(30), 7232–7242. doi: 10.1002/chem.201805985.
21. Hugel, H., (2005). Symmetry rules. *Chemistry in Australia, 72*(3), 14–16. doi: 10.2555/0314-4240.72.3.1642.
22. Nohira, H., & Nohira, T., (2012). Quantization of chemical reaction: The dynamic correlation diagram method free from noncrossing rule. *Journal of Theoretical & Computational Chemistry, 11*(2), 379–389. doi: 10.1142/S0219633612500253.
23. Langrand, C., Eastes, R. E., & Cheymol, N., (2000). Cycloadditions experiments [4+2] and [2+2] in microchemistry. Correlation diagrams use for setting up Woodward-Hoffmann rules. Dewar-Zimmerman rules. *Actualite Chimique,* (6), 28–38.
24. Heravi, M. M., Ghanbarian, M., Zadsirjan, V., & Alimadadi, J. B., (2019). Recent advances in the applications of Wittig reaction in the total synthesis of natural products

containing lactone, pyrone, and lactam as a scaffold. *Monatshefte fuer Chemie, 150*(8), 1365–1407. doi: 10.1007/s00706-019-02465-9.

25. Muller, C., Cokoja, M., & Kuhn, F. E., (2014). Modern variants of Wittig, Peterson, and Tebbe protocols. In: van Leeuwen, P. W. N. M., (ed.), *Science of Synthesis, C-1 Building Blocks in Organic Synthesis* (Georg Thieme Verlag) (Vol. 2, pp. 1–29).
26. Staudinger, H., (1907). Contribution to our knowledge of the ketenes. First paper. Diphenylketene. *Justus Liebigs Annalen der Chemie, 356*(1, 2), 51–123. doi: 10.1002/jlac.19073560106.
27. Cossio, F. P., Lecea, B., Cuevas, C., Mielgo, A., & Palomo, C., (1993). A novel entry for the asymmetric Staudinger reaction: Experimental and computational studies on the formation of β-lactams through [2+2] cycloaddition reaction of ketenes to imines. *Anales de Quimica, 89*(1), 119–122.
28. Tidwell, T. T., (2008). Hugo (Ugo) Schiff, Schiff bases, and a century of β-lactam synthesis. *Angewandte Chemie, International Edition, 47*(6), 1016–1020. doi: 10.1002/anie.200702965.
29. Holder, R. W., Graf, N. A., Duesler, E., & Moss, J. C., (1983). Gem-alkylcyclopentadienes. 2. Secondary deuterium kinetic isotope effect study of the cycloaddition of diphenylketene and 5,5-dimethylcyclopentadiene. *Journal of the American Chemical Society, 105*(9), 2929–2931. doi: 10.1021/ja00347a086.
30. Pasto, D. J., & Warren, S. E., (1982). Cycloaddition of substituted allenes with 1,1-dichloro-2,2-difluoroethene. A model for the two-step, diradical-intermediate cycloaddition of allenes. *Journal of the American Chemical Society, 104*(13), 3670–3676. doi: 10.1021/ja00377a021.
31. Baldwin, J. E., & Kapecki, J. A., (1969). Kinetic isotope effects on the (2+2) cycloadditions of diphenylketene with α- and β-deuteriostyrene. *Journal of the American Chemical Society, 91*(11), 3106, 3107. doi: 10.1021/ja01039a059.
32. Lemal, D. M., (2017). Pathways for concerted [2 + 2] cycloaddition to cumulenes. *Journal of Organic Chemistry, 82*(24), 13012–13019. doi: 10.1021/acs.joc.7b01911.
33. Marcus, R. A., (1995). Global potential energy contour plots for chemical reactions. Stepwise vs concerted 2 + 2 cycloaddition. *Journal of the American Chemical Society, 117*(16), 4683–4690. doi: 10.1021/ja00121a024.
34. Taggi, A. E., Hafez, A. M., Wack, H., Young, B., Ferraris, D., & Lectka, T., (2002). The Development of the first catalyzed reaction of ketenes and imines: Catalytic, asymmetric synthesis of β-lactams. *Journal of the American Chemical Society, 124*(23), 6626–6635. doi: 10.1021/ja0258226.
35. Jin, J. H., Zhao, J., Yang, W. L., & Deng, W. P., (2019). Asymmetric synthesis of spirooxindole β-lactams via isothiourea-catalyzed mannich/lactamization reaction of aryl acetic acids with isatin-derived ketimines. *Advanced Synthesis & Catalysis, 361*(7), 1592–1596. doi: 10.1002/adsc.201801621.
36. Hodous, B. L., & Fu, G. C., (2002). Enantioselective Staudinger synthesis of β-lactams catalyzed by a planar-chiral nucleophile. *Journal of the American Chemical Society, 124*(8), 1578, 1579. doi: 10.1021/ja012427r.
37. Lee, E. C., Hodous, B. L., Bergin, E., Shih, C., & Fu, G. C., (2005). Catalytic asymmetric Staudinger reactions to form β-lactams: An unanticipated dependence of diastereoselectivity on the choice of the nitrogen substituent. *Journal of the American Chemical Society, 127*(33), 11586, 11587. doi: 10.1021/ja052058p.

38. Duguet, N., Campbell, C. D., Slawin, A. M. Z., & Smith, A. D., (2008). *N*-Heterocyclic carbene catalyzed β-lactam synthesis. *Organic & Biomolecular Chemistry, 6*(6), 1108–1113. doi: 10.1039/b800857b.
39. Zhang, Y. R., He, L., Wu, X., Shao, P. L., & Ye, S., (2008). Chiral *N*-heterocyclic carbene catalyzed Staudinger reaction of ketenes with imines: Highly enantioselective synthesis of N-Boc β-lactams. *Organic Letters, 10*(2), 277–280. doi: 10.1021/ol702759b.
40. Kirichok, A. A., Shton, I. O., Pishel, I. M., Zozulya, S. A., Borysko, P. O., Kubyshkin, V., Zaporozhets, O. A., et al., (2018). Synthesis of multifunctional spirocyclic azetidines and their application in drug discovery. *Chemistry – A European Journal, 24*(21), 5444–5449. doi: 10.1002/chem.201800193.
41. Sheehan, J. C., Buhle, E. L., Corey, E. J., Laubach, G. D., & Ryan, J. J., (1950). Total synthesis of a 5-phenylpenicillin: Methyl 5-phenyl-(2-carbomethoxyethyl)penicillinate. *Journal of the American Chemical Society, 72*, 3828, 3829.
42. Zhang, P., Wang, P., & Li, Y., (2007). Synthesis of new 3-(3-*t*-butoxyl)succinimidyl monocyclic β-lactams. *Chinese Journal of Organic Chemistry, 27*(8), 1031–1033.
43. Zhang, P., Liu, N., Wang, L. Z., & Li, Y., (2009). Synthesis of 3-(3*S*-*t*-butyl)succinimidyl-β-lactams and the stereoselectivity of the reaction. *Gaodeng Xuexiao Huaxue Xuebao, 30*, 90–94.
44. Kanwar, S., Saluja, A., Khurana, J. P. S., & Sharma, S. D., (2001). Stereocontrolled synthesis of β-lactams via Staudinger reaction between phenoxyketenes and chiral imines. *Journal of the Indian Chemical Society, 78*(3), 137–141.
45. Jiao, L., Liang, Y., & Xu, J., (2006). Origin of the relative stereoselectivity of the β-lactam formation in the Staudinger reaction. *Journal of the American Chemical Society, 128*(18), 6060–6069. doi: 10.1021/ja056711k.
46. Kurteva, V., & Alexandrova, M., (2019). Constrained 1-phenylethyl amine analogues as chiral auxiliaries in stereoselective *trans*-β-lactam formation via Staudinger cycloaddition. *Journal of Heterocyclic Chemistry, 56*(3), 930–937. doi: 10.1002/jhet.3471.
47. Alcaide, B., Polanco, C., & Sierra, M. A., (1998). Alkyne-$Co_2(CO)_6$ complexes in the synthesis of fused tricyclic β-lactam and azetidine systems. *Journal of Organic Chemistry, 63*(20), 6786–6796. doi: 10.1021/jo980114h.
48. Ren, X. F., & Turos, E., (1994). Synthesis of novel β-lactam core structures related to the penam and penem antibiotics. *Journal of Organic Chemistry, 59*(20), 5858–5861. doi: 10.1021/jo00099a007.
49. Leon, F., Rivera, D. G., & Wessjohann, L. A., (2008). Multiple multicomponent macrocyclizations including bifunctional building blocks (MiBs) based on Staudinger and Passerini three-component reactions. *Journal of Organic Chemistry, 73*(5), 1762–1767. doi: 10.1021/jo7022125.
50. Palomo, C., Oiarbide, M., Esnal, A., Landa, A., Miranda, J. I., & Linden, A., (1998). Practical synthesis of α-amino acid *N*-carboxy anhydrides of polyhydroxylated α-amino acids from β-lactam frameworks. Model studies toward the synthesis of directly linked peptidyl nucleoside antibiotics. *Journal of Organic Chemistry, 63*(17), 5838–5846.
51. Palomo, C., Oiarbide, M., Landa, A., Esnal, A., & Linden, A., (2001). A β-lactam-based stereoselective access to β,γ-dihydroxy α-amino acid-derived peptides with either α,β-like or unlike configurations. *Journal of Organic Chemistry, 66*(12), 4180–4186. doi: 10.1021/jo001786m.

52. Firestone, R. A., Maciejewicz, N. S., & Christensen, B. G., (1974). Total synthesis of β-lactam antibiotics. VI. 3-Arylephalosporins. *Journal of Organic Chemistry, 39*(23), 3384–3387. doi: 10.1021/jo00937a017.
53. Decuyper, L., Franceus, J., Dhaene, S., Debruyne, M., Vandoorne, K., Piens, N., Dewitte, G., et al., (2018). Chemoenzymatic approach toward the synthesis of 3-*O*-(α/β)-glucosylated 3-hydroxy-β-lactams. *ACS Omega, 3*(11), 15235–15245. doi: 10.1021/acsomega.8b01969.
54. Zhang, P., Li, W. H., & Li, Y., (2004). Synthesis of α-bromo-β-lactam derivatives of 1,5-benzothiazepines under microwave irradition. *Chinese Journal of Organic Chemistry, 24*, 334–337.
55. D'hooghe, M., Mollet, K., De Vreese, R., Jonckers, T. H. M., Dams, G., & De Kimpe, N., (2012). Design, synthesis, and antiviral evaluation of purine-β-lactam and purine-aminopropanol hybrids. *Journal of Medicinal Chemistry, 55*(11), 5637–5641. doi: 10.1021/jm300383k.
56. Hagmann, W. K., Kissinger, A. L., Shah, S. K., Finke, P. E., Dorn, C. P., Brause, K. A., Ashe, B. M., et al., (1993). Orally active β-lactam inhibitors of human leukocyte elastase. 2. Effect of *C*-4 substitution. *Journal of Medicinal Chemistry, 36*(6), 771–777. doi: 10.1021/jm00058a015.
57. Arshad, M. F., Al-Otaibi, F., Kumar, S., Nagarethinam, S., Elkerdasy, A., & Upmanyu, N., (2017). Discovery of novel β-lactam amalgamated 4-methylthiazole-5-carboxylic acid derivatives as potential antimicrobial agents. *Acta Poloniae Pharmaceutica-Drug Research, 74*(6), 1699–1709.
58. Yadav, R. N., Banik, I., & Banik, B. K., (2018). Microwave-induced synthesis of enantiopure β-lactams from *L*-glyceraldehyde. *Journal of the Indian Chemical Society, 95*(11), 1393–1395.
59. Valiullina, Z. R., Galeeva, A. M., Selezneva, N. K., & Miftakhov, M. S., (2018). Synthesis of β-lactam and anomalous minor products in the (*i*-Pr)$_2$NEt-promoted reaction of *N*-chloroglycine methyl ester derivative with dichloroacetyl chloride. *Russian Journal of Organic Chemistry, 54*(10), 1559–1561. doi: 10.1134/S1070428018100196.
60. Saini, P., Bari, S. S., Sahoo, S. C., Khullar, S., Mandal, S. K., & Bhalla, A., (2019). Stereoselective synthesis and characterization of novel *trans*-4-(thiophenyl)pyrazolyl-β-lactams and their *C*-3 functionalization. *Tetrahedron, 75*(33), 4591–4601. doi: doi.org/10.1016/j.tet.2019.07.001.
61. O'Boyle, N. M., Carr, M., Greene, L. M., Bergin, O., Nathwani, S. M., McCabe, T., Lloyd, D. G., et al., (2010). Synthesis and evaluation of azetidinone analogues of combretastatin α-4 as tubulin targeting agents. *Journal of Medicinal Chemistry, 53*(24), 8569–8584. doi: 10.1021/jm101115u.
62. Mozaffari, A., Jarrahpour, A., Alborz, M., & Turos, E., (2019). One-pot multicomponent synthesis of β-lactams via in situ generated imines. *ChemistrySelect, 4*(19), 5950–5953. doi: 10.1002/slct.201900385.
63. Gummidi, L., Kerru, N., Ibeji, C. U., & Singh, P., (2019). Crystal structure and DFT studies of (*E*)-1-(4-fluorophenyl)-3-(1*H*-indol-1-yl)-4-styrylazetidin-2-one. *Journal of Molecular Structure, 1187*, 50–58. doi: 10.1016/j.molstruc.2019.03.053.
64. Uzir, M. H., (2016). Enantioselective Synthesis. In: Drioli, E., & Giorno, L., (eds.), *Encyclopedia of Membranes* (pp. 700–700). Springer Berlin Heidelberg: Berlin, Heidelberg.

65. Xu, J., Yuan, S., Peng, J., Miao, M., Chen, Z., & Ren, H., (2017). Enantioselective [2+2] annulation of simple aldehydes with isatin-derived ketimines via oxidative *N*-heterocyclic carbene catalysis. *Chemical Communications (Cambridge, United Kingdom), 53*(24), 3430–3433. doi: 10.1039/c7cc01232b.
66. Zhang, H. M., Gao, Z. H., & Ye, S., (2014). Bifunctional *N*-heterocyclic carbene-catalyzed highly enantioselective synthesis of spirocyclic oxindolo-β-lactams. *Organic Letters, 16*(11), 3079–3081. doi: 10.1021/ol501205v.
67. Taggi, A. E., Hafez, A. M., Wack, H., Young, B., Drury, W. J. III., & Lectka, T., (2000). Catalytic, asymmetric synthesis of β-lactams. *Journal of the American Chemical Society, 122*(32), 7831, 7832. doi: 10.1021/ja001754g.
68. Mukaiyama, T., Usui, M., Shimada, E., & Saigo, K., (1975). Convenient method for the synthesis of carboxylic esters. *Chemistry Letters,* (10), 1045–1048. doi: 10.1246/cl.1975.1045.
69. Novosjolova, I., (2013). The Mukaiyama reagent: An efficient condensation agent. *Synlett, 24*(1), 135–136. doi: 10.1055/s-0032-1317530.
70. Huang, H., Iwasawa, N., & Mukaiyama, T., (1984). A convenient method for the construction of β-lactam compounds from β-amino acids using 2-chloro-1-methylpyridinium iodide as condensing reagent. *Chemistry Letters,* (8), 1465–1466. doi: 10.1246/cl.1984.1465.
71. Poeylaut-Palena, A. A., & Mata, E. G., (2009). Cross metathesis on solid support. Novel strategy for the generation of β-lactam libraries based on a versatile and multidetachable olefin linker. *Journal of Combinatorial Chemistry, 11*(5), 791–794. doi: 10.1021/cc900072z.
72. Mendez, L., Testero, S. A., & Mata, E. G., (2007). Versatile and efficient solid-supported synthesis of C3-anchored monocyclic β-lactam derivatives. *Journal of Combinatorial Chemistry, 9*(2), 189–192. doi: 10.1021/cc060165p.
73. Mendez, L., & Mata, E. G., (2010). Synthesis of multicyclic β-Lactam derivatives via solid-phase-generated ketenes. *Journal of Combinatorial Chemistry, 12*(6), 810–813. doi: 10.1021/cc100140r.
74. Zarei, M., & Jarrahpour, A., (2011). A mild and efficient route to 2-azetidinones using the cyanuric chloride-DMF complex. *Synlett,* (17), 2572–2576. doi: 10.1055/s-0030–1289517.
75. Bose, A. K., Manhas, M. S., Van, D. V. J. M., Bari, S. S., Wagle, D. R., Hegde, V. R., & Krishnan, L., (1985). Studies on lactams. 73. Enantiospecific synthesis of β-lactams via cycloaddition. *Tetrahedron Letters, 26*(1), 33–36. doi: 10.1016/S0040-4039(00)98458-2.
76. Bananezhad, B., & Islami, M. R., (2017). Stereoselective synthesis of 3-(5-benzoyl-1-methyl-1*H*-pyrrol-2-yl)-2-azetidinone derivatives via an in situ generated ketene. *Synlett, 28*(12), 1453–1456. doi: 10.1055/s-0036-1558974.
77. Wu, G. G., (2000). A concise asymmetric synthesis of a β-lactam-based cholesterol absorption inhibitor. *Organic Process Research & Development, 4*(4), 298–300. doi: 10.1021/op990196r.
78. Gilman, H., & Speeter, M., (1943). Reformatskii reaction with benzalaniline. *Journal of the American Chemical Society, 65*, 2255, 2256.
79. Troisi, L., Granito, C., & Pindinelli, E., (2010). Novel and recent synthesis and applications of β-lactams. *Topics in Heterocyclic Chemistry, 22*(heterocyclic scaffolds I), 101–209. doi: 10.1007/7081_2009_12.

80. Mandal, B., Ghosh, P., & Basu, B., (2010). Recent approaches toward solid phase synthesis of β-lactams. *Topics in Heterocyclic Chemistry, 22*(heterocyclic scaffolds I), 261–311. doi: 10.1007/7081_2009_9.
81. Morrill, L. C., Smith, S. M., Slawin, A. M. Z., & Smith, A. D., (2014). Isothiourea-mediated asymmetric functionalization of 3-alkenoic acids. *Journal of Organic Chemistry, 79*(4), 1640–1655. doi: 10.1021/jo402591v.
82. Palomo, C., Cossio, F. P., Arrieta, A., Odriozola, J. M., Oiarbide, M., & Ontoria, J. M., (1989). The Reformatskii type reaction of Gilman and Speeter in the preparation of valuable β-lactams in carbapenem synthesis: Scope and synthetic utility. *Journal of Organic Chemistry, 54*(24), 5736–5745. doi: 10.1021/jo00285a021.
83. Seitz, J. D., Wang, T., Vineberg, J. G., Honda, T., & Ojima, I., (2018). Synthesis of a next-generation taxoid by rapid methylation amenable for ^{11}C-labeling. *Journal of Organic Chemistry, 83*(5), 2847–2857. doi: 10.1021/acs.joc.7b03284.
84. Drazic, T., Roje, M., Jurin, M., & Pescitelli, G., (2016). Synthesis, separation and absolute configuration determination by ECD spectroscopy and TDDFT calculations of 3-amino-β-lactams and derived guanidines. *European Journal of Organic Chemistry,* (24), 4189–4199. doi: 10.1002/ejoc.201600641.
85. Ojima, I., & Habus, I., (1990). Asymmetric synthesis of β-lactams by chiral ester enolate – imine condensation. *Tetrahedron Letters, 31*(30), 4289–4292 doi: 10.1016/S0040-4039(00)97603-2.
86. Ojima, I., Park, Y. H., Sun, C. M., Brigaud, T., & Zhao, M., (1992). New and efficient routes to norstatine and its analogs with high enantiomeric purity by β-lactam synthon method. *Tetrahedron Letters, 33*(39), 5737–5740. doi: 10.1016/0040-4039(92)89019-9.
87. Shimizu, M., Teramoto, Y., & Fujisawa, T., (1995). Creation of chirality in the reaction of the chiral ester enolate-imine condensation leading to the stereodivergent synthesis of β-lactams. *Tetrahedron Letters, 36*(5), 729–732. doi: 10.1016/0040-4039(94)02327-8.
88. Fujioka, H., Yamanaka, T., Matsunaga, N., Fuji, M., & Kita, Y., (1992). Asymmetric synthesis of carbapenems using chiral acetals: Synthesis of a key intermediate to (+)-PS-5. *Synlett,* (1), 35–36. doi: 10.1055/s-1992-21256.
89. Tomioka, K., Ahmed, H. M., Kambara, T., Fujieda, H., Hayashi, S., Nomura, Y., Kanai, M., & Koga, K., (1999). Catalytic asymmetric reaction of lithium ester enolates with imines. *Chemical Communications (Cambridge),* (8), 715–716. doi: 10.1039/a901424a.
90. Fujieda, H., Kanai, M., Kambara, T., Iida, A., & Tomioka, K., (1997). A ternary complex reagent for an asymmetric reaction of lithium ester enolates with imines. *Journal of the American Chemical Society, 119*(8), 2060, 2061. doi: 10.1021/ja963581u.
91. Kambara, T., & Tomioka, K., (2000). Controlling factors in chiral bisoxazoline-catalyzed asymmetric lithium ester enolate-imine condensation producing a β-lactam. *Chemical & Pharmaceutical Bulletin, 48*(10), 1577–1580. doi: 10.1248/cpb.48.1577.
92. Hussein, M. A., Iida, A., & Tomioka, K., (1999). Studies aimed at enhancement of reactivity and enantioselectivity of a lithium ester enolate using a chiral tridentate lithium amide. *Tetrahedron, 55*(37), 11219–11228. doi: 10.1016/S0040-4020(99)00644-4.
93. Evans, C. D., Mahon, M. F., Andrews, P. C., Muir, J., & Bull, S. D., (2011). Intramolecular ester enolate-imine cyclization reactions for the asymmetric synthesis of polycyclic β-lactams and cyclic β-amino acid derivatives. *Organic Letters, 13*(23), 6276–6279. doi: 10.1021/ol202750u.
94. Michel, K., Froehlich, R., & Wuerthwein, E. U., (2009). Diastereo- and enantioselective synthesis of functionalized β-lactams from oxiranecarbaldimines and lithium ester

enolates. *European Journal of Organic Chemistry,* (32), 5653–5665. doi: 10.1002/ejoc.200900864.
95. Schunk, S., & Enders, D., (2002). Solid-phase synthesis of monocyclic β-lactam derivatives. *Journal of Organic Chemistry, 67*(23), 8034–8042. doi: 10.1021/jo0261552.
96. Schunk, S., & Enders, D., (2000). Solid-phase synthesis of β-lactams via the ester enolate-imine condensation route. *Organic Letters, 2*(7), 907–910. doi: 10.1021/ol0055465.
97. Hovestad, N. J., Ford, A., Jastrzebski, J. T. B. H., & Van, K. G., (2000). Functionalized carbosilane dendritic species as soluble supports in organic synthesis. *Journal of Organic Chemistry, 65*(20), 6338–6344. doi: 10.1021/jo991726k.
98. Abraham, C. J., Paull, D. H., Dogo-Isonagie, C., & Lectka, T., (2009). Diastereoselective synthesis of *trans*-β-lactams using a simple multifunctional catalyst. *Synlett,* (10), 1651–1654. doi: 10.1055/s-0029-1217348.
99. Del, R. E., Lopez, R., Menendez, M. I., Sordo, T. L., & Ruiz-Lopez, M. F., (1998). Theoretical study of ester enolate-imine condensation route to β-lactams. *Journal of Computational Chemistry, 19*(16), 1826–1833. doi: 10.1002/(SICI)1096-987X(199812)19:16<1826::AID-JCC4>3.0.CO;2-N.
100. Panunzio, M., Cozzi, P. G., Kretz, C. M., Bandini, E., & Martelli, G., (1996). Syntheses of β-lactam antibiotics via ester enolate-imine condensation routes. *Topics in Heterocyclic Systems: Synthesis, Reactions and Properties, 1*, 119–140.
101. Benaglia, M., Cinquini, M., & Cozzi, F., (2000). The *S*-thioester enolate/imine condensation: A shortcut to β-lactams. *European Journal of Organic Chemistry,* (4), 563–572. doi: 10.1002/(SICI)1099-0690(200002)2000:4<563::AID-EJOC563>3.0.CO;2-M.
102. Cainelli, G., & Panunzio, M., (1991). Total synthesis of non-classical β-lactam antibiotics by the ester enolate-imine condensation route. *Farmaco, 46*(1, Suppl.), 177–190.
103. Brown, M. J., (1989). Literature review of the ester enolate imine condensation. *Heterocycles, 29*(11), 2225–2244. doi: 10.3987/REV-89-407.
104. Hart, D. J., & Ha, D. C., (1989). The ester enolate-imine condensation route to β-lactams. *Chemical Reviews (Washington, DC, United States), 89*(7), 1447–1465. doi: 10.1021/cr00097a003.
105. Brown, S., Jordan, A. M., Lawrence, N. J., Pritchard, R. G., & Mcgown, A. T., (1998). A convenient synthesis of the paclitaxel side-chain via a diastereoselective Staudinger reaction. *Tetrahedron Letters, 39*(21). doi: 10.1016/S0040–4039(98)00548-6.
106. Jayaraman, M., Deshmukh, A. R., & Bhawal, B. M., (1996). Application of (+)-(1*S*,2*S*)-2-amino-1-phenylpropane-1,3-diol in the formal total synthesis of carbapenems, novel 4-cyano-β-lactams and β-hydroxy aspartates. *Tetrahedron, 52*(27), 8989–9004. doi: 10.1016/0040-4020(96)00463-2.
107. Khasanov, A. B., Ramirez-Weinhouse, M. M., Webb, T. R., & Thiruvazhi, M., (2004). Novel asymmetric approach to proline-derived spiro-β-lactams. *Journal of Organic Chemistry, 69*(17), 5766–5769. doi: 10.1021/jo049430o.
108. Marx, D., & Menendez, M. I., (2018). Torquoselective mechanochemical activation of the Staudinger reaction to form β-lactams. *Journal of Organic Chemistry, 83*(4), 2438–2441. doi: 10.1021/acs.joc.7b03036.
109. Rai, A., Singh, P. K., Shukla, P., & Rai, V. K., (2016). Carbocation catalyzed carboxylic acid activation in Staudinger reaction for stereoselective synthesis of β-lactams. *Tetrahedron Letters, 57*(46), 5084–5088. doi: 10.1016/j.tetlet.2016.10.012.

110. Tukulula, M., Louw, S., Njoroge, M., & Chibale, K., (2020). Synthesis and *in vitro* antiprotozoan evaluation of 4-/8-aminoquinoline-based lactams and tetrazoles. *Molecules, 25*(24), 5941/1–5941/13. doi: 10.3390/molecules25245941.
111. Wang, Y., Liang, Y., Jiao, L., Du, D. M., & Xu, J., (2006). Do reaction conditions affect the stereoselectivity in the Staudinger reaction? *Journal of Organic Chemistry, 71*(18), 6983–6990. doi: 10.1021/jo0611521.
112. Zarei, M., (2017). β-Lactam preparation via Staudinger reaction with activated dimethylsulfoxide. *Journal of Heterocyclic Chemistry, 54*(2), 1161–1166. doi: 10.1002/jhet.2685.
113. Yang, H., Li, H., Wei, G., & Jiang, Z., (2021). Photoredox catalytic phosphite-mediated deoxygenation of α-diketones enables Wolff rearrangement and Staudinger synthesis of β-lactams. *Angewandte Chemie, International Edition, 60*(36), 19696–19700. doi: 10.1002/anie.202107080.
114. Nakahara, K., Yamaguchi, K., Yoshitake, Y., Yamaguchi, T., & Harano, K., (2009). Chemical reactivity of dihydropyrazine derivatives. Cycloaddition behavior toward ketenes. *Chemical & Pharmaceutical Bulletin, 57*(8), 846–852. doi: 10.1248/cpb.57.846.
115. Yang, Z., Li, S., Zhang, Z., & Xu, J., (2014). Base-switched annuloselectivity in the reactions of ethyl malonyl chloride and imines. *Organic & Biomolecular Chemistry, 12*(48), 9822–9830. doi: 10.1039/C4OB01454E.
116. Smith, S. R., Douglas, J., Prevet, H., Shapland, P., Slawin, A. M. Z., & Smith, A. D., (2014). Isothiourea-catalyzed asymmetric synthesis of β-lactams and β-amino esters from arylacetic acid derivatives and *N*-sulfonylaldimines. *Journal of Organic Chemistry, 79*(4), 1626–1639. doi: 10.1021/jo402590m.
117. Kadam, P., Karpoormath, R., Omondi, B., Chenia, H., Ramjugernath, D., & Koorbanally, N. A., (2015). Stereo-selective synthesis, structural and antibacterial studies of novel glycosylated $\beta^{2,3}$-amino acid analogues. *Medicinal Chemistry Research, 24*(8), 3174–3193. doi: 10.1007/s00044-015-1370-4.
118. Shi, J., Linden, A., & Heimgartner, H., (2013). Reactions of acid chlorides/ketenes with 2-substituted 4,5-dihydro-4,4-dimethyl-1,3-thiazoles: Formation of penam derivatives. *Helvetica Chimica Acta, 96*(8), 1462–1481. doi: 10.1002/hlca.201300165.
119. Thi, H. D., Thuy, G. L. N., Catak, S., Van, S. V., Nguyen, T. V., D'Hooghe, M., (2018). Use of 3-hydroxy-4-(trifluoromethyl)azetidin-2-ones as building blocks for the preparation of trifluoromethyl-containing aminopropanes, 1,3-oxazinan-2-ones, aziridines, and 1,4-dioxan-2-ones. *Synthesis, 50*(7), 1439–1456. doi: 10.1055/s-0036-1591537.
120. Berry, S., Bari, S. S., Banik, B. K., & Bhalla, A., (2017). Stereoselective synthesis of novel monocyclic *trans*-3-halogenated-4-pyrazolyl-β-lactams: Potential synthons and promising biologically active agents. *Synthetic Communications, 47*(23), 2239–2246. doi: 10.1080/00397911.2017.1371759.
121. Brandi, A., Cicchi, S., & Cordero, F. M., (2008). Novel syntheses of azetidines and azetidinones. *Chemical Reviews (Washington, DC, United States), 108*(9), 3988–4035. doi: 10.1021/cr800325e.
122. Coquerel, Y., & Rodriguez, J., (2015). The Wolff rearrangement: Tactics, strategies and recent applications in organic synthesis. In: Rojas, C. M., (ed.), *Molecular Rearrangements in Organic Synthesis* (pp. 59–84).
123. Candeias, N. R., Trindade, A. F., Gois, P. M. P., & Afonso, C. A. M., (2014). The Wolff rearrangement. In: Knochel, P., & Molander, G. A., (eds.), *Comprehensive Organic Synthesis*, (2nd edn., Vol. 3, pp. 944–991).

124. Kumar, R. R., & Balasubramanian, M., (2009). Wolff rearrangement. In: Li, J. J., (ed.), *Name Reactions for Homologations* (pp. 257–273).
125. Kirmse, W., (2002). 100 years of the Wolff rearrangement. *European Journal of Organic Chemistry,* (14), 2193–2256. doi: 10.1002/1099-0690(200207)2002:14<2193::AID-EJOC2193>3.0.CO;2-D.
126. Wolff, L., (1902). Über diazoanhydride. *Justus Liebigs Annalen der Chemie, 325*(2), 129–195. doi: 10.1002/jlac.19023250202.
127. Lowe, G., & Ridley, D. D., (1973). Synthesis of β-lactams by photolytic Wolff rearrangement. *Journal of the Chemical Society, Chemical Communications,* (10), 328–329. doi: 10.1039/C39730000328.
128. Lowe, G., & Ridley, D. D., (1973). Synthesis of β-lactams by photolytic Wolff rearrangement. *Journal of the Chemical Society, Perkin Transactions 1: Organic and Bio-Organic Chemistry,* (18), 2024–2029. doi: 10.1039/P19730002024.
129. Corrie, J. E. T., Hlubucek, J. R., & Lowe, G., (1977). Synthesis of a cephalosporin analog. *Journal of the Chemical Society, Perkin Transactions 1: Organic and Bio-Organic Chemistry,* (12), 1421–1425. doi: 10.1039/P19770001421.
130. Tomioka, H., Kondo, M., & Izawa, Y., (1981). Substituent effects on the product distribution in diazo amide photochemistry. Role of ground-state conformational populations. *Journal of Organic Chemistry, 46*(6), 1090–1094. doi: 10.1021/jo00319a010.
131. Podlech, J., (1996). Stereoselective synthesis of aminoalkyl-substituted β-lactams via cycloaddition of ketenes generated from α-amino acids. *Synlett,* (6), 582–584. doi: 10.1055/s-1996–5469.
132. Podlech, J., & Linder, M. R., (1997). Cycloadditions of ketenes generated in the Wolff rearrangement. Stereoselective synthesis of aminoalkyl-substituted β-lactams from α-amino acids. *Journal of Organic Chemistry, 62*(17), 5873–5883. doi: 10.1021/jo9705069.
133. Allen, A. D., Godoy, J., Fu, N., Nagy, M., Spadaro, S., Tidwell, T. T., & Vukovic, S., (2008). Spiro-aziridine and bislactam formation from bisketene-imine cycloadditions. *Journal of the American Chemical Society, 130*(8), 2386–2387. doi: 10.1021/ja077623y.
134. Vaske, Y. S. M., Mahoney, M. E., Konopelski, J. P., Rogow, D. L., & McDonald, W. J., (2010). Enantiomerically pure trans-β-lactams from α-amino acids via compact fluorescent light (CFL) continuous-flow photolysis. *Journal of the American Chemical Society, 132*(32), 11379–11385. doi: 10.1021/ja1050023.
135. Linder, M. R., & Podlech, J., (1999). Synthesis of peptidomimetics containing a β-lactam moiety using peptidic diazoketones and imines in a Staudinger reaction. *Organic Letters, 1*(6), 869–871. doi: 10.1021/ol9908171.
136. Lawlor, M. D., Lee, T. W., & Danheiser, R. L., (2000). Rhodium-catalyzed rearrangement of α-diazo thiol esters to thio-substituted ketenes. Application in the synthesis of cyclobutanones, cyclobutenones, and β-lactams. *Journal of Organic Chemistry, 65*(14), 4375–4384. doi: 10.1021/jo000227c.
137. Chen, L., Wang, K., Shao, Y., & Sun, J., (2019). Stereoselective synthesis of fully substituted β-lactams via metal-organo relay catalysis. *Organic Letters, 21*(10), 3804–3807. doi: 10.1021/acs.orglett.9b01255.
138. Chen, L., Zhang, L., Shao, Y., Xu, G., Zhang, X., Tang, S., & Sun, J., (2019). Rhodium-catalyzed *C=N* bond formation through a rebound hydrolysis mechanism and application in β-lactam synthesis. *Organic Letters, 21*(11), 4124–4127. doi: 10.1021/acs.orglett.9b01312.

139. Liu, M., Chen, Y., & Fu, N., (2013). Convenient synthesis of adamantyl-substituted lactams via uncatalyzed Staudinger reaction. *Synthetic Communications, 43*(7), 1055–1062. doi: 10.1080/00397911.2011.622061.
140. Musio, B., Mariani, F., Sliwinski, E. P., Kabeshov, M. A., Odajima, H., & Ley, S. V., (2016). Combination of enabling technologies to improve and describe the stereoselectivity of Wolff-Staudinger cascade reaction. *Synthesis, 48*(20), 3515–3526. doi: 10.1055/s-0035-1562579.
141. Liu, M., Wang, J. A., Yuan, X., Jiang, R., & Fu, N., (2017). One-pot synthesis of *trans*-β-lactams from ferrocenylketene generated by thermal Wolff rearrangement. *Synthetic Communications, 47*(24), 2369–2377. doi: 10.1080/00397911.2017.1378358.
142. Huang, Z., Wang, C., Tokunaga, E., Sumii, Y., & Shibata, N., (2015). Stereoselective synthesis of β-lactam-triflones under catalyst-free conditions. *Organic Letters, 17*(22), 5610–5613. doi: 10.1021/acs.orglett.5b02827.
143. Synofzik, J., Dar'in, D., Novikov, M. S., Kantin, G., Bakulina, O., & Krasavin, M., (2019). α-Acyl-α-diazoacetates in transition-metal-free β-lactam synthesis. *Journal of Organic Chemistry, 84*(18), 12101–12110. doi: 10.1021/acs.joc.9b02030.
144. Minuto, F., Lambruschini, C., & Basso, A., (2021). Ketene 3-component Staudinger reaction (K-3CSR) to β-lactams: A new entry in the class of photoinduced multicomponent reactions. *European Journal of Organic Chemistry,* (22), 3270–3273. doi: 10.1002/ejoc.202100577.
145. Synofzik, J., Bakulina, O., Balabas, O., Dar'in, D., & Krasavin, M., (2020). Catalyst-free synthesis of diastereomerically pure 3-sulfonylazetidin-2-ones via microwave-assisted tandem Wolff rearrangement-Staudinger cycloaddition. *Synthesis, 52*(20), 3029–3035. doi: 10.1055/s-0040-1707193.
146. Synofzik, J., Bakulina, O., Dar'in, D., Kantin, G., & Krasavin, M., (2020). Dialkyl diazomalonates in transition-metal-free, thermally promoted, diastereoselective Wolff β-lactam synthesis. *Synlett, 31*(13), 1273–1276. doi: 10.1055/s-0040–1707811.
147. Taubinger, A. A., Fenske, D., & Podlech, J., (2008). Synthesis of β,β'-diamino acids from α-amino acid-derived β-lactams by ring opening with nucleophiles. Utilization in the synthesis of peptidomimetics. *Tetrahedron, 64*(37), 8659–8667. doi: 10.1016/j.tet.2008.07.006.
148. Shellhamer, D. F., Alexander, K. L., Bunting, S. A., Elwin, S. L., Licata, C. J., Milligan, J. C., Robinson, R. D., et al., (2015). Improved synthetic utility of a sluggish electrophile: Reaction of chlorosulfonyl isocyanate with unreactive and reactive alkenes. *Synthesis, 47*(13), 1944–1950. doi: 10.1055/s-0034–1380553.
149. Ishiguro, M., Tanaka, R., Namikawa, K., Nasu, T., Inoue, H., Nakatsuka, T., Oyama, Y., & Imajo, S., (1997). 5,6-*cis*-Penems: Broad-spectrum *anti*-methicillin-resistant *Staphylococcus aureus* β-lactam antibiotics. *Journal of Medicinal Chemistry, 40*(14), 2126–2132. doi: 10.1021/jm9703348.
150. Clauss, K., (1969). Reaction of chlorosulfonyl isocyanate with olefins. *Justus Liebigs Annalen der Chemie, 722*, 110–121. doi: 10.1002/jlac.19697220111.
151. Shellhamer, D. F., Davenport, K. J., Hassler, D. M., Hickle, K. R., Thorpe, J. J., Vandenbroek, D. J., Heasley, V. L., et al., (2010). Reaction of chlorosulfonyl isocyanate with fluorosubstituted alkenes: Evidence of a concerted pathway. *Journal of Organic Chemistry, 75*(22), 7913–7916. doi: 10.1021/jo101240s.
152. Shellhamer, D. F., Bunting, S. A., Hickle, K. R., Horn, P. C., Milligan, J. C., Shipowick, D. E., Smith, L. B., et al., (2013). Kinetic studies on the reaction of chlorosulfonyl

isocyanate with monofluoroalkenes: Experimental evidence for both stepwise and concerted mechanisms and a pre-equilibrium complex on the reaction pathway. *Journal of Organic Chemistry, 78*(2), 246–252. doi: 10.1021/jo3016488.
153. Atmaca, U., Daryadel, S., Taslimi, P., Celik, M., & Guelcin, I., (2019). Synthesis of β-amino acid derivatives and their inhibitory profiles against some metabolic enzymes. *Archiv der Pharmazie (Weinheim, Germany), 352*(12), 1900200/1–1900200/9. doi: 10.1002/ardp.201900200.
154. Pirkle, W. H., Bowen, W. E., & Vuong, D. V., (1994). Liquid chromatographic separation of the enantiomers of cyclic β-amino esters as their *N*-3,5-dinitrobenzoyl derivatives. *Journal of Chromatography A, 676*(2), 297–302. doi: 10.1016/0021-9673(94)80429-X.
155. Black, T. H., Olson, J. T., & Abt, D. C., (1992). A short synthesis of γ-lactams via the spontaneous ring expansion of β-lactams. *Synthetic Communications, 22*(18), 2729–2733. doi: 10.1080/00397919208021675.
156. Pinazzi, C. P., Noireaux, P., & Reyx, D., (1974). Functionalization of polyalkadienes. Study of the molecular models obtainable from isoprene polymers by utilization of chlorosulfonylisocyanate. *Makromolekulare Chemie, 175*(10), 2849–2863. doi: 10.1002/macp.1974.021751006.
157. Moriconi, E. J., & Meyer, W. C., (1971). Reaction of dienes with chlorosulfonyl isocyanate. *Journal of Organic Chemistry, 36*(19), 2841–2849. doi: 10.1021/jo00818a025.
158. Haug, T., Lohse, F., Metzger, K., & Batzer, H., (1968). Preparations and reactions of β-lactams. *Helvetica Chimica Acta, 51*(8), 2069–2089. doi: 10.1002/hlca.19680510828.
159. Moriconi, E. J., & Meyer, W. C., (1968). 1,2- And 1,4-cycloaddition of chlorosulfonyl isocyanate to dienes. *Tetrahedron Letters,* (35), 3823–3827. doi: 10.1016/S0040-4039(01)99111-7.
160. Friedrich, H. J., (1971). Cycloaddition of chlorosulfonyl isocyanate to stilbene. *Tetrahedron Letters,* (31), 2981–2984. doi: 10.1016/S0040-4039(01)97041-8.
161. Moriconi, E. J., & Kelly, J. F., (1968). The stereospecific cycloaddition of chlorosulfonyl isocyanate to *cis*- and *trans*-β-methylstyrene and *cis*- and *trans*-3-hexene. *Tetrahedron Letters,* (12), 1435–1439. doi: 10.1016/S0040-4039(01)98973-7.
162. Paquette, L. A., Wyvratt, M. J., & Allen, G. R. Jr., (1970). Unsaturated heterocyclic systems. LXX. Stereochemistry of the thermal fragmentation of β-lactams. Comparison with the pyrolysis of 1-azetines. *Journal of the American Chemical Society, 92*(6), 1763–1765. doi: 10.1021/ja00709a060.
163. Liang, G. J., & Chen, A. Q., (2011). Synthesis and crystal structure of 4-oxo-3-azatricyclo[5.3.1.02,5]undecane-9-carboxylic acid methyl ester. *Gaodeng Xuexiao Huaxue Xuebao, 32*(2), 300–305.
164. Dener, J. M., Fantauzzi, P. P., Kshirsagar, T. A., Kelly, D. E., & Wolfe, A. B., (2001). Large-scale syntheses of Fmoc-protected non-proteogenic amino acids: Useful building blocks for combinatorial libraries. *Organic Process Research & Development, 5*(4), 445–449. doi: 10.1021/op010204f.
165. Hauser, F. M., & Ellenberger, S. R., (1987). A superior procedure for the preparation of 2-azetidinones from volatile olefins. *Synthesis,* (3), 324–324. doi: 10.1055/s-1987-27936.
166. Lee, M. R., Stahl, S. S., & Gellman, S. H., (2008). Synthesis of β-lactams bearing functionalized side chains from a readily available precursor. *Organic Letters, 10*(22), 5317–5319. doi: 10.1021/ol802274x.
167. Cheung, L. L. W., & Yudin, A. K., (2010). Synthesis of highly substituted cyclobutane fused-ring systems from *N*-vinyl β-lactams through a one-pot domino process. *Chemistry – A European Journal, 16*(13), 4100–4109. doi: 10.1002/chem.200902748.

168. Barrett, A. G. M., Betts, M. J., & Fenwick, A., (1985). Acyl and sulfonyl isocyanates in β-lactam synthesis. *Journal of Organic Chemistry, 50*(2), 169–175. doi: 10.1021/jo00202a006.
169. Williams, C. I., & Whitehead, M. A., (1999). Semi-empirical study of isocyanate geometries, and β-lactam formation through alkene-isocyanate cyclo-addition reactions. *Journal of Molecular Structure: Theochem, 491*, 93–101. doi: 10.1016/S0166-1280(99)00092-5.
170. Cossio, F. P., Roa, G., Lecea, B., & Ugalde, J. M., (1995). Substituent and solvent effects in the [2 + 2] cycloaddition reaction between olefins and isocyanates. *Journal of the American Chemical Society, 117*(49), 12306–12313. doi: 10.1021/ja00154a033.
171. Charif, I. E., Mekelleche, S. M., & Ahmed, A. B., (2003). Quantum study of cycloaddition of olefins with isocyanates. *Journal de la Societe Algerienne de Chimie, 13*(1), 75–83.
172. Cossio, F. P., Lecea, B., Lopez, X., Roa, G., Arrieta, A., & Ugalde, J. M., (1993). An *ab initio* study on the mechanism of the alkene-isocyanate cycloaddition reaction to form β-lactams. *Journal of the Chemical Society, Chemical Communications*, (18), 1450–1452. doi: 10.1039/C39930001450.
173. Freitag, D., Drees, M., Goutal, S., Strassner, T., & Metz, P., (2005). Synthetic and computational studies on intramolecular [2 + 2] sulfonyl isocyanate-olefin cycloadditions. *Tetrahedron, 61*(23), 5615–5621. doi: 10.1016/j.tet.2005.03.075.
174. Furman, B., Kaluza, Z., Stencel, A., Grzeszczyk, B., & Chmielewski, M., (2007). β-Lactams from carbohydrates. *Topics in Heterocyclic Chemistry, 7* (Heterocycles from carbohydrate precursors), 101–132. doi: 10.1007/7081_2006_046.
175. Furman, B., Borsuk, K., Kaluza, Z., Lysek, R., & Chmielewski, M., (2004). Stereochemistry of [2+2] cycloaddition of chlorosulfonyl isocyanate to olefins. *Current Organic Chemistry, 8*(6), 463–473. doi: 10.2174/1385272043485792.
176. Arrieta, A., Lecea, B., & Cossio, F. P., (2010). Computational studies on the synthesis of β-lactams via [2+2] thermal cycloadditions. *Topics in Heterocyclic Chemistry, 22*(heterocyclic scaffolds I), 313–347. doi: 10.1007/7081_2009_10.
177. Barrett, A. G. M., Mortier, J., Sabat, M., & Sturgess, M. A., (1988). Iron(II) vinylidenes and chromium carbene complexes in β-lactam synthesis. *Organometallics 7*(12), 2553–2561. doi: 10.1021/om00102a022.
178. Barrett, A. G. M., & Sturgess, M. A., (1987). Cationic iron vinylidene complexes in bicyclic β-lactam synthesis. *Journal of Organic Chemistry, 52*(17), 3940, 3941. doi: 10.1021/jo00226a047.
179. Balijepalli, A. S., McNeely, J. H., Hamoud, A., & Grinstaff, M. W., (2020). Guidelines for β-lactam synthesis: Glycal protecting groups dictate stereoelectronics and [2+2] cycloaddition kinetics. *Journal of Organic Chemistry, 85*(19), 12044–12057. doi: 10.1021/acs.joc.0c00510.
180. Borsuk, K., Kazimierski, A., Solecka, J., Urbanczyk-Lipkowska, Z., & Chmielewski, M., (2002). Stereocontrolled formation of oxacephams from carbohydrates. *Carbohydrate Research, 337*(21–23), 2005–2015. doi: 10.1016/S0008-6215(02)00139-8.
181. Cierpucha, M., Solecka, J., Frelek, J., Szczukiewicz, P., & Chmielewski, M., (2004). Synthesis, biological, and chiroptical activity of 3-phenyl-clavams. *Bioorganic & Medicinal Chemistry, 12*(2), 405–416. doi: 10.1016/j.bmc.2003.10.043.
182. Kardos, M., Kiss, L., & Fueloep, F., (2015). Stereocontrolled synthesis of difunctionalized azetidinones and $\beta^{2,3}$-amino acid derivatives from cyclodienes by ring-opening and cross-metathesis reactions. *Asian Journal of Organic Chemistry, 4*(10), 1155–1159. doi: 10.1002/ajoc.201500286.

183. Malpass, J. R., & Tweddle, N. J., (1972). Control over cyclization paths in the reaction of chlorosulfonyl isocyanate with olefins. *Journal of the Chemical Society, Chemical Communications,* (22), 1244–1245. doi: 10.1039/c39720001244.
184. Moriconi, E. J., Kelly, J. F., & Salomone, R. A., (1968). Olefinic intermediates in the reaction of 1,1-dimethyl-1,1,2-trimethyl and 1,1,2,2-tetramethylcyclopropanes with chlorosulfonyl isocyanate. *Journal of Organic Chemistry, 33*(9), 3448–3452. doi: 10.1021/jo01273a019.
185. Shellhamer, D. F., Brady, D. L., Flores, F. V., & Perry, M. C., (2020). Reaction of *p*-toluenesulfonyl isocyanate with electron-rich alkenes and monofluoroalkenes. *Journal of Undergraduate Chemistry Research, 19*(1), 10–13.
186. Miyamoto, H., Urbanczyk-Lipkowska, Z., & Toda, F., (2000). Enantioselective photoreaction of amides to β-lactams in the solid state. *Trends in Organic Chemistry, 8*, 93–100.
187. Pasto, D. J., Chen, A. F. T., Ciurdaru, G., & Paquette, L. A., (1973). Uniparticulate electrophilic addition to alkenylidenecyclopropanes. *Journal of Organic Chemistry, 38*(5), 1015–1026. doi: 10.1021/jo00945a033.
188. Kinugasa, M., & Hashimoto, S., (1972). The reactions of copper(I) phenylacetylide with nitrones. *Journal of the Chemical Society, Chemical Communications,* 466, 467. doi: 10.1039/C39720000466.
189. Stecko, S., Furman, B., & Chmielewski, M., (2014). Kinugasa reaction: An 'ugly duckling' of β-lactam chemistry. *Tetrahedron, 70*(43), 7817–7844. doi: 10.1016/j.tet.2014.06.024.
190. Chigrinova, M., MacKenzie, D. A., Sherratt, A. R., Cheung, L. L. W., & Pezacki, J. P., (2015). Kinugasa reactions in water: From green chemistry to bioorthogonal labeling. *Molecules, 20*(5), 6959–6969. doi: 10.3390/molecules20046959.
191. Khangarot, R. K., & Kaliappan, K. P., (2013). Kinugasa reaction: A direct one-pot route to highly functionalized β-lactams. *European Journal of Organic Chemistry,* (34), 7664–7677. doi: 10.1002/ejoc.201300597.
192. Aranda, M. T., Perez-Faginas, P., & Muniz, R. G., (2013). An update on the synthesis of β-lactams. *Advances in Organic Synthesis, 6*, 296–354. doi: 10.2174/9781608050291113060008.
193. Mandal, B., & Basu, B., (2013). Synthesis of β-lactams through alkyne-nitrone cycloadditions. *Topics in Heterocyclic Chemistry, 30*(β-lactams), 85–110. doi: 10.1007/7081_2012_85.
194. Aranda, M. T., Perez-Faginas, P., & Gonzalez-Muniz, R., (2009). An update on the synthesis of β-lactams. *Current Organic Synthesis, 6*(3), 325–341. doi: 10.2174/157017909788921899.
195. Marco-Contelles, J., (2004). β-Lactam synthesis by the Kinugasa reaction. *Angewandte Chemie, International Edition, 43*(17), 2198–2200. doi: 10.1002/anie.200301730.
196. Pal, R., Ghosh, S. C., Chandra, K., & Basak, A., (2007). Synthesis of β-lactams using the Kinugasa reaction. *Synlett,* (15), 2321–2330. doi: 10.1055/s-2007-986637.
197. Shintani, R., & Fu, G. C., (2003). Catalytic enantioselective synthesis of β-lactams: Intramolecular Kinugasa reactions and interception of an intermediate in the reaction cascade. *Angewandte Chemie, International Edition, 42*(34), 4082–4085. doi: 10.1002/anie.200352103.
198. Santoro, S., Liao, R. Z., Marcelli, T., Hammar, P., & Himo, F., (2015). Theoretical study of mechanism and stereoselectivity of catalytic Kinugasa reaction. *Journal of Organic Chemistry, 80*(5), 2649–2660. doi: 10.1021/jo502838p.

199. Ye, M. C., Zhou, J., & Tang, Y., (2006). Trisoxazoline/Cu(II)-promoted Kinugasa reaction. Enantioselective synthesis of β-lactams. *Journal of Organic Chemistry, 71*(9), 3576–3582. doi: 10.1021/jo0602874.
200. Ding, L. K., & Irwin, W. J., (1976). *Cis*- and *trans*-azetidin-2-ones from nitrones and copper acetylide. *Journal of the Chemical Society, Perkin Transactions 1: Organic and Bio-Organic Chemistry,* (22), 2382–2386. doi: 10.1039/P19760002382.
201. Malig, T. C., Yu, D., & Hein, J. E., (2018). A revised mechanism for the Kinugasa reaction. *Journal of the American Chemical Society, 140*(29), 9167–9173. doi: 10.1021/jacs.8b04635.
202. Mames, A., Stecko, S., Mikolajczyk, P., Soluch, M., Furman, B., & Chmielewski, M., (2010). Direct, catalytic synthesis of carbapenams via cycloaddition/rearrangement cascade reaction: Unexpected acetylenes' structure effect. *Journal of Organic Chemistry, 75*(22), 7580–7587. doi: 10.1021/jo101355h.
203. Kabala, K., Grzeszczyk, B., Stecko, S., Furman, B., & Chmielewski, M., (2015). Approach to monobactams and nocardicins via diastereoselective Kinugasa reaction. *Journal of Organic Chemistry, 80*(24), 12038–12046. doi: 10.1021/acs.joc.5b01979.
204. Mucha, L., Parda, K., Staszewska-Krajewska, O., Stecko, S., Ulikowski, A., Frelek, J., Suszczynska, A., et al., (2016). Diastereoselective synthesis of β-lactams via Kinugasa reaction of acyclic chiral nitrones. *Tetrahedron: Asymmetry, 27*(1), 12–21. doi: 10.1016/j.tetasy.2015.11.006.
205. Kutaszewicz, R., Grzeszczyk, B., Gorecki, M., Staszewska-Krajewska, O., Furman, B., & Chmielewski, M., (2019). Bypassing the stereoselectivity issue: Transformations of Kinugasa adducts from chiral alkynes and achiral open-chain nitrones. *Organic & Biomolecular Chemistry, 17*(25), 6251–6268. doi: 10.1039/c9ob00940j.
206. Wolosewicz, K., Michalak, M., Adamek, J., & Furman, B., (2016). Studies on the enantioselective Kinugasa reaction: Efficient synthesis of β-lactams catalyzed by *N*-PINAP/CuX complexes. *European Journal of Organic Chemistry,* (12), 2212–2219. doi: 10.1002/ejoc.201600050.
207. Kumar, Y., Singh, P., & Bhargava, G., (2016). Cu(I) mediated Kinugasa reactions of α,β-unsaturated nitrones: A facile, diastereoselective route to 3-(hydroxy/bromo)methyl-1-aryl-4-(-styryl)azetidin-2-ones. *New Journal of Chemistry, 40*(10), 8216–8219. doi: 10.1039/C6NJ01747A.
208. Kowalski, M. K., Mloston, G., Obijalska, E., Linden, A., & Heimgartner, H., (2016). First application of fluorinated nitrones for the synthesis of fluoroalkylated β-lactams via the Kinugasa reaction. *Tetrahedron, 72*(35), 5305–5313. doi: 10.1016/j.tet.2016.06.073.
209. Takayama, Y., Ishii, T., Ohmiya, H., Iwai, T., Schwarzer, M. C., Mori, S., Taniguchi, T., et al., (2017). Asymmetric synthesis of β-lactams through copper-catalyzed alkyne-nitrone coupling with a prolinol-phosphine chiral ligand. *Chemistry – A European Journal, 23*(35), 8400–8404. doi: 10.1002/chem.201702070.
210. Imai, K., Takayama, Y., Murayama, H., Ohmiya, H., Shimizu, Y., & Sawamura, M., (2019). Asymmetric synthesis of α-alkylidene-β-lactams through copper catalysis with a prolinol-phosphine chiral ligand. *Organic Letters, 21*(6), 1717–1721. doi: 10.1021/acs.orglett.9b00276.
211. Kowalski, M. K., Mloston, G., Obijalska, E., & Heimgartner, H., (2017). Application of diethyl ethynylphosphonate to the synthesis of 3-phosphonylated β-lactams via the Kinugasa reaction. *ARKIVOC (Gainesville, FL, United States),* (2), 59–67. doi: 10.3998/ark.5550190.p009.660.

212. Popik, O., Grzeszczyk, B., Staszewska-Krajewska, O., Furman, B., & Chmielewski, M., (2020). Synthesis of β-lactams via diastereoselective, intramolecular Kinugasa reactions. *Organic & Biomolecular Chemistry, 18*(15), 2852–2860. doi: 10.1039/d0ob00228c.
213. Shu, T., Zhao, L., Li, S., Chen, X. Y., Von, E. C., Rissanen, K., & Enders, D., (2018). Asymmetric synthesis of spirocyclic β-lactams through copper-catalyzed Kinugasa/Michael domino reactions. *Angewandte Chemie, International Edition, 57*(34), 10985–10988. doi: 10.1002/anie.201806931.
214. Hosseini, A., & Schreiner, P. R., (2019). Synthesis of exclusively 4-substituted β-lactams through the Kinugasa reaction utilizing calcium carbide. *Organic Letters, 21*(10), 3746–3749. doi: 10.1021/acs.orglett.9b01192.
215. Ahn, C., Kennington, J. W. Jr., & DeShong, P., (1994). A new approach to the synthesis of monocyclic β-lactam derivatives. *Journal of Organic Chemistry, 59*(21), 6282–6286. doi: 10.1021/jo00100a033.
216. Wang, Y., Zheng, Z., & Zhang, L., (2014). Ruthenium-catalyzed oxidative transformations of terminal alkynes to ketenes by using tethered sulfoxides. Access to β-lactams and cyclobutanones. *Angewandte Chemie, International Edition, 53*(36), 9572–9576. doi: 10.1002/anie.201403796.
217. Kim, I., Roh, S. W., Lee, D. G., & Lee, C., (2014). Rhodium-catalyzed oxygenative [2 + 2] cycloaddition of terminal alkynes and imines for the synthesis of β-lactams. *Organic Letters, 16*(9), 2482–2485. doi: 10.1021/ol500856z.
218. Basak, A., Ghosh, S. C., Bhowmich, T., Das, A. K., & Bertolasi, V., (2002). An asymmetric synthesis of β-lactams: On the use of chiral oxazolidones in the Kinugasa reaction. *Tetrahedron Letters, 43*(31), 5499–5501. doi: 10.1016/S0040-4039(02)00974-7.
219. Basak, A., & Pal, R., (2005). Synthesis of β-lactam nucleoside chimera via Kinugasa reaction and evaluation of their antibacterial activity. *Bioorganic & Medicinal Chemistry Letters, 15*(8), 2015–2018. doi: 10.1016/j.bmcl.2005.02.064.
220. Pal, R., & Basak, A., (2006). A novel synthesis of β-lactam fused cyclic enediynes by intramolecular Kinugasa reaction. *Chemical Communications (Cambridge, United Kingdom),* (28), 2992–2994. doi: 10.1039/B605743H.
221. Basak, A., Pal, R., & Das, S., (2009). Synthesis of β-lactam fused enediynes by intramolecular Kinugasa reaction: Comparison of reactivity with monocyclic analogues. *International Journal of Chemistry (Toronto, ON, Canada), 1*(1), 63–74. doi: 10.5539/ijc.v1n1p63.
222. Stecko, S., Mames, A., Furman, B., & Chmielewski, M., (2008). Diastereoselective synthesis of carbapenams via Kinugasa reaction. *Journal of Organic Chemistry, 73*(18), 7402–7404. doi: 10.1021/jo801212q.
223. Khangarot, R. K., & Kaliappan, K. P., (2011). A stereoselective synthesis of sugar-derived chiral β-lactams. *European Journal of Organic Chemistry,* (30), 6117–6127, S6117/1–S6117/28. doi: 10.1002/ejoc.201100953.
224. Stecko, S., Mames, A., Furman, B., & Chmielewski, M., (2009). Asymmetric Kinugasa reaction of cyclic nitrones and nonracemic acetylenes. *Journal of Organic Chemistry, 74*(8), 3094–3100. doi: 10.1021/jo900121x.
225. Grzeszczyk, B., Polawska, K., Shaker, Y. M., Stecko, S., Mames, A., Woznica, M., Chmielewski, M., & Furman, B., (2012). Asymmetric Kinugasa reaction involving six-membered cyclic nitrones. *Tetrahedron, 68*(52), 10633–10639. doi: 10.1016/j.tet.2012.09.031.

226. Zhang, X., Hsung, R. P., Li, H., Zhang, Y., Johnson, W. L., & Figueroa, R., (2008). A highly stereoselective synthesis of chiral α-amino-β-lactams via the Kinugasa reaction employing ynamides. *Organic Letters, 10*(16), 3477–3479. doi: 10.1021/ol801257j.
227. Piotrowska, D. G., Bujnowicz, A., Wroblewski, A. E., & Glowacka, I. E., (2015). A new approach to the synthesis of 4-phosphonylated β-lactams. *Synlett, 26*(3), 375–379. doi: 10.1055/s-0034–1379506.
228. Hussein, M., Nasr El, D. A., Fares, F., Dorcet, V., Hachem, A., & Gree, R., (2016). A new direct synthesis of α-methylene- and α-alkylidene-β-lactams. *Tetrahedron Letters, 57*(18), 1990–1993. doi: 10.1016/j.tetlet.2016.03.083.
229. Nasr El, D. A., Gree, D., Roisnel, T., Caytan, E., Hachem, A., & Gree, R., (2016). Synthesis of fluorine-containing exoalkylidene β-lactams. *European Journal of Organic Chemistry,* (3), 556–561. doi: 10.1002/ejoc.201501347.
230. Michalak, M., Stodulski, M., Stecko, S., Woznica, M., Staszewska-Krajewska, O., Kalicki, P., Furman, B., et al., (2012). Synthesis of *N*,4-diaryl substituted β-lactams via Kinugasa cycloaddition/rearrangement reaction. *Tetrahedron, 68*(52), 10806–10817. doi: 10.1016/j.tet.2011.11.007.
231. Gao, Z., Zhang, H., Qin, L., & He, K., (2020). Progress in the synthesis of ezetimibe. *Zhongguo Yiyao Gongye Zazhi, 51*(8), 21–38. doi: 10.16522/j.cnki.cjph.2020.08.002.
232. Lo, M. M. C., & Fu, G. C., (2002). Cu(I)/bis(azaferrocene)-catalyzed enantioselective synthesis of β-lactams via couplings of alkynes with nitrones. *Journal of the American Chemical Society, 124*(17), 4572, 4573. doi: 10.1021/ja025833z.
233. Coyne, A. G., Mueller-Bunz, H., & Guiry, P. J., (2007). The asymmetric synthesis of β-lactams: HETPHOX/Cu(I) mediated synthesis via the Kinugasa reaction. *Tetrahedron: Asymmetry, 18*(2), 199–207. doi: 10.1016/j.tetasy.2007.01.006.
234. Baeza, B., Casarrubios, L., & Sierra, M. A., (2013). Towards a general synthesis of 3-metal-substituted β-lactams. *Chemistry – A European Journal, 19*(35), 11536–11540. doi: 10.1002/chem.201301114.
235. Basak, A., Chandra, K., Pal, R., & Ghosh, S. C., (2007). Kinugasa reaction under click chemistry conditions. *Synlett,* (10), 1585–1588. doi: 10.1055/s-2007-980383.
236. Basak, A., & Ghosh, S. C., (2004). L-Proline-mediated one-pot synthesis of 3-exomethylene β-lactams via Kinugasa reaction. *Synlett,* (9), 1637–1639. doi: 10.1055/s-2004-829097.
237. Chen, J. H., Liao, S. H., Sun, X. L., Shen, Q., & Tang, Y., (2012). Tris(oxazoline)/copper-catalyzed coupling of alkynes with nitrones: A highly enantioselective access to β-lactams. *Tetrahedron, 68*(25), 5042–5045. doi: 10.1016/j.tet.2012.04.049.
238. Maciejko, M., Stecko, S., Staszewska-Krajewska, O., Jurczak, M., Furman, B., & Chmielewski, M., (2012). An entry to the carbapenem antibiotic scaffold via the asymmetric Kinugasa reaction. *Synthesis, 44*(18), 2825–2839. doi: 10.1055/s-0032-1316732.
239. Soluch, M., Grzeszczyk, B., Staszewska-Krajewska, O., Chmielewski, M., & Furman, B., (2016). Synthesis of thienamycin methyl ester from 2-deoxy-D-ribose via Kinugasa reaction. *Journal of Antibiotics, 69*(3), 164–168. doi: 10.1038/ja.2015.108.
240. Qi, J., Wei, F., Huang, S., Tung, C. H., & Xu, Z., (2021). Copper(I)-catalyzed asymmetric interrupted Kinugasa reaction: Synthesis of α-thiofunctional chiral β-lactams. *Angewandte Chemie, International Edition, 60*(9), 4561–4565. doi: 10.1002/anie.202013450.
241. Qi, J., Wei, F., Tung, C. H., & Xu, Z., (2021). Modular synthesis of α-quaternary chiral β-lactams by a synergistic copper/palladium-catalyzed multicomponent reaction. *Angewandte Chemie, International Edition, 60*(25), 13814–13818. doi: 10.1002/anie.202100601.

242. Saito, T., Kikuchi, T., Tanabe, H., Yahiro, J., & Otani, T., (2009). Enantioselective synthesis of β-lactams via the indabox-Cu(II)-catalyzed Kinugasa reaction. *Tetrahedron Letters, 50*(35), 4969–4972. doi: 10.1016/j.tetlet.2009.06.050.
243. Xu, C., Yang, Y., Wu, Y., He, F., He, H., Deng, P., & Zhou, H., (2020). Development of TsDPEN based imine-containing ligands for the copper-catalyzed asymmetric Kinugasa reaction. *RSC Advances, 10*(31), 18107–18114. doi: 10.1039/D0RA03276J.
244. Zarei, M., (2020). $CuFe_2O_4$ nanoparticles catalyze the reaction of alkynes and nitrones for the synthesis of 2-azetidinones. *New Journal of Chemistry, 44*(40), 17341–17345. doi: 10.1039/d0nj02660c.
245. McKay, C. S., Kennedy, D. C., & Pezacki, J. P., (2009). Studies of multicomponent Kinugasa reactions in aqueous media. *Tetrahedron Letters, 50*(17), 1893–1896. doi: 10.1016/j.tetlet.2009.02.035.
246. Santoro, S., & Himo, F., (2021). Mechanism of the Kinugasa reaction revisited. *Journal of Organic Chemistry, 86*(15), 10665–10671. doi: 10.1021/acs.joc.1c01351.
247. Yoshimura, T., Takuwa, M., Tomohara, K., Uyama, M., Hayashi, K., Yang, P., Hyakutake, R., et al., (2012). Protonation-assisted conjugate addition of axially chiral enolates: Asymmetric synthesis of multisubstituted β-lactams from α-amino acids. *Chemistry – A European Journal, 18*(48), 15330–15336. doi: 10.1002/chem.201201339.
248. Hyakutake, R., Yoshimura, T., Ueda, Y., Hayashi, K., Furuta, T., & Kawabata, T., (2018). Asymmetric synthesis of β-lactams by intramolecular conjugate addition of serine and cysteine derivatives via memory of chirality. *Heterocycles, 97*(2, Spec. Issue), 1128–1147. doi: 10.3987/com-18-s(t)95.
249. Clayden, J., Watson, D. W., Helliwell, M., & Chambers, M., (2003). β-Lactams or γ-lactams by 4-*exo-trig* or 5-*endo-trig* anionic cyclisation of lithiated acrylamide derivatives. *Chemical Communications (Cambridge, United Kingdom),* (20), 2582, 2583. doi: 10.1039/B308029C.
250. Martinez-Cuezva, A., Lopez-Leonardo, C., Bautista, D., Alajarin, M., & Berna, J., (2016). Stereocontrolled synthesis of β-lactams within [2]-rotaxanes: Showcasing the chemical consequences of the mechanical bond. *Journal of the American Chemical Society, 138*(28), 8726–8729. doi: 10.1021/jacs.6b05581.
251. Martinez-Cuezva, A., Bautista, D., Alajarin, M., & Berna, J., (2018). Enantioselective formation of 2-azetidinones by ring-assisted cyclization of interlocked *N*-(α-methyl) benzyl fumaramides. *Angewandte Chemie, International Edition, 57*(22), 6563–6567. doi: 10.1002/anie.201803187.
252. Martinez-Cuezva, A., Lopez-Leonardo, C., Alajarin, M., & Berna, J., (2019). Stereocontrol in the synthesis of β-lactams arising from the interlocked structure of benzylfumaramide-based hydrogen-bonded [2]rotaxanes. *Synlett, 30*(8), 893–902. doi: 10.1055/s-0037-1611705.
253. Takabatake, T., Yoneda, T., Otsuka, J., Kagawa, N., & Toyota, M., (2019). Artificial intelligence designed drug synthesis: One-pot preparation of trans β-lactams and application to cholesterol absorption inhibitor SCH 47949 synthesis. *Tetrahedron Letters, 60*(34), 150942/1–150942/4. doi: 10.1016/j.tetlet.2019.07.033.
254. Panunzio, M., Bandini, E., D'Aurizio, A., Xia, Z., & Mu, X., (2008). Synthesis of venlafaxine from azadiene via a *hetero*-Diels-Alder approach: New microwave-assisted transketalization and hydroxymethylation reactions. *Synthesis,* (11), 1753–1756. doi: 10.1055/s-2008-1072581.

255. Mu, X., Xia, Z., Wang, C., Panunzio, M., & Zeng, L., (2009). *Hetero* Diels-Alder reaction versus intramolecular cyclization of 2-aza-butadiene in the synthesis of venlafaxine. *Youji Huaxue, 29*(1), 61–65.
256. Bongini, A., Panunzio, M., & Venturini, A., (2010). Silyloxyazadienes: One intermediate and two competitive pericyclic reactions. *Physical Chemistry Chemical Physics, 12*(19), 5067–5073. doi: 10.1039/b925351c.
257. Komjati, B., Szokol, B., Kovats, B., Kegye, P., & Nagy, J., (2018). DFT study of stereoselective ketene-imine cycloadditions, evaluation of possible solvent effects with IEF-PCM. *Periodica Polytechnica, Chemical Engineering, 62*(4), 503–509. doi: 10.3311/PPch.12527.
258. Bongini, A., Xia, Z., & Panunzio, M., (2007). An *ab initio* study of substituent effects on the electrocyclization of silyloxyazadienes. *European Journal of Organic Chemistry,* (21), 3533–3538, S3533/1–S3533/24. doi: 10.1002/ejoc.200700036.
259. Suarez, D., & Sordo, T. L., (1997). Theoretical study of the zwittazido cleavage of 4-azido-2-pyrrolinones: The role of solvent and substituents. *Journal of the American Chemical Society, 119*(43), 10291–10301. doi: 10.1021/ja970517j.
260. Bongini, A., Panunzio, M., Tamanini, E., Martelli, G., Vicennati, P., & Monari, M., (2003). Lewis acid-catalyzed electrocyclization of 2-aza-1,3-butadienes to *NH*-β-lactams. *Tetrahedron: Asymmetry, 14*(8), 993–998. doi: 10.1016/S0957-4166(03)00162-9.
261. Arrieta, A., Cossio, F. P., & Lecea, B., (2000). New insights on the origins of the stereocontrol of the Staudinger reaction: [2 + 2] cycloaddition between ketenes and *N*-silylimines. *Journal of Organic Chemistry, 65*(25), 8458–8464. doi: 10.1021/jo0007736.
262. Long, S., Monari, M., Panunzio, M., Bandini, E., D'Aurizio, A., & Venturini, A., (2011). *Hetero*-Diels-Alder (HDA) strategy for the preparation of 6-aryl- and heteroaryl-substituted piperidin-2-one scaffolds: Experimental and theoretical studies. *European Journal of Organic Chemistry,* (31), 6218–6225, S6218/1–S6218/28. doi: 10.1002/ejoc.201100930.
263. Wolfe, S., Ro, S., & Shi, Z., (2001). Generation, electrocyclic ring opening, and unprecedented conversion of a 3-acylaminoazetinone into *cis*-3,4-disubstituted azetidinones. *Canadian Journal of Chemistry, 79*(8), 1259–1271. doi: 10.1139/v01-102.
264. Vallavoju, N., Sreenithya, A., Ayitou, A. J. L., Jockusch, S., Sunoj, R. B., & Sivaguru, J., (2016). Photoreactions with a twist: Atropisomerism-driven divergent reactivity of enones with *UV* and visible light. *Chemistry – A European Journal, 22*(32), 11339–11348. doi: 10.1002/chem.201601509.
265. Toda, F., Miyamoto, H., Takeda, K., Matsugawa, R., & Maruyama, N., (1993). Some enantioselective photocyclization reactions in inclusion crystals with optically active host compounds. *Journal of Organic Chemistry, 58*(23), 6208–6211. doi: 10.1021/jo00075a013.
266. Le Blanc, S., Pete, J. P., & Piva, O., (1992). New access to spiranic β-lactams. *Tetrahedron Letters, 33*(15), 1993–1996. doi: 10.1016/0040-4039(92)88122-L.
267. Toda, F., Miyamoto, H., Inoue, M., Yasaka, S., & Matijasic, I., (2000). Enantioselective photocyclization of amides to β-lactam derivatives in inclusion crystals with an optically active host. *Journal of Organic Chemistry, 65*(9), 2728–2732. doi: 10.1021/jo991832m.
268. Sivasubramanian, K., Kaanumalle, L. S., Uppili, S., & Ramamurthy, V., (2007). Value of zeolites in asymmetric induction during photocyclization of pyridones, cyclohexadienones and naphthalenones. *Organic & Biomolecular Chemistry, 5*(10), 1569–1576. doi: 10.1039/b702572f.

269. Wu, L. C., Cheer, C. J., Olovsson, G., Scheffer, J. R., Trotter, J., Wang, S. L., & Liao, F. L., (1997). Crystal engineering for absolute asymmetric synthesis through the use of *meta*-substituted aryl groups. *Tetrahedron Letters, 38*(18), 3135–3138. doi: 10.1016/S0040-4039(97)00615-1.
270. Kumarasamy, E., Jesuraj, J. L., Omlid, J. N., Ugrinov, A., & Sivaguru, J., (2011). Light-induced enantiospecific 4π ring closure of axially chiral 2-pyridones: Enthalpic and entropic effects promoted by *H*-bonding. *Journal of the American Chemical Society, 133*(43), 17106–17109. doi: 10.1021/ja203087a.
271. Somekawa, K., Oda, H., & Shimo, T., (1991). Intramolecular photochemistry of 3-alkynyloxy-2-pyridones and isomerization of the photoadduct to tricyclic β-lactam. *Chemistry Letters,* (12), 2077, 2078. doi: 10.1246/cl.1991.2077.
272. Kametani, T., Mochizuki, T., & Honda, T., (1982). Synthetic studies on β-lactam antibiotics: Conversion of 2-pyridone into azetidin-2-one. *Heterocycles, 19*(1), 89–90. doi: 10.3987/R-1982–01–0089.
273. Begley, W. J., Lowe, G., Cheetham, A. K., & Newsam, J. M., (1981). Novel β-lactams derived from the photoisomer of *N*-benzyloxycarbonylmethylene-2-pyridone. *Journal of the Chemical Society, Perkin Transactions 1: Organic and Bio-Organic Chemistry,* (9), 2620–2624.
274. Tanaka, K., Fujiwara, T., & Urbanczyk-Lipkowska, Z., (2002). Highly enantioselective photocyclization of 1-alkyl-2-pyridones to β-lactams in inclusion crystals with optically active host compounds. *Organic Letters, 4*(19), 3255–3257. doi: 10.1021/ol026497u.
275. Shailaja, J., Karthikeyan, S., & Ramamurthy, V., (2002). Cyclodextrin mediated solvent-free enantioselective photocyclization of *N*-alkylpyridones. *Tetrahedron Letters, 43*(51), 9335–9339. doi: 10.1016/S0040-4039(02)02338-9.
276. Kurita, J., Yoneda, T., Kakusawa, N., & Tsuchiya, T., (1990). Studies on seven-membered heterocycles. XXXI. Synthesis of 1,4-oxazepinones and 1,4-diazepinones from 2-pyridones and their conversion into fully unsaturated 1,4-oxazepines and 1,4-diazepines. *Chemical & Pharmaceutical Bulletin, 38*(11), 2911–2918. doi: 10.1248/cpb.38.2911.
277. Toda, F., & Tanaka, K., (1988). Enantioselective photoconversion of pyridones into β-lactam derivatives in inclusion complexes with optically active host compounds. *Tetrahedron Letters, 29*(34), 4299–4302. doi: 10.1016/S0040-4039(00)80480-3.
278. Haaf, K., & Rüchardt, C., (1990). A new approach to enantiomerically pure β-lactams from α-amino acids by applying the isonitrile-nitrile rearrangement. *Chemische Berichte, 123*(3), 635–638. doi: 10.1002/cber.19901230333.
279. Ihara, M., Haga, Y., Yonekura, M., Ohsawa, T., Fukumoto, K., & Kametani, T., (1983). Synthesis of β-lactam antibiotics by the sulfeno-cycloamination. *Journal of the American Chemical Society, 105*(25), 7345–7352. doi: 10.1021/ja00363a023.
280. Miller, M. J., (1986). Hydroxamate approach to the synthesis of β-lactam antibiotics. *Accounts of Chemical Research, 19*(2), 49–56. doi: 10.1021/ar00122a004.
281. Okawara, T., Matsuda, T., Noguchi, Y., & Furukawa, M., (1982). A one-pot synthesis of β-lactams by the reaction of β-haloacyl chlorides with α-amino acids. *Chemical & Pharmaceutical Bulletin, 30*(5), 1574–1578. doi: 10.1248/cpb.30.1574.
282. Nakatsuka, S., Tanino, H., & Kishi, Y., (1975). Biogenetic-type synthesis of penicillin-cephalosporin antibiotics. II. Oxidative cyclization route to β-lactam thiazoline derivatives. *Journal of the American Chemical Society, 97*(17), 5010–5012. doi: 10.1021/ja00850a043.

283. Biloski, A. J., Wood, R. D., & Ganem, B., (1982). A new β-lactam synthesis. *Journal of the American Chemical Society, 104*(11), 3233–3235. doi: 10.1021/ja00375a057.
284. Adam, W., Groer, P., Humpf, H. U., & Saha-Moeller, C. R., (2000). Synthesis of optically active α-methylene β-lactams through lipase-catalyzed kinetic resolution. *Journal of Organic Chemistry, 65*(16), 4919–4922. doi: 10.1021/jo0003089.
285. Zhang, Q., Qiu, Z. Y., Lan, Z. Y., & Zheng, J. N., (2008). Synthesis of aztreonam. *Zhongguo Xinyao Zazhi, 17*(5), 393–395.
286. Manchand, P. S., Luk, K. C., Belica, P. S., Choudhry, S. C., Wei, C. C., & Soukup, M., (1988). A novel synthesis of the monobactam antibiotic carumonam. *Journal of Organic Chemistry, 53*(23), 5507–5512. doi: 10.1021/jo00258a020.
287. Molina, L., Perani, A., Infante, M. R., Manresa, M. A., Maugras, M., Achilefu, S., Steve, M. J., & Selve, C., (1995). Bioactive surfactants containing A β-lactam group: Synthesis and properties. *Journal of the Chemical Society, Chemical Communications,* (12), 1279, 1280. doi: 10.1039/C39950001279.
288. Hughes, D. L., (1992). *The Mitsunobu Reaction* (Vol. 42). In Organic Reactions (Hoboken, NJ, United States), Paquette, L. A. Ed., Wiley.
289. Mattingly, P. G., Kerwin, J. F. Jr., & Miller, M. J., (1979). A facile synthesis of substituted *N*-hydroxy-2-azetidinones. A biogenetic type β-lactam synthesis. *Journal of the American Chemical Society, 101*(14), 3983–3985. doi: 10.1021/ja00508a056.
290. Durham, T. B., & Miller, M. J., (2002). Conversion of glucuronic acid glycosides to novel bicyclic β-lactams. *Organic Letters, 4*(1), 135–138. doi: 10.1021/ol017026v.
291. Misner, J. W., Fisher, J. W., Gardner, J. P., Pedersen, S. W., Trinkle, K. L., Jackson, B. G., & Zhang, T. Y., (2003). Enantioselective synthesis of the carbacephem antibiotic loracarbef via mitsunobu and Dieckmann cyclization from an unnatural amino acid. *Tetrahedron Letters, 44*(32), 5991–5993. doi: 10.1016/S0040-4039(03)01483-7.
292. Rzasa, R. M., Shea, H. A., & Romo, D., (1998). Total synthesis of the novel, immunosuppressive agent (-)-Pateamine A from *Mycale sp.* Employing a β-lactam-based macrocyclization. *Journal of the American Chemical Society, 120*(3), 591, 592. doi: 10.1021/ja973549f.
293. Lotz, B. T., & Miller, M. J., (1993). Diastereoselective synthesis of the carbacephem framework. *Journal of Organic Chemistry, 58*(3), 618–625. doi: 10.1021/jo00055a013.
294. Swaren, P., Massova, I., Bellettini, J. R., Bulychev, A., Maveyraud, L., Kotra, L. P., Miller, M. J., et al., (1999). Elucidation of mechanism of inhibition and *x*-ray structure of the TEM-1 β-lactamase from *Escherichia coli* inhibited by A N-sulfonyloxy-β-lactam. *Journal of the American Chemical Society, 121*(23), 5353–5359. doi: 10.1021/ja990400q.
295. Ageno, G., Banfi, L., Cascio, G., Guanti, G., Manghisi, E., Riva, R., & Rocca, V., (1995). Enantiospecific and diastereoselective synthesis of 4,4-disubstituted 3-amino-2-azetidinones, starting from δ-serine. *Tetrahedron, 51*(29), 8121–8134. doi: 10.1016/0040-4020(95)00427-A.
296. Meloni, M. M., & Taddei, M., (2001). Solid-phase synthesis of β-lactams via the Miller hydroxamate approach. *Organic Letters, 3*(3), 337–340. doi: 10.1021/ol006779z.
297. Frlan, R., Hrast, M., & Gobec, S., (2019). Application of the *N*-dibenzyl protective group in the preparation of β-lactam pseudopeptides. *Molecules, 24*(7), 1261–1275. doi: 10.3390/molecules24071261.
298. Kamimura, A., Morita, R., Matsuura, K., Mitsudera, H., & Shirai, M., (2003). A convenient stereoselective synthesis of β-lactams from β-hydroxy-α-thioalkylesters prepared from

Michael/Aldol tandem reaction or stereoselective addition of thiols to the Baylis-Hillman adducts. *Tetrahedron, 59*(50), 9931–9938. doi: 10.1016/j.tet.2003.10.035.
299. Malachowski, W. P., Tie, C., Wang, K., & Broadrup, R. L., (2002). The synthesis of azapeptidomimetic β-lactam molecules as potential protease inhibitors. *Journal of Organic Chemistry, 67*(25), 8962–8969. doi: 10.1021/jo026280d.
300. Zegrocka, O., Abramski, W., Urbanczyk-Lipkowska, Z., & Chmielewski, M., (1993). Formation of β-lactams fused to the pyranoid ring via the Mitsunobu reaction. *Carbohydrate Research, 307*(1, 2), 33–43. doi: 10.1016/S0008-6215(98)00017-2.
301. Bonache, M. A., Cativiela, C., Garcia-Lopez, M. T., & Gonzalez-Muniz, R., (2006). β-Lactams derived from phenylalanine and homologues: Effects of the distance between the aromatic rings and the α-stereogenic reactive center on the memory of chirality. *Tetrahedron Letters, 47*(33), 5883–5887. doi: 10.1016/j.tetlet.2006.06.057.
302. Gerona-Navarro, G., Garcia-Lopez, M. T., & Gonzalez-Muniz, R., (2002). General approach for the stereocontrolled construction of the β-lactam ring in amino acid-derived 4-alkyl-4-carboxy-2-azetidinones. *Journal of Organic Chemistry, 67*(11), 3953–3956. doi: 10.1021/jo025571j.
303. Perez-Faginas, P., O'Reilly, F., O'Byrne, A., Garcia-Aparicio, C., Martin-Martinez, M., Perez De, V. M. J., Garcia-Lopez, M. T., & Gonzalez-Muniz, R., (2007). Exceptional stereoselectivity in the synthesis of 1,3,4-trisubstituted 4-carboxy β-lactam derivatives from amino acids. *Organic Letters, 9*(8), 1593–1596. doi: 10.1021/ol070533d.
304. Qian, X., Zheng, B., Burke, B., Saindane, M. T., & Kronenthal, D. R., (2002). A stereoselective synthesis of BMS-262084, an azetidinone-based tryptase inhibitor. *Journal of Organic Chemistry, 67*(11), 3595–3600. doi: 10.1021/jo010757o.
305. Santa, Z., Parkanyi, L., Nemeth, I., Nagy, J., & Nyitrai, J., (2001). Synthesis of pure methyl [(2*S*,3*R*,α*R*)-1-(3-bromo-4-methoxyphenyl)-3-(α-acetoxy)ethyl-4-oxoazetidin-2-carboxylate] and its enantiomer. *Tetrahedron: Asymmetry, 12*(1), 89–94. doi: 10.1016/S0957-4166(01)00003-9.
306. Bose, A. K., Ghosh-Mazumdar, B. N., & Chatterjee, B. G., (1960). Ease of cyclization to the β-lactam ring. *Journal of the American Chemical Society, 82*, 2382–2386. doi: 10.1021/ja01494a066.
307. Sheehan, J. C., & Bose, A. K., (1950). A new synthesis of β-lactams. *Journal of the American Chemical Society, 72*(11), 5158–5161. doi: 10.1021/ja01167a099.
308. Pedroni, J., Boghi, M., Saget, T., & Cramer, N., (2014). Access to β-lactams by enantioselective palladium(0)-catalyzed C(sp^3)-H alkylation. *Angewandte Chemie, International Edition, 53*(34), 9064–9067. doi: 10.1002/anie.201405508.
309. Kugelman, M., Gala, D., Jaret, R. S., Nyce, P. L., & McPhail, A. T., (1990). Synthesis of azetidinones from *L*-threonine: Formation of unusually stable bicyclic hemiketals and *cis*-azetidinones. *Synlett,* (7), 431, 432. doi: 10.1055/s-1990-21119.
310. Hanessian, S., Bedeschi, A., Battistini, C., & Mongelli, N., (1985). A new synthetic strategy for the penems. Total synthesis of (5*R*,6*S*,8*R*)-6-(α-hydroxyethyl)-2-(hydroxymethyl)penem-3-carboxylic acid. *Journal of the American Chemical Society, 107*(5), 1438, 1439. doi: 10.1021/ja00291a069.
311. Jacobi, P. A., Murphree, S., Rupprecht, F., & Zheng, W., (1996). Formal total syntheses of the β-lactam antibiotics thienamycin and PS-5. *Journal of Organic Chemistry, 61*(7), 2413–2427. doi: 10.1021/jo952092u.
312. Nilsson, B. M., Ringdahl, B., & Hacksell, U., (1990). β-Lactam analogs of oxotremorine. 3- And 4-methyl-substituted 2-azetidinones. *Journal of Medicinal Chemistry, 33*(2), 580–584. doi: 10.1021/jm00164a018.

313. Turner, J. J., Sikkema, F. D., Filippov, D. V., Van, D. M. G. A., & Van, B. J. H., (2001). Synthesis of β-lactams via ring opening of a serine derived aziridine. *Synlett,* (11), 1727–1730. doi: 10.1055/s-2001-18090.
314. Aydin, B. O., Yurtoglu, E., Arcelik, N., Sahin, E., Secen, H., & Altundas, R., (2019). Asymmetric synthesis of aminocyclooctenecarbonitriles: Cyclooctene β-lactam and hydroxyaminocyclooctene carboxylic precursors. *Tetrahedron Letters, 60*(3), 284–287. doi: 10.1016/j.tetlet.2018.12.034.
315. Kang, S. H., Kim, M., & Ryu, D. H., (2003). A stereoselective synthesis of a key intermediate to 1β-methylcarbapenem via aziridine ring-opening reaction. *Synlett,* (8), 1149, 1150. doi: 10.1055/s-2003–39910.
316. Palomo, C., Aizpurua, J. M., Balentova, E., Jimenez, A., Oyarbide, J., Fratila, R. M., & Miranda, J. I., (2007). Synthesis of β-lactam scaffolds for ditopic peptidomimetics. *Organic Letters, 9*(1), 101–104. doi: 10.1021/ol0626241.
317. Wang, X., Meng, F., Wang, Y., Han, Z., Chen, Y. J., Liu, L., Wang, Z., & Ding, K., (2012). Aromatic spiroketal bisphosphine ligands: Palladium-catalyzed asymmetric allylic amination of racemic Morita-Baylis-Hillman adducts. *Angewandte Chemie, International Edition, 51*(37), 9276–9282. doi: 10.1002/anie.201204925.
318. Zhou, P., Liu, Y., Zhou, L., Zhu, K., Feng, K., Zhang, H., Liang, Y., et al., (2016). Potent antitumor activities and structure basis of the chiral β-lactam bridged analogue of combretastatin α-4 binding to tubulin. *Journal of Medicinal Chemistry, 59*(22), 10329–10334. doi: 10.1021/acs.jmedchem.6b01268.
319. Shimizu, M., Kume, K., & Fujisawa, T., (1995). Switchover of the diastereoselectivity induced by a simple selection of titanium(IV) halides in the addition reaction of silyl ketene acetals to a chiral imine. *Tetrahedron Letters, 36*(29), 5227–5230. doi: 10.1016/0040-4039(95)00978-L.
320. Tatsuta, K., Takahashi, M., Tanaka, N., & Chikauchi, K., (2000). Novel synthesis of (+)-4-acetoxy-3-hydroxyethyl-2-azetidinone from carbohydrate. A formal total synthesis of (+)-thienamycin. *Journal of Antibiotics, 53*(10), 1231–1234. doi: 10.7164/antibiotics.53.1231.
321. Momoi, Y., Okuyama, K. I., Toya, H., Sugimoto, K., Okano, K., & Tokuyama, H., (2014). Total synthesis of (-)-haouamine B pentaacetate and structural revision of haouamine B. *Angewandte Chemie, International Edition, 53*(48), 13215–13219. doi: 10.1002/anie.201407686.
322. Paquette, L. A., Rothhaar, R. R., Isaac, M., Rogers, L. M., & Rogers, R. D., (1998). Diastereo- and enantiodifferentiation in indium-promoted allylations of 2,3-azetidinediones in water. Definition of long-range stereocontrol elements on π-facial selectivity for β-lactam synthesis. *Journal of Organic Chemistry, 63*(16), 5463–5472. doi: 10.1021/jo980372e.
323. Allmendinger, T., Rihs, G., & Wetter, H., (1988). Synthesis of an optically active penem intermediate by asymmetric amidoalkylation. *Helvetica Chimica Acta, 71*(2), 395–403. doi: 10.1002/hlca.19880710213.
324. Mukaiyama, T., Matsueda, R., & Suzuki, M., (1970). Peptide synthesis via the oxidation-reduction condensation by the use of 2,2'-dipyridyldisulfide as an oxidant. *Tetrahedron Letters,* (22), 1901–1904. doi: 10.1016/S0040-4039(01)98113-4.
325. Kobayashi, S., Limori, T., Izawa, T., & Ohno, M., (1981). $Ph_3P-(PyS)_2-CH_3CN$ as an excellent condensing system for β-lactam formation from β-amino acids. *Journal of the American Chemical Society, 103*(9), 2406–2408. doi: 10.1021/ja00399a044.

326. Ham, W. H., Oh, C. Y., Lee, Y. S., & Jeong, J. H., (2000). A new synthesis of a key intermediate of β-lactam antibiotics via diastereoselective alkylation of β-hydroxy ester. *Journal of Organic Chemistry, 65*(24), 8372–8374. doi: 10.1021/jo000467s.
327. Miyachi, N., Kanda, F., & Shibasaki, M., (1989). Use of copper(I) trifluoromethanesulfonate in β-lactam synthesis. *Journal of Organic Chemistry, 54*(15), 3511–3513. doi: 10.1021/jo00276a001.
328. Palomo, C., Aizpurua, J. M., Urchegui, R., Iturburu, M., Ochoa De, R. A., & Cuevas, C., (1991). A convenient method for β-lactam formation from β-amino acids using phenyl phosphorodichloridate reagent. *Journal of Organic Chemistry, 56*(6), 2244–2247. doi: 10.1021/jo00006a053.
329. Chopade, M. U., Patil, H. S., Nikalje, M. D., Chopade, A. U., & Gaikwad, S., (2018). One pot facile synthesis of anti-microbial β-lactam derivatives catalyzed by $Fe(acac)_3$. *Chemical Methodologies, 3*(3), 368–382. doi: 10.22034/chemm.2018.151216.1099.
330. Dailler, D., Rocaboy, R., & Baudoin, O., (2017). Synthesis of β-lactams by palladium(0)-catalyzed *C*(sp^3)-*H* carbamoylation. *Angewandte Chemie, International Edition, 56*(25), 7218–7222. doi: 10.1002/anie.201703109.
331. Zhou, T., Jiang, M. X., Yang, X., Yue, Q., Han, Y. Q., Ding, Y., & Shi, B. F., (2020). Synthesis of chiral β-lactams by Pd-catalyzed enantioselective amidation of methylene *C*(sp^3)-*H* bonds. *Chinese Journal of Chemistry, 38*(3), 242–246. doi: 10.1002/cjoc.201900533.
332. Zhang, Q., Chen, K., Rao, W., Zhang, Y., Chen, F. J., & Shi, B. F., (2013). Stereoselective synthesis of chiral α-amino-β-lactams through palladium(II)-catalyzed sequential arylation/amidation of *C*(sp^3)-*H* bonds. *Angewandte Chemie, International Edition, 52*(51), 13588–13592. doi: 10.1002/anie.201306625.
333. Tong, H. R., Zheng, W., Lv, X., He, G., Liu, P., & Chen, G., (2020). Asymmetric synthesis of β-lactam via palladium-catalyzed enantioselective intramolecular *C*(sp^3)-*H* amidation. *ACS Catalysis, 10*(1), 114–120. doi: 10.1021/acscatal.9b04768.
334. Ling, P. X., Fang, S. L., Yin, X. S., Zhang, Q., Chen, K., & Shi, B. F., (2017). Palladium-catalyzed sequential monoarylation/amidation of *C*(sp^3)-*H* bonds: Stereoselective synthesis of α-amino-β-lactams and *anti*-α,β-diamino acid. *Chemical Communications (Cambridge, United Kingdom), 53*(47), 6351–6354. doi: 10.1039/c7cc02426f.
335. Zhang, S. J., Sun, W. W., Cao, P., Dong, X. P., Liu, J. K., & Wu, B., (2016). Stereoselective synthesis of diazabicyclic β-lactams through intramolecular amination of unactivated *C*(sp^3)-*H* bonds of carboxamides by palladium catalysis. *Journal of Organic Chemistry, 81*(3), 956–968. doi: 10.1021/acs.joc.5b02532.
336. Sun, W. W., Cao, P., Mei, R. Q., Li, Y., Ma, Y. L., & Wu, B., (2014). Palladium-catalyzed unactivated *C*(sp^3)-*H* bond activation and intramolecular amination of carboxamides: A new approach to β-lactams. *Organic Letters, 16*(2), 480–483. doi: 10.1021/ol403364k.
337. Nozawa-Kumada, K., Saga, S., Matsuzawa, Y., Hayashi, M., Shigeno, M., & Kondo, Y., (2020). Copper-catalyzed oxidative benzylic *C*(sp^3)-*H* cyclization for the synthesis of β-lactams. *Chemistry – A European Journal, 26*(20), 4496–4499. doi: 10.1002/chem.201905777.
338. Wu, X., Zhao, Y., Zhang, G., & Ge, H., (2014). Copper-catalyzed site-selective intramolecular amidation of unactivated *C*(sp^3)-*H* bonds. *Angewandte Chemie, International Edition, 53*(14), 3706–3710. doi: 10.1002/anie.201311263.
339. Wang, Z., Ni, J., Kuninobu, Y., & Kanai, M., (2014). Copper-catalyzed intramolecular *C*(sp^3)-*H* and *C*(sp^2)-*H* amidation by oxidative cyclization. *Angewandte Chemie, International Edition, 53*(13), 3496–3499. doi: 10.1002/anie.201311105.

340. Aihara, Y., & Chatani, N., (2016). Nickel-catalyzed reaction of *C-H* bonds in amides with I_2: Ortho-iodination via the cleavage of *C*(sp^2)-*H* bonds and oxidative cyclization to β-lactams via the cleavage of *C*(sp^3)-*H* bonds. *ACS Catalysis, 6*(7), 4323–4329. doi: 10.1021/acscatal.6b00964.
341. Wu, X., Zhao, Y., & Ge, H., (2014). Nickel-catalyzed site-selective amidation of unactivated *C*(sp^3)-*H* bonds. *Chemistry – A European Journal, 20*(31), 9530–9533. doi: 10.1002/chem.201403356.
342. Wu, X., Yang, K., Zhao, Y., Sun, H., Li, G., & Ge, H., (2015). Cobalt-catalyzed site-selective intra- and intermolecular dehydrogenative amination of unactivated sp^3 carbons. *Nature Communications, 6*, 6462/1–6462/10. doi: 10.1038/ncomms7462.
343. Cho, I., Jia, Z. J., & Arnold, F. H., (2019). Site-selective enzymatic *C-H* amidation for synthesis of diverse lactams. *Science (Washington, DC, United States), 354*(6440), 575–578. doi: 10.1126/science.aaw9068.
344. Cho, I., Jia, Z. J., & Arnold, F. H., (2019). Genetically tunable enzymatic *C-H* amidation for lactam synthesis. *ChemRxiv,* 1–11.
345. Zhao, Q., & Li, C., (2008). Preference of β-lactam formation in Cu(I)-catalyzed intramolecular coupling of amides with vinyl bromides. *Organic Letters 10*(18), 4037–4040. doi: 10.1021/ol801545a.
346. Zhang, Q., Chen, K., & Shi, B. F., (2014). Recent progress in the synthesis of functionalized β-lactams through transition-metal-catalyzed *C*(sp^3)-*H* amidation. *Synlett, 25*(14), 1941–1945. doi: 10.1055/s-0035-1560178.
347. Deyrup, J. A., (1983). Aziridines. *Chemistry of Heterocyclic Compounds (Chichester, United Kingdom), 42*(Small Ring Heterocycl., Pt. 1), 1–214.
348. Righi, G., & Bonini, C., (1999). Metal halide opening of oxirane and aziridine ring: Recent methodologies and applications to the synthesis of natural products. *Recent Research Developments in Organic Chemistry, 3*(Pt. 2), 343–356.
349. Kametani, T., & Honda, T., (1986). Application of aziridines to the synthesis of natural products. *Advances in Heterocyclic Chemistry, 39*, 181–236. doi: 10.1016/S0065-2725(08)60765-5.
350. Deyrup, J. A., & Clough, S. C., (1969). New route to β-lactams. *Journal of the American Chemical Society, 91*(16), 4590, 4591. doi: 10.1021/ja01044a070.
351. Deyrup, J. A., & Clough, S. C., (1974). Interconversions of aziridine carboxylates and β-lactams. *Journal of Organic Chemistry, 39*(7), 902–907. doi: 10.1021/jo00921a009.
352. Sharma, S. D., Kanwar, S., & Rajpoot, S., (2006). Aziridines as templates: A general strategy for the stereospecific synthesis of 2-azetidinones. *Journal of Heterocyclic Chemistry, 43*(1), 11–19. doi: 10.1002/jhet.5570430103.
353. Huang, L., Zhao, W., Staples, R. J., & Wulff, W. D., (2013). Multifaceted interception of 2-chloro-2-oxoacetic anhydrides: A catalytic asymmetric synthesis of β-lactams. *Chemical Science, 4*(2), 622–628. doi: 10.1039/C2SC21240D.
354. Alper, H., Perera, C. P., & Ahmed, F. R., (1981). A novel synthesis of β-lactams. *Journal of the American Chemical Society, 103*(5), 1289–1291. doi: 10.1021/ja00395a082.
355. Alper, H., Urso, F., & Smith, D. J. H., (1983). Regiospecific metal-catalyzed ring expansion of aziridines to β-lactams. *Journal of the American Chemical Society, 105*(22), 6737, 6738. doi: 10.1021/ja00360a045.
356. Lu, S. M., & Alper, H., (2004). Carbonylative ring expansion of aziridines to β-lactams with rhodium-complexed dendrimers on a resin. *Journal of Organic Chemistry, 69*(10), 3558–3561. doi: 10.1021/jo030353r.

357. Ardura, D., Lopez, R., & Sordo, T. L., (2006). A theoretical study of rhodium(I) catalyzed carbonylative ring expansion of aziridines to β-lactams: Crucial activation of the breaking *C-N* bond by hyperconjugation. *Journal of Organic Chemistry, 71*(19), 7315–7321. doi: 10.1021/jo0610356.
358. Alper, H., & Hamel, N., (1987). Regiospecific synthesis of α-methylene-β-lactams by a homogeneous palladium catalyzed ring expansion-carbonylation reaction. *Tetrahedron Letters, 28*(28), 3237–3240. doi: 10.1016/S0040-4039(00)95481-9.
359. Ganem, B., (1988). A one-pot conversion of an aziridine to a β-lactam using nickel tetracarbonyl. *Chemtracts: Organic Chemistry, 1*(6), 447–448.
360. Chamchaang, W., & Pinhas, A. R., (1988). A one-pot conversion of an aziridine to a β-lactam using nickel tetracarbonyl. *Journal of the Chemical Society, Chemical Communications,* (11), 710–711. doi: 10.1039/C39880000710.
361. Chamchaang, W., & Pinhas, A. R., (1990). The conversion of an aziridine to a β-lactam. *Journal of Organic Chemistry, 55*(9), 2943–2950. doi: 10.1021/jo00296a070.
362. Tanner, D., & Somfai, P., (1993). Palladium-catalyzed transformation of a chiral vinylaziridine to a β-lactam. An enantioselective route to the carbapenem (+)-PS-5. *Bioorganic & Medicinal Chemistry Letters, 3*(11), 2415–2418. doi: 10.1016/S0960-894X(01)80967-7.
363. Ley, S. V., & Middleton, B., (1998). A new route to functionalized π-allyltricarbonyliron lactam complexes from aziridines and their use in stereoselective synthesis and oxidative conversion to β-lactams. *Chemical Communications (Cambridge),* (18), 1995, 1996. doi: 10.1039/a806236f.
364. Davoli, P., Moretti, I., Prati, F., & Alper, H., (1999). Carbonylation of silylated hydroxymethyl aziridines to β-lactams. *Journal of Organic Chemistry, 64*(2), 518–521. doi: 10.1021/jo981568h.
365. Davoli, P., & Prati, F., (2000). A novel approach to a precursor of the carbapenem antibiotic PS-5 via aziridine stereospecific carbonylation. *Heterocycles, 53*(11), 2379–2389. doi: 10.3987/COM-00-8974.
366. Davoli, P., Forni, A., Moretti, I., Prati, F., & Torre, G., (2001). On the effect of ring substituents in the carbonylation of aziridines. *Tetrahedron, 57*(9), 1801–1812. doi: 10.1016/S0040-4020(00)01152-2.
367. Ardura, D., & Lopez, R., (2007). A theoretical investigation of the $Co(CO)_4$-catalyzed carbonylative ring expansion of *N*-benzoyl-2-methylaziridine to β-lactams: Reaction mechanism and effect of substituent at the aziridine Cα atom. *Journal of Organic Chemistry, 72*(9), 3259–3267. doi: 10.1021/jo0625249.
368. Piens, N., & D'hooghe, M., (2017). Carbonylation of aziridines as a powerful tool for the synthesis of functionalized β-lactams. *European Journal of Organic Chemistry,* (40), 5943–5960. doi: 10.1002/ejoc.201700698.
369. Yan, D. M., Crudden, C. M., Chen, J. R., & Xiao, W. J., (2019). A career in catalysis: Howard Alper. *ACS Catalysis, 9*(7), 6467–6483. doi: 10.1021/acscatal.9b01789.
370. Piens, N., Van, H. K., Vogt, D., & D'hooghe, M., (2017). Cobalt carbonyl-catalyzed carbonylation of functionalized aziridines to versatile β-lactam building blocks. *Organic & Biomolecular Chemistry, 15*(22), 4816–4821. doi: 10.1039/C7OB00832E.
371. Mahadevan, V., Getzler, Y. D. Y. L., & Coates, G. W., (2002). [Lewis acid]+[$Co(CO)_4$]- complexes: A versatile class of catalysts for carbonylative ring expansion of epoxides and aziridines. *Angewandte Chemie, International Edition, 41*(15), 2781–2784. doi: 10.1002/1521-3773(20020802)41:15<2781::AID-ANIE2781>3.0.CO;2-S.

372. Piotti, M. E., & Alper, H., (1996). Inversion of stereochemistry in the $Co_2(CO)_8$-catalyzed carbonylation of aziridines to β-lactams. The first synthesis of highly strained *trans*-bicyclic β-lactams. *Journal of the American Chemical Society, 118*(1), 111–116. doi: 10.1021/ja9531586.
373. Davoli, P., Spaggiari, A., Ciamaroni, E., Forni, A., Torre, G., & Prati, F., (2004). Synthesis of *N*-bridgehead fused bicyclic β-lactams through organometal-mediated transformations of 1,2-dialkenylaziridines. *Heterocycles, 63*(11), 2495–2514. doi: 10.3987/COM-04-10166.
374. Fontana, F., Tron, G. C., Barbero, N., Ferrini, S., Thomas, S. P., & Aggarwal, V. K., (2010). Stereoselective synthesis of *trans*-β-lactams by palladium-catalyzed carbonylation of vinyl aziridines. *Chemical Communications (Cambridge, United Kingdom), 46*(2), 267–269. doi: 10.1039/B920564K.
375. Calet, S., Urso, F., & Alper, H., (1989). Enantiospecific and stereospecific rhodium(I)-catalyzed carbonylation and ring expansion of aziridines. Asymmetric synthesis of β-lactams and the kinetic resolution of aziridines. *Journal of the American Chemical Society, 111*(3), 931–934. doi: 10.1021/ja00185a023.
376. Decamps, S., Sevaille, L., Ongeri, S., & Crousse, B., (2014). Access to novel functionalized trifluoromethyl β-lactams by ring expansion of aziridines. *Organic & Biomolecular Chemistry, 12*(33), 6345–6348. doi: 10.1039/C4OB01262C.
377. Tanner, D., & Somfai, P., (1987). An aziridine route to chiral β-lactams. A novel entry to (+)-thienamycin. *Tetrahedron Letters, 28*(11), 1211–1214. doi: 10.1016/S0040-4039(00)95328-0.
378. Konev, A. S., Novikov, M. S., Khlebnikov, A. F., & Tehrani, K. A., (2007). A simple route to side-chain fluorinated β-lactams from ring-fluorinated aziridines. *Journal of Fluorine Chemistry, 128*(2), 114–119. doi: 10.1016/j.jfluchem.2006.10.013.
379. Chiba, K., Mori, M., & Ban, Y., (1985). A novel synthesis of (±)-3-aminonocardicinic acid. *Tetrahedron, 41*(2), 387–392. doi: 10.1016/S0040-4020(01)96430-0.
380. Li, W., Liu, C., Zhang, H., Ye, K., Zhang, G., Zhang, W., Duan, Z., You, S., & Lei, A., (2014). Palladium-catalyzed oxidative carbonylation of *N*-allylamines for the synthesis of β-lactams. *Angewandte Chemie, International Edition, 53*(9), 2443–2446. doi: 10.1002/anie.201309081.
381. Chirila, A., Van, V. K. M., Paul, N. D., & De Bruin, B., (2018). [Co(MeTAA)] metalloradical catalytic route to ketenes via carbonylation of carbene radicals. *European Journal of Inorganic Chemistry,* (20, 21), 2251–2258. doi: 10.1002/ejic.201800101.
382. Zhang, Z., Liu, Y., Ling, L., Li, Y., Dong, Y., Gong, M., Zhao, X., et al., (2011). Pd-catalyzed carbonylation of diazo compounds at atmospheric pressure: A catalytic approach to ketenes. *Journal of the American Chemical Society, 133*(12), 4330–4341. doi: 10.1021/ja107351d.
383. Xie, P., Qian, B., Huang, H., & Xia, C., (2012). Stereoselective synthesis of *trans* β-lactams via palladium/*N*-heterocyclic carbene-catalyzed carbonylative [2+2] cycloaddition. *Tetrahedron Letters, 53*(13), 1613–1616. doi: 10.1016/j.tetlet.2012.01.073.
384. Li, L. L., Ding, D., Song, J., Han, Z. Y., & Gong, L. Z., (2019). Catalytic generation of C1 ammonium enolates from halides and CO for asymmetric cascade reactions. *Angewandte Chemie, International Edition, 58*(23), 7647–7651. doi: 10.1002/anie.201901501.
385. Hsiao, Y., & Hegedus, L. S., (1997). Synthesis of optically active imidazolines, azapenams, dioxocyclams, and bis-dioxocyclams. *Journal of Organic Chemistry, 62*(11), 3586–3591. doi: 10.1021/jo962343e.

386. Ugi, I., & Steinbrückner, C., (1960). Über ein neues kondensations-prinzip. *Angewandte Chemie, 71*(7, 8), 267, 268. doi: 10.1002/ange.19600720709.
387. Dyker, G., (2000). Amino acid derivatives by multicomponent reactions. In: Schmalz, H. G., (ed.), *Organic Synthesis Highlights IV* (pp. 53–57).
388. Krasavin, M., (2012). Amine (Imine) component surrogates in the Ugi reaction and related isocyanide-based multicomponent reactions. In: Nenajdenko, V., (ed.), *Isocyanide Chemistry* (pp. 195–231).
389. Rahmati, A., & Pashmforoush, N., (2015). Synthesis of various heterocyclic compounds via multi-component reactions in water. *Journal of the Iranian Chemical Society, 12*(6), 993–1036. doi: 10.1007/s13738-014-0562-z.
390. Hulme, C., & Nixey, T., (2003). Rapid assembly of molecular diversity via exploitation of isocyanide-based multi-component reactions. *Current Opinion in Drug Discovery & Development, 6*(6), 921–929.
391. Ugi, I., & Steinbrückner, C., (1961). Isonitriles. IX. α-Addition of immonium ions and carboxylate anions to isonitriles. *Chemische Berichte, 94*, 2802–2814. doi: 10.1002/cber.19610941032.
392. Ugi, I., (1982). From isocyanides via four-component condensations to antibiotic syntheses. *Angewandte Chemie International Edition in English, 21*(11), 810–819. doi: 10.1002/anie.198208101.
393. Isenring, H. P., & Hofheinz, W., (1981). A simple two-step synthesis of diphenylmethyl esters of 2-oxo-1-azetidineacetic acids. *Synthesis,* (5), 385–387. doi: 10.1055/s-1981-29461.
394. Kehagia, K., & Ugi, I. K., (1995). The synthesis of 4-acetoxy-azetidine-2-ones as key intermediates for β-lactams. *Tetrahedron, 51*(35), 9523–9530. doi: 10.1016/0040-4020(95)00542-G.
395. Gedey, S., Van, D. E. J., & Fueloep, F., (2002). Liquid-phase combinatorial synthesis of alicyclic β-lactams via Ugi four-component reaction. *Organic Letters, 4*(11), 1967–1969. doi: 10.1021/ol025986r.
396. Rainoldi, G., Lesma, G., Picozzi, C., Lo Presti, L., & Silvani, A., (2018). One step access to oxindole-based β-lactams through Ugi four-center three-component reaction. *RSC Advances, 8*(61), 34903–34910. doi: 10.1039/c8ra08165d.
397. Pitlik, J., & Townsend, C. A., (1997). Solution-phase synthesis of a combinatorial monocyclic β-lactam library: Potential protease inhibitors. *Bioorganic & Medicinal Chemistry Letters, 7*(24), 3129–3134. doi: 10.1016/S0960-894X(97)10170-6.
398. Kanizsai, I., Gyonfalvi, S., Szakonyi, Z., Sillanpaeae, R., & Fueloep, F., (2007). Synthesis of bi- and tricyclic β-lactam libraries in aqueous medium. *Green Chemistry, 9*(4), 357–360. doi: 10.1039/B613117D.
399. Gedey, S., Van, D. E. J., & Fueloep, F., (2004). Synthesis of alicyclic β-lactams via the Ugi reaction on a solid support. *Letters in Organic Chemistry, 1*(3), 215–220. doi: 10.2174/1570178043401153.
400. Basso, A., Banfi, L., Riva, R., & Guanti, G., (2004). U-4C-3CR versus U-5C-4CR and stereochemical outcomes using suitable bicyclic β-amino acid derivatives as bifunctional components in the Ugi reaction. *Tetrahedron Letters, 45*(3), 587–590. doi: 10.1016/j.tetlet.2003.10.193.
401. Kanizsai, I., Szakonyi, Z., Sillanpaeae, R., & Fueloep, F., (2006). A comparative study of the multicomponent Ugi reactions of an oxabicycloheptene-based β-amino acid in water and in methanol. *Tetrahedron Letters, 47*(51), 9113–9116. doi: 10.1016/j.tetlet.2006.10.069.

402. Pirrung, M. C., & Das, S. K., (2004). β-Lactam synthesis by Ugi reaction of β-keto acids in aqueous solution. *Synlett,* (8), 1425–1427. doi: 10.1055/s-2004-825606.
403. Vishwanatha, T. M., Narendra, N., & Sureshbabu, V. V., (2011). Synthesis of β-lactam peptidomimetics through Ugi MCR: First application of chiral $N^{β}$-Fmoc amino alkyl isonitriles in MCRs. *Tetrahedron Letters, 52*(43), 5620–5624. doi: 10.1016/j.tetlet.2011.08.090.
404. Aviles, E., Prudhomme, J., Le Roch, K. G., & Rodriguez, A. D., (2015). Structures, semisyntheses, and absolute configurations of the antiplasmodial α-substituted β-lactam monamphilectines B and C from the sponge svenzea flava. *Tetrahedron, 71*(3), 487–494. doi: 10.1016/j.tet.2014.11.060.
405. Aviles, E., & Rodriguez, A. D., (2010). Monamphilectine A, a potent antimalarial β-lactam from marine sponge *Hymeniacidon sp*: Isolation, structure, semisynthesis, and bioactivity. *Organic Letters, 12*(22), 5290–5293. doi: 10.1021/ol102351z.
406. Szakonyi, Z., Sillanpaa, R., & Fulop, F., (2010). Synthesis of conformationally constrained tricyclic β-lactam enantiomers through Ugi four-center three-component reactions of a monoterpene-based β-amino acid. *Molecular Diversity, 14*(1), 59–65. doi: 10.1007/s11030-009–9143-y.
407. Gedey, S., Vainiotalo, P., Zupko, I., De Witte, P. A. M., & Fueloep, F., (2003). Liquid-phase synthesis of mixture-based bicyclic β-lactam libraries. *Journal of Heterocyclic Chemistry, 40*(6), 951–956. doi: 10.1002/jhet.5570400601.
408. Cheibas, C., Cordier, M., Li, Y., & El Kaim, L., (2019). A Ugi straightforward access to bis-β-lactam derivatives. *European Journal of Organic Chemistry,* (27), 4457–4463. doi: 10.1002/ejoc.201900678.
409. Zeng, X. H., Wang, H. M., Yan, Y. M., Wu, L., & Ding, M. W., (2014). One-pot regioselective synthesis of β-lactams by a tandem Ugi 4CC/SN cyclization. *Tetrahedron, 70*(23), 3647–3652. doi: 10.1016/j.tet.2014.04.033.
410. Corredor, M., Garrido, M., Bujons, J., Orzaez, M., Perez-Paya, E., Alfonso, I., & Messeguer, A., (2015). Efficient synthesis of conformationally restricted apoptosis inhibitors bearing a triazole moiety. *Chemistry – A European Journal, 21*(40), 14122–14128. doi: 10.1002/chem.201502380.
411. Ghabraie, E., Balalaie, S., Mehrparvar, S., & Rominger, F., (2014). Synthesis of functionalized β-lactams and pyrrolidine-2,5-diones through a metal-free sequential Ugi-4CR/cyclization reaction. *Journal of Organic Chemistry, 79*(17), 7926–7934. doi: 10.1021/jo5010422.
412. Garrido, M., Corredor, M., Orzaez, M., Alfonso, I., & Messeguer, A., (2016). Regioselective synthesis of a family of β-lactams bearing a triazole moiety as potential apoptosis inhibitors. *ChemistryOpen, 5*(5), 485–494. doi: 10.1002/open.201600052.
413. Ramanivas, T., Parameshwar, M., Gayatri, G., Nanubolu, J. B., & Srivastava, A. K., (2017). Asymmetric synthesis of functionalized 2,5-pyrrolidinediones and β-lactams through diastereospecific cycloisomerization/rearrangement of chiral ethanolamine-derived Ugi adducts. *European Journal of Organic Chemistry,* (16), 2245–2257. doi: 10.1002/ejoc.201700031.
414. Zidan, A., Garrec, J., Cordier, M., El-Naggar, A. M., Abd El-Sattar, N. E. A., Ali, A. K., Hassan, M. A., & El Kaim, L., (2017). β-Lactam synthesis through diodomethane addition to amide dianions. *Angewandte Chemie, International Edition, 56*(40), 12179–12183. doi: 10.1002/anie.201706315.
415. Li, Z., Sharma, U. K., Liu, Z., Sharma, N., Harvey, J. N., & Van, D. E. E. V., (2015). Diversity-oriented synthesis of β-lactams and γ-lactams by post-Ugi nucleophilic

cyclization: Lewis acids as regioselective switch. *European Journal of Organic Chemistry,* (18), 3957–3962. doi: 10.1002/ejoc.201500270.
416. Ghoshal, A., Kumar, A., Yugandhar, D., Sona, C., Kuriakose, S., Nagesh, K., Rashid, M., et al., (2018). Identification of novel β-lactams and pyrrolidinone derivatives as selective histamine-3 receptor (H3R) modulators as possible anti-obesity agents. *European Journal of Medicinal Chemistry, 152*, 148–159. doi: 10.1016/j.ejmech.2018.04.020.
417. Alcaide, B., Almendros, P., Aragoncillo, C., Callejo, R., Pilar, R. M., & Rosario, T. M., (2015). Investigation of the Passerini and Ugi reactions in β-lactam aldehydes. Synthetic applications. *Organic & Biomolecular Chemistry, 13*(5), 1387–1394. doi: 10.1039/C4OB02289K.
418. Blackie, M. A. L., Feng, T. S., Smith, P. J., & Chibale, K., (2016). Synthesis of novel β-lactams and *in vitro* evaluation against the human malaria parasite *Plasmodium falciparum. ARKIVOC (Gainesville, FL, United States),* (3), 214–235. doi: 10.3998/ark.5550190.p009.465.
419. Fan, L., Adams, A. M., Polisar, J. G., & Ganem, B., (2008). Studies on the chemistry and reactivity of α-substituted ketones in isonitrile-based multicomponent reactions. *Journal of Organic Chemistry, 73*(24), 9720–9726. doi: 10.1021/jo8019708.
420. Gao, X., Shan, C., Chen, Z., Liu, Y., Zhao, X., Zhang, A., Yu, P., et al., (2018). One-pot synthesis of β-lactams by the Ugi and Michael addition cascade reaction. *Organic & Biomolecular Chemistry, 16*(33), 6096–6105. doi: 10.1039/C8OB01176A.
421. Neochoritis, C. G., Stotani, S., Mishra, B., & Doemling, A., (2015). Efficient isocyanide-less isocyanide-based multicomponent reactions. *Organic Letters, 17*(8), 2002–2005. doi: 10.1021/acs.orglett.5b00759.
422. Xu, J., (2012). Synthesis of β-lactams with π electron-withdrawing substituents. *Tetrahedron, 68*(52), 10696–10747. doi: 10.1016/j.tet.2012.04.007.
423. Johnson, P. Y., Schmuff, N. R., & Hatch, C. E. III., (1975). Ring contraction reactions of 2-aminopyrazolidine-3-ones. New synthesis of mono- and bicyclo-β-lactams. *Tetrahedron Letters,* (47), 4089–4090. doi: 10.1016/S0040-4039(00)91182-1.
424. Johnson, P. Y., & Hatch, C. E. III., (1975). Synthesis and photolysis of 2-acylpyrazolidin-3-ones. Model for the photochemical syntheses of 6-azapenicillin isomers. *Journal of Organic Chemistry, 40*(7), 909–915. doi: 10.1021/jo00895a021.
425. Johnson, P. Y., & Hatch, C. E. III., (1975). Photochemical formation of spiro and bicyclo 1-acylaminoazetidin-2-ones. Models for the syntheses of penicillin-like systems. II. *Journal of Organic Chemistry, 40*(24), 3502–3510. doi: 10.1021/jo00912a008.
426. White, J. D., Perri, S. T., & Toske, S. G., (1992). Nitrosative deamination of 1-aminoazetidin-2-ones. An entry to *N*-unsubstituted β-lactams. *Tetrahedron Letters, 33*(4), 433–436. doi: 10.1016/S0040-4039(00)93960-1.
427. Stork, G., & Szajewski, R. P., (1974). Carboxy β-lactams by photochemical ring contraction. *Journal of the American Chemical Society, 96*(18), 5787–5791. doi: 10.1021/ja00825a015.
428. Lawton, G., Moody, C. J., & Pearson, C. J., (1984). Synthesis of aza-β-lactams by photochemical ring contraction. *Journal of the Chemical Society, Chemical Communications,* (12), 754–756. doi: 10.1039/C39840000754.
429. Moody, C. J., Pearson, C. J., & Lawton, G., (1985). Regioselectivity in the photochemical ring contraction of 4-diazopyrazolidine-3,5-diones to give aza-β-lactams. *Tetrahedron Letters, 26*(26), 3167–3170. doi: 10.1016/S0040-4039(00)98647-7.
430. Moore, H. W., & Arnold, M. J., (1983). Photolysis of 4-diazopyrrolidine-2,3-diones. A new synthetic route to mono- and bicyclic β-lactams. *Journal of Organic Chemistry, 48*(19), 3365–3367. doi: 10.1021/jo00167a056.

431. Black, D. S. C., & Boscacci, A. B., (1974). Photorearrangement of 3-oxo-1-pyrroline 1-oxides. *Journal of the Chemical Society, Chemical Communications,* (4), 129, 130. doi: 10.1039/c39740000129.
432. Sato, M., Ogasawara, H., Takayama, K., & Kaneko, C., (1987). Synthesis of 1,3-dioxin-4-ones and their use in synthesis. Part XIV. A novel synthetic method of lactams from 1,3-dioxin-4-ones via intramolecular ketene trapping. *Heterocycles, 26*(10), 2611–2614. doi: 10.3987/R-1987–10–2611.
433. Hirokami, S., Hirai, Y., Nagata, M., Yamazaki, T., & Date, T., (1979). Photochemical reactions of 4-pyrimidones. Structures and properties of the β-lactams formed. *Journal of Organic Chemistry, 44*(13), 2083–2087. doi: 10.1021/jo01327a008.
434. Johnson, M. R., Fazio, M. J., Ward, D. L., & Sousa, L. R., (1983). Synthesis of β-lactams by the photochemical extrusion of sulfur dioxide from 1,1-dioxo-4-thiazolidinones. *Journal of Organic Chemistry, 48*(4), 494–499. doi: 10.1021/jo00152a017.
435. Procter, G., Nally, J., & Ordsmith, N. H. R., (1995). β-Lactams from tetrahydro-1,2-oxazine-3,6-diones, and a labeling study of the product stereochemistry. *Tetrahedron, 51*(47), 12837–12842. doi: 10.1016/0040-4020(95)00737-S.
436. White, J. D., & Toske, S. G., (1993). Photochemically mediated ring contraction of pyrazolidin-3-ones to β-lactams. *Bioorganic & Medicinal Chemistry Letters, 3*(11), 2383–2388. doi: 10.1016/S0960-894X(01)80960-4.
437. Bender, D. R., Bjeldanes, L. F., Knapp, D. R., & Rapoport, H., (1975). Rearrangement of pyruvates to malonates. Synthesis of β-lactams. *Journal of Organic Chemistry, 40*(9), 1264–1269. doi: 10.1021/jo00897a018.
438. Bender, D., Rapoport, H., & Bordner, J., (1975). Stereochemistry of β-lactams derived from α-keto-γ-lactams by ring contraction. X-ray analysis and differential behavior with shift reagents of difunctional β-lactams. *Journal of Organic Chemistry, 40*(22), 3208–3213. doi: 10.1021/jo00910a009.
439. Bender, D. R., Brennan, J., & Rapoport, H., (1978). Periodate oxidation of α-keto γ-lactams. Enol oxidation and β-lactam formation. Mechanism of periodate hydroxylation reactions. *Journal of Organic Chemistry, 43*(17), 3354–3362. doi: 10.1021/jo00411a020.
440. Sheradsky, T., & Zbaida, D., (1978). The desulfurization of mesoionic thiazol-4-ones; stereospecific formation of β-lactams. *Tetrahedron Letters,* (23), 2037–2040. doi: 10.1016/S0040–4039(01)94743–4.
441. Minami, T., & Agawa, T., (1974). Reactions of sulfur diimides with phenyl and chloroketenes. *Journal of Organic Chemistry, 39*(9), 1210–1215. doi: 10.1021/jo00923a010.
442. Molchanov, A. P., & Tran, T. Q., (2012). Regioselective cycloaddition of *C*-aryl- and *C*-carbamoylnitrones to methyl 2-benzylidenecyclopropanecarboxylate. *Russian Journal of Organic Chemistry, 48*(10), 1283–1288. doi: 10.1134/S1070428012100041.
443. Molchanov, A. P., Tran, T. Q., Stepakov, A. V., Starova, G. L., & Kostikov, R. R., (2014). Regioselective cycloaddition of *C*-carbamoylnitrones to methyl (*E*)-2-(2-phenylcyclopropylidene)acetate and methyl (*E*)-2-methylidene-3-phenylcyclopropane-1-carboxylate. *Russian Journal of Organic Chemistry, 50*(1), 78–82. doi: 10.1134/S1070428014010151.
444. Paschetta, V., Cordero, F., Paugam, R., Ollivier, J., Brandi, A., & Salaun, J., (2001). Spirocyclopropane-isoxazolidines, octahydropyrindin-4-ones and β-lactams with biomimetic ethylene release from the electrophilic substitution of a 1,1-ethyleneallylzinc complex. *Synlett,* (8), 1233–1236. doi: 10.1055/s-2001-16057.

445. Cordero, F. M., Pisaneschi, F., Goti, A., Ollivier, J., Salauen, J., & Brandi, A., (2000). New synthesis of β-lactams by ethylene extrusion from spirocyclopropane isoxazolidines. *Journal of the American Chemical Society, 122*(33), 8075, 8076. doi: 10.1021/ja000108e.
446. Cordero, F. M., Pisaneschi, F., Salvati, M., Paschetta, V., Ollivier, J., Salauen, J., & Brandi, A., (2003). Selective ring contraction of 5-spirocyclopropane isoxazolidines mediated by acids. *Journal of Organic Chemistry, 68*(8), 3271–3280. doi: 10.1021/jo034003g.
447. Diethelm, S., & Carreira, E. M., (2013). Total synthesis of (±)-gelsemoxonine. *Journal of the American Chemical Society, 135*(23), 8500–8503. doi: 10.1021/ja403823n.
448. Diethelm, S., & Carreira, E. M., (2015). Total synthesis of gelsemoxonine through a spirocyclopropane isoxazolidine ring contraction. *Journal of the American Chemical Society, 137*(18), 6084–6096. doi: 10.1021/jacs.5b02574.
449. Diethelm, S., Schoenebeck, F., & Carreira, E. M., (2014). Mechanistic insight into the spirocyclopropane isoxazolidine ring contraction. *Organic Letters, 16*(3), 960–963. doi: 10.1021/ol403693t.
450. Cordero, F. M., Salvati, M., Pisaneschi, F., & Brandi, A., (2004). Novel prospects of the acidic thermal rearrangement of spiro[cyclopropane-1,5'-isoxazolidines] to β-lactams. *European Journal of Organic Chemistry*, (10), 2205–2213. doi: 10.1002/ejoc.200300595.
451. Wei, C., Zhu, J. F., Zhang, J. Q., Deng, Q., & Mo, D. L., (2019). Synthesis of spirofluorenyl-β-lactams through cycloaddition and ring contraction from *N*-aryl fluorenone nitrones and methylenecyclopropanes. *Advanced Synthesis & Catalysis, 361*(17), 3965–3973. doi: 10.1002/adsc.201900523.
452. Padwa, A., Koehler, K., & Rodriguez, A., (1981). Nitrone cycloaddition. New approach to β-lactams. *Journal of the American Chemical Society, 103*(16), 4974, 4975. doi: 10.1021/ja00406a065.
453. Padwa, A., Koehler, K. F., & Rodriguez, A., (1984). New synthesis of β-lactams based on nitrone cycloaddition to nitroalkenes. *Journal of Organic Chemistry, 49*(2), 282–288. doi: 10.1021/jo00176a013.
454. Jasinski, R., (2015). A new mechanistic insight on β-lactam systems formation from 5-nitroisoxazolidines. *RSC Advances, 5*(62), 50070–50072. doi: 10.1039/C5RA07663C.
455. Aurich, H. G., & Ruiz, Q. J. L., (1994). Effects of configuration and *N*-substitution on the formation of β-lactams from bicyclic cyano-substituted isoxazolidines. *Tetrahedron, 50*(13), 3943–3950. doi: 10.1016/S0040-4020(01)89669-1.
456. Di Nunno, L., & Scilimati, A., (1993). A new synthesis of β-phenylamino thioesters and β-lactams via base-induced ring-opening of 2-phenyl-3-aryl-5-phenylthioisoxazolidines. *Tetrahedron, 49*(47), 10965–10976. doi: 10.1016/S0040-4020(01)80250-7.
457. Purrington, S. T., & Sheu, K. W., (1992). β-Lactams from 5,5-difluoroisoxazolidines. *Tetrahedron Letters, 33*(23), 3289–3292. doi: 10.1016/S0040-4039(00)92069-0.
458. Baldwin, S. W., & Aube, J., (1987). Asymmetric synthesis with chiral hydroxylamines. Synthesis of optically pure 4-substituted azetidinones. *Tetrahedron Letters, 28*(2), 179–182. doi: 10.1016/S0040-4039(00)95680-6.
459. Baran, P. S., Shenvi, R. A., & Mitsos, C. A., (2005). A remarkable ring contraction en route to the chartelline alkaloids. *Angewandte Chemie, International Edition, 44*(24), 3714–3717. doi: 10.1002/anie.200500522.
460. Easton, C. J., (1985). Syntheses of β-lactams by ring contraction of isothiazolidinones. *Journal of the Chemical Society, Perkin Transactions 1: Organic and Bio-Organic Chemistry*, (1), 153–157.

461. Giordano, R. C., Ribeiro, M. P. A., & Giordano, R. L. C., (2006). Kinetics of beta-lactam antibiotics synthesis by penicillin *G* acylase (PGA) from the viewpoint of the industrial enzymatic reactor optimization. *Biotechnology Advances, 24*(1), 27–41. doi: 10.1016/j.biotechadv.2005.05.003.
462. Becka, S., Stepanek, V., Vyasarayani, R. W., Grulich, M., Marsalek, J., Plhackova, K., Dobisova, M., et al., (2014). Penicillin *G* acylase from *Achromobacter* sp. CCM 4824: An efficient biocatalyst for syntheses of beta-lactam antibiotics under conditions employed in large-scale processes. *Applied Microbiology and Biotechnology, 98*(3), 1195–1203. doi: 10.1007/s00253–013–4945–3.
463. Pan, X., Li, A., Peng, Z., Ji, X., Chu, J., & He, B., (2020). Efficient synthesis of β-lactam antibiotics with in situ product removal by a newly isolated penicillin G acylase. *Bioorganic Chemistry, 99*, 103765.
464. Ulijn, R. V., Martin, L. D., Halling, P. J., Moore, B. D., & Janssen, A. E. M., (2002). Enzymatic synthesis of β-lactam antibiotics via direct condensation. *Journal of Biotechnology, 99*(3), 215–222. doi: 10.1016/S0168-1656(02)00211-0.
465. Koreishi, M., Tani, K., Ise, Y., Imanaka, H., Imamura, K., & Nakanishi, K., (2007). Enzymatic synthesis of β-lactam antibiotics and *N*-fatty-acylated amino compounds by the acyl-transfer reaction catalyzed by penicillin V acylase from *Streptomyces* mobaraensis. *Bioscience, Biotechnology, and Biochemistry, 71*(6), 1582–1586. doi: 10.1271/bbb.70052.
466. Van, L. L. M., De Vroom, E., Van, R. F., & Sheldon, R., (1999). Enzymatic synthesis of β-lactam antibiotics using penicillin-*G* acylase in frozen media. *FEBS Letters, 456*(1), 89–92. doi: 10.1016/S0014-5793(99)00939-4.
467. Kurochkina, V. B., (1999). Enzymatic synthesis of β-lactam antibiotics. II. Aminocephalosporins. *Antibiotiki i Khimioterapiya, 44*(8), 6–11.
468. Kurochkina, V. B., & Nys, P. S., (1999). Enzymatic synthesis of beta-lactam antibiotics. I. Cefazolin. *Antibiotiki i Khimioterapiya, 44*(5), 12–16.
469. Hamed, R. B., Gomez-Castellanos, J. R., Henry, L., Warhaut, S., Claridge, T. D. W., & Schofield, C. J., (2019). Biocatalytic production of bicyclic β-lactams with three contiguous chiral centers using engineered crotonases. *Communications Chemistry, 2*(1), 1–10. doi: 10.1038/s42004-018-0106-z.
470. Bachmann, B. O., Li, R., & Townsend, C. A., (1998). β-Lactam synthetase: A new biosynthetic enzyme. *Proceedings of the National Academy of Sciences of the United States of America, 95*(16), 9082–9086. doi: 10.1073/pnas.95.16.9082.
471. Marahiel, M. A., Stachelhaus, T., & Mootz, H. D., (1997). Modular peptide synthetases involved in nonribosomal peptide synthesis. *Chemical Reviews (Washington, DC, United States), 97*(7), 2651–2674. doi: 10.1021/cr960029e.
472. Nys, P. S., & Bartoshevich, Y. E., (1994). Role of biocatalysis in development and improvement of processes for production of beta-lactam antibiotics in Russia. *Antibiotiki i Khimioterapiya, 39*(8), 3–14.
473. Kurochkina, V. B., & Sklyarenko, A. V., (2005). Enzymatic synthesis of beta-lactam antibiotics. Analytical review. *Antibiotiki i Khimioterapiya, 50*(5, 6), 39–58.
474. Qu, F., Li, D., Wang, L., & Yi, B., (2012). Progress in enzymatic synthesis of β-lactam antibiotics with α-amino acid ester hydrolase. *Zhongguo Yiyao Gongye Zazhi, 43*(5), 381–385.
475. Pan, X., Li, A., Peng, Z., Ji, X., Chu, J., & He, B., (2020). Efficient synthesis of β-lactam antibiotics with *in situ* product removal by a newly isolated penicillin *G* acylase. *Bioorganic Chemistry, 99*, 103765/1–103765/9. doi: 10.1016/j.bioorg.2020.103765.

476. Goncalves, L. R. B., Ferreira, A. L. O., Fernandez-Lafuente, R., Guisan, J. M., Giordano, R. C., & Giordano, R. L. C., (2008). Influence of mass transfer limitations on the enzymatic synthesis of β-lactam antibiotics catalyzed by penicillin *G* acylase immobilized on glyoxyl-agarose. *Bioprocess and Biosystems Engineering, 31*(5), 411–418. doi: 10.1007/s00449-007-0176-2.
477. Wong, P. K., Madhavarao, M., Marten, D. F., & Rosenblum, M., (1977). A new synthesis of β-lactams. *Journal of the American Chemical Society, 99*(8), 2823, 2824. doi: 10.1021/ja00450a082.
478. Berryhill, S. R., & Rosenblum, M., (1980). Use of organoiron complexes in β-lactam synthesis. Preparation of 2-methylcarbopenam. *Journal of Organic Chemistry, 45*(10), 1984–1986. doi: 10.1021/jo01298a046.
479. Berryhill, S. R., Price, T., Rosenblum, M., (1983). β-Lactam synthesis using organoiron intermediates. Preparation of 3-carbomethoxycarbapenam. *Journal of Organic Chemistry, 48*(2), 158–162. doi: 10.1021/jo00150a004.
480. Liebeskind, L. S., Welker, M. E., & Fengl, R. W., (1986). Transformations of chiral iron complexes used in organic synthesis. Reactions of η^5-$CpFe(PPh_3)(CO)COCH_3$ and related species leading to a mild, stereospecific synthesis of β-lactams. *Journal of the American Chemical Society, 108*(20), 6328–6343. doi: 10.1021/ja00280a034.
481. Schulz, T., Weismann, D., Wallbaum, L., Guthardt, R., Thie, C., Leibold, M., Bruhn, C., & Siemeling, U., (2015). New stable and persistent acyclic diaminocarbenes. *Chemistry – A European Journal, 21*(40), 14107–14121. doi: 10.1002/chem.201502315.
482. Wallbaum, L., Schulz, T., Bruhn, C., & Siemeling, U., (2019). The reaction of a particularly electrophilic acyclic diaminocarbene with carbon monoxide: Formation of β- and γ-lactams. *Zeitschrift fuer Naturforschung, B: A Journal of Chemical Sciences, 74*(2), 221–226. doi: 10.1515/znb-2018–0255.
483. Worgull, D., Dickmeiss, G., Jensen, K. L., Franke, P. T., Holub, N., & Jorgensen, K. A., (2011). Optically active bicyclic *N*-heterocycles by organocatalytic asymmetric Michael addition/cyclization sequences. *Chemistry – A European Journal, 17*(15), 4076–4080. doi: 10.1002/chem.201100233.
484. D'annibale, A., Pesce, A., Resta, S., & Trogolo, C., (1997). Manganese(III)-promoted free radical cyclizations of enamides leading to β-lactams. *Tetrahedron, 53*(38), 13129–13138. doi: 10.1016/S0040–4020(97)00835–1.
485. Attenni, B., Cerreti, A., D'Annibale, A., Resta, S., & Trogolo, C., (1998). Mn(III)-promoted sulfur-directed 4-*exo-trig* radical cyclization of enamides to β-lactams. *Tetrahedron, 54*(39), 12029–12038. doi: 10.1016/S0040-4020(98)83055-X.
486. D'Annibale, A., Nanni, D., Trogolo, C., & Umani, F., (2000). Diastereoselectivity in Mn(III)-promoted 4-*exo-trig* cyclization of enamides to β-lactams. *Organic Letters, 2*(3), 401, 402. doi: 10.1021/ol991363a.
487. D'Annibale, A., Pesce, A., Resta, S., & Trogolo, C., (1997). Ceric ammonium nitrate promoted free radical cyclization reactions leading to β-lactams. *Tetrahedron Letters, 38*(10), 1829–1832. doi: 10.1016/S0040–4039(97)00162–7.
488. Tarui, A., Kawashima, N., Sato, K., Omote, M., Miwa, Y., Minami, H., & Ando, A., (2010). Simple, chemoselective, and diastereoselective Reformatsky-type synthesis of α-bromo-α-fluoro-β-lactams. *Tetrahedron Letters, 51*(15), 2000–2003. doi: 10.1016/j.tetlet.2010.02.023.
489. Isoda, M., Sato, K., Funakoshi, M., Omura, K., Tarui, A., Omote, M., & Ando, A., (2015). Diastereoselective synthesis of *Syn*-β-lactams using Rh-catalyzed reductive

Mannich-type reaction of α,β-unsaturated esters. *Journal of Organic Chemistry, 80*(16), 8398–8405. doi: 10.1021/acs.joc.5b01233.
490. Cossio, F. P., & Palomo, C., (1985). Reagents and synthetic methods. 53. Stereoselective annelation of trimethylsiloxyacetic acids and imines into 3-hydroxy-β-lactams. *Tetrahedron Letters, 26*(35), 4239–4242. doi: 10.1016/S0040-4039(00)99002-6.
491. Overman, L. E., & Osawa, T., (1985). A convenient synthesis of 4-unsubstituted β-lactams. *Journal of the American Chemical Society, 107*(6), 1698–1701. doi: 10.1021/ja00292a040.
492. Toyofuku, M., Fujiwara, S. I., Shin-ike, T., Kuniyasu, H., & Kambe, N., (2005). Palladium-catalyzed intramolecular selenocarbamoylation of alkynes with carbamoselenoates: Formation of α-alkylidene-β-lactam framework. *Journal of the American Chemical Society, 127*(27), 9706–9707. doi: 10.1021/ja052175k.
493. Kong, W. J., Liu, Y. J., Xu, H., Chen, Y. Q., Dai, H. X., & Yu, J. Q., (2016). Pd-catalyzed α-selective *C-H* functionalization of olefins: En route to 4-imino-β-lactams. *Journal of the American Chemical Society, 138*(7), 2146–2149. doi: 10.1021/jacs.5b13353.
494. El Bakri, Y., Marmouzi, I., El Jemli, M., Anouar, E. H., Karthikeyan, S., Harmaoui, A., Faouzi, M. E. A., et al., (2019). Synthesis, biological activity and molecular modeling of a new series of condensed 1,2,4-triazoles. *Bioorganic Chemistry, 92*, 103193/1–103193/13. doi: 10.1016/j.bioorg.2019.103193.
495. Moore, H. W., Hernandez, L., & Sing, A., (1976). Chlorocyanoketene. A new β-lactam synthesis. *Journal of the American Chemical Society, 98*(12), 3728–3730. doi: 10.1021/ja00428a068.
496. Moore, H. W., Hernandez, L. Jr., Kunert, D. M., Mercer, F., & Sing, A., (1981). A new synthetic route to 2-azetidinones. Ring contraction of 4-azido-2-pyrrolinones to 3-cyano-2-azetidinones. *Journal of the American Chemical Society, 103*(7), 1769–1777. doi: 10.1021/ja00397a031.
497. Zhang, L., Ma, L., Zhou, H., Yao, J., Li, X., & Qiu, G., (2018). Synthesis of α-methylene-β-lactams enabled by base-promoted intramolecular 1,2-addition of *N*-propiolamide and *C-C* bond migrating cleavage of aziridine. *Organic Letters, 20*(8), 2407–2411. doi: 10.1021/acs.orglett.8b00742.
498. Zhu, L., Xiong, Y., & Li, C., (2015). Synthesis of α-methylene-β-lactams via PPh_3-catalyzed umpolung cyclization of propiolamides. *Journal of Organic Chemistry, 80*(1), 628–633. doi: 10.1021/jo502265a.
499. Wasserman, H. H., & Glazer, E., (1975). Conversion of amino acids to β-lactam derivatives via cyclopropanone. *Journal of Organic Chemistry, 40*(10), 1505, 1506. doi: 10.1021/jo00898a031.
500. He, M., & Bode, J. W., (2008). Enantioselective, *NHC*-catalyzed bicyclo-β-lactam formation via direct annulations of enals and unsaturated *N*-sulfonyl ketimines. *Journal of the American Chemical Society, 130*(2), 418,419. doi: 10.1021/ja0778592.
501. Lu, H., & Li, C., (2006). General and highly efficient synthesis of 2-alkylideneazetidines and β-lactams via copper-catalyzed intramolecular *N*-vinylation. *Organic Letters, 8*(23), 5365–5367. doi: 10.1021/ol062274i.
502. Faltracco, M., Sukowski, V., Van, D. M., Hamlin, T. A., Bickelhaupt, F. M., & Ruijter, E., (2020). Diastereoselective synthesis of β-lactams by ligand-controlled stereodivergent intramolecular Tsuji-Trost allylation. *Journal of Organic Chemistry, 85*(15), 9566–9584. doi: 10.1021/acs.joc.0c00575.
503. Van, D. S. F. H., Kleijn, H., Jastrzebski, J. T. B. H., & Van, K. G., (1989). The syntheses of β-lactams from zinc enolates of *N,N*-disubstituted α-amino acid esters and imines:

Substituent and solvent effects. *Tetrahedron Letters, 30*(6), 765–768. doi: 10.1016/S0040-4039(01)80304-X.
504. Georg, G. I., Cheruvallath, Z. S., Himes, R. H., Mejillano, M. R., & Burke, C. T., (1992). Synthesis of biologically active taxol analogs with modified phenylisoserine side chains. *Journal of Medicinal Chemistry, 35*(22), 4230–4237. doi: 10.1021/jm00100a031.
505. Sekine, A., Hori, K., Ohashi, Y., Yagi, M., & Toda, F., (1989). X-Ray structural studies of chiral β-lactam formation from an achiral oxo amide using the chiral crystal environment. *Journal of the American Chemical Society, 111*(2), 697–699. doi: 10.1021/ja00184a045.
506. Vessally, E., Babazadeh, M., Hosseinian, A., Edjlali, L., & Sreerama, L., (2018). Recent advances in synthesis of functionalized β-lactams through cyclization of *N*-propargyl amine/amide derivatives. *Current Organic Chemistry, 22*(2), 199–205. doi: 10.2174/1385272821666170519113904.

CHAPTER 4

Reactions of β-Lactams

4.1 BRIEF INTRODUCTION

Being the most used antibiotics, β-lactams have very broad biological and medicinal applications, as indicated in so many literature mentioned in Chapter 1 of this book, as well as the enormous amounts of uncited references. Therefore, it is unnecessary to duplicate the biological and medicinal applications of β-lactams in this chapter again. It is perceivable that β-lactams will be continually used on the battlefield against infectious diseases in the future.

On the other hand, the chemical applications of β-lactams are also very broad, as demonstrated in the syntheses where β-lactams are directly applied as the precursors or starting materials. However, the synthetic applications of β-lactams focusing on the modification of the functional groups on the side chains of β-lactams, e.g., β-lactams-based semi-syntheses, will not be discussed here, owing to the enormous amounts of functional groups and their associated reactivities. For example, a C=C double bond on the side chain could undergo hydroxylation, halogenation, epoxidation, dihydroxylation, aminohydroxylation, oxidation, ozonolysis, etc. Therefore, it is impossible to include all these synthetic modifications of β-lactam derivatives in this Chapter, and only those modifications on the β-lactam rings will be collected. These modifications may include the functionalization at N1, substitution on C3 or C4, reduction of the C2-carbonyl group, and cleavage of N1-C2, C2-C3, C3-C4, and N1-C4 bonds.

The Chemistry and Biology of Beta-Lactams. Zerong Wang, PhD

4.2 CLEAVAGE OF THE AMIDE BOND OF Β-LACTAMS

The most common synthetic application using β-lactams as the precursors or substrates is the cleavage of the N1-C2 bond of the β-lactam moiety, affording β-amino acid derivatives. As mentioned in Chapter 2, the antibiotic activities of β-lactams are primarily originating from the relative instability of the four-membered ring due to its ring strain. Therefore, β-lactams can be easily opened during their exposure to nucleophiles, affording a variety of derivatives of β-amino acids depending on the nature of nucleophiles.

4.2.1 HYDROLYSIS OF Β-LACTAMS

When the nucleophile is water or hydroxide, β-amino acids are formed in most cases in the hydrolysis of β-lactams. For example, treatment of 0.1 mol of 4-ethylazetidin-2-one in 10 mL ethanol with a stoichiometric amount of $Ba(OH)_2 \cdot 8H_2O$ in 50 mL of water at 50°C for 10 hours yielded 93% of 3-aminopentanoic acid [1]. In another example, the cycloaddition of isobutene with sulfurisocyanatidic chloride (i.e., chlorosulfonyl isocyanate (CSI)) yields 2,2-dimethyl-4-oxoazetidine-1-sulfonyl chloride and (3-methylbut-3-enoyl)sulfamoyl chloride, in a ratio of 7:3. Then, 400 g of 2,2-dimethyl-4-oxoazetidine-1-sulfonyl chloride was stirred in 2 L water at 60–65°C for 40 minutes and neutralized with concentrated NaOH. After the water was evaporated to dryness, the residue was extracted with acetone. Removal of acetone followed by distillation gave 135 g of 3-amino-3-methylbutanoic acid, as shown in Scheme 4.1 [2]. Cycloaddition of CSI with allenes, such as 2,4-dimethylpenta-2,3-diene and 2-methylpenta-2,3-diene, gives both 3-alkylidene-β-lactams and unsaturated amides. In this case, the solution of a stoichiometrically equal amount of allenes in anhydrous ether was added to an ice-bath cooled ether solution of $ClSO_2NCO$, until the completion of the reaction as monitored by IR. For the case of 2,4-dimethylpenta-2,3-diene, 67% and 28% of the expected 2,2-dimethyl-4-oxo-3-(propan-2-ylidene)azetidine-1-sulfonyl chloride and 3-methyl-2-(prop-1-en-2-yl)but-2-enamide were obtained. However, for the case of 2-methylpenta-2,3-diene, 37% and 25% of 3-ethylidene-2,2-dimethyl-4-oxoazetidine-1-sulfonyl chloride and 2-(prop-1-en-2-yl)but-2-enamide were obtained, where the 3-alkylidene-β-lactam is a mixture of *cis*- and *trans*-isomers with a ratio of 13:87, and the amide is a 29:71 mixture of *cis*- and *trans*-isomers (Scheme 4.2). Removal of the chlorosulfonyl group was accomplished by the addition of the solution of 3-alkylidene-β-lactams in 20% excess pyridine to an acetone solution containing 2 equivalents of thiophenol, as illustrated in Scheme 4 2. After that, the 3-alkylidene-β-lactams were hydrolyzed with concentrated HCl to afford 3-amino-3-methyl-2-isopropylidenebutanoic acid hydrochloride and 3-amino-2-ethylidene-3-methylbutanoic acid hydrochloride, respectively. The existing double bond can be further hydrogenated to give the corresponding saturated β-amino acids [3]. In another synthesis, the reaction of 2-(bromomethyl)-2-ethylbutanoyl chloride with *O*-benzylhydroxylamine in pyridine afforded 1-(benzyloxy)-3,3-diethylazetidin-2-one. Subsequent reaction of this β-lactam with HCl at room temperature gave 2,2-diethyl-3-aminopropanoic acid [4]. Similarly, [2+2]-cycloaddition of cyclohexene with *N*-benzyl isocyanate in a sealed tube in an atmosphere of dry CO_2 at ~80°C gave 7-benzoyl-7-azabicyclo[4.2.0]octan-8-one. Subsequent acid hydrolysis of the lactam ring led to 2-aminocyclohexane-1-carboxylic acid [5].

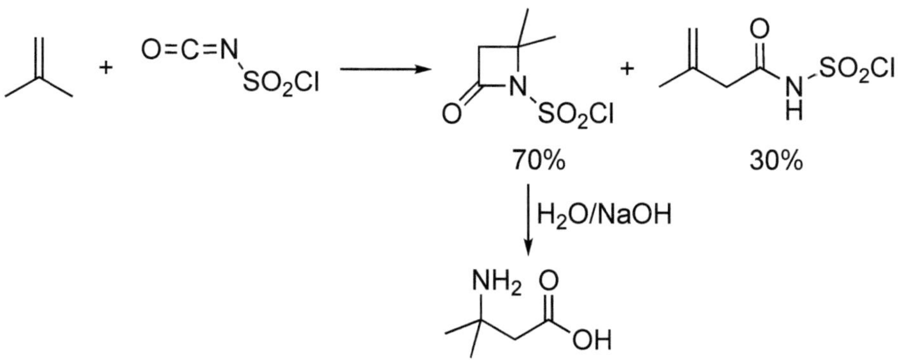

SCHEME 4.1 Preparation of 3-amino-3-methylbutanoic acid from the reaction between isobutene and sulfurisocyanatidic chloride

In addition, hydrolysis of the resulting β-lactams obtained from the Staudinger reaction allows the preparation of *N*-substituted β-amino acids directly. For example, oxidative cross-condensation of two different amines in the presence of an oxidation agent (O_2 in this case) allows the generation of imines of diverse scaffolds, which are then allowed to undergo [2+2]-cycloaddition with ketene to give β-lactams. Subsequent hydrolysis of the β-lactams with 10% HCl leads to the *N*-substituted β-amino acids. Specifically, when 0.321 g of benzylamines (3.0 mmol), 28.2 mg of 2,4,6-trihydroxybenzoic acid monohydrate (0.15 mmol, 5 mol.%), 0.775 g of octan-1-amine (6.0 mmol) and 1.5 mL of toluene were added to a 20 mL two-neck flask equipped

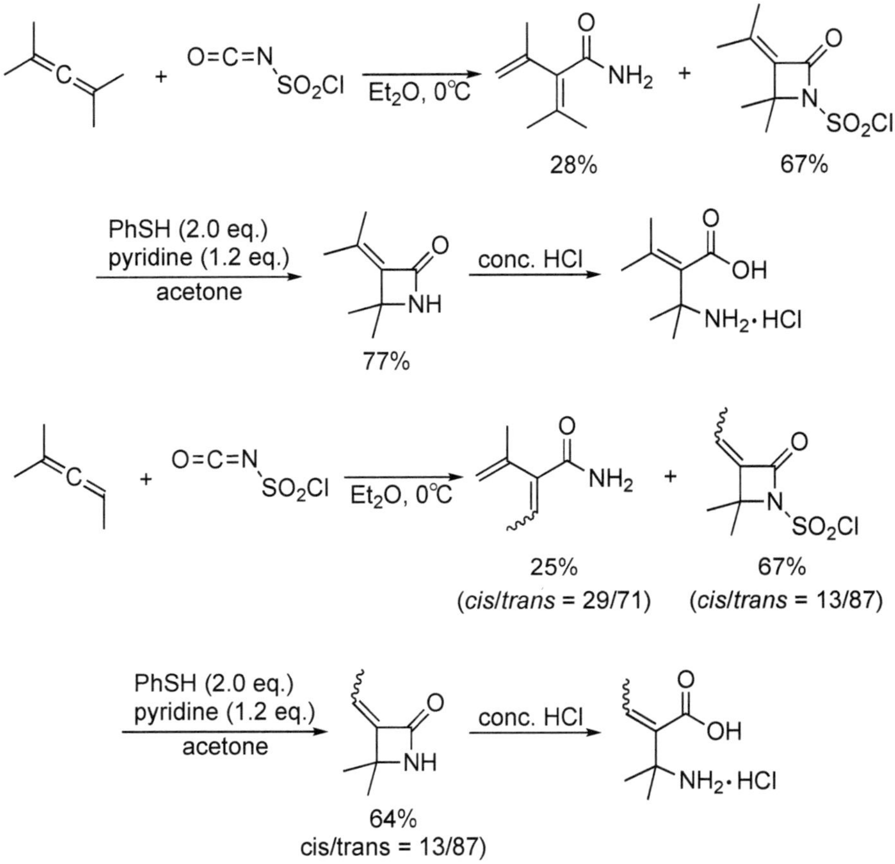

SCHEME 4.2 The synthesis of 2-(2-aminopropan-2-yl)-3-methylbut-2-enoic acid and 2-(2-aminopropan-2-yl)but-2-enoic acid

with an O_2 balloon at room temperature, the mixture was heated at 90°C under stirring for 2 hours and then cooled to room temperature. After that, 300 mg of 4Å molecular sieves and 0.627 mL Et_3N (4.5 mmol) were added, followed by a dropwise addition of a solution of 0.488 g of 2-methoxyacetyl chloride (4.5 mmol) in 3.0 mL CH_3CN at 0°C. Work-up of this reaction mixture afforded 58% of (3*R*,4*S*)-3-methoxy-1-octyl-4-phenylazetidin-2-one. When 86.8 mg of this β-lactam in 1 mL CH_3CN was mixed with 1.0 mL 10% HCl, the mixture was heated at 100°C for 24 hours to give 92% of (2*R*,3*S*)-2-methoxy-3-(octylamino)-3-phenylpropanoic acid hydrochloride, as illustrated in Scheme 4.3 [6].

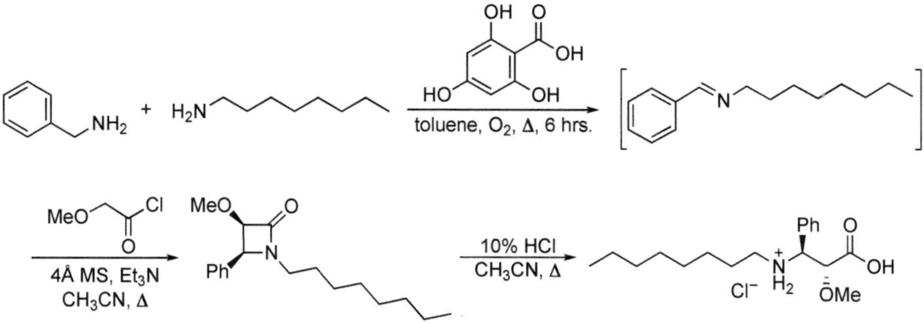

SCHEME 4.3 Preparation of (2*R*,3*S*)-2-methoxy-3-(octylamino)-3-phenylpropanoic acid

Additional examples about the acidic hydrolysis of β-lactams into β-amino acid derivatives include the hydrolysis of (2a*R*,7b*R*)-1,2a,3,7b-tetrahydro-2*H*-indeno[1,2-*b*]azet-2-one with 18% HCl to afford 87% of (1*R*,2*R*)-1-amino-2,3-dihydro-1*H*-indene-2-carboxylic acid hydrochloride [7], refluxing of (1*S*,12*R*)-13-azabicyclo[10.2.0]tetradecan-14-one or (1*R*,12*R*)-13-azabicyclo[10.2.0]tetradecan-14-one in 18% HCl for 2 hours to afford (1*S*,2*R*)-2-aminocyclododecane-1-carboxylic acid or (1*R*,2*R*)-2-aminocyclododecane-1-carboxylic acid, respectively [8], hydrolysis of (1*R*,2*R*,5*S*,6*R*,7*R*)-6,8,8-trimethyl-3-azatricyclo[5.1.1.02,5]nonan-4-one in 15% HCl at room temperature for 2 hours to afford 91% of (1*R*,2*R*,3*S*,4*R*,5*R*)-2-amino-4,6,6-trimethylbicyclo[3.1.1]heptane-3-carboxylic acid hydrochloride, for which the yield could be increased to 94% with 18% HCl [9], and conversion of dimethyl (4-oxoazetidin-2-yl)phosphonate into a mixture of 3-amino-3-(dimethoxyphosphoryl)propanoic acid and 3-amino-3-(hydroxy(methoxy)phosphoryl)propanoic acid with trifluoroacetic acid or dilute HCl [10]. For comparison, basic hydrolysis of 0.5 M of 1-methyl-2-azetidinone or 1-benyl-4-phenyl-2-azetidinone in 85% ethanol using equimolar NaOH (0.5 M) revealed a second-order kinetics [11]. Basic hydrolysis of *tert*-butyl (1*R*,2*R*,5*S*,6*R*,7*R*)-6,8,8-trimethyl-4-oxo-3-azatricyclo[5.1.1.0^{2,5}]nonane-3-carboxylate with LiOH in aqueous THF at 25°C for 7 hours yielded 91% of (1*R*,2*R*,3*S*,4*R*,5*R*)-2-((*tert*-butoxycarbonyl)amino)-4,6,6-trimethylbicyclo[3.1.1]-heptane-3-carboxylic acid [9].

An interesting aspect of the hydrolysis of β-lactams appears in the case of spiro-β-lactams. The Staudinger reaction between (*E*)-*N*-benzyl-1-phenylmethanimine and ketene (i.e., ((1*R*,4*R*)-bicyclo[2.2.1]hept-5-en-2-ylidene)methanone) prepared from (1*R*,4*R*)-bicyclo[2.2.1]hept-5-ene-2-carboxylic acid afforded a mixture of four stereoisomers, as shown in

Scheme 4.4 [12]. These stereoisomers include (1'*R*,3*S*,4*R*,4'*R*)-1-benzyl-4-phenylspiro[azetidine-3,2'-bicyclo[2.2.1]heptan]-5'-en-2-one, (1'*R*,3*R*,4*S*,4'*R*)-1-benzyl-4-phenylspiro[azetidine-3,2'-bicyclo[2.2.1]heptan]-5'-en-2-one, (1'*R*,3*R*,4*R*,4'*R*)-1-benzyl-4-phenylspiro[azetidine-3,2'-bicyclo[2.2.1]heptan]-5'-en-2-one and (1'*R*,3*S*,4*S*,4'*R*)-1-benzyl-4-phenylspiro[azetidine-3,2'-bicyclo[2.2.1]heptan]-5'-en-2-one. However, hydrolysis of (1'*R*,3*S*,4*R*,4'*R*)-1-benzyl-4-phenylspiro[azetidine-3,2'-bicyclo[2.2.1]heptan]-5'-en-2-one with either base (e.g., 5% NaOH in THF at 65°C for 6 hours, 32% NaOH in dioxane at 100°C for 10 hours, LiOH in H_2O/THF at 65°C for 6 hours, Cs_2CO_3 in H_2O/dioxane at 50°C for 1 day, and even $NaOCH_3$ in MeOH at 50°C for 1 day) or acid (e.g., 10 *N* HCl in dioxane at room temperature for 72 hours) was unsuccessful, and the β-lactam was fully recovered. In contrast, only 15% of this β-lactam was recovered when it was hydrolyzed with 10 *N* HCl in dioxane at 50°C for 4 hours. While no starting material could be retained when it was hydrolyzed with 10 *N* HCl in dioxane at 100°C for 10 hours or hydrolyzed with 32% HBr in acetic acid at room temperature for 16 hours, no identifiable product could be obtained either. Similarly, (1'*R*,3*R*,4*S*,4'*R*)-1-benzyl-4-phenylspiro[azetidine-3,2'-bicyclo[2.2.1]heptan]-5'-en-2-one was fully recovered when it was treated with Dowex 50X8 in THF at 60°C for 7 hours, 57% HI in acetic acid at room temperature for 24 hours and 50% H_2SO_4 in acetic acid at room temperature for 72 hours. It also decomposed completely when it was treated with 10 *N* HCl in dioxane at 100°C for 24 hours. Interestingly, 73% of (1*R*,2*R*,4*R*)-2-((*S*)-(benzylamino)(phenyl)methyl)bicyclo[2.2.1]hept-5-ene-2-carboxylic acid was obtained when this β-lactam was hydrolyzed with 32% HBr in acetic acid at room temperature for 16 hours. Likewise, hydrolysis of (1'*S*,3*R*,4*S*,4'*R*)-1-benzyl-4-phenylspiro[azetidine-3,2'-bicyclo[2.2.1]heptan]-2-one with 32% HBr under the same condition afforded 71% of (1*S*,2*R*,4*R*)-2-((*S*)-(benzylamino)(phenyl)methyl)bicyclo[2.2.1]heptane-2-carboxylic acid (Scheme 4.4). In contrast, (1'*S*,3*R*,4*R*,4'*R*)-1-benzyl-4-phenylspiro[azetidine-3,2'-bicyclo[2.2.1]heptan]-2-one was fully recovered when it was treated with 10% HCl in dioxane at 100°C for 24 hours, 32% HBr in acetic acid at room temperature for 48 hours or lipase from Candida *antarctica* in toluene at 60°C for 48 hours. The reason for the difference in stability of these β-lactams upon hydrolysis is unclear.

Comparison of alkaline solvolysis of β-lactams with those linear amides indicates the existence of a greater sensitivity to polar effects for β-lactams, and the rate-limiting step is the attack of hydroxide ion at the carbonyl carbon of the amide group at room temperature, whereas deamination of the amino acid appears to be the rate-limiting step for the hydrolysis of β-lactams at 80°C. Also, the steric effect from the substituent at position 3 of

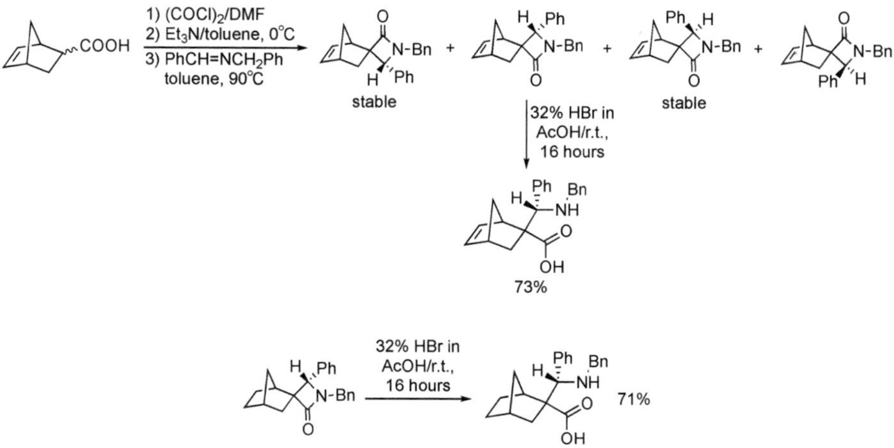

SCHEME 4.4 Preparation of (1'*S*,3*R*,4*S*,4'*R*)-1-benzyl-4-phenylspiro[azetidine-3,2'-bicyclo [2.2.1]heptan]-2-one and its acidic hydrolysis into (1*S*,2*R*,4*R*)-2-((*S*)-(benzylamino)(phenyl) methyl)bicyclo[2.2.1]heptane-2-carboxylic acid

2-azetidinones greatly impacts the rate of hydrolysis. For example, hydrolysis of 3,3-dimethyl β-lactam is about 25 times slower than that of 3-methyl β-lactam. Similarly, hydrolysis of 1-alkyl substituted 2-azetadinones is about an order of magnitude slower with respect to the unsubstituted compound. In addition, position 1 is influenced by polar effects to a greater degree than position 3 [13].

In addition to acidic/basic hydrolysis of β-lactams to afford β-amino acids, conversion of β-lactams into β-amino acids also occurs under enzymatic catalysis, as shown in the enantioselective hydrolysis of 3,4-disubstituted β-lactams to prepare a key side-chain intermediate of taxol [14]. One obvious advantage of using enzymatic hydrolysis of β-lactams is that enantiomerically pure β-amino acids can be achieved even though a racemic mixture of β-lactams is used, which is often the case when the β-lactams are synthesized from the Staudinger reaction. For example, a solvent-free hydrolysis of 6-azabicyclo[3.2.0]hept-3-en-7-one in the presence of *Candida antarctica* lipase B (CAL-B) afforded (1*R*,2*S*)-2-aminocyclopent-3-ene-1-carboxylic acid and unhydrolyzed (1*S*,5*R*)-6-azabicyclo[3.2.0]hept-3-en-7-one with greater than 200 of enantioselectivity for the β-amino acid. Similarly, racemic (3*R*)-3-(*tert*-butyl)-6-azabicyclo[3.2.0]heptan-7-ones and (1*S*,6*R*)-3-azatricyclo[4.2.1.0^{2,5}]non-7-en-4-ones have been hydrolyzed under the same condition to (1*R*,2*S*,4*R*)-2-amino-4-(*tert*-butyl)cyclopentane-1-carboxylic acid and (1*R*,2*R*,3*S*,4*S*)-3-aminobicyclo[2.2.1]hept-5-ene-2-carboxylic acid, respectively, with the stereoisomer untouched, as shown in Scheme 4.5

[15]. Under a similar condition, racemic 13-azabicyclo[10.2.0]tetradecan-14-one has been hydrolyzed in a mixture of water and isopropyl ether into (1*R*,2*S*)-2-aminocyclododecane-1-carboxylic acid with CAL-B, leaving (1*S*,12*R*)-13-azabicyclo[10.2.0]tetradecan-14-one unchanged (Scheme 4.5) [8]. It should be pointed out that the absolute configuration and conformation of these cyclic β-lactams have been determined by the combination of vibrational circular dichroism (VCD) spectroscopy and quantum chemical calculations at the *ab initio* (DFT) level of theory [16]. Enzymatic hydrolysis of dimethyl (4-oxoazetidin-2-yl)phosphonate in the presence of lipases or β-lactamase afforded 3-amino-3-phosphonopropanoic acid of only 25% ee [10]. More examples of enzymatic hydrolysis of β-lactams can be found in the recent review [17].

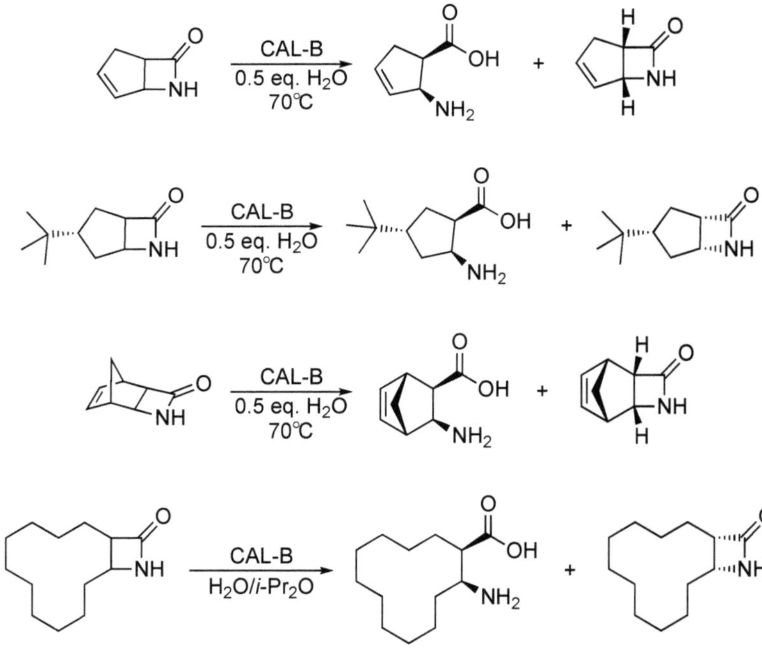

SCHEME 4.5 Examples of the enzymatic hydrolysis of β-lactams

4.2.2 *ALCOHOLYSIS OF* Β*-LACTAMS*

When the nucleophile is alcohol, ring-opening of the β-lactams leads to the formation of β-amino esters. This process is generally known as the alcoholysis of β-lactams. Biochemically, the resistance of β-lactams from bacteria has largely been exerted by means of β-lactamases, via the attack of the β-lactam antibiotics from their serine residue [18] or threonine residue [19], which in fact belong to alcohol. Examples of the alcoholysis of β-lactams include the conversion of (1*R*,2*R*,5*S*,6*R*,7*R*)-6,8,8-trimethyl-3-azatricyclo[5.1.1.0^{2,5}]nonan-4-one with 10% HCl in EtOH at room temperature for 1 hours to afford 86% of ethyl (1*R*,2*R*,3*S*,4*R*,5*R*)-2-amino-4,6,6-trimethylbicyclo[3.1.1]heptane-3-carboxylate [9], conversion of (1*R*,2*R*,5*S*,7*R*)-8,8-dimethyl-3-azatricyclo[5.1.1.0^{2,5}]nonan-4-one into 89% of ethyl (1*R*,2*R*,3*S*,5*R*)-2-amino-6,6-dimethylbicyclo[3.1.1]heptane-3-carboxylate hydrochloride with HCl in EtOH; treatment of *tert*-butyl (1*R*,2*R*,5*S*,7*R*)-2,8,8-trimethyl-4-oxo-3-azatricyclo[5.1.1.0^{2,5}]nonane-3-carboxylate with sodium methoxide in MeOH to give 92% of methyl (1*R*,2*R*,3*S*,5*R*)-2-((*tert*-butoxycarbonyl)amino)-2,6,6-trimethylbicyclo[3.1.1]heptane-3-carboxylate (Scheme 4.6) [9]. Similarly, heating (1*S*,2*S*,5*R*,7*S*)-2,8,8-trimethyl-3-azatricyclo[5.1.1.0^{2,5}]nonan-4-one in EtOH in the presence of 22% HCl afforded 60% of ethyl (1*S*,2*S*,3*R*,5*S*)-2-amino-2,6,6-trimethylbicyclo[3.1.1]heptane-3-carboxylate [20]. Under a very similar condition, (1*R*,6*S*)-7-azabicyclo[4.2.0]oct-3-en-8-one, (1*R*,6*S*)-7-azabicyclo[4.2.0]oct-4-en-8-one and (1*R*,5*S*)-6-azabicyclo[3.2.0]hept-3-en-7-one have been treated with HCl in EtOH at 0°C and converted into ethyl (1*R*,6*S*)-6-aminocyclohex-3-ene-1-carboxylate, ethyl (1*R*,2*S*)-2-aminocyclohex-3-ene-1-carboxylate and ethyl (1*R*,2*S*)-2-aminocyclopent-3-ene-1-carboxylate, respectively. These β-amino esters have been applied in the preparation of carbocyclic nucleoside analogs [21].

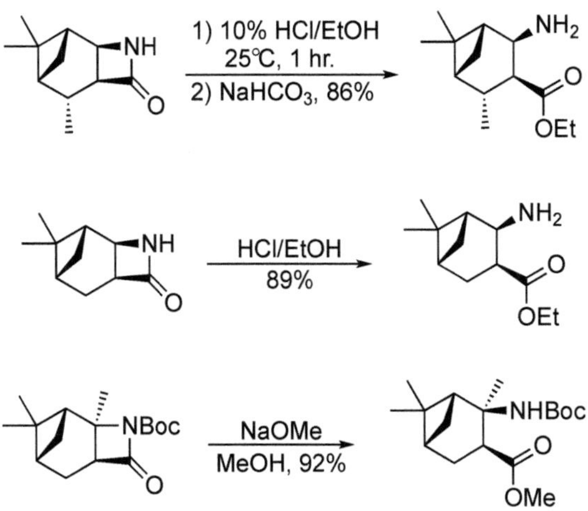

SCHEME 4.6 Examples of the alcoholysis of β-lactams

In another example, the addition of (*E*)-penta-1,3-diene solution in ether to the ether solution of chlorosulfonyl isocyanate (CSI) at -10°C to 0°C gave (*S*,*E*)-2-oxo-4-(prop-1-en-1-yl)azetidine-1-sulfonyl chloride, which was reduced with sodium sulfite in aqueous ether at 0 to 5°C to yield (*S*,*E*)-4-(prop-1-en-1-yl) azetidin-2-one. The solution of this β-lactam in dry methanol was injected with dry HCl to give methyl (*S*,*E*)-3-aminohex-4-enoate hydrochloride. Subsequent transformations of this β-amino acid methyl ester led to *N*-benzoyldaunosamine, i.e., *N*-((2*S*,3*S*,4*S*)-3,6-dihydroxy-2-methyltetrahydro-2*H*-pyran-4-yl)benzamide or *N*-((2*R*,3*R*,4*S*)-3,6-dihydroxy-2-methyltetrahydro-2*H*-pyran-4-yl)benzamide, as shown in Scheme 4.7 [22, 23]. Similarly, treatment of a solution of 2-methyl-4-oxo-2-(prop-1-en-2-yl)azetidine-1-sulfonyl chloride obtained from the cycloaddition between 2,3-dimethylbuta-1,3-diene and CSI in ether with a saturated solution of NaOH in CH_3OH at 5–20°C gave methyl 3-((methoxysulfonyl) amino)-3,4-dimethylpent-4-enoate [24]. Also, (1*S*,5*R*)-6-azabicyclo[3.2.0] hept-3-en-7-one prepared in 50–65% yield from cyclopentadiene and CSI by an improved procedure was treated with HCl in methanol to give afford methyl (1*S*,2*R*)-2-aminocyclopent-3-ene-1-carboxylate hydrochloride, which was then neutralized with aqueous Na_2CO_3 to give methyl (1*S*,2*R*)-2-aminocyclopent-3-ene-1-carboxylate, and reduced with $LiAlH_4$ in ether to generate ((1*S*,2*R*)-2-aminocyclopent-3-en-1-yl)methanol, a key intermediate in the total synthesis of (±)-naphthyridinomycin [25]. Furthermore, treatment of *tert*-butyl (2*S*,3*R*)-3-acetoxy-2-((2*S*,3*S*,4*R*,5*R*)-3,4-bis(benzyloxy)-5-((benzyloxy)methyl)tetrahydrofuran-2-yl)-4-oxoazetidine-1-carboxylate in methanol with 5.0 equivalents of Et_3N and 3.0 equivalents of DMAP at room temperature yielded 94% of methyl (2*R*,3*R*)-3-((2*S*,3*S*,4*R*,5*R*)-3,4-bis(benzyloxy)-5-((benzyloxy)methyl) tetrahydrofuran-2-yl)-3-((*tert*-butoxycarbonyl)amino)-2-hydroxypropanoate [26]. In addition, dimethyl (4-oxoazetidin-2-yl)phosphonate was converted into butyl 3-amino-3-(dimethoxyphosphoryl)propanoate in BuOH/$SOCl_2$ solution with HCl or in BuOH/H_2SO_4/CH_2Cl_2, or methyl 3-amino-3-(hydroxy(methoxy) phosphoryl)propanoate in MeOH with gaseous HCl [10].

In addition to general usage of HCl, azide has been reported to catalyze the methanolysis of a series of trifluoromethylated β-lactams. For example, (3*S*,4*R*)-3-methyl-1-tosyl-4-(trifluoromethyl)azetidin-2-one, (3*S*,4*R*)-3-ethyl-1-tosyl-4-(trifluoromethyl)azetidin-2-one, (3*S*,4*R*)-3-(*tert*-butyl)-1-tosyl-4-(trifluoromethyl)azetidin-2-one, and (3*S*,4*R*)-3-benzyl-1-tosyl-4-(trifluoromethyl)azetidin-2-one have been treated with sodium azide and methanol in DMF at room temperature to afford α-alkyl-β-trifluoromethyl-β-amino methyl esters in good to excellent yields [27].

Another important example of the reaction between alcohol and β-lactam can be found in the total synthesis of cryptophycins, which are macrocyclic cytotoxins produced by cyanobacteria, exhibiting potent tumor-selective antitumor activities by binding microtubules and disrupting cellular mitosis, particularly effective for limiting multidrug-resistant cancer cell lines. In this total synthesis, $Bu_4N^+CN^-$ initiated intramolecular reaction in CH_2Cl_2 between hydroxyl group and the nearby β-lactam moiety converted (3*R*,4*S*,*E*)-8-(((*R*)-3-(4-methoxyphenyl)-1-oxo-1-(2-oxoazetidin-1-yl)propan-2-yl)amino)-3-methyl-8-oxo-octa-1,6-dien-4-yl (*S*)-2-hydroxy-4-methylpentanoate into (3*S*,10*R*,16*S*,*E*)-16-((*R*)-but-3-en-2-yl)-3-isobutyl-10-(4-methoxybenzyl)-1,4-dioxa-8,11-diazacyclohexadec-13-ene-2,5,9,12-tetraone, in 68% yield. Subsequent transformations completed the synthesis of cryptophycin-24 (Arenastatin A), i.e., (3*S*,10*R*,16*S*,*E*)-3-isobutyl-10-(4-methoxybenzyl)-16-((*S*)-1-((2*R*,3*R*)-3-phenyloxiran-2-yl)ethyl)-1,4-dioxa-8,11-diazacyclohexadec-13-ene-2,5,9,12-tetraone, as shown in Scheme 4.8 [28].

One example of cyanide-triggered transesterification allows the ring expansion of β-lactam, as shown in Scheme 4.9. In this case, treatment of diethyl (5*R*,6*S*)-6-(2-iodoethyl)-7-oxo-1-azabicyclo[3.2.0]heptane-2,2-dicarboxylate with sodium methoxide or sodium cyanide in methanol led to the generation of 3,3-diethyl 7-methyl (7*S*,7a*R*)-hexahydro-3*H*-pyrrolizine-3,3,7-tricarboxylate. In comparison, treatment of diethyl (5*R*,6*S*)-6-(2-iodoethyl)-7-oxo-1-azabicyclo[3.2.0]heptane-2,2-dicarboxylate in acetic acid,(1*S*,7a*R*)-1-carboxy-5,5-bis(ethoxycarbonyl)octahydropyrrolizin-4-ium iodide was formed slowly. However, treatment of this compound in trifluoroacetic acid, the same iodide was generated quickly [29]. Extension of this sodium cyanide triggered ring expansion of β-lactam allows the transformation of 3-vinyl-β-lactams into methyl (*R*)-*N*-substituted pyrrolidine-3-carboxylates in methanol [30].

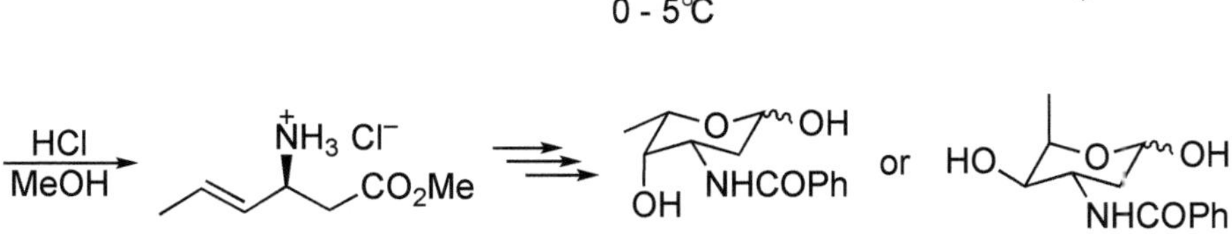

SCHEME 4.7 Formation of (*S*,*E*)-4-(prop-1-en-1-yl)azetidin-2-one from the reaction between (*E*)-penta-1,3-diene and CSI, and its methanolysis

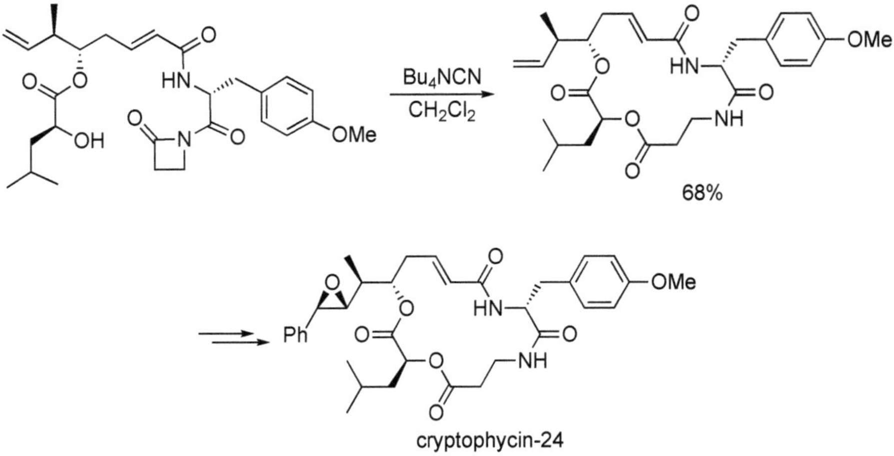

SCHEME 4.8 Cyanide promoted intramolecular alcoholysis of (3*R*,4*S*,*E*)-8-(((*R*)-3-(4-methoxyphenyl)-1-oxo-1-(2-oxoazetidin-1-yl)propan-2-yl)amino)-3-methyl-8-oxoocta-1,6-dien-4-yl (*S*)-2-hydroxy-4-methylpentanoate as a key step in the synthesis of cryptophycin-24

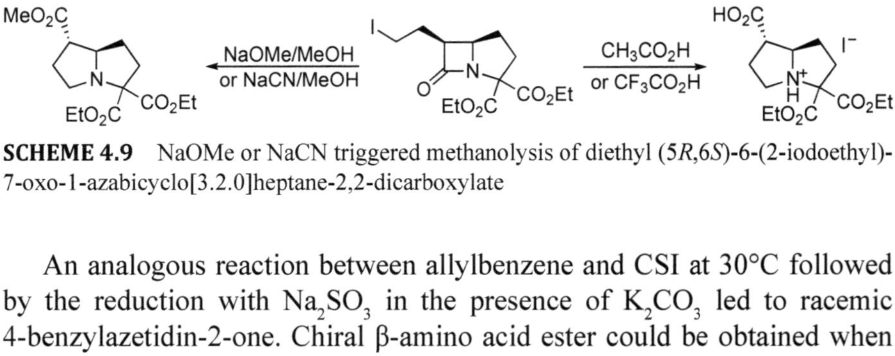

SCHEME 4.9 NaOMe or NaCN triggered methanolysis of diethyl (5*R*,6*S*)-6-(2-iodoethyl)-7-oxo-1-azabicyclo[3.2.0]heptane-2,2-dicarboxylate

An analogous reaction between allylbenzene and CSI at 30°C followed by the reduction with Na_2SO_3 in the presence of K_2CO_3 led to racemic 4-benzylazetidin-2-one. Chiral β-amino acid ester could be obtained when this racemic β-lactam was hydrolyzed in the presence of a lipase. For example, when this β-lactam was hydrolyzed in the diisopropyl ether in the presence of CAL-B (known as Novozyme 435) and 0.2 *M* methanol at 45°C for 3 hours, the conversion rate of the reaction was 40%, and a mixture of methyl (*S*)-3-amino-4-phenylbutanoate and (*R*)-3-amino-4-phenylbutanoic acid was obtained, with 81% and 59% *ee*, respectively. When the temperature was raised to 55°C, the hydrolysis of this β-lactam in the presence of 0.4 *M* MeOH resulted in 81% *ee* of (*S*)-3-amino-4-phenylbutanoate, but only 7% *ee* of (*R*)-3-amino-4-phenylbutanoic acid, as shown in Scheme 4.10. Further reaction of 4-benzylazetidin-2-one with paraformaldehyde gave 4-benzyl-1-(hydroxymethyl)azetidin-2-one, which can also be resolved by similar hydrolysis in the presence of different lipases [31].

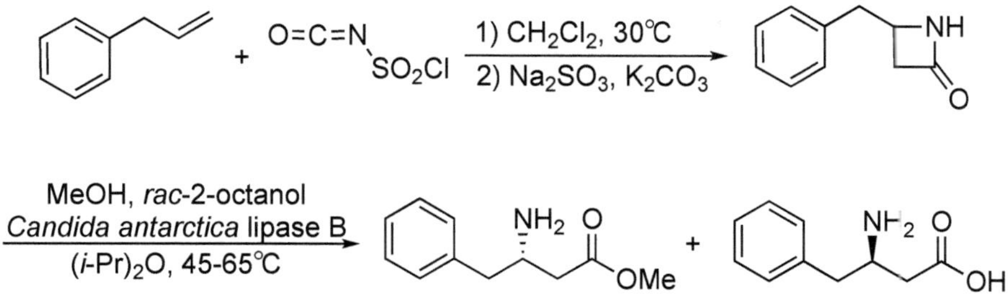

SCHEME 4.10 Enzymatic methanolysis of 4-benzylazetidin-2-one to afford methyl (*S*)-3-amino-4-phenylbutanoate

Another example of enzymatic alcoholysis can be found in the conversion of racemic benzyl 2-oxo-3-phenylazetidine-1-carboxylate into allyl (*R*)-3-(((benzyloxy)carbonyl)amino)-2-phenylpropanoate from the reaction of this β-lactam with allyl alcohol in the presence of an organic base and CAL-B in the organic solvent, with benzyl (*S*)-2-oxo-3-phenylazetidine-1-carboxylate untouched. In this transformation, the tested solvents include toluene, THF, methyl *t*-butyl ether, and acetonitrile. The organic bases used are pyridine (0.5 equivalent), DIPEA (0.5 eq.), DBU (0.5 eq.), and DBN (0.1 ~ 0.5 eq.), among these bases, DBN works the best [32].

4.2.3 AMINOLYSIS OF Β-LACTAMS

When the nucleophile is an amine, β-amino amide would be the expected product from the reaction between amine and β-lactam. This is generally known as aminolysis of β-lactam. For example, aminolysis of *tert*-butyl 2-benzyl-4-oxoazetidine-1-carboxylate (either one of the resolved enantiomers) with 25% NH_3 in methanol yielded *tert*-butyl (*S*)-(4-amino-4-oxo-1-phenylbutan-2-yl)carbamate and *tert*-butyl (*R*)-(4-amino-4-oxo-1-phenylbutan-2-yl)carbamate [31]. For a specific example, when (*R*)-2-((benzyloxy)carbamoyl)hexyl methanesulfonate in THF was treated with K_2CO_3 in the presence of a phase transfer catalyst, (*R*)-1-(benzyloxy)-3-butylazetidin-2-one was formed. This β-lactam was then allowed to react with (*S*)-*N*-(5-fluoropyridin-2-yl)pyrrolidine-2-carboxamide in THF in the presence of 2-ethylhexanoic acid, to yield (*S*)-1-((*R*)-2-(((benzyloxy)amino)methyl)hexanoyl)-*N*-(5-fluoropyridin-2-yl)pyrrolidine-2-carboxamide. Further formylation with formic acid/Ac_2O in isopropyl acetate afforded (*S*)-1-((*R*)-2-((*N*-(benzyloxy)formamido)methyl)hexanoyl)-*N*-(5-fluoropyridin-2-yl)pyrrolidine-2-carboxamide (Scheme 4.11). This compound is the key intermediate to making *in vitro* peptide deformylase inhibitor, which also shows a good *in vivo* antibacterial activity [33].

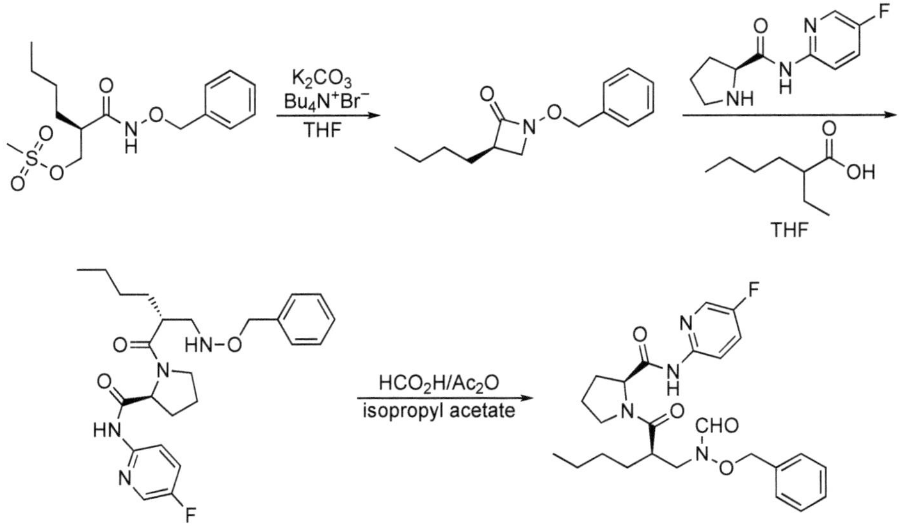

SCHEME 4.11 Aminolysis of (*R*)-1-(benzyloxy)-3-butylazetidin-2-one with (*S*)-*N*-(5-fluoropyridin-2-yl)pyrrolidine-2-carboxamide

It has been shown that the Zn^{2+}-tris(hydroxymethyl)aminomethane (Tris) system has a great catalytic effect on the hydrolysis and aminolysis of some β-lactam antibiotics. Among the tested β-lactam antibiotics including imipenem, SCH 29482, aztreonam, and nocardicin A in aqueous Tris solution at 35.0°C in the presence of a metal ion (e.g., Zn^{2+}, Cd^{2+}, Co^{2+}, Cu^{2+}, Ni^{2+}, and Mn^{2+}), the decomposition kinetics of these β-lactams indicated that Tris and metal ion (Cd^{2+} and Zn^{2+}) exerted a great catalytic effect on the hydrolysis and aminolysis of imipenem and SCH 29482. Such catalytic effect arises from a ternary complex in which the metal ion plays a double role by: (a) placing the antibiotic and Tris in the right position for the reaction; and (b) lowering the pKa of Tris' hydroxyl group to generate a strong nucleophile [34].

Meropenem, i.e., (4*R*,5*S*,6*S*)-3-(((3*S*,5*S*)-5-(dimethylcarbamoyl)pyrrolidin-3-yl)thio)-6-((*R*)-1-hydroxyethyl)-4-methyl-7-oxo-1-azabicyclo[3.2.0]hept-2-ene-2-carboxylic acid, is a novel 1β-methylcarbapenem antibiotic exhibiting a potent antibacterial activity against a wide range of Gram-positive and Gram-negative bacteria, with additional stability against renal dehydropeptidase-I. Although it shows a good stability in aqueous solution due to the effect of 1β-Me group against the hydrolysis of the β-lactam ring, intermolecular aminolysis of the β-lactam does occur to afford the dimer products, i.e., (4*R*,5*S*,6*S*)-3-(((3*S*,5*S*)-1-((2*S*,3*R*)-2-((2*S*,3*R*)-5-carboxy-4-(((3*S*,5*S*)-5-(dimethylcarbamoyl)pyrrolidin-3-yl)thio)-3-methyl-3,4-dihydro-2*H*-pyrrol-2-yl)-3-hydroxybutanoyl)-5-(dimethylcarbamoyl)-pyrrolidin-3-yl)thio)-6-((*R*)-1-hydroxyethyl)-4-methyl-7-oxo-1-azabicyclo[3.2.0]-hept-2-ene-2-carboxylic acid (involving migration of double bond) and (4*R*,5*S*,6*S*)-3-(((3*S*,5*S*)-1-((2*S*,3*R*)-2-((2*S*,3*R*)-5-carboxy-4-(((3*S*,5*S*)-5-(dimethylcarbamoyl)-pyrrolidin-3-yl)thio)-3-methyl-2,3-dihydro-1*H*-pyrrol-2-yl)-3-hydroxybutanoyl)-5-(dimethylcarbamoyl)pyrrolidin-3-yl) thio)-6-((*R*)-1-hydroxyethyl)-4-methyl-7-oxo-1-azabicyclo[3.2.0]hept-2-ene-2-carboxylic acid, as illustrated in Scheme 4.12 [35].

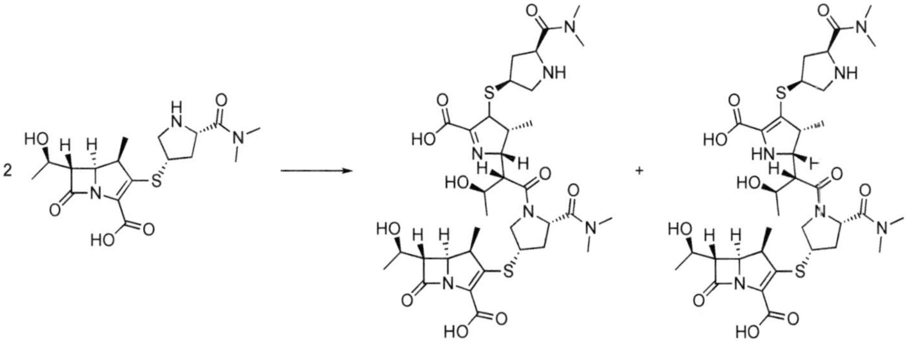

SCHEME 4.12 Self aminolysis of meropenem

Aminolysis of cephalosporin type β-lactams with a 3'-group that can be cleaved during the aminolysis indicates that the opening of the β-lactam occurs prior to the loss of the carbamic acid anion from position 3.' For example, when (6*R*,7*S*)-3-((carbamoyloxy)methyl)-7-methoxy-8-oxo-7-(2-(thiophen-2-yl)acetamido)-5-thia-1-azabicyclo[4.2.0]oct-2-ene-2-carboxylic acid (i.e., cefoxitin) was aminolyzed in liquid ammonia at –50°C, (*R*)-2-((*S*)-2-amino-1-methoxy-2-oxo-1-(2-(thiophen-2-yl)acetamido) ethyl)-5-((carbamoyloxy)methyl)-3,6-dihydro-2*H*-1,3-thiazine-4-carboxylate was formed first, which then underwent elimination at the 3'-position to yield (*R*)-2-((*S*)-2-amino-1-methoxy-2-oxo-1-(2-(thiophen-2-yl) acetamido)ethyl)-5-methylene-5,6-dihydro-2*H*-1,3-thiazine-4-carboxylate, as shown in Scheme 4.13. Similarly, treatment of (6*R*,7*S*)-7-(5-amino-5-carboxypentanamido)-3-((carbamoyloxy)methyl)-7-methoxy-8-oxo-5-thia-1-azabicyclo[4.2.0]oct-2-ene-2-carboxylic acid (i.e., cephamycin C) in liquid ammonia yielded (2*R*)-2-((1*S*)-2-amino-1-(5-amino-5-

carboxylatopentanamido)-1-methoxy-2-oxoethyl)-5-((carbamoyloxy) methyl)-3,6-dihydro-2*H*-1,3-thiazine-4-carboxylate and subsequently (2*R*)-2-((1*S*)-2-amino-1-(5-amino-5-carboxylatopentanamido)-1-methoxy-2-oxoethyl)-5-methylene-5,6-dihydro-2*H*-1,3-thiazine-4-carboxylate (Scheme 4.13) [36].

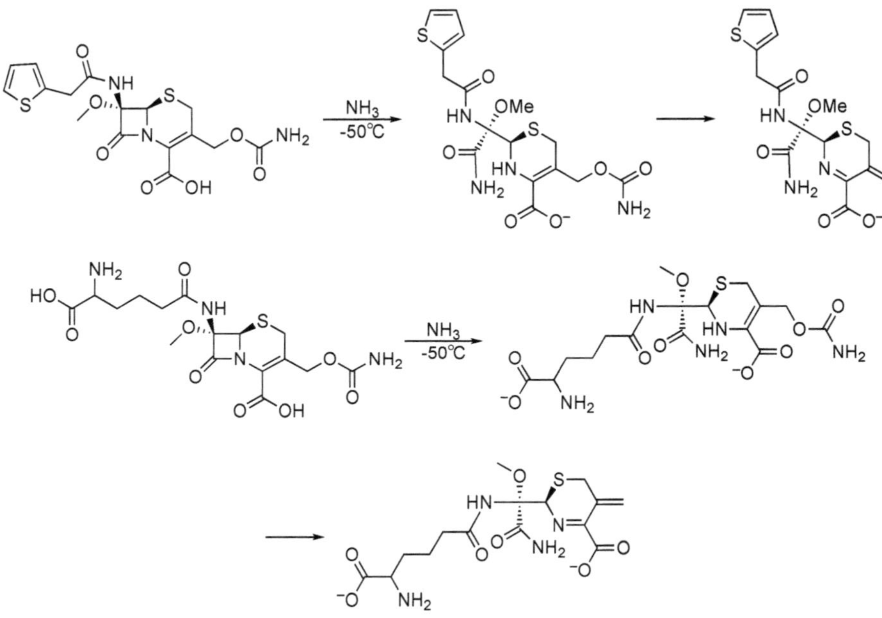

SCHEME 4.13 Aminolysis of cefoxitin and cephamycin C in liquid NH_3

Aminolysis of (2*S*,5*R*,6*R*)-3,3-dimethyl-7-oxo-6-(2-phenylacetamido)-4-thia-1-azabicyclo[3.2.0]heptane-2-carboxylic acid (i.e., benzylpenicillin) with hydroxyamine yielded (2*R*,4*S*)-2-((*R*)-2-(hydroxyamino)-2-oxo-1-(2-phenylacetamido)ethyl)-5,5-dimethylthiazolidine-4-carboxylic acid. The kinetic study only shows a first-order dependence on the amine concentration. The rate enhancement is greater than 10^6 with respect to the reaction between other primary amines and benzylpenicillin predicted from a Bronsted plot. This is compatible with the rate-limiting formation of the tetrahedral intermediate due to a rapid intramolecular general acid-catalyzed breakdown of the intermediate. In comparison, the rate enhancement for the aminolysis of 8-oxo-3-(pyridin-1-ium-1-ylmethyl)-7-(2-(thiophen-2-yl)acetamido)-5-thia-1-azabicyclo[4.2.0]oct-2-ene-2-carboxylate (i.e., cephaloridine) with hydroxylamine is greater than 10^4, indicating that the fission of β-lactam *C-N* bond and expulsion of the leaving group at C3' are not concerted [37]. Kinetic study on the aminolysis of (2S,5R,6R)-6-amino-3,3-dimethyl-7-oxo-4-thia-1-azabicyclo[3.2.0]heptane-2-carboxylic acid (i.e., 6-APA) and monobactam (2S,3S)-3-((Z)-2-(2-ammoniothiazol-4-yl)-2-(((2-carboxypropan-2-yl)oxy) imino)acetamido)-2-methyl-4-oxoazetidine-1-sulfonate (i.e., aztreonam) by propylamine or ethanolamine reveals that aztreonam is slightly more reactive than 6-APA, although it is generally assumed that the amide bond should be less activated in the monobactams. When ethanolamine is applied as the nucleophile, the rate law has a kinetic term proportional to $[RNH_2]$ $[RNH_3^+]$. It is assumed that this aminolysis of β-lactams proceeds through mechanisms in which either a water molecule or a second amine molecule may act as a bifunctional catalyst, assisting proton transfer from the attacking amine molecule to the leaving amino group [38]. More experimental data on aminolysis can be found in a review [39].

Quantum mechanical treatment of both ammonolysis (NH_3 as the nucleophile) and aminolysis (CH_3NH_2 as the nucleophile) of 2-azetidinone at MP2/6-31G**, B3LYP/6-31G** and G2(MP2,SVP) levels of theory reveals two different mechanistic routes for these two processes: a concerted pathway through a 1,2 addition of the H-NRH bond to the amide *C-N* bond, and a stepwise mechanism through several tetrahedral intermediates. For the stepwise mechanism, the *syn*-periplanar orientation of the attacking NH_2R with respect to the lone pair of the amidic nitrogen atom is more favored than the *anti*-periplanar orientation. The G2(MP2,SVP) level of calculation predicts that the nonconcerted route is the favored mechanism, whereas MP2/6-31G**SCRF level of computation indicates that the electrostatic effect of solvent tends to stabilize the stepwise transition structures in the aminolysis of β-lactam [40].

Also, quantum mechanical treatment of CH_3NH_2-assisted aminolysis of 2-azetidinone at MP2/6-31G**, B3LYP/6-31G** and G2(MP2,SVP) levels of theory revealed the existence of potential concerted pathway and two different stepwise routes involving several tetrahedral intermediates. One nonconcerted route involving an *N* inversion of a tetrahedral intermediate is the most favored mechanism, and the formation of the first antiperiplanar intermediate is the rate-determining step. Through a low energy barrier inversion process, this intermediate isomerizes to a *syn*-periplanar structure that, in turn, connects with a transition state for the ring-opening and *syn* *H*-transfer process. The computed Gibbs energy profile based on 2-azetidinone indicates a clear kinetic preference of stepwise mechanism for the aminolysis of monocyclic β-lactams, whereas for more rigid systems, such as benzylpenicillin, the largely impeded *N*-inversion process may favor a concerted or the nonconcerted *anti*-periplanar mechanism [41]. Further treatment of the aminolysis of 7-oxo-4-thia-1-azabicyclo[3.2.0]heptane-2-carboxylic acid by methylamine in water at the B3LYP/6-31+G* level of theory indicates that the concerted mechanism is favored, where the electrostatic effect of the carboxylate group plays an important role due to the enhancement of the solute-solvent interaction. In contrast, the stepwise mechanism is favored for this reaction in the gas phase [42]. More calculations on the aminolysis of β-lactams with different computational methods (e.g., KINBETA, KINUNI, KINUNI-MW) can be found in the literature [43–45].

A particular type of β-lactams, i.e., N-aroyl β-lactams, belong to imides containing both *exo*- and endocyclic acyl moieties. Aminolysis of this type β-lactams also give the ring-opened products. However, the rate law for the aminolysis of these β-lactams is dominated by a first-order dependence on the amine concentration in its free base form, indicating an uncatalyzed aminolysis reaction. Kinetic data also indicate a concerted mechanism for the aminolysis of these β-lactams [46].

4.2.4 HYDRAZINOLYSIS OF Β*-LACTAMS*

The phthaloyl group is often applied to protect the primary amino group in organic synthesis and cleaved by hydrazinolysis. Therefore, the phthaloyl group has been used in the synthesis of β-lactam containing additional amino groups [47–49]. For example, when 17.7 g of 2-phenyl-5,6-dihydro-4H-1,3-thiazine (0.1 mol) in 70 mL benzene and 22.4 g of phthaloylglycyl chloride (i.e., 2-(1,3-dioxoisoindolin-2-yl)acetyl chloride, 0.1 mol) in 110 mL dry benzene were combined at room temperature, a salt-like sparingly soluble precipitate fell out. Then, to this mixture was added 13.6 mL absolute triethylamine (0.1 mol) in 70 mL of benzene under vigorous stirring, and the resulting mixture was heated at 80°C for 2.5 hours to give 70.5% of 2-(8-oxo-6-phenyl-5-thia-1-azabicyclo[4.2.0]octan-7-yl)isoindoline-1,3-dione. Then, 18.2 g of this β-lactam (50 mmol) suspended in 250 mL dioxane was treated with 4.8 g 80% hydrazine hydrate (75 mmol) in 15 mL of dioxane under heating for 3 hours to give 7-amino-6-phenyl-5-thia-1-azabicyclo[4.2.0]octan-8-one (Scheme 4.14) [50]. Although this reaction did not show the hydrazinolysis of β-lactam, the further study indicates that the phthaloyl protecting group is more reactive than the β-lactam moiety during hydrolysis and hydrazinolysis. For example, treatment of the mixture of (2*R*,3*R*,4*S*,5*S*)-2-(((1*E*,2*E*)-3-phenylallylidene) amino)tetrahydro-2*H*-pyran-3,4,5-triyl triacetate and 2-(1,3-dioxoisoindolin-2-yl)acetyl chloride (phthaloylglycyl chloride) in CH_2Cl_2 at -10°C with Et_3N yielded (2*R*,3*R*,4*S*,5*S*)-2-(3-(1,3-dioxoisoindolin-2-yl)-2-oxo-4-((*E*)-styryl) azetidin-1-yl)tetrahydro-2*H*-pyran-3,4,5-triyl triacetate. Removal of the phthaloyl group by refluxing this β-lactam in ethanol in the presence of methylhydrazine and subsequent acidification with 5 *N* HCl afforded 2-oxo-4-((*E*)-styryl)-1-((2*R*,3*R*,4*S*,5*S*)-3,4,5-triacetoxytetrahydro-2*H*-pyran-2-yl) azetidin-3-aminium chloride. In addition, (2*R*,3*R*,4*S*,5*S*)-2-(3-(1,3-dioxoisoindolin-2-yl)-2-oxo-4-((*E*)-styryl)azetidin-1-yl)tetrahydro-2*H*-pyran-3,4,5-triyl triacetate was converted into (2*R*,3*R*,4*S*,5*S*)-2-(3-(1,3-dioxoisoindolin-2-yl)-2-formyl-4-oxoazetidin-1-yl)tetrahydro-2*H*-pyran-3,4,5-triyl triacetate via ozonolysis, subsequent reduction by $LiH(O\text{-}t\text{-}Bu)_3$ and deprotection of the phthaloyl group yielded 2-(hydroxymethyl)-4-oxo-1-((2*R*,3*R*,4*S*,5*S*)-3,4,5-triacetoxytetrahydro-2*H*-pyran-2-yl)azetidin-3-aminium chloride (Scheme 4.15) [51]. In both cases, the β-lactam moiety survives under the condition to remove the phthaloyl group. Similar strategy has been applied to the syntheses of (3*S*,4*R*)-3-amino-4-(2-methoxyphenyl)-1-(4-methoxyphenyl)azetidin-2-one, (3*S*,4*R*)-3-amino-4-(2-fluorophenyl)-1-(4-methoxyphenyl)azetidin-2-one and (3*R*,4*S*)-3-amino-1-(4-methoxyphenyl)-4-((*E*)-2-methoxystyryl)azetidin-2-one via the hydrazinolysis of the corresponding phthaloyl protected β-lactams in methanol, respectively [52]. Also, treatment of benzyl (2*S*,4*S*,5*S*,6*S*)-6-(1,3-dioxoisoindolin-2-yl)-3,3-dimethyl-7-oxo-4-thia-1-azabicyclo[3.2.0] heptane-2-carboxylate 4-oxide with *p*-TsOH allowed the Morin ring expansion to form benzyl (6*S*,7*S*)-7-(1,3-dioxoisoindolin-2-yl)-3-methyl-8-oxo-5-thia-1-azabicyclo[4.2.0]oct-2-ene-2-carboxylate, which was then deprotected with hydrazine hydrate in DMF to give benzyl (6*S*,7*S*)-7-amino-3-methyl-8-oxo-5-thia-1-azabicyclo[4.2.0]oct-2-ene-2-carboxylate (Scheme 4.16) [53].

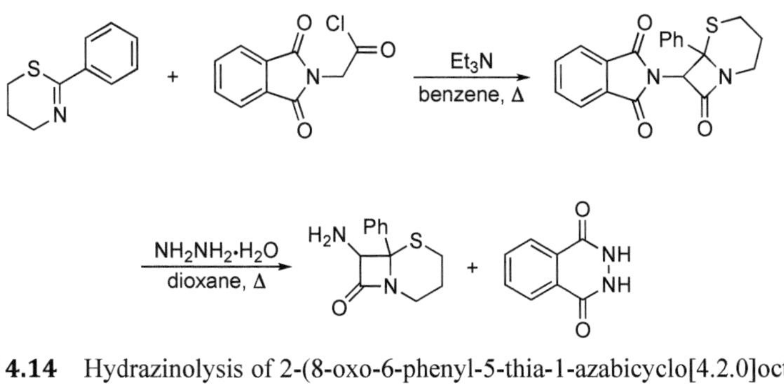

SCHEME 4.14 Hydrazinolysis of 2-(8-oxo-6-phenyl-5-thia-1-azabicyclo[4.2.0]octan-7-yl) isoindoline-1,3-dione into 7-amino-6-phenyl-5-thia-1-azabicyclo[4.2.0]octan-8-one

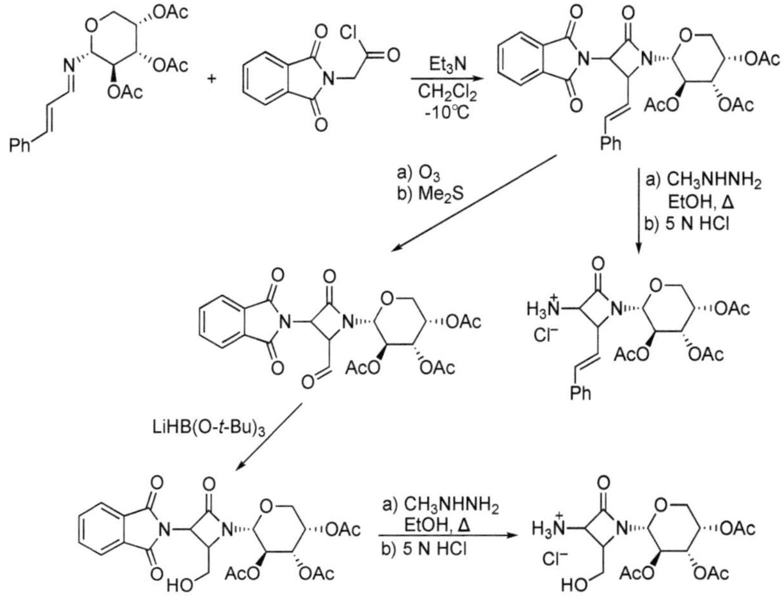

SCHEME 4.15 Deprotection of the phathaloyl group with methylhydrazine

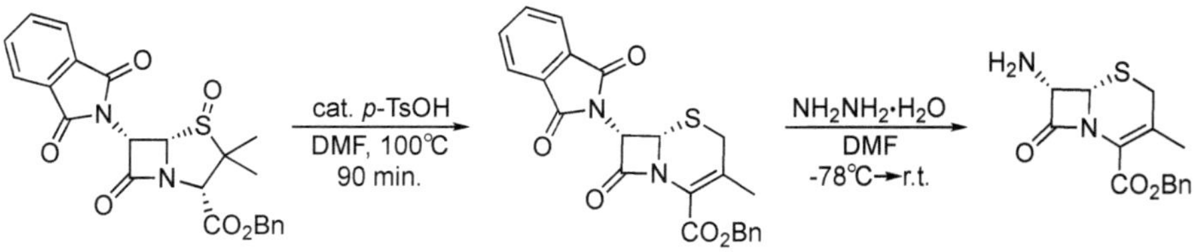

SCHEME 4.16 Hydrazinolysis of phthaloyl group to afford benzyl (6*S*,7*S*)-7-amino-3-methyl-8-oxo-5-thia-1-azabicyclo[4.2.0]oct-2-ene-2-carboxylate

Similarly, when (2*S*,5*R*,6*R*)-*N*-benzyl-6-(1,3-dioxoisoindolin-2-yl)-3,3-dimethyl-7-oxo-4-thia-1-azabicyclo[3.2.0]heptane-2-carboxamide was treated with Na_2S/H_2O in THF at 0°C, 46% of 2-(((2*S*,5*R*,6*R*)-2-(benzylcarbamoyl)-3,3-dimethyl-7-oxo-4-thia-1-azabi-cyclo-[3.2.0]heptan-6-yl)carbamoyl)benzoic acid was obtained, along with 3% of (2*R*,4*S*)-4-(benzylcarbamoyl)-2-((1,3-dioxoisoindolin-2-yl)methyl)-5,5-dimethylthiazolidin-3-ium chloride (that was obtained from the mother liquid) (Scheme 4.17) [54].

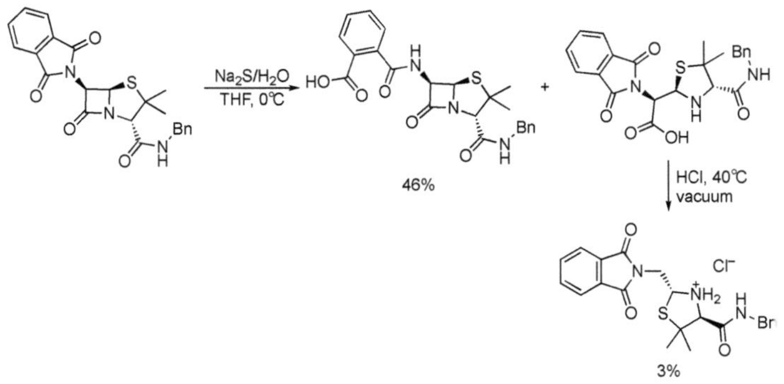

SCHEME 4.17 Cleavage of the phthaloyl group with Na_2S

Similarly, when (2*S*,5*R*,6*R*)-*N*-benzyl-6-(1,3-dioxoisoindolin-2-yl)-3,3-dimethyl-7-oxo-4-thia-1-azabicyclo[3.2.0]heptane-2-carboxamide was treated with methylhydrazine in CH_2Cl_2 at -75°C, the hydrazine also attacked the phthaloyl group. When the reaction mixture was brought up to room temperature, (2*S*,5*R*,6*R*)-6-amino-*N*-benzyl-3,3-dimethyl-7-oxo-4-thia-1-azabicyclo[3.2.0]heptane-2-carboxamide was obtained after removal of the phthaloyl protecting group. In comparison, when the same molecule was treated with hydrazine under the same condition, although the phthaloyl group was half cleaved, no (2*S*,5*R*,6*R*)-6-amino-*N*-benzyl-3,3-dimethyl-7-oxo-4-thia-1-azabicyclo[3.2.0]heptane-2-carboxamide was obtained when the reaction mixture was brought up to the room temperature, as shown in Scheme 4.18 [54].

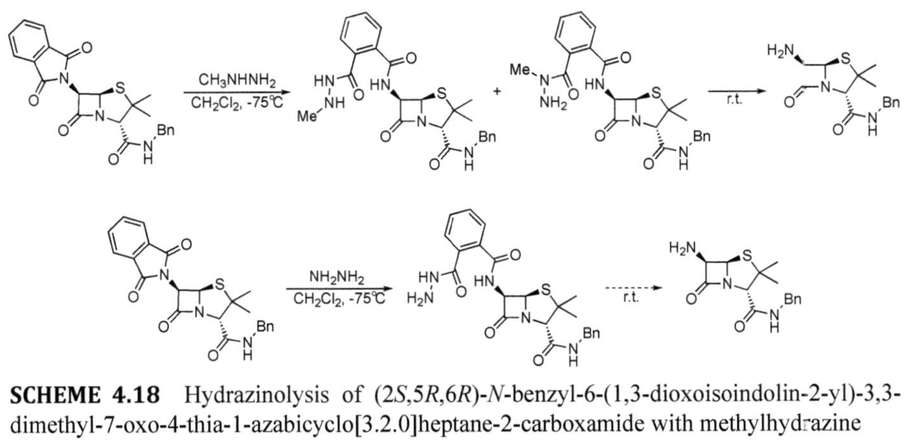

SCHEME 4.18 Hydrazinolysis of (2*S*,5*R*,6*R*)-*N*-benzyl-6-(1,3-dioxoisoindolin-2-yl)-3,3-dimethyl-7-oxo-4-thia-1-azabicyclo[3.2.0]heptane-2-carboxamide with methylhydrazine

However, when hydrazine was added to the toluene solution of (2*S*,5*R*,6*R*)-*N*-benzyl-6-(1,3-dioxoisoindolin-2-yl)-3,3-dimethyl-7-oxo-4-thia-1-azabicyclo[3.2.0]heptane-2-carboxamide, HPLC-MS analysis immediately detected (2*S*,5*R*,6*R*)-*N*-benzyl-6-(2-(hydrazinecarbonyl) benzamido)-3,3-dimethyl-7-oxo-4-thia-1-azabicyclo[3.2.0]heptane-2-carboxamide, (2*R*,4*S*)-*N*-benzyl-2-((*R*)-1-(1,3-dioxoisoindolin-2-yl)-2-hydrazineyl-2-oxoethyl)-5,5-dimethylthiazolidine-4-carboxamide as well as (2*R*,4*S*)-*N*-benzyl-2-((*R*)-1-(2-(hydrazinecarbonyl)benzamido)-2-hydrazineyl-2-oxoethyl)-5,5-dimethylthiazolidine-4-carboxamide. The latter has been obtained in 24% yield, which upon heating in alcohol, underwent decomposition to afford 36% of (2*R*,4*S*)-2-((*R*)-1-amino-2-hydrazineyl-2-oxoethyl)-*N*-benzyl-5,5-dimethylthiazolidine-4-carboxamide and 45% of (2*R*,4*S*)-*N*-benzyl-2-((*R*)-1-(1,3-dioxoisoindolin-2-yl)-2-hydrazineyl-2-oxoethyl)-5,5-dimethylthiazolidine-4-carboxamide (after ring closure), as shown in Scheme 4.19 [54].

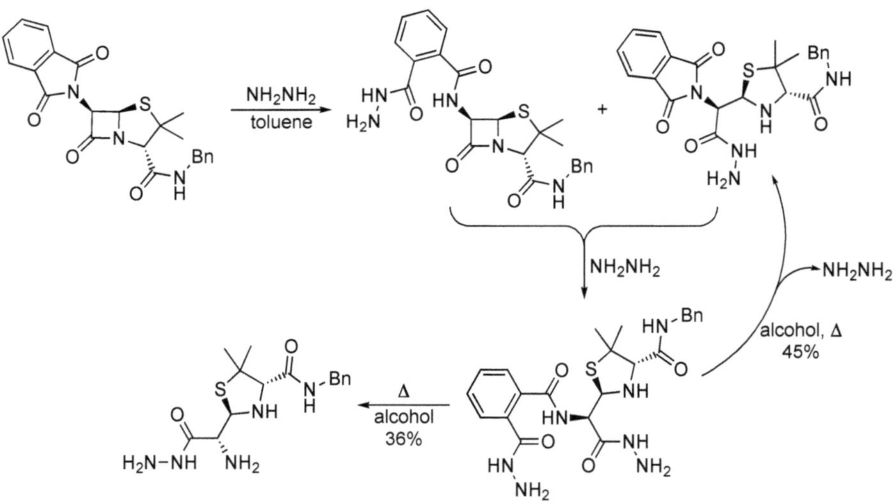

SCHEME 4.19 Cleavage of β-lactam with hydrazine

4.2.5 *THIOLYSIS OF* B*-LACTAMS*

Compared to the hydrazinolysis of β-lactams, the thiolysis of β-lactams has more practical usage. Thiol is normally more nucleophilic than hydrazine, and particularly many β-lactamases react with β-lactam antibiotics via their sulfhydryl groups from the cysteine residues [55–57]. Specifically, it is said that serum and cellular proteins could form adducts with amoxicillin, a process potentially causing the development of allergic reactions to this antibiotic. Treatment of amoxicillin by several thiol-containing compounds, including dithiothreitol, *N*-acetyl-*L*-cysteine, and glutathione, led to the detection of amoxicillin diketopiperazine. While the formation of diketopiperazine from the reaction between human serum albumin and amoxicillin is not associated with *L*-cysteine, the effect of thiol is catalytic and can render complete amoxicillin conversion. Also, those β-lactam antibiotics with an amino group in the antibiotic lateral chain would react with a thiol to give diketopiperazine, such as amoxicillin and ampicillin, whereas other β-lactam antibiotics such as cefaclor and benzylpenicillin do not undergo this route to yield diketopiperazine. A general reaction between amoxicillin or ampicillin and thiol is illustrated in Scheme 4.20 [58], where (2*R*,4*S*)-2-((2*R*,5*R*)-5-(4-hydroxyphenyl)-3,6-dioxopiperazin-2-yl)-5,5-dimethylthiazolidine-4-carboxylic acid and (2*R*,4*S*)-2-((2*R*,5*R*)-3,6-dioxo-5-phenylpiperazin-2-yl)-5,5-dimethylthiazolidine-4-carboxylic acid are formed from amoxicillin and ampicillin, respectively.

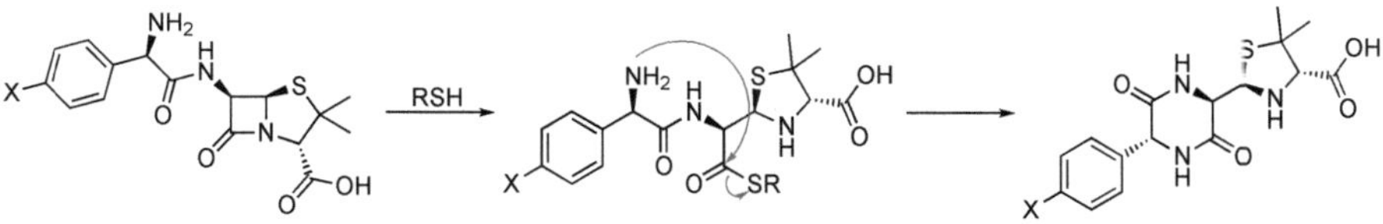

SCHEME 4.20 The mechanism for the conversion of amoxicillin and ampicillin into the corresponding diketopiperazines via thiolysis

Recently, a reactivity-guided drug discovery approach for electrophilic natural products used methyl (4-bromobenzoyl)cysteinate to react with a series of enone, β-lactam and β-lactone-based electrophilic natural products (e.g., parthenolide, andrographolide, wortmannin, penicillin G, salinosporamide), so that the natural products can be easily identified by mass spectrometry due to the unique characteristic of bromine isotopes as well as NMR spectroscopy. The reaction between penicillin G and the protected cysteine is illustrated in Scheme 4.21 for the thiolysis of β-lactam. However, the purification and NMR analysis of the direct thiolysis product, i.e., (2*R*,4*S*)-2-((*S*)-2-(((*R*)-2-(4-bromobenzamido)-3-methoxy-3-oxopropyl)thio)-2-oxo-1-(2-phenylacetamido)ethyl)-5,5-dimethylthiazolidine-4-carboxylic acid were very difficult, so that this direct product was acetylated with acetic anhydride in pyridine to yield (2*R*,4*S*)-3-acetyl-2-((*S*)-2-(((*R*)-2-(4-bromobenzamido)-3-methoxy-3-oxopropyl) thio)-2-oxo-1-(2-phenylacetamido)-ethyl)-5,5-dimethylthiazolidine-4-carboxylic acid. Interestingly, the epoxide-based electrophilic natural products (triptolide, epoxomicin, eponemycin, cyclomarin, salinamide) preferably react with thiophenol rather than cysteine [59].

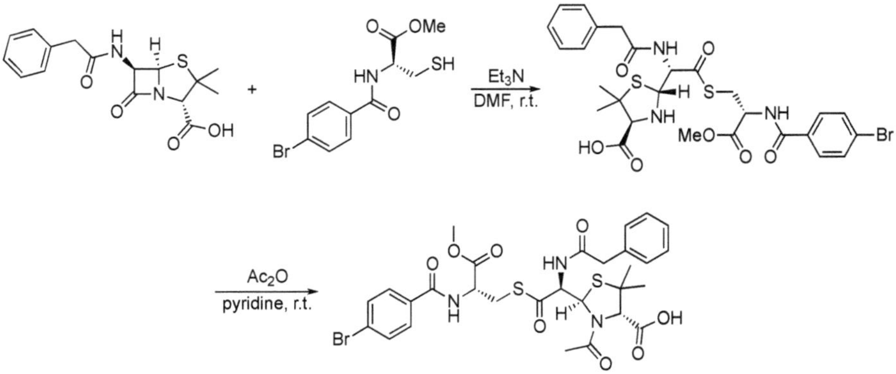

SCHEME 4.21 The reaction between methyl (4-bromobenzoyl)cysteinate and penicillin G

HPLC and ^{1}H NMR investigation on the thiol catalyzed hydrolysis of benzylpenicillin indicated the formation of a thioester intermediate, and the solvent water acted as a general acid catalyst in the breakdown of the tetrahedral intermediate, with a solvent kinetic isotope effect of 2.2–2.4. The catalytically reactive form of the thiol is actually the thiolate anion, and the breakdown of the tetrahedral intermediate is the rate-limiting step [60]. Interestingly, a similar study on the thiol-catalyzed hydrolysis of cephalosporins with HPLC and ^{1}H NMR spectroscopy revealed a low solvent kinetic isotopic effect (ca. 1.1), indicating that water probably does not act as a general acid catalyst to facilitate the breakdown of the tetrahedral intermediate by protonating the departing amine [61].

It should be pointed out that even though the β-lactam is normally reactive towards thiols, under special situations the β-lactam has been untouched when another reactive group presents on the β-lactam. For example, the quinidine catalyzed reaction between 4-oxoazetidin-2-yl formate and 2-methylbenzenethiol afforded 89% of (*S*)-4-(*o*-tolylthio)azetidin-2-one with 49% *ee*, rather than the thiolysis product *S*-(*o*-tolyl) 3-amino-3-(formyloxy) propanethioate. Similarly, the reaction between 4-oxoazetidin-2-yl formate and 2-mercaptobenzaldehyde yielded 85% of (2a*S*,8*R*)-8-hydroxy-2,2a-dihydro-1*H*,8*H*-azeto[2,1-*b*]benzo[*e*][1,3]thiazin-1-one (38% *ee*), as shown in Scheme 4.22 [62].

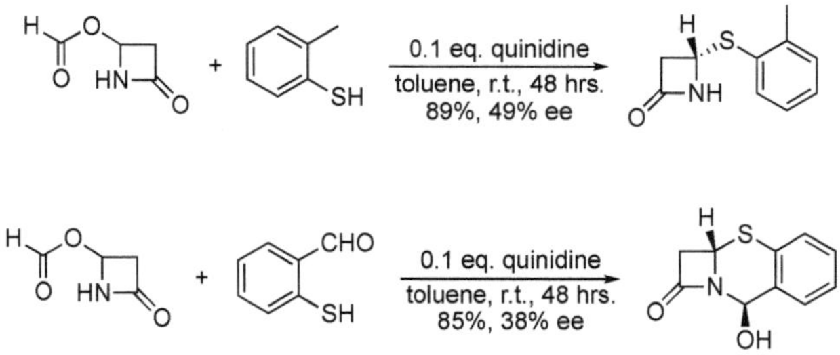

SCHEME 4.22 Quinidine catalyzed reactions between 4-oxoazetidin-2-yl formate and benzenethiols

Theoretical study on the reaction between (5*R*)-6-amino-4-thia-1-azabicyclo[3.2.0]heptan-7-one and 2-mercaptoethanol or 2-mercaptoethylamine with density functional theory (B3LYP/6-31+G* level) reveals that formation of the tetrahedral intermediate is the rate-determining step of the process. Particularly, the reaction involving 2-mercaptoethylamine is a general acid-catalyzed process due to the presence of an effective proton donor group, that reduced the energy of formation for the tetrahedral intermediate from 12.04 kcal/mol with 2-mercaptoethanol to 4.51 kcal/mol with 2-mercaptoethylamine [63]. In comparison, semi-empirical calculation of the reaction between 2-mercaptoethanol (or 2-mercaptoethylamine) and penicillin (or cephalosporine) indicates that the rate-limiting step in the reaction with penicillin is the cleavage of the tetrahedral intermediate, which is consistent with an intramolecular acid-catalyzed process for the thiolysis by 2-mercaptoethylamine; whereas the rate-determining step in the reaction of cephalosporine is the formation of the tetrahedral intermediate, due to the presence of an appropriate leaving group at position 3' of cephalosporine [64].

4.2.6 *CLEAVAGE OF* Β*-LACTAM BY HYDRIDE*

The simple β-lactam can be converted into azetidine, 1,3-amino alcohol, or 3-amino aldehyde by reducing the amide bond with hydride. For example, treatment of (3*S*,4*S*)-1-benzyl-3-(benzyloxy)-4-(2-chloropropan-2-yl)azetidin-2-one in Et_2O at 0°C with 5 equivalents of $LiAlH_4$ yielded 62% of (*S*)-2-((*R*)-1-benzyl-3,3-dimethylaziridin-2-yl)-2-(benzyloxy)ethan-1-ol and 4% of (2*S*,3*R*)-1-benzyl-3-(benzyloxy)-2-(2-chloropropan-2-yl)azetidine. The former is created via the intramolecular cyclization of (2*S*,3*S*)-3-(benzylamino)-2-(benzyloxy)-4-chloro-4-methylpentan-1-ol, the 1,3-amino alcohol intermediate. For comparison, when (3*S*,4*S*)-3-(benzyloxy)-

1-(*tert*-butyl)-4-(2-chloropropan-2-yl)azetidin-2-one was treated under the same condition, 52% of (3*S*,4*S*)-4-(benzyloxy)-*N*-(*tert*-butyl)-2,2-dimethyltetrahydrofuran-3-amine and 3% of (2*S*,3*R*)-3-(benzyloxy)-1-(*tert*-butyl)-2-(2-chloropropan-2-yl)azetidine were obtained. In this case, the intramolecular cyclization of 1,3-amino alcohol intermediate, i.e., (2*S*,3*S*)-2-(benzyloxy)-3-(*tert*-butylamino)-4-chloro-4-methylpentan-1-ol, to form the aziridine type product is very challenging due to the presence of a bulky *tert*-butyl group on the nitrogen atom. Therefore, intramolecular cyclization with terminal hydroxyl group takes place to give the tetrahydrofuran derivative, as shown in Scheme 4.23 [65].

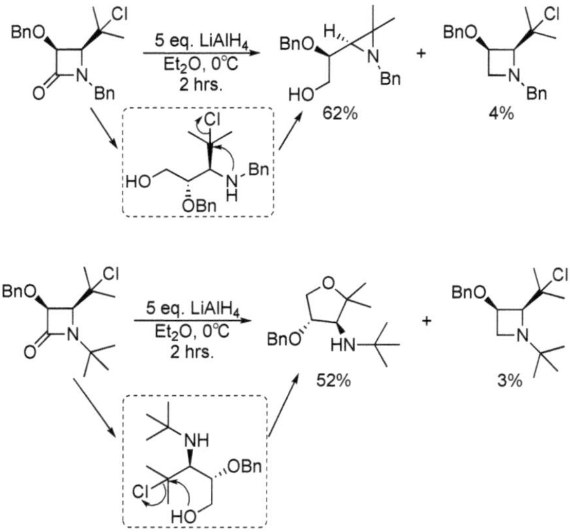

SCHEME 4.23 The impact of the *N*-substituent on the reduction of β-lactam with $LiAlH_4$

In another example, the Staudinger reaction between ketenes and Schiff bases yielded a series of β-lactams (55–89% yields), which were then reduced by 2.0 equivalents of $LiAlH_4$ to afford 32–75% of 1,3-amino alcohols. Specifically, when (*E*)-*N*-isopropyl-1-phenylmethanimine and 2-(benzyloxy)acetyl chloride in CH_2Cl_2 were treated with 3.0 equivalents of Et_3N at room temperature for 15 hours, 85% of (3*S*,4*R*)-3-(benzyloxy)-1-isopropyl-4-phenylazetidin-2-one was obtained. Upon treatment with 2.0 equivalents of $LiAlH_4$ in Et_2O for 3 hours, 42% of (2*S*,3*R*)-2-(benzyloxy)-3-(isopropylamino)-3-phenylpropan-1-ol was yielded [66].

The reactivity of β-lactam towards a variety of hydride-reducing reagents has been demonstrated by (*S*)-2-((*tert*-butyldimethylsilyl) oxy)-2-((2*R*,3*R*)-3-methoxy-1-(4-methoxyphenyl)-4-oxoazetidin-2-yl) acetonitrile, which contains both β-lactam and cyano moieties that can be reduced by the hydride. For example, when this β-lactam was treated in MeOH at room temperature by 7 equivalents of $NaBH_4$ in the presence of 1.0 equivalent of $NiCl_2$, the cyano group was reduced to amino functionality and the β-lactam is unchanged, affording 100% of (3*R*,4*R*)-4-((*R*)-2-amino-1-((*tert*-butyldimethylsilyl)oxy)ethyl)-3-methoxy-1-(4-methoxyphenyl)azetidin-2-one. In comparison, when it was treated with 3.0 equivalents of $LiBH_4$ in Et_2O at room temperature, the β-lactam rather than the cyano group was reduced, yielding 82% of (2*S*,3*R*,4*R*)-2-((*tert*-butyldimethylsilyl)oxy)-5-hydroxy-4-methoxy-3-((4-methoxyphenyl) amino)pentanenitrile. (2*R*,3*R*,4*R*)-5-Amino-4-((*tert*-butyldimethylsilyl) oxy)-2-methoxy-3-((4-methoxyphenyl)amino)pentan-1-ol can then be obtained by either reducing the cyano group with 9.0 equivalents of $NaBH_4$ in the presence of 1.0 equivalent of $NiCl_2$ in MeOH or reducing the β-lactam moiety with 3.0 equivalent of $LiBH_4$ in Et_2O at room temperature (Scheme 4.24). On the other hand, when this β-lactam was treated with a mild reducing reagent, such as 6.0 equivalents of DIBAL-H in THF at -78°C, the β-lactam moiety was converted into aldehyde and amine, affording 50% of (2*S*,3*R*,4*R*)-2-((*tert*-butyldimethylsilyl)oxy)-4-methoxy-3-((4-methoxyphenyl)amino)-5-oxopentanenitrile. When the same β-lactam was reduced with AlH_2Cl in Et_2O at room temperature, 48% of (*S*)-2-((*tert*-butyldimethylsilyl)oxy)-2-((2*R*,3*S*)-3-methoxy-1-(4-methoxyphenyl) azetidin-2-yl)acetonitrile was obtained (Scheme 4.24) [67].

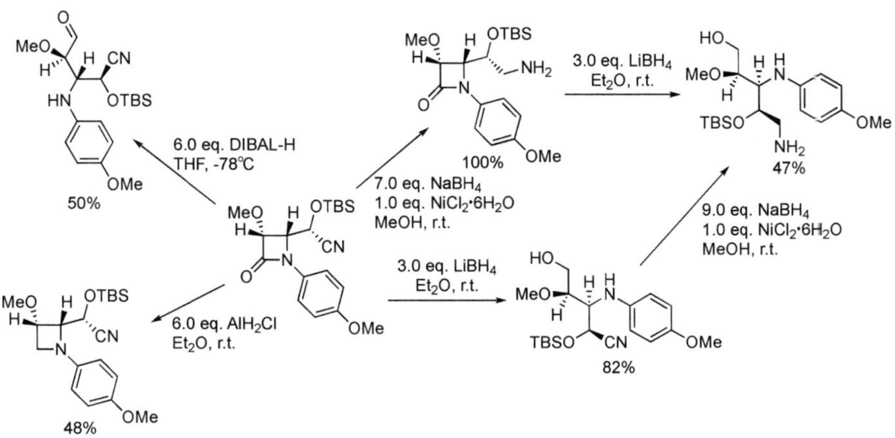

SCHEME 4.24 Reduction of (*S*)-2-((*tert*-butyldimethylsilyl)oxy)-2-((2*R*,3*R*)-3-methoxy-1-(4-methoxyphenyl)-4-oxoazetidin-2-yl)acetonitrile by various reducing agents

Similarly, a series of *cis*- and *trans*-β-lactams have been treated in Et_2O with monochloroalane (AlH_2Cl) that was generated *in situ* from $LiAlH_4/AlCl_3$ to give azetidines. In addition, the reactivity between the resulting azetidines and $AlEt_2Cl$ has been further investigated, indicating that azetidines with 4-methoxyphenyl or 2-furyl group at C2 and a benzyl or allyl substituent at nitrogen atom efficiently reacted with $AlEt_2Cl$ to stereoselectively give olefins through a fragmentation process; whereas acetal or thioacetal azetidines gave pyrrolidines as the sole products in a stereo-controlled manner. Specifically, treatment of (3*S*,4*R*)-1-benzyl-4-(4-methoxyphenyl)-3-phenoxyazetidin-2-one or (3*R*,4*S*)-1-benzyl-4-(4-methoxyphenyl)-3-vinylazetidin-2-one with AlH_2Cl in Et_2O, each yielded 91% of (2*R*,3*R*)-1-benzyl-2-(4-methoxyphenyl)-3-phenoxyazetidine or (2*S*,3*S*)-1-benzyl-2-(4-methoxyphenyl)-3-vinyl-azetidine, respectively. Further treatment of these azetidines in CH_2Cl_2 with $AlEt_2Cl$ afforded 57% of (*Z*)-1-methoxy-4-(2-phenoxyvinyl)benzene and 76% of (*E*)-1-(buta-1,3-dien-1-yl)-4-methoxybenzene, accordingly (Scheme 4.25) [68]. Moreover, when (2*S*,3*S*)-1-(4-methoxyphenyl)-4-oxo-3-phenylazetidine-2-carbaldehyde was treated with ethane-1,2-diol in the presence of acidic catalyst (*p*-TsOH), 94% of (3*S*,4*S*)-4-(1,3-dioxolan-2-yl)-1-(4-methoxyphenyl)-3-phenylazetidin-2-one was obtained. Likewise, when this β-lactam was reduced with monochloroalane in CH_2Cl_2, 85% of (2*S*,3*R*)-2-(1,3-dioxolan-2-yl)-1-(4-methoxyphenyl)-3-phenylazetidine was yielded. Further treatment of this compound with $AlEt_2Cl$ in CH_2Cl_2, 35% of (4a*S*,7*R*,7a*S*)-5-(4-methoxyphenyl)-7-phenylhexahydro-5*H*-[1,4]dioxino[2,3-*b*]pyrrole was obtained. In contrast, when (2*S*,3*R*)-1-benzyl-2-(bis(phenylthio)methyl)-3-phenoxyazetidine was treated with 2.0 equivalents of $AlEt_2Cl$ under refluxing condition, 72% of 1-benzyl-3-(phenylthio)-1*H*-pyrrole was yielded (Scheme 4.25) [68].

In another example, reduction of (3*R*,4*S*)-1-isopropyl-4-phenyl-3-(*p*-tolyl)azetidin-2-one with $LiAlH_4$ in Et_2O was performed at 0°C to refluxing temperature for 3 hours, 52% of (2*S*,3*S*)-3-(isopropylamino)-3-phenyl-2-(*p*-tolyl)propan-1-ol was obtained. This product is possibly formed by the reduction of the carbonyl to hydroxyl, followed by ring-opening and further reduction of the terminal aldehyde group to the hydroxyl group. In contrast, reduction of (3*R*,4*S*)-1-isopropyl-4-phenyl-3-(*p*-tolyl)azetidin-2-one with AlH_3, formed *in situ* by 3.0 equivalent of $LiAlH_4$ and $AlCl_3$ at 0°C for 30 minutes, followed by refluxing for 4 hours, resulted in the formation of (2*S*,3*S*)-1-isopropyl-2-phenyl-3-(*p*-tolyl)-azetidine, in 80% of yield. In this case, the carbonyl group is completely converted into the methylene group, without an accompanying ring-opening product (Scheme 4.26) [69].

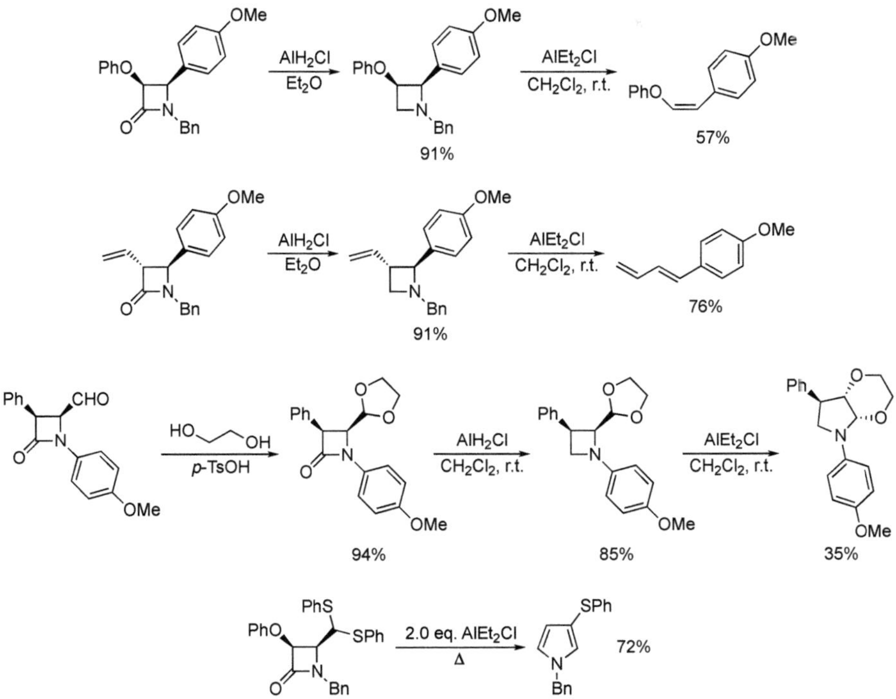

SCHEME 4.25 Treatment of β-lactams with diethylaluminum chloride

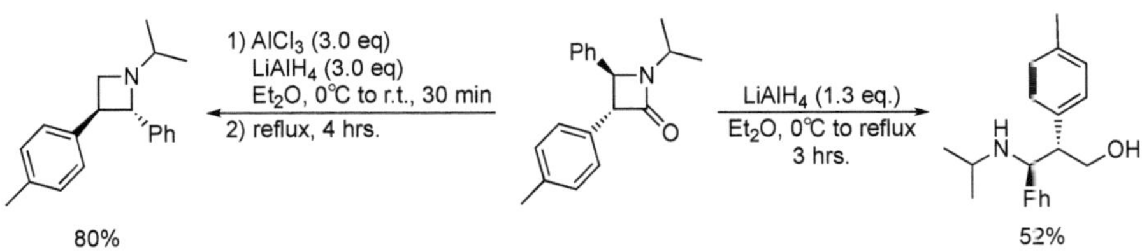

SCHEME 4.26 Reduction of (3*R*,4*S*)-1-isopropyl-4-phenyl-3-(*p*-tolyl)azetidin-2-one with $LiAlH_4$ with or without the presence of $AlCl_3$

In a very similar manner, reduction of (3*S*,4*S*)-3-methoxy-1-(4-methoxyphenyl)-4-(trifluoromethyl)azetidin-2-one with 2.0 equivalent of $LiAlH_4$ in Et_2O afforded 77% of (2*S*,3*S*)-4,4,4-trifluoro-2-methoxy-3-((4-methoxyphenyl)-amino)butan-1-ol. In contrast, reduction of the same substrate with 3.0 equivalents of aluminum chloride dihydride (AlH_2Cl) in Et_2O at room temperature yielded 85% of (2*S*,3*R*)-3-methoxy-1-(4-methoxyphenyl)-2-(trifluoromethyl)azetidine. Further treatment of this compound with 2.0 equivalents of $Me_3O^+BF_4^-$ in CH_2Cl_2 at room temperature, followed by 4.0 equivalents of sodium acetate in refluxing CH_3CN gave 86%

of (2*S*,3*S*)-4,4,4-trifluoro-2-methoxy-3-((4-methoxy-phenyl)(methyl)amino) butyl acetate, as illustrated in Scheme 4.27. Likewise, treatment of (3*S*,4*S*)-3-benzyloxy-1-(4-methoxyphenyl)-4-(trifluoromethyl)azetidin-2-one with AlH_2Cl, and then $Me_3O^+BF_4^-$, the reaction intermediate was further treated with 4.0 equivalents of *tert*-butylamine in refluxing CH_3CN instead of sodium acetate to afford 78% of (2*R*,3*S*)-2-(benzyloxy)-N^1-(*tert*-butyl)-4,4,4-trifluoro-N^3-(4-methoxyphenyl)-N^3-methylbutane-1,3-diamine [70]. More examples for the reduction of β-lactams with the combination of $LiAlH_4$ and $AlCl_3$ can be found in the preparation of azetidines with a sugar appendage [71], and synthesis of stereo-defined aziridines and azetidines starting from the reduction of 4-(1- or 2-haloalkyl)azetidin-2-ones (X = Cl, n = 0; X = Br, n = 1) with $LiAlH_4$, involving the 1,2-fission of the β-lactam moiety [65].

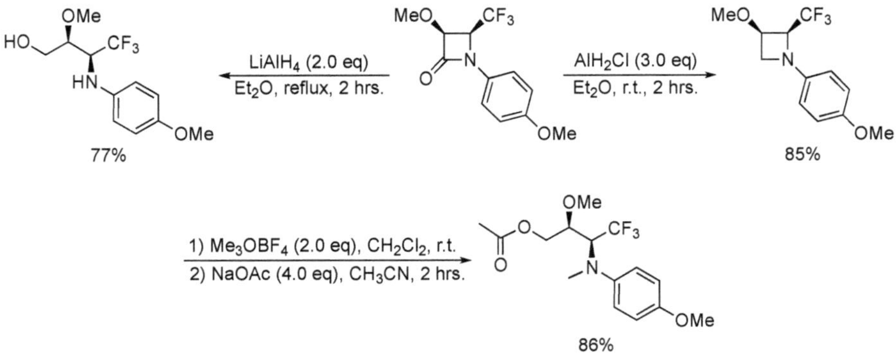

SCHEME 4.27 Reduction of (3*S*,4*S*)-3-methoxy-1-(4-methoxyphenyl)-4-(trifluoromethyl) azetidin-2-one with $LiAlH_4$ and AlH_2Cl yielding different products

4.2.7 MISCELLANEOUS REACTION RELATING TO THE CLEAVAGE OF AMIDE BONDS

In addition to the above reaction examples that cleave the amide bonds of β-lactams to give a variety of β-amino acids, esters, thioesters, amides, etc., there are reactions involving β-lactams of additional nucleophilic functionality that participates in a cascade reaction after the amide bond is cleaved. For example, treatment of (3*R*,4*S*)-4-((2*R*,4*S*)-4-((*tert*-butyldimethylsilyl)oxy) piperidin-2-yl)-3-methoxy-1-(*p*-tolyl)azetidin-2-one with sodium methoxide in methanol, should afford the alcoholysis product of methyl (2*R*,3*S*)-3-((2*R*,4*S*)-4-((*tert*-butyldimethylsilyl)oxy)piperidin-2-yl)-2-methoxy-3-(*p*-tolylamino) propanoate. However, intramolecular lactamization between the piperidine nitrogen and methyl ester moiety occurred to afford quantitative yield of

(2*R*,7*S*,8a*R*)-7-((*tert*-butyldimethylsilyl)oxy)-2-methoxy-1-(*p*-tolylamino) hexahydroindolizin-3(2*H*)-one (Scheme 4.28) [72].

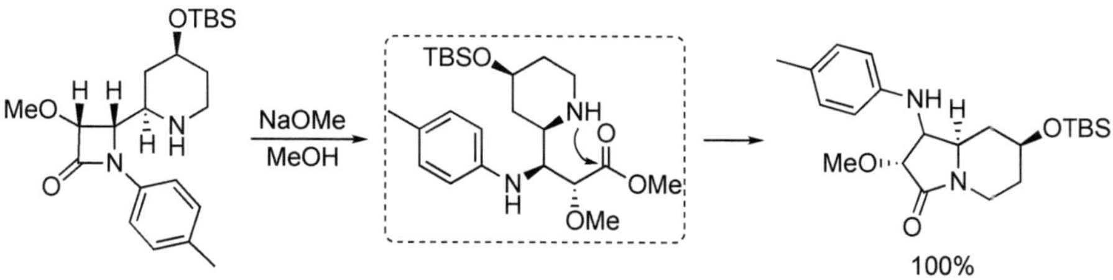

SCHEME 4.28 The mechanism for the transformation of (3*R*,4*S*)-4-((2*R*,4*S*)-4-((*tert*-butyldimethylsilyl)oxy)piperidin-2-yl)-3-methoxy-1-(*p*-tolyl)azetidin-2-one into (2*R*,7*S*,8a*R*)-7-((*tert*-butyldimethylsilyl)oxy)-2-methoxy-1-(*p*-tolylamino)hexahydroindolizin-3(2*H*)-one with NaOMe

In another example, when the benzyloxycarbonyl group within 1-(*tert*-butyl) 2-methyl 2-(3-(((benzyloxy)carbonyl)amino)propyl)-4-oxoazetidine-1,2-dicarboxylate was removed via hydrogenation, the newly exposed amino group intramolecularly attacks the β-lactam carbonyl group to give 81–83% of methyl 4-((*tert*-butoxycarbonyl)amino)-2-oxoazepane-4-carboxylate (Scheme 4.29), whereas the lactamization product of *tert*-butyl 2,5-dioxo-1,6-diazaspiro[3.5]nonane-1-carboxylate arising from the reaction between the newly exposed amino group and methyl ester did not occur due to the high ring strain of the β-lactam moiety [73].

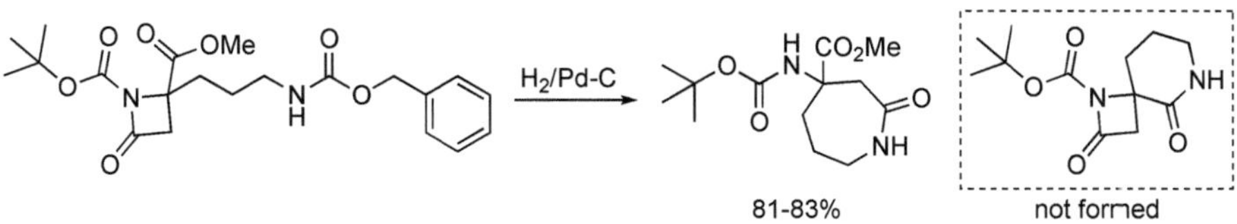

SCHEME 4.29 Hydrogenation of 1-(*tert*-butyl) 2-methyl 2-(3-(((benzyloxy)carbonyl) amino)propyl)-4-oxoazetidine-1,2-dicarboxylate, yielding methyl 4-((*tert*-butoxycarbonyl) amino)-2-oxoazepane-4-carboxylate

Compared to the base-initiated ring expansion, acid is also effective to promote the ring expansion of β-lactam. For example, when (*S*)-2-(benzylamino)-2-((2*S*,3*R*)-3-methoxy-1-(4-methoxyphenyl)-4-oxoazetidin-2-yl)acetonitrile was stirred in MeOH in the presence of sulfuric acid at room temperature for 63 hours, 66% of (2*S*,3*S*,4*R*)-1-benzyl-4-methoxy-3-((4-methoxyphenyl)amino)-5-oxopyrrolidine-2-carbonitrile was obtained, as shown in Scheme 4.30 [74]. Clearly, this product is formed from the intramolecular lactamization between the nitrogen atom connecting to the benzyl group and methyl ester group.

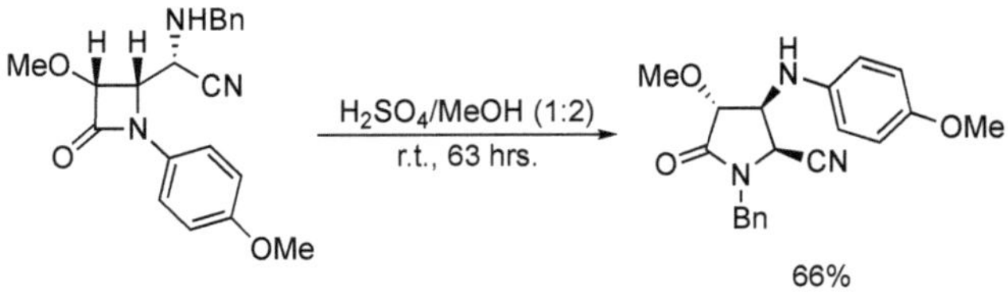

SCHEME 4.30 Conversion of (*S*)-2-(benzylamino)-2-((2*S*,3*R*)-3-methoxy-1-(4-methoxyphenyl)-4-oxoazetidin-2-yl)acetonitrile into (2*S*,3*S*,4*R*)-1-benzyl-4-methoxy-3-((4-methoxyphenyl)amino)-5-oxopyrrolidine-2-carbonitrile

The ring-opening of β-lactam via the cleavage of N1-C2 bond also occurs even when the nucleophile is an aromatic ring. For example, treatment of 1-(4-methoxyphenyl)azetidin-2-one with 2.0 equivalents of trifluoromethanesulfonic acid (i.e., triflic acid) in CH_2Cl_2 at 0–18°C afforded 94% of 6-methoxy-2,3-dihydroquinolin-4(1*H*)-one. Similarly, 1-(2-methoxyphenyl)azetidin-2-one was transformed into 95% of 8-methoxy-2,3-dihydroquinolin-4(1*H*)-one (Scheme 4.31). However, the presence of a strong electron-withdrawing group on the phenyl ring such as the nitro group reduces the nucleophilicity of the benzene ring, leading to the failure of a similar transformation. The presence of hydroxyl group also results in the failure of this reaction, although the reason is unclear [75].

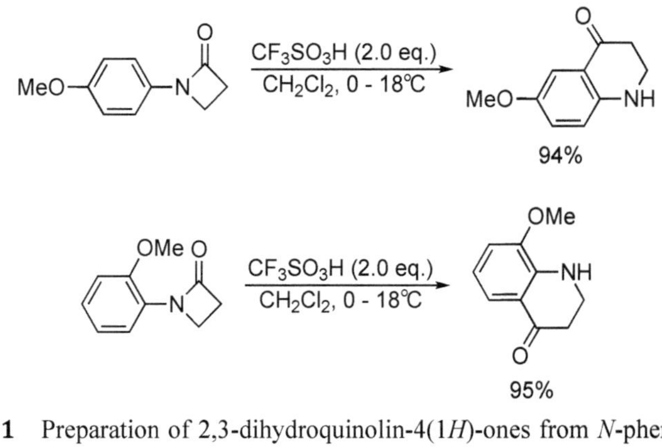

SCHEME 4.31 Preparation of 2,3-dihydroquinolin-4(1*H*)-ones from *N*-phenyl azetidin-2-ones

It should be mentioned that a group of β-lactams with a strong electron-withdrawing group on the nitrogen atom can be activated by a simple Lewis base, such as DMAP or 4-(pyrrolidin-1-yl)pyridine (PPY), to temporarily form nucleophilic nitrogenous species that undergoes addition reaction with the aldehyde group of ethyl 2-oxoacetate, as illustrated in Scheme 4.32. The model reaction between 1-((4-nitrophenyl) sulfonyl)-4-phenylazetidin-2-one and ethyl 2-oxoacetate has been carried out in $CHCl_3$, CH_3CN, THF, DMSO or DMF in the presence of 10 mol% of PPY, that affords ethyl 3-((4-nitrophenyl)sulfonyl)-6-oxo-4-phenyl-1,3-oxazinane-2-carboxylate. The results indicate that the highest diastereoselectivity can be achieved in DMSO, whereas the highest yield of the product can be obtained in DMF. In the tested reactions, 1-((4-nitrophenyl)sulfonyl)-4-phenylazetidin-2-one, 4-phenyl-1-tosylazetidin-2-one, and 1-(methylsulfonyl)-4-phenylazetidin-2-one all afforded the expected products, whereas *tert*-butyl 2-oxo-4-phenylazetidine-1-carboxylate and 4-phenyl-1-(2,2,2-trifluoroacetyl)azetidin-2-one failed in this transformation (Scheme 4.32). Following a similar mechanism, the dimer of β-lactam can also form as a side product, as illustrated in the conversion of 4-phenyl-1-tosylazetidin-2-one into (*Z*)-4-methyl-*N*-(2-(6-oxo-4-phenyl-3-tosyl-1,3-oxazinan-2-ylidene)-1-phenylethyl)benzene sulfonamide [76].

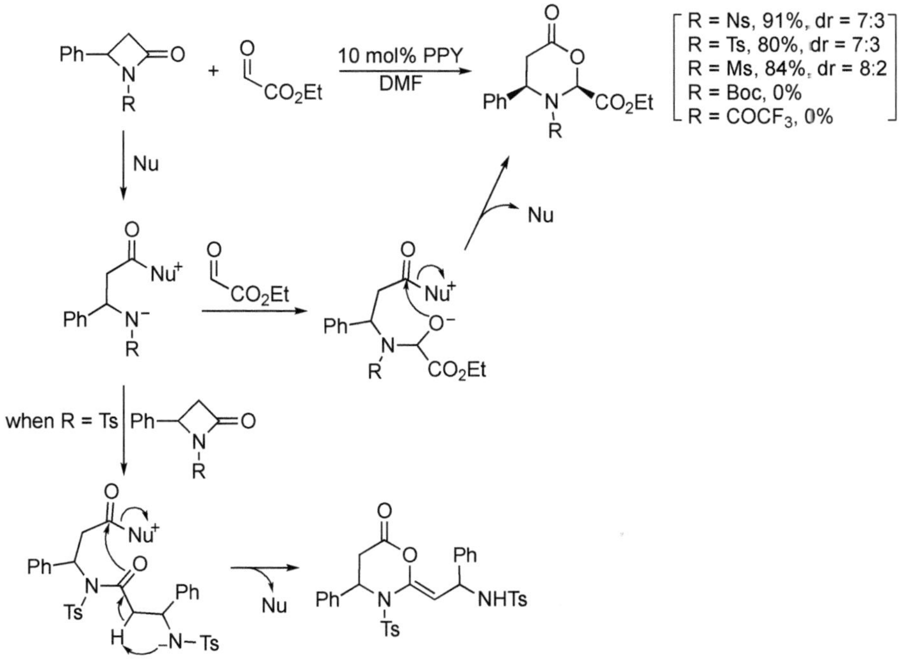

SCHEME 4.32 The base-catalyzed reaction between 4-phenyl-1-acylazetidin-2-ones and ethyl 2-oxoacetate that affords ethyl 6-oxo-4-phenyl-3-acyl-1,3-oxazinane-2-carboxylates

4.3 PREPARATION OF POLYAMIDES

It will not be surprising that β-lactam shares a common property of lactam undergoing ring-opening polymerization to give polyamides, involving the cleavage of N1-C2 bond. However, due to the high ring strain, β-lactams demonstrate much higher reactivity for ring-opening polymerization than other lactams. For example, the polymerization of pyrrolidone, the five-membered lactam, is prevented even by the introduction of a single methyl group under anionic polymerization conditions, whereas the tendency toward ring opening is so strong in β-lactam that the anionic polyaddition with the formation of high molecular weight poly(β-amide)s proceeds even when all the hydrogen atoms on the α- and β-carbon atoms are substituted, although the polymerization rate decreases with the increasing number of substituents. In addition, the equilibrium between polyamides and lactams observed in five-, six-, and seven-membered rings is negligible in the case of β-lactams [77]. The biggest feature of polyamides from β-lactams is that these polyamides can avoid proteolytic degradation [78]. Anyway, the polymerization of β-lactams can be initiated under both acidic or basic conditions, proceeding via cationic or anionic ring-opening polymerization. The resulting polymers have been called nylon-3, analogous to the name of nylon-6 from the polymerization of caprolactam [79–82]. The cationic ring-opening polymerization of β-lactams proceeds as a stepwise polyaddition. When stoichiometric quantities of acid (e.g., acetic acid) and β-lactams are used, the resulting mixture contains the acetylamino acid together with large quantities of the oligomeric compounds containing two, three, or four repeating units. This condition is best known to form poly(β-amino acid)s with a medium degree of polymerization (up to 50) when the polymerization is carried out at 100 to 150°C with less than a stoichiometric amount of carboxylic acid, where the average molecular weight of the polymer is controlled by the quantity of carboxylic acid applied. In contrast, poly(β-amino acid)s of high molecular weight can be accessed via anionic ring-opening polymerization, which proceeds rapidly even at room temperature, with complete consumption of the lactam present. Therefore, it is ideal to perform this polymerization in a solvent in order to dissipate the heat of the reaction. Particularly, the anionic ring-opening polymerization can be initiated immediately by the addition of a small quantity of *N*-acyl-β-lactam, leading to poly(β-amino acid)s of narrow molecular weight distribution. The number of chain initiation corresponds to the concentration of the added *N*-acyl-β-lactam, the resulting polymers are similar in structure to natural silk, with molecular weight as high as that of fibroin (200,000–300,000 g/mol) [77].

For example, benzyl (*S*)-4-oxoazetidine-2-carboxylate prepared from *L*-aspartic acid was subject to acyl transfer reaction by refluxing in 2-methoxyethan-1-ol at 121°C in the presence of tetrabutoxytitanium [$Ti(OBu)_4$] to give 2-methoxyethyl (*S*)-4-oxoazetidine-2-carboxylate. The polymerization of this β-lactam was performed in CH_2Cl_2 at a concentration of 18% w/w in the presence of sodium 2-oxopyrrolidin-1-ide initiator, to yield 76% of poly[2-methoxyethyl β-*L*-aspartate], as shown in Scheme 4.33. Also, poly[2-(2-methoxyethoxy)ethyl β-*L*-aspartate] was prepared from the anionic ring-opening polymerization of 2-(2-methoxyethoxy)ethyl (*S*)-4-oxoazetidine-2-carboxylate by a similar approach in 67% yield [83]. The formed polymers were estimated to be greater than 2×10^5 g/mol in molecular weights, and demonstrated conformation resembling the well-known α-helix in polypeptides. A chiral poly(isobutyl(2*S*,3*R*)-3-benzyloxyaspartate) with a molecular weight of 543 kDa and 230 kDa (estimated with from viscometric measurement and gel-permeation chromatography respectively) was synthesized from the ring-opening polymerization of isobutyl (2*R*,3*R*)-3-(benzyloxy)-4-oxoazetidine-2-carboxylate, that was prepared in multistep after the initial cycloaddition between 2-(benzyloxy)acetyl chloride and (*S*,*E*)-1-(2,2-dimethyl-1,3-dioxolan-4-yl)-*N*-(4-methoxyphenyl)methanimine [84]. Similarly, benzyl (*S*)-4-oxoazetidine-2-carboxylate has been converted into 2-(2-(2-methoxyethoxy)ethoxy)ethyl (*S*)-4-oxoazetidine-2-carboxylate, 2,5,8,11-tetraoxatridecan-13-yl (*S*)-4-oxoazetidine-2-carboxylate and isobutyl (*S*)-4-oxoazetidine-2-carboxylate in the presence of 2-(2-(2-methoxyethoxy)ethoxy)ethan-1-ol, 2,5,8,11-tetraoxatridecan-13-ol and 2-methyl-propan-1-ol, respectively. The anionic ring-opening polymerizations of these monomers have been initiated with metal-amido complexes to afford block copoly(β-peptides) [85].

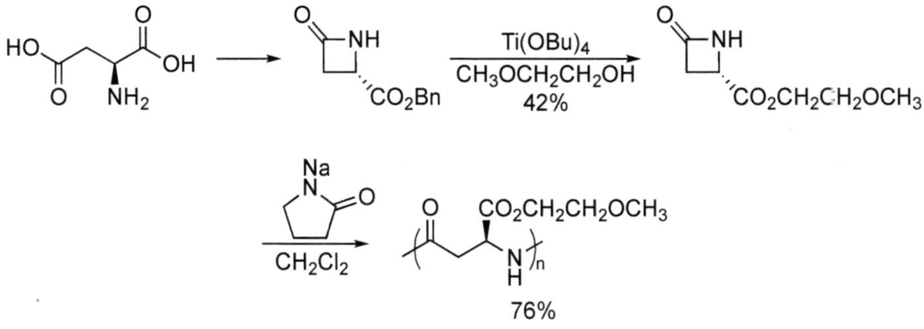

SCHEME 4.33 Preparation of 2-methoxyethyl (*S*)-4-oxoazetidine-2-carboxylate and its based initiated polymerization

These poly(β-peptide)s have attracted quite a lot of attention as novel advanced materials due to their capability of taking helical conformations stabilized by intramolecular hydrogen bonds, which have demonstrated some unusual properties, such as piezoelectricity, thermochromicity, and perm-selectivity. The poly(β-peptide)s prepared from the anionic ring-opening polymerization of benzyl (*S*)-4-oxoazetidine-2-carboxylate (triggered by *tert*-butylmagnesium chloride) were then suspended in a large excess of alcohol in the presence of titanium tetrabutoxide for transesterification to achieve the ideal materials [82].

In addition to the preparation of poly(β-asparate) as outlined above, poly(β^3-homolysine) has been prepared from the anionic ring-opening polymerization of *tert*-butyl (*S*)-(4-(4-oxoazetidin-2-yl)butyl)carbamate followed by the removal of the Boc protecting group as shown in Scheme 4.34. In order to make this β-lactam, N^2-(((9*H*-fluoren-9-yl)methoxy)carbonyl)-N^6-(*tert*-butoxycarbonyl)-*L*-lysine was initially treated with isobutyl carbonochloridate (IBC-Cl) to form (9*H*-fluoren-9-yl)methyl *tert*-butyl (6-chloro-6-oxohexane-1,5-diyl)(*S*)-dicarbamate, which was then allowed to react with diazomethane in THF to yield (9*H*-fluoren-9-yl)methyl *tert*-butyl (7-diazo-6-oxoheptane-1,5-diyl)(*S*)-dicarbamate. Subsequently, silver

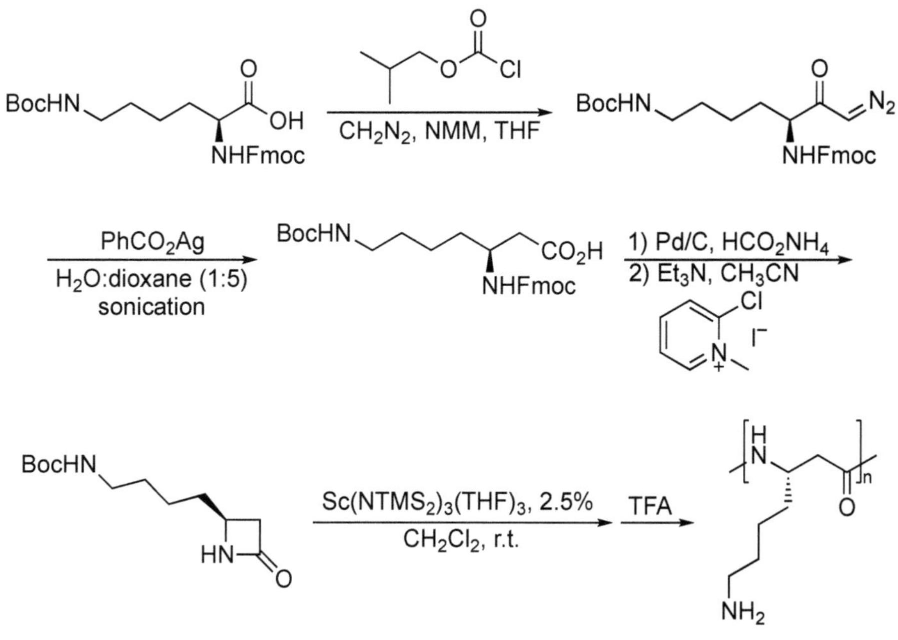

SCHEME 4.34 The approach to making poly(β^3-homolysine)

benzoate triggered the Arndt-Eistert reaction led to (*S*)-3-((((9*H*-fluoren-9-yl) methoxy)carbonyl)amino)-7-((*tert*-butoxycarbonyl)amino)heptanoic acid, from which the Fmoc protecting group was removed via hydrogenation. The resulting (*S*)-3-amino-7-((*tert*-butoxycarbonyl)amino)heptanoic acid was converted into the β-lactam with 2-chloro-1-methylpyridin-1-ium iodide. The polymerization of this β-lactam has been catalyzed with $Sc(NTMS_2)_3(THF)_3$. It is said that poly($β^3$-homolysine) as well as poly(*D*-lysine) are more effective gene delivery agents than the analogous oligomers or polymers of *L*-Lys for cultured cells [86].

Nylon-3 homo- and copolymers have been prepared via lithium bis(trimethylsilyl)amide initiated ring-opening polymerization of β-lactams (e.g., (1*S*,8*R*)-9-azabicyclo[6.2.0]decan-10-one, (1*S*,6*R*)-7-azabicyclo[4.2.0] octan-8-one, and *tert*-butyl (((2*S*,3*S*)-2-methyl-4-oxoazetidin-3-yl)methyl) carbamate). Meanwhile, the use of an *N*-acyl-β-lactam as a co-initiator in the polymerization process allows the placement of a specific functional group, borne by the *N*-acyl-β-lactam, at the *N*-terminus of each polymer chain. The *N*-acyl-β-lactam has been selected from the following list: (*S*)-2-((2,2-dimethyl-4-oxoazetidin-3-yl)methyl)isoindoline-1,3-dione, (*S*)-*N*-((2,2-dimethyl-4-oxoazetidin-3-yl)methyl)acetamide,(*S*)-4-(*tert*-butyl)-*N*-((2,2-dimethyl-4-oxoazetidin-3-yl)methyl)benzamide, (*S*)-*N*-((2,2-dimethyl-4-oxoazetidin-3-yl)methyl)octanamide, *N*-(((2*S*,3*S*)-2-methyl-4-oxoazetidin-3-yl)methyl)octanamide and (*S*)-*N*-((2,2-dimethyl-4-oxoazetidin-3-yl)methyl)-2-(tritylthio)acetamide [79]. The copolymer arising from copolymerization of (1*S*,6*R*)-7-azabicyclo[4.2.0]octan-8-one and *tert*-butyl (((2*S*,3*S*)-2-methyl-4-oxoazetidin-3-yl)methyl)carbamate upon deprotection has been applied as the mimicry of antimicrobial host-defense peptides [87]. This strategy has been applied to make functionally diverse nylon-3 copolymers [80].

To conduct the anionic ring-opening polymerization of β-lactams with living characteristics, the polymerization of 3,3-dimethyl-2-azetidinone or 4,4-dimethyl-2-azetidinone has been performed at 25°C in *N*,*N*-dimethylacetamide containing 5 wt.% of lithium chloride. The resulting polyamides have a narrow molecular weight distribution as indicated by gel-permeation chromatography. Particularly, the number average molecular weights estimated from ^{1}H NMR spectroscopy were nearly the same as the expected values according to the ratio of the consumed lactam to the corresponding *N*-benzoyl derivative used as an activator [88, 89]. Moreover, ring-opening polymerization with a strong organic base such as 1-*tert*-butyl-4,4,4-tris(dimethylamino)-2,2-bis[tris(dimethylamino)phosphoranyliden amino]-2λ^5,4λ^5-catenadi(phosphazene) (*t*-BuN=P[N=P(NMe$_2$)$_3$]$_3$) in *N,N*-dimethylacetamide in the presence of LiCl allows the preparation of linear and crystalline poly(β-peptide)s without metal residue [81].

Different from Bestian's preliminary work for the anionic ring-opening polymerization, lithium bis(trimethylsilyl)amide has been used to initiate the anionic ring-opening polymerizations of a series of β-lactams, including *tert*-butyl (((2*S*,3*S*)-2-methyl-4-oxoazetidin-3-yl)methyl)carbamate, *tert*-butyl (*S*)-((2,2-dimethyl-4-oxoazetidin-3-yl)methyl)carbamate, 2-(((2*S*,3*S*)-2-methyl-4-oxoazetidin-3-yl)methyl)isoindoline-1,3-dione and (*S*)-2-((2,2-dimethyl-4-oxoazetidin-3-yl)methyl)isoindoline-1,3-dione, all prepared from the cycloaddition of CSI with the corresponding alkenes, i.e., *tert*-butyl (*E*)-but-2-en-1-ylcarbamate, *tert*-butyl (3-methylbut-2-en-1-yl)carbamate, (*E*)-2-(but-2-en-1-yl)isoindoline-1,3-dione and 2-(3-methylbut-2-en-1-yl) isoindoline-1,3-dione, as well as (1*S*,6*R*)-7-azabicyclo[4.2.0]octan-8-one. However, (1*S*,8*R*)-9-azabicyclo[6.2.0]decan-10-one and (1*S*,12*R*)-13-azabicyclo[10.2.0]tetradecan-14-one do not work at all in THF, possibly due to the high steric hindrance. Also, the possible mechanism for polymerization has been discussed [78]. As pointed out by Bestian, the presence of *N*-acyl-β-lactam facilitates the polymerization, the unsuccessful polymerizations of the above seven β-lactams do take place smoothly in the presence of 10% co-initiator of 1-benzoyl-4-phenyl-azetidin-2-one or (1*S*,8*R*)-9-benzoyl-9-azabicyclo-[6.2.0]decan-10-one, affording good yields of polymers. The resulting polymers from (1*S*,6*R*)-7-azabicyclo[4.2.0]octan-8-one do not dissolve in common organic solvents, such as THF, CH_2Cl_2, DMF or DMSO, whereas the polymers arising from (1*S*,8*R*)-9-azabicyclo[6.2.0]decan-10-one and (1*S*,12*R*)-13-azabicyclo[10.2.0]tetradecan-14-one are quite soluble in THF or CH_2Cl_2, possibly due to the large cycloalkyl rings that hinder the formation of intramolecular hydrogen bonding, thereby promoting polymer solubility. In addition, the polymerization also takes place in the presence of a small amount of acyl chloride or carboxylic acid anhydride under the anionic polymerization condition [78].

A carbohydrate-based polyamide has been prepared by means of anionic ring-opening polymerization of (3*S*,4*R*)-3-methoxy-4-((1*S*,2*R*)-1,2,3-trimethoxypropyl)azetidin-2-one, which was prepared from the key intermediate of (2*S*,3*R*,4*S*,5*R*)-3-amino-2,4,5,6-tetramethoxyhexanoic acid. This intermediate was prepared from the starting material of (4a*R*,6*S*,7*S*,8*S*,8a*S*)-8-azido-6-methoxy-2-phenylhexahydropyrano[3,2-*d*][1,3]dioxin-7-ol in seven steps. The anionic ring-opening polymerization was performed in CH_2Cl_2 using potassium *tert*-butoxide as the initiator, as outlined in Scheme 4.35 [90].

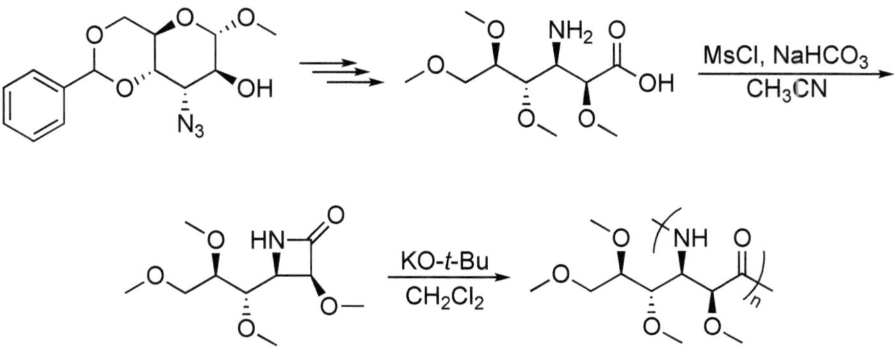

SCHEME 4.35 Preparation of (3*S*,4*R*)-3-methoxy-4-((1*S*,2*R*)-1,2,3-trimethoxypropyl) azetidin-2-one and its base triggered polymerization

In addition, the anionic ring-opening polymerization of β-lactam has been applied to make enantiopure poly(amido-saccharide)s (PASs) with a defined molecular weight and narrow dispersity, with the main chain structure composed of pyranose rings linked to an amide with α-stereochemistry through the 1- and 2-positions of the pyranose rings. The monomer of (1*S*,3*R*,4*S*,5*R*,6*R*)-4,5-bis(benzyloxy)-3-((benzyloxy) methyl)-2-oxa-8-azabicyclo[4.2.0]octan-7-one was synthesized in one-step from the cycloaddition of benzyl-protected δ-glucal, i.e., (2*R*,3*S*,4*R*)-3,4-bis(benzyloxy)-2-((benzyloxy)methyl)-3,4-dihydro-2*H*-pyran with CSI in toluene at -55°C, as illustrated in Scheme 4.36. The subsequent polymerization was then initiated with lithium bis(trimethylsilyl)amide (LiHMDS) in the presence of *p*-*tert*-butyl benzoyl chloride as a co-initiator. The degree of polymerization ranging from 25 to > 120 as well as the corresponding yields varied depending on the amount of initiator and co-initiator applied. The benzyl protecting groups are removed in liquid ammonia by sodium reduction [91]. The formed PASs forms a left-handed helical structure. In contrast, the cycloaddition of (2*R*,3*S*,4*S*)-3,4-bis(benzyloxy)-2-((benzyloxy) methyl)-3,4-dihydro-2*H*-pyran with $ClSO_2NCO$ gave (1*R*,3*R*,4*S*,5*S*,6*S*)-4,5-bis-(benzyloxy)-3-((benzyloxy)methyl)-2-oxa-8-azabicyclo[4.2.0] octan-7-one, and its corresponding polymer forms a right-handed helical structure [92].

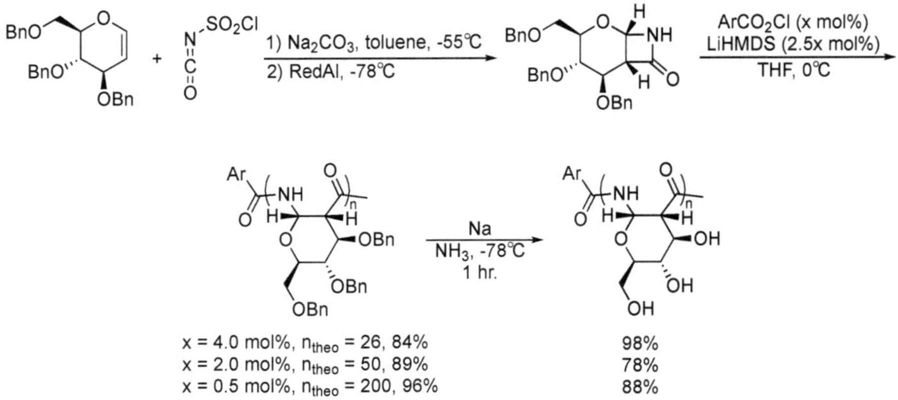

SCHEME 4.36 Preparation of (1*S*,3*R*,4*S*,5*R*,6*R*)-4,5-bis(benzyloxy)-3-((benzyloxy) methyl)-2-oxa-8-azabicyclo[4.2.0]octan-7-one and its ring-opening polymerization

Recently, 3,4-dibenzyl glucal, i.e., ((2*R*,3*S*,4*R*)-3,4-bis(benzyloxy)-3,4-dihydro-2*H*-pyran-2-yl)methanol was converted into *tert*-butyl (((2*R*,3*S*,4*R*)-3,4-bis(benzyloxy)-3,4-dihydro-2*H*-pyran-2-yl)methyl)((2-nitrophenyl) sulfonyl)carbamate, which was then allowed to undergo cycloaddition with room temperature stable trichloroacetyl isocyanate (TCAI) in CH_3CN for 8–10 days to afford *tert*-butyl (((1*S*,3*R*,4*S*,5*R*,6*R*)-4,5-bis(benzyloxy)-7-oxo-2-oxa-8-azabicyclo[4.2.0]octan-3-yl)methyl)((2-nitrophenyl)sulfonyl) carbamate. Removal of 2-nitrophenylsulfonyl protecting group with K_2CO_3/PhSH yielded *tert*-butyl (((1*S*,3*R*,4*S*,5*R*,6*R*)-4,5-bis(benzyloxy)-7-oxo-2-oxa-8-azabicyclo[4.2.0]octan-3-yl)methyl)carbamate. The ring-opening polymerization of this β-lactam was initiated with LiHMDS in the presence of perfluorophenyl 2-(benzylthio)acetate in THF to give polypeptide as shown in Scheme 4.37 [93]. The monomer/initiator ratio has been controlled in the range from 12 to 50, and the number averaged molecular weight of the resulting polypeptides varies from 4.1 kDa to 26.6 kDa. Upon the deprotection of the benzyl protecting group, the polymers demonstrated mucoadhesive properties. Particularly, the cycloaddition between tri-*O*-octyl-*D*-glucal, i.e., (2*R*,3*S*,4*R*)-3,4-bis(octyloxy)-2-((octyloxy)methyl)-3,4-dihydro-2*H*-pyran and CSI yielded (1*S*,3*R*,4*S*,5*R*,6*R*)-4,5-bis(octyloxy)-3-((octyloxy) methyl)-2-oxa-8-azabicyclo[4.2.0]octan-7-one, which upon ring-opening polymerization, afforded poly(β-peptide)s demonstrating thermotropic liquid crystalline phases [94].

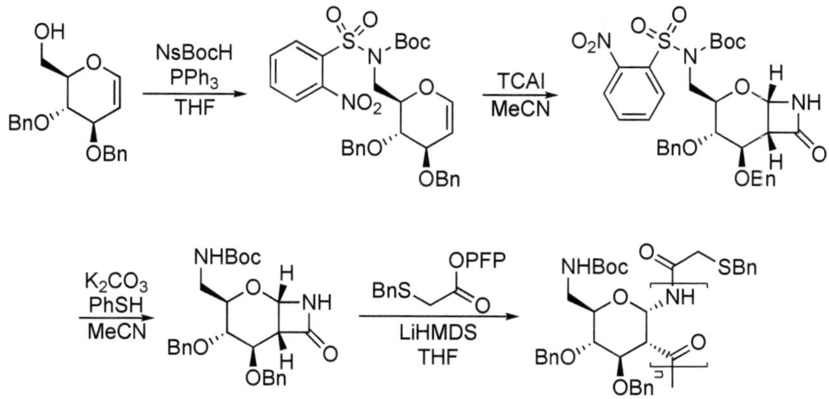

SCHEME 4.37 The total synthesis of *tert*-butyl (((1*S*,3*R*,4*S*,5*R*,6*R*)-4,5-bis(benzyloxy)-7-oxo-2-oxa-8-azabicyclo[4.2.0]octan-3-yl)methyl)carbamate and its ring-opening polymerization

In a very similar manner, the glycal of maltose, i.e., (2*R*,3*S*,4*R*)-4-(benzyloxy)-2-((benzyloxy)methyl)-3-(((2*R*,3*R*,4*S*,5*R*,6*R*)-3,4,5-tris(benzyloxy)-6-((benzyloxy)methyl)tetrahydro-2*H*-pyran-2-yl)oxy)-3,4-dihydro-2*H*-pyran was converted into the corresponding β-lactam in cycloaddition with $ClSO_2NCO$, i.e., (1*S*,3*R*,4*S*,5*R*,6*R*)-5-(benzyloxy)-3-((benzyloxy)methyl)-4-(((2*R*,3*R*,4*S*,5*R*,6*R*)-3,4,5-tris(benzyloxy)-6-((benzyloxy)methyl)tetrahydro-2*H*-pyran-2-yl)oxy)-2-oxa-8-azabicyclo[4.2.0]octan-7-one. The LiHMDS initiated ring-opening polymerization of this β-lactam in THF in the presence of 3,3-dimethylbutanoyl chloride leads to PASs of high molecular weight (up to 31500 g/mol) with narrow dispersity (PDI < 1.1) [95]. The LiHMDS initiated ring-opening polymerization of (1*R*,3*R*,4*R*,5*R*,6*S*)-4,5-bis(benzyloxy)-3-((benzyloxy)methyl)-2-oxa-8-azabicyclo[4.2.0]octan-7-one in the presence of perfluorophenyl 6-palmitamidohexanoate or perfluorophenyl N^2,N^6-dipalmitoyl-*L*-lysinate as a co-initiator leads to enantiopure poly(amido-saccharide)s, which inhibits the formation of biofilm. Similarly, the random copolymers formed from (1*R*,3*R*,4*R*,5*R*,6*S*)-4,5-bis(benzyloxy)-3-((benzyloxy)methyl)-2-oxa-8-azabicyclo[4.2.0]octan-7-one and (1*R*,3*R*,4*S*,5*R*,6*S*)-4,5-bis(benzyloxy)-3-((benzyloxy)methyl)-2-oxa-8-azabicyclo[4.2.0]octan-7-one under a similar condition behave analogously [96].

On the other hand, β-Lactam has been well known to interact with enzymes by connecting to the nucleophilic group of enzymes along with the opening of the β-lactam ring. Inversely, enzymes can also trigger the polymerization of β-lactams. The most commonly used biocatalyst in polymerization reactions is the immobilized CAL-B on acrylic resin, commercially available as N435 [97], possibly due to its stability at elevated temperature and its acceptance of various substrates, as demonstrated in the synthesis of poly(β-alanine) with an average degree of polymerization up to 8 due to the limited solubility of polyamide in the reaction medium [98]. An atomistic model for the formation of polyamide from CAL-B catalyzed polymerization of β-lactam has recently been reported [99]. This enzyme has been applied to promote the copolymerization of β-lactam and ε-caprolactone (ε-CL), as shown in Scheme 4.38. The feed ratio of β-lactam and ε-caprolactone has been set as 1:3, 1:1, and 3:1, in order to determine the differences in the structure of the resulting copolymers. The results reveal that the copolymer, i.e., poly(ε-CL-co-β-lactam) is an alternating and random copolymer, containing short oligo(ε-CL) or oligo(β-lactam) segments. Particularly, the alternating copolymer is the most abundant structure from the copolymerization with an equal amount of β-lactam and ε-caprolactone; random copolymer containing short oligo(ε-CL) attached to the alternating repeating units is observed as the dominant structure from the copolymerization of 3:1 ε-CL/β-lactam mixture; whereas the dominant random copolymer resulting from the copolymerization of 1:3 ε-CL/β-lactam mixture contains short oligo(β-lactam) attaching to the alternating repeating units. Also, a yield of up to 50% can be achieved from the copolymerization of 1:1 ε-CL/β-lactam, and deviation from this ratio leads to a lower yield of copolymerization. The highest degree of polymerization is measured for 12, for the copolymerization of 1:1 ε-CL/β-lactam [100].

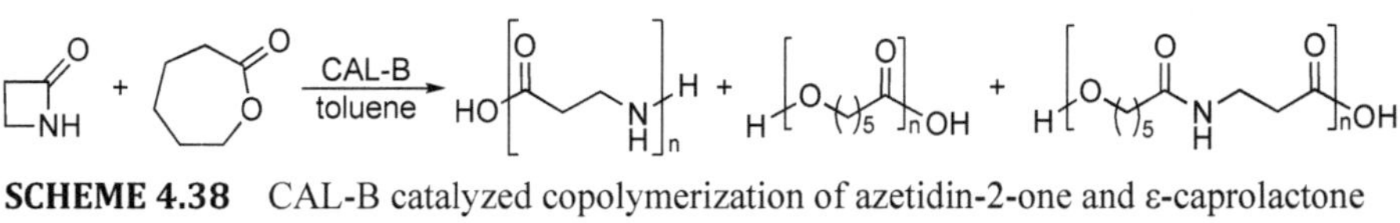

SCHEME 4.38 CAL-B catalyzed copolymerization of azetidin-2-one and ε-caprolactone

Recently, *N*-thiocarboxyanhydrides have been proved as complementary substrates for the preparation of poly-β-amides, when the polymerization of β-lactams fails. For example, LiHMDS catalyzed polymerization of benzyl (*S*)-4-oxoazetidine-2-carboxylate in the presence of acyl chloride failed to give the expected product of 20-mer but led to a recovery of only

35.5% of a compound that has little benzyl ester groups, although several other reports have indicated the polymerization of this monomer under this condition. However, simple primary amine-initiated polymerization of 1,3-thiazinane-2,6-dione gave 96.3% of the expected polymer with benzyl ester groups as the sidechains. This polymerization condition has several advantages with respect to the polymerization of β-lactams, such as open-vessel reaction, application of undried solvent, mild polymerization condition, compatibility with the base labile group, and controllable dual functionalization [101].

4.4 APPLICATION IN THE SYNTHESIS OF UNNATURAL Α-AMINO ACID DERIVATIVES AND RELATING PEPTIDES

These practices involve the cleavage of C2-C3 bond of β-lactams under special conditions, as cleavage of the regular C-C bond does not occur smoothly. One of such modifications is the conversion of 3-hydroxy-β-lactams into *N*-carboxyanhydride (NCA) derivatives of α-amino acids by means of oxidation, from which other molecules arising from the ring-opening of NCAs can be formed (interested readers please see volume 3 of this book series). One of the early examples of such transformation is the oxidation of 3-hydroxy-β-lactams by means of $Me_2S{\cdot}Br_2/Et_3N$, affording azetidin-2,3-diones in nearly quantitative yields. The resulting azetidin-2,3-diones can be easily converted into α-amino acid NCAs when treated with *m*-chloroperbenzoic acid (mCPBA) in CH_2Cl_2. Subsequent treatment of the NCAs with nucleophiles leads to the derivatives of α-amino acids. For example, treatment of 3-hydroxy-1,4-bis(4-methoxyphenyl) azetidin-2-one with $Me_2S{\cdot}Br_2/Et_3N$ in CH_2Cl_2 at room temperature for 1 hour affords 1,4-bis(4-methoxyphenyl)azetidin-2,3-dione. Oxidation of this azetidin-2,3-dione with mCPBA in CH_2Cl_2 at -20°C leads to 3,4-bis(4-methoxyphenyl)oxazolidin-2,5-dione, a process known as Baeyer-Villiger oxidation. However, in contrast to the normal Baeyer-Villiger oxidation where the most substituted carbon atom (possibly with higher electron density) migrates, the carbonyl group migrates in this case. Treatment of this NCA in pyridine with *t*-butylamine at room temperature for 3.5 hours gave *N*-(*tert*-butyl)-2-(4-methoxyphenyl)-2-((4-methoxyphenyl)amino) acetamide, in an overall yield of 70% from 1,4-bis(4-methoxyphenyl) azetidin-2,3-dione, as outlined in Scheme 4.39 [102].

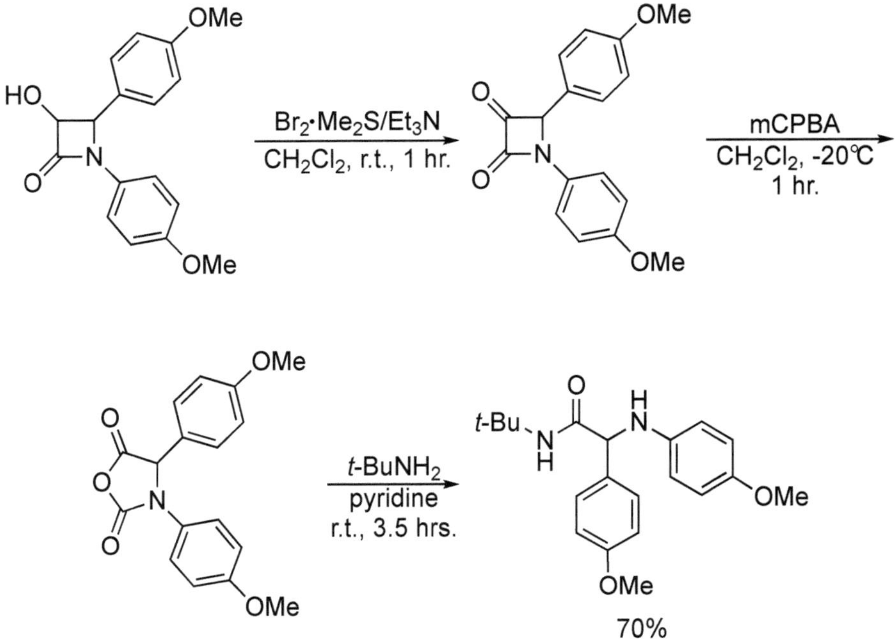

SCHEME 4.39 Transformation of 3-hydroxy-1,4-bis(4-methoxyphenyl)azetidin-2-one into *N*-(*tert*-butyl)-2-(4-methoxyphenyl)-2-((4-methoxyphenyl)amino)acetamide

In another example, treatment of 2-(benzyloxy)acetyl chloride with Et_3N in CH_2Cl_2 at -78°C generates 1-benzyloxy-ketene *in situ*, which undergoes the Staudinger reaction with *tert*-butyl ((2*S*,3*R*,*Z*)-4-(benzylimino)-1-((*tert*-butyldiphenylsilyl)oxy)-3-methoxybutan-2-yl) carbamate to yield *tert*-butyl ((1*R*,2*S*)-1-((2*S*,3*R*)-1-benzyl-3-(benzyloxy)-4-oxoazetidin-2-yl)-3-((*tert*-butyldiphenylsilyl)-oxy)-1-methoxypropan-2-yl)carbamate. Catalytic hydrogenation of the resulting β-lactam using ammonium formate as hydrogen source in refluxing isopropanol yielded 95% of *tert*-butyl ((1*R*,2*S*)-1-((2*R*,3*R*)-1-benzyl-3-hydroxy-4-oxoazetidin-2-yl)-3-((*tert*-butyldiphenylsilyl)oxy)-1-methoxypropan-2-yl)carbamate. Deprotection of the *tert*-butyldiphenylsilyl group with tetrabutylammonium fluoride (TBAF) and subsequent protection with 2,2-dimethoxypropane in refluxing benzene in the presence of a catalytic amount of *p*-toluenesulfonic acid yielded 75% of *tert*-butyl (*S*)-4-((*R*)-((2*R*,3*R*)-1-benzyl-3-hydroxy-4-oxoazetidin-2-yl) (methoxy)methyl)-2,2-dimethyloxazolidine-3-carboxylate. Treatment of this 3-hydroxy-β-lactam with P_2O_5 in DMSO afforded *tert*-butyl

(*S*)-4-((*R*)-((*S*)-1-benzyl-3,4-dioxoazetidin-2-yl)(methoxy)methyl)-2,2-dimethyloxazolidine-3-carboxylate, which was then oxidized with mCPBA in CH_2Cl_2, giving *tert*-butyl (*S*)-4-((*R*)-((*S*)-3-benzyl-2,5-dioxooxazolidin-4-yl)(methoxy)methyl)-2,2-dimethyloxazolidine-3-carboxylate. The resulting NCA was then refluxed in methanol for 2 hours to yield 80% of *tert*-butyl (*S*)-4-((1*S*,2*S*)-2-(benzylamino)-1,3-dimethoxy-3-oxopropyl)-2,2-dimethyloxazolidine-3-carboxylate, as illustrated in Scheme 4.40 [103]. Similarly, refluxing of (*S*)-3-benzyl-4-(4-methoxybenzyl)oxazolidin-2,5-dione in methanol gave 89% of methyl (*S*)-2-(benzylamino)-3-(4-methoxyphenyl)propanoate [104].

Similarly, (3*R*,4*R*)-4-((*R*)-((*tert*-butyldimethylsilyl)oxy)(phenyl)methyl)-3-hydroxy-1-(4-methoxyphenyl)azetidin-2-one was oxidized with CrO_3/pyridine to give (*S*)-4-((*R*)-((*tert*-butyldimethylsilyl)oxy)(phenyl)methyl)-1-(4-methoxyphenyl)azetidin-2,3-dione, which was further oxidized with mCPBA in CH_2Cl_2 at -40°C to give (*S*)-4-((*R*)-((*tert*-butyldimethylsilyl)oxy)(phenyl)methyl)-3-(4-methoxyphenyl)-oxazolidin-2,5-dione. Refluxing of this NCA in methanol afforded methyl (2*S*,3*R*)-3-((*tert*-butyldimethylsilyl)oxy)-2-((4-methoxyphenyl)amino)-3-phenylpropanoate, in an overall yield of 80%. Reduction of this α-amino ester with $LiBH_4$ in THF led to 76% of (2*R*,3*R*)-3-((*tert*-butyldimethylsilyl)oxy)-2-((4-methoxyphenyl)amino)-3-phenylpropan-1-ol. Likewise, the reaction of *tert*-butyl ((*S*)-1-((*R*)-3-(4-methoxyphenyl)-2,5-dioxooxazolidin-4-yl)ethyl)carbamate with methyl *L*-alaninate generated dipeptide of methyl ((2*R*,3*S*)-3-((*tert*-butoxycarbonyl)amino)-2-((4-methoxyphenyl)amino)butanoyl)-*L*-alaninate (Scheme 4.41) [105]. Similarly, (*R*)-3-benzyl-4-((*S*)-1-((*tert*-butyldimethylsilyl)oxy)ethyl)-4-methyloxazolidine-2,5-dione obtained from mCPBA oxidation in CH_2Cl_2 at -40°C (in 90% yield) was allowed to react with methyl *L*-phenylalaninate in the presence of 1 equivalent of KCN for 24 hours to afford 87% of methyl ((2*R*,3*S*)-2-(benzylamino)-3-((*tert*-butyldimethylsilyl)oxy)-2-methylbutanoyl)-*L*-phenylalaninate. In contrast, only a trace amount of this dipeptide was formed in the absence of KCN. A similar situation was observed during the coupling between this NCA and methyl *L*-valinate, where 90% of methyl ((2*R*,3*S*)-2-(benzylamino)-3-((*tert*-butyldimethylsilyl)oxy)-2-methylbutanoyl)-*L*-valinate was obtained, whereas no such dipeptide formed without KCN in presence [106].

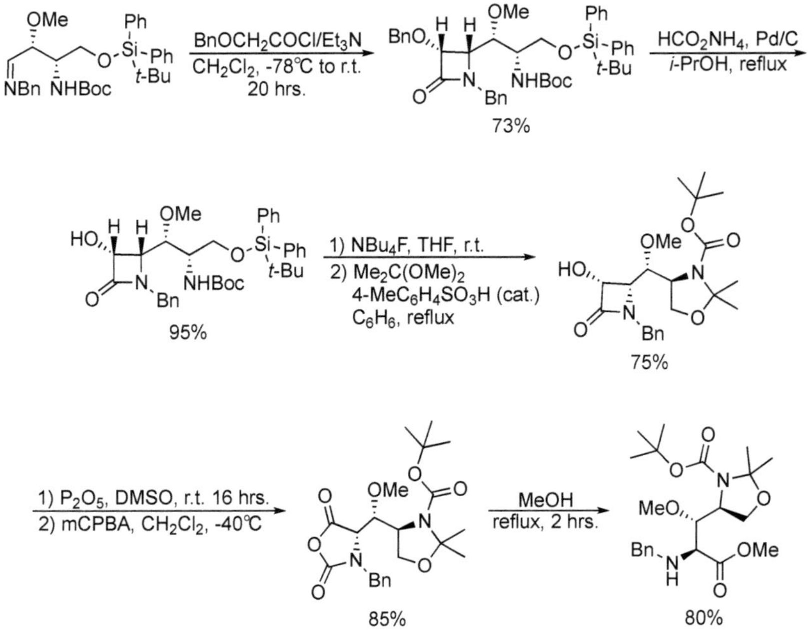

SCHEME 4.40 Synthesis of *tert*-butyl (*S*)-4-((*R*)-((2*R*,3*R*)-1-benzyl-3-hydroxy-4-oxoazetidin-2-yl)(methoxy)methyl)-2,2-dimethyloxazolidine-3-carboxylate and its oxidation into tert-butyl (*S*)-4-((1*S*,2*S*)-2-(benzylamino)-1,3-dimethoxy-3-oxopropyl)-2,2-dimethyloxazolidine-3-carboxylate

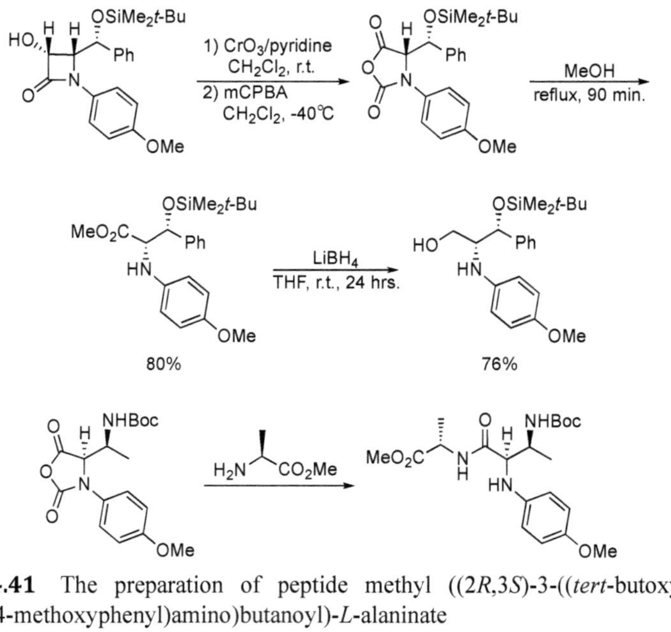

SCHEME 4.41 The preparation of peptide methyl ((2*R*,3*S*)-3-((*tert*-butoxycarbonyl)amino)-2-((4-methoxyphenyl)amino)butanoyl)-*L*-alaninate

Following this strategy, a tripeptide has been prepared as illustrated in Scheme 4.42. In this approach, the β-lactam was obtained from the Staudinger reaction by addition of the solution of (benzyloxy)acetyl chloride (2.21 mL) in 10 mL of CH_2Cl_2 to a solution of (*R*,*E*)-*N*-benzyl-2-((*tert*-butyldiphenylsilyl) oxy)-3-methylbutan-1-imine and Et_3N in CH_2Cl_2, and directly converted into the corresponding NCA without removal of the benzyl protecting group, by means of the oxidation with NaOCl in a catalytic amount of TEMPO in a KH_2PO_4/K_2HPO_4 buffer (pH = 6.9). The resulting NCA of (*S*)-3-benzyl-4-((*R*)-1-((*tert*-butyldiphenylsilyl)oxy)-2-methylpropyl)oxazolidine-2,5-dione was allowed to react with methyl *L*-leucinate to give dipeptide of methyl ((2*S*,3*R*)-2-(benzylamino)-3-((*tert*-butyl-diphenylsilyl)oxy)-4-methylpentanoyl)-*L*-leucinate. After deprotection of the *N*-benzyl group by catalytic hydrogenation in EtOH, the dipeptide of methyl ((2*S*,3*R*)-2-amino-3-((*tert*-butyldiphenylsilyl) oxy)-4-methylpentanoyl)-*L*-leucinate was then coupled with another NCA obtained in the same way like the previous one, i.e., (*S*)-3-benzyl-4-((*R*)-((*tert*-butyldimethylsilyl)oxy)(phenyl)methyl)oxazolidin 2,5-dione, in DMF in the presence of NaN_3 to give 70% of the tripeptide, i.e., methyl ((2*S*,3*R*)-2-(2-(benzylamino)-3-((*tert*-butyldimethylsilyl)-oxy)-3-phenylpropanamido)-3-((*tert*-butyldiphenylsilyl)oxy)-4-methylpentanoyl)-*L*-leucinate [107].

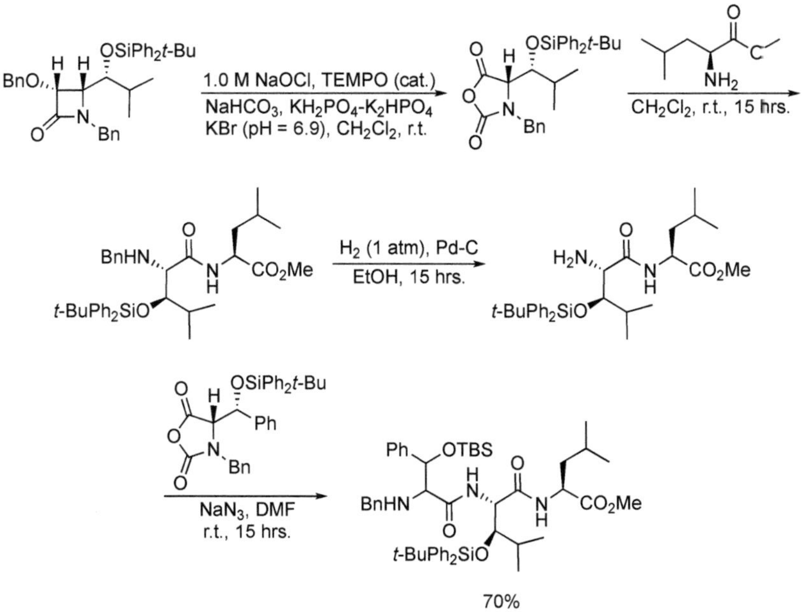

SCHEME 4.42 β-Lactam based synthesis of tripeptide methyl ((2*S*,3*R*)-2-(2-(benzylamino)-3-((*tert*-butyldimethylsilyl)oxy)-3-phenylpropanamido)-3-((*tert*-butyldiphenylsilyl) oxy)-4-methylpentanoyl)-*L*-leucinate

Similarly, α-hydroxy-β-lactams with a quaternary carbon at C4 that are formed by the standard Staudinger reaction are converted into α,α-dialkyl α-amino acid NCAs in a one-pot manner. These NCAs are then coupled with another amino acid to form conformationally restricted dipeptide segments, as illustrated in the preparation of ((*S*)-2-amino-2-methyl-4-phenylbutanoyl)-*L*-valine in 71% yield [108]. This approach has been applied to make peptidyl nucleoside analogs as illustrated in Scheme 4.43, where the acetone-protected *L*-threose was initially converted to imine (*E*)-*N*-benzyl-1-((4*S*,5*S*)-5-((benzyloxy)methyl)-2,2-dimethyl-1,3-dioxolan-4-yl)methanimine, which then underwent the Staudinger reaction with 2-(benzyloxy)acetyl chloride and Et_3N to afford (3*R*,4*S*)-1-benzyl-3-(benzyloxy)-4-((4*S*,5*S*)-5-((benzyloxy)methyl)-2,2-dimethyl-1,3-dioxolan-4-yl)azetidin-2-one. Oxidation of this β-lactam with perchloric acid at room temperature for 2–4 hours, followed by $NaIO_4$ oxidation in aqueous acetone at 0°C for 2 hours, led to the corresponding NCA of (*S*)-3-benzyl-4-((4*S*,5*S*)-5-((benzyloxy)methyl)-2,2-dimethyl-1,3-dioxolan-4-yl)oxazolidine-2,5-dione. Coupling with benzyl *L*-leucinate in CH_2Cl_2 at room temperature for 24 hours yielded benzyl ((*S*)-2-(benzylamino)-2-((4*S*,5*S*)-5-((benzyloxy)methyl)-2,2-dimethyl-1,3-dioxolan-4-yl)acetyl)-*L*-leucinate, as displayed in Scheme 4.43 [109].

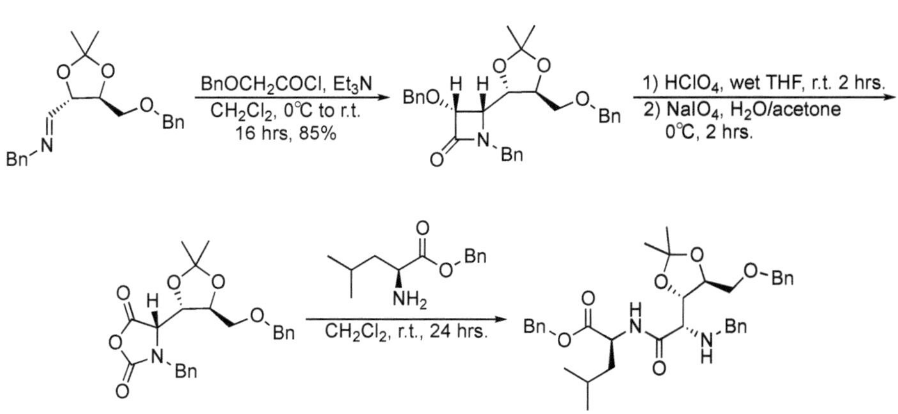

SCHEME 4.43 *L*-Threose-based synthesis of dipeptide benzyl ((*S*)-2-(benzylamino)-2-((4*S*,5*S*)-5-((benzyloxy)methyl)-2,2-dimethyl-1,3-dioxolan-4-yl)acetyl)-*L*-leucinate

The efficiency of this approach to creating α-amino acid NCA has been illustrated in the preparation of even more complicated peptidyl glycoside, as demonstrated in Scheme 4.44, where the generation of NCA from 3-hydroxy-β-lactam has been repeatedly applied. In this synthesis, it is unclear how one of the three benzyl protecting groups in (3*R*,4*S*)-1-benzyl-3-(benzyloxy)-4-((4*S*,5*S*)-5-((benzyloxy)methyl)-2,2-dimethyl-1,3-dioxolan-4-yl) azetidin-2-one obtained in Scheme 4.43 was selectively removed during the catalytic hydrogenation, affording 95% of (3*R*,4*S*)-1-benzyl-3-(benzyloxy)-4-((4*S*,5*S*)-5-(hydroxymethyl)-2,2-dimethyl-1,3-dioxolan-4-yl)azetidin-2-one. Treatment of this β-lactam with 4-nitrophenyl chloroformate followed by acyl substitution with NH_4Cl led to 90% of ((4*S*,5*S*)-5-((2*S*,3*R*)-1-benzyl-3-(benzyloxy)-4-oxoazetidin-2-yl)-2,2-dimethyl-1,3-dioxolan-4-yl)methylcarbamate, which was then catalytically hydrogenated to yield 3-hydroxy-β-lactam. Subsequent oxidation with NaOCl in combination with TEMPO yielded ((4*S*,5*S*)-5-((*S*)-3-benzyl-2,5-dioxooxazolidin-4-yl)-2,2-dimethyl-1,3-dioxolan-4-yl)methylcarbamate. In a very similar manner, (*E*)-*N*-benzyl-1-((3a*R*,4*R*,6*R*,6a*R*)-6-methoxy-2,2-dimethyltetrahydrofuro [3,4-*d*][1,3]dioxol-4-yl)-methanimine was converted into a NCA of (*R*)-3-benzyl-4-((3a*R*,4*R*,6*R*,6a*R*)-6-methoxy-2,2-dimethyltetrahydrofuro [3,4-*d*][1,3]dioxol-4-yl)oxazolidine-2,5-dione as shown in Scheme 4.44. Reaction of this NCA with methanol followed by catalytic hydrogenation led to unnatural α-amino acid of methyl (*R*)-2-amino-2-((3a*R*,4*R*,6*R*,6a*R*)-6-methoxy-2,2-dimethyltetrahydrofuro[3,4-*d*][1,3]dioxol-4-yl)acetate. Coupling of this unnatural α-amino acid with the NCA prepared in the first part in Scheme 4.44 in CH_2Cl_2 afforded the peptidyl glycoside, i.e., methyl (*R*)-2-((*S*)-2-(benzylamino)-2-((4*S*,5*S*)-5-((carbamoyloxy)methyl)-2,2-dimethyl-1,3-dioxolan-4-yl)acetamido)-2-((3a*R*,4*R*,6*R*,6a*R*)-6-methoxy-2,2-dimethyltetrahydrofuro[3,4-*d*][1,3]dioxol-4-yl)acetate [110].

Likewise, coupling of (*S*)-3-benzyl-4-((*S*)-1-((*tert*-butyldimethylsilyl) oxy)ethyl)oxazolidine-2,5-dione prepared in the same strategy with *t*-butyl glycinate in CH_2Cl_2 at room temperature afforded 75% of *tert*-butyl *N*-benzyl-*O*-(*tert*-butyldimethylsilyl)-*L*-allothreonylglycinate, which was further coupled with ((benzyloxy)carbonyl)-*L*-leucine in the presence of DCC and HOBt to yield *tert*-butyl *N*-(((benzyloxy)carbonyl)-*L*-leucyl)-*O*-(*tert*-butyldimethylsilyl)-*L*-allothreonylglycinate [111].

The practical application of this strategy has been demonstrated in the synthesis of enalapril, an angiotensin-converting enzyme (ACE) inhibitor, which belongs to a family of peptides in controlling hypertension and congestive heart failure, including enalapril, quinapril, trandolapril, and moexipril, etc. In this total synthesis, treatment of methyl (*S*)-2-(((*E*)-((*S*)-2,2-dimethyl-1,3-dioxolan-4-yl)methylene)amino) propanoate in CH_2Cl_2 at -78°C with 2-(benzyloxy)acetyl chloride and Et_3N yielded 75% of methyl (*S*)-2-((2*S*,3*R*)-3-(benzyloxy)-2-((*S*)-2,2-

dimethyl-1,3-dioxolan-4-yl)-4-oxoazetidin-1-yl)propanoate, with excellent diastereoselectivity. Acidic hydrolysis of the acetal protecting group, followed by $NaIO_4$ diol cleavage led to methyl (*S*)-2-((2*R*,3*R*)-3-

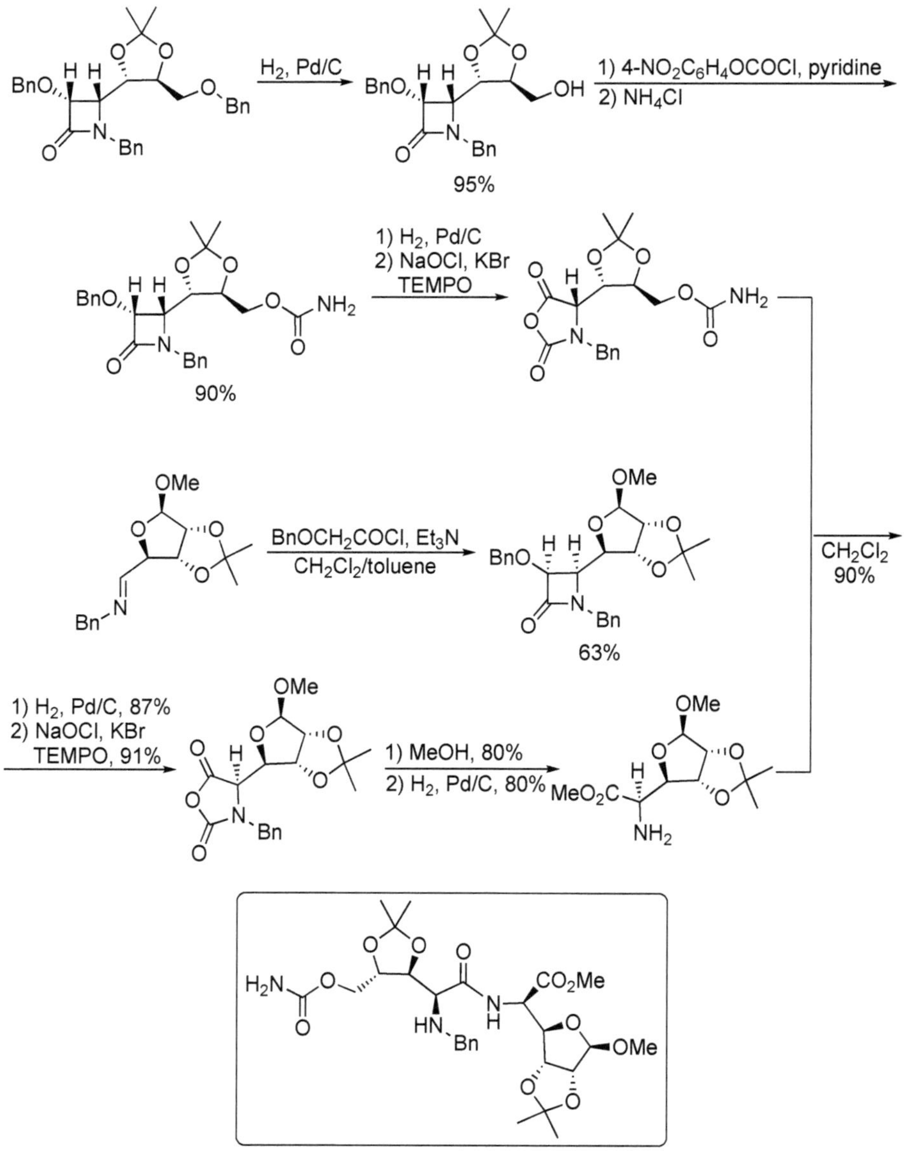

SCHEME 4.44 The synthesis of methyl (*R*)-2-((*S*)-2-(benzylamino)-2-((4*S*,5*S*)-5-((carbamoyloxy)methyl)-2,2-dimethyl-1,3-dioxolan-4-yl)acetamido)-2-((3a*R*,4*R*,6*R*,6a*R*)-6-methoxy-2,2-dimethyltetrahydrofuro[3,4-*d*][1,3]dioxol-4-yl)acetate

(benzyloxy)-2-formyl-4-oxoazetidin-1-yl)propanoate. Subsequent Wittig reaction and catalytic hydrogenation afforded the expected 3-hydroxy-β-lactam of methyl (*S*)-2-((3*R*,4*S*)-3-hydroxy-2-oxo-4-phenethylazetidin-1-yl)propanoate. Basic saponification of the methyl ester group allowed the subsequent coupling with benzyl *L*-prolinate in the presence of EDC, giving benzyl ((*S*)-2-((3*R*,4*S*)-3-hydroxy-2-oxo-4-phenethylazetidin-1-yl)propanoyl)-*L*-prolinate. Then, oxidation with NaOCl in CH_2Cl_2 in combination with TEMPO created NCA of benzyl ((*S*)-2-((*S*)-2,5-dioxo-4-phenethyloxazolidin-3-yl)propanoyl)-*L*-prolinate. Esterification in ethanol cleaved the NCA ring and following catalytic hydrogenation removed the benzyl group. Finally, mixing the product with maleic acid in EtOAc led to (*S*)-*N*-((*S*)-1-((*S*)-2-carboxypyrrolidin-1-yl)-1-oxopropan-2-yl)-1-ethoxy-1-oxo-4-phenylbutan-2-aminium (*Z*)-3-carboxyacrylate, the complex of enalapril and maleic acid, as outlined in Scheme 4.45 [112].

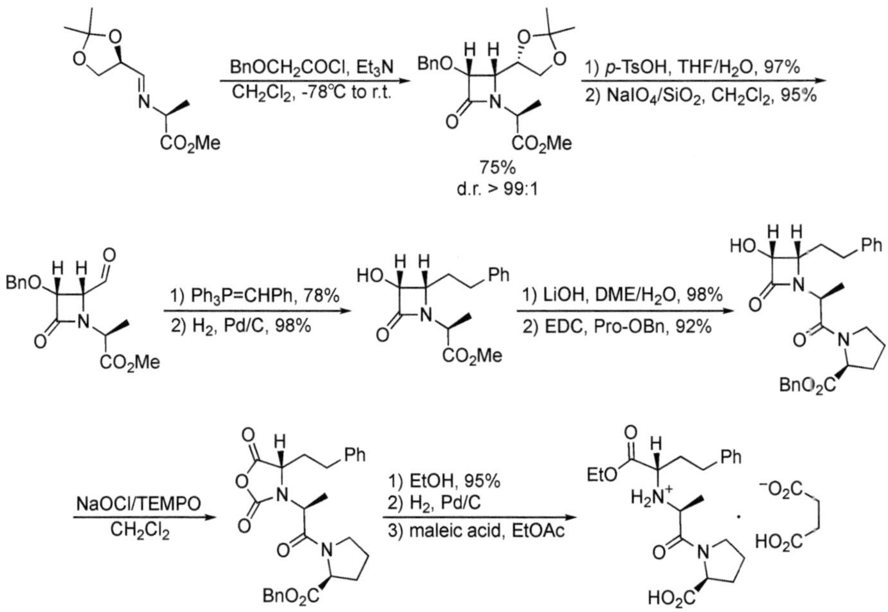

SCHEME 4.45 The total synthesis of enalapril

It should be emphasized that the major application of α-amino acid NCAs is to form polypeptides via ring-opening polymerization of the corresponding NCAs, as mentioned in Volume 3 of this book series. Such application is therefore not repeated here anymore.

4.5 NUCLEOPHILIC SUBSTITUTION ON C3 OR C4

The electronic characters of C3 and C4 in β-lactams are different, as C3 is adjacent to the electron-withdrawing carbonyl group at C2, whereas C4 is connecting to N1, which could be considered as an electron-donating group. For this reason, C3 can be deprotonated to carbanion and function as a nucleophile, whereas C4 would be better attached with a leaving group to function as an electrophile. For example, treatment of (3*R*,4*R*)-3-(benzyloxy)-1-((*S*)-1-phenylethyl)-4-(trifluoromethyl)azetidin-2-one with LiHMDS in THF at -78°C removed a proton at C3, affording the expected carbanion, which underwent S_N2 reaction with methyl iodide to yield (3*R*,4*R*)-3-(benzyloxy)-3-methyl-1-((*S*)-1-phenylethyl)-4-(trifluoromethyl)azetidin-2-one (Scheme 4.46). Interestingly, in the absence of electrophile (MeI in this case), the *in situ* generated carbanion underwent either [1,2]-Wittig rearrangement or [2,3]-Wittig rearrangement to afford 60–65% of (3*R*,4*R*)-3-benzyl-3-hydroxy-1-((*S*)-1-phenylethyl)-4-(trifluoromethyl)azetidin-2-one and (3*R*,4*R*)-3-hydroxy-1-((*S*)-1-phenylethyl)-3-(*o*-tolyl)-4-(trifluoromethyl)azetidin-2-one, in a ratio of 4:6 [113].

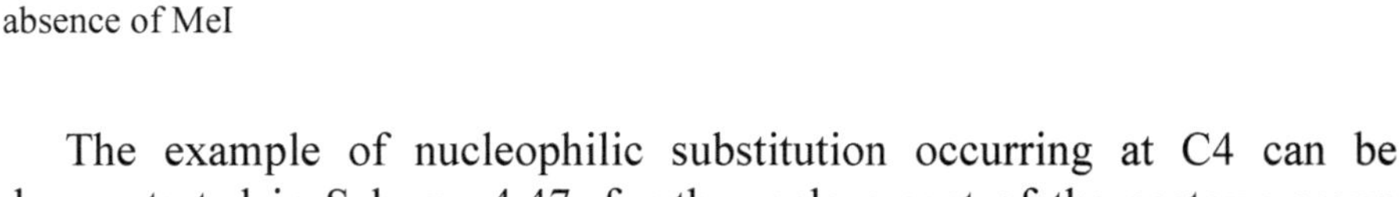

SCHEME 4.46 Different reaction paths during the treatment of (3*R*,4*R*)-3-(benzyloxy)-1-((*S*)-1-phenylethyl)-4-(trifluoromethyl)azetidin-2-one with LiHMDS in the presence or absence of MeI

The example of nucleophilic substitution occurring at C4 can be demonstrated in Scheme 4.47, for the replacement of the acetoxy group with a nitrogenous nucleophile. Treatment of *N*-(benzyloxy)benzamide in anhydrous ethanol with sodium ethoxide allows the formation of sodium benzoyl(benzyloxy)amide. Subsequent addition of 1.05 equivalents of (2*R*,3*R*)-3-((*R*)-1-((*tert*-butyldimethylsilyl)oxy)ethyl)-4-oxoazetidin-2-yl acetate affords 61% of *N*-(benzyloxy)-*N*-((2*R*,3*S*)-3-((*R*)-1-((*tert*-butyldimethylsilyl)oxy)ethyl)-4-oxoazetidin-2-yl)benzamide. The replacement of the acetoxy group with this sodium amide may be an S_N2 substitution. Accordingly, an inversion of the configuration at C4 of the β-lactam is expected. The actual result displayed in Scheme 4.47 is contradicting to S_N2 substitution. The authors have realized this issue and affirmed the verification of this stereochemistry by NMR homo decoupling experiments. Thus, the exact reaction scheme has been used from the original paper. A similar result in stereochemistry has been shown in the second reaction in Scheme 4.47.

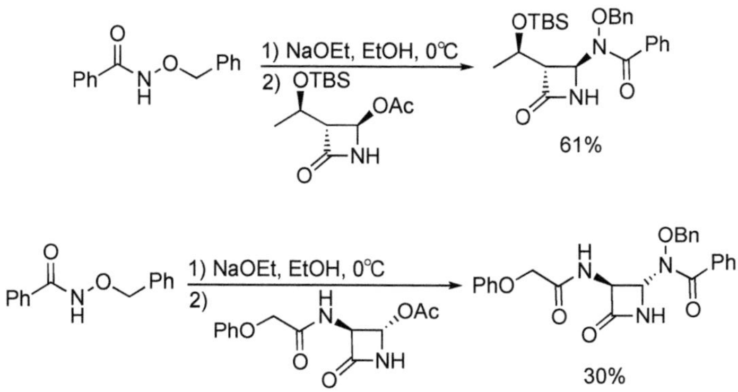

SCHEME 4.47 Substitution of 4-acetoxyl group of the β-lactam with an amide nucleophile

4.6 CLEAVAGE OF C3-C4 BOND

In fact, the bond between C3 and C4 is a C-C bond, which normally does not break. However, in a few cases, C3-C4 bond does participate in the cationic rearrangement. One occasion occurs for the β-lactams with a ketal functionality at the C3 carbon and a nucleophilic sp^2 (imino or carbonyl) group at C4, of which the C3 ketal group would stabilize the emerging positive charge on the ketal carbon in the presence of Lewis acid, leading to pyrazine-2,3-dione or dihydro-1,4-oxazine derivatives, as illustrated in Scheme 4.48. When X = O, and $R_1 = R_2$ = H, treatment of 3,3-dimethoxy-1-(4-methoxyphenyl)-4-oxoazetidine-2-carbaldehyde with $SnCl_2$ yielded 75% of 4-(4-methoxyphenyl)-2*H*-1,4-oxazine-2,3(4*H*)-dione. This transformation is proposed to proceed in two paths, both involving the cleavage of C3-C4 bond, and the hydrolysis of ketal to a carbonyl group. When X = NC_6H_4-OMe, and $R_1 = R_2$ = Me, treatment of 3,3-dimethoxy-1-(4-methoxyphenyl)-4-(1-((4-methoxyphenyl)imino)ethyl)-4-methylazetidin-2-one under this condition afforded 99% of 1,4-bis(4-methoxyphenyl)-5,6-dimethyl-1,4-dihydropyrazine-2,3-dione [115]. In contrast, when two methyl groups exist at C3, instead of two methoxy groups, treatment of the corresponding β-lactam, such as 4-(dimethoxymethyl)-1-(4-methoxyphenyl)-3,3,4-trimethylazetidin-2-one, with $SnCl_2$ does not give any rearranged product, but only 1-(4-methoxyphenyl)-2,3,3-trimethyl-4-oxoazetidine-2-carbaldehyde. Further treatment of this aldehyde with sulfuric acid promotes the cleavage of C3-C4 bond of β-lactam, affording 1-(4-methoxyphenyl)-3,3,5-trimethylpyrrolidine-2,4-dione, as illustrated in Scheme 4.49 [115].

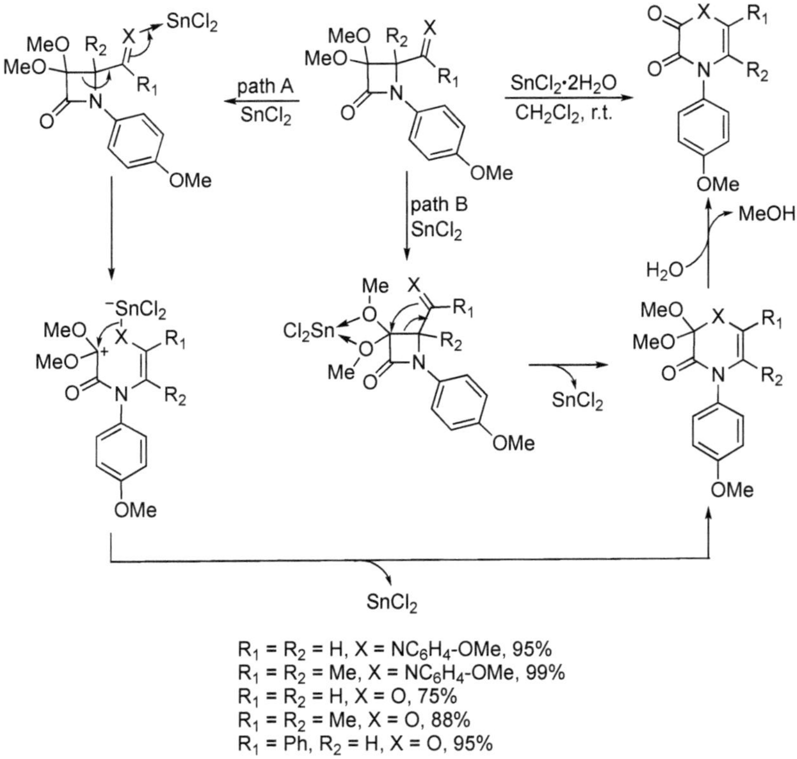

SCHEME 4.48 Mechanism for the conversion of 4-substituted-3,3-dimethoxy-1-(4-methoxyphenyl)azetidin-2-ones into 4-(4-methoxyphenyl)-6-substituted-2*H*-1,4-oxazine-2,3(4*H*)-diones

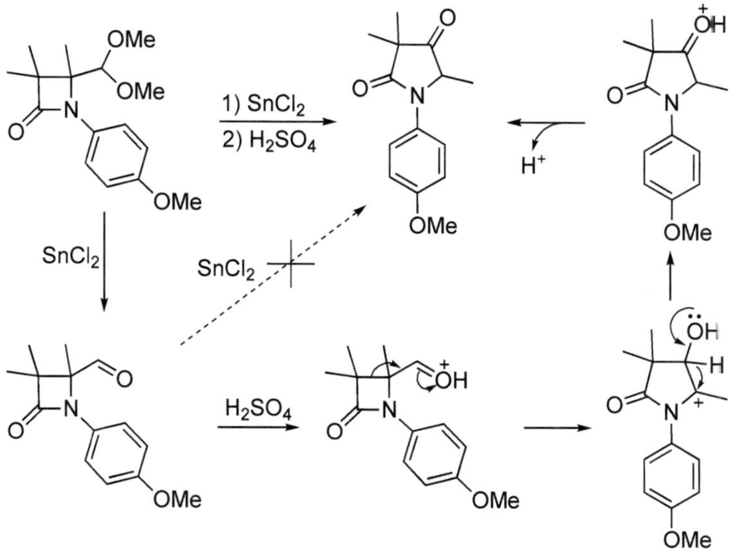

SCHEME 4.49 The mechanism for the transformation of 4-(dimethoxymethyl)-1-(4-methoxyphenyl)-3,3,4-trimethylazetidin-2-one into 1-(4-methoxyphenyl)-3,3,5-trimethylpyrrolidine-2,4-dione

Another example involving the cleavage, of C3-C4 of β-lactam is illustrated in Scheme 4.50, for a series of rearrangements of the β-lactams into substituted 3,4-dihydropyridin-2(1*H*)-ones. The plausible mechanism involves the cleavage of C3-C4 bond as demonstrated by the rearrangement of (*R*)-3-methoxy-4,4-di((*E*)-styryl)-1-(*p*-tolyl)azetidin-2-one, affording (*Z*)-(2-(((1*E*,4*E*)-1,5-diphenylpenta-1,4-dien-3-ide-3-yl)(*p*-tolyl)amino)-2-oxoethylidene)(methyl)oxonium, where its resonance structure undergoes ring-closure to form (3*R*,4*R*)-3-methoxy-4-phenyl-6-((*E*)-styryl)-1-(*p*-tolyl)-3,4-dihydropyridin-2(1*H*)-one in 94% yield. For this rearrangement, the substituent at C3 is critical. When X is an electron-withdrawing group, such as chloro, phthalimido, or *p*-nitrophenyl group, no rearrangement occurs. However, when X is an electron-donating group, such as NH_2, OMe, or SMe, substituted 3,4-dihydropyridin-2(1*H*)-ones are obtained in excellent yields [116].

4.7 CLEAVAGE OF N1-C4 BOND

Compared to the cleavage of N1-C2 bond, the cleavage of N1-C4 β-lactams is relatively challenging and uncommon. One example is the

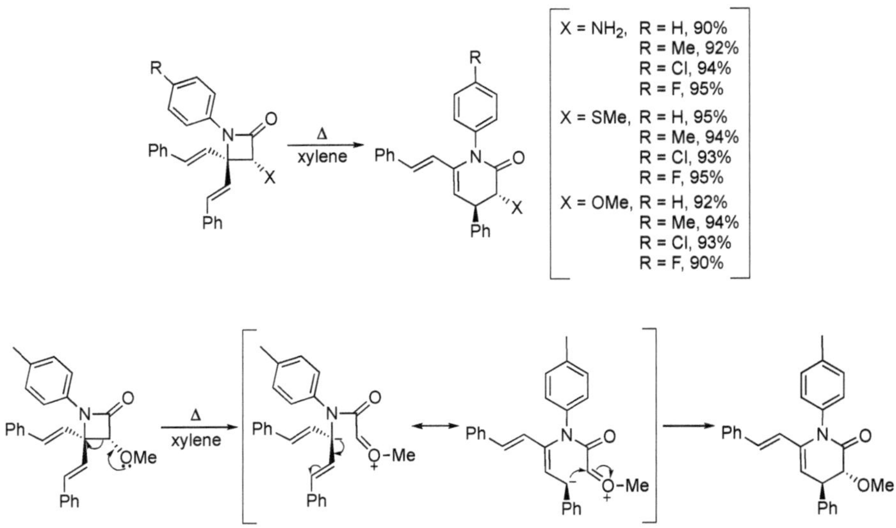

SCHEME 4.50 The mechanism for the transformation of a series of (*R*)-3-methoxy-4,4-di((*E*)-styryl)-1-(*p*-tolyl)azetidin-2-one and its analogs into 1,3-disubstituted-4-phenyl-6-((*E*)-styryl)-3,4-dihydropyridin-2(1*H*)-ones

reaction between (3*S*,4*R*)-3-((*R*)-1-((*tert*-butyldimethylsilyl)oxy)ethyl)-4-((*R*)-phenylsulfinyl)azetidin-2-one and (benzyloxy)trimethylsilane in CH_3CN at 0°C, in the presence of a catalytic amount of trimethylsilyl trifluoromethanesulfonate (TMSOTf), affording 82% of *N*-((1*R*,2*R*,3*R*)-1-(benzyloxy)-3-((*tert*-butyldimethylsilyl)oxy)-2-cyanobutyl)-acetamide and *N*-((1*S*,2*R*,3*R*)-1-(benzyloxy)-3-((*tert*-butyldimethylsilyl)oxy)-2-cyanobutyl)acetamide, in a ratio of 7:1 (Scheme 4.51). The reaction of the same β-lactam with methoxytrimethylsilane, isopropoxytrimethylsilane or (cyclohexyloxy)trimethylsilane also occurs under a similar condition, yielding the corresponding products with the best stereoselectivity for (cyclohexyloxy)trimethylsilane. Similar reactions proceed equally well with 4-alkoxyazetidin-2-ones in the presence of a catalytic amount of TMSOTf in solvents containing one cyano group, such as acetonitrile, propionitrile, butyronitrile, and benzonitrile [117]. In this case, the nitrile becomes an acylating agent. The mechanism of this reaction is outlined in the conversion of (3*R*,4*R*)-3-((*R*)-1-((*tert*-butyldimethylsilyl)oxy) ethyl)-4-methoxyazetidin-2-one into a mixture of *N*-((1*R*,2*R*,3*R*)-3-((*tert*-butyldimethylsilyl)oxy)-2-cyano-1-methoxybutyl)-benzamide and *N*-((1*S*,2*R*,3*R*)-3-((*tert*-butyldimethylsilyl)oxy)-2-cyano-1-methoxybutyl) benzamide, in 81% yield, in *anti*/*syn* ratio of 15:1 (Scheme 4.52) [117].

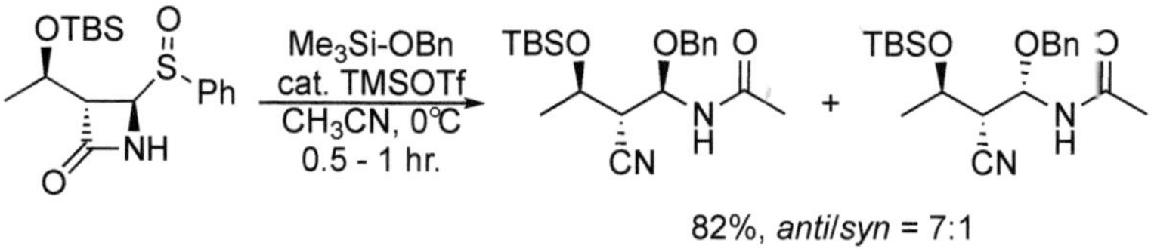

SCHEME 4.51 TMSOTf catalyzed reaction between Me_3SiOBn and (3*S*,4*R*)-3-((*R*)-1-((*tert*-butyldimethylsilyl)oxy)ethyl)-4-((*R*)-phenylsulfinyl)azetidin-2-one

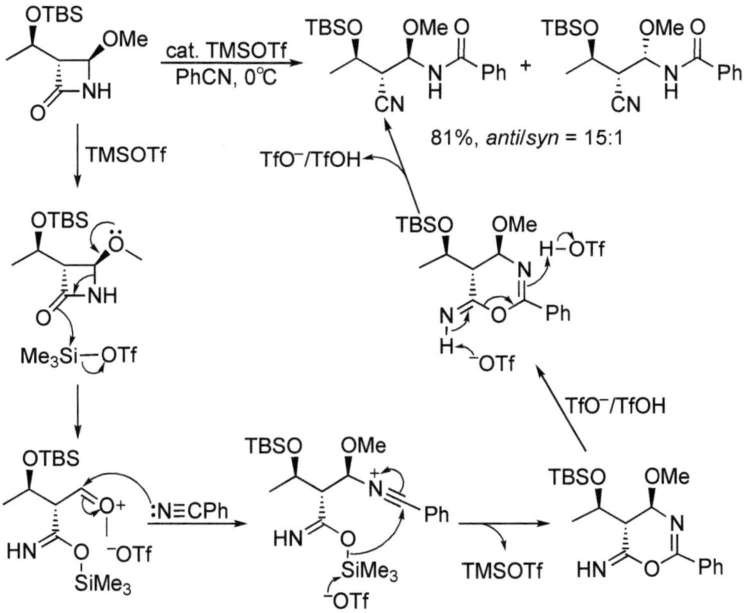

SCHEME 4.52 The mechanism for TMSOTf catalyzed reaction of (3*R*,4*R*)-3-((*R*)-1-((*tert*-butyldimethylsilyl)oxy)ethyl)-4-methoxyazetidin-2-one in benzonitrile

The cyanide catalyzed ring-expansion of β-lactam bearing a formyl group at C4 also involves the cleavage of the N1-C4 bond, as shown in Scheme 4.53 for the one-pot transformation of (2*R*,3*R*)-3-methoxy-1-(4-methoxyphenyl)-4-oxoazetidine-2-carbaldehyde into 50% of (*R*)-3-methoxy-1-(4-methoxyphenyl)-pyrrolidine-2,5-dione. In this transformation, treatment of the β-lactam with 4-methoxyaniline in refluxing acetonitrile gave (3*R*,4*S*)-3-methoxy-1-(4-methoxyphenyl)-4-((*E*)-((4-methoxyphenyl)imino)methyl)azetidin-2-one. Subsequent addition of 20 mol% of tetrabutylammonium cyanide triggered the ring-expansion of this β-lactam to (*R*)-3-methoxy-1-(4-methoxyphenyl)-5-((4-methoxyphenyl) imino)pyrrolidin-2-one. Hydrolysis of the imine lead to the final substituted succinimide in a total yield of 55% [118].

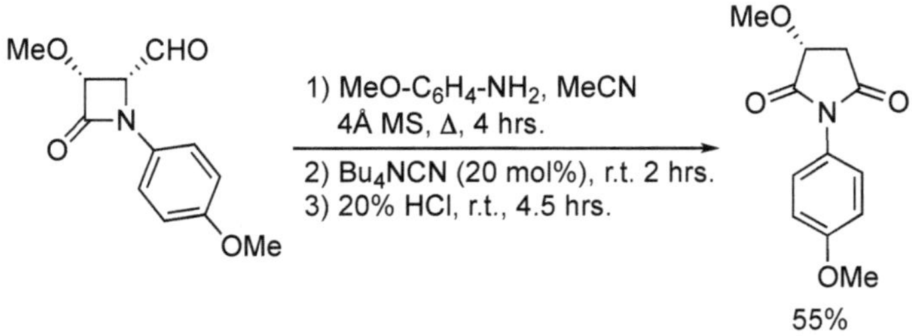

SCHEME 4.53 Transformation of (2*R*,3*R*)-3-methoxy-1-(4-methoxyphenyl)-4-oxoazetidine-2-carbaldehyde into (*R*)-3-methoxy-1-(4-methoxyphenyl)pyrrolidine-2,5-dione

Another example of the cleavage of the N1-C4 bond of β-lactams has been demonstrated in a series of studies on the reactions between a number of aromatic or aliphatic aldehydes and ε-amido-allylindiums generated *in situ* from *N*-toluenesulfonyl-4-vinylazetidin-2-ones in the presence of indium iodide (InI) and a catalytic amount of $Pd(PPh_3)_4$ to afford (3-Z)-2,6-*anti*-enediols, with effective 1,5-stereo-control [119–121]. For example, treatment of (3*R*,4*S*)-4-((*E*)-prop-1-en-1-yl)-1-tosyl-3-((triisopropylsilyl)oxy)azetidin-2-one with 3.0 equivalents of indium iodide and 2.0 equivalents of ligand *N*-methylimidazole (NMI) in THF/EtOH (9:1) in the presence of $Pd(PPh_3)_4$, followed by addition of benzaldehyde afforded 83% of (2*R*,5*S*,6*S*,*Z*)-6-hydroxy-5-methyl-6-phenyl-*N*-tosyl-2-((triisopropylsilyl)oxy)hex-3-enamide and 15% of (2*R*,5*R*,6*R*,*E*)-6-hydroxy-5-methyl-6-phenyl-*N*-tosyl-2-((triisopropylsilyl)oxy)hex-3-enamide. The former was measured with high diastereoselectivity, whereas the latter was a mixture of four isomers. Under this reaction condition, treatment of (3*R*,4*R*)-4-((*Z*)-prop-1-en-1-yl)-1-tosyl-3-((triisopropylsilyl)oxy)azetidin-2-one with $Pd(PPh_3)_4$ and indium iodide also gave similar results as shown in Scheme 4.54. It is assumed that the $Pd(PPh_3)_4$ facilitates the initial cleavage of the N1-C4 bond to form the palladium allylic complex, which was converted into the indium complex. Allylic addition to benzaldehyde allows the expected products as illustrated in Scheme 4.54. Starting from either β-lactam, the favored *trans*-indium complex will react with benzaldehyde, whereas the *cis*-indium complex does not react with aldehyde due to 1,3-diaxial strain in the transition state, which explains the observed stereoselectivity. In this reaction, the presence of ligand NMI dramatically increases the reaction rate, the yield of the major product as well as the conversion rate [120]. However, the substitution of the ligand NMI with Et_3N reverses the stereoselectivity, as displayed in Scheme 4.55, where a mixture of (2*R*,5*R*,6*R*,*E*)-6-hydroxy-5-methyl-6-phenyl-*N*-tosyl-2-((triisopropylsilyl)oxy)hex-3-enamide and (2*R*,5*R*,6*S*,*E*)-6-hydroxy-5-methyl-6-phenyl-*N*-tosyl-2-((triisopropylsilyl)oxy)hex-3-enamide was obtained in a good yield, with the former as the predominant product [121].

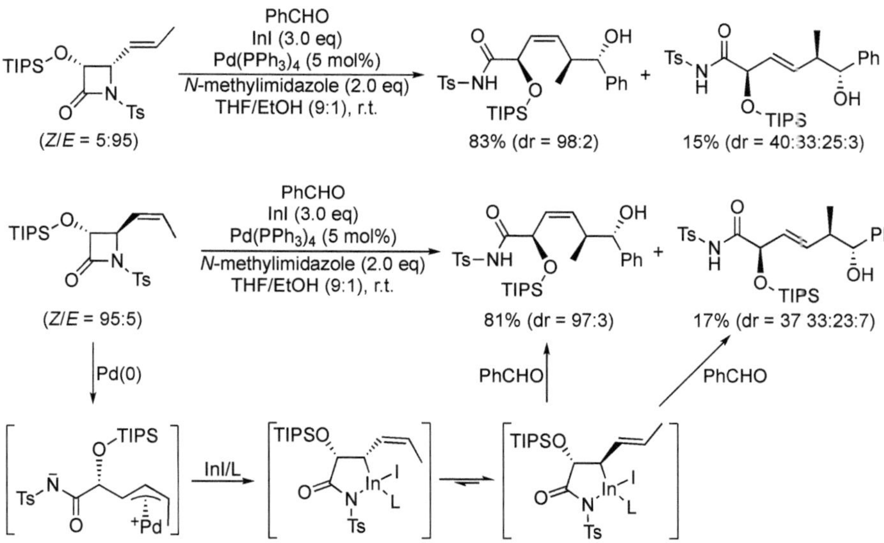

SCHEME 4.54 The mechanism for the transformation of (3*R*,4*R*)-4-((*Z*)-prop-1-en-1-yl)-1-tosyl-3-((triisopropylsilyl)oxy)azetidin-2-one into (2*R*,5*S*,6*S*,*Z*)-6-hydroxy-5-methyl-6-phenyl-*N*-tosyl-2-((triisopropylsilyl)oxy)hex-3-enamide and its stereoisomers

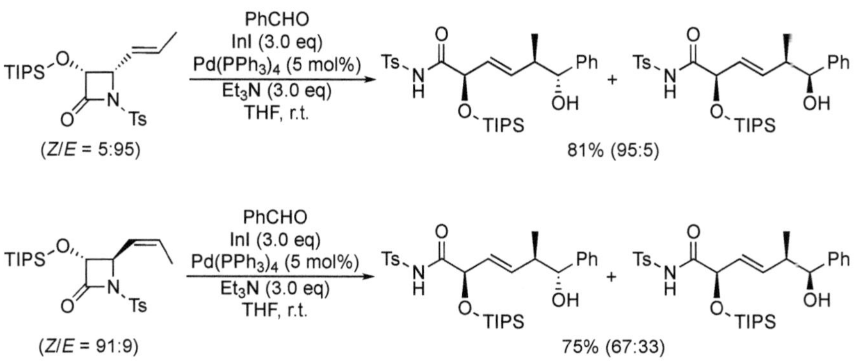

SCHEME 4.55 The reactions of (3*R*,4*S*)-4-((*E*)-prop-1-en-1-yl)-1-tosyl-3-((triisopropylsilyl)oxy)azetidin-2-one and (3*R*,4*R*)-4-((*Z*)-prop-1-en-1-yl)-1-tosyl-3-((triisopropylsilyl)oxy)azetidin-2-one with benzaldehyde in the presence of Et_3N and indium iodide

Extension of this protocol to β-lactams with a terminal alkyne group at C4 allows the preparation of allene derivatives, as demonstrated in Scheme 4.56, in which (3*R*,4*S*)-4-ethynyl-1-tosyl-3-((triisopropylsilyl)oxy)azetidin-2-one was converted into 60% of (2*R*,6*S*)-6-hydroxy-6-phenyl-*N*-tosyl-2-((triisopropylsilyl)oxy)hexa-3,4-dienamide and (2*R*,6*R*)-6-hydroxy-6-phenyl-*N*-tosyl-2-((triisopropylsilyl)oxy)hexa-3,4-dienamide, in a ratio of 73:27 in the presence of 2.0 equivalents of NMI. In contrast, when the ligand was changed to pyridine, the stereoselectivity was inversed, affording 53% of these mixtures, but in a ratio of 19:81 [122].

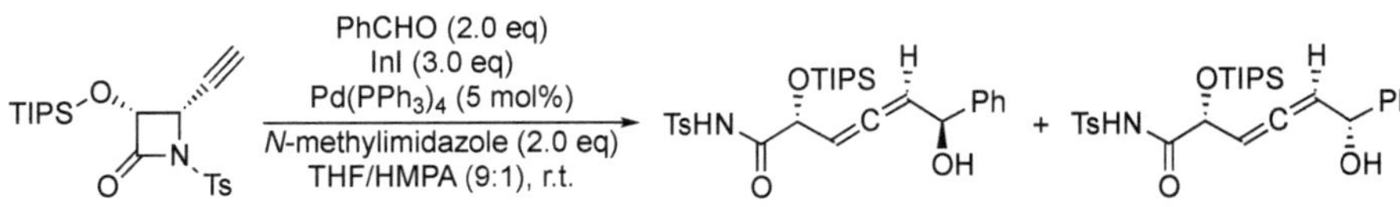

SCHEME 4.56 The reaction between (3*R*,4*S*)-4-ethynyl-1-tosyl-3-((triisopropylsilyl)oxy)azetidin-2-one and benzaldehyde in the presence of indium iodide and NMI

4.8 FUNCTION AT N1

There are many examples to introduce an alkyl group or acyl group to N1 by treatment of the β-lactams with weak bases such as K_2CO_3 followed by alkylating agents or acylating agents [123], so that they are not presented here. Examples of *N*-acyl-β-lactams can be found in literature [28, 79, 124], whereas *N*-alkyl-β-lactams and *N*-aryl-β-lactams are readily accessible from the Staudinger reaction and other synthetic methods described in Chapter 3. Examples of the alkenylation at N1 can be found in palladium-catalyzed coupling with a vinyl halide. In order to make carbapenem skeleton, (3*S*,4*R*)-4-((*Z*)-3-bromo-4-((*tert*-butyldimethylsilyl)oxy)but-2-en-1-yl)-3-((*R*)-1-((*tert*-butyldimethylsilyl)oxy)ethyl)azetidin-2-one has been synthesized as the model substrate to screen a suitable catalyst and accompanying ligand. Among the tested ligands, including tri-*o*-tolylphosphane, BINAP, (oxybis(2,1-phenylene))bis(diphenylphosphane) (abbreviated as DPEphos), (S)-(+)-2-(diphenylphosphino)-2'-methoxy-1,1'-binaphthyl (i.e., (S)-MOP), DPEphos is superior to others. When the toluene solution containing the substrate, $Pd(OAc)_2$ and ligand as well as 2.0 equivalents of

Cs_2CO_3 was heated in toluene at 80°C for 6 hours, 96% of (5*R*,6*S*)-6-((*R*)-1-((*tert*-butyldimethylsilyl)oxy)ethyl)-2-(((*tert*-butyldimethylsilyl)oxy) methyl)-1-azabicyclo[3.2.0]hept-2-en-7-one was obtained (Scheme 4.57). In contrast, when an electron-withdrawing group, such as an ester, was directly attached to the vinyl group, this reaction condition failed. More reactive vinyl iodide is needed in order to obtain a high yield of the product. It was found that when a toluene solution of ethyl (*S*,*Z*)-4-((2*R*,3*S*)-3-((*R*)-1-((*tert*-butyldimethylsilyl)oxy)ethyl)-4-oxoazetidin-2-yl)-2-iodopent-2-enoate, $Pd(OAc)_2$, and DPEphos was heated at 100°C for 2 minutes and was then added to a suspension of K_2CO_3 in toluene, the resulting mixture was heated at 100°C for 22 hours, ethyl (4*S*,5*R*,6*S*)-6-((*R*)-1-((*tert*-butyldimethylsilyl)oxy)ethyl)-4-methyl-7-oxo-1-azabicyclo[3.2.0]hept-2-ene-2-carboxylate can be obtained in a good yield (Scheme 4.57) [125]. A similar condition using CuI catalyzed coupling also works, giving excellent yield of carbapenem skeletons, as shown in Scheme 4.58, where ethyl (*E*)-4-((2*R*,3*S*)-3-((*R*)-1-((*tert*-butyldimethylsilyl)oxy)ethyl)-4-oxoazetidin-2-yl)-2,3-diiodobut-2-enoate was converted into 94% of ethyl (5*R*,6*S*)-6-((*R*)-1-((*tert*-butyldimethylsilyl) oxy)ethyl)-3-iodo-7-oxo-1-azabicyclo[3.2.0]hept-2-ene-2-carboxylate in toluene at 40°C, in the presence of CuI, and ligand bipyridine, base K_3PO_4 and 1.0 equivalent of water as additive [126].

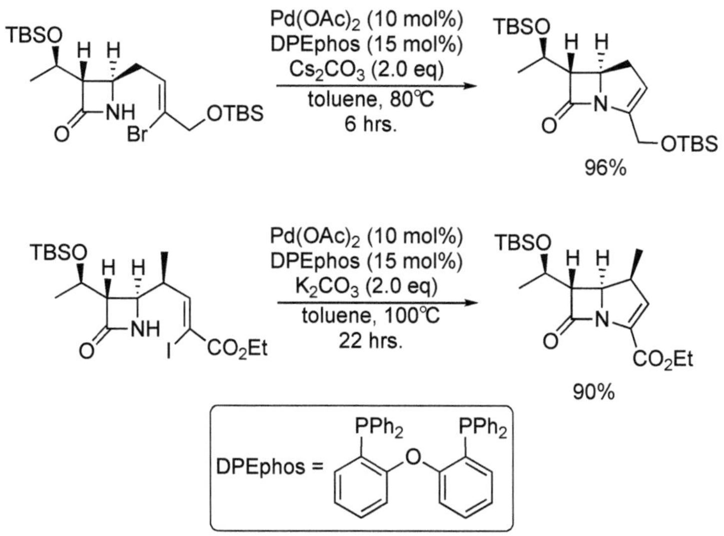

SCHEME 4.57 $Pd(OAc)_2$ catalyzed intramolecular *N*-alkenylation of β-lactams

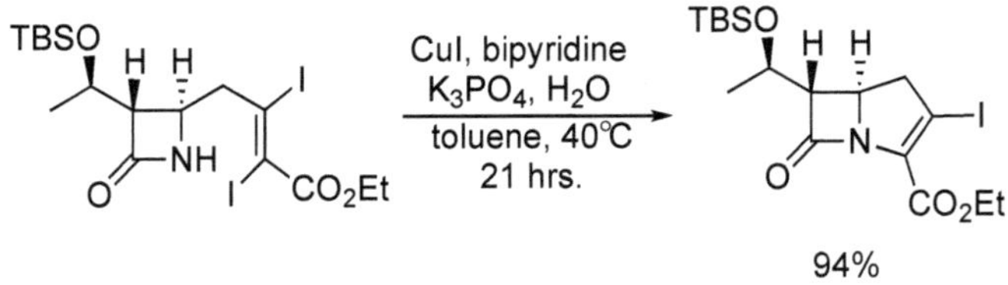

SCHEME 4.58 CuI catalyzed intramolecular *N*-alkenylation of ethyl (*E*)-4-((2*R*,3*S*)-3-((*R*)-1-((*tert*-butyldimethylsilyl)oxy)ethyl)-4-oxoazetidin-2-yl)-2,3-diiodobut-2-enoate

In addition, *aza*-Wacker cyclization condition also allows the coupling between N1 and vinyl functionality. For example, in 1.0 atm pressure of oxygen, (*S*)-*N*-allyl-*N*-(benzyloxy)-4-oxoazetidine-2-carboxamide was oxidatively converted into (*S*)-4-(benzyloxy)-2-methylene-1,4-diazabicyclo[4.2.0]octane-5,8-dione in refluxing acetonitrile, in the presence of 20 mol% of $Pd(OAc)_2$ and 40 mol% of pyridine (as the ligand). In the absence of oxygen, the oxidizing agent is needed for the reaction to proceed smoothly. As an example, when 1.6 equivalents of benzoquinone was applied as the oxidant, (*S*,*E*)-*N*-(benzyloxy)-*N*-(but-2-en-1-yl)-4-oxoazetidine-2-carboxamide in refluxing acetonitrile was transformed into 60% of (2*S*,6*S*)-4-(benzyloxy)-2-vinyl-1,4-diazabicyclo[4.2.0] octane-5,8-dione in the presence of 10 mol% $Pd(OAc)_2$ and 1.5 equivalents of acetic acid (Scheme 59) [127].

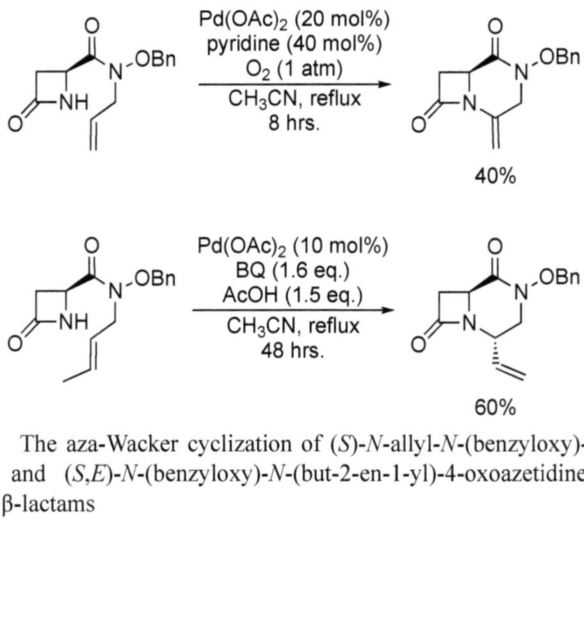

SCHEME 4.59 The aza-Wacker cyclization of (*S*)-*N*-allyl-*N*-(benzyloxy)-4-oxoazetidine-2-carboxamide and (*S*,*E*)-*N*-(benzyloxy)-*N*-(but-2-en-1-yl)-4-oxoazetidine-2-carboxamide affording fused β-lactams

More examples of the transformations of β-lactams into a variety of organic compounds have been collected in several pieces of literature [128–132].

KEYWORDS

- **hydrolysis of β-lactam**
- **alcoholysis of β-lactam**
- **aminolysis of β-lactam**
- **hydrazinolysis of β-lactam**
- **thiolysis of β-lactam**
- **Staudinger reaction**
- **ring-opening polymerization**
- **unnatural amino acid**

REFERENCES

1. Haug, T., Lohse, F., Metzger, K., & Batzer, H., (1968). Preparations and reactions of β-lactams. *Helvetica Chimica Acta, 51*(8), 2069–2089. doi: 10.1002/hlca.19680510828.
2. Graf, R., (1963). Reactions with *N*-carbonylsulfamoyl chloride. III. Reactions with olefins and aldehydes-concerning β-lactams. *Justus Liebigs Annalen der Chemie, 661*, 111–157. doi: 10.1002/jlac.19636610109.
3. Moriconi, E. J., & Kelly, J. F., (1968). The reaction of chlorosulfonyl isocyanates with allenes and olefins. *Journal of Organic Chemistry, 33*(8), 3036–3046. doi: 10.1021/jo01272a005.
4. Nicolaus, B. J. R., Bellasio, E., Pagani, G., & Testa, E., (1963). Syntheses and properties of *N*-alkoxy- and *N*-hydroxyazetidin-2-ones. *Gazzetta Chimica Italiana, 93*(6), 618–634.
5. Arbuzov, B. A., & Zobova, N. N., (1967). Synthesis of *N*-benzoylazetidin-2-ones and *N*-benzoylazacyclo-3-butene-2-ones by the action of benzoyl isocyanate on olefins and acetylenic compounds. *Dokl. Akad. Nauk SSSR, 172*(4), 845–848.
6. Yamamoto, Y., Kodama, S., Nishimura, R., Nomoto, A., Ueshima, M., & Ogawa, A., (2021). One-pot construction of diverse β-lactam scaffolds via the green oxidation of amines and its application to the diastereoselective synthesis of β-amino acids. *Journal of Organic Chemistry, 86*(17), 11571–11582. doi: 10.1021/acs.joc.1c01128.
7. Evans, C. D., Mahon, M. F., Andrews, P. C., Muir, J., & Bull, S. D., (2011). Intramolecular ester enolate-imine cyclization reactions for the asymmetric synthesis of polycyclic β-lactams and cyclic β-amino acid derivatives. *Organic Letters, 13*(23), 6276–6279. doi: 10.1021/ol202750u.
8. Forro, E., & Fueloep, F., (2007). Do lipases also catalyze the ring cleavage of inactivated cyclic *trans*-β-lactams? *Tetrahedron: Asymmetry, 17*(23), 3193–3196. doi: 10.1016/j.tetasy.2006.11.030.

9. Szakonyi, Z., & Fulop, F., (2011). Monoterpene-based chiral β-amino acid derivatives prepared from natural sources: Syntheses and applications. *Amino Acids, 41*(3), 597–608. doi: 10.1007/s00726-011-0891-5.
10. Malinowska, B., (2011). Selective opening of the β-lactam ring in *O,O*-dimethyl 4-oxoazetidin-2-ylphosphonate. *Przemysl Chemiczny, 90*(5), 907–909.
11. Holley, R. W., & Holley, A. D., (1949). Synthesis and reactivity of some 1-alkyl-2-azetidinones (*N*-alkyl-β-lactams). *Journal of the American Chemical Society, 71*(6), 2124–2129. doi: 10.1021/ja01174a061.
12. Cremonesi, G., Dalla, C. P., Forni, A., & La Rosa, C., (2013). Stereoselective synthesis of constrained norbornane-derived spiro-β-lactams. *Tetrahedron, 69*(3), 1175–1182. doi: 10.1016/j.tet.2012.11.048.
13. Washkuhn, R. J., & Robinson, J. R., (1971). Relative substituent effects on alkaline solvolysis of β-lactams (2-azetidinones) and amides. *Journal of Pharmaceutical Sciences, 60*(8), 1168–1175. doi: 10.1002/jps.2600600811.
14. Galla, Z., Beke, F., Forro, E., & Fulop, F., (2016). Enantioselective hydrolysis of 3,4-disubstituted β-lactams. An efficient enzymatic method for the preparation of a key taxol side-chain intermediate. *Journal of Molecular Catalysis B: Enzymatic, 123*, 107–112. doi: 10.1016/j.molcatb.2015.11.011.
15. Forro, E., & Fulop, F., (2008). Vapor-assisted enzymatic hydrolysis of β-lactams in a solvent-free system. *Tetrahedron: Asymmetry, 19*(8), 1005–1009. doi: 10.1016/j.tetasy.2008.04.003.
16. Vass, E., Hollosi, M., Forro, E., & Fulop, F., (2006). VCD spectroscopic investigation of enantiopure cyclic β-lactams obtained through lipolase-catalyzed enantioselective ring-opening reaction. *Chirality, 18*(9), 733–740. doi: 10.1002/chir.20312.
17. Forro, E., & Fulop, F., (2012). Recent lipase-catalyzed hydrolytic approaches to pharmacologically important β- and γ-amino acids. *Current Medicinal Chemistry, 19*(36), 6178–6187. doi: 10.2174/092986731120906 6178.
18. Medeiros, A. A., (1997). Beta-lactamases: Quality and resistance. *Clinical Microbiology and Infection: The Official Publication of the European Society of Clinical Microbiology and Infectious Diseases, 3*(Suppl. 4), 4S2-4S9.
19. Toomer, C. A., Lambert, P. A., Sansom, C. E., & Schwalbe, C. H. W., (1990). Active-site modeling of class I β-lactamases. *Biochemical Society Transactions, 18*(5), 921, 922. doi: 10.1042/bst0180921.
20. Dhar, P., Chan, P., Cohen, D. T., Khawam, F., Gibbons, S., Snyder-Leiby, T., Dickstein, E., et al., (2014). Synthesis, antimicrobial evaluation, and structure-activity relationship of α-pinene derivatives. *Journal of Agricultural and Food Chemistry, 62*(16), 3548–3552. doi: 10.1021/jf403586t.
21. Mario, R. A., & Lorand, K., (2019). Alicyclic β- and γ-amino acids: Useful scaffolds for the stereocontrolled access to amino acid-based carbocyclic nucleoside analogs. *Molecules (Basel, Switzerland), 24*(1), 161/1–161/15. doi: 10.3390/molecules24010161.
22. Hauser, F. M., & Rhee, R. P., (1981). A brief total synthesis of *N*-benzoyl-*D,L*-daunosamine. *Journal of Organic Chemistry, 46*(1), 227, 228. doi: 10.1021/jo00314a066.
23. Hauser, F. M., Rhee, R. P., & Ellenberger, S. R., (1984). Total synthesis of optically active *N*-benzoyldaunosamine from an azetidinone. *Journal of Organic Chemistry, 49*(12), 2236–2240. doi: 10.1021/jo00186a031.
24. Moriconi, E. J., & Meyer, W. C., (1971). Reaction of dienes with chlorosulfonyl isocyanate. *Journal of Organic Chemistry, 36*(19), 2841–2849. doi: 10.1021/jo00818a025.

25. Evans, D. A., & Biller, S. A., (1985). The total synthesis of (±)-naphthyridinomycin. I. Preparation of a key tricyclic lactam intermediate. *Tetrahedron Letters, 26*(16), 1907–1910. doi: 10.1016/S0040–4039(00)98338–2.
26. Dondoni, A., Massi, A., Sabbatini, S., & Bertolasi, V., (2004). Three-component Staudinger-type stereoselective synthesis of *C*-glycosyl-β-lactams and their use as precursors for *C*-glycosyl isoserines and dipeptides. A polymer-assisted solution-phase approach. *Advanced Synthesis & Catalysis, 346*(11), 1355–1360. doi: 10.1002/adsc.200404100.
27. Petrik, V., Roeschenthaler, G. V., & Cahard, D., (2011). Diastereoselective synthesis of *trans*-trifluoromethyl-β-lactams and α-alkyl-β-trifluoromethyl-β-amino esters. *Tetrahedron, 67*(18), 3254–3259. doi: 10.1016/j.tet.2011.03.001.
28. Vidya, R., Eggen, M., Nair, S. K., Georg, G. I., & Himes, R. H., (2003). Synthesis of cryptophycins via an *N*-acyl-β-lactam macrolactonization. *Journal of Organic Chemistry, 68*(25), 9687–9693. doi: 10.1021/jo0302197.
29. Cavagna, F., Linkies, A., Pietsch, H., & Reuschling, D., (1980). Pyrrolizidines by rearrangement of β-lactams. *Angewandte Chemie International Edition in English, 19*(2), 129, 130. doi: 10.1002/anie.198001291.
30. Bose, A. K., Krishnan, L., Wagle, D. R., & Manhas, M. S., (1986). Studies on lactams. Part 77. A novel chemical transformation of 3-vinyl-4-substituted-2-azetidinones. *Tetrahedron Letters, 27*(49), 5955–5958. doi: 10.1016/S0040-4039(00)85371-X.
31. Li, X. G., & Kanerva, L. T., (2006). Lipase-involved strategy to the enantiomers of 4-benzyl-β-lactam as a key intermediate in the preparation of β-phenylalanine derivatives. *Advanced Synthesis & Catalysis, 348*(1+2), 197–205. doi: 10.1002/adsc.200505253.
32. Gianolio, E., Mohan, R., & Berkessel, A., (2016). Enantiopure *N*-benzyloxycarbonyl-β^2-amino acid allyl esters from racemic β-lactams by dynamic kinetic resolution using *Candida antarctica* lipase B. *Advanced Synthesis & Catalysis, 358*(1), 30–33. doi: 10.1002/adsc.201500820.
33. Jiang, X., Prasad, K., Prashad, M., Slade, J., Repic, O., & Blacklock, T. J., (2006). β-Amino amides from β-lactams: Application to the formal synthesis of a peptide-deformylase Inhibitor. *Synlett,* (18), 3179–3181. doi: 10.1055/s-2006-951499.
34. Mendez, R., Alemany, T., & Martin-Villacorta, J., (1992). Catalysis of hydrolysis and aminolysis of non-classical β-lactam antibiotics by metal ions and metal chelates. *Chemical & Pharmaceutical Bulletin, 40*(12), 3228–3233. doi: 10.1248/cpb.40.3228.
35. Takeuchi, Y., Sunagawa, M., Isobe, Y., Hamazume, Y., & Noguchi, T., (1995). Stability of a 1^{β}-methylcarbapenem antibiotic, meropenem (SM-7338) in aqueous solution. *Chemical & Pharmaceutical Bulletin, 43*(4), 689–692. doi: 10.1248/cpb.43.689.
36. Grabowski, E. J. J., Douglas, A. W., & Smith, G. B., (1985). Ammonolysis of cephamycins: Carbon-13 NMR characterization of the intermediates from β-lactam ring cleavage prior to loss of the 3'-group. *Journal of the American Chemical Society, 107*(1), 267, 268. doi: 10.1021/ja00287a057.
37. Llinas, A., & Page, M. I., (2004). Intramolecular general acid catalysis in the aminolysis of β-lactam antibiotics. *Organic & Biomolecular Chemistry, 2*(5), 651–654. doi: 10.1039/B313900J.
38. Diaz, N., Suarez, D., Sordo, T. L., Mendez, R., & Villacorta, J. M., (2003). A combined theoretical and experimental research project into the aminolysis of β-lactam antibiotics:

The importance of bifunctional catalysis. *European Journal of Organic Chemistry,* (21), 4161–4172. doi: 10.1002/ejoc.200300418.

39. Lopez, R., Menendez, M. I., Diaz, N., Suarez, D., Campomanes, P., & Sordo, T. L., (2000). Theoretical studies of hydrolysis, alcoholysis and aminolysis of β-lactams. *Recent Research Developments in Physical Chemistry, 4*(Pt. 1), 157–173.
40. Diaz, N., Suarez, D., & Sordo, T. L., (1999). Ammonolysis and aminolysis of β-lactams: A theoretical study. *Chemistry: A European Journal, 5*(3), 1045–1054. doi: 10.1002/(SICI)1521-3765(19990301)5:3<1045::AID-CHEM1045>3.0.CO;2-M.
41. Diaz, N., Suarez, D., & Sordo, T. L., (1999). Importance of a synperiplanar stepwise mechanism through neutral intermediates in the aminolysis of monocyclic β-lactams: A theoretical analysis. *Journal of Organic Chemistry, 64*(25), 9144–9152. doi: 10.1021/jo9910942.
42. Diaz, N., Suarez, D., & Sordo, T. L., (2001). Theoretical study of amine-assisted aminolysis of penicillins – the kinetic role of the carboxylate group. *European Journal of Organic Chemistry,* (4), 793–801. doi: 10.1002/1099-0690(200102)2001:4<793::AID-EJOC793>3.0.CO;2-Z.
43. Lujan, A., Gonzalez, J. L., Del MarCanedo, M., & Grande, C., (1993). Determination of the optimized individual kinetic constants of the aminolysis reaction in basic medium of β-lactam antibiotics by a numerical computational method. *Journal of Pharmaceutical Sciences, 82*(11), 1167–1171. doi: 10.1002/jps.2600821121.
44. Grande, C., Canedo, M. D. M., & Gonzalez, J. L., (1994). Kinetic treatment of basic aminolysis of beta-lactam antibiotics by a univariant optimization computational method. *International Journal of Chemistry (Calcutta, India), 5*(2), 77–85.
45. Grande, M. C., Gonzalez Hernandez, J. L., (1994). New computational treatment of kinetic data on the basic aminolysis of beta-lactam antibiotics by a joint uni- and multivariate optimization methods. *International Journal of Chemistry (Calcutta, India), 5*(3), 107–116.
46. Tsang, W. Y., Ahmed, N., & Page, M. I., (2007). The aminolysis of *N*-aroyl beta-lactams occurs by a concerted mechanism. *Organic & Biomolecular Chemistry, 5*(3), 485–493.
47. Kukolja, S., & Lammert, S. R., (1975). Azetidinone antibiotics. XIV. Removal of a phthaloyl protective group from acid and base sensitive compounds. *Journal of the American Chemical Society, 97*(19), 5582, 5583. doi: 10.1021/ja00852a044.
48. Misner, J. W., Fisher, J. W., Gardner, J. P., Pedersen, S. W., Trinkle, K. L., Jackson, B. G., & Zhang, T. Y., (2003). Enantioselective synthesis of the carbacephem antibiotic loracarbef via Mitsunobu and Dieckmann cyclization from an unnatural amino acid. *Tetrahedron Letters, 44*(32), 5991–5993. doi: 10.1016/S0040-4039(03)01483-7.
49. Lagerlund, I., (1976). Strained heterocyclic compounds. 10. Introduction of an amide side chain in place of halogen in 7-halo-8-oxo-1-azabicyclo[4.2.0]octane. *Acta Chemica Scandinavica, Series B: Organic Chemistry and Biochemistry, B30*(4), 318–322.
50. Paul, L., Polczynski, P., & Hilgetag, G., (1967). β-Lactams. V. Preparation of some condensed β-lactams. *Chemische Berichte, 100*(8), 2761–2765. doi: 10.1002/cber.19671000840.
51. Khalil, N. S. A. M., (2005). Synthesis of novel α-*L*-arabinopyranosides of β-lactams with potential antimicrobial activity. *Nucleosides, Nucleotides & Nucleic Acids, 24*(9), 1277–1287. doi: 10.1080/15257770500230285.
52. Del Buttero, P., Molteni, G., & Papagni, A., (2003). Stereoselective synthesis of 3-amino-4-substituted-2-azetidinones via [2+2] cycloadditions of tricarbonyl(η^6-arene)

chromium(0)complexed imines. *Tetrahedron: Asymmetry, 14*(24), 3949–3953. doi: 10.1016/j.tetasy.2003.09.046.

53. Fekner, T., Baldwin, J. E., Adlington, R. M., Jones, T. W., Prout, C. K., & Schofield, C. J., (2000). Syntheses of (6*S*)-cephalosporins from 6-aminopenicillanic acid. *Tetrahedron, 56*(33), 6053–6074. doi: 10.1016/S0040-4020(00)00486-5.
54. Ceric, H., & Sindler-Kulyk, M., (2009). Hydrazinolysis study of phthalimido- and phthalisoimido-penicillin amide derivatives. *ARKIVOC (Gainesville, FL, United States),* (7), 237–246. doi: 10.3998/ark.5550190.0010.723.
55. Bhattacharjee, N., Triboulet, S., Dubee, V., Fonvielle, M., Edoo, Z., Hugonnet, J. E., Etheve-Quelquejeu, M., et al., (2019). Negative impact of carbapenem methylation on the reactivity of β-lactams for cysteine acylation as revealed by quantum calculations and kinetic analyses. *Antimicrobial Agents and Chemotherapy, 63*(4), e02039-18/1–e02039-18/12. doi: 10.1128/AAC.02039–18.
56. Bhattacharjee, N., Field, M. J., Simorre, J. P., Arthur, M., & Bougault, C. M., (2016). Hybrid potential simulation of the acylation of *Enterococcus faecium L,D*-transpeptidase by carbapenems. *Journal of Physical Chemistry B, 120*(21), 4767–4781. doi: 10.1021/acs.jpcb.6b02836.
57. Knap, A. K., & Pratt, R. F., (1987). Inactivation of the thiol RTEM-1 β-lactamase by 6-β-bromopenicillanic acid. Identity of the primary active-site nucleophile. *Biochemical Journal, 247*(1), 29–33. doi: 10.1042/bj2470029.
58. Pajares, M. A., Zimmerman, T., Sanchez-Gomez, F. J., Ariza, A., Torres, M. J., Blanca, M., Canada, F. J., Montanez, M. I., & Perez-Sala, D., (2020). Amoxicillin inactivation by thiol-catalyzed cyclization reduces protein haptenation and antibacterial potency. *Frontiers in Pharmacology, 11*, 189/1–189/16. doi: 10.3389/fphar.2020.00189.
59. Castro-Falcon, G., Hahn, D., Reimer, D., & Hughes, C. C., (2016). Thiol probes to detect electrophilic natural products based on their mechanism of action. *ACS Chemical Biology, 11*(8), 2328–2336. doi: 10.1021/acschembio.5b00924.
60. Llinás, A., Donoso, J., Vilanova, B., Frau, J., Muñoz, F., & Page, M. I., (2000). Thiol-catalyzed hydrolysis of benzylpenicillin. *Journal of the Chemical Society, Perkin Transactions 2: Physical Organic Chemistry,* 1521–1525. doi: 10.1039/b001091j.
61. Llinas, A., Vilanova, B., & Page, M. I., (2004). Thiol-catalyzed hydrolysis of cephalosporins and possible rate-limiting amine anion expulsion. *Journal of Physical Organic Chemistry, 17*(6, 7), 521–528. doi: 10.1002/poc.760.
62. Koziol, A., Altieri, E., Furman, B., Solecka, J., & Chmielewski, M., (2011). Quinidine catalyzed reaction between 4-formyloxyazetidin-2-one and some thiophenols, thiols and alcohols. *ARKIVOC (Gainesville, FL, United States),* (4), 37–53. doi: 10.3998/ark.5550190.0012.405.
63. Garcías, R. C., Coll, M., Donoso, J., & Muñoz, F., (2006). Density functional theory study of the thiolysis reaction in penicillins. *Journal of Molecular Structure: Theochem, 773*, 29–34. doi: 10.1016/j.theochem.2006.07.001.
64. Garcias, R. C., Coll, M., Donoso, J., Vilanova, B., & Munoz, F., (2005). Theoretical study of thiolysis in penicillins and cephalosporines. *International Journal of Chemical Kinetics, 37*(7), 434–443. doi: 10.1002/kin.20082.
65. Van, B. W., Dejaegher, Y., Van, L. R., & De Kimpe, N., (2006). Reduction of 4-(haloalkyl) azetidin-2-ones with $LiAlH_4$ as a powerful method for the synthesis of stereo-defined aziridines and azetidines. *Organic Letters, 8*(6), 1101–1104. doi: 10.1021/ol053015g.

66. D'hooghe, M., Dekeukeleire, S., Mollet, K., Lategan, C., Smith, P. J., Chibale, K., & De Kimpe, N., (2009). Synthesis of novel 2-alkoxy-3-amino-3-arylpropan-1-ols and 5-alkoxy-4-aryl-1,3-oxazinanes with antimalarial activity. *Journal of Medicinal Chemistry, 52*(13), 4058–4062. doi: 10.1021/jm9002632.
67. Alcaide, B., Almendros, P., Cabrero, G., & Ruiz, M. P., (2007). Stereocontrolled access to orthogonally protected *anti,anti*-4-aminopiperidine-3,5-diols through chemoselective reduction of enantiopure beta-lactam cyanohydrins. *The Journal of Organic Chemistry, 72*(21), 7980–7991. doi: 10.1021/jo701452a.
68. Alcaide, B., Almendros, P., Aragoncillo, C., & Salgado, N. R., (1999). Novel diethylaluminum chloride promoted reactions of the azetidine ring: Efficient and stereocontrolled entry to functionalized olefins, pyrrolidines, and pyrroles. *Journal of Organic Chemistry, 64*(26), 9596–9604. doi: 10.1021/jo991203c.
69. Koch, V., Lorion, M. M., Barde, E., Braese, S., & Cossy, J., (2019). Cobalt-catalyzed α-arylation of substituted α-halogeno β-lactams. *Organic Letters, 21*(16), 6241–6244. doi: 10.1021/acs.orglett.9b02122.
70. Hang, D. T., Decuyper, L., Mollet, K., Kenis, S., De Kimpe, N., Van, N. T., & D'hooghe, M., (2016). Synthesis of trifluoromethylated azetidines, aminopropanes, 1,3-oxazinanes, and 1,3-oxazinan-2-ones starting from 4-trifluoromethyl-β-lactam building blocks. *Synlett, 27*(7), 1100–1105. doi: 10.1055/s-0035-1561316.
71. Sanap, S. P., Ghosh, S., Jabgunde, A. M., Pinjari, R. V., Gejji, S. P., Singh, S., Chopade, B. A., & Dhavale, D. D., (2010). Synthesis, computational study and glycosidase inhibitory activity of polyhydroxylated conidine alkaloids – a bicyclic imino-sugar. *Organic & Biomolecular Chemistry, 8*(14), 3307–3315. doi: 10.1039/c004690f.
72. Alcaide, B., Almendros, P., Alonso, J. M., & Aly, M. F., (2003). Useful dual Diels-Alder behavior of 2-azetidinone-tethered aryl imines as azadienophiles or azadienes: A β-lactam-based stereocontrolled access to optically pure, highly functionalized indolizidine systems. *Chemistry – A European Journal, 9*(14), 3415–3426. doi: 10.1002/chem.200304712.
73. Nunez-Villanueva, D., Bonache, M. A., Infantes, L., Garcia-Lopez, M. T., Martin-Martinez, M., & Gonzalez-Muniz, R., (2011). Quaternary α,α-2-oxoazepane α-amino acids: Synthesis from ornithine-derived β-lactams and incorporation into model dipeptides. *Journal of Organic Chemistry, 76*(16), 6592–6603. doi: 10.1021/jo200894d.
74. Alcaide, B., Almendros, P., Cabrero, G., & Ruiz, M. P., (2012). Stereoselective cyanation of 4-formyl and 4-imino-β-lactams: Application to the synthesis of polyfunctionalized γ-lactams. *Tetrahedron, 68*(52), 10761–10768. doi: 10.1016/j.tet.2012.02.062.
75. Lange, J., Bissember, A. C., Banwell, M. G., & Cade, I. A., (2011). Synthesis of 2,3-dihydro-4(1*h*)-quinolones and the corresponding 4(1*H*)-quinolones via low-temperature fries rearrangement of *N*-arylazetidin-2-ones. *Australian Journal of Chemistry, 64*(4), 454–470. doi: 10.1071/CH10465.
76. Barbier, V., Marrot, J., Couty, F., & David, O. R. P., (2016). β-Lactams as formal dipoles through amide-bond activation. *European Journal of Organic Chemistry,* (3), 549–555. doi: 10.1002/ejoc.201501342.
77. Bestian, H., (1968). Poly-β-amides. *Angewandte Chemie, International Edition in English, 7*(4), 278–285. doi: 10.1002/anie.196802781.
78. Zhang, J., Kissounko, D. A., Lee, S. E., Gellman, S. H., & Stahl, S. S., (2009). Access to poly-β-peptides with functionalized side chains and end groups via controlled

ring-opening polymerization of β-lactams. *Journal of the American Chemical Society, 131*(4), 1589–1597. doi: 10.1021/ja8069192.

79. Zhang, J., Markiewicz, M. J., Mowery, B. P., Weisblum, B., Stahl, S. S., & Gellman, S. H., (2012). *C*-Terminal functionalization of nylon-3 polymers: Effects of *C*-terminal groups on antibacterial and hemolytic activities. *Biomacromolecules, 13*(2), 323–331. doi: 10.1021/bm2013058.
80. Zhang, J., Markiewicz, M. J., Weisblum, B., Stahl, S. S., & Gellman, S. H., (2012). Functionally diverse nylon-3 copolymers from readily accessible β-lactams. *ACS Macro Letters, 1*(6), 714–717. doi: 10.1021/mz300172y.
81. Yang, H., Zhao, J., Yan, M., Pispas, S., & Zhang, G., (2011). Nylon 3 synthesized by ring opening polymerization with a metal-free catalyst. *Polymer Chemistry, 2*(12), 2888–2892. doi: 10.1039/c1py00334h.
82. Morillo, M., De Ilarduya, A. M., & Munoz-Guerra, S., (2002). Synthesis of poly(α-alkyl β,*L*-aspartate)s by transesterification. *Macromolecular Rapid Communications, 23*(14), 849–852. doi: 10.1002/1521-3927(20021001)23:14<849::AID-MARC849>3.0.CO;2-Y.
83. López-Carrasquero, F., Martínez De, I. A., & Muñoz-Guerra, S., (1994). Poly(β-*L*-aspartate)s containing ethylene oxide units in the side chain: Synthesis and structural studies. *Polymer Journal (Tokyo, Japan), 26*(6), 694–704. doi: 10.1295/polymj.26.694.
84. Garcia-Martin, M. D. G., De Paz, M. V., & Galbis, J. A., (1997). Synthesis of poly[isobutyl (2*S*,3*R*)-3-benzyloxyaspartate]. *Macromolecular Chemistry and Physics, 198*(1), 219–227. doi: 10.1002/macp.1997.021980116.
85. Cheng, J., & Deming, T. J., (2001). Controlled polymerization of β-lactams using metal-amido complexes: Synthesis of block copoly(β-peptides). *Journal of the American Chemical Society, 123*(38), 9457, 9458. doi: 10.1021/ja0110022.
86. Eldred, S. E., Pancost, M. R., Otte, K. M., Rozema, D., Stahl, S. S., & Gellman, S. H., (2005). Effects of side chain configuration and backbone spacing on the gene delivery properties of lysine-derived cationic polymers. *Bioconjugate Chemistry, 16*(3), 694–699. doi: 10.1021/bc050017c.
87. Mowery, B. P., Lee, S. E., Kissounko, D. A., Epand, R. F., Epand, R. M., Weisblum, B., Stahl, S. S., & Gellman, S. H., (2007). Mimicry of antimicrobial host-defense peptides by random copolymers. *Journal of the American Chemical Society, 129*(50), 15474–15476. doi: 10.1021/ja077288d.
88. Hashimoto, K., Hotta, K., Okada, M., & Nagata, S., (1995). Synthesis of monodisperse polyamides by living anionic polymerization of β-lactams in the mixture of *N,N*-dimethylacetamide and lithium chloride. *Journal of Polymer Science, Part A: Polymer Chemistry, 33*(12), 1995–1999. doi: 10.1002/pola.1995.080331206.
89. Hashimoto, K., Oi, T., Yasuda, J., Hotta, K., & Okada, M., (1997). Molecular weight distribution of polyamides obtained in anionic polymerization of methyl-substituted β-lactams and aminolysis of their *N*-benzoyl lactams. *Journal of Polymer Science, Part A: Polymer Chemistry, 35*(9), 1831–1838. doi: 10.1002/(SICI)1099-0518(19970715)35:9<1831::AID-POLA24>3.0.CO;2-5.
90. García-Martín, M. D. G., De Paz Báñez, M. V., & Galbis, J. A., (2000). Preparation of 3-amino-3-deoxy-2,4,5,6-tetra-*O*-methyl-*D*-altronic acid hydrochloride, a precursor in the preparation of a chiral β-polyamide (nylon 3 analog). *Journal of Carbohydrate Chemistry, 19*(7), 805–815. doi: 10.1080/07328300008544119.

91. Dane, E. L., & Grinstaff, M. W., (2012). Poly-amido-saccharides: Synthesis via anionic polymerization of a β-lactam sugar monomer. *Journal of the American Chemical Society, 134*(39), 16255–16264. doi: 10.1021/ja305900r.
92. Xiao, R., Dane, E. L., Zeng, J., McKnight, C. J., & Grinstaff, M. W., (2017). Synthesis of altrose poly-amido-saccharides with β-*N*-(1→2)-*D*-amide linkages: A right-handed helical conformation engineered in at the monomer level. *Journal of the American Chemical Society, 139*(40), 14217–14223. doi: 10.1021/jacs.7b07405.
93. Balijepalli, A. S., Sabatelle, R. C., Chen, M., Suki, B., & Grinstaff, M. W., (2020). A synthetic bioinspired carbohydrate polymer with mucoadhesive properties. *Angewandte Chemie, International Edition, 59*(2), 704–710. doi: 10.1002/anie.201911720.
94. Ghobril, C., Heinrich, B., Dane, E. L., & Grinstaff, M. W., (2014). Synthesis of hydrophobic carbohydrate polymers and their formation of thermotropic liquid crystalline phases. *ACS Macro Letters, 3*(4), 359–363. doi: 10.1021/mz5000703.
95. Xiao, R., Zeng, J., & Grinstaff, M. W., (2018). Biologically active branched polysaccharide mimetics: Synthesis via ring-opening polymerization of a maltose-based β-lactam. *ACS Macro Letters, 7*(7), 772–777. doi: 10.1021/acsmacrolett.8b00302.
96. Dane, E. L., Grinstaff, M. W., Ballok, A. E., & O'Toole, G. A., (2014). Synthesis of bioinspired carbohydrate amphiphiles that promote and inhibit biofilms. *Chemical Science, 5*(2), 551–557. doi: 10.1039/C3SC52777H.
97. Schwab, L. W., Baum, I., Fels, G., & Loos, K., (2010). Mechanistic insight in the enzymatic ring-opening polymerization of β-propiolactam. *ACS Symposium Series, 1043*(green polymer chemistry: Biocatalysis and biomaterials), 265–278. doi: 10.1021/bk-2010-1043.ch019.
98. Schwab, L. W., Kroon, R., Schouten, A. J., & Loos, K., (2008). Enzyme-catalyzed ring-opening polymerization of unsubstituted β-lactam. *Macromolecular Rapid Communications, 29*(10), 794–797. doi: 10.1002/marc.200800117.
99. Baum, I., Elsasser, B., Schwab, L. W., Loos, K., & Fels, G., (2011). Atomistic model for the polyamide formation from β-lactam catalyzed by Candida Antarctica lipase B. *ACS Catalysis, 1*(4), 323–336. doi: 10.1021/cs1000398.
100. Stavila, E., Alberda, V. E. G. O. R., Woortman, A. J. J., & Loos, K., (2014). Lipase-catalyzed ring-opening copolymerization of ε-caprolactone and β-lactam. *Biomacromolecules, 15*(1), 234–241. doi: 10.1021/bm401514k.
101. Zhou, M., Xiao, X., Cong, Z., Wu, Y., Zhang, W., Ma, P., Chen, S., et al., (2020). Water-insensitive synthesis of poly-β-peptides with defined architecture. *Angewandte Chemie, International Edition, 59*(18), 7240–7244. doi: 10.1002/anie.202001697.
102. Cossio, F. P., Lopez, C., Oiarbide, M., Palomo, C., Aparicio, D., & Rubiales, G., (1988). Reagents and synthetic methods. 69. Synthetic utility of azetidine-2,3-diones. A new approach to 3-hydroxyethyl-β-lactams and α-amino acid derivatives. *Tetrahedron Letters, 29*(25), 3133–3136. doi: 10.1016/0040-4039(88)85105-0.
103. Palomo, C., Aizpurua, J. M., Cabre, F., Garcia, J. M., & Odriozola, J. M., (1994). Synthesis of β-alkylserine *N*-carboxyanhydrides through β-lactams via cycloaddition reaction of alkoxyketenes to chiral α-alkoxyaldehyde-derived imines. *Tetrahedron Letters, 35*(17), 2721–2724. doi: 10.1016/S0040–4039(00)77015–8.
104. Palomo, C., Aizpurua, J. M., Ganboa, I., Maneiro, E., & Odriozola, B., (1994). A β-lactam framework as a β-alanyl dication equivalent: New synthesis of α-amino acid *N*-carboxy anhydrides (NCAs) derived from β-substituted alanines. *Journal of the Chemical Society, Chemical Communications,* (12), 1505–1507. doi: 10.1039/C39940001505.

105. Palomo, C., Aizpurua, J. M., Ganboa, I., Carreaux, F., Cuevas, C., Maneiro, E., & Ontoria, J. M., (1994). New synthesis of α-amino acid *N*-carboxy anhydrides through Baeyer-Villiger oxidation of α-keto β-lactams. *Journal of Organic Chemistry, 59*(11), 3123–3130. doi: 10.1021/jo00090a033.
106. Palomo, C., Aizpurua, J. M., Urchegui, R., & Garcia, J. M., (1995). A facile access to peptides containing δ-α-methyl β-alkylserines by coupling of α-branched leuchs anhydrides with α-amino esters. *Journal of the Chemical Society, Chemical Communications,* (22), 2327, 2328. doi: 10.1039/C39950002327.
107. Palomo, C., Aizpurua, J. M., Cuevas, C., Urchegui, R., & Linden, A., (1996). Generation of threonine- and azathreonine *N*-carboxy anhydrides from α-hydroxy β-lactams promoted by 2,2,6,6-tetramethylpiperidinyl-1-oxyl (TEMPO) in combination with sodium hypochlorite. *Journal of Organic Chemistry, 61*(13), 4400–4404. doi: 10.1021/jo952178n.
108. Palomo, C., Aizpurua, J. M., Ganboa, I., Odriozola, B., Urchegui, R., & Görls, H., (1996). Concise synthesis of α-alkyl α-amino acids and their incorporation into peptides via β-lactam-derived α-amino acid *N*-carboxy anhydrides. *Chemical Communications (Cambridge, United Kingdom),* (11), 1269, 1270. doi: 10.1039/CC9960001269.
109. Palomo, C., Oiarbide, M., & Esnal, A., (1997). Synthetic studies towards peptidyl nucleoside antibiotics: First synthesis of a polyoxamic acid derivative enabling direct coupling with α-amino acid esters. *Chemical Communications (Cambridge), 1997*(7), 691, 692. doi: 10.1039/a700815e.
110. Palomo, C., Oiarbide, M., Esnal, A., Landa, A., Miranda, J. I., & Linden, A., (1998). Practical synthesis of α-amino acid *N*-carboxy anhydrides of polyhydroxylated α-amino acids from β-lactam frameworks. Model studies toward the synthesis of directly linked peptidyl nucleoside antibiotics. *Journal of Organic Chemistry, 63*(17), 5838–5846. doi: 10.1021/jo980354x.
111. Palomo, C., Oiarbide, M., Ganboa, I., & Miranda, J. I., (2001). A concise synthesis of α-amino acid *N*-carboxy anhydrides of (2*S*,3*S*)-β-substituted serines. *Tetrahedron Letters, 42*(51), 8955–8957. doi: 10.1016/S0040-4039(01)01766-X.
112. Palomo, C., Ganboa, I., Oiarbide, M., Sciano, G. T., & Miranda, J. I., (2002). A β-lactam route to short peptide segments related to angiotensin-converting enzyme (ACE) inhibitors. *ARKIVOC (Gainesville, FL, United States),* (5), 8–16. doi: 10.3998/ark.5550190.0003.502.
113. Garbi, A., Allain, L., Chorki, F., Ourevitch, M., Crousse, B., Bonnet-Delpon, D., Nakai, T., & Begue, J. P., (2001). Novel [1,2]- and [2,3]-Wittig rearrangements of α-benzyloxy β-CF_3-β-lactam enolates. *Organic Letters, 3*(16), 2529–2531. doi: 10.1021/ol016198p.
114. Majewski, M. W., Miller, P. A., Oliver, A. G., & Miller, M. J., (2017). Alternate "drug" delivery utilizing β-lactam cores: Syntheses and biological evaluation of β-lactams bearing isocyanate precursors. *Journal of Organic Chemistry, 82*(1), 737–744. doi: 10.1021/acs.joc.6b02272.
115. Alcaide, B., Martin-Cantalejo, Y., Rodriguez-Lopez, J., & Sierra, M. A., (1993). New reactivity patterns of the β-lactam ring: Tandem *C*3-*C*4 bond breakage-rearrangement of 4-acyl- or 4-imino-3,3-dimethoxy-2-azetidinones promoted by stannous chloride ($SnCl_2$-$2H_2O$). *Journal of Organic Chemistry, 58*(17), 4767–4770. doi: 10.1021/jo00069a054.
116. Singh, P., Singh, P., Kumar, K., Kumar, V., Mahajan, M. P., & Bisetty, K., (2012). Synthetic studies on the role of substituents at *C*-3 position on *C*3-*C*4 bond cleavage

of β-lactam ring: Convenient route for diastereoselective synthesis of pyridin-2-ones. *Heterocycles, 86*(2), 1301–1322. doi: 10.3987/COM-12-S(N)83.

117. Kita, Y., Shibata, N., Yoshida, N., Kawano, N., & Matsumoto, K., (1994). An unprecedented cleavage of the β-lactam ring: Stereoselective synthesis of chiral β-amido cyanides. *Journal of Organic Chemistry, 59*(5), 938, 939. doi: 10.1021/jo00084a002.
118. Alcaide, B., Almendros, P., Cabrero, G., & Ruiz, M. P., (2005). Organocatalytic ring expansion of β-lactams to γ-lactams through a novel *N*1-*C*4 bond cleavage. Direct synthesis of enantiopure succinimide derivatives. *Organic Letters, 7*(18), 3981–3984. doi: 10.1021/ol051504a.
119. Klimczak, U. K., & Zambron, B. K., (2015). Effective 1,5-stereocontrol in Pd(0)/InI promoted reactions of chiral *N*-Ts-4-vinylazetidin-2-ones with aldehydes. An efficient entry into nonracemic *semi*-protected (3*Z*)-2,6-*anti*-enediols. *Chemical Communications (Cambridge, United Kingdom), 51*(31), 6796–6799. doi: 10.1039/c5cc01485a.
120. Plata, P., Klimczak, U., & Zambron, B. K., (2018). Acyclic remote 1,5- and 1,4,5-stereocontrol in the catalytic stereoselective reactions of β-lactams with aldehydes: The effect of the *N*-methylimidazole ligand. *Journal of Organic Chemistry, 83*(23), 14527–14552. doi: 10.1021/acs.joc.8b02333.
121. Domin, S., Plata, P., & Zambron, B. K., (2019). Diastereoselectivity switch in the Pd(0)/InI-mediated reactions of β-lactams with aldehydes: An entry into nonracemic semiprotected (3*E*)-2,6-ENEDIOLS. *Journal of Organic Chemistry, 84*(19), 12268–12280. doi: 10.1021/acs.joc.9b01471.
122. Domin, S., Kedzierski, J., & Zambron, B. K., (2019). Remote 1,5-stereoselectivity control by an *N*-ligand switch in the Pd(0)/InI-promoted reactions of 4-ethynyl-β-lactams with aldehydes. *Organic Letters, 21*(11), 3904–3908. doi: 10.1021/acs.orglett.9b00891.
123. Alajarín, M., Molina, P., Vidal, Á., & Tovar, F., (1998). The first *aza*-Wittig reaction of the β-lactam carbonyl group. *Synlett,* (11), 1288–1290. doi: 10.1055/s-1998-1920.
124. Sato, S., Inokuma, T., Otsubo, N., Burton, D. R., & Barbas, C. F., (2013). Chemically programmed antibodies as HIV-1 attachment inhibitors. *ACS Medicinal Chemistry Letters, 4*(5), 460–465. doi: 10.1021/ml400097z.
125. Kozawa, Y., & Mori, M., (2003). Synthesis of 3-alkoxycarbonyl-1β-methylcarbapenem by using the palladium-catalyzed *C-N* bond-forming reaction between vinyl halide and β-lactam nitrogen. *Journal of Organic Chemistry, 68*(8), 3064–3067. doi: 10.1021/jo020584i.
126. Jiang, B., Tian, H., Huang, Z. G., & Xu, M., (2008). Successive copper(I)-catalyzed cross-couplings in one pot: A novel and efficient starting point for synthesis of carbapenems. *Organic Letters, 10*(13), 2737–2740. doi: 10.1021/ol800845r.
127. Jobbins, M. O., & Miller, M. J., (2014). Syntheses of hydroxamic acid-containing bicyclic β-lactams via palladium-catalyzed oxidative amidation of alkenes. *Journal of Organic Chemistry, 79*(4), 1620–1625. doi: 10.1021/jo402544p.
128. Alcaide, B., & Almendros, P., (2004). β-Lactams as versatile synthetic intermediates for the preparation of heterocycles of biological interest. *Current Medicinal Chemistry, 11*(14), 1921–1949. doi: 10.2174/0929867043364856.
129. Kamath, A., & Ojima, I., (2012). Advances in the chemistry of β-lactam and its medicinal applications. *Tetrahedron, 68*(52), 10640–10664. doi: 10.1016/j.tet.2012.07.090.
130. Alcaide, B., Almendros, P., & Aragoncillo, C., (2007). β-Lactams: Versatile building blocks for the stereoselective synthesis of non-β-lactam products. *Chemical Reviews (Washington, DC, United States), 107*(11), 4437–4492. doi: 10.1021/cr0307300.

131. Nagpal, R., Bhalla, J., & Bari, S. S., (2019). A comprehensive review on C-3 functionalization of β-lactams. *Current Organic Synthesis, 16*(1), 3–16. doi: 10.2174/1570179415666181116103341.
132. Ombito, J. O., & Singh, G. S., (2019). Recent progress in chemistry of β-lactams. *Mini-Reviews in Organic Chemistry, 16*(6), 544–567. doi: 10.2174/1570193X15666180914165303.

Index

B

C

D

F

G

K

L

M

O

U

V